T0328625

ERGONOMICS

ERGONOMICS

HOW TO DESIGN FOR EASE AND EFFICIENCY

THIRD EDITION

KATRIN E. KROEMER ELBERT

HENRIKE B. KROEMER

ANNE D. KROEMER HOFFMAN

ACADEMIC PRESS

An imprint of Elsevier

Academic Press is an imprint of Elsevier
125 London Wall, London EC2Y 5AS, United Kingdom
525 B Street, Suite 1650, San Diego, CA 92101, United States
50 Hampshire Street, 5th Floor, Cambridge, MA 02139, United States
The Boulevard, Langford Lane, Kidlington, Oxford OX5 1GB, United Kingdom

Notices
Knowledge and best practice in this field are constantly changing. As new research and experience broaden our understanding, changes in research methods, professional practices, or medical treatment may become necessary.

Practitioners and researchers must always rely on their own experience and knowledge in evaluating and using any information, methods, compounds, or experiments described herein. In using such information or methods they should be mindful of their own safety and the safety of others, including parties for whom they have a professional responsibility.

To the fullest extent of the law, neither the Publisher nor the authors, contributors, or editors, assume any liability for any injury and/or damage to persons or property as a matter of products liability, negligence or otherwise, or from any use or operation of any methods, products, instructions, or ideas contained in the material herein.

Library of Congress Cataloging-in-Publication Data
A catalog record for this book is available from the Library of Congress

British Library Cataloguing-in-Publication Data
A catalogue record for this book is available from the British Library

ISBN: 978-0-12-813296-8

For information on all Academic Press publications visit our website at https://www.elsevier.com/books-and-journals

Working together
to grow libraries in
developing countries

www.elsevier.com • www.bookaid.org

Publisher: Matthew Deans
Acquisition Editor: Brian Guerin
Editorial Project Manager: Thomas Van Der Ploeg
Production Project Manager: Vijay Bharath. R
Designer: Mark Rogers

Typeset by Thomson Digital

Contents

Section II Design Applications

Chapter 8   Ergonomic Models, Methods, Measurements

Chapter 9   Designing to Fit the Moving Body

Chapter 10   The Computer Workplace

Chapter 11   The Individual Within an Organization

Section III Further Information

Preface to the Third Edition

What do we know about the human body and mind at work? Given what we know, how then should we design the work task, tools, the interface with the machine, and work procedures so that the human can perform safely, efficiently, and with satisfaction--perhaps even enjoy working?

These challenges are the main themes of this book. The solutions are the WHY and HOW of ergonomics/human engineering.

These first lines in the 1994 first edition of Ergonomics–How To Design For Ease And Efficiency appeared again in the second edition of 2001, amended 2003. They remain the motto of this third edition of our book.

We have updated and refreshed content to reflect and/or clarify current approaches to mind and body, including more recent findings, research, applications, and equipment. In particular, we've substantially updated the chapters on the computer workplace (Chapter 10) and designing for special populations (Chapter 14) and added a chapter on the individual within an organization (Chapter 11). Further, we have made improvements to the organization and flow of the information in the chapters. For ease of reading, we avoided interrupting the text with names and dates of other authors but have used a small marker to refer to further "Notes" at the end of each chapter for the interested reader.

We have left out some older references which, in the earlier editions, supported ergonomic findings and recommendations that are now well established and generally accepted. Of course, we continue to provide the complete references for all previously and newly included material, so the third part of the book remains a substantive source of past and present ergonomic information.

We remain grateful to all who gave suggestions for improvements. Our principal influence is our father, Professor Emeritus Karl Heinrich Eberhard Kroemer, who is the original primary author of this text and remains our inspiration for all things ergonomic. His passion for relevant and practical research, rigorous engineering, and dissemination of information is reflected in this book. He has provided significant updates, insight, and guidance, and is our role model for methodical technical work with an eye for improving the human work condition.

We would like to hear from you! Please, tell us what we did well, what we should do better in the next edition, and remind us of any past knowledge not to be forgotten. We can be reached at katrin.ergonomics@gmail.com.

Katrin E. Kroemer Elbert, PhD
Henrike B. Kroemer, PhD, LP
Anne D. Kroemer Hoffman, MBA

Using This Book

In this book, we discuss interactions of the human with work tasks, task equipment and the work environment.

Our intent in exploring the interactions is to build a knowledge-based understanding so that we can:

- utilize human abilities,
- amplify human capabilities,
- facilitate human efficiency,
- avoid overloading or underloading, and, we hope,
- enjoy doing our tasks.

This book comprises physical, psychological, biomechanical, and social features of the human as they concern individual or group relations to job, equipment, task environment, and organization. The authors include administrative concerns, consider economic constraints, and reflect aspects of today's ever-evolving technology. This book is comprehensive of all facets of human factors/ergonomics.

The writing and layout of this book rely on the concept of providing basic background knowledge and building from there to detailed information that allows the reader to delve further into special and novel applications. The person new to the field, the undergraduate and the graduate student will all gain a broad and thorough understanding of "Human-Technology Systems."

The authors acknowledge that, although the physical aspects of humans are changing only slowly, requirements for design are ever changing as the ergonomic database is ever expanding. To that end, we include general references to international standards and organizations to allow the interested reader to seek the latest information and requirements, which are updating more frequently than this text.

Further, we intend this book to be useful to any interested reader and try to present issues relevant to all humans. We use the pronouns "he" and "she" interchangeably when referring to gender non-specific features of human factors. While we understand that "gender" may be used to describe masculinity/femininity, much of the literature we review assumes gender refers to biological aspects of maleness/femaleness.

Ergonomic understanding will benefit established professionals, both specialists and generalists, concerned with people's performance and well-being at work. Among these are the designer, engineer, architect, industrial hygienist, industrial physician, industrial psychologist, occupational nurse, and manager. Of course, all of us are interested in humanizing work, making it more safe, efficient, and satisfying.

EASY READING

To avoid breaking the flow of reading, we have placed small markers in the text: they indicate where references or explanations are in order. These appear in a separate "Notes" section at the end of each chapter, which the reader may skip or consult.

THREE SECTIONS OF THIS BOOK

Section I: "The Ergonomic Knowledge Base" consists of Chapters 1–7. Here, we explore the properties of the human body and mind in the task environment. The focus is on human dimensions, capabilities and limitations; that is, the human factors to be considered in designing for ease and efficiency. For everybody's convenience, we use traditional disciplinary divisions into anatomy, physiology and psychology, whereas, as we all know, the human functions holistically and synergistically. The chapters contain references to overlapping information covered in other chapters.

Section II: "Design Applications" contains Chapters 8–15. Here, we discuss the design of tasks, equipment and environment based on the knowledge base developed in Section I about human attributes, strengths and limitations.

Section III: "Further Information" at the end of the book, includes a list of all **References** mentioned in the text (including previous editions), an extensive **Glossary** with concise descriptions and definitions, and a detailed **Index** that refers the reader to specific pages.

THREE WAYS TO USE THIS BOOK

1. Read straight through from beginning to the end, as a university course, and work on the "Challenges" listed at the end of each chapter.
2. Select a chapter of particular interest, absorb the background information, and proceed to design applications that make use of that information.
3. Start with the index, pick a topic of interest, and look up the information in the book section(s) that the index cites.

Introducing Ergonomics and Human Factors Engineering

WHAT ERGONOMICS IS

Ergonomics is the use of scientific principles, methods, and data drawn from a variety of disciplines for the development of systems in which people play a significant role. The field of applications extends from a single person using a simple tool to a complex multi-person socio-technical organization.

Ergonomic specialists rely on the understanding that systems are meant to benefit people, whether they are consumers, production workers, system operators or maintenance crews. This user-centered design philosophy acknowledges human variability as a design parameter. The resultant design features utilize human capabilities, consider human limitations and have built-in safeguards to avoid or reduce the effects of human error or of system failure.

Ergonomics is neutral: it takes no sides, neither employers' nor workers'. It is not for or against progress. It is not a philosophy, but a scientific discipline and practical technology.

WHAT ERGONOMICS DOES

Ergonomics focuses on the human as the most important component of our technological systems. This explains the early terms "human factors engineering," "man-machine systems," and the recent "human-systems integration." The aim is to assure that all human-made tools, devices, equipment, machines, environments and their organizations advance, directly or indirectly, the well-being of humans and their performance.

Accordingly, ergonomics has three related tasks:

1. Study, research, and experimentation to determine specific human traits and characteristics that need to be known for engineering design.
2. Apply this knowledge in the design of tools, machines, shelter, environments, work tasks and procedures so that they fit and accommodate the human.
3. Observe the actual performance of humans and their equipment in the real or a simulated environment, assess the suitability of the designed human-machine system and determine whether improvements are possible.

HOW ERGONOMICS CAME ABOUT

Using objects found in the environment as tools is an ancient activity. Pieces of stone, bone, and wood must have been selected for their fit to the hand and their suitability as cutters, scrapers, pounders, and missiles. Purposeful shaping of these tools naturally came next, followed by assembling raw materials to make, for example, protective clothing and shelters. These fundamental "ergonomic" activities grew into individual and communal skills and then into processes for creating finished products from diverse materials.

As human society grew more complex, organizational and managerial challenges developed. Training workers and soldiers, for example, became necessary, together with forming and controlling their behavior. Major projects, such as building the pyramids of ancient Egypt, assembling armies for warfare, sheltering the inhabitants of cities and supplying them with food and water, required sophisticated knowledge of human needs and desires; careful planning and complex logistics had to be mastered. The aims and means of training became sophisticated as well: Roman soldiers, for example, underwent well-organized exercises and conditioning until they could perform military exercises with ease: "drying the legions" of the Roman Empire relied on the principle of adapting and improving the physiological capabilities of the recruits to meet challenging physical requirements; when they no longer showed sweat on the skin, they were "dry" and fit.

Evolution of Disciplines

Artists, military officers, employers and athletes apparently were always interested in body build and physical performance. Specialized "medicine men" and "herb women" treated illnesses and injuries. About 400 BC, Hippocrates (often considered the father of western medicine) described a scheme of four body types: the "moist" type was believed to be dominated by black gall, the "dry" type by yellow gall, the "cold" type by slime, and the "warm" type by blood. These "humors" had been in some balance to keep an individual healthy. Knowledge of medicine, anatomy and anthropology began to assemble.

Over the centuries, more exact information accumulated into specialized disciplines. In the 15th–17th centuries, gifted people such as Leonardo da Vinci (1452–1519) and Alfonzo Giovanni Borelli (1608–1679) could still master all existing knowledge of anatomy, physiology and equipment design; these individuals were artist, scientist, and engineer in one. Da Vinci, along with being an exceptional artist, understood nature, mechanics, anatomy, physics, architecture, and weaponry. His designs for machines included a bicycle, helicopter, and an airplane. Borelli's book *de Motu Animalium*, published in 1680, applied the modern principle of scientific investigation by continuing Galileo's custom of testing hypotheses against observation. Borelli extended rigorous analytical methods to biology (he can be called the "father of biomechanics"); he was the first to understand that the levers of the human musculoskeletal system favor quick motion rather than force.

In the 18th century, the sciences of anatomy and physiology diversified and amassed specific detailed knowledge. Psychology began to develop as a distinct field of study. Well into the 19th century, the sciences tended to be oriented toward theories: the stereotype is the scientist in a white coat who devotes his life to research in the laboratory. But increasing industrialization with its employment of human workers, focused interest on "applied" aspects of the "pure" sciences.[1] In the early 1800s, in France, Lavoisier, Duchenne, Amar, and Dunod researched energy capabilities of the human body. Marey developed methods to describe human motions at work. Bedaux made studies to determine work and payment systems before Taylor and the Gilbreths did similar work in the United States in the early 1900s. In England, the Industrial Fatigue Research Board considered theoretical and practical aspects of the human at work. In Italy, Mosso constructed dynamometers and ergometers to research fatigue. In Scandinavia, Johannsson and Tigerstedt developed the scientific disciplines of work physiology. In 1913, Rubner founded a Work Physiology Institute in Germany.

In the United States, Benedict and Cathcard described the efficiencies of muscular work in 1913. The Harvard Fatigue Laboratory was established in 1927.

In the first half of the 20th century, applied physiology and psychology were well advanced and widely recognized, both in their theoretical research "to study human characteristics" and in the application of this knowledge "for the appropriate design of the living and work environment." Two distinct approaches to studying human characteristics had developed: one concerned chiefly with physiological and physical properties of the human, the other interested mainly in psychological and social traits. Although there was much overlap between these approaches, the physical and physiological aspects were studied mainly in Europe and the psychological and social aspects and North America.

Directions in Europe

Based on a broad fundament of anatomical, anthropological and physiological research, "work" physiology assumed great importance in Europe, particularly in the hunger years associated with the First World War. Marginal living conditions stimulated research on such topics as the minimal nutrition required to perform certain activities; the consumption of energy while doing agricultural, industrial, military and household tasks; the relationships between energy consumption and heart rate; the assessment of muscular capabilities; suitable body postures at work; the design of equipment and work stations to fit the human body. Another development in the 1920s was psycho-technology, which involved testing individuals for their ability to perform physical and mental work, their vigilance and attention, their ability to carry mental workload, their behavior as drivers of vehicles, their ability to read road signs, and related topics.

Directions in North America

Most psychologists around 1900 were strictly scientific and deliberately avoided studying problems that strayed outside the boundaries of pure research. Some investigators, however, pursued practical concerns, such as sending and receiving Morse code, measuring perception and attention at work, using psychology in advertising, and promoting industrial efficiency.

A particularly important step was the development of intelligence testing,[2] used to screen military recruits during the First World War and, later, to select industrial workers for their mental capabilities which certain jobs required. The concept of industrial psychology won acceptance.

Some of the best-known, most puzzling findings in industrial psychology resulted from experiments at the Hawthorne Works near Chicago in the mid-1920s. The experiments were designed to assess relationships between efficiency and lighting in workrooms where electrical equipment was produced. The bizarre finding was that the workers' productivity increased whenever the illumination was changed, regardless of increase or decrease; apparently, productivity responded to the attention paid to the workers by the researchers. This phenomenon became known as the observer or Hawthorne Effect.[3]

Industrial psychology[4] divided into special branches, including personnel psychology, organizational behavior, industrial relations, and engineering psychology. Under the pressures of the Second World War, the "human factor" as part of a "man-machine system" became a major concern. Technological development lead to machines and systems that put higher demands on the attention, endurance, and strength of individuals and teams than many could muster. For example: operators had to observe radar screens over periods of many hours, with the intent of detecting and distinguishing some blips from others. In high-performance aircraft, the pilot had to endure forceful accelerations, for instance in tight turns and in steep dives or climbs. In these cases, the pilot might be unable to operate hand controls properly and could even black out. Crew members had to fit into tiny tank and aircraft cockpits, and (from the 1960s on) into spacecraft, which required that small persons be selected from expert aviators. Stressful conditions make it difficult to maintain morale and performance. New tasks and machinery generate new needs to consider human physique and psychology, purposefully, and knowingly, in the design of jobs, equipment, and environments.

Names for the Discipline: "Ergonomics" and "Human Factors"

Early on, in Europe and North America, anthropologists, physiologists, psychologists, sociologists, statisticians, and engineers used various terms to describe their activities of studying the human and applying the information obtained in design, selection, and training.

In January of 1950, British researchers met in Cambridge, England, to select a name for the new society to represent their activities. Among others, the term "ergonomic" was proposed.[5] In late 1949, KFH Murrell had derived that word from the Greek terms *ergon*, indicating work and effort, and *nomos*, meaning law or usage; apparently, he re-invented a word already used by W. Jastrzebowski in Poland nearly a hundred years earlier. That term was neutral; it implied no priority among contributing disciplines; it was easily remembered and recognized and could be used in any language. "Ergonomics" was formally accepted as the name of the new society at its council meeting in early 1950.[6]

In the United States, a group convened in 1956 to establish a formal society. They selected "human factors" instead of "ergonomics." Often, the word "engineering" is added or substituted to indicate applications, as in "human (factors) engineering". In 1992, the Human Factors Society renamed itself the Human Factors and Ergonomics Society[7] HFES with (in 2016) about 4500 individual members.[8]

The International Ergonomics Association IEA has 54 member societies (in 2016[9]) with 28 in Europe, 13 in Asia, 9 in the Americas and 4 in Africa and the Oceania regions.

There has been some discussion of whether human factors differs from ergonomics—whether one relies more heavily on psychology or on physiology, or is more theoretical or practical than the other. Today, the two terms are usually considered synonymous: the Canadian Society uses "human factors" in its English name, and "ergonomie" in its French version.

TODAY'S ERGONOMIC KNOWLEDGE BASE

The development of space travel first forced humans to fold themselves into minuscule capsules; now people live and function in near weightlessness for months. Today, many professional activities involve long hours using computers of one kind or another as the primary work tools. Yet, heavy physical work still persists in many jobs in industry, in commercial fishing, in agriculture and forestry—some hard work is even new, such as in airline baggage handling.

The field of ergonomics/human factors continues to grow and to change, driven by new technologies and by the resulting new tasks for people. Classic sciences still provide fundamental information about human beings (Fig. 1). The anthropological basis consists of Anatomy, describing the build of the human body; Orthopedics, concerned with the skeletal system; Physiology,

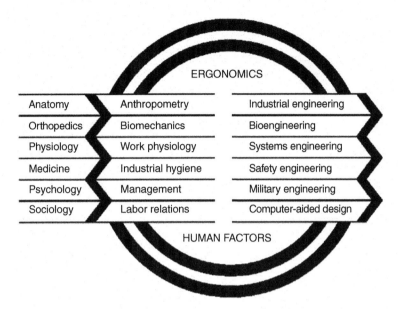

FIGURE 1   Origins, developments, and applications of ergonomics/human factors.

dealing with the functions and activities of living body, including the physical and chemical processes involved; Medicine, concerned with injuries and illnesses and their prevention and healing; Psychology, the science of mind and behavior; and Sociology, concerned with the development, structure, interaction, and behavior of individuals or groups. Of course, physics, chemistry, statistics, and mathematical modeling supply methodology and knowledge.

Several applied disciplines developed from these basic sciences; these include primarily Anthropometry, the measuring and description of the physical dimensions of the human body; Biomechanics (Anthromechanics), describing the physical behavior of the body in mechanical terms; Industrial Hygiene, concerned with the control of occupational health hazards; Industrial Psychology, discussing people's attitudes and behavior at work; Management, dealing with and coordinating the intentions of the employer and the employees; and Work Physiology, applying physical knowledge and measuring techniques to the body at work. Of course, associated disciplines such as Labor Relations have developed which are also part of, or contribute to, or overlap with, ergonomics.

Their topical areas overlap and intertwine; their research produces practical applications in such fields as Industrial Engineering, also called Integrated Engineering, concerned with the interactions among people, machinery, and energies; Bioengineering, working to replace worn or damaged body parts; Systems Engineering, in which the human is an important component of the overall work unit; Safety Engineering and Industrial Hygiene, which focus on the well-being of humans; and Military Engineering, which relies on the human as soldier or operator. As computers have become integrated into all aspects of modern day life, computer-aided design is used in all aspects of engineering. Naturally, other application disciplines also rely on ergonomic knowledge and data, such as oceanographic, aeronautical, and astronautical engineering.

GOALS OF ERGONOMICS

The (US) National Research Council asserted in 1983[10] that "design begins with an understanding of the user's role in overall system performance and that systems exist to serve their users, whether they are consumers, system operators, production workers, or maintenance crews. This user-oriented design philosophy acknowledges human variability as a design parameter. The resultant designs incorporate features that take advantage of unique human capabilities as well as built-in safeguards." The human is never indentured to the system but is the beneficiary, the passenger, the participant, the operator, the supervisor, the controller, and the decision maker.

There is a hierarchy of goals in ergonomics. The most essential and basic task is to generate *tolerable* working conditions that do not pose unavoidable dangers to human life or health. When this basic requirement is assured, the next goal is to generate *acceptable* conditions to which the people involved can voluntarily agree. The final goal is to generate *optimal* conditions that are so well adapted to human characteristics, capabilities, and desires that physical, mental, and social well-being is achieved.

ERGONOMICS DEFINED

Ergonomics (human factors, human engineering, human factors engineering) is the study of human characteristics for the appropriate design of the living and work environment. Its fundamental aim is that all man-made tools, devices, equipment, machinery, and environments advance, directly or indirectly, well-being and performance of people. As more knowledge about humans becomes available, as novel opportunities develop to apply human capabilities in modern systems, and as new needs arise for protecting the person from outside events, ergonomics changes and expands.

NOTES

1 For more details, see McFarland (1946), Chapanis, Garner, and Morgan (1949), Lehmann (1953, 1962), Floyd and Welford (1954), Woodson (1954), Brouha (1960, 1967), Grandjean (1963), and Scherrer (1967).

2 Gould (1981) provides a partly amusing, partly disturbing account of the early years of such testing.

3 Roethlisberger and Dickson (1943), Parsons (1974, 1990), and Jones (1990).

4 The term first appeared as a misprint of "individual" psychology, as Muchinsky (1983), p. 12.

5 Edholm and Murrell (1974), Monod and Valentin (1979), and Koradecka (2000).

6 The original proposal for the name included two alternative suggestions. One was "Ergonomic Society." Note that there is no "s" following Ergonomic; the final "s" apparently slipped into the ballot and has made deriving an adjective or adverb difficult. The other alternative was "Human Research Society."

7 Christensen, Topmiller and Gill (1988), Kroemer (1993a,b), and Meister (1999).

8 Accessed 7 July 2016 from https://hfes.org.

9 Accessed 21 June 2016 from https://iea.cc/.

10 National Research Council, Committee on Human Factors (1983), Research needs for human factors, pp. 2–3, Washington, DC: National Academy Press.

The Ergonomic Knowledge Base

Size and Mobility of the Human Body

Overview

"How tall are you? What's your weight?" For us to design things that fit people, we must know the dimensions of their bodies; however, body builds can vary considerably among individuals. Measurements of body sizes are available for some populations, but for many people on Earth only estimates exist. If we have suitable statistical descriptors and know the relationships among various body data, we can calculate probable body dimensions; if not, we must measure. Understanding the properties, capabilities, and limitations of the body allows us to design equipment and tools that utilize and enhance human strengths.

1.1 HUMANS SPREADING OVER THE EARTH

Development of humans

The human species appears to have grown like a big bush: some branches at first develop but then dry up, while others grow more and more twigs, some of which may vanish while others flourish. We can trace the development of the human race by fossils and by reconstructing mitochondrial DNA over several million years in Africa, for hundreds of thousands of years in Europe and Asia, and for some 20,000 years in the Americas.

Homo in Africa

Current paleoanthropology suggests that *Australopithecine* was a predecessor (or possibly close evolutionary cousin) of the genus *Homo* about 3 million years ago in Africa, where *Homo erectus* then developed. One humanoid branch started about 250,000 years ago and remained in Africa. Another branch developed 60,000 or 70,000 years later. Some of its members stayed in Africa, others spread into Eurasia. Eventually, *Homo sapiens* spread all over the Earth.[1]

Neanderthals and Cro-Magnons

Remains of anatomically modern humans who lived 130,000–180,000 years ago have been found in South Africa and in the Levant. About 150,000 years ago, the *Neanderthals* emerged, primarily in central Europe. They apparently were stocky, heavy-set, and cold-adapted with a brain as big as our current one. For several thousand years, they existed side-by-side with *Cro-Magnons* but then vanished about 30,000 years ago. The Cro-Magnons grew into *Homo sapiens*.

Popular notions about the different appearances of Cro-Magnons and Neanderthals are mostly based on conjecture, often in the style of a commercial Hollywood-type movie. For example, there is no indication that the Cro-Magnons were dark skinned, or the Neanderthals light. Furthermore, there is no evidence of violent struggles for superiority between the two races.[2]

Humans spread from Africa

From Africa, *Homo sapiens* spread over the Earth. Roughly 50,000 years ago, Australia was settled by early humans who arrived from eastern Indonesia. Their descendants became the Aboriginal population. Most of the current inhabitants of Indonesia, the Philippines, and parts of Southeast Asia may be descendants of a population that emigrated from modern-day Taiwan about 4000–6000 years ago.[3]

Emigrants from Asia in the Americas

Waves of peoples, the earliest around 20,000 years ago, crossed what was then the Bering land bridge from East Asia to modern-day Alaska. Some moved into the areas of today's Canada and the United States, others followed the pacific coastal areas into

South America. Their descendants populated the entire hemisphere, becoming the ancestors of North, Central, and South American native peoples.

After its old history of Neanderthals and Cro-Magnons, the European population was reconstituted twice fairly recently: around 8000 years ago by people from the Near East and then 2000 years later by Indo-Europeans from southern Russia.

Europe

Thus, the human stock with its many current branches appears African in origin and about a quarter-million years old. Today, the number of people is growing fast; "population explosions" are occurring in some parts of the Earth. The total number of humans was about 10^9 (1000 million, or 1 billion) around 1800. In 1900, about 1.7 billion people lived on Earth. The second billion was reached by 1930. The third billion was present in 1960, the fifth in 1987. In 1998, about 5.8 billion people lived on Earth, including 4.6 billion in developing countries. In 2017, the total human population on Earth has been estimated to be 7.5 billion. If current birth and death rates continue, about 10 billion people will live on Earth in 2050; projections beyond this time vary widely.

Population growth

Emigration from certain areas and immigration to others are on a much smaller scale than population growth but can be locally of great importance. In North America, for example, during the last few centuries, waves of immigrants from certain geographical areas have been changing the composition of the inhabitant population, replacing most native peoples with Europeans. In today's United States, the influx of Cubans and Haitians is strongly felt in Florida, the arrival of Central and South Americans affects mostly southwestern states, and Asians are very evident along the Pacific coast.

Local population changes

1.2 ANTHROPOLOGY AND ANTHROPOMETRY

Anthropology, the study of mankind, was primarily philosophical and aesthetic in nature until about the middle of the 19th century. Size, proportions, and appearance of the human body have always been of interest to artists, warriors, and physicians. In the mid-1800s, Adolphe Quetelet applied statistics to anthropological data of body measurements. This was the beginning of modern anthropometry, the measurement and description of the human body. By the end of the 19th century, anthropometry was a widely applied scientific discipline, used both in measuring the bones of early people and in assessing the body sizes and proportions of contemporaries. A new offspring, biomechanics (related to structure and movement), emerged. Today, engineers have become highly interested in the application of anthropometric and anthromechanical information.[4]

Measurement of the human body

Unification of measuring methods became necessary and was achieved primarily by anthropologists who convened 1906 in Monaco and 1912 in Geneva. They established landmarks on the body, mostly on bones, to and from which to take measurements. In 1914, Rudolf Martin published his *Lehrbuch der Anthropologie*. The several editions of this authoritative textbook shaped the discipline for decades. Beginning in the 1960s, increasing engineering needs for anthropometric information, newly developing measuring techniques, and advanced statistical considerations stimulated the need for updated standardization. Since the 1980s, the International Standardization Organization (ISO) has established conforming anthropometric techniques worldwide.[5]

Standardized measurements

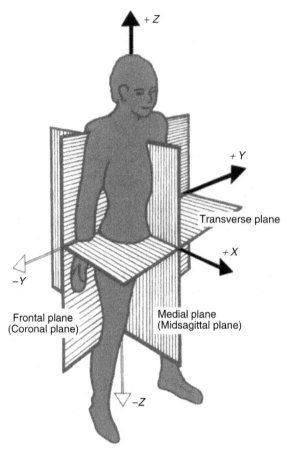

FIGURE 1.1   **Reference planes in anthropometry.** Note the directions of x, y, and z—see text.

1.2.1 Measurement Techniques

Most body measurements are defined by the two endpoints of the distance measured. For example, forearm length is the elbow-to-fingertip distance; stature (height) starts at the floor on which the subject stands and extends to the highest point on the skull.

Reference planes

Fig. 1.1 shows the three standard reference planes: the medial (mid-sagittal), the frontal (or coronal), and the transverse planes, all at 90 degrees to each other and usually set to meet in the center of mass of the whole body. A Cartesian coordinate system is seldom employed in anthropometry but used routinely in anthromechanics (see Chapters 2 and 6)[6]; the convention is to set the +x axis pointing forward from the subject, +y to the subject's left, and +z upward. The origin of the axes may be moved to any point of the body, such as to a hand, if appropriate for modeling.

Body length

The classic reference posture of a person standing upright is similar to that shown in Fig. 1.1 but with the heels of the feet kept together, called "anatomical position." Fig. 1.2 lists other descriptive terms.

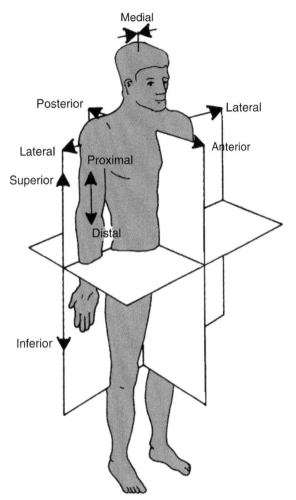

FIGURE 1.2 **Terminology used in anthropometry.** *Source: Adapted from Kroemer, Kroemer, and Kroemer (2010).*

For measurement of stature, the subjects may assume one of four customary positions: standing naturally upright; standing stretched to maximum height; leaning against a wall with the back flattened and buttocks, shoulders, and back of the head touching the wall; or lying on the back. Lying supine results in the largest measure; it is mostly used with infants who are unable to stand. The difference between measures when the standing subject either stretches or just stands upright can easily be 2 cm or more. This demonstrates that standardization is needed to assure uniform postures and comparable results.[7]

1.2.2 Body Posture during Measurements

For most measurements, the subject stands upright with body segments in line with each other or at right angles. When measurements are taken on a seated person, the flat

Body posture

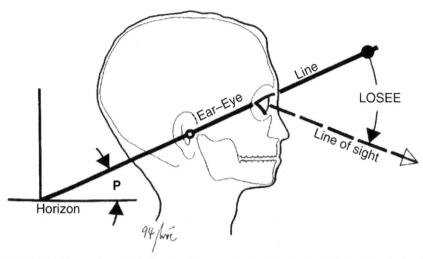

FIGURE 1.3　**The Ear–Eye Line describes head posture in the lateral view.** The E–E Line is also the reference for the angle of the line of sight, LOSEE, in the medial plane.

and horizontal surfaces of feet and foot support are so arranged that the thighs are horizontal, the lower legs vertical, and the feet flat on their horizontal support. The subject is nude, or nearly so, and does not wear shoes.

Head posture　　If measurements include the head, it is positioned upright ("erect"), with the pupils on the same horizontal level. The traditional term for this posture was "head in the Frankfurt Plane," but an easier description is by the Ear–Eye (E–E) Line: it passes through the right ear hole and the outside juncture of the lids of the right eye. When the head is upright, the E–E Line is angled by about 11 degrees above the horizon, as shown in Fig. 1.3.

1.2.3　Body Measures

> *Height* is a straight-line, point-to-point vertical measurement.
> *Breadth* is a straight-line, point-to-point horizontal measurement running across the body or a segment.
> *Depth* is a straight-line, point-to-point horizontal measurement running fore-aft the body.
> *Distance* is a straight-line, point-to-point measurement between landmarks on the body.
> *Curvature* is a point-to-point measurement following a contour; this measurement is usually neither closed nor circular.
> *Circumference* is a closed measurement that follows a body contour; hence this measurement is usually not circular.
> *Reach* is a point-to-point measurement following the long axis of the arm or leg.
> All measurements are taken on people maintaining prescribed erect standing or sitting stances.

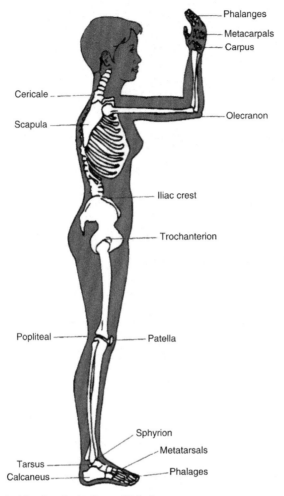

FIGURE 1.4 Anatomical landmarks in the sagittal view.

Figs. 1.4 and 1.5 illustrate anatomical landmarks on the human body which serve as starting and endpoints for body measures.

1.2.4 Classical Measuring Techniques

The *Morant technique* used a set of measurement grids, usually attached to the inside corner of two vertical walls meeting at right angles. The subject was placed in front of the grids, and projections of the body onto the grids were used to determine anthropometric values. Related box-like jigs with grids are still in some use for determining foot dimensions.

Grid technique

In the classical method, the anthropometrist's hands guide special measuring instruments to the bony landmarks on the body of the subject. The largest instrument is the *anthropometer*, a graduated rod with a sliding edge at right angle, used to measure long

Hand-held measuring devices

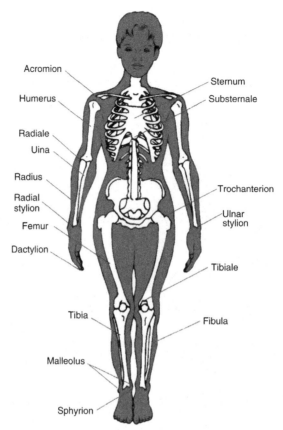

FIGURE 1.5   **Anatomical landmarks in the frontal view.**

straight distances. The *spreading caliper* consists of two curved branches joined in a hinge, used to determine the distance between the tips of the branches. A small *sliding caliper* serves to take short measurements, such as finger thickness or finger length. Thickness of skinfolds, a measure of body fat, is determined with a *skinfold caliper*. A *cone* serves to measure the diameter around which fingers can close. Circumferences and curvatures are measured with a *tape*.

Simple but clumsy

Measuring body dimensions with these traditional instruments is simple in principle, but requires experience and skill and consumes a great deal of time. Each measurement and tool must be selected in advance, and what was not measured in the test session remains as an unknown. A major shortcoming of the classical technique is that most body measures appear unrelated to each other in space. For example, as one looks at a subject from the side, stature, eye height, and shoulder height are located in different yet undefined frontal planes.

1.2.5 New Measurement Techniques

Photography

Photographs can record all three-dimensional (3D) aspects of the human body. They allow the recording of practically infinite numbers of measurements, taken from the recording at one's convenience. However, photographs also have drawbacks: the body is depicted in two dimensions; a scale may be difficult to establish; parallax distortions occur; and bony landmarks under the skin cannot be palpated on the photograph. For these and other reasons, two-dimensional photographic and video anthropometry have not been widely used.

An electromechanical probe[8] can be placed on body landmarks and their locations can be registered in three dimensions. Computerized data storage and processing allows 3D identification of body points and of their changes by motion.

Placing a probe

A laser can be used as a distance-measuring device to determine the shape of irregular bodies. Measurements are taken by either rotating the body to be measured, or by rotating the sending and receiving units of the laser device around the body. Markers may be placed on points of the body surface so that the laser can recognize them, for example to indicate the location of a bone landmark. Scanning[9] is fast and collects vast amounts of measurements. Computerized data storage and processing allows detailed 3D descriptions of the body and of changes due to motion, training, or aging.

Body scanning

1.3 AVAILABLE ANTHROPOMETRIC INFORMATION

In the past, interest in the body build of populations other than one's own group was based mostly on curiosity and general wish to know; however, as industry and marketing began to reach around the globe, body size became a matter of practical interest to designers and engineers. Around 1970, first compilations of worldwide anthropometric information appeared.[10] Since then, an increasing number of publications describe national populations. Modern measuring technology and sophisticated statistics allow anthropometric assessment of populations to be taken rapidly and accurately.

Anthropometric sourcebooks

Table 1.1 shows 1990 estimates of stature of the Earth's population, divided into regional groups. Table 1.2 contains measured data on specific population samples. These data are included only to demonstrate the variability that may be encountered, and not to reflect current populations: in many cases the surveys were done decades ago and included only small groups.[11]

Earth's population

1.3.1 Variability

Anthropometric data show considerable variability stemming from four sources: variability in the measurement itself, variability within the individual, variability between individuals, and secular variability.

1.3.1.1 *Measurement Variability*

Varying degrees of care can be exercised in selecting population samples, using measurement instruments, storing the measured data, and applying statistical treatments. Depending on the care taken during measurements, the resulting information may be quite variable. An exceptionally large standard deviation (see statistics later in this chapter) of a data set can be a warning signal.

Poor data

TABLE 1.1   Estimates of Average Stature (in cm) in 20 Regions of the Earth

Region	Females	Males
Africa		
North	161	169
West	153	167
Southeastern	157	168
America, South/Central		
Native Indians	148	162
European and African extraction	162	175
America, North	165	179
Asia		
North	159	169
Southeast	153	163
Near East	157	168
Australia		
European extraction	167	177
China		
South	152	166
Europe		
Northern	169	181
Central	166	177
Eastern	163	175
Southeast	162	173
France	163	177
Iberian Peninsula	160	171
India		
Northern	154	167
Southern	150	162
Japan	159	172

Source: Adapted from Jürgens, Aune, and Pieper (1990).

1.3.1.2 Intraindividual Variability

Changes over time

An individual's body size changes from youth to age, and it depends on nutrition, physical exercise, and health as well as genetics. Such changes in the same individual become apparent in *longitudinal* studies, in which an individual is observed over years and decades. Most (but not all) such changes with age follow the scheme shown in Fig. 1.6. During childhood and adolescence, body dimensions such as stature increase rapidly. From early 20s into the 50s, little change occurs and in general, stature remain almost steady. From the sixth decade on, many dimensions decline, while others—such as weight or bone circumference—often increase.

1.3.1.3 Interindividual Variability

People differ

Individuals differ from each other in body proportions and in specific measurements such as arm length, weight, and height. Data describing a population sample are usually collected in a *cross-sectional* study, in which every subject is measured at about the same moment in time. This means that people of different ages, nutrition, fitness, etc.,

TABLE 1.2 Measured Heights and Weights of Adults: Averages (with Standard Deviations when Available)

Measured where, group (Date of publication)	Sample size	Stature (mm)	Weight (kg)
Algeria: Females (1990)	666	1576 (56)	61 (13)
Brazil: Males (1988)	3076	1699 (67)	nda
Cameroon: Urban Females (2006)	1156	1620	64
Urban males, 35–44 years old (2006)	558	1721	75
China: Females (Taiwan) (2002)	about 600	1572 (53)	52 (7)
Males (Taiwan) (2002)	about 600	1705 (59)	67 (9)
France: Females (2006)	5510	1625 (71)	62 (12)
Males (2006)	3986	1756 (77)	77 (13)
Germany: Female army applicants (2006)	301	1674	64
Male army applicants (2006)	1036	1795	75
Great Britain: Females (2001)	3870	1611	68
Males (2001)	3233	1746	81
India: Females (1997)	251	1523 (66)	50 (10)
Males (1997)	710	1650 (70)	57 (11)
East-Ctr. India male farm workers (2002)	300	1638 (56)	57 (7)
East India male farm workers (1997)	134	1621 (58)	54 (7)
South India male workers (1992)	128	1607 (60)	57 (5)
North-E. (Assam) Male farm workers (2016)	130	1628 (46)	55 (7)
Iran: Female students (1997)	74	1597 (58)	56 (10)
Male students (1997)	105	1725 (58)	66 (10)
Ireland: Males (1991)	164	1731 (58)	74 (9)
Italy: Females (1991)	753	1610 (64)	58 (8)
Females (2002)	386	1611 (62)	58 (9)
Males (1991)	913	1733 (71)	75 (10)
Males (2002)	410	1736 (67)	73 (11)
Japan: Females (1990)	240	1584 (50)	54 (6)
Males (1990)	248	1688 (55)	66 (8)
Netherlands: Females 18–65 years old (2002)	691	1679 (75)	73 (16)
Males 18–65 years old (2002)	564	1813 (90)	84 (16)
Russia: Female herders (ethnic Asians)	246	1588 (55)	nda
Female students (Russians)	207	1637 (57)	61 (8)
Female students (Uzbeks)	164	1578 (49)	56 (7)
Fem. factory workers (Russians)	205	1606 (53)	61 (8)
Fem. factory workers (Uzbeks)	301	1580 (54)	58 (9)
Male students (Russians)	166	1757 (56)	71 (9)
Male students (Uzbeks)	150	1700 (52)	65 (7)
Male factory workers (Russians)	192	1736 (61)	72 (10)
Male factory workers (ethnic mix)	150	1700 (59)	68 (8)
Male farm mechanics (Asians)	520	1704 (58)	64 (8)
Male coal miners (Russians)	150	1801 (61)	nda
Male construction workers (Russians) (all 1999)	150	1707 (69)	nda

(Continued)

I. THE ERGONOMIC KNOWLEDGE BASE

TABLE 1.2   Measured Heights and Weights of Adults: Averages (with Standard Deviations when Available) (*cont.*)

Measured where, group (Date of publication)	Sample size	Stature (mm)	Weight (kg)
Saudi Arabia: Males (1986)	1440	1675 (61)	nda
Singapore: Male pilot trainees (1995)	832	1685 (53)	nda
Sri Lanka: Females (1985)	287	1523 (59)	nda
Males (1985)	435	1639 (63)	nda
Thailand: Females (1991)	250	1512 (48)	nda
Females (1991)	711	1540 (50)	nda
Males (1991)	250	1607 (20)	nda
Males (1991)	1478	1654 (59)	nda
Turkey: Male soldiers (1991)	5108	1702 (60)	63 (7)
U.S.A.: Females (2004)	about 3800	1625	75
Males (2004)	about 3800	1762	87
Female army soldiers (1989)	2208	1629 (64)	62 (8)
Female army soldiers (2014)	1986	1626 (64)	67 (11)
Male army soldiers (1989)	1774	1756 (67)	76 (11)
Male army soldiers (2009)	1475	1760 (nda)	84 (nda)
Male Army soldiers (2014)	4082	1755 (67)	85 (14)
Canadian and US Females 18–26 years old	1264	1640 (73)	69 (18)
(2002)	1127	1778 (79)	86 (18)
Canadian and US Males 18–65 years old	30	1559 (61)	49
(2002)	41	1646 (60)	59
Vietnamese females living in the USA (1993)			
Vietnamese males living in the USA (1993)			

Abbreviation: *nda*, no data available.

Source: Adapted from Kroemer (2017) who listed all sources except Patel et al. (2016) for North-East (Assam) Indian farm workers.

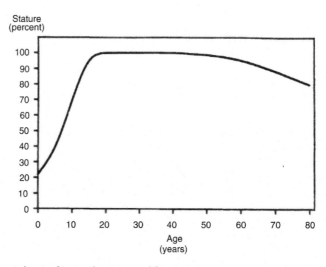

FIGURE 1.6   Approximate changes in stature with age.

are included in the sample set. Most textbooks, including this one, contain data gathered in cross-sectional studies.

1.3.1.4 Secular Variations

Factual and anecdotal evidence shows that people in present day are larger, on average, than their ancestors; yet, reliable anthropometric information on this development is available only for about the last 100 years. During the latter half of the 20th century, stature increased in North America and in Europe by about 1 cm per decade, on average. The probable reason was that improved nutrition and hygiene have allowed individuals to achieve more of their genetically determined body size potential. If this explanation is correct, then the rate of increase should slowly taper off until a final body size is reached: indeed, for stature[12] in the United States, this seems to happening. Also during the latter half of the 20th century, body weight increased by about 2 kg per decade, but since about 2010, that increase has risen dramatically and has generated concerns about obesity in many countries.

Trends taller and fatter

Most reliable data depicting secular trends are from military surveys. However, soldiers constitute a selected sample of the general population: for example, they exclude people older than about 50 years as well as people who are unusual in their body dimensions, such as extremely short or tall. Further, only fairly healthy individuals are included.[13]

Military surveys

1.3.2 Population Samples

The Human Biometry Data Bank at the University René Descartes in Paris contains extensive anthropometric information on Europeans. Publications of the World Health Organization (WHO) and ISO standards contain some international body size data. Unfortunately, sufficient information on most populations on Earth is missing. Available anthropometric data, such as listed in Table 1.2, are usually limited to a few dimensions, commonly stature and weight, often measured on small population samples.

Civilians' body sizes

Body dimensions of soldiers have long been of interest for a variety of reasons, among them to provide uniforms, armor, and equipment. Armies have personnel willing to and capable of performing body measurements on large numbers of soldiers, available on command. Hence, anthropometric information about soldiers has a long history and is rather complete. For example, the US military anthropometric data bank contains the data of about 100 surveys from many nations, though most on US military personnel.

Body sizes of soldiers

Among the US military services, the Army is the largest and anthropometrically least biased sample of the total US adult population. Therefore, the body dimensions of the Army often serve as useful estimates for the general North American adult population. However, the user of this information must realize that the military consists of relatively young and healthy individuals. Thus, civilian data are likely to differ somewhat; only the soldiers' heads, hands, and feet are considered similar to civilians' dimensions.[14] It is of some interest to note that even the military data show strong increases in body weight from 1988 to 2012,[15] accompanied by enlarged trunk depth measures and associated larger circumferences especially of the trunk and thighs—apparently, the worldwide obesity trend pertains both to soldiers and civilians.

US Army body sizes

Table 1.3 provides anthropometric data excerpted from surveys done on Chinese in Taiwan measured between 1996 and 2000; on Russians in Moscow measured between

Body sizes in China, Russia and United States

TABLE 1.3 Common Body Measurements and Their Applications

Dimensions, applications	Population	Females mean (SD)	Males mean (SD)
1. Stature A main measure for comparing population samples. Reference for the minimal height of overhead obstructions; add to height to allow for clearance, head covering, shoes, stride	Chinese (Taiwan) Russians (Moscow) US Army soldiers	1572 (53) 1637 (57) 1629 (64)	1705 (59) 1757 (56) 1756 (69)
2. Eye height, standing Origin of the visual field of a standing person. Reference for the location of visual obstructions and of visual targets such as displays; consider slump and motion	Chinese (Taiwan) Russians (Moscow) US Army soldiers	nda 1526 (57) 1520 (62)	nda 1637 (55) 1642 (67)
3. Shoulder height (acromion), standing Starting point for arm length measurements; near the center of rotation of the upper arm. Reference point for hand reaches; consider slump and motion	Chinese (Taiwan) Russians (Moscow) US Army soldiers	1285 (50) 1334 (54) 1335 (58)	1396 (53) 1440 (54) 1441 (63)
4. Elbow height, standing Reference for height and distance of the work area of the hand and the location of controls and fixtures; consider slump and motion	Chinese (Taiwan) Russians (Moscow) US Army soldiers	978 (38) 1010 (42) 1004 (45)	1059 (40) 1083 (48) 1083 (50)
5. Hip height (trochanter), standing Traditional anthropometric measure, indicator of leg length and the height of the hip joint. Used for comparing population samples	Chinese (Taiwan) Russians (Moscow) US Army soldiers	802 (41) nda 845 (45)	860 (48) nda 901 (49)
6. Knuckle height, standing Reference for low locations of controls, handles, and handrails; consider slump and motion of the standing person	Chinese (Taiwan) Russians (Moscow) US Army soldiers	708 (33) 731 (34) nda	757 (32) 773 (39) nda
7. Fingertip height (dactylion), standing Reference for the lowest locations of controls, handles, and handrails; consider slump and motion of the standing person	Chinese (Taiwan) Russians (Moscow) US Army soldiers	618 (32) 635 (32) 613 (35)	659 (30) 668 (37) 654 (37)
8. Sitting height Reference for the minimal height of overhead obstructions; add to height to allow for clearance, head covering, trunk motion of the seated person	Chinese (Taiwan) Russians (Moscow) US Army soldiers	846 (32) 859 (32) 857 (33)	910 (30) 912 (32) 918 (36)
9. Sitting eye height Origin of the visual field of a seated person. Reference point for the location of visual targets such as displays; consider slump and motion	Chinese (Taiwan) Russians (Moscow) US Army soldiers	732 (31) 742 (29) 748 (30)	791 (29) 790 (33) 805 (33)
10. Sitting shoulder height (acromion) Starting point for arm length measurements; near the center of rotation of the upper arm. Reference for hand reaches; consider slump and motion	Chinese (Taiwan) Russians (Moscow) US Army soldiers	561 (27) nda 563 (29)	602 (26) nda 603 (31)
11. Sitting elbow height Reference for the height of an armrest, of the work area of the hand and of keyboard and controls; consider slump and motion of the seated person	Chinese (Taiwan) Russians (Moscow) US Army soldiers	252 (25) 236 (24) 232 (26)	264 (24) 243 (25) 245 (29)
12. Sitting thigh height (clearance) Reference for the minimal clearance needed between seat pan and the underside of a table or desk; add clearance for clothing and motions	Chinese (Taiwan) Russians (Moscow) US Army soldiers	nda 148 (14) 168 (14)	nda 151 (18) 181 (16)
13. Sitting knee height Traditional anthropometric measure for lower leg length. Reference for the minimal clearance below the underside of a table or desk; add shoe height	Chinese (Taiwan) Russians (Moscow) US Army soldiers	471 (24 527 (24) 511 (27)	521 (29) 562 (25) 554 (28)
14. Sitting popliteal height Reference for the height of a seat; add shoe height	Chinese (Taiwan) Russians (Moscow) US Army soldiers	379 (18) 423 (23) 388 (24)	411 (19) 468 (24) 430 (25)

TABLE 1.3 Common Body Measurements and Their Applications (*cont.*)

Dimensions, applications	Population	Females mean (SD)	Males mean (SD)
15. Shoulder-elbow length	Chinese (Taiwan)	309 (18)	338 (19)
Traditional anthropometric measure for comparing population samples	Russians (Moscow)	nda	nda
	US Army soldiers	334 (17)	364 (18)
16. Elbow-fingertip length	Chinese (Taiwan)	384 (27)	427 (27)
Traditional anthropometric measure. Reference for fingertip reach when moving	Russians (Moscow)	nda	nda
the forearm in the elbow	US Army soldiers	440 (23)	480 (23)
17. Overhead grip reach, sitting	Chinese (Taiwan)	1105 (44)	1208 (49)
Reference for the height of overhead controls operated by a seated person.	Russians (Moscow)	1169 (46)	1276 (47)
Consider ease of motion, reach, and finger/hand/arm strength	US Army soldiers	1196 (62)	1303 (68)
18. Overhead grip reach, standing	Chinese (Taiwan)	1831 (67)	2002 (79)
Reference for the height of overhead controls operated by a standing person.	Russians (Moscow)	nda	nda
Add shoe height. Consider ease of motion, reach, and strength	US Army soldiers	1968 (98)	2141 (104)
19. Forward grip reach	Chinese (Taiwan)	651 (33)	710 (36)
Reference for forward reach distance. Consider ease of motion, reach, and finger/	Russians (Moscow)	702 (37)	759 (38)
hand/arm strength	US Army soldiers	693 (43)	757 (44)
20. Arm length, vertical	Chinese (Taiwan)	669 (31)	738 (33)
A traditional measure for comparing population samples. Reference for the	Russians (Moscow)	nda	nda
location of controls very low on the operator's side. Consider ease of motion,	US Army soldiers	722 (37)	787 (39)
reach, strength			
21. Downward grip reach	Chinese (Taiwan)	nda	nda
Reference for the location of controls low on the side of the operator. Consider	Russians (Moscow)	nda	nda
ease of motion, reach, and finger/hand/arm strength	US Army soldiers	607 (30)	663 (32)
22. Chest depth	Chinese (Taiwan)	213 (19)	217 (19)
A traditional measure for comparing population samples. Reference for the clearance	Russians (Moscow)	242 (21)	245 (20)
between seat backrest and the location of obstructions in front of the trunk	US Army soldiers	247 (27)	254 (26)
23. Abdominal depth, sitting	Chinese (Taiwan)	nda	nda
A traditional measure for comparing population samples. Reference for the clearance	Russians (Moscow)	nda	nda
between seat backrest and the location of obstructions in front of the trunk	US Army soldiers	223 (32)	255 (37)
24. Buttock-knee depth, sitting	Chinese (Taiwan)	530 (26)	558 (31)
Reference for the clearance between seat backrest and the location of obstructions	Russians (Moscow)	584 (29)	610 (30)
in front of the knees	US Army soldiers	591 (33)	618 (31)
25. Buttock-popliteal depth, sitting	Chinese (Taiwan)	nda	nda
Reference for the depth of a seat.	Russians (Moscow)	496 (29)	517 (26)
	US Army soldiers	485 (29)	503 (27)
26. Shoulder breadth (biacromial)	Chinese (Taiwan)	324 (25)	369 (28)
A traditional measure for comparing population samples. Indicator of the distance	Russians (Moscow)	360 (16)	397 (25)
between the centers of rotation of the two upper arms	US Army soldiers	365 (18)	416 (19)
27. Shoulder breadth (bideltoid)	Chinese (Taiwan)	406 (24)	460 (23)
Reference for the lateral clearance required near shoulder level. Add space for	Russians (Moscow)	412 (21)	458 (23)
ease of motion and tool use	US Army soldiers	450 (29)	510 (33)
28. Hip breadth, sitting	Chinese (Taiwan)	353 (23)	360 (27)
Reference for seat width. Add space for clothing and ease of motion	Russians (Moscow)	372 (23)	362 (23)
	US Army soldiers	399 (33)	379 (30)

(Continued)

I. THE ERGONOMIC KNOWLEDGE BASE

TABLE 1.3   Common Body Measurements and Their Applications (*cont.*)

Dimensions, applications	Population	Females mean (SD)	Males mean (SD)
29. Span A traditional measure for comparing population samples. Reference for sideway reach	Chinese (Taiwan) Russians (Moscow) US Army soldiers	1571 (62) 1640 (75) 1660 (83)	1738 (69) 1782 (68) 1814 (85)
30. Elbow span (arms akimbo) Reference for the lateral space needed at upper body level for ease of motion and tool use	Chinese (Taiwan) Russians (Moscow) US Army soldiers	801 (39) 870 (38) nda	894 (45) 935 (37) nda
31. Head length (depth) A traditional measure for comparing population samples. Reference for headgear size	Chinese (Taiwan) Russians (Moscow) US Army soldiers	187 (6) nda 190 (7)	197 (7) nda 196 (7)
32. Head breadth A traditional measure for comparing population samples. Reference for headgear size	Chinese (Taiwan) Russians (Moscow) US Army soldiers	161 (9) nda 148 (5)	167 (8) nda 154 (6)
33. Hand length A traditional measure for comparing population samples. Reference for hand tool and gear size. Consider manipulations, gloves, tool use	Chinese (Taiwan) Russians (Moscow) US Army soldiers	167 (8) 168 (8) 181 (10)	183 (10) 188 (9) 193 (10)
34. Hand breadth A traditional measure for comparing population samples. Reference for hand tool and gear size, and for an opening through which a hand must fit. Consider gloves, tool use	Chinese (Taiwan) Russians (Moscow) US Army soldiers	75 (4) 76 (3) 78 (4)	86 (5) 87 (5) 88 (4)
35. Foot length A traditional measure for comparing population samples. Reference for shoe and pedal size	Chinese (Taiwan) Russians (Moscow) US Army soldiers	nda 239 (11) 246 (12)	nda 266 (12) 271 (13)
36. Foot breadth A traditional measure for comparing population samples. Reference for shoe size, spacing of pedals	Chinese (Taiwan) Russians (Moscow) US Army soldiers	nda 88 (4) 93 (5)	nda 97 (6) 102 (5)
37. Weight (in kg) A traditional measure for comparing population samples. Reference for body size, clothing, strength, health, etc. Add weight for clothing and equipment worn on the body	Chinese (Taiwan) Russians (Moscow) US Army soldiers	52 (7) 60 (7) 68 (11)	67 (9) 71 (9) 86 (14)

Dimensions correspond to numbering in Figs. 1.7–1.9. All measures represent the mean (with standard deviation in parentheses) in *mm*, except weight in *kg*. Abbreviation: *nda*, no data available.
Gordon et al. (2014) provide exact definitions of the measurements.
Sources: Chinese (Taiwan), 25–34 years of age, measured between 1996 and 2000: Wang, Wang and Lin (2002). Russians, students in Moscow, 18–22 years of age, measured between 1984 and 1986: Strokina and Pakhomova (1999). US Army soldiers, 4082 males and 1986 females, 17–58 years of age, measured 2010, 2011, and 2012: Gordon et al. (2014).

1984 and 1986; and on US Army soldiers measured between 2010 and 2012. Figs. 1.7–1.9 depict the tabulated body dimensions.

1.4 ANTHROPOMETRIC STATISTICS

> The mean *m* (often called *average*), standard deviation *S*, and sample size *n* completely describe a normally distributed (Gaussian, bell-shaped) data set.

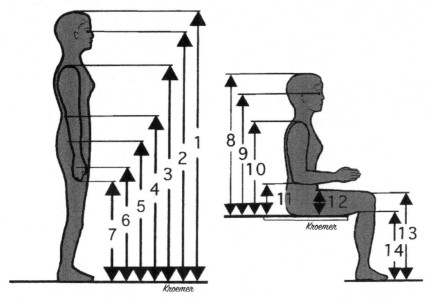

FIGURE 1.7   Measurements of heights; numbered as in Table 1.3.

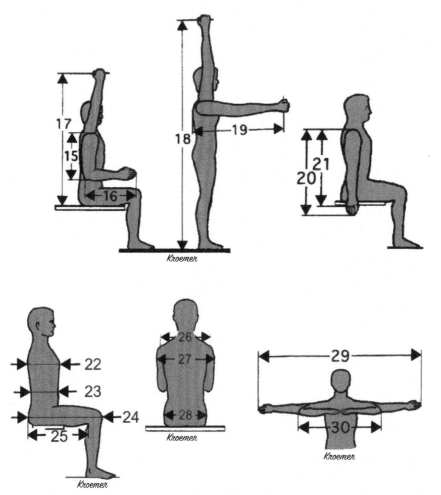

FIGURE 1.8   Measurements of reaches, heights, depths, breadths, and spans; numbered as in Table 1.3.

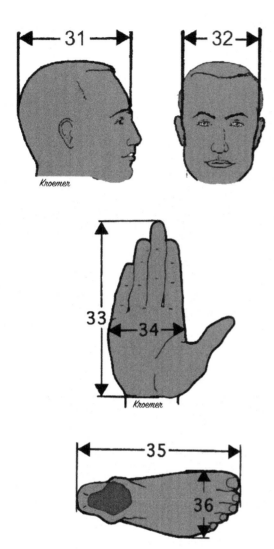

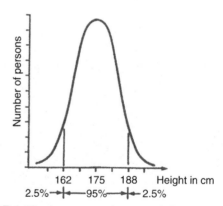

FIGURE 1.9 Measurements taken on head, hand, and foot; numbered as in Table 1.3.

FIGURE 1.10 A representative normal distribution of stature measurements: 95% of all data points lie between 162 and 188 cm.

Normal distributions

Conveniently, anthropometric data usually appear in a reasonably normal distribution, where all data points fall along a bell-shaped contour, symmetrical to the left and right of the average. (Unfortunately, there are some important exceptions: weight and muscle strength data are usually not normally distributed.) Fig. 1.10 sketches such a normal (or Gaussian) distribution where the measured heights appear in bell shape on both sides of the mean value, here about 175 cm.

Easy statistics

Most anthropometric data sets can be easily described by simple (parametric) statistics, listed in Table 1.4. The data cluster at the center of the set, symmetrically left and

TABLE 1.4 Statistical Formulas of Particular Use in Anthropometry

Measures of central tendency		
Mean, average m	$m = \Sigma x/n$ (where n is the number of data points)	E#1
Median	Middle value of values in numerical order	
Mode	Most often occurring value	
Measures of variability		
Range	$x_{max} - x_{min}$	E#2
Standard deviation S	$S = (variance)^{1/2} = [\Sigma(x - m)^2/n]^{1/2}$	E#3
Coefficient of variation CV	$CV = S/m$ (in percent: $CV = 100S/m$ %)	E#4
Standard error of the mean SE	$SE = S/n^{1/2}$	E#5
Symmetry (skewness) β_1	$\beta_1 = \Sigma(x - m)^3/nS^3$	E#6
Peakedness (kurtosis) β_2	$B_2 = \Sigma(x - m)^4/nS^4$	E#7
Percentile value location in a data distribution		
Percentile p	$p = m + kS$ (k from Table 1.5)	E#8
Sampling		
Sample size n *(for m)*	$n \geq (1.96\ CV)^2$ (CV in %)	E#9a
Sample size n (for 5th/95th p)	$n \geq (1.96\ CV)^2 (1.534)^2$ (CV in %)	E#9b
Measures of relations between variables x and y		
Correlation coefficient r between two samples x and y	$r = S_{xy}/(S_xS_y)$ $r = \Sigma[(x - m_x)(y - m_y)]/[\Sigma(x - m_x)^2\ \Sigma(y - m_y)^2]^{1/2}$	E#10
Coefficient of determination R	$R = r^2$	E#11
Regression	$y = a + b\,x$ $b = r\,S_y/S_x$	E#12 E#13
Ratio scaling		
Scaling factor E	$E = d_x/D_x = d_y/D_y$	E#14
Combining data sets		
Covariance COV	$COV_{x,y} = r_{xy}\,S_x\,S_y$	E#15
Sum of two mean values	$m_z = m_x + m_y$	E#16
Standard deviation of the sum	$S_z = \left(S_x^2 + S_y^2 + 2rS_xS_y\right)^{1/2}$	E#17
Difference of two mean values	$m_z = m_x - m_y$	E#18
Standard deviation of the difference	$S_z = \left(S_x^2 + S_y^2 - 2rS_xS_y\right)^{1/2}$	E#19
Factor k in a combined set	$k_z = n_x k_x + n_y k_y$	E#20
Percentile p in a combined set	$p_z = n_x p_x + n_y p_y$	E#21

right of the 50th percentile, which coincides with the mean m (the arithmetic average). The standard deviation S (or SD) describes the variability (spread) of all collected data; the sample size is n. Skewness β_1 is dimensionless; if its value is zero, the set of data is symmetrically distributed. Kurtosis β_2 is also dimensionless; if its value is 3, the data are normally distributed. (That value is set to zero in some other formulas.)

No "average person"

It's easy to calculate the average in a set of data, but that mean value is just one of the statistical descriptors of the distribution.[16] One cannot "design for the average" because that design would be too big for half the people and too small for the other half—consider the mean headaches caused by a doorway at mean stature. Furthermore, practically nobody is average in many or all features. People differ, and ergonomic design must accommodate variability.

Check the CV

One simple way to check on data diversity is to divide the standard deviation of the data in question by their mean to get the coefficient of variation (*CV*)—see E#4 in Table 1.4. For most body dimensions the *CV* is between 3% and 10%; larger values are suspect and should prompt a thorough examination of the data. (However, in most strength data the *CV* is between 10% and 85%.)

Use percentiles

The location of every datum in a distribution of data can be described by its percentile value. For example: in Fig. 1.10, the 2.5th percentile is at 162 cm, so 2.5% of all data are smaller while 97.5% are larger; the range from 162 to 188 cm includes 95% of all data. Accommodating specific body dimensions, or a range of body sizes, is best done by selecting percentile points in sets of anthropometric data. Only judicious use of percentiles avoids the fallacy of "designing for the average."

PERCENTILES ARE USEFUL IN SEVERAL WAYS

First, they help to establish the portion of a user population that will be included in (or excluded from) a specific design solution. For example: a certain product may need to fit everybody who is larger than 5th percentile or smaller than 95th percentile in a specified dimension, such as grip size or arm reach. So, the central 90% of all users will be accommodated, but the 5% having values smaller than 5th percentile and the 5% having values larger than 95th percentile will not be fitted. This "5 to 95" strategy is widely used.

Second, percentiles help to select subjects for fit trials. For example: if the product needs to be tested, individuals having 5th or 95th percentile values (or any other chosen values) in the critical dimensions are selected to participate in use trials.

Third, any body dimension, design value, or subject score can be exactly located in the general distribution. For example: a certain foot length can be identified as a given percentile value of that dimension; a given seat height can be describe as fitting a distinct percentile value of lower leg length (popliteal height); a test score can be explained as being a discrete percentile value on the distribution of scores.

Finally, the use of percentiles helps in the selection of people who can use a given product. For example: if a cockpit of an airplane is designed to fit 10th–90th percentiles, one can select cockpit crews whose body measures are above the 10th and below the 90th percentile in the critical design dimensions.

Often, one wants to know a given percentile value, say the 5th percentile. There are two ways to determine this. One, check a graph of the data distribution, such as shown

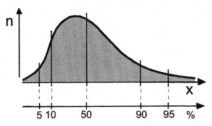

FIGURE 1.11 Skewed distribution sectioned into percentile portions.

in Fig. 1.11 and measure, count, estimate, or eyeball the critical value. This works whether the distribution is normal, skewed, binomial, or in any other form. If the data are normally distributed, there is a second, easy (and exact) approach: calculate. Calculation of a percentile values p uses Eq. (E#8) (see Table 1.4):

$$p = m + kS \tag{1.1}$$

with m the mean and S the standard deviation; take factor k from Table 1.5.

1.5 USING ANTHROPOMETRIC DATA

Ergonomists need to use relevant and reliable body dimensions to design equipment and tools that utilize and enhance human strengths. Some specific examples follow.[17]

EXAMPLE: ARM LENGTH

The task: Calculate 95p shoulder-to-fingertip length.

The solution: You know the mean lower arm (LA) link length (with the hand) to be 443 mm with a standard deviation of 23 mm. The mean upper arm (UA) link length is 336 mm and its standard deviation is 17 mm.

You are tempted to calculate both 95th percentile values and add them. But you may not use the sum of the two 95p lengths because this would disregard their correlation. Therefore, you must calculate the sum of the mean values first:

$m_A = m_{LA} + m_{UA} = 443 + 336 = 779$ mm (Equation E#16 in Table 1.4)

Next, you calculate standard deviation of the new mean, using an assumed correlation coefficient of 0.4:

$SD_A = \left[23^2 + 17^2 + 2 \times 0.4 \times 23 \times 17\right]^{1/2} = 33.6$ mm (E#17)

Now you can calculate the 95p total arm length:

$A_{95} = 779$ mm $+ 1.64 \times 33.6$ mm $= 834$ mm (E#8, with $k = 1.64$ from Table 1.5)

The result: The 95p shoulder-to-fingertip length is calculated to be 834 mm.

I. THE ERGONOMIC KNOWLEDGE BASE

TABLE 1.5   Percentile Values and Associated k Factors

Below mean				Above mean			
Percentile	Factor k	Percentile	Factor k	Percentile	Factor k	Percentile	Factor k
0.001	−4.25	24	−0.71	50 (mean)	0	77	0.74
0.01	−3.72	25	−0.67	51	0.03	78	0.77
0.1	−3.09	26	−0.64	52	0.05	79	0.81
0.5	−2.58	27	−0.61	53	0.08	80	0.84
1	−2.33	28	−0.58	54	0.10	81	0.88
2	−2.05	29	−0.55	55	0.13	82	0.92
2.5	−1.96	30	−0.52	56	0.15	83	0.95
3	−1.88	31	−0.50	57	0.18	84	0.99
4	−1.75	32	−0.47	58	0.20	85	1.04
5	−1.64	33	−0.44	59	0.23	86	1.08
6	−1.55	34	−0.41	60	0.25	87	1.13
7	−1.48	35	−0.39	61	0.28	88	1.18
8	−1.41	36	−0.36	62	0.31	89	1.23
9	−1.34	37	−0.33	63	0.33	90	1.28
10	−1.28	38	−0.31	64	0.36	91	1.34
11	−1.23	39	−0.28	65	0.39	92	1.41
12	−1.18	40	−0.25	66	0.41	93	1.48
13	−1.13	41	−0.23	67	0.44	94	1.55
14	−1.08	42	−0.20	68	0.47	95	1.64
15	−1.04	43	−0.18	69	0.50	96	1.75
16	−0.99	44	−0.15	70	0.52	97	1.88
17	−0.95	45	−0.13	71	0.55	97.5	1.96
18	−0.92	46	−0.10	72	0.58	98	2.05
19	−0.88	47	−0.08	73	0.61	99	2.33
20	−0.84	48	−0.05	74	0.64	99.5	2.58
21	−0.81	49	−0.03	75	0.67	99.9	3.09
22	−0.77	50 (mean)	0	76	0.71	99.99	3.72
23	−0.74					99.999	4.26

Any percentile value p (in a normal distribution of data) can be calculated from the mean m and the standard deviation S, $p = m + k S$. (This is E#8 in Table 1.4.) Note that k is negative for percentile values below the mean.

EXAMPLE: TORSO MASS

The task: Determine the mass of the female torso at 75th percentile.

The solution: From a recent publication, you take it that the mass of torso and head combined has an average of 35.8 kg and a standard deviation of 5.2 kg. The estimated mass of the head, measured separately, has a mean of 5.8 kg and a standard deviation of 1.2 kg. You assume the correlation between head and torso masses to be 0.1.

The mean torso mass is the difference between the average values of torso and head masses:

$m_{torso} = 35.8\,kg - 5.8\ kg = 30.0\ kg$ (Equation E#18 in Table 1.4)

The standard deviation of the mean torso mass results from:

$SD_{torso} = \left[5.2^2 + 1.2^2 - 2 \times 0.1 \times 5.2 \times 1.2\right]^{1/2}\ kg = 5.2\ kg$ (E#19)

The mass of a 75th percentile torso is:

$mass_{torso}\,75p = 30.0\ kg + 0.67 \times 5.2\ kg = 33.5\ kg$ (E#8, with $k = 0.67$ from Table 1.5)

The result: The mass of the 75th percentile torso is calculated to be 33.5 kg.

1.6 BODY PROPORTIONS

We often judge the human body by how its components "fit together." European images of a beautiful person are affected by esthetic codes, canons, and rules founded on often ancient (Egyptian, Greek, Roman) beauty concepts of the human body. Leonardo da Vinci's drawing of the body within a frame of graduated circles and squares has been adopted, in simplified form, as the emblem of the US Human Factors and Ergonomics Society.

The body beautiful

Categorizing body builds into different types is called somatotyping (from the Greek *soma* for body). In the 1940s, William Herbert Sheldon established a system of three body types, intended to describe (male) body proportions. Sheldon rated each person's appearance in proportions of ectomorphic, endomorphic, and mesomorphic components: long/lean/slim, strong/sturdy/large, and stocky/solid/muscular body builds, respectively. This typology was originally based on intuitive assessment, not on actual body measurements; these were introduced into the system later by Sheldon's disciples. In 1967, Heath and Carter standardized the procedure using Sheldon's terms: their body typology has been widely employed. Unfortunately, these and other attempts at somatotyping have not provided reliable predictors of human performance in technological systems and, hence, are of little value for ergonomists.[18]

Types of body build

1.6.1 Body Image

Everyone has a mental picture of the physical appearance of the own body. This body image may affect lifestyle behaviors, in particular when we have a distorted view of our own appearance: a common associate of some weight and eating problems, especially in

How we see ourselves

societies that scorn the obese. For example, a person suffering from *anorexia nervosa* may appear emaciated yet still complain of looking and feeling "fat." Individuals who engage intensely in athletics requiring rigid weight or body shape limitations, such as ballet dancers, gymnasts, and runners, are prone to exhibit body image disturbances and may exercise compulsively and in excess.

ASKING ABOUT BODY SIZE INSTEAD OF MEASURING?

Self-reporting is notoriously inaccurate: women and men tend to underreport their weight,[19] short people tend to overestimate their stature while heavy people often underestimate their weight.

1.6.2 "Desirable" Body Weight

Healthy body weight?

Adipose (fat-containing) tissue is a normal part of the human body: it stores fat for use as energy under high metabolic demands. However, excess of such fatty tissue is an obvious indicator of obesity. People who are severely overweight have a higher risk of health problems and of early death than their slimmer contemporaries do: the more overweight, the higher the risk. Causes for obesity may be both genetic and behavioral: they include too much caloric intake, too little physical activity, and metabolic and endocrine malfunctions.

1.6.3 Body Mass Index

BMI

It is a matter of judgment (since there are no natural cutoff points) to establish quantitative definitions of normal body weight, of emaciation or underweight compared to overweight and obesity. Adolphe Quetelet (1796–1874) calculated a Body Mass Index (BMI) by dividing a person's body weight (in kilograms) by the square of body height (stature, in meters). Quetelet's BMI procedure has seen widespread use since the 1990s. The WHO classifies BMI values below 18.5 as "underweight," above 25 as "overweight," and BMI values above 30 as "obese."

Using the metric system, to calculate BMI, divide weight (kg) by squared height (m).
$$BMI = (\text{weight [kg]})/(\text{height [m]})^2$$
Example: if height = 1.65 m and weight = 68 kg, then BMI = $68/(1.65)^2 = 25$
Using the old US system, to calculate BMI, divide weight (lbs) by squared height (in), then multiply with a conversion factor of 703.
$$BMI = 703 \times (\text{weight [lb]})/(\text{height [in]})^2$$
Example: if height = 5'5" = 65 in and weight = 150 lbs, then BMI = $[150/(65)^2] \times 703 = 25$

Among the troublesome issues with the concept of BMI[20] is that body composition varies among individuals of the same age, height, and weight. Age is an important moderator of the BMI-to-fatness relationship; at equal BMIs, older individuals are fatter than younger adults. BMI varies strongly with age in children. As women generally have smaller bones and less muscle tissue than men, one might expect that women's BMIs would be less than those of men for any given percentile; however, this generalization actually applies only below the 75th percentile of the BMI distribution. In the upper quarter of the BMI distribution, women's BMIs are generally higher than men's. (See also Chapters 3 and 14 for additional considerations of BMI.)[21]

Troubling BMI issues

1.7 DEALING WITH STATISTICS

Some groups of anthropometric data show strong covariation with each other: as one increases, another (or several others) increases as well—this happens, for example, among weight and trunk circumferences. Conversely, as one measure increases, others may decrease: as we advance into old age, often body heights decrease (see Fig. 1.6). Many body dimensions show little correlation with each other. Fig. 1.12 shows examples of how values of one variable can scatter with respect to another variable; the relationships may be positive or negative, strong or weak. The correlation coefficient, r, (formally, the *Pearson product–moment correlation coefficient*; E#10 in Table 1.4) numerically expresses the linear relationship between two variables.

Relations among body dimensions

Human stature does not correlate well with weight (or head length, or waist circumference, and other measures), whereas other body dimensions are closely related, such as eye height with stature. Table 1.6A lists selected correlation coefficients r among body dimensions of US Army personnel, male and female.[22] Unfortunately, similar tables for other nationalities and for civilians seem not to exist.

Anthropometric correlations

The data in Table 1.6B show that even in the military, a group of individuals selected for their youthful fitness, there is little correlation between stature and weight: r is only around 0.5. This points to a basic problem with the BMI and with other body indices which rely on the presumption of a robust positive relation between body weight and stature—which does not exist. Unfortunately, some people still believe in an ill-conceived 1966 design scheme which presented all body heights, all body breadths, and all segment lengths as fixed percentages of stature. For instance, hip breadth was said to be 19.1% of height—misleading nonsense, of course, because hip breadth relates well only to a few trunk circumferences but varies widely among individuals and between males and females as groups. Furthermore, what useful gadget could be designed for a fixed "average" hip breadth?

Lacking co-variation

The linear relationship between a dependent variable y and an independent variable x is a bivariate regression as expressed by the Eq. ($y = a + bx$) where a is the intercept and b the slope (E#12 in Table 1.4). Note that often a linear relationship between x and y is assumed but not verified.

Regression

The coefficient of determination, R, is simply the square of the correlation coefficient r between the two variables used in a bivariate regression equation (or among more variables in multiple regression equations). R measures the strength of association by

Determination

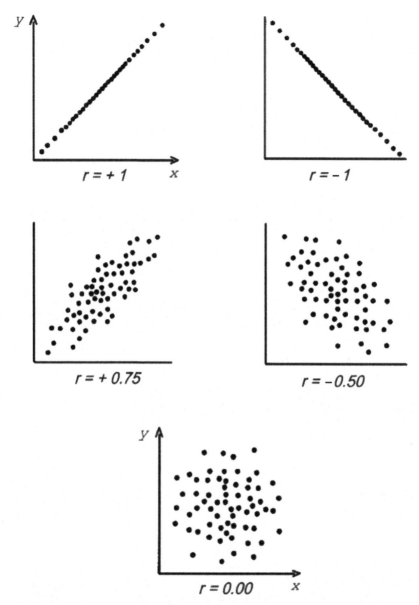

FIGURE 1.12   Scatter diagrams and associated correlations.

stating the proportion of variation in the dependent variable y which is predictable from the independent variable x.

"0.7 Conven-
tion"   It is common practice in engineering anthropometry (in fact, in ergonomics altogether) to require a correlation coefficient r of at least 0.7 as a basis for design decisions that involve related body dimensions. The reason for this" 0.7 convention" is that one should be able to explain at least 50% of the variance of the predicted value from of the predictor

TABLE 1.6 Correlations between Anthropometric Data on US Soldiers

	1 Age	2 W	3 Stat	4 OFR	5 WH	6 CH	7 SH	8 PH	9 SC	10 CC
1 Age [302]		0.219	0.041	0.017	0.044	−0.055	0.066	−0.074	0.155	0.193
2 Weight [125]	0.195		0.529	0.493	0.491	0.370	0.422	0.242	0.845	0.806
3 Stature [100]	−0.021	0.546		0.928	0.848	0.840	0.755	0.808	0.377	0.222
4 Overhead fingertip reach [84]	−0.013	0.525	0.937		0.704	0.905	0.554	0.868	0.384	0.199
5 Wrist height, standing [128]	0.028	0.527	0.856	0.749		0.625	0.754	0.587	0.300	0.255
6 Crotch height [39]	0.090	0.351	0.852	0.890	0.673		0.330	0.915	0.267	0.093
7 Sitting height [94]	0.026	0.447	0.741	0.578	0.692	0.347		0.343	0.285	0.202
8 Popliteal height, sitting [87]	−0.094	0.341	0.852	0.883	0.673	0.924	0.383		0.188	0.023
9 Shoulder circumference [91]	0.122	0.861	0.399	0.413	0.334	0.250	0.326	0.256		0.808
10 Chest circumference [34]	0.279	0.873	0.312	0.308	0.357	0.135	0.287	0.137	0.859	
11 Waist circumference [115]	0.364	0.849	0.276	0.251	0.343	0.060	0.298	0.074	0.703	0.839
12 Buttock circumference [24]	0.190	0.935	0.401	0.380	0.412	0.204	0.373	0.191	0.781	0.815
13 Span [99]	−0.016	0.497	0.815	0.908	0.535	0.840	0.398	0.844	0.445	0.281
14 Biacromial breadth [11]	0.034	0.496	0.487	0.506	0.295	0.370	0.407	0.394	0.633	0.419
15 Hip breadth, standing [66]	0.209	0.831	0.453	0.416	0.457	0.239	0.464	0.224	0.672	0.727
16 Head circumference [62]	0.125	0.508	0.342	0.312	0.302	0.224	0.303	0.240	0.433	0.421
17 Head length [63]	−0.002	0.371	0.346	0.315	0.295	0.260	0.302	0.268	0.295	0.271
18 Head breadth [61]	0.198	0.320	0.114	0.098	0.112	0.034	0.128	0.035	0.303	0.311
19 Hand length [60]	0.032	0.453	0.650	0.724	0.464	0.676	0.300	0.679	0.372	0.242
20 Foot length [52]	−0.012	0.512	0.700	0.734	0.537	0.687	0.383	0.697	0.409	0.299

(Continued)

I. THE ERGONOMIC KNOWLEDGE BASE

TABLE 1.6   Correlations between Anthropometric Data on US Soldiers (*cont.*)

	11 WC	12 BC	13 Sp	14 BB	15 HiB	16 HC	17 HeL	18 HeB	19 HaL	20 FL
1 Age [302]	0.299	0.258	0.011	0.025	0.283	0.073	0.027	0.044	0.044	0.026
2 Weight [125]	**0.767**	**0.897**	0.438	0.440	**0.778**	0.428	0.329	0.420	0.430	0.493
3 Stature [100]	0.167	0.361	**0.787**	0.505	0.372	0.348	0.354	0.124	0.637	0.673
4 Overhead fingertip reach [84]	0.132	0.313	**0.907**	0.535	0.294	0.337	0.345	0.095	**0.737**	**0.732**
5 Wrist height, standing [128]	0.217	0.363	0.453	0.303	0.397	0.250	0.261	0.403	0.403	0.468
6 Crotch height [39]	0.061	0.185	**0.870**	0.418	0.146	0.287	0.302	0.043	**0.706**	**0.703**
7 Sitting height [94]	0.142	0.351	0.336	0.384	0.438	0.246	0.255	0.159	0.256	0.330
8 Popliteal height, sitting [87]	−0.031	0.063	**0.840**	0.420	0.051	0.241	0.271	0.020	0.685	0.671
9 Shoulder circumference [91]	0.697	**0.726**	0.395	0.574	0.601	0.353	0.264	0.261	0.355	0.379
10 Chest circumference [34]	**0.781**	**0.707**	0.167	0.304	0.603	0.393	0.191	0.246	0.186	0.288
11 Waist circumference [115]		**0.738**	0.109	0.214	0.673	0.223	0.117	0.229	0.127	0.170
12 Buttock circumference [24]	**0.859**		0.258	0.327	**0.915**	0.313	0.226	0.220	0.258	0.323
13 Span [99]	0.201	0.352		0.565	0.203	0.345	0.338	0.083	**0.827**	**0.775**
14 Biacromial breadth [11]	0.311	0.411	0.575		0.294	0.287	0.259	0.152	0.441	0.456
15 Hip breadth, standing [66]	**0.799**	**0.902**	0.355	0.404		0.232	0.160	0.196	0.180	0.250
16 Head circumference [62]	0.376	0.427	0.320	0.301	0.364		**0.824**	0.497	0.342	0.360
17 Head length [63]	0.222	0.301	0.304	0.235	0.259	**0.820**		0.131	0.337	0.339
18 Head breadth [61]	0.277	0.268	0.131	0.180	0.235	0.541	0.120		0.082	0.113
19 Hand length [60]	0.166	0.320	**0.810**	0.433	0.298	0.330	0.306	0.137		**0.825**
20 Foot length [52]	0.220	0.390	**0.766**	0.445	0.377	0.333	0.304	0.161	**0.806**	

Values for women are listed above the diagonal, for men below. Values larger than 0.7 are bolded. Pairs of data that correlate above 0.700 are similar in the male and female groups; also, correlations below 0.300 are similar for both genders.

variable: This requires $R = r^2$ to be at least 0.5, so r is at least 0.7075. (Note that r depends on the sample size n).

1.8 HOW TO OBTAIN MISSING DATA

In terms of anthropometry, Europe and North America long had the best-measured populations on earth,[23] now joined by Russians and Taiwanese—see Table 1.2. Yet even there, most civilian populations were not assessed comprehensively, and current data on subgroups are sparse. Lucky is the ergonomist who is particularly interested in body sizes of Italians visiting swimming beaches,[24] Portuguese and American children,[25] Irish workers,[26] Indian vehicle drivers,[27] American farmers,[28] pregnant American women,[29] American wheelchair users,[30] or of hand sizes.[31] But in many cases, the exact body dimensions needed for a design are neither available in the literature nor on the internet.

Occasionally, one can simply use the data of a population of known dimensions, if one has good reason to assume that this population is similar to the one on which data are missing. Yet one has to consider the underlying assumption: for example, are Taiwanese similar in size to all Chinese? Do Japanese males have body proportions that are similar to those of their American contemporaries?[32] Are the body sizes of Russian students in Moscow similar to those of students in Tashkent? Can American women's body sizes be treated as small men's sizes[33] for equipment design purposes?

For a quick check, it may suffice to take measures on a few coworkers to get a rough estimate. How much effort to put into the data depends on the importance of the application and the risk and consequences of being wrong. The most thorough approach is to actually measure a sufficiently large and carefully selected, statistically appropriate sample of the population to be fitted. The literature provides help in understanding important aspects of anthropometric surveys[34] such as sampling composite populations, and, of course, of measuring procedures. Getting to understand such issues quickly leads to the conclusion that planning and conducting an anthropometric survey is a complex and time-consuming task which should be done by qualified anthropometrists and trained specialists.

Say you are designing student desks and want to know how tall 10-year old girls who attend schools in Spain are. It is impossible to measure them all, so the true value of that population statistic (in this example, the girls' mean height) cannot be observed and remains unknown. However, it is possible to *estimate* the true value, with some degree of certainty, from measures taken on a random sample from the whole population of interest. In order to trust the sample statistics, the sampling procedure needs to be carefully planned and executed: this is often a daunting, time-consuming, and expensive task.

To determine the required sample size, one starts with a "worst case" scenario: select the dimension that is expected to show the largest CV (waist circumference, for example is highly variable); then decide whether the data should describe low/high percentile values (often the 5th/95th percentiles) or the mean value of the distribution. The sample size needed to obtain information at the tails of distributions is larger than needed to estimate the center of the distribution. When the precision is 1% and the confidence level is 95% (as often desired), the minimal sample sizes[35] for every subgroup

Estimate or calculate?

Measure?

Sampling

Sample size

are $n \geq (1.96CV)^2 (1.534)^2$ for estimation of 5th/95th percentiles, or $n \geq (1.96CV)^2$ to estimate mean values (E#9 in Table 1.4) with the CV in percent.

Predict?
A rather challenging task is the prediction of future body dimensions, needed when equipment must be designed for use in decades to come. In the 1960s, for example, NASA was concerned about the body sizes of astronauts by the end of the century. Fig. 1.13 shows predicted values for male NASA flight crews in terms of stature; predictions were also made for female astronauts. For both groups, stature was assumed to increase linearly over the years, but that "secular" growth has slowed since the turn of the century. Note that at that time, all astronauts were fairly small and short so that they could fit into the tiny cockpits of the early space vehicles.

Clothing sizing in the USA is typical for use, misuse, and nonuse of correlations.

Sizing of clothes for men is a fairly well organized and standardized procedure. Most men's jacket sizes run from 38 to 56, meaning that they should fit men with chest circumferences between 38 and 56 in, with increments of 1 or 2 in. So, chest circumference is the primary "predictor variable" for other size variables, such as coat length, shoulder width, and sleeve length. Similarly, slacks are ordered by waist circumference and shirts by neck circumference.

In men's shirts, a given neck circumference usually is associated with a given chest circumference, while sleeve length may vary by 1 or 2 in increments. This is an attempt to cover various body dimensions with a few shirt sizes, but it has obvious shortcomings: if a person needs a large neck size, the corresponding shirt also usually comes with ballooning chest and waist circumferences, which the wearer may not need.

In contrast, women's clothes are not well standardized. There was one ill-defined prototype size 12 (based mostly on 1941 measurements), from which larger and smaller sizes were derived in nonstandard manners, as deemed suitable by each manufacturer. Hence, a woman well fitted by clothes of size 10 made by one producer may need a size 12, or 8, in clothing tailored by another company. There remains a plethora of nonstandard sizing schemes (mis)labeled junior, misses, mature, petite, tall, and/or plus sizing. In recent years, so-called vanity sizing or size inflation has complicated women's clothing: ready-to-wear clothing of the same nominal size may now fit a larger body. This makes the wearer "feel better" about (or more likely to purchase) the size being considered.

The wide diversion of body dimensions has long been demonstrated but is often overlooked: Table 1.7 lists that in a sample of 4063 men, 1055 had a "middling" stature between 38th and 62nd percentile, this 26% spread being indeed a very generous approximation of "middle." Selecting those who were also "middling" in other body measures showed that after the fourth iteration less than 1% of the remaining men were "in the middle." Similarly, Table 1.8 shows that trying to accommodate the center range

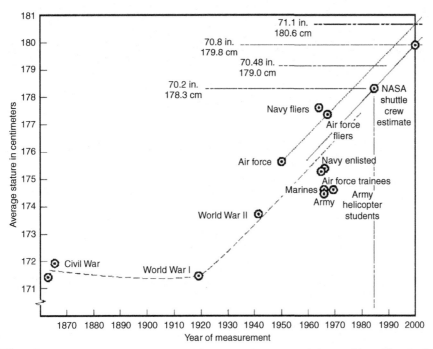

FIGURE 1.13 **Predictions of average stature of NASA male personnel.** *Source: Adapted from Roebuck et al. (1988).*

TABLE 1.7 How many Individuals are "Approximately Middle" in Body Dimensions?

Step	Sample size	Remaining in the "approximately middle" sample	
1	Of the original 4063 men	1055 were between 38th and 62nd percentile in Stature	26%
2	Of the remaining 1055 men	302 were between 38th and 62nd percentile in Chest Circumference	7%
3	Of the remaining 302 men	143 were between 38th and 62nd percentile in Arm (sleeve) Length	4%
4	Of the remaining 143 men	73 were between 38th and 62nd percentile in Crotch Height	2%
5	Of the remaining 73 men	28 were between 38th and 62nd percentile in Torso Circumference	<1%
6	Of the remaining 28 men	12 were between 38th and 62nd percentile in Hip Circumference	<0.3%
7	Of the remaining 12 men	6 were between 38th and 62nd percentile in Neck Circumference	<0.2%
8	Of the remaining 6 men	3 between 38th and 62nd percentile in Waist Circumference	<0.1%
9	Of the remaining 3 men	2 were between 38th and 62nd percentile in Thigh Circumference	<0.1%
10	Of the remaining 2 men	None were between 38th and 62nd percentile in Crotch Length	0%

Source: Adapted from Daniels (1952).

TABLE 1.8　How Many Users would a "5th to 95th Percentile" Specification Accommodate?

5th to 95th percentile requirement in:	Percentage of user population accommodated
Stature	90%
and chest circumference	82%
and waist length	78%
and shoulder breadth	71%
and waist circumference	67%

Source: Adapted from Gordon et al. (1997).

of 90% of users, 5th to 95h percentile, may exclude every third user after checking the sixth body dimension.

PHANTOMS, GHOSTS, AND THE "AVERAGE PERSON"

Several misleadingly simple body proportion templates have been used in the past. Some presume, falsely, that all body dimensions (lengths, breadths, depths, and circumferences) are in given fixed proportions (percentages) of one body dimension, generally stature. Other templates pretend that individuals exist whose body segments are all of the same percentile value. Among these fallacious single-percentile constructs are ghostly figures that consist of all 5th percentile or of solely 95th percentile body segments; the most prominent computational illusion is the all-50th-percentile phantom, the "average person." Of course, designs for these figments do not fit many actual users.

1.9 "FITTING" DESIGN PROCEDURES

Fitting which to what?

Information about body size is needed when an object must be made to fit the human body. Examples include tool handles to hold, protective equipment and clothing to wear, chairs to sit on, displays to look at, and workstations in general. Several of these applications are discussed elsewhere in this book; see, for example, the sections on hand tools in Chapter 9 and on computer workstations in Chapter 10.

Fitting how?

Two primary fitting methods are often used. One approach is to select two percentile points to cut out ranges from the whole data set—say, to make a glove size fit hands[36] from 5th to 25th percentile size. Another approach uses just one cutoff percentile: a large one to assure that even big individuals can pass through an escape opening, or small percentile value that prevents even small users to reach into the danger zone of a stamping press.

Functional anthropometry

Traditional anthropometric data describe the body in a fixed, upright stance. In everyday activities, however, we bend and stretch, we move the body and its segments. This means that the designer must convert static information into "functional" (or dynamic) anthropometry in order to accommodate the body during sequences of motions to achieve desired postures and perform various activities. These define zones of safety,

TABLE 1.9 Guidelines for the Conversion of Standard Measuring Postures to Functional Stances and Motions

Functional stance and motion	Suggested conversion
Slumped standing or sitting	Deduct 5%–10% from relevant height measurements
Relaxed trunk	Add 5%–10% to trunk circumferences and depths
Wearing shoes	Add approximately 25 mm to relevant standing and sitting heights; more for "high heels"
Wearing light clothing	Add about 5% to relevant dimensions
Wearing heavy clothing	Add 15% or more to relevant dimensions. (Note that heavy clothing may strongly reduce mobility)
Extended reaches	Add 10% or more to relevant reach measures for strong motions of the trunk
Use of hand tools	Center of handle is at about 40% of hand length, measured from the wrist
Forward bending of head, neck, and trunk	Ear-Eye Line declines to near horizontal
Comfortable seat height	Subtract up to 10% from popliteal height

Source: Adapted from Kroemer, Kroemer, and Kroemer (2010).

convenience, expediency, of minimally required or largest covered space. Table 1.9 helps in the conversion of static data into dynamic information.

Size and location of "zones of comfort, of convenience or expediency" depend on the person and the task; preferences may include previous customs and developed skills, ease of posture and motion, speed and accuracy of task performance, minimal energy expenditure, and personal safety. Preferred working areas of the hand are approximately within partial spheres around the elbow or shoulder, as in Fig. 1.14. The fastest and most controlled hand motions are essentially flat and horizontal, with the forearm about horizontal and the upper arm about vertical; the joint movements primarily occurring

Workspaces of hands and feet

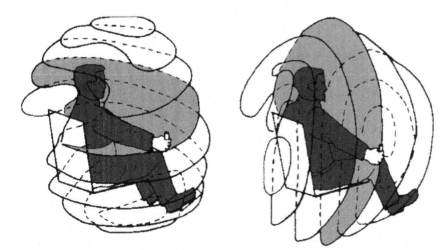

FIGURE 1.14 **Envelopes of horizontal and vertical hand reaches.** *Source: Adapted from Ignazi et al. (1982).*

in the shoulder by abduction and adduction. These hand motions are also within view and focus of the eyes, as illustrated in Fig. 1.15.[37]

Fitting all, "small-to-tall"

The "min-max strategy" is often successfully applied (instead of the useless average person concept) when workstations, tools, and tasks must be designed to fit small and large operators, and everybody in-between. The *minimal* dimensions derive from the smallest operators' needs to see, reach, apply force, etc. However, these measures may not accommodate large individuals who can reach farther and need more open space for their bodies: their needs establish the *maximal* boundaries.

The "min-max strategy" requires, of course, that the designer first decide what the important user and use features are; then, to determine the minimal and the maximal values (e.g., of body size, or reach, or strength) that shall be accommodated.

One size does not fit all

The best solution, if feasible, is to provide workplaces and tools and other work objects that can be adjusted to various dimensions. Adjustability between minima and maxima usually allows good fit but may be expensive. The other good solution is to have the items in different sizes so that each user can select what fits best individually. Often, a compromise is attempted (e.g., in automobiles) when the overall space is made large to allow everybody to fit in while the interior components have some adjustment features to better accommodate small users. However, the "one size fits all" design concept often does not.

Work at elbow height

The location of the work surface is usually the first design concern. Its height depends on the physical work to be performed, on the dimensions of the work piece itself, and possibly on the need to observe the work being done. As a general rule, the manipulation itself should be performed at about the height of the operator's elbows when the upper arms hang down alongside the trunk or are slightly elevated forward and sideways, as shown in Fig. 1.16.

Using the US soldier data in Table 1.3 for elbow heights of standing individuals, one can calculate 93 cm for a 5th percentile standing female operator; it is 108 cm for a 95th percentile standing female. For standing male operators, the respective elbow heights are 100 and 117 cm. These minima and maxima bracket hand workspace heights that apply to most standing operators.

Standing at work

As can be noted from Fig. 1.16, working heights get lower when the operator does not stand erect but bends, however, the thickness of soles and heels on shoes worn can offset that "slump." If the work piece is large, and the manipulation is performed on its upper part, the support surface (bench height) must be lowered to allow the hands to be at elbow level. Work that demands close visual observation requires a fairly short viewing distance. In that case, particularly if the manipulation needs relatively little force and energy, the work area may be well above elbow height. (In turn, this may require support for the elevated hands and forearms).

If the workstation is fully enclosed, as shown in Fig. 1.17 (same as Fig. 9.10), standing is made a bit comfortable by some "toe space" at the bottom of the workstation which allows the operator to step in closely to the work. This space should be high enough to accommodate individuals wearing thick soles, but shallow enough so that one does not

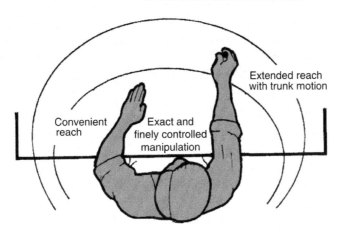

FIGURE 1.15 **Areas for exact, convenient, and extended manipulation.** *Source: Adapted from Proctor and Van Zandt (1994).*

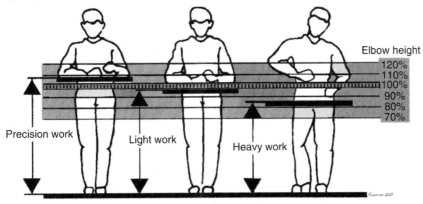

FIGURE 1.16 **Elbow height is often a good measure to determine the work height.** *Source: Adapted from Kroemer, Kroemer, and Kroemer (2010).*

FIGURE 1.17 Enclosing a workstation can make it difficult to sit or stand.

hit the edge of the foot space cutout with the instep of the foot. Thus, a toe space depth of about 10 cm and a height also of about 10 cm should be appropriate.

Sitting at work

For a sitting operator, the elbow height is appropriately measured from the seat surface, not the floor: the data in Table 1.3 yield 5th and 95th percentile elbow heights of sitting female soldiers of 19 and 27 cm, respectively; 20 and 29 cm for males (note how similar the values are). However, the work surface can be lowered only until it touches the upper side of the thighs: Table 1.3 supplies a thigh height of about 21 cm above the seat height for 95th percentile male operators. This represents the necessary height of the open space underneath the working surface to accommodate the legs of most seated operators: the required 21 cm are in conflict with the about 19 cm minimal elbow height just determined—this is even without considering the thickness of the work surface. The solution is to provide adjustment in work surface height, not only in seat height. (Another way to determine the needed height of the leg space is to use the "knee height," also given in Table 1.3, plus some allowance for shoe heels).

Legroom

The width of the legroom (which is not shown in Fig. 1.16) should exceed the hip width of the widest operator. The depth of the legroom should go beyond the largest distance from the front of the belly to the kneecaps. This is not a dimension customarily measured by anthropometrists; it can be estimated from the difference between buttock-knee depth and abdominal depth, both listed in Table 1.3. The deeper the leg space, the more the feet can be extended forward. The height of the work seat should be adjustable to fit individuals with long and short lower legs, as discussed in more detail in Chapter 10 on office design.

Sit–stand station

Occasionally we are called upon to design a workstation at which the operator can either sit or stand, as shown in Fig. 1.18. This task combines the "min-max" space requirements for sitting and standing. For sitting at a stand-up workstation, there must be a very tall and stable chair. This also requires a support surface for the feet: attaching a small board or bar to the chair reduces the stability of the chair. Also, such a foot board or bar provides little support surface for the feet which must be kept in place by muscle tension instead of being able to move to different positions—incorporating a solid support surface into the workstation is a better solution.

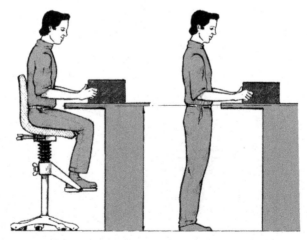

FIGURE 1.18   **Sitting or standing at work.**

An even better solution, though more expensive, is to make the workstation adjustable in height so that it provides a lower work surface when the operator sits and a higher surface when the operator stands. An additional advantage is that the height adjustment can be tailored to the individual operator, whether sitting or standing.

1.9.1 Safe Distances

Body dimensions, often hand or arm lengths, determine the "safe" distances by which the operator's body (usually a finger) must stay clear of a danger point such as the edge of a hazardous gadget—an electric arc, a saw, a press mold, or pinch point. Body measures determine the straight-line distance between the danger point and the barrier (wall, safety guard, enclosure of an opening) beyond which the operator's body cannot proceed toward the hazard.

For hand safety, either the size of an opening, which allows only a finger to penetrate, can determine the safe distance; or the location and size of a barrier, which by the arm can overreach, may establish the minimal distance. In the first case, the safe distance would be the length of the longest possible finger, with a safety margin; in the second case, the longest hand/arm reach would determine the distance. Many variations are possible, some shown in Fig. 1.19. For foot safety, the most likely barrier is at the ankle, so that only the toes and distal part of the foot can penetrate further toward the danger point; or the whole leg may have to be considered, probably restrained in the hip area.

Finger and foot safety

In most cases, the most protruding (usually the longest) involved body segment should be considered under the given conditions of barrier and mobility. A further safety margin, often at least 10%, should be added to those body dimensions. Certain conditions, such as holding an object that, if entrapped, might pull the hand toward the hazard point, give reason to extend the safety distance further.

Safety margins

1.10 DESIGN STEPS

"Good"

For proper consideration of the different sizes and proportions of the human body, identify the critical dimension(s) to be fitted. Decide on the smallest percentile values (not necessarily the 5th) and the largest values (not always the 95th) that must be accommodated.

"Caution"

Two (or more) body dimensions may be used to create an imaginary nth-percentile person; stature and weight are often used but many other dimensions are not highly correlated with either stature or weight; therefore these two variables are often not good predictors (Table 1.6).

"Bad"

Constructing single-percentile phantoms (such as the 5th, the 50th, a.k.a. "average", or the 95th percentile "person") is nonsense.[38]

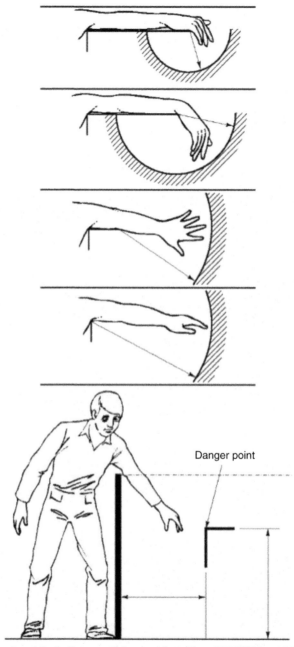

FIGURE 1.19   **Examples of "safe distances."** *Source: Adapted from DIN 31001 and BS 5304.*

The design process can be broken down into four steps:

1. Select those anthropometric measures that directly relate to defined design dimensions. Examples are hand length related to handle size; shoulder and hip breadth related to escape-hatch diameter; head length and breadth related to helmet size; eye height related to the heights of windows and displays; knee height and hip breadth related to the leg room in a console.
2. For each of these pairings, determine whether the design must fit only one given percentile of the body dimension, or a range along that body dimension. For example, the escape hatch must be fitted to the largest extreme values of shoulder breadth and hip breadth, considering clothing and equipment worn; handle size of pliers is probably selected to fit a smallish hand; the leg room of a console must accommodate the tallest knee heights; the height of a seat should be adjustable to fit individuals with short and with long lower legs.
3. Combine all selected design values in a careful drawing, mock-up, or computer model to ascertain that they are compatible. For example, the required legroom clearance height, needed for sitting individuals with long lower legs, may be very close to the height of the working surface, determined from elbow height.
4. Determine whether one design will fit all users. If not, several sizes or adjustment capability must be provided to fit all users. For example, one large bed size fits all sleepers; gloves and shoes must come in different sizes; seat heights are adjustable.

The ultimate test of a design is through its actual use. This can be simulated with test subjects who represent the min-max values selected by the designer.

Fit check

The following story, slightly paraphrased, appeared in the *Washington Post* of May 25, 1984: The Navy has adopted new flight training standards that will require its aviators, as a whole, to have longer arms and shorter legs. These standards will exclude 73% of all college-age women and 13% of the college-age men, according to a military spokesman. He said the new standards were devised because some aviation candidates could not reach rudder pedals or see over instrument panels. Some taller pilots were so tightly wedged that their helmets bumped the aircrafts' canopies. "We found out that manufacturers are still building airplanes the way they want, but God is not making people to fit them." Previously, 39% of the female applicants and 7% of the men were ineligible to become aviation candidates because of their size.

Six years later, in 1990, Buckle and coworkers checked the cockpit dimensions of aircraft used in many countries in respect to eight critical body dimensions of pilots, including eye height, hand and leg sizes, and reaches. In many cases, the fit was marginal, at best. For example, based on eye height, 13% of the British male and 73% of the female pilot candidates would have to be excluded from being crew members.

1.11 CHAPTER SUMMARY

Anthropometry and anthromechanics provide the basic descriptors needed to design "fitting" equipment and work procedures.

Information about body size is available in a rather complete manner only for military populations. Soldiers are a select sample of the general population. They are relatively young and generally healthy, and do not include individuals of unusual body dimensions. Furthermore, until recently, there were not many female soldiers. Thus, estimates of the general civilian populations derived from military data have been rather unreliable in the past.

In North America, recent surveys of the US Army have remedied many of the problems associated with missing anthropometric information. The US Army is fairly similar to the general population in its composition, including females. Thus, Army information in this book can be used to approximate body dimensions of the general population of the United States. Statistical treatment of the measured data allows us to establish correlations among them and to predict, using this knowledge, data that were not actually measured.

Statistics applied to anthropometric data provide a wealth of information. For example, the standard deviation divided by the mean is a measure of the variability of the recorded data. Large variability could indicate either a truly diverse sample, or problems in data acquisition or treatment. Much variability stems from the fact that most surveys are performed as cross-sections of the population, meaning that at one moment in time measurements are taken on individuals of different ages. While this reflects interindividual differences, it does not inform us about intraindividual changes occurring with aging or with changes in health.

In recent decades, body sizes have been expanding in many areas of the Earth. In particular, body weight increased but growth in many length dimensions has also occurred, such as in stature and leg length. Such information is important for the design of closely fitting equipment, especially clothing.

Use of statistics on body size and strength has often been faulty. It is indefensible and inexcusable to design tools, equipment, or workstations for a phantom called "average person." Instead, ranges of body dimensions must be considered, and the ways in which certain body dimensions change with respect to each other. For example, it is inadequate although common practice to establish height–weight indices (such as in desired body weight) when in fact the correlation between height and weight is very low. Instead, fairly simple rules apply which consider variations in body dimensions. For ergonomic design of equipment, a four-step procedure guides proper "sizing" of designed products.

1.12 CHALLENGES

How are the body dimensions of soldiers different from those of civilians?

If such differences exist, how important would they be for the design of special tools, equipment, workstations, or work procedures?

Are there rules that describe how to derive "functional" body dimensions from those measured in static, standardized postures?

How do secular changes in body dimensions affect the design of technical products?

When is designing for "fit from the 5th to the 95th percentile" appropriate and when is it not?

Are women equivalent to small men, anthropometrically?

Are children equivalent to small men, anthropometrically?

Which procedures are appropriate to select fair subsamples from a general population for measurement purposes?

How can correlations among body dimensions reduce the needed number of measurements taken on a population sample?

How do you know if your anthropometric data are good enough to make design decisions?

How precisely must body dimensions be known to suffice for specific design purposes?

When might it be important to account for cultural differences in user expectations and not just physical differences?

What are some of the impacts of using two-dimensional (i.e., in-plane) data for a three-dimensional world?

When might you need to have difference reference points for anthropometric data?

How often should anthropometric data be updated, and for how long is our historic data still valid?

NOTES

1   The information regarding human evolution and movement is ever changing, and the interested reader should check for updated findings.

2   Classifications of people into races started with Carl Linnaeus' four-race taxonomy of 1735: Caucasians (Europeans, "white"), Mongolians (East Asians, "yellow"), Ethiopians (sub-Saharan Africans, "black") and Native Americans ("red"), to which the anthropologist Johann Friedrich Blumenbach (1752–1840) added the "race" of Malayans ("brown" Southeast Asians and Pacific Islanders) around 1780. These classifications relied primarily on superficial differences in skin color, hair form, and eye lid shape and have no bearing on ergonomics.

3   Marco Polo (1254–1324) traveled from Europe to Asia, staying in China for 20 years. There he found people much advanced of Europe in mathematics, technology, and civilized amenities. By 1292 he returned to Italy, was taken prisoner in the war between Venice and Genoa, and (in detention) began dictating his reminiscences of his travels. This manuscript, circulating from about 1298 onwards, was immensely popular (although largely disbelieved) and stimulated much interest in the study of other countries.

4   About 500 BC, the philosopher and physician Alcmaeon of Croton deliberately and carefully dissected human cadavers. He noted the difference between arteries and veins and found the sense organs connected to the brain by nerves. (The word anatomy stems from the Greek "to cut up.") In Rome, the physician Galen (Claudius Galenus, about 129–199) could dissect only animals, which occasionally misled him when transferring his observations to the human body. In 1316, Mondino de Luzzi, professor at the medical school of Bologna, Italy, published the first book devoted entirely to anatomy. In 1543, the anatomist Andreas Vesalius upset many traditional but false Egyptian and Greek notions about anatomy of the human body in his book De Corporis Humani Fabrica, concerning the structure of the human body. His book contained careful illustrations of anatomical features, drawn by a student of Titian (Asimov 1989).

5   The International Standardization Organization (ISO) has established conforming anthropometric techniques worldwide: see, for example, ISO 7250.

6   Some reference systems reverse the directions.

7   Specific terminology and measuring conventions have been described in English by Hertzberg (1968); Garrett and Kennedy, 1971; Roebuck, Kroemer, and Thomson (1975); NASA/Webb (1978); Roebuck (1995); recently by Bradtmiller (2015); Gordon et al. (2014); Gordon et al. (1989); ISO 15535; Paquette, Gordon, and Bradtmiller (2009).

8   Paquet and Feathers (2004).

9   Gordon et al. (2014); Paquette et al. (2009).

10   Hertzberg (1968); Garrett and Kennedy (1971); NASA/Webb (1978).

11   Depending on the research need, more recent data may be available from the World Health Organization (WHO) or other governmental agencies.

12   Greiner and Gordon (1990).

13   Most of today's soldiers are too big to fit into medieval body armor.

14   McConville, Robinette, and Churchill (1981).

15   Gordon et al. (1989); Gordon et al. (2014).

16   Bellera, Foster, and Hanley (2012); Gordon et al. (1989); Blackwell and Bradtmiller (2012); ISO 15535; Roebuck (1995); Sokal and Rohlf (2011); Zar (2010).

17   Anyone who flies commercial airlines realizes that airline seats have become smaller and more tightly packed just as passengers have gotten wider and taller. Ergonomics has a vital role to play here: not just for comfort for the flying public but also for safety and egress in an emergency. Design of seats, seat layout, restraints, crew training, and evacuation procedures must consider both physiology and psychology. Valid anthropometric data that represent the passenger population and testing that considers emotional state and cognitive abilities of stressed passengers are needed.

18   Hippocrates developed, about 400 BC, a scheme that included four body types, supposedly determined by their fluids: the "moist" type was dominated by black gall and the "dry" by yellow gall; the "cold" type was governed by slime, the "warm" by blood. In 1921, the psychiatrist Ernst Kretschmer described a system of three body types intended to relate body build to personality traits: the asthenic, pyknic, and athletic body builds. His "athletic" type referred to character traits, not sports performance capabilities.

19   Bowman and Delucca (1992).

20   Cronk and Roche (1982); Andres (1984); Heymsfield, Allison, Heshka, and Pierson (1995); Siervogel, Roche, Guo, Mukherjee, and Chumlea (1991); Williamson (1995).

21   People commonly say they weigh too much for their height, or jokingly say they are undertall for their weight. In fact, in general there is little correlation between stature and weight.

22   Older correlation tables for soldiers are in publications by NASA/Webb (1978); Kroemer, Kroemer, and Kroemer (1997); Roebuck et al. (1975).

23   Peebles and Norris (1998); Kroemer, Kroemer, and Kroemer (2010).

24   Coniglio et al. (1991).

25   Froufe, Ferreira, and Rebelo (2002); Lueder and Rice (2007).

26   Gallwey and Fitzgibbon (1991).

27   Kulkarniet al. (2012).

28   Casey (1989).

29   Culver and Viano (1990).

30   Paquet and Feathers (2004).

31   Hsiao, Whitestone, Kau, and Hildreth (2015); Kroemer (2016); Mirmohammadi et al. (2016).

32   They do not, as Nakanishi and Nethery (1999) showed.

33   No, as stated by Fullenkamp, Robinette, and Daanen (2008).

34   Gordon et al. (1989); Gordon et al. (2014); ISO 15535; Paquette et al. (2009).

35   Bellera et al. (2012); Gordon, Blackwell, Bradtmiller (2012); ISO 15535; Sokal and Rohlf (2011); Zar (2010).

36   Hsiao et al. (2015); Robinette and Vetch (2016).

37   The literature contains some ill-defined but overly precise recommendations for location and size of work spaces for the hands, such as for men within a "radius of 394 mm from the shoulder" and for women within a "356 mm radius" (Nicholson, 1991).

38   Daniels (1952).

Bones, Muscles, and Strength of the Human Body

Overview

The human musculoskeletal system is a physiological entity; however, a biomechanical construct can well explain essentials of its functioning and many results of its functions. In this anthromechanical view, long bones and their joints make up the structural links. Muscles are like linear motors that exert pull forces between connected links. This generates moments within the musculoskeletal system which result in moments or forces that can be exerted to outside objects (such as tools or equipment) in order to perform work tasks. The specifics of body articulations and of muscles decisively determine the functioning of this structure. The hands are of particular interest because they perform most of the work; it is also especially important to understand the capabilities and limitations of the spinal column. Such mechanical interpretation of the physical system enables assessment of internal strains and facilitates designing work tasks appropriate to human capabilities.

2.1 UNDERSTANDING THE HUMAN BODY

The function of the human body has long been of interest. Leonardo da Vinci (1452–1519) and Giovanni Alfonso Borelli (1608–1679) combined mechanical with anatomical and physiological explanations to describe the functioning of the human body. Following Borelli's approach, we usually model the human body as built on a structure of solid bone links which articulate in joints of various degrees of freedom; muscles that bridge the articulations move (or keep in position) the body segments (with attributed volume and mass properties). The Webers (1836) relied on Borelli's model in their discussion of the mechanics of the legs, Harless (1860) and von Meyer (1863) used it for their considerations of body mass properties, as did Braune and Fischer (1889) in their analysis of the biomechanics of a gunfiring infantryman. In 1873 von Meyer modeled body segments as ellipsoids and spheres. Dempster refined and expanded this biomechanical model in the 1950s. The Simons and Gardner model of 1960 still depicted body segments as uniform geometric shapes: cylinders for the appendages, neck, and torso, and a sphere for the head. Using equations developed by Barter in 1957, inertial parameters can be computed for the geometric forms and the moment for the total body of inertia. This elementary work still is the basis for much of the present biodynamic modeling.[1]

Body as a mechanical system
Treating the human body as a mechanical system[2] entails gross simplifications, such as disregarding mental functions. Still, many components of the body may be well considered in terms of technical analogies, such as:

- articulations: joints and bearing surfaces
- bones: structural members, lever arms, central axes
- contours: surfaces of geometric bodies
- flesh: volumes, masses
- linings of articulations/joints: lubricated sliding surfaces
- muscles: motors, dampers, or locks
- nerves: control circuits, feedforward and feedback conduit
- neural control: feedforward and feedback of signals
- organs: generators or consumers of energy
- tendons: cables transmitting muscle forces
- tendon sheaths: sliding enclosures, pulleys

2.2 THE SKELETAL SYSTEM

2.2.1 Bones

The human skeleton is composed of 206 bones[3] which articulate with other bones in their intermediate joints and are held in position by connective tissues.[4]

Biomechanically, the main function of human skeletal bone is to provide an internal framework for many parts of the body, see Fig. 2.1. The bones connect at one end or both ends in body joints. They are the lever arms on which muscles pull.

Bone framework

In early childhood, when mineralization is low, bone is soft and flexible. Mature bone is firm and hard, and thus can resist high strain while retaining certain elastic properties. Bones of the elderly are more mineralized and therefore stiffer. With aging, bones change their geometry, similar to pipes getting wider in diameter but thinner in their walls. In mechanical terms, osteoporosis in the elderly means hollowing of bones and decreasing bone mass with thinned walls getting more brittle although the moment of inertia remains about the same.

Young and old bone[5]

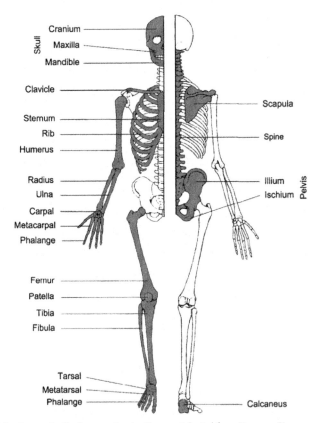

FIGURE 2.1 **Major bones in the human body.** *Source: Adapted from Kroemer, Kroemer, and Kroemer (2010).*

Use it or lose it

Bone cells are nourished through blood vessels. The flow of blood helps to continuously resorb and rebuild bone material throughout one's life. Physical exercise strengthens bone (and muscle) through local strain which encourages growth, whereas disuse encourages resorption (per Wolff's law of bone remodeling): however, overstrain can cause structural damage in terms of micro- or macro-fractures.

Joint locations, link lengths

The joint location for center-of-rotation for simple articulations can easily be determined for hinge joints in fingers and elbows. However, the center-of-rotation is more difficult to assess for complex joints with several degrees of freedom, such as in the hip or in the shoulder. Once the joint locations are established, the distance between adjacent joint centers defines the straight-line link length of a body segment. Combining segment links appropriately establishes the solid framework of the body. Table 2.1 contains information on the relations between bone lengths and link lengths.

"Stick man"

The human skeletal system is often simplified into a relatively small number of straight-line links (representing long bones) and joints (representing major articulations) as shown in Fig. 2.2. Data on link lengths, body segment masses, and on joint mobilities, for example, are needed together with appropriate terminology and computational algorithms so that such anthromechanical models[6] can represent the human body realistically. Fig. 2.3 depicts conventions used to describe motions, forces, and torques at body joints.

Body center of mass

For anthromechanical calculations, the mass of a body segment, even of the whole body, is often treated as concentrated in one point in the body, where its physical characteristics respond in the same way as if distributed throughout the body. The location of the center of mass (also called center of gravity) shifts with body posture, with muscular contractions, food and fluid ingestion or excretion, and respiration. Table 2.2 lists relative locations of mass centers.

Body volume

Archimedes' principle allows us to calculate body volume, v: the person or a body member is immersed in a container filled with water, and the displaced water corresponds to the volume. Other volume information can be obtained from cadaver dissection or from model calculations.

Body weight

Weight w is the force which a mass m exerts on earth in the direction of gravitational acceleration $g = 9.81$ ms^{-2}. Body weight $w = m\,g$ can be measured easily with a variety of scales.

Body mass

Mass m is volume times density: $m = v\,d$.

TABLE 2.1   Ratios (in %) of Link Length to Bone Length

Segment ratio	Mean (%)	Standard deviation (%)
Thigh link/femur length	93.3	0.9
Shank link/tibia length	107.8	1.8
Upper arm link/humerus length	89.4	1.6
Forearm link/ulna length	98.7	2.7
Forearm link/radius length	107.1	3.5

Source: Data from NASA/Webb (1978) and Dempster et al. (1964).

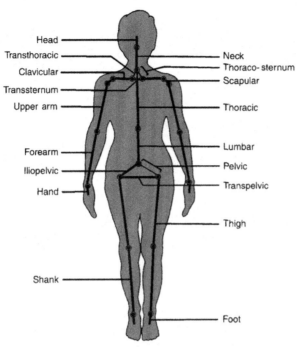

FIGURE 2.2 **A simple anthromechanical model of the human skeleton.** *Source: Adapted from NASA/Webb (1978).*

Density d is mass per unit volume v: $d = m/v = w/(g\,v)$. The specific density d_s is the ratio of d to the density of water d_{H2O}: $ds = d/d_{H2O}$.

Body density

The human body's density is very close to that of water—we almost float. The human body is not homogeneous throughout; its density varies depending on cavities, water content, fat tissue, bone components, etc. Still, in many cases it is sufficient to assume that either the body segment considered or even the whole body, is of constant (average) density.

Basic structural components such as skin, muscle, and bone are relatively constant in percentage from individual to individual. Fat, however, varies in percentage of total mass (or of weight) throughout the body and among individuals. There are several techniques to determine body fat, including skinfold measures: in selected areas of the body, the fold thickness of skin with its underlayment of fat tissue is measured with a special caliper. Knowing the fat content, we can express body mass as lean body mass plus fat mass.

Lean body

As new technologies mature and become available to the ergonomist, other methods of estimating body segments and relating them to movements allow for more accurate or detailed measurements.[7]

2.2.2 Connective Tissues

Several types of tissues connect parts of the body:[8]

- Tendons are strong yet elastic elongations of muscles, connecting them with bones.
- Ligaments connect bones and provide capsules around joints.

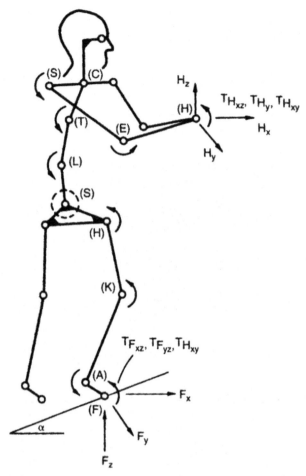

FIGURE 2.3 **Terminology to describe motions, forces, and torques at body joints.** *Source: Adapted from Kroemer, Kroemer, and Kroemer (1990), (2010).*

- Cartilage is a translucent, viscoelastic, flexible material capable of rapid growth, which serves as articulation surfaces at the joints, at the ends of the ribs, and as discs between the vertebrae.
- Muscles are the organs generating force and movement between bone linkages.

2.2.3 Joints

Joint designs

The mobility of body joints is determined by many factors: the shape of the bones at their articulations, the encapsulation of the joint by ligaments, the supply of cartilaginous membranes, the provision of discs or volar plates, and the action of muscles. Fig. 2.4 describes the basic types of bony articulations; several human joints have combinations of these features.

TABLE 2.2 Locations of the Centers of Mass of Body Segments, in Percent of Segment Length Measured From its Proximal End

Segment	Harless (1860)	Braune & Fischer (1889)	Fischer (1906)	Dempster (1955)	Clauser, McConville & Young (1969)
	(n=2)	(n=3)	(n=1)	(n=8)	(n=13)
Head[a]	36.2	—	—	43.3	46.6
Trunk[b]	44.88[b]	—	—	—	38.0[b]
Total arm	—	—	44.6	—	41.3
Upper arm	—	47.0	45.0	43.6	51.3
Forearm and hand	—	47.2	46.2	67.7	62.6
Forearm[b]	42.0[b]	42.2[b]	—	43.0[b]	39.0[b]
Hand[b]	39.7[b]	—	—	49.4[b]	18.0[b]
Total leg[b]	—	—	41.2[b]	43.3[b]	38.2[b]
Thigh[b]	48.9[b]	44.0[b]	43.6[b]	43.3[b]	37.2[b]
Shank and foot	—	52.4	53.7	43.7	47.5
Shank	43.3	42.0	43.3	43.3	37.1
Foot	44.4	44.4	—	42.9	44.9
Total body[c]	58.6	—	—	—	58.8

[a]Percent of head length, measured from the crown down.
[b]The values on these lines are not comparable with others because the investigators used differing definitions for segment lengths.
[c]Percent of stature, measured from the floor up.
Source: Adapted from Roebuck et al. (1975), based on data from Clauser, McConville, and Young (1969).

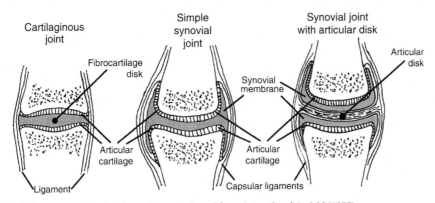

FIGURE 2.4 Types of body joints. *Source: Adapted from Astrand and Rodahl (1977).*

Our thumbs and fingers are able to perform complex finely controlled motions in their articulations within the wrist and digits. Our arms give us long reaches, and our shoulder and elbow joints provide extensive mobility. Legs provide us with powerful abilities to move: these long body members rotate about their articulations with the trunk at the hip joints, which provide wide-ranging angular freedom.

Digit, arm, leg mobility

Simpler but large angular motions occur in the knee joints. The ankle articulation allows three-axes excursions which are small but important for subtle balance of the standing body. The spine has rather complex mobility, as discussed in some detail below.

Types of joints

Some bony joints have no remaining mobility, such as the seams in the skull of an adult; others have very limited mobility, such as the connections of the ribs to the sternum. Hip and shoulder joints have pan-like bone structures inside which the knob-shaped proximal bones of upper arm and thigh can rotate. The technical analog for the shoulder or hip is a ball joint that can move about three axes of rotation (three "degrees of freedom"): the upper arm and leg can (1) rotate fore-aft as well as (2) left-right and they can (3) twist. Other joints have two degrees of freedom, such as the knuckles of the hand digits or the ill-defined wrist joint (discussed later), where the hand may be bent in flexion and extension and can deviate laterally. (The capability to perform the twisting pronation/supination is located in the forearm, not in the wrist.) The elbow and knees have the simplest joints, hinges, with only one axis of rotation, so the forearm and lower leg can simply swing forth and back.

Joint lubrication

Synovial fluid present in a joint provides lubrication and thereby facilitates movement. For example, while a person is running, the cartilage in the knee joints can show an increase in thickness of about 10%, caused relatively quickly by synovial fluid seeping into the cartilage from the underlying bone marrow cavity. Similarly, fluid seeps into the spinal discs (which are composed of fibrous cartilage) when they are not compressed, for example while lying down to sleep. This makes them more pliable directly after getting up than during the day, when they are "squeezed out" by the load of body masses and their accelerations. (This is demonstrated by noting that one is tallest immediately after getting up compared to after a day's effort.) Synovial fluid also facilitates smooth gliding of the flexor and extensor tendons of the hand's digits inside their sheaths.

Motion ranges

The terms *mobility* or *flexibility* indicate the range of motion that can be achieved in a body articulation. Figs. 2.5 and 2.6 depict common mobility measurements. The actual range depends on the design of the joint, on training (use), age, and gender. It is properly measured by an angle from a known reference position such as the so-called neutral position, which is located somewhere within the range, or as the enclosed angle between the smallest and largest excursions achieved by adjacent body segments about their common joint. Table 2.3 presents an overview of mobilities measured on American students in 1982 and 1983. Of the 32 measurements taken, 24 showed significantly more mobility in females than males, while men were more flexible only in ankle flexion and wrist abduction. The differences were small throughout, however.[9]

2.2.3.1 Artificial Joints

Joints may fail

Natural joints may fail as a result of disease, trauma, or long-term wear and tear. If conservative medical treatment does not succeed, joints may be replaced with artificial, man-made devices, such as is frequently done in hips and knees. In the United States alone, about a million hip and knee joints are implanted annually, predominantly in elderly patients. Joint replacement typically relieves pain while restoring function and mobility to the patient.

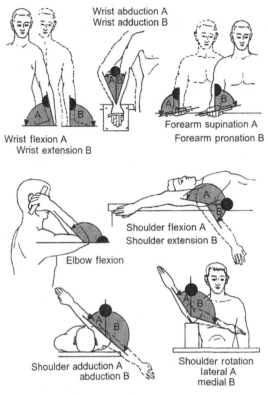

FIGURE 2.5 Hand and arm joint displacements.

The degeneration of cartilage in the hip or knee due to trauma or arthritis is the primary indication for replacement of the articulating surfaces with artificial joints. If needed, all articulating surfaces are replaced: in the hip, typically the head of the femur (thigh bone) is removed and substituted by a spherical metallic or ceramic ball on a stem. At the same time, the acetabular socket is resurfaced with a liner, generally ceramic or plastic. In the knee, the articulating surfaces on the bottom of the femur are commonly replaced with metal, and articulating surfaces at the top of the tibia (shinbone) and on the patella (kneecap) are resurfaced with plastic.[10]

Hip and knee replacements

A single-component, molded plastic integral hinge usually can replace phalangeal finger joints. This simple artificial joint is successful because of the low loads carried by the joints and the minimal debris generated by wear.

Finger joint replacement

The artificial joint is designed to yield near-normal range of motion to facilitate return to typical daily activities. Of course, it must also be suitable for implantation: since the body provides a corrosive and warm environment, the implant material must be nontoxic, of low reactivity, generate minimal wear, and maintain its structural strength. Finally, the design of the device should consider the possibility of salvage (revision surgery): ideally, sufficient bone and soft tissue should remain to allow for replacement of the implant or fusion of the joint if needed.

Design requirements

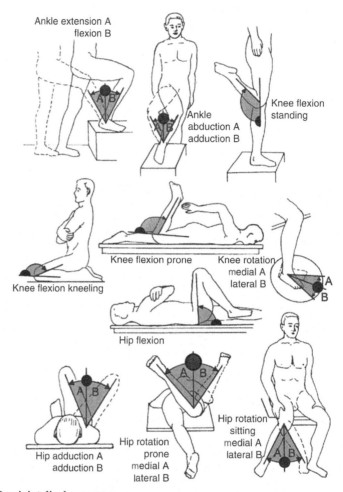

FIGURE 2.6 Leg joint displacements.

In engineering terms, *strain* is the result or the effect of *stress*: stress is the system input (the cause), strain the output (the effect). In the 1930s, the psychologist Hans Selye introduced the concept of stressors causing stress (distress, if excessive). It is confusing to use the term "stress" when it may mean either the cause or the result. (See Chapters 4 and 11 for more on "job stress.") To avoid confusion, in this text the engineering terminology is used: stress (defined in engineering terms as force applied over an area) produces strain (defined in engineering terms as a relative change in length).

TABLE 2.3 Comparison of Mobility (in Degrees) of Female and Male American Students

Joint	Movement	5th percentile		50th percentile		95th percentile		Difference[a]
		Female	Male	Female	Male	Female	Male	
Neck	Ventral flexion	34.0	25.0	51.5	43.0	69.0	60.0	fem + 8.5
	Dorsal flexion	47.5	38.0	70.5	56.5	93.5	74.0	fem + 14.0
	Right rotation	67.0	56.0	81.0	74.0	95.0	85.0	fem + 7.0
	Left rotation	64.0	67.5	77.0	77.0	90.0	85.0	none
Shoulder	Flexion	169.5	161.0	184.5	178.0	199.5	193.5	fem + 6.5
	Extension	47.0	41.5	66.0	57.5	85.0	76.0	fem + 8.5
	Adduction	37.5	36.0	52.5	50.5	67.5	63.0	NS
	Abduction	106.0	106.0	122.5	123.5	139.0	140.0	NS
	Medial rotation	94.0	68.5	110.5	95.0	127.0	114.0	fem + 15.5
	Lateral rotation	19.5	16.0	37.0	31.5	54.5	46.0	fem + 5.5
Elbow	Flexion	135.5	122.51	148.0	138.0	160.5	150.0	fem + 10.0
Wrist	Supination	87.0	86.0	108.5	107.5	130.0	135.0	NS
	Pronation	63.0	42.5	81.0	65.0	99.0	86.5	fem + 16.0
	Extension	56.5	47.0	72.0	62.0	87.5	76.0	fem + 10.0
	Flexion	53.5	50.5	71.5	67.5	89.5	85.0	fem + 4.0
	Adduction	16.5	14.0	26.5	22.0	36.5	30.0	fem + 4.5
	Abduction	19.0	22.0	28.0	30.5	37.0	40.0	male + 2.5
Hip	Flexion	103.0	95.0	125.0	109.5	147.0	130.0	fem + 15.5
	Adduction	27.0	15.5	38.5	26.0	50.0	39.0	fem + 12.5
	Abduction	47.0	38.0	66.0	59.0	85.0	81.0	fem + 7.0
	Medial rotation (prone)	30.5	30.5	44.5	46.0	58.5	62.5	NS
	Lateral rotation (prone)	29.0	21.5	45.5	33.0	62.0	46.0	fem + 12.5
	Medial rotation (sitting)	20.5	18.0	32.0	28.0	43.5	43.0	fem + 4.0
	Lateral rotation (sitting)	20.5	18.0	33.0	26.5	45.5	37.0	fem + 6.5
Knee	Flexion (standing)	99.5	87.0	113.5	103.5	127.5	122.0	fem + 10.0
	Flexion (prone)	116.0	99.5	130.0	117.0	144.0	130.0	fem + 13.0
	Medial rotation	18.5	14.5	31.5	23.0	44.5	35.0	fem + 8.5
	Lateral rotation	28.5	21.0	43.5	33.5	58.5	48.0	fem + 10.0
Ankle	Flexion	13.0	18.0	23.0	29.0	33.0	34.0	male + 6.0
	Extension	30.5	21.0	41.0	35.5	51.5	51.5	fem + 5.5
	Adduction	13.0	15.0	23.5	25.0	34.0	38.0	NS
	Abduction	11.5	11.0	24.0	19.0	36.5	30.0	fem + 5.0

Abbreviation: *NS*, not significant.

[a]*Differences are listed only if significant ($\alpha < 0.5$).*

Sources: Adapted from Kroemer, Kroemer, and Kroemer (1990) with data from Houy (1982) and Staff (1983).

Measurement of actual loads

To design ergonomic interventions for joint replacement recipients, we seek to understand the motions and loading of the joints. Loading of the joint may be estimated indirectly using traditional techniques of correlating limb position, motion velocity and acceleration, and force-sensor readings to determine the balance of forces across the joint. Numerical modeling has focused on nonlinear optimization of the forces in the tendons and ligaments at a joint along with electromyographic information of muscle activity to predict torque and force distribution during various activities.

To measure forces directly at the hip joint, for example, a specially instrumented total hip replacement with a three-axis load cell in the neck of the metal component has been used.[11] The loads measured during daily activities can then be sent telemetrically to recorders and analyzed. Other, noninvasive techniques show the effect of various activities.[12] The use of more realistic motion simulation and loading leads to improvements in understanding the capabilities of patients and improved designs for functioning.

2.2.4 The Spinal Column

Spine structure

The spinal column is a complex stack[13] of 24 bones, each called a vertebra, which make it the only "solid" structure in the human torso that keeps the rib cage from sagging into the pelvis. It consists of 7 cervical, 12 thoracic, and 5 lumbar vertebrae. The topmost vertebra is the *atlas* because it bears the skull of the head. The bottom region (*sacrum*) of the spine consists of a fused group of rudimentary vertebrae, which then terminates at the tailbone (*coccyx*). The sacrum articulates slightly via the sacroiliac joints with the bones of the pelvic girdle (see Fig. 2.7).

Spine curvature

The stack of vertebrae, seen from the side, naturally has two forward bends (*lordoses*) in the cervical and lumbar sections, and one backward bend (*kyphosis*) in the chest area (Fig. 2.7). Only in the front and rear views is the spinal column in a straight line; a lateral distortion is called *scoliosis*.

Spinal vertebra

Fig. 2.8 is a schematic sketch of a vertebra: its bone forms a solid main body and behind it a ring-like opening (*foramen*) that protects the spinal cord which runs up and down. It also has openings (*foramina*) for sensory and motor nerve branches (roots) to the left and right. The bone structure has five protrusions: the *spinous process* (SP in Figure 2.8) points backward and the two *transverse processes* (TP) point left and right. These protrusions act as lever arms for attaching ligaments and muscles which stabilize or bend the spinal column by their coordinated pulls. Two other protrusions, the *superior articulation processes*, point upward, ending with rounded surfaces that fit into their counterparts of the next vertebra. These are the *facet* joints (F).

Spine joints

Fig. 2.9 schematically depicts the five lumbar vertebrae (L1 through L5) atop the sacrum, seen from the right. The flat main body of each spinal vertebra sits on top of an *intervertebral disc*, a tough fibrocartilage cushion enclosed by a strong fibrous ring. The disc, filled with viscous synovial fluid, is an elastic pad that absorbs shocks and allows the vertebrae above and below to make small changes in their angles of tilt and twist. However, the small angular displacements add up over the stack of vertebrae and discs, giving the whole spinal column considerable ability to twist and bend, both fore-aft and sideways (left-right). The heavy lines in the figure indicate load transmitting surfaces, whereas the discs provide elasticity and displacement, the facet joints between the inferior and superior

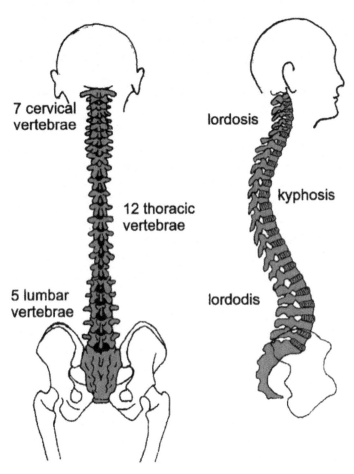

FIGURE 2.7   The spinal column.

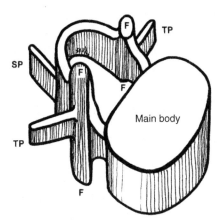

FIGURE 2.8   A simple model of basic mechanical structures of a vertebra.

I. THE ERGONOMIC KNOWLEDGE BASE

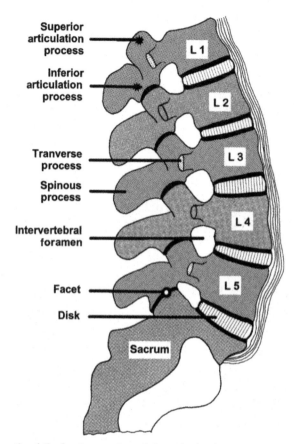

FIGURE 2.9 Schematic of the lumbar section of the spinal column, with bearing surfaces drawn in heavy lines; spine is seen from the right.

processes have no cushioning and hence provide little displacement but strong load bearing capability. The sketch shows the six bearing surfaces with the vertebrae above and below: two interfaces are at the elastic discs at the bottom and top of the main body; four bony articulations are at the facets of the two superior and two inferior surfaces.

Spine functions   Being the only solid (albeit elastic) structure between the shoulders and the pelvis, the spinal column keeps them at distance. Its top vertebra (the atlas) supports the head. Its thoracic vertebrae are linked elastically through connective tissues with bones of the shoulder and with the ribs. Below, the lumbar section atop the sacrum ties elastically to the pelvis. At its posterior side, the stack of vertebrae provides the spinal canal, a "bone tunnel" which protects the spinal cord that carries signals between the brain and sectors of the body.

Spine flexibility   The spinal column transfers compressive and shear forces as well as bending and twisting moments between the bones of the pelvis and of the shoulder and head. This complex rod[14] is held in delicate balance by ligaments connecting the vertebrae and by muscles that pull at the rear and the sides of the spinal column. Longitudinal muscles of

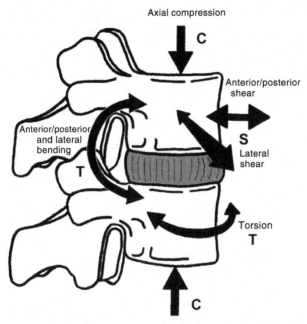

FIGURE 2.10 **Forces and torques acting on the spinal column.** *Source: Adapted from Marras (2008) and Kroemer, Kroemer, and Kroemer (2010).*

the trunk also both stabilize and load the spine. Even though only small displacements are possible between adjacent bones of the spine, the head and neck can twist and can bend, fore-aft and sideways, within fairly narrow limits by using the upper section of the spinal column. The larger movements of the trunk, in bending as well as twisting, occur mostly in the lower parts of the spine, especially in the lumbar section.

Loads on the spine are depicted in Fig. 2.10. Compression C is a major load. Yet, owing to the naturally slanted arrangement of load-bearing surfaces at the discs, particularly if the spine is bent, the spine is also subjected to shear S. Furthermore, the spine must withstand torques T, coming with bending both fore/aft and sideways, and with axial twisting. A healthy spine is capable of withstanding considerable loading yet flexible enough to allow motions. There is, however, a trade-off between load carried and flexibility. If there is no external load on the spine, only its anatomical structure (joints, ligaments, and muscles) restricts its mobility. Applying load to the spinal column reduces its mobility until, under heavy load, the range of possible postures become limited. The ability of the spine to transmit large forces, mostly in compression, is remarkable. Yet, overloading can cause damage, often to the spinal disks in the lumbar region, which is frequently the location of discomfort, pain, and injury because it must transmit substantial forces and torques to and from the upper body. Low back pain (LBP) is common, especially among people who handle loads (see Chapter 13).

The traditional model of the spine has been a straight column, as depicted in Fig. 2.11, with axial compression force transmitted along a straight line. If that so-called thrust line stays within the stack of vertebrae, this simplification allows a unique description of its

Spine strains

Stability under load

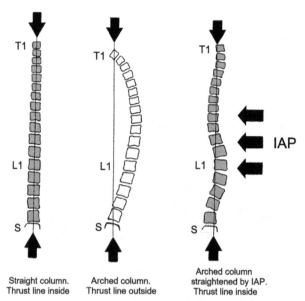

FIGURE 2.11 **Models of the spinal column under longitudinal compression: as a straight column, arched and stabilized by intraabdominal pressure (IAP).** *Source: Adapted from Aspden (1988).*

geometry and strain under the applied load. If the spinal column arches, its load-bearing capability depends on the curvature: transverse forces may keep the arch straight enough to remain stable. Such stabilizing forces may come from ligaments and muscles, helped by intra-abdominal pressure (*IAP*). If the thrust line falls outside the arch, tensile force must keep the arch stable within its possible position range, or it buckles.

Pressure within the abdominal cavity regularly increases when we are ready to lift a heavy weight or do a similar hard effort. Fig. 2.12 illustrates that IAP generates pressure vectors P against the retaining walls; the pressure is of the same magnitude in all directions. Forces P can stiffen the spinal column against buckling; they can also reduce the spinal compression force C that results from the pull force M of longitudinal trunk muscles as well as from the load due to the weight of body segment masses and of the external load.

The gel-like round core of the intervertebral disc, the *nucleus pulposus*, is the main load-bearing and load-transmitting element. It is kept in place by surrounding layers of elastic material, the *annulus fibrosus*. The disc with its surrounding ligaments allows relative movements of adjacent vertebrae, and they form a physiological shock absorber. When not functioning properly because of injury or deterioration, the cartilaginous end plates of the vertebrae and possibly also the facet joints may sustain unsuitable strains. Severe displacement of its cartilaginous components and/or serious dislocation of the bony structures of the spinal column may squeeze the spinal cord as it runs through its spinal tunnel or impinge on nerve branches (roots) of the cord as they emanate laterally between vertebrae.

The nucleus pulposus has no blood supply of its own but is nourished through exchange of tissue fluid, which circulates through the disc as a result of osmotic forces, gravitation, and the pumping effects of body movements on the spinal column. Thus,

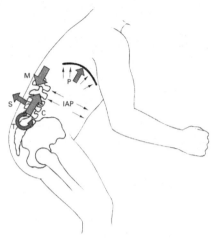

FIGURE 2.12   Intraabdominal pressure (IAP) and its resulting force vectors (P) may reduce the spinal compressive force (C) which is produced by trunk muscle pull (M). Shear force (S) and torque (T) also load the spine.

the nourishment fluid flow improves with activity and is reduced by immobilization. Tissue fluid circulation is needed to provide water, solutes, glycosaminoglycans, protein, and collagen. If this proper balance is not achieved, fissures can develop in the annulus through which nuclear material may penetrate and herniate peripheral areas.

The spinal column, especially in its discs, is often the location of injury, discomfort, and pain, because it must endure substantial strains due to internal and external stresses. For example, when standing or sitting, impacts and vibrations from the lower body are transmitted primarily through the spinal column into the upper body. Conversely, forces and impacts induced through the upper body, such as when working with the hands, especially when handling loads, are transmitted downward through the spinal column to floor or seat supporting the body. (See the discussion of material handling in Chapter 13.) Thus, the spinal column must absorb and dissipate a great deal of energy, whether it is transmitted to the body from the outside or generated inside by muscles for exertion of work to the outside.

Low back pain (LBP) is just that: a painful sensation of disorder in the low back area. LBP may stem from a large number of possible sources,[15] many believed to be basically associated with time-related changes in the spinal column and in its supporting ligaments and muscles. Degeneration increases as one gets older, apparently resulting from a combination of repetitive trauma and the normal aging processes. Strong activity demands may trigger the occurrence of various low back symptoms; however, except in cases of acute injuries, the causes of LBP often remain unclear. Classification of LBP is difficult, and different clinicians frequently diagnose it differently. Furthermore, many people who show objective signs of spinal degeneration (as diagnosed by spinal imaging, such as X-rays, CT scans, MRI scans, myelograms, and discograms) do not feel any pain. (Generally speaking, clinicians— and engineers—treat functional problems, not imaging results, so asympomatic low back disorders likely require no intervention.)[16]

Back pain

Why LBP?

Skeletal adjustments for pregnancy

In pregnant women, increasing bulk at the abdomen shifts the center of mass forward. To counteract this, the mother-to-be can lean back slightly, which curves the lumbar spine more. Increasing the lumbar curve can create complications: vertebrae are more likely to slip against each other, which can cause back pain, even injury. The female spine has several features that help to prevent problems. In women, the lordotic curve in the lower back mainly includes three vertebrae, which distributes strain over a wider area than in men, where it spans only two vertebrae. Furthermore, the facet joints are larger in women than in men and the joints are oriented at a slightly different angle, which makes them able to resist higher force and better brace the vertebrae against slipping.

2.2.5 The Hands

Bones and joints

Fig. 2.13 shows a top (dorsal) view of the bones and joints of the right hand.[17] Its 27 bones provide the solid mechanical structures with various joints designs and varying

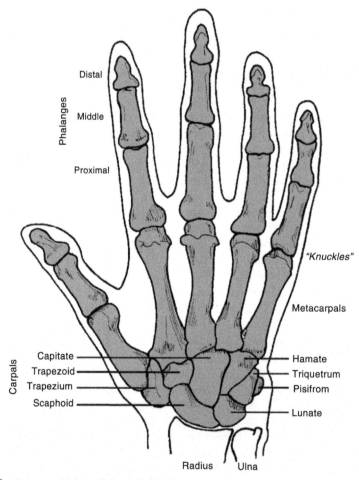

FIGURE 2.13   **Bones and joints of the right hand.**

mobility against each other. The basic structure of the five digits is similar except that the thumb has three distal bones while the fingers have four bones.

As we examine mobility, starting at the tips of the digits, we find that all joints of the *phalanges* are simple one-axis single joints, whereas the "knuckles" (*metacarpal-phalangeal joints*) provide considerable mobility in their two-axes articulations. Only very small displacement is possible at the carpal-metacarpal joints of the four fingers and among the eight carpal bones; however, the thumb demonstrates exceptional mobility so that it can oppose any of the other digits. At the base of the hand, two carpal bones (*scaphoid* and *lunate*) interface with the distal portions of the two forearm bones, radius and ulna. That gliding wrist joint gives the hand wide-ranging mobility in two axes: bending up–down and left–right. The ability to twist is actually located within the forearm. Combined, the various articulations give the hand its complex dexterity.

Joint designs and encapsulations establish mobility limits but the actual motions are affected by the pull of muscles (see below). Some of the muscles that bend (flex) and straighten (extend) the hand are located within the hand; these so-called *intrinsic* muscles control most of the finer aspects of manipulations. However, other muscles are in the forearm (away from the hand, therefore called *extrinsic*) from where they produce most of the powerful digit activities. The contractile forces generated by the muscles are transmitted to the hand by pull on tendons that cross the wrist. Fig. 2.14 highlights the extensor tendons at the back of the hand and Fig. 2.15 shows the flexor tendons on the palmar side. Fig. 2.16 sketches a cross-section at the base of the hand: it shows that the extensor tendons lie at the back side of the hand while the flexor tendons pass at the

Hand mobility

Motion control

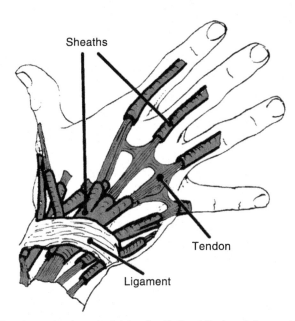

FIGURE 2.14 **Digit extensor tendons straighten the digits of the hand.** *Source: Adapted from Kroemer and Kroemer (2001) and Putz-Anderson (1988).*

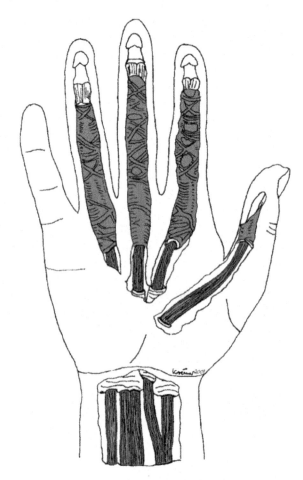

FIGURE 2.15 **Digit flexor tendons bend the digits of the hand.** *Source: Adapted from Kroemer and Kroemer (2001).*

palmar side through a narrow opening, the *carpal tunnel* formed by carpal bones and a transverse ligament.

Tendon sheaths   Tendons are kept in place by synovial sheaths which also provide lubrication for their gliding. These tubular membranes have complex designs, depending on their locations and purposes. Reenforced by ring-like or crosswise ligaments where needed, sheaths supply attachments to bones and "pulleys" at which the tendons pull to stabilize or articulate sections of the digits, as sketched (much simplified) in Fig. 2.17.

2.3 MUSCLE

Skeletal muscles are the "motors" that propel and stabilize the parts of the skeletal system, hence body segments and the whole body.[18]

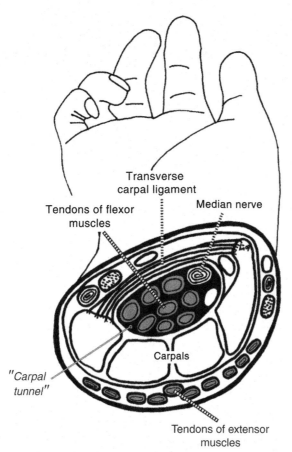

FIGURE 2.16 Cross-section of the right hand, distal from the wrist. Carpal bones and the transverse carpal ligament form the "carpal tunnel" through which pass digit flexor tendons and the median nerve. *Source: Adapted from Kroemer and Kroemer (2001).*

The human body has three types of muscle, which together comprise about 40% of the weight of the body. *Cardiac muscle* brings about contractions of the heart to pump blood. *Smooth muscle* constricts blood vessels. *Skeletal (voluntary) muscle*, under control of the somatic nervous system (see Chapter 4), maintains postural balance and moves body segments.

The human body has about 600 or 700 skeletal muscles (depending on how one counts), identified by Latin names. (The Greek terms for muscle, *mys* or *myo*, are often used as prefixes.) Most skeletal muscles are formed like a spindle, with a large "belly" between slender ends. The proximal end (toward the center of the body) of the muscle is called the *origin*; the distal end is called the *insertion*. Commonly, tendons extend from the muscle ends to bones. These connectors can be long: for example, the flexor tendons between muscle in the forearm and phalanges in the hand can measure nearly 20 cm.

Types of muscle

Skeletal muscles

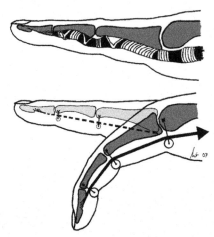

FIGURE 2.17   **Ring-like and cruciform ligaments keep the flexor tendon attached to finger bones and, by providing "pulleys," allow it to bend the finger by pulling.** *Source: Adapted from Kroemer and Kroemer (2001).*

"Body motors"

Skeletal muscles connect two body members, usually across one common joint but in some cases across two or more joints (as in the hands). Their pulling creates torques (moments) around body joints which serve as pivots.

Muscle tension

Muscles perform their functions by temporarily and often quickly developing internal lengthwise tension. This is habitually called contraction, but in fact tensioning a muscle does not always shorten it; overtly, it may stay at the same length, called *isometric*, or can even be lengthened by an overwhelming external force.

Co-contraction

The usual arrangement of skeletal muscles is in a "functional pair" of counteracting muscles: one flexes, the other muscle extends (as in the upper arm, where the biceps and triceps move the forearm). The active muscle is called *agonist* (or *protagonist*), its opponent *antagonist*. Their simultaneous activation controls the speed of body segment motion and the strength of exertion.

> Careless use of the term *contraction* can lead to confusion[19] because the muscle may not actually shorten. During a "static contraction," muscle length remains more or less the same, isometric. In an "eccentric contraction," the muscle is actually lengthened. Muscular contraction can produce tension in the muscle only to the extent that resistance against the shortening exists. Therefore, the event of a "contraction" does not necessarily imply a forceful effort of the muscle. To avoid misleading implications and contradictions in terms, it is often better to use the terms *activation*, *effort*, *exertion* instead of contraction, and *tense* or *tighten* for contracting.

2.3.1 Architecture of Skeletal Muscle

Muscle "bundles"

Skeletal muscle has a complex structure: as shown in Fig. 2.18, every muscle actually consists of bundles of (10–150) muscle fibers. A membrane, the sarcolemma, wraps the

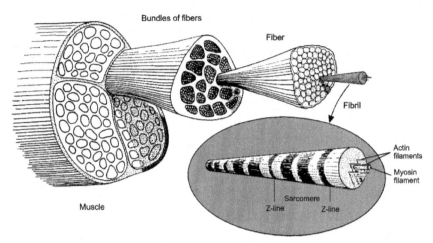

Bundles of fibers

Fiber

Fibril

Actin filaments

Myosin filament

Sarcomere

Muscle

Z-line   Z-line

FIGURE 2.18 **Major components of muscle.** *Source: Adapted from Kroemer, Kroemer, and Kroemer (2010), Wilmore et al. (2008), and Astrand, Rodahl, and Dahl (2003).*

fiber bundle which also contains mitochondria, the "energy factories of the muscle." Muscle is pervaded by nerves and blood vessels. A muscle usually has several 100,000 parallel fibers in its middle (the muscle belly), and tapers off toward its ends (the origin and insertion).

The muscle fiber is cylindrical, with a diameter of 10^5 to 10^6 Ångstrom (1 Å = 10^{-10} m) and up to 500 mm long. Each fiber is a single large cell with several hundred cell nuclei located at regular intervals near its surface. Every muscle fiber, in turn, consists of fibrils also packed into bundles wrapped by connective tissue. These myofibrils are mostly parallel to each other by the hundreds or thousands.

Viewed via a microscope, series of lighter and darker stripes appear across each fibril. This striping has led to the name *striated muscle*. One distinct kind of stripe indicates the locations of a z-disc (from the German word *zwischen*, between), a dense membrane crossing the fibril. The part between two adjacent z-discs is called the sarcomere; see Fig. 2.19. That distance is about 250 Å when the muscle is at rest. Thus, 1 mm of muscle length can contain many thousands of sarcomeres end-to-end.

The z-discs contain important transverse parts of a network of tubules, sacs, cisterns, and channels which permeate fibers and hence the whole muscle. This is the "control and plumbing" network of the muscle, into which blood transports chemical messages as well as oxygen and energy carriers, and from which it removes metabolic byproducts (see also the discussion on metabolism in Chapter 3).

Fibrils consist of bundles of filaments, rod-shaped polymerized protein molecules which attach end-to-end and lie side-by-side. There are two kinds of filaments, the thicker one (about 150 Å in diameter) consisting mainly of *myosin* and the thinner one (90 Å) of *actin*. A myofibril contains usually 10–2500 tightly packed myosin filaments and about double as many actin filaments; the actin rods extend from both sides of each z-disc.

The actin and myosin filaments constitute the contracting microstructure of the muscle. Lengthwise, several actin strands wrap around each of the myosin in form of a

Fibers and fibrils

Striation, sarcomere

Control and plumbing

Bundles of filaments

Muscle contraction

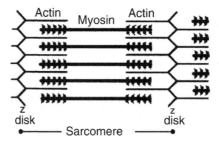

FIGURE 2.19   **Schematic of a sarcomere with the myofilaments actin and myosin.** *Source: Adapted from Kroemer, Kroemer, and Kroemer (2010).*

double helix. When a muscle is at rest, its actins and myosins are separated by troponin and tropomoysin proteins. For a contraction, these pull away, which allows actin strands to connect with myosin at so-called cross-bridges (which resemble tiny golf club heads), serving as temporary attachments by which the actins ratchet themselves along the myosin rods. Fig. 2.20 sketches actin and myosin rods within shortened, relaxed, and stretched muscle; the length of the sarcomere changes accordingly.

Muscle length

From its resting length, a muscle may actively contract or be passively stretched, as illustrated in Fig. 2.20. As a general rule, sarcomeres (and hence the total muscle) can contract to about 60% of resting length, or may be stretched to about 160% without damage. At about double resting length, muscle fibers, tendons, or tendon attachments are likely to experience damage.

Muscle actions

Muscle may remain relaxed, may be extended by an external force, or it may shorten (contract). Table 2.4 lists terms that have been applied to various conditions of muscle

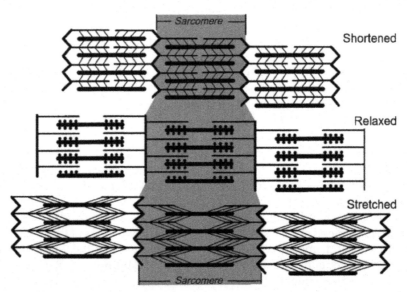

FIGURE 2.20   **Schematic of sarcomeres in shortened, relaxed, and stretched muscle.** *Source: Adapted from Kroemer, Kroemer, and Kroemer (2010).*

TABLE 2.4 Terms Describing Muscle Efforts

Term	Meaning	Same as	Opposite of
Concentric	Muscle shortens against resistance		Eccentric
Eccentric	Lengthening of a resisting muscle by external force		Concentric
Isoacceleration	Acceleration remains constant		Acceleration changes
Isoforce	Muscle force (tension) remains constant	Isokinetic, isotonic	Muscle force (tension) varies
Isoinertial	Muscle moves a constant mass		
Isokinematic	Velocity of muscle shortening (or of lengthening) remains constant		
Isokinetic	Muscle force (tension) remains constant	Isoforce, isotonic	
Isometric	Muscle length remains constant	Static	Dynamic
Isotonic	Muscle force (tension) remains constant	Isoforce, isokinetic	Muscle force (tension) varies
Static	Muscle length remains constant	Isometric	Dynamic

action. However, muscle may not actually behave in the manner implied by these wordings. For example, an "isometric" contraction does involve some muscle fiber shortening which is largely compensated for by simultaneous tendon elongation; so the body limbs involved may not show any displacement. Most traditional descriptors of muscle strength exertion (iso-inertial, -kinetic, -metric, -tonic) are simplistic, even misleading,[20] because they do not accurately describe the actual physiologic events. Furthermore, these terms may be misleadingly applied to strength testing devices. Of all listed terms, only the physical descriptors eccentric, concentric, static, and dynamic are not ambiguous.

2.3.2 Control of Muscle Contraction

Electrical impulses, generated in the brain or spinal cord and transmitted along the motor nerves, control the activity of muscles.[21] In a relaxed muscle, connections between actin and myosin filaments are chemically neutralized. An incoming nervous signal initiates filament activities: they establish cross-bridges between actins and myosins. This may result in:

Nervous signals

- shortening of the muscle (concentric contraction), or
- keeping the muscle at the same length (isometric or static contraction), or
- lengthening of the muscle (eccentric contraction),

Depending on whether or not the muscle overcomes a counteracting external resistance.

When nervous stimulation stops, muscular activities end as well.

2.3.2.1 The Motor Unit

The *motor endplate* is the contact area between the axon end of a motor nerve (alpha-motoneuron) and the sarcolemma of muscle. Each motor nerve has many (usually

Innervation

hundreds or thousands) of such myoneural junctions and hence innervates the activities of that many muscle fibers. The fibers controlled by the same signal are called a *motor unit*. However, the fibers of one unit usually do not lie close together but are distributed throughout the muscle; therefore, "firing" one motor unit does not cause a strong contraction in one spot but brings about rather weak contractions throughout the muscle.

Innervation ratio

Motor units can be classified by their innervation ratios, meaning the number of fibers innervated by one neuron. Muscles used for finely controlled actions, such as rotation of the eyeball, have a ratio such as 1:7, whereas muscles for gross activities generally have ratios around 1:1000 or even smaller.

Slow but enduring fiber

Another classification of motor units describes their types of muscle fibers. Type I fiber is short and appears red because it is penetrated by many capillaries, which provide good blood supply and oxygen storage: hence it resists fatigue. A contraction is produced by a relatively low action potential, but takes relatively a long time (60–120 ms) to peak; therefore, Type I is called a slow fiber.

Fast but fatiguing fiber

Type II fibers appear light in comparison because they are not profusely capillarized. They are less resistant to fatigue but perform better under anaerobic conditions. They require high action potentials and produce fast twitches (15–50 ms) to peak. Type II fibers are further subdivided according to their supply with capillaries.

Note that the times just given for slow or fast reactions have been measured in isolated muscle preparations, usually taken from cats, and with artificial stimulation. In the human muscle, distinct groupings of muscle fibril groupings are not as prevalent, and the actual behavior of the stimulated muscle depends on many factors, such as external resistance and fatigue—discussed below.

2.3.3 Activation of the Motor Unit

Muscle fibers are enclosed by a semipermeable membrane, the *sarcolemma*. Among other purposes it serves to receive and distribute muscle action signals. When a muscle is at rest, sodium (positive) and some potassium (positive) ions accumulate on the outside of the sarcolemma, while (negative) chlorine ions are on the inside. This establishes a polarized membrane, with a transmural electrical potential of nearly 100 mV.

Axon action impulse

An action impulse arriving from a motoneuron must be strong enough to depolarize the sarcolemma by at least 40 mV. If this threshold is not achieved, the motor unit does not react; if achieved, the motor unit contracts completely. This "all-or-none principle" governs motor unit contraction.

Potential propagation

After sufficiently strong depolarization, the permeability of the sarcolemma increases so that positive ions can neutralize negative ions. This local depolarization, called endplate potential, propagates at a speed of up to 5 m/s along the membrane. The wave, acting like an electric current, causes hydrolysis of water molecules which releases hydrogen and hydroxyl ions, which, in turn, split off a phosphate group from adenosine triphosphate (ATP), the main energy store at the muscle.

Splitting ATP frees energy

Decomposition of ATP causes formation of adenosine diphosphate (ADP) and of phosphoric acid. This reaction liberates energy, about 11 kcal per mole of ATP. This is the primary source of energy for muscular contraction (see more on energy release in Chapter 3).[22]

A single contraction of a motor unit is called a *twitch*: after an action impulse arrives at the motor unit, there is an initial *latency period*, typically about 10 ms in a Type II fiber, during which no perceptible muscle tension or change in muscle length takes place. Then follows the *contraction period* which usually takes up to 40 ms. During this period, energy from the decomposition of ATP allows actins to pull along the myosin filaments, achieving muscle contraction. Next, the actin–myosin cross-bridges disengage and the muscle returns to its resting length during the *relaxation period*. During this time of no contraction, typically about 40 ms, ADP picks up a phosphate group to resynthesize ATP. The final *recovery* phase lasts another 40 ms or so, if no contraction signal arrives. During this time, aerobic metabolism oxidizes glucose and glycogen, ATP and phosphocreatine are regenerated in the mitochondria. A twitch time of about 130 ms is typical for isolated Type II fiber but can last more than 200 ms in skeletal muscle and may be as short as 10 ms in ocular muscle.

OVERHEAD WORK[23]

An everyday example of muscle fatigue is the pain we experience during overhead work, as shown in Fig. 2.21. After a fairly short time, we feel severe discomfort in shoulder muscles that keep the arms elevated, and often in neck muscles that keep the head tilted back, and we find it impossible to continue even though nerve impulses still arrive at the neuromuscular junctions and action potentials continue to spread over muscle fibers.

FIGURE 2.21 **Fatiguing overhead work.** *Source: Adapted from Nordin, Andersson, and Pope (1997).*

Twitch summa-
tion
Still looking at a single motor unit, we see *summation* (also called *superposition*) of twitches when they are initiated so quickly after each other that a contraction is not yet completely released by the time the next stimulus arrives. In this case, the new contraction builds on a level higher than if the fiber were completely relaxed, and accordingly, higher contractile tension is generated in the muscle. This "staircase" effect takes place when excitation impulses arrive at frequencies of 10 or more per second.

When more than about 30 stimuli per second arrive at a muscle, successive contractions fuse together, which causes a maintained contraction called *tetanus*. Such superposition of twitches generates muscle tension two or three times larger than a single twitch; a full tetanus may have tension that is five times stronger than in a single twitch.

Rate coding
As just discussed, in a single motor unit the frequency of contractions and the strength of contraction are controlled by the frequency of the exciting nervous signals. This is called *rate coding*. However, the fibers belonging to one motor unit are generally in various locations within the muscle. Therefore, activation of one motor unit brings about a weak contraction throughout the muscle.

Recruitment
coding
If more than one alpha-motoneuron excite their motor units to contract at the same time, *recruitment coding* takes place. The larger the number of motor units contracting simultaneously, the higher the contractile strength exerted by the total muscle. Thus, nervous control of muscle strength (more on this below) follows a complex pattern of rate and recruitment coding.

2.3.4 Muscle Fatigue

If there is insufficient time for relaxation and recovery before the next contraction, we experience the feeling of *muscle fatigue*, which signals that muscle is becoming unable to continue or repeat an effort. The reasons are variable and complex: they commonly relate to lacking energy delivery to the involved muscle, to accumulation of metabolic byproducts, to overexertion of contraction mechanisms, even to events within the nervous system. The occurrence of such fatigue depends on type and intensity of the effort, on the fiber type of the involved muscles, and on individual fitness, training, and motivation.

Fatigue relieved
by rest
Muscular fatigue is overcome by rest, during which accumulated metabolic by-products are removed and ATP and phosphocreatine are rebuilt.

Endurance
Fatigue of a single motor unit, or of a whole muscle, depends on the frequency and intensity of muscular contraction and on the length of time it is maintained. The more strength exertion is required of a given muscle, the shorter the period through which this strength can be maintained. Fig. 2.22 shows schematically this relationship between strength exertion and endurance during a static, isometric, effort.

> Because of muscle fatigue, maximal strength can be maintained for only a few seconds. A practical rule is that 50% of strength is available for about 1 minute; less than 20% can be applied continuously for a long period of time.

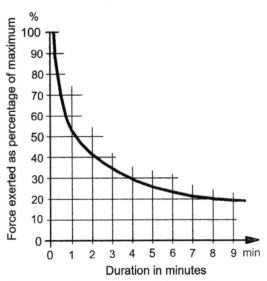

FIGURE 2.22   Relation between strength and endurance during an isometric effort.

2.3.4.1 Length–Strength Relationships

When contraction stimuli do not exist and no external force is applied, the muscle relaxes at its resting length. At other lengths, the isometric force that the muscle is able to generate depends on its length: hence, the length–strength relationship is of interest.

If there is no external resistance, nervous stimulation causes the muscle to contract to its smallest possible length, which is about 60% of resting length. In this condition, the actin fibrils are completely curled around the myosin fibrils, so the z-lines are as close as possible. This is the shortest feasible length of the involved sarcomeres; therefore, at that position, the *active* contractile force of the muscle is zero. As external loads stretch the muscle longer and longer, the actin proteins can establish suitable cross-bridges with the myosin rods; so, the active contractile tension increases.   Active contractile force

Like a rubber band, the muscle shows *passive* resistance to being stretched. If an external force lengthens a muscle beyond its resting length, its sarcomeres become increasingly elongated. This slides the actin and myosin fibrils along each other even when they counter with a contraction effort. At a muscle stretch of 120%–130% of resting length, the cross-bridges between the actin and myosin rods overlap optimally to generate a contractile force. But if the sarcomeres are elongated further, the overlap gets smaller and smaller until, at about 160% of resting length, so little overlap remains that no active contractile force can be developed. With further elongation, the passive stretch resistance increases strongly until the point of damage to (or even breakage of) muscle or tendon or attachment.   Passive resistance to stretch

Accordingly, active contractile force developed within a muscle is zero at approximately 60% resting length, becomes about 90% at resting length, goes to maximum at about 120%–130% of resting length, and then falls back to zero at about 160% resting length. Above resting length, the tension in the muscle is the sum of active   Summation of active and passive tensions

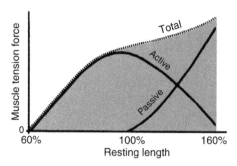

FIGURE 2.23   Active, passive, and total static tension (force) within a muscle at different lengths.

and passive components (see Fig. 2.23), which explains why we stretch ("preload") muscles for a strong force exertion, such as in bringing the arm behind the shoulder before throwing.

To assess muscle strength by considering transmission of muscular effort within the biomechanic links of the body. The arm–hand system is a typical example, sketched (much simplified) in Fig. 2.24.

The distal end of the upper arm link articulates in the elbow joint around the proximal end of the forearm link. The flexor muscle connects both links with its origin near the proximal end of the upper arm link[24] and its insertion on the forearm link at a relatively short distance m from the elbow. The flexor muscle generates a force M, which pulls on the forearm at the angle α (which changes with elbow angle). Hence, $M\ sin\ \alpha$ is the component of M that is perpendicular to the forearm; it generates a torque T about the elbow. The magnitude of T depends on force M, the lever arm m and the pull angle α:

$$T = m\ M\ (\sin \alpha) \tag{E#2.1}$$

Torque T transforms into hand force H, which is perpendicular to the forearm link at its lever arm h, according to lengths of the lever arms m and h:

$$T = h\ H \tag{E#2.2}$$

Solving for hand force H yields

$$H = T\ /\ h = m\ M\ (\sin \alpha)\ /\ h \tag{E#2.3}$$

If α, m, h, and H are measured, one can compute the muscle force M:

$$M = h\ H\ /\ [m\ (\sin \alpha)] \tag{E#2.4}$$

Muscle tension during motion

The summation of active and passive tension within the muscle also appears during motion, when muscle length changes. The higher the velocity of muscle contraction, the faster actin and myosin filaments slide by each other and the less time is available for the cross-bridges to develop and hold. Accordingly, muscle tension during motion (either concentric or eccentric) is lower than static tension.

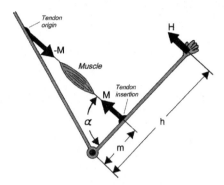

FIGURE 2.24 **Schematic of a flexor muscle exerting pull force *M* between origin and insertion at bone links.** *Source: Adapted from Kroemer, Kroemer, and Kroemer (2010).*

2.4 VOLUNTARY STRENGTH

Voluntary muscle strength is defined as the maximal force (or tension stress: force divided by cross-section) that the muscle can develop voluntarily between its origin and insertion.

The contraction characteristics of a muscle depend on its types of fibers as well as their numbers and orientations. Muscles fibers may be at various directions relative to the long axis of the muscle, like in a feather; such arrangements are called *penniforms*. In the *fusiform* arrangement, fibrils and fibers lie parallel to the center line of the muscle. In each case, the strength capability of a muscle depends on its number of fibers side-by-side, which determines its cross-sectional thickness. Thus, assuming equal tension, a thicker muscle can develop more force than a thinner muscle.

Strength training increases the thickness of fibers, but probably not their number. Endurance training also increases capillary density and mitochondrial volume. Either kind of training also improves the overall coordination of motor unit activation in the central nervous system (CNS, see also Chapter 4) because "muscles are the slaves of their motoneurons."[25]

Muscular contraction results from activation of motor units; both rate and recruitment coding control the cooperative effort of the participating units. High frequency of excitatory nerve signals and concurrent stimulation of many motor units generates large muscle strength. However, it appears that one cannot voluntarily activate more than about two-thirds of all fibers of a muscle at once. Apparently, this limitation assures that structural tensile capacity normally is not exceeded—although this can occur as a result of a reflex, which might damage or even tear a muscle or tendon.

Body motion, or lack thereof, is a major factor which determines the force, torque, work, power, or impulse transmitted from the body to an external object. When the body does not move, muscles maintain their length. In physiological terms, this called an *isometric* condition. In physics terms, this is called a static condition: all forces acting within the system are in equilibrium, as Newton's first law requires.

Inherent muscle strength

Strength training

Muscle activation

Static and dynamic strength

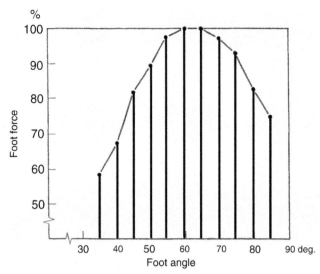

FIGURE 2.25 **Results of separate static foot strength measurements.** The convenience line drawn to connect the individual measurements should not be misread: the measurements were NOT done during a continuous motion.

In everyday activities, muscle strength is generally exerted while the involved body segments are in motion: unfortunately, measurement of dynamic strength is difficult to do because of a lack of suitable measuring devices[26] and because the measurement conditions are difficult to control. In contrast, static exertions are easy to measure and easy to control. Therefore, most strength data available in the literature report static (isometric) forces, as illustrated in Fig. 2.25 which shows, schematically, measurements of static foot force, taken separately at various degrees of foot angle.

NOTE OF CAUTION

Connecting measured isometric strength values with a continuous line, habitually done in the literature, can be misleading: the connecting curve (as in Fig. 2.25) seems to imply motion when in fact strength was exerted and measured in separate static positions, independently, one after the other, with a break after each exertion.

Force units

In the international system of measurements SI (Système International), the unit for force measurement is the Newton (N). Other units are occasionally used: the kilogram-force (also called kilopond, kp) of 9.81 N, or the pound-force of 4.45 N. The units gram and pound are measures of mass which, multiplied with 9.81 m/s^2 acceleration, exert that weight force on a scale on Earth.

Torque (also called moment) is the product of force and its lever arm (distance) to the articulation about which it acts; the direction of the force must be at a right angle to its lever arm. In kinesiology, the lever arm is often called the mechanical advantage.

Forces as well as torques are vectors: described both by magnitude and by direction.

Within a muscle, filament contraction in the longitudinal direction of the muscle fiber generates tension (in units of force per cross-sectional area). The tensions in each filament combine to a resultant tension of the muscle, with its magnitude proportional to the cross-sectional thickness of the muscle. Estimates of muscle tensions range from 16 to 61 N/cm^2: an oft-used general value called *specific (skeletal muscle) tension*[27] is 30 N/cm^2. So, if the cross-sectional area is known (such as from MRI scans or from cadaver measurements), one can estimate the force of the muscle. However, it is worth keeping in mind that this calculation relies on assumptions about cross-section and specific tension values—since no technique directly measures the muscle force in the living human.

Muscle tension

2.4.1 Regulation of Strength Exertion

The conceptual model in Fig. 2.26 helps to understand steps and factors in the exertion of voluntary strength. Feedforward commands from the CNS stimulate muscular actions

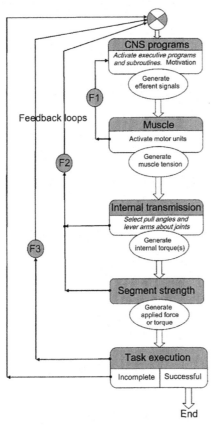

FIGURE 2.26 Schematic of the generation and control of muscle effort. *Source: Adapted from Kroemer, Kroemer, and Kroemer (1997).*

aimed to execute a given task. During every step of the ensuing activities, feedback loops report on the status of activities, so that new commands can adjust the actions in order to achieve the task efficiently and safely.

Feedforward

The CNS initiates activities by calling up an "executive program" (*engram*) which, innate or learned, exists for all normal muscular activities, such as walking, using pliers, or lifting objects. "Subroutines" modify that general program to make it appropriate for the specific case, such as walking downstairs, holding a tool securely, lifting with caution. Another modifier is motivation, which determines how much of the structurally possible strength a person will exert under the given conditions. Table 2.5 contains a listing of circumstances that may increase or decrease someone's willingness to exert.

As displayed in Fig. 2.26, these complex interactions result in excitation commands transmitted to the motor units involved where the signals trigger muscle contractions. The actual tension developed in the muscle depends on the rate and frequency of the signals received, and on the size of the muscle and the number of its activated motor units.

The muscle force acts on body members according to the existing biomechanical conditions, in particular depending on pull angles and lever arms. These modify the output of the muscle effort. (These conditions, specifically pull angles and lever arms, remain essentially constant in a static effort but they change over the course of dynamic activities.) The internal transmission generates torque(s) around body joints (e.g., from the upper arm over elbow and wrist to the hand) until the final output arrives at the body segment which executes the intended action (e.g., the hand holding a tool).

Feedback

Fig. 2.26 also shows several feedback loops which serve to monitor the muscular exertions, body segment positions, and motions. The first feedback loop, F1, is a reflex-like arc that originates at receptors (*proprioceptors*, see Chapter 4) in the joints signaling location, in the tendons indicating changes in muscle tension, and in the muscles indicating their lengths. These signals influence the signal generator in the spinal cord quickly.

Two additional feedback loops, F2 and F3, originate at *exteroceptors* and pass their signals through a "comparator" and then reach the CNS. Loop F2 starts at receptors sending kinesthetic signals, reporting on events related to touch, pressure, and body position. When lifting an object, for example, body position and motion are monitored continuously along with the sensations of pressure in the hand and of the object rubbing against one's body. Feedback loop F3 provides information on sounds and vision related to the effort to the comparator. For example, this may be the sounds or movements generated in equipment by the exertion of strength; it may be the pointer of an instrument that indicates the force applied; or it may be the experimenter or coach giving feedback and exhortation to the subject, depending on the status of the effort (see Table 2.5).

The comparator in the last two feedback loops compares the actual conditions reported by the body sensors to those conditions expected according to the feedforward signals. If actual and expected conditions differ, new commands are generated to eliminate discrepancies.

TABLE 2.5 Factors Likely to Increase (+) or Decrease (−) Muscular Performance

Circumstances	Likely effect
Feedback of results to subject	+
Instructions on how to exert	+
Arousal of ego involvement/aspiration	+
Stimulating drugs	+
Subject's outcry, startling noise	+
Hypnotic commands to perform strongly	+
Setting of goals, incentives	+
Competition, contest	?
Spectators	?
Verbal encouragement	?
Fear of injury	−

Source: Adapted from Kroemer, Kroemer, and Kroemer (1990).

2.4.2 Measuring Strength

Ethical considerations preclude subjecting people to tests that can damage their muscles, tendons, or any other part of their body. Thus, tests of human body strength do not probe structural integrity; instead, the test subject assesses what magnitude of effort is tolerable or suitable under the experimental conditions—that assessment determines the person's willingness to exercise that effort. This fact is expressed in the term *maximal voluntary effort* (MVE) or *maximum voluntary contraction* (MVC).

Consequently, all human strength tests are self-controlled by the test subject. A person's MVE depends on that individual's situational motivation as well as on experience and skill, on fitness and training status of the musculoskeletal system (see also Table 2.5). Further, the outcome of strength tests depends on the kind and placement of the measuring apparatus used.

Instruction also affects the magnitude of the MVE. The classical approach is the experimenter telling the subject to exert the maximal force possible either in a single maximal voluntary effort or in repeated MVEs. Instructions ask explicitly for MVEs that follow the subject's self-assessment of safety, suitability, or avoidance of fatigue: typical for this approach are "psychophysical tests" regarding lifting or lowering loads (more in Chapter 13) where the subject receives the instruction to imagine that the task would have to be done continuously for, say, 6 hours.

Considering the feedforward section in Fig. 2.26, it becomes apparent that current technology does not provide suitable means to measure the executive programs, the subroutines, or the effects of will or motivation on the signals generated in the CNS; presently, only very general information can be gleaned from an electroencephalogram (EEG).

Efferent excitation impulses along motor nerves to muscles can be recorded in an electromyogram (EMG).[28] Interpretation of an EMG relies on several basic premises. One assumption is that the signals relate to the same motor units. Another assumption

Maximal voluntary effort

Measurement opportunities

is that the amplitude of the electrical excitation complies with the magnitude of muscle exertion. In an isometric effort, there usually is a nearly linear increase in EMG intensity from rest to maximal voluntary muscle exertion; but in dynamic muscle use, the EMG–force relationship is complex and difficult to establish. Another complication comes with exertion time, when rate and recruitment coding of motor unit excitation often vary; this is particularly the case of muscle "fatigue" when the EMG generally shows a shift toward lower frequencies, usually together with an increase in amplitude. Being such complex technique to assess complex muscle strength, electromyography is currently not viable to measure muscle strength.[29]

Presently (in 2018), no instruments are available to directly measure the tensions within muscle filaments, fibrils, fibers, muscle, or groups of muscles in the living human. The conditions of internal transmission such as mechanical advantages and pull angles are difficult and often practically impossible to record and control. Yet, description of the body posture, and of involved segment positions during static exertion help to understand how muscle tension is conveyed within the body to the point of exertion.

Signals along the feedback loops shown in Fig. 2.26 are observable but, with existing technology, do not provide quantitative information about the magnitude of muscular exertion.

2.4.2.1 Practical Assessment of Human Strength

Muscles in the human body exert force at a lever arm; by this internal transmission, they develop torque around a body joint. This fact leads to the following definition: body segment strength is the maximal torque that a muscle (or a group of muscles) can develop about a skeletal articulation.

However, the outcome of interest is often just the resulting force applied to an object outside: a push exerted with the hand to a workpiece or with a foot to a pedal. For such practical purposes, we want to know the force vector available at the body's "business end" and we don't really care about internal "gear ratios" or other anthromechanic specifics.

Therefore, purely for practical reasons, body strength is commonly measured as the static force (occasionally, torque) applied by a body segment (usually hand or foot) to an instrument placed and used in standardized manner. Consequently, three statements describe common strength measurements:

1. Body strength is what an instrument measures.
2. Body strength depends on physical conditions as well as motivation.
3. Body strength is measured in static efforts.

Measuring instruments

Newton's second law states that force is mass multiplied by acceleration, $f = m \times a$. Force is not a basic physical unit but a derived one, which establishes necessary design principles for devices to measure force or torque. Newton's third law establishes that force action requires an opposing equally strong reaction force. This means that instruments must be firmly attached to solid structures that provide reaction resistance, or that they must be held between two parts of the body; for example, pinch force gages are held between opposing hand digit tips, usually between thumb and a finger.

Most grip strength gages measure the flexion strength of finger phalanges against the "heel of the palm," particularly the muscular pad at the base of the thumb; commonly,

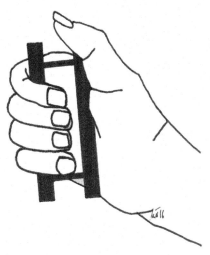

FIGURE 2.27   Basic features of a hand grip strength gage.

one surface of the instrument is placed under the bent middle phalanges of the fingers while the opposing surface of the gage rests against the *thenar eminence* (the group of muscles on the palm of the human hand at the base of the thumb), as sketched in Fig. 2.27. The amount for executable force depends decidedly on the size of the device (especially on the distance between its opposing surfaces), on the shape, and texture of the surfaces of the two handles, on the overall arrangement within the hand, and on how the test subject holds hand and arm. Thus, the several kinds of pinch and grip "dynamometers" (most are actually static rather than "dynamic") on the market are likely to yield different strength readings.[30]

Besides pinch and grip gages, a large variety of devices is available, or can be built. Basically, they consist of a device that defines (1) the body part (often hand) which exerts strength, (2) the coupling of the body part with the device, and (3) the reaction to the strength exertion; the device either remaining in place for static testing or moving in a specific manner for dynamic testing. The device incorporates a sensor which reacts to the applied exertion and transmits a signal for recording the exerted force or torque.

Obviously, the outcome of strength tests not only reflects the individual's physiological capacities and motivation but also depends, often strongly, on situational conditions such as:

Situational conditions

- the body segment used for the exertion (i.e., one foot or both feet, hand(s), or shoulder)
- body posture that allows the use of strong muscles at advantageous leverage
- ability to brace the body against a support structure including friction at the shoes
- suitable "coupling" at the interface between body and device
- body motion or lack thereof
- skill and experience

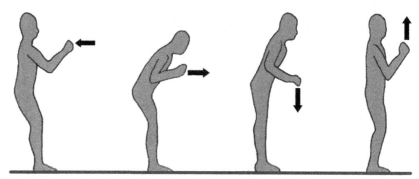

FIGURE 2.28   **Body stability determines maximal hand forces.**

Body posture

The effects of standing stances on the magnitude of possible hand force are illustrated in Fig. 2.28. Body postures determine vertical forces, horizontal push and pull depend on stance and body weight;[31] in each case, the need to maintain postural stability limits reaction forces and hence the hand force exertions. If the situation allows bracing the body against solid structures, force exertion can be markedly facilitated (see Chapter 9).

Static efforts

As discussed earlier, static (isometric) muscular exertion yields the highest force (tension) measurements, mostly because involved myofilaments can establish solid cross-bridges which they cannot do well in concentric or eccentric motions. Furthermore, the static condition allows strict experimental control during strength tests. The majority of information about human muscle strength reflects static data.

Variables in experiments

In an experiment, *independent* variables are those that are purposely manipulated in order to observe resulting changes in *dependent* variables, such as strength exertions. In a static strength test, muscle length (body posture) is set to a fixed value and hence there is no body motion relative to the measuring instrument; thus, the only dependent variable of importance is force. (For simplicity, *time* is disregarded as a variable.) In contrast, any dynamic test condition involves motion (changes in biomechanics and muscle length) and hence introduces the variables of displacement, velocity, acceleration, and moment of inertia. These variables, either dependent or independent, are rather difficult to control,[32] even with sophisticated equipment; therefore, information on human dynamic strength is limited to specific applications.

Isoinertial tests

The one case in which dynamic exertions are commonly measured is *isoinertial* tests, where the handled mass (weight) is a constant value. Movement of a constant mass (as in lifting) may be either a controlled independent or a dependent variable. Displacement and its derivatives (velocity, acceleration) can be dependent or independent variables. Force applied is likely to be a dependent value, according to Newton's second law (force equals mass times acceleration).

Free dynamic tests

Athletic performance and many common daily activities are classified as "free dynamic" activities. The subject freely choses which variables to control or to make dependent: this applies to force, torque, impulse, and displacement and its time derivatives. Even if mass and repetition are controlled, overall, these actions are unregulated and out of systematic control, but realistic.

2.4.2.2 *The Strength Test Protocol*

In many cases, the human factors engineer needs data on human strength under operational conditions which have not been reported in the literature or by professional organizations. In such case, specific strength testing may be necessary. This must be done in an organized and planned manner to assure validity and reliability.

After choosing the type of strength test to be done, and while selecting the measurement techniques and the measurement devices, an experimental protocol[33] must be devised. The following protocol is based on the "Caldwell regimen" and concerns the selection of subjects, their information, and protection; the control of the experimental conditions; the use, calibration,[34] and maintenance of the measurement devices; and (usually) the avoidance of training and fatigue effects.

Regarding the selection of subjects, care must be taken that they are in fact a representative sample of the population of interest. Regarding the management of the experimental conditions, the control of motivational aspects is particularly important. Other than for sports and medical function testing, the experimenter should not exhort the test subjects (see Table 2.5). The strength measure (force or torque) is the result of a maximal voluntary isometric muscular effort. The strength measure has vector qualities and therefore should be described by both magnitude and direction.

1. Measure static strength according to the following conditions:
   a. Static strength is assessed during a steady exertion sustained for 4 seconds.
   b. The transient periods of 1 second each, before and after the steady exertion, are disregarded.
   c. The strength datum is the mean score recorded during the first 3 seconds of the steady exertion.
2. Treat each subject in the same manner, as follows:
   a. The person should be informed about the test purpose and procedures.
   b. Instructions should be kept factual and not include emotional appeals.
   c. The subject should be told to "increase to maximal exertion (without jerk) in about 1 second and then maintain this effort during a 4-second count." (Special conditions may dictate using a different procedure.)
   d. Inform the subject about his/her performance in qualitative, noncomparative, positive terms. Do not give instantaneous feedback during the exertion.
   e. Rewards, goal setting, competition, spectators, fear, noise, etc. can affect the subject's motivation and performance and, therefore, should be avoided.
3. Provide a minimal rest period of 2 minutes between related efforts; more if symptoms of fatigue are apparent.
4. Describe the conditions existing during strength testing:
   a. Body parts and muscles chiefly used
   b. Body position (or movement)
   c. Body support/reaction force available
   d. Coupling of the subject to the measuring device (to describe location of the strength vector)
   e. Strength measuring and recording device

5. Describe the subjects:
   a. Population and sample selection, sample size
   b. Current health: medical examination and questionnaire recommended
   c. Gender
   d. Age
   e. Anthropometry (at a minimum, height and weight)
   f. Training and experience related to the strength testing
6. Report statistical measures of the experimental results:
   a. Number of data collected
   b. Minimum and maximum
   c. Median and mode
   d. Mean and standard deviation for normally distributed data points; for
      nonnormal distribution, lower and upper percentile values such a 1st, 5th, 10th,
      25th, 75th, 90th, 95th, 99th

Adhering to this protocol outline facilitates comparison between similarly acquired strength data; naturally it can be modified to suit the needs of the experiment.

2.5 DESIGNING FOR BODY STRENGTH

The ergonomist considering human strength should consider a number of factors.

Is the exertion by hand, by foot, or with the whole body? For each, specific design information is available (see Chapters 9 and 13). If it is possible to choose the body segment, the selection should aim to achieve the safest, least strenuous, and most efficient performance. For example, compared to hand movements over the same distance, foot motions consume more energy, are less accurate and slower, but are stronger.

Is a maximal or a minimal strength exertion the critical design factor? The expected strength exertions range, obviously, from the measured minimum to the maximum. "Average" user strength is a statistical phantom and usually has no design value. The structural strength of the object must be high above the highest perceivable strength application that even the strongest operator will not break a handle or a pedal. However, the minimum operational strength requirement is determined by the weakest operator who can still achieve the desired result, so that a door handle or brake pedal can be successfully operated or a heavy object be moved.

Is it static or dynamic use? If static, information about isometric strength capabilities applies. If dynamic, static values provide some guidance for slow motions although they are on the high side for concentric movement and low for eccentric movement. For strenuous dynamic activities, additional considerations include physical (circulatory, respiratory, metabolic) endurance capabilities of the operator and prevailing environmental conditions. Physiologic and ergonomic texts[35] provide related information.

What strength data? Measured strength data are often not normally distributed yet are routinely treated, statistically, as if they were and are reported in terms of means (averages) and standard deviations. This dubious computational procedure is not of great practical concern, however, because usually the data points of design interest are the extremes and not the central values. Often the 5th and 95th percentile values are selected. These can generally be easily determined, if not by calculation then by estimation.

EXAMPLE: VALVE OPERATING FORCE

The task: As assistant manager in a chemical plant you want to make sure that even "weak" operators can close the hand-operated emergency shut-off levers in the pipelines; on the other extreme, the valves must be structurally so solid that even brute strength applied does not break them.

The solution: Using a simulated valve handle, you measure the operating strengths of a random sample of plant employees. The results are:

- For weaker operators, the average pull force of the weaker hand was 116 N with a standard deviation (SD) of 37 N.
- For brute strength, the mean push force of both hands combined on the lever was 331 N; the SD was 173 N.

Based on this information you might decide to select, for breaking strength, a force near the 100th percentile. That calculation (with $k = 3$, from Table 1.5) uses (331 N + 3 × 173 N) and results in 850 N as the value to design for structural strength of the valve assembly. The "weak" operating force should be near the 5th percentile, at 55 N (=116 N − 1.65 × 37 N).

The result: When you inspect the plot of the raw measurements, you see that their distribution is distinctly non-normal, with the bulk of the strength measurements clustered at the low end of the scale and a long, thin tail of results at the high end. This convinces you not to rely on percentiles calculated from means and standard deviations, but rather to estimate the cutoff points from of the curve. Doing so, you decide that the operating requirement should be set to 50 N but the valve should be this designed to withstand a force of 1000 N.

Depending on the specific situation, you may also want to include factor of safety in the final strength recommendation.

Similar to the situational conditions listed previously, the quality and quantity of the force (or torque) that the body can transmit to an outside object depends on mechanical and physical conditions, especially on the:

- body segment employed, for example, hand or foot
- type of body object attachment, such as a simple touch or an encompassing grasp
- coupling type (i.e., friction and/or interlocking)
- direction of force (or torque) vector
- static or dynamic exertion
- needs for control and caution in task execution

Consideration and proper selections of these conditions are critical tasks for the designer and ergonomist.

2.6 CHAPTER SUMMARY

Anthromechanics and anthropometry (see Chapter 1) provide the basic descriptors needed to design "fitting" equipment and work procedures.

The human body is often modeled as a rigid skeleton with joints which allow angular displacements. The body members have mass properties, and are moved by muscles. The long bones of the segments establish the lever arms at which muscles attach. Body articulations are of different kinds, and some wear out and can be replaced by artificial joints.

Muscles convert chemically stored energy into mechanically useful force and work. The arrangement of muscles within the human body is rather complex. Usually, agonistic muscle is opposed by antagonistic muscle; these pairings of muscle groups regulate motion and strength exertion. Complexity of anthromechanical conditions poses interesting challenges for modeling the human body. Furthermore, voluntary and involuntary control of muscular exertions, fatigue, posture, motion characteristics, and motivation strongly influence the muscular output.

Use of statistics on body strength has often been faulty. It is indefensible to design tools, equipment, or workstations for a phantom called "average person." Instead, ranges of capabilities must be considered: Often, the weakest user limits the operational energy and the strongest exertion determines the structural strength of the equipment.

Understanding the musculo-mechanical conditions is hampered by insufficient definition of dynamic circumstances and by the difficulty of controlling them in experiments. Therefore, isometric (or static) muscle efforts are fairly well researched, while little systematic knowledge exists about dynamic capabilities. Newly developed procedures to test body strength, both statically and dynamically, should provide both theoretical insights and practical information to be used in the ergonomic design of work tasks and equipment.

The force or torque that the body can transmit to an outside object depends on mechanical and physical conditions, especially on

- whether the exertion is static or dynamic,
- the body segment that is employed (often hand or foot), and
- the coupling between the body and the object.

Consideration and proper selections of these conditions are critical tasks for the designer and ergonomist.

2.7 CHALLENGES

Can the "staircase effect" be generated by external electrical stimulation of muscles? If so, might there be a danger of overexertion?

Is external electrical stimulation of muscle contraction useful for building stronger muscles?

How does cocontraction of agonistic and antagonistic muscles about a joint influence the loading of that joint?

Could wearing a tight belt around the waist increase intraabdominal pressure? To what, if any, effect?

Which tissues in the trunk determine the posture of the spinal column?

Regarding the forces at the spinal disc, what is the impact of giving up the assumption of a straight vertebral column in favor of a column with kyphosis and lordosis?

What are the factors limiting a person's exertion of strength? What are some practical means to affect (or control) a person's motivation for strength generation?

What can the investigator do to increase or decrease muscular effort of a test subject? How can introducing competition, spectators, or verbal encouragement help or hurt?

Is it possible to develop "conversion algorithms" to determine dynamic muscle strength from static strength, and vice versa?

NOTES

1 Easterby, Kroemer, and Chaffin (1982); Kroemer (2007); Kroemer et al. (1988); Oezkaya et al. (2012); Peterson and Bronzino (2015).

2 This is occasionally called "iatromechanics."

3 Most people have 206 bones in their body but the number of the sesamoid bones in the hand and foot varies among individuals.

4 In 1680, Giovanni Alfonso Borelli's book "De Motu Animalium" (about the motion of animals) was published posthumously: it explained muscular action in mechanical terms, describing the actions of muscles and bones in terms of a system of forces and levers. In 1721, Jean Bernoulli wrote his "Physiomechanicae Dissertatio de Motu Musculorum", a physiomechanical study of the motion of muscles. In 1799, Charles-Augustin de Coulomb published his "Memoire sur la Force des Hommes", discussing human body strength.

5 Chaffin, Andersson, and Martin (2006); Kroemer et al. (2010); Kumar (2008); Nordin and Frankel (1989); Oezkaya et al. (2012).

6 Chandler, Clauser, McConville, Reynolds, and Young (1975); Kroemer (1973); NASA (1978); McConville et al. (1980); NASA since 1989; Roebuck (1995); Kroemer et al. (2010); Kumar (2008); Gordon et al. (2014).

7 Some specific examples for body segment computation can be found in, for example, the use of DEXA and other techniques in Durkin and Dowling (2003) and the use of 3D photogrammetry as described by Peyer, Morris, and Sellers (2015).

8 The physician Marie Francois Xavier Bichat (1771–1802) performed many postmortems during which he observed that the various organs were built of a mixture of simpler structures. Basically, these are flat and delicately thin, and in his 1800 book "Treatise on Membranes", Bichat called them tissues. (Asimov 1989)

9 It seems intuitive that more "flexibility" in body joints should indicate better physical performance and reduced risk of injury. Yet, in comparison with untrained individuals, athletes in the following categories have been found to be less flexible: soccer players, runners, those participating in sports for 5 years or longer, even ballet dancers in some hip movements. (Burton, 1991)

10 Routinely successful total hip replacement started in the 1960s after it had been attempted for about a century. Today, at least 90% of patients with hip and knee replacements are reported to be pleased with their new joints and continue to function well for decades after surgery.

11 As described by Bergmann et al. (2001).

12 See, for example, by Giarmatzis et al. (2015).

13 Karpandji (1988); Panjabi et al. (1992) describe the geometry of vertebrae in detail. Hukins and Meakin (2000), Meakin et al. (1996), Chaffin et al. (2006), Marras (2008), and Vieira (2008) discuss biomechanics of the spine.

14 Yettram and Jackman (1981); Aspden (1988); Meakin et al. (1996) discuss Euler buckling of the spine. Marras (2008) presents an overview.

15 Marras (2008), Marras et al. (2016), Vieira (2008).

16 Low back problems have been diagnosed among Egyptians 5000 years ago and were discussed in 1713 by Bernadino Ramazzini. The problem is not confined to humans; dogs and other quadrupeds can suffer from back pain as well.

17 Walji (2008) illustrates the musculoskeletal components of hand and arm.

18 The physician Galen (Claudius Galenus, about 129–199) studied human physiology, first at a gladiator school in Pergamon and then in Rome. He identified muscles and showed that they worked in groups. He also demonstrated the importance of the spinal cord by cutting it in various positions in animals and noting the extent of the resulting paralysis. (Asimov 1989)

19 Cavanagh (1988).

20 Kumar (2004), pp. 4–7.

21 The physiologist Albrecht von Haller (1708–1777) demonstrated that muscles could be made to contract by a stimulus transmitted to them by a nerve connected with the spinal cord and brain. In 1780, the anatomist Luigi Galvani (1737–1798) observed that muscles of dissected frog legs twitched when struck by an electrical spark.

22 In death, the ADP complex disintegrates and firmly bonds the myosin cross-bridges to the actin molecules, causing *rigor mortis*.

23 More in Kroemer et al. (2010).

24 This is a simplification; in reality, the biceps has two heads, one of which reaches beyond the shoulder joint.

25 Basmajian and Deluca (1985), p. 431.

26 Applying hand (or foot) force to a measuring device attached to a lever which rotates at constant speed (a device often falsely called isokinetic) does not change the lengths of involved muscles in isokinematic (nor in isokinetic) manner.

27 Enoka (1988).

28 Merlett et al. (2004), Sommerich and Marras (2004).

29 In 1666, Franceso Redi thought that the electric shock generated by the ray fish was muscular in origin. This was demonstrated by Luiggi Galvani in 1791 by dipolarizing the muscles in frogs' legs. His book "De Viribus Electricitatis" (about the features of electricity) is a milestone in the development of neurophysiology. During the 19th century it was popular to stimulate muscles through the skin by applying an external electrical current in the belief that this would perform miraculous cures on a wide variety of ailments. (as discussed by Basmajian and De Luca (1985), p 1–6)

30 Imran (2008). Feathers et al. (2015) provide extensive listings of data on grip, hand, and finger strength.

31 Andres (2004); see also Chapter 9

32 Systematic listings of experimental variables in dynamic and static efforts by Kroemer et al. (1990); Gallagher et al. (2004).

33 Based on the "Caldwell regimen", Caldwell et al. (1974); Gallagher et al. (2004).

34 You cannot measure if you cannot calibrate.

35 Such as by Astrand et al. (2004); Hall (2016); Kroemer et al. (2010); Kumar (2004); Kumar (2008); Wilmore et al. (2008); Winter (2009).

How the Body Does Its Work

Overview

In many respects, one may compare the way in which the body generates energy with the functioning of a combustion engine: fuel (food) is combusted, for which oxygen must be present. The combustion yields energy that moves parts mechanically. The fueling and cooling system (blood vessels) moves supplies (oxygen, carbohydrates, and fat derivatives) to the combustion sites (the muscles and other organs) and removes combustion by-products (lactic acid, carbon dioxide, water, and heat) for dissipation at the surfaces of the skin and lungs.

3.1 INTRODUCTION[1]

Control systems

Body processes[2] are governed by several complex and overlapping control systems: the central nervous system (CNS), the hormonal system, and the limbic system. Control centers (in the brain and spinal cord: see Chapter 4) rely on feedback from various body parts and provide feedforward signals according to general (autonomic, innate, learned) principles and according to situation-dependent (voluntary, motivational) rules.

Supply systems

The *respiratory system* provides oxygen for energy metabolism and dissipates metabolic by-products. The *circulatory system* carries oxygen from the lungs to cells that consume oxygen. The circulatory system also brings "fuel," that is, derivatives of carbohydrates and fats to the cells and removes metabolic by-products from the combustion sites. The *metabolic system* supports the chemical processes in the body, particularly those that yield energy.

3.2 THE RESPIRATORY SYSTEM

The respiratory system absorbs oxygen and dispels carbon dioxide, water, and heat. It interacts closely with the circulatory system, which provides the means of transport of those essentials as well as the nutrients that nourish the body. Fig. 3.1 schematically indicates the interaction between the two systems.

The respiratory system moves air to and from the lungs, where part of the oxygen contained in the inhaled air is absorbed into the bloodstream; it also removes carbon dioxide, water, and heat from the blood into the air to be exhaled. Between 2×10^6 and 6×10^6 alveoli provide an adult with about 70–90 m^2 of exchange surfaces in the lungs.

Lung actions

Air exchange is brought about by the pumping action of the thorax. The diaphragm separating the chest cavity from the abdomen descends about 10 cm when the abdominal muscles relax. Muscles connecting the ribs contract and raise the ribs. When the dimensions of the rib cage and its included thoracic cavity increase both toward the outside and in the direction of the abdomen, air is sucked into the lungs. When the inspiratory muscles relax, the lung tissue, thoracic wall, and abdomen elastically recoil to their resting positions without involving expiratory muscles. The recoil expels air from the lungs. When ventilation needs are very high, as in heavy physical work, the recoil forces

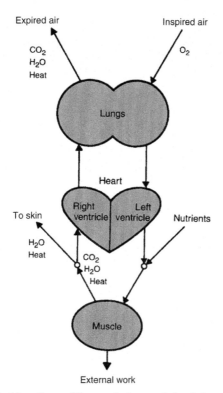

FIGURE 3.1   **The interrelated functions of the respiratory and circulatory systems.**

are augmented by activities of expiratory (intercostal) muscles, and contraction of the muscles in the abdominal wall further assists expiration.

The mucus-covered surfaces in the nose, mouth, and throat adjust the temperature of the inward-flowing air to body temperature, moisten or dry the air, and cleanse it of particles. In a normal climate (see Chapter 6), about 10% of the total heat loss of the body, whether at rest or work, occurs in the respiratory tract. The percentage increases to about 25% at outside temperatures of about $-30°C$. In a cold environment, heating and humidifying the inspired air cools the mucosa; during expiration, some of the heat and water is recovered by condensation from the air to be exhaled (hence the "runny nose" in the cold). Altogether, the energy required for breathing is relatively small, amounting to only about 2% of the total oxygen uptake of the body at rest and increasing to not more than 10% during heavy exercise.

Heat exchanges

3.2.1 Respiratory Volumes

The volume of air exchanged in the lungs depends on the requirements associated with the work performed. When the respiratory muscles are relaxed, there is still air left in the lungs. A forced maximal expiration reduces this amount of air to the *residual capacity* (or *residual volume*); see Fig. 3.2. A maximal inspiration adds the volume called *vital*

Lung capacity

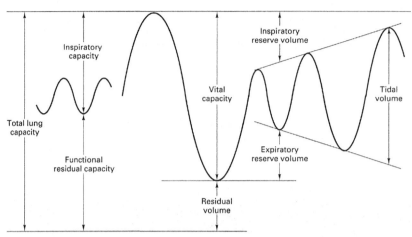

FIGURE 3.2   Respiratory volumes. *Source: Adapted from Kroemer, Kroemer, and Kroemer (1990) and Kroemer et al. (2010).*

capacity. Both volumes together are the *total lung capacity*. Only the so-called tidal volume is moved, leaving both an inspiratory and an expiratory *reserve volume* within the vital capacity during rest or submaximal work.

Vital capacity and other respiratory volumes are usually measured with the help of a spirometer. The obtained results depend on age, training, sex, body size, and body position of the subject. The total lung volume of highly trained, tall young males is between 7 and 8 liters (L) and their vital capacity up to 6 L. Women have lung volumes about 10% smaller, and untrained persons have volumes of about 60–80% of their athletic peers.

Minute volume

The movement of gas in and out of the lungs is called *pulmonary ventilation*. It is calculated by multiplying the frequency of breathing by the expired tidal volume. This is called the (respiratory expired) *minute volume*. The respiratory system is able to increase the volume of air that it moves and the amount of oxygen that it absorbs by large multiples. The minute volume can be increased from about 5 to 100 L/min or more, an increase in air volume by a factor of 20 or more. Though not exactly linearly related to minute volume, the oxygen consumption shows a similar increase.

Breathing frequency

At rest, one breathes 10–20 times every minute. In light exercise, primarily the tidal volume is increased, while with heavier work the respiratory frequency quickly increases up to about 45 breaths/min, together with the increasing tidal volume. This indicates that breathing frequency, which can be measured easily, is not a reliable indicator of the heaviness of work performed.

3.3 THE CIRCULATORY SYSTEM

Transport systems

The circulatory system carries oxygen from the lungs to the cells, where nutritional materials, also brought by circulation from the digestive tract, are metabolized. Metabolic by-products (CO_2, heat, and water) are dissipated by circulation. The circulatory and respiratory systems are closely interrelated, as shown in Fig. 3.1.

Water is the largest component of the body by weight, making up about 60% of body weight in men and about 50% in women. In slim individuals, the percentage of total water is higher than in obese persons, since adipose tissue contains very little water. The relation between water and lean (fat-free) body mass is rather constant in "normal" adults, about 72%. Small changes in hydration are normal, typically on the order of 2%, for example as occurs during the menstrual cycle in women.

Depending on age, gender, and training, approximately 10% of the total fluid volume consists of blood. Volumes of 4–4.5 L of blood in women and 5–6 L in men are normal. The specific heat of blood is 3.85 Joules (J) (0.92 cal) per gram. (In comparison, the specific heat of water is 4.18 J/g, the specific heat of fat is 1.67 J/g, the specific heat of atmospheric air is 1.01 J/g).

Of the total blood volume, about 55% is plasma, which is mostly water. The remaining 45% consists of formed elements (solids), predominantly red cells (*erythrocytes*), white cells (*leukocytes*), and platelets (*thrombocytes*). The percentage of red-cell volume in the total blood volume is called the hematocrit or packed cell volume (PCV).[3]

Blood is classified into four groups according to the content of certain antigens and antibodies: O, A, B, and AB. The importance of this classification lies primarily in the incompatibility reactions between the types in blood transfusions. Another classification categorizes blood according to its Rh (originally Rhesus) factor.[4]

Blood carries dissolved materials—particularly oxygen and nutritive materials—as well as hormones, enzymes, salts, and vitamins. It removes waste products, including dissolved carbon dioxide and heat.

Red blood cells transport oxygen. The oxygen attaches to hemoglobin, an iron-containing protein molecule of the red blood cell. Each molecule of hemoglobin contains four atoms of iron, which combine loosely and reversibly with four molecules of oxygen. Hemoglobin molecules can react simultaneously with oxygen and carbon dioxide. Hemoglobin has high affinity for carbon monoxide (CO), which takes up space otherwise occupied by oxygen; this property explains the toxicity of CO.

Marginal notes: Body water · Blood · Blood groups · Blood functions · Oxygen transport

---

The body can lose up to 15% of its blood volume without dramatic effects. At 20% loss, blood pressure is reduced and pulse and breathing are affected; gravely so at 30% loss, where heart rate is increased as well. At 40% loss, death is imminent without a transfusion.

3.3.1 Architecture of the Circulatory System

The circulatory system is nominally divided into two subsystems: the *systemic* and the *pulmonary* circuits, each powered by one-half of the heart (so it can be considered a "double pump"). The left side of the heart supplies the systemic section, which branches from the arteries through the arterioles and capillaries to the metabolizing organ (muscle, for example); from there, the branches combine again from venules to veins that lead to the heart's right side. The pulmonary system starts at the right ventricle, which

Marginal note: Two subsystems

powers the blood flow through the pulmonary artery, the lungs, and the pulmonary vein to the left side of the heart.

Heart chambers

Each half of the heart has an antechamber (*atrium*) and a chamber (*ventricle*). The left and right atria receive blood from the veins, which is then flushed through a valve into the ventricles.[5]

The heart is a pump

The heart is a hollow muscle which contains blood; it produces blood flow via contraction and with the aid of valves. The ventricle of the heart is filled through the valve-controlled opening from the atrium. The heart muscle contracts (called *systole*), and when the internal pressure is equal to the pressure in the aorta, the aortic valve opens and the blood is ejected from the heart into the systemic circulation. Continuing contraction of the heart increases the pressure further, since less volume of blood can escape from the aorta than the heart presses into it. Part of the excess volume is kept in the aorta and its large branches, which act as an elastic pressure vessel ("windkessel"). Then, the aortic valve closes with the beginning of the relaxation (*diastole*) of the heart, while the elastic properties of the aortic walls propel the stored blood into the arterial tree, where elastic blood vessels smooth out the waves of blood volume. At rest, about half the volume in the ventricle (the *stroke volume*) is ejected, while the other half (the *residual volume*) remains in the heart. During exercise, the heart ejects a larger portion of the volume it contains and increases its contraction frequency. When a great deal of blood is required but cannot be supplied, such as during very strenuous physical work with small muscle groups or during maintained isometric contractions, the heart rate can become very high.

Heart rate

At a heart rate of 75 beats/min, the diastole takes less than 0.5 second and the systole just over 0.3 second; at a heart rate of 150 beats/min, the periods are close to 0.2 second each. Hence, an increase in heart rate occurs mainly by shortening the duration of the diastole.

Specialized cardiac cells (the sinoatrial nodes) serve as natural "pacemakers," determining the frequency of contractions by propagating stimuli to other cells of the heart muscle. The heart has its own intrinsic control system, which operates, at (individually different) 50–70 beats/min in the absence of external influences. Changes in heart action stem from the CNS.

The events in the right heart are similar to those in the left, but the pressure in the pulmonary artery is only about one-fifth of those during systole in the left heart.

ECG or EKG

Myocardial action potentials are recorded in the electrocardiogram (ECG or EKG), sketched in Fig. 3.3. The different waves have been given alphabetic identifiers: the P wave is associated with the electrical stimulation of the atrium, while the Q, R, S, and T waves are associated with ventricular events. The ECG is employed chiefly for clinical diagnoses; however, with appropriate apparatus, it can be used for counting and recording the heart rate. Fig. 3.3 shows the electrical, pressure, and sound events during a contraction–relaxation cycle of the heart.[6]

3.3.2 Pathways of Blood

Changes in minute volume

Since only a finite blood volume exists in the body, cardiac output can be changed by two factors: the frequency of contraction (heart rate) and the pressure generated by each contraction in the blood. Both determine the so-called (cardiac) *minute volume*. The cardiac output of an adult at rest is around 5 L/min. During strenuous exercise, this level

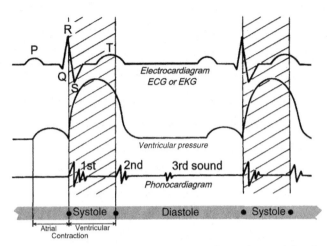

FIGURE 3.3 Scheme of the electrocardiogram, the pressure fluctuation, and the phonogram of the heart with its three sounds.

might be raised by a factor of 5 to about 25 L/min, while a well-trained athlete may reach up to 35 L/min.

A healthy heart can pump much more blood through the body than is usually needed. Hence, a circulatory limitation is more likely to lie in the transporting capability of the vascular portions of the circulatory system than in the heart itself. As mentioned, the arterial section of the vascular system (upstream of the metabolizing organ) has relatively strong elastic walls that act as a pressure vessel ("windkessel"), thus transmitting pressure waves far into the body, though with much loss of pressure along the way. At the arterioles of the consumer organ, the blood pressure is reduced to approximately one-third its value at the heart's aorta.

Limited transport capability

As blood seeps through the consumer organ (such as a muscle) via capillaries, the pressure differential from the arterial side to the venous side maintains the transport of blood through the *capillary bed*. Here, the exchanges of oxygen, nutrients, and metabolic by-products between the working tissue and the blood take place. If a lack of oxygen or the accumulation of metabolites requires high blood flow, smooth muscles that encircle the fine blood vessels remain relaxed, allowing the pathways to remain open. The large cross-sectional opening reduces blood flow velocity and blood pressure, allowing nutrients and oxygen to enter the extracellular space of the tissue, and permitting the blood to accept metabolic by-products from the tissue.

Capillary bed

Constriction of the capillary bed by tightening the encircling smooth muscle reduces local blood flow so that other organs in more need of blood may be better supplied. Such compression of the capillary bed can also occur if the striated muscle itself contracts strongly, at more than about 20% of its maximal capability. If this contraction is maintained, the muscle hinders or shuts off its own blood supply and cannot continue the contraction. Thus, sustained strong static contraction is self-limiting (see the discussion of muscle endurance in Chapter 2). A typical example of muscles cutting off their own blood flow is overhead work, wherein muscles must keep the arms elevated (such as in

Restricting blood flow in capillaries

Fig. 2.21 in Chapter 2). After a fairly short time, one must let them hang down to allow muscle relaxation and renewed blood flow.

> The designer of equipment and work tasks should be careful not to require sustained muscular contractions, such as keeping the body in a fixed position or grasping a handle tightly. Instead, the work should permit frequent changes in muscle tension, best achieved by allowing movement.

Large veins

The venous portion of the systemic system has a large cross section and provides low flow resistance; only about one-tenth of the total pressure loss occurs here. Valves are built into the venous system, allowing blood to flow only toward the right ventricle.

Swollen ankles

From the field of physics, Pascal's Law states that the static pressure in a column of fluid depends on the height of the column. However, the hydrostatic pressure in, for example, the feet of a standing person is not as large as is expected from physics, because the valves in the veins of the extremities modify the value: in a standing person, the arterial pressure in the feet may be only about 100 mmHg higher than in the head. Nevertheless, blood, water, and other body fluids in the lower extremities are pooled there, leading to a well-known increase in volume of the lower extremities ("swollen ankles"), particularly when one stands or sits still over extended periods of time.

3.3.3 Regulation of Circulation

Blood supply priorities

If the concentration of metabolites in a muscle increases, smooth muscles encircling blood vessels will relax, allowing more blood to flow. At the same time, signals from the CNS can trigger the constriction of other less important vessels supplying blood to organs. This leads to quick redistribution of the blood supply, which favors skeletal muscles over the digestive system (prioritization of muscles over digestion). However, even in heavy exercise, the systemic blood flow is so controlled that the arterial blood pressure is sufficient for an adequate blood supply to the brain, heart, and other vital organs. To accomplish this, neural vasoconstrictive commands can override local dilatory control. For example, the temperature-regulating center in the hypothalamus can affect vasodilation in the skin if this is needed to maintain a suitable body temperature, even if it means a reduction of blood flow to the working muscles (prioritization of skin over muscles).

Heart output

Circulation at the arterial side of a consumer organ is regulated both by local control and by impulses from the CNS, the latter having overriding power. The heart increases its output through a higher heartbeat frequency and higher blood pressure. At the venous side of the circulation, the constriction of veins, combined with the pumping action of dynamically working muscles and forced respiratory movements, facilitates the return of blood to the heart. These venous and pulmonary actions make increased cardiac output possible, because the heart cannot pump more blood than it receives.

Heart rate related to oxygen uptake

Overall, the heart rate follows oxygen consumption and hence energy production of the dynamically working muscle in a linear fashion from moderate to rather heavy work. However, the heart rate at a given level of oxygen intake is higher when the work is performed with the arms than with the legs. This reflects the use of different muscles and muscle masses with different lever arms to perform the work. Smaller muscles doing the same external work as larger muscles are more strained and require more oxygen. Also, static (isometric) muscle contraction increases the heart rate, apparently because the body tries to bring blood to the tensed muscles.

Work in a hot environment causes a higher heart rate than work at a moderate temperature (as discussed in Chapter 6). Also, emotions such as nervousness, apprehension, and fear can affect the heart rate, especially at rest and during light work.

3.4 THE METABOLIC SYSTEM

The term *homeostasis* characterizes the remarkable internal stability of human body functions: Core temperature, fluid volume, blood pH, and many other functions of the body stay nearly the same even under extremes of our natural environment.

Homeostasis

Over time, the human body also maintains a balance between energy input and output. The input is determined by nutrients, from which chemically stored energy is liberated during the metabolic processes within the body. The output is mostly heat and work. Work is measured in terms of physically useful energy, that is, energy transmitted to outside objects. The amount of such external work performed strains individuals differently, depending on their physique and training. There is close interaction between the metabolic, circulatory, and respiratory systems, as sketched in Fig. 3.4.

THE HUMAN ENERGY MACHINE

There is a useful analogy between the human body and a combustion engine[7]: in the cylinder of the engine, an explosive combustion of a fuel–air mixture transforms chemically stored energy into physical kinetic energy and heat. The kinetic energy moves the pistons of the engine, and gears transfer their motion to the wheels of the car. The engine must be cooled to prevent overheating. Waste products are expelled.

This whole process can work only in the presence of oxygen and when there is fuel in the tank. In the "human machine," muscles are analogous to the cylinders and pistons while bones and joints are equivalent to the gears. Nutrients (mostly carbohydrates and fats) are the fuels that are oxidized to yield energy. As the muscle machine works, heat and metabolic by-products are generated and must be expelled.

3.4.1 Metabolism and Work

The term "metabolism" refers to all chemical processes in the living body. Used in a narrower sense here, it describes the (overall) energy-yielding processes.

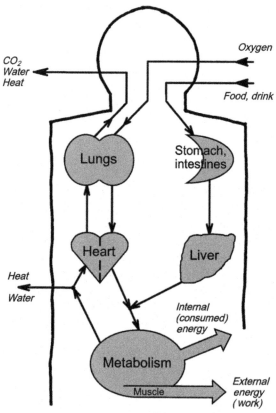

FIGURE 3.4 Interactions among energy inputs, metabolism, and outputs of the human body. *Source: Adapted from Kroemer, Kroemer, and Kroemer (1990) and Kroemer et al. (2010).*

Input equals output?   The balance between energy input I (via nutrients) and outputs can be expressed by the equation:

$$I = M = H + W + S \tag{3.1}$$

where M is the metabolic energy generated, which is divided into the heat H that must be dispelled to the outside, the work W, and the change in energy storage S in the body. (A gain in S is counted as positive, a loss as negative).

The units of energy (or work) are Joules (J) or calories (cal) with $4.2\,\text{J} = 1\,\text{cal}$. More precisely,

$$1\,\text{J} = 1\,\text{N m} = 0.2389\,\text{cal} = 10^7\,\text{ergs} = 0.948\,\text{BTU}\,10^{-3} = 0.7376\,\text{ft lb}$$

In many countries, the kilocalorie ($1\,\text{kcal} = 1\,\text{Cal} = 1000\,\text{cal}$) is used to measure the energy content of foodstuffs.

The units of power are $1\,\text{kcal/h} = 1.163\,\text{W}$, with $1\,\text{W} = 1\,\text{J/s}$.

Assuming no change in energy storage and also that there are no other heat gains or losses to the environment, one can simplify the energy–balance equation to

$$I = H + W \qquad (3.2)$$

Human energy efficiency e (work efficiency) is defined as the ratio of work performed and energy input:

$$e\,(\text{in}\%) = 100\,\frac{W}{I} = 100\,\frac{W}{M} \qquad (3.3)$$

In everyday activities, only about 5% or less of the energy input is converted into "work"—that is, energy usefully transmitted to outside objects. Under favorable circumstances, highly trained athletes may attain perhaps 25%. The remainder, by far the largest portion of the input, is finally converted into heat.

"Good heaters" but "bad workers"

Work (in the physical sense) is done by the skeletal muscles, which move body segments against external (or internal) resistances. For this, the muscle is able, in its mitochondria, to convert chemical energy into physical work or energy. From resting, it can increase its energy generation up to 50-fold. Such enormous variation in metabolic rate not only requires quickly adapting supplies of nutrients and oxygen to the muscle but also generates large amounts of waste products (mostly heat, carbon dioxide, and water) that must be removed. Thus, during the performance of physical work, the ability to maintain the internal equilibrium of the body is largely dependent on the circulatory and respiratory functions that serve the muscles involved. Among these functions, the control of body temperature is of particular importance. This function interacts with the external environment, particularly the surrounding temperature and humidity, as discussed in Chapter 6 in more detail.

Converting chemical energy into physical work

3.4.2 Energy Transformation in the Body

In living organisms, such as the human body, energy transformation involves chemical reactions that either liberate energy, most often as heat, or require energy. The first kind of reaction, called catabolism, is *exergonic* (or *exothermic*). The opposite kind of reaction requires energy input; called anabolism, it is *endergonic* (or *endothermic*). Generally, the breakage of molecular bonds is exergonic, while the formation of bonds is endergonic.

Depending on the molecular combinations, bond breakage releases different amounts of energies. Often, reactions do not simply go from the most complex to the most broken-down state, but achieve the process in steps, with intermediate and temporarily incomplete stages.

Energy is supplied to the body in food or drink. Energy is contained in specific chemical compounds that are changed in the course of digestion and then, during the following assimilation, reassembled to different molecular combinations. Their energy content can be released for use by the body; the ergonomist is particularly interested in energy use by muscles to perform physical work.

Energy in food and drink

3.4.3 The Energy Pathways

Ingestion of food and drink

In the mouth, chewing destroys the structure of the food mechanically, and saliva starts its chemical breakdown. Saliva is 99.5% water (a solvent) and 0.5% salts, enzymes, and other chemicals. The enzyme *lysozyme* destroys bacteria, thus protecting the mucous membranes from infection and the teeth from decay. Another enzyme, *salivary amylase*, breaks down starch.

During swallowing, breathing stops and the epiglottis closes for a second or two so that a blob of food (*bolus*) can avoid the windpipe (*trachea*) and slide down the gullet (*esophagus*). Liquids need only about 1 second, but solid food takes up to 8 seconds to glide to the stomach.[8]

Stomach actions

The stomach generates gentle waves, two to four per minute, which mix the bolus with gastric juice. The stomach's gastric juices start breaking up proteins chemically but do little to break down fats and carbohydrates. Carbohydrate-rich foods leave the stomach within 2 hours, while protein-containing foods are slower; fatty foods stay up to 6 hours.

Digestion and absorption

The contents of the stomach empty into the top of the small intestine, called the *duodenum* (Latin for "12 fingers" which reflects the approximate length of that section of intestine). Here, true chemical digestion occurs through the breakup of large complex molecules into smaller ones. These can be transported across cell membranes and absorbed into blood and lymph. It takes the food 3–5 hours to move through the small intestine, a tube about 3 cm in diameter and 7 m in length. During this time, about 90% of all the nutrients are extracted. Then in the large intestine, final processing is completed, and solid waste is eliminated as feces.

Assimilation

The digested and absorbed foodstuffs are reassembled ("assimilated") into new molecules that can be used for body growth and repair or can be easily degraded to release or store energy.

Altogether, it takes 5–12 hours after eating a meal to extract its nutrients and to pass it through the digestive tract.

IF YOU STOP SMOKING, WILL YOU GAIN WEIGHT?

Nicotine prolongs the time during which food stays in the stomach before being sent to the intestines—and a full stomach turns off the appetite. Nicotine also dulls the desire for sweets and lessens the activity of the enzyme that regulates the storage of fat. In addition, nicotine speeds up the metabolism rate by increasing the resting metabolic rate (see the discussion later in this chapter) and, possibly, the thermic effects of exercise or work. When one stops smoking, the nicotine content of the body is greatly reduced. Consequently, sweet food becomes more attractive, the stomach is emptied faster, and fat is stored in the cells more easily (especially in subcutaneous fat). Eight out of 10 people gain a bit of weight after stopping smoking, but about 20% gain more than 15 kg.[9] Using nicotine substitutes (in the form of nicotine gum or a patch), medications (to reduce cravings), and behavioral modification techniques can facilitate smoking cessation while minimizing relapse and weight gain.

3.4.4 Foodstuffs

Our food is a mixture of organic compounds (foodstuffs), water, salts, minerals, vitamins, and of fibrous material (mostly cellulose). Energy is contained in the primary foodstuffs—*carbohydrates, fats, proteins*—and in alcohol.[10]

> The nutritionally usable energy amounts per gram are, on average: 4.2 kcal (18 kJ) from carbohydrates, 4.5 kcal (19 kJ) from protein, about 7 kcal (30 kJ) from alcohol, and 9.5 kcal (40 kJ) from fat.

3.4.4.1 *Carbohydrates*

Carbohydrates range from small to rather large molecules, and most are composed of only three chemical elements: carbon (C), oxygen (O), and hydrogen (H). (The ratio of H to O usually is 2 to 1—as in water—hence the name carbohydrate, meaning "watered carbon.") Carbohydrates are digested by breaking the bonds between monosaccharides so that the compounds are reduced to simple sugars that can be absorbed through the walls of the intestines into the bloodstream.

The blood transports the monosaccharide *glucose* ($C_6H_{12}O_6$) to the liver. From there, glucose is sent either to the CNS (the brain and spinal cord) or to the muscles for direct use. Glucose, the primary (but not the largest) source of energy, is extensively used by the CNS and it provides the quick energy required for muscular actions.

Glucose

The liver can change glucose into a long-chain molecule called *glycogen*, a polysaccharide ($C_6H_{10}O_5$)$_x$. Glycogen is deposited in the liver and stored as energy near skeletal muscle. When the glycogen storage areas are filled, the liver converts glucose to fat and stores it in adipose tissue. Here, excess energy input becomes felt and seen. So, we can "become fat" without eating fat.[11,12]

Glycogen

3.4.4.2 *Fat*

Fat is the carrier of vitamins A, D, E, and K in food. It is also the major energy source for the body. Fat is a triglyceride: a molecule formed by joining a glycerol nucleus to three fatty-acid radicals. Unsaturated fat has double bonds between adjacent carbon atoms; hence, the compound is not saturated with all the hydrogen atoms it could accommodate. Most plant fats are polyunsaturated and therefore liquid at room temperature, while most animal fats are saturated and thus solid at room temperature.

When fat is digested, the bonds linking the glycerol with the three fatty acids are broken. Digestion takes place primarily in the small intestine, where the glycerol and fatty-acid molecules can pass through cell membranes. The water-repellent fatty acids are absorbed into the lymph, which finally empties into the bloodstream, which is already carrying the water-soluble glycerol.

Digestion

Regulated by the liver, fat is usually transported for storage to adipose tissue from where it will be used when needed for energy (see below). Fat also cushions vital organs (such as the heart, liver, spinal cord, brain, and the eyeballs) against impact. A layer of

Storage

fat under the skin insulates the body against heat transfer to and from the environment. As an illustration, note the appearance of swimmers, who usually have more "rounded" bodies than do "lanky" long-distance runners.

3.4.4.3 Protein

Protein is the third major food component. It consists of chains of amino acids joined together by peptide bonds. Many such bonds exist; hence, proteins come in a variety of types and sizes. Digestion breaks the protein bonds into amino acids, which are then absorbed into the bloodstream. The amino acids are transported to the liver, which disperses some to cells throughout the body, to be rebuilt into new proteins. However, most amino acids become enzymes: organic catalysts that control the chemical reactions between molecules without being consumed themselves. Still others become hemoglobin (the carrier of oxygen in the blood), antibodies, hormones, or collagen. Usually, the body employs protein for these important functions, but it can use its amino acids for energy if other energy carriers are not available.

3.4.5 Stored Energy

At the level of the cell, glucose and glycogen are the primary and first-used sources of energy, particularly in the CNS and at muscles. Under normal conditions of nutrition and exertion, body fat accounts for most of the stored energy reserves.

Fat in the body

On average, about 16% of body weight is fat in a young man. The percentage increases to 22% by middle age and is even more if the man is obese. Young women average about 22% of body weight in form of fat, a percentage that usually rises to some 35% by middle age. Ranges of 12–20% in men and 20–30% in women are considered normal. Athletes generally have lower percentages of body fat—usually about 15% in men and 20% in women. However, the ideal body fat percentage for an athlete varies by the requirements of the sport involved. For example, body builders, triathletes, and gymnasts tend to have much lower body fat percentages than shot putters and heavyweight-boxers.

Obesity has been defined as fat content of more than 25% in men and greater than 33% in women. More specifically, measures of overweight and obesity can be defined as follows: a body mass index (BMI, see Chapter 1) over 25 is considered to represent being overweight, while a BMI of 30 or higher is considered to represent being obese. (While BMI does not measure body fat directly, it is strongly correlated with more direct measures of body fat obtained from skinfold thickness measurements, bioelectric impedance, underwater weighing and other methods. From a morbidity perspective, BMI is strongly correlated with various adverse health outcomes consistent with the more direct measure of body fat percentage.)

However, lower body fat is not always better. The total body fat consists of subcutaneous storage fat and essential body fat. Maintaining an excessively low storage fat percentage results in decreased protection of essential internal organs in the torso; maintaining an excessively low essential fat percentage compromises normal functioning of bodily organs and the nervous system. The percentage of essential body fat for women is greater than that for men due the fat requirements in women for childbearing and various other hormonal functioning. In general, a BMI below 18.5 is considered to represent being underweight.

Assuming 15% fat in a 60-kg person as a low value, body fat amounts to 9 kg. Twenty-five percent of a 100-kg person, or 36% of a 70-kg person, means approximately 25 kg of body fat. Given that each gram of fat yields about 9.5 kcal, the percentages entail an energy storage in the form of fat of about 85,500 Cal for a slender, lightweight person, and nearly 240,000 Cal for a heavy person.

Most of us have about 400 grams of glycogen stored near the muscles, about 100 grams in the liver, and some in the bloodstream. With an energy value of about 4.2 kcal/g, we have only some 2200 Cal as energy available from glycogen. About the same amount is present in the form of glucose, which is not directly stored by the body.

Glycogen in the body

3.4.6 Energy Release

We generally have an enormous amount of energy stored in our bodies in form of fat, but extraction of the energy from it involves a relatively complex chemical processes that takes some time. The energy in glucose (stored as glycogen) is more easily liberated than the energy from fat.

The utilization of the energy present in the human body is achieved by *catabolism*, a destructive metabolic process in which organic molecules are broken down, releasing their internal bond energies. Overall, this is accomplished in the human body by *aerobic* metabolism, that is, with oxygen required and involved—but there are some *anaerobic* phases within the process (meaning they do not require the presence of oxygen).

Catabolism

Glucose can be oxidized according to the formula:

Glucose and glycogen use

$$C_6H_{12}O_6 + 6O_2 \rightarrow 6CO_2 + 6H_2O + \text{Energy} \tag{3.4}$$

This means that one molecule of glucose combines with six molecules of oxygen, generating six molecules each of carbon dioxide and water, while energy (about 690 kcal/mol) is released.

Glycogen is the storage form of glucose. It is a large branched polymer of glucose residues that can be broken down easily to yield glucose molecules when energy is needed. Glucose and glycogen molecules can be broken into fragments which oxidize each other. This process is anaerobic and yields much less energy than aerobic reactions

Fat catabolism takes place in anaerobic steps, except for the last step, which is aerobic. First, fat is split into glycerol and fatty acids. Glycerol is used in a manner similar to glucose, while the fatty acid is reformed to acetic acid, which enters the Krebs cycle and, finally, becomes oxidized to carbon dioxide and water. The energy yield is 2340 kcal/mol, more than three times that released from glucose, but the complex fat breakdown takes longer.

Fat use

3.4.7 Energy by ATP–ADP Conversion

Living muscle cells store "quick-release" energy in the form of the molecular compound *adenosine triphosphate* (ATP). Hydrolysis can easily and quickly break down its phosphate bonds to *adenosine diphosphate* (ADP). This anaerobic reaction releases energy:

$$ATP + H_2 \rightarrow ADP + Energy\,(\text{output}) \tag{3.5a}$$

While the ATP available in the mitochondria of the muscle can provide energy for about two seconds, ATP must be resynthesized for continuous operation. This is done through *creatine phosphate* (CP), which transfers a phosphate molecule to the ADP. Energy must be supplied for this endergonic reaction to occur:

$$ADP + CP + Energy\,(\text{input}) \rightarrow ATP + H_2O \tag{3.5b}$$

The energy needed for the rebuilding of ATP from ADP is liberated in the breakdown of complex molecules to simpler ones, ultimately to CO_2 and H_2O. First glucose is used, then glycogen, and finally fats (and possibly even proteins). This "combustion of foodstuffs" is the ultimate source of the body's energy, keeping the ATP–ADP conversion going.

3.4.8 Muscular Work

The mitochondria are the "cellular power factories" of muscle (as mentioned earlier in this chapter). They provide chemically stored energy in the form of ATP and release it, as just described, so that muscle can contract, thereby converting chemical into mechanical energy.

The first few seconds

At the very beginning of muscular effort, breaking the phosphate bond of ATP releases "quick energy" for muscular contraction. However, the contracting muscle consumes its local supply of ATP in about 2 seconds.

The first 10 seconds

The next source of immediate energy is CP, which transfers a phosphate molecule to the just-created molecule of ADP, turning the ADP back into ATP. (Note that this cycle of converting ATP into ADP and back to ATP does not require the presence of oxygen.) Since a human has three to five times more ADP than ATP, there is enough energy available for a muscle to perform up to about 10 seconds of high activity.

After 10 seconds

After about 10 seconds of ATP–ADP–ATP reactions, energy must be supplied to sustain the reformation of ATP. Now, the energy absorbed from the foodstuffs comes into play: glucose (then glycogen, then fat, then finally even protein) is broken down, releasing energy for the recreation of ATP.

Given that only a few seconds have elapsed since the activity started, there simply has not been enough time to use oxygen in the energy-conversion process. Thus, while releasing energy, the breakdown of glucose (generating carbon dioxide and water) is not complete, but other metabolic by-products are generated also, particularly lactic acid. If this metabolic by-product is not resynthesized within about a minute in the presence of oxygen, the muscles simply cannot continue to work further.

The foregoing anaerobic energy release relies primarily on the breakdown of glucose, although some glycogen is also involved. This is why glucose is called the primary, most easily accessible, and most metabolized energy carrier for the body.

If the physical activity continues, it must be performed at a level at which oxygen is sufficiently available to maintain the energy-conversion processes. Hence, the energy generated by a quick burst of maximal effort cannot be maintained at a high level for extended periods of time.

After minutes and longer

To allow long-lasting work, the energy demanded from the muscles must be so low that the oxygen supply at the mitochondria level allows continuous aerobic energy conversion, without the generation of metabolic by-products that would lead to fatigue (see below) and force termination of the work.

Continual work

3.4.8.1 Aerobic and Anaerobic Metabolism

Without oxygen, a molecule of glucose yields two molecules of ATP. With oxygen, the glucose energy yield is 36 molecules of ATP. Even richer in energy is the fat molecule *palmitate*, which yields about 130 molecules of ATP.

Because the energy yield is so much more efficient under aerobic conditions, in which no metabolic by-products (that can cause fatigue and exhaustion) are generated, one is able to keep up a fairly high-energy expenditure, as long as ATP is replaced as quickly as it is used up and no metabolite such as lactic acid is developed.

However, if a very heavy expenditure of energy is required over long periods of time, such as in heavy physical work or during a marathon run, the interacting metabolic system and the oxygen-supplying circulatory system might become overtaxed.

In our everyday activities, we usually manage to conform our energy output with the body's abilities to develop energy under sufficient supply of oxygen. If needed, we simply take a break; while one is resting, accumulated metabolic by-products are resynthesized, and the metabolic, circulatory, and respiratory systems return to their normal states.

Many of the single intermediate steps in the metabolic reactions are, in fact, anaerobic; but eventually, oxygen must be provided. Thus, overall, sustained energy use in the human body is aerobic.

We may now refine the earlier used analogy of the "Human Energy Machine":

- The "muscle motor" runs on the energy derived from the breakdown of ATP into ADP.
- The "ATP battery" is recharged by rebuilding ATP from ADP with the help of energy released in the combustion of glucose, glycogen, and fatty and amino acids.

3.4.9 Energy Use and Body Weight

The balance between energy input, output, and storage follows from Equation (3.1): if the input exceeds the output, storage (body fat and hence weight) is increased;

conversely, weight decreases if the input is less than the output. Approximately 7000–8000 kcal make a difference of 1 kg in body weight.

Apparently, the body tries to maintain a given energy storage, and hence body weight, regulated by a "set-point" principle: to keep the system in homeostasis, a deviation from the "set" internal state generates a signal that initiates the responses necessary to counter the perturbing influence. If the body weight declines, adjustments in food intake usually occur via an increased appetite, and changes in whole-body metabolism favor regaining the lost weight. Thus, even if one voluntarily reduces the food intake, the body tries to extract enough energy from the reduced food intake to maintain the old body weight. Similar coordinated counteracting adjustments in intake and energy expenditure also occur when one's body weight is elevated from the normally maintained level.

The set point can be adjusted voluntarily (typically by a change in dietary or exercise habits) or through internally regulated mechanisms (such as fever, CNS lesions, hormones, and toxins).

3.5 ENERGY REQUIREMENTS AT REST AND AT WORK

Basal metabolism

A minimal amount of energy is needed to keep the body functioning, even if it is just resting. This *basal metabolism* is strictly measured on a person who has fasted for 12 hours and has no protein intake for at least 2 days, with complete physical rest in a neutral ambient temperature. Under these conditions, the basal metabolic values depend primarily on one's age, gender, height, and weight. Altogether, there is relatively little interindividual variation; hence a commonly accepted figure for basal metabolism is 1 kcal (or 4.2 kJ)/kg/h, or 4.9 kJ/min for a person of 70 kg.

Resting metabolism

The highly controlled conditions under which basal metabolism is measured are rather difficult to accomplish in most practical applications. Therefore, one usually simply measures the *resting metabolism* before the working day, with the subject as well at rest as possible. Depending on the given conditions, resting metabolism is 10–15% higher than basal metabolism. The mean resting metabolism rate (RMR) for 162-cm (5'4") man is 1400 kcal/day, while for a 193-cm (6'4") man, it is 2230 kcal/day. The mean RMR for a 155-cm (5'1") woman is 1240 kcal/day and for a 180-cm (5'11") woman is 1570 kcal/day. In addition to mass and gender, age is another variable in RMR, though age-related decline can be largely counteracted with exercise.

Work metabolism

The increase in metabolism from resting to working is called *work metabolism*. This increase above the resting level represents the amount of energy needed to perform the work.

Oxygen debt

At the start of physical work, oxygen uptake follows the demand sluggishly. As Fig. 3.5 shows, after a slow onset, oxygen intake rises rapidly and then gradually approaches the level at which the oxygen requirements of the body are met.

Thus, there is a discrepancy between oxygen demand and available oxygen during the first section of physical work: the energy liberation in the body is largely anaerobic during this time. The resulting *oxygen deficit* must be repaid at some later time, usually during rest after work. The amount of the deficit depends on the kind of work performed and on

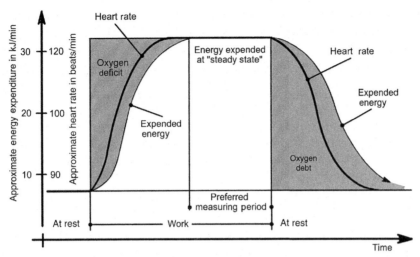

FIGURE 3.5 Scheme of energy liberation, energy expenditures, and heart rate at steady state work. *Source: Adapted from Kroemer et al. (2010).*

the person, but the *oxygen debt* repaid is approximately twice as large as the oxygen deficit incurred. Of course, given the close interaction between the circulatory and the metabolic systems, the heart rate reacts similarly; but as it increases faster at the start of work than oxygen uptake does, it also falls back more quickly to the resting level.[13]

If the workload stays below about 50% of the worker's maximal oxygen uptake, then oxygen consumption, heart rate, and cardiac output can achieve and can stay on the required supply level. This condition of stabilized functions at work is called the steady state. Obviously, a well-trained person can attain this equilibrium level between demands and supply even at a relative high workload in a few minutes, while an ill-trained person would be unable to attain a steady state at that high requirement level, but could be in equilibrium at a lower level of demand. Of course, the oxygen uptake, as well as the heart rate, should be measured during the steady-state period.

Steady state

3.5.1 Fatigue

Fatigue is operationally defined as a "reduced muscular ability to continue an existing effort."

Fatigue is best researched in regard to maintained static (isometric) muscle contraction. If the effort exceeds about 15% of a maximal voluntary contraction (see Fig. 2.22 in Chapter 2), intramuscular pressure reduces blood flow through muscle tissue—may even cut it off in a maximal effort—in spite of a reflex increase in systolic blood pressure. Insufficient blood flow brings about depletion of energy sources available (especially ATP and CP) at the muscle, accumulation of potassium ions, and depletion of extracellular sodium in the extracellular fluid. Combined with an intracellular accumulation of phosphate (from the degradation of ATP), these biochemical events perturb the coupling between nervous excitation and muscle-fiber contraction. This uncoupling between CNS control and muscle

Causes of fatigue

action signals the onset of fatigue. Formation of lactic acid (a by-product of the energy conversion process) also occurs but probably is not the primary reason for fatigue. Also, the increase in the number of positive hydrogen ions resulting from anaerobic metabolism causes a drop in intramuscular pH, which then inhibits enzymatic reactions, notably those in the breakdown of ATP—all causing fatigue.[14]

If the energetic work demands exceed about half the person's maximal oxygen uptake, anaerobic energy-yielding metabolic processes play increasing roles. The onset of the feeling of fatigue usually coincides with the depletion of glycogen deposits at the working muscles, a decrease in blood glucose, and an increase in blood lactate. These events impede especially the processes associated with formation and detachment of sarcomere cross-bridges. Furthermore, so-called central fatigue may occur in the nervous control system, associated with one's sense of effort and motivation; however, as often occurs in competitive sports, a new emotional drive to perform can temporarily overcome fatigue.

Rest periods

The benefit of feeling fatigue is prevention of serious damage. When severe exercise enforces anaerobic metabolic processes, it brings about a continuously growing oxygen deficit with the consequence that a balance between demand and supply cannot be achieved; no steady state exists, and the work requirements exceed capacity levels—as sketched in Fig. 3.6. The resulting fatigue can be counteracted by the insertion of rest periods. Given the same ratio of "total resting time" to "total working time," many short rest periods have more recovery value than a few long rest periods.

Tips to avoid fatigue:

- Increase fitness and skill.
- Allow short bursts of dynamic work and avoid long periods of static effort.
- Keep energetic work and muscle demands low.
- Take many short rest pauses; this is better than taking a few long breaks.

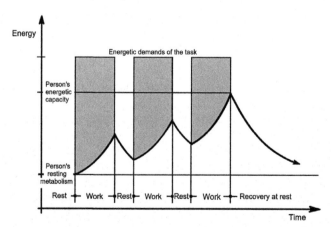

FIGURE 3.6 **Metabolic reactions to the attempt of doing work that exceeds one's capacity even when interspersed with rest periods.** *Source: Adapted from Kroemer et al. (2010).*

The incidence of fatigue indicates the need to improve the task conditions; the proper ergonomic approach is to design out any work requirements that generate fatigue (such as shown in Fig. 2.21 for overhead work). The human engineering solution is fitting the task to the human.

3.5.2 Assessing Energy Expenditures

The ability to perform physical work is different from person to person and depends on gender, age, body size, health, motivation, and the environment (see Fig. 3.7).

3.5.2.1 Energy Requirements of a Task

To match a person's work capacity with the requirements of a job, one needs to know the individual's energy capacity and how much the job demands from this capacity. In order to measure an individual's capacity, one makes the person perform a known amount of work (usually on a bicycle ergometer or a treadmill) and measures the subject's reactions. To measure the energy requirements of a work task, one lets a "standard person" perform the job and again measures the person's reactions to the task. (Since "standard people" do not exist, one simply measures the reaction of various workers actually doing the job, assuming that they represent the "normal.")

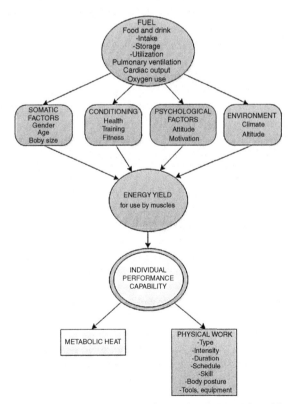

FIGURE 3.7 **Determiners of individual physical work capacity.** *Source: Adapted from Kroemer et al. (2010).*

TABLE 3.1　Oxygen Needed, Respiratory Quotient RQ, and Energy Released in Nutrient Metabolism

	Carbohydrate	Fat	Protein	Average[a]
O_2 consumed (L/g)	0.83	2.02	0.79	na
RQ or RER	1.00	0.71	0.80	na
Energy yield (kJ/g)	18	40	19	na
Energy yield (kJ/L O_2)	21.2	19.7	18.9	**21**
Energy yield (kcal/L O_2)	5.05	4.69	4.49	**5**

[a]Assumes the construct of a "normal" adult on a "normal" diet doing "normal" work.

O₂ uptake

A person's oxygen consumption (and CO_2 release, as per Equation (3.4)) while performing work is a measure of his or her metabolic energy production. The instruments used for this purpose rely on the principle that the difference in O_2 and CO_2 contents between the exhaled and inhaled air indicates the oxygen absorbed and carbon dioxide released in the lungs.

RQ or RER

The *respiratory exchange quotient* (RQ, also called *respiratory exchange rate*, RER) compares the carbon dioxide expired with the oxygen consumed. One gram of carbohydrate needs 0.83 L of oxygen to be metabolized and releases the same volume of carbon dioxide—see Equation (3.4). Hence, the RQ is 1 (unit). The energy released is 18 kJ/g, equivalent to 21.2 kJ (5.05 kcal) per liter of oxygen. Table 3.1 shows these relationships for carbohydrates, fat, and protein conversion.

Caloric value of O₂ uptake

Assuming an overall "average" caloric value of 5 kcal/L-O_2, one can calculate the energy conversion occurring in the body from the volume of oxygen consumed. Given observation periods of five or more minutes, this is a reliable assessment of the metabolic processes, that is, of the effort of the body while performing a task.

Standardized tests

Assessments of human energy capabilities use various techniques of measuring oxygen consumption in standardized tests with normalized external work, done mostly on bicycle ergometers, treadmills, or steps. The selection of this equipment is based not so much on theoretical considerations as on availability and ease of use.

3.5.2.2 Measuring Oxygen Uptake

Measuring the oxygen consumed over a sufficiently long period of time (such as 5 minutes) is a practical way to assess the metabolic processes. (A physician or physiologist should supervise this test.) The method is called "indirect calorimetry" since it does not measure the expenditures of energy directly.

Traditionally, indirect calorimetry has been performed by collecting all exhaled air during a certain observation period in airtight (Douglas) bags. The volume of the exhaled air (easily 100 L/min in heavy work) is then measured and analyzed for oxygen and carbon dioxide as needed for the determination of the RQ. This requires a rather elaborate air-collecting system, which mostly limits the procedure to the laboratory.

A major improvement in this technique has been to divert only a known percentage of the exhaled air into a small collection bag, which means that only a relatively small device has to be carried by the subject. Still, in both collection techniques, the subject must wear a face mask with valves and a nose clip, which can become quite uncomfortable and hinders speaking.

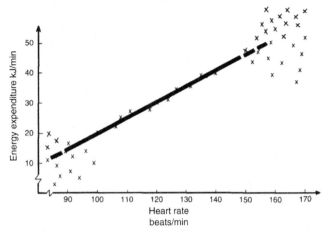

FIGURE 3.8   Schematic representation of the relationships between energy expenditure and heart rate.

Significant advances have been made through the use of instantaneously reacting sensors that can be placed into the air flow of the exhaled air, allowing a breath-by-breath analysis without air collection and without a constraining face mask.

The circulatory and metabolic processes interact closely because nutrients and oxygen must be brought to the muscle or other metabolizing organ and, at the same time, metabolic by-products must be removed from it for proper functioning. Therefore, the heart rate (as a primary indicator of circulatory functions) and oxygen consumption (representing the metabolic conversion taking place in the body) have a linear and reliable relationship in the range between light and heavy work, shown in Fig. 3.8. If one knows this relationship, one often can simply substitute heart rate assessments for measurements of metabolic processes, particularly oxygen uptake. This is a very attractive shortcut, since heart rate measurements can be performed easily.

Heart rate

3.5.2.3 *Counting the Heart Rate*

The simplest way to count the heart rate is to feel the pulse in an artery with a fingertip (called *palpation*), often in the wrist or perhaps in the neck; or one may listen to the sound of the beating heart with a stethoscope. All the measurer needs to do is count the number of heartbeats over a given period of time (such as 15 seconds) and, from this, calculate an average heart rate per minute. More refined techniques use deformations of tissue due to changes in filling of the imbedded blood vessels as indicators of blood pulses. These *plethys-mographic* methods range from measuring the change in the volume of tissues mechanically (e.g., in a finger) to using photoelectric techniques that react to changes in the transmissibility of light depending on the blood filling (such as of the ear lobe). Other common techniques rely on electric signals associated with the pumping actions of the heart sensed by electrodes placed on the chest (ECG).

The use of heart rate has a major advantage over oxygen consumption as an indicator of metabolic processes (and work load): the heart rate responds faster to work demands and hence more easily indicates quick changes in body functions due to changes in work requirements.

Reliability The reliability of these techniques of assessing metabolic processes is limited primarily by the intra- and inter-individual relationships between circulatory and metabolic functions. Statistically speaking, the regression line (shown in Fig. 3.8) relating heart rate to oxygen uptake (aerobic energy production) is different in slope and y-axis intersect from person to person and from task to task. Accordingly, the scatter of the data around the line, indicated by the coefficient of correlation, is also variable. The correlation is low at light loads, under which the heart rate is barely elevated, and circulatory functions can be easily influenced by psychological events (excitement, fear, etc.) that may be completely independent of the task proper. With very heavy work, the oxygen–heart rate relationship may also fall apart, for example, when cardiovascular capacities may be exhausted before metabolic or muscular limits are reached. The presence of heat load also influences the oxygen–heart-rate relationship.

3.5.2.4 Subjective Rating of Perceived Effort

Rating perceived efforts Humans are able to make absolute and relative judgments about the strain generated in the body by a given work task. This correlation between effort and its psychologically perceived intensity probably has been used as long as people have sought to express their preference for one type of work over another. Based on the relationships between the intensity of a physical stimulus and its perceptual sensation (Fechner's and Weber's Laws), in the 1960s, Borg[15] began to develop formal techniques to rate the perceived exertion associated with different efforts using the power function:

$$P = e + f(I - g)^n \tag{3.5}$$

where P is the intensity of the perceived effort, I is the intensity of the physical stimulus (such as, the work performed), and e, f, g, and n are the constants specific to the task.

Asking a person to rate the perceived effort (RPE) of standardized work-loads provides a procedure (similar to the methods previously described) to assess an individual's capability to perform stressful work.

3.5.2.5 Rating Scales

RPE on a formal "light" to "hard" scale can be used to measure the intensity of a task. In 1960, Borg developed a category scale, modified in 1985, which ranges from 6 to 20 (to match heart rates from 60 to 200 beats/min). Every second number is anchored by a verbal expression:

The 1960 Borg RPE Scale (Modified 1985)

6—No exertion at all

7—Extremely light

8

9—Very light

10

11—Light

12

13—Somewhat hard

14

15—Hard

16

17—Very hard

18

19—Extremely hard

20—Maximal exertion

In 1980, Borg proposed his "General Scale" which he claimed was a category scale with ratio properties that yields ratios and levels and allows comparisons but still retains the same correlation (of about 0.88) with the heart rate as the RPE scale, particularly if large muscles are involved in the effort. Normally, the scale (called the CR-10 scale) has 10 intensity levels, but one might rate the intensity of a sensation such as an ache or a pain higher than 10.

The Borg General (CR-10) Scale

0—Nothing at all

0.5—Extremely weak (just noticeable)

1—Very weak

2—Weak (light)

3—Moderate

4—Somewhat strong

5—Strong (heavy)

6

7—Very strong

8

9

10—Extremely strong (almost maximal)

11 or higher—The individual's maximum

Borg's instructions for the use of the CR-10 scale (slightly modified from his wording): While the subject looks at the rating scale, the experimenter says, "I will not ask you to specify the feeling, but do select a number that most accurately corresponds to your perception of [specific symptoms]. If you don't feel anything, for example, if there is no [symptom], you answer zero: nothing at all. If you start feeling something, just about noticeable, you answer 0.5: extremely weak, just noticeable. If you have an extremely strong feeling of [symptom], you answer 10: extremely strong, almost maximal. This would be the absolute strongest you have ever experienced. The more you feel—the stronger the feeling—the higher the number which you choose. Keep in mind that there are no wrong numbers; be honest, and do not overestimate or underestimate your ratings. Do not think of any other sensation than the one I ask you about. Do you have any questions?"

Let the subject get well acquainted with the rating scale before the test. During the test, let the subject do the ratings toward the end of every work period, about 30 seconds before stopping or changing the workload. If the test must be stopped before the scheduled end of the work period, let the subject rate the feeling at the moment of stoppage.

3.5.3 Estimating Energy Expenditures for Specific Work Functions

Numerous measurements of energy requirements at work have resulted in tabulations of energy expenditures for a range of body postures and for many professional or athletic activities: see Tables 3.2–3.4.

Instead of using such general tables, one can compose the total energetic cost of given work activities by adding together the energy costs of the work elements that, combined, make up this activity. If one knows both the time spent in a given element of a certain activity and its metabolic cost per time unit, one can simply calculate the energy requirements of this element by multiplying its unit metabolic cost by its duration.

An example of daily energy expenditure:

- For a person resting (sleeping) 8 h/day, at an energetic cost of approximately 5.1 kJ/min, the total energy cost is calculated to be 2448 kJ (5.1 kJ/min × 60 min/h × 8 h).
- If the person then does 6 hours of light work while sitting, at 7.4 kJ/min, this adds another 2664 kJ to the energy expenditure.
- With an additional 6 hours of light work done while standing (8.9 kJ/min), and further with 4 hours of walking (11.0 kJ/min), the total expenditure during the full 24-hour day would come to about 11,000 kJ (approximately 2600 kcal).

TABLE 3.2 Energy Consumption (to be Added to Basal Metabolism) at Various Activities

	kJ/min
LYING DOWN, SITTING, STANDING	
Resting while lying down	0.2
Resting when sitting	0.4
Sitting with light work	2.5
Standing still and relaxed	2.0
Standing with light work	4.0
WALKING WITHOUT LOAD	
Walking on horizontal smooth surface at 2 km/h	7.6
Walking on horizontal smooth surface at 3 km/h	10.8
Walking on horizontal smooth surface at 4 km/h	14.1
Walking on horizontal smooth surface at 5 km/h	18.0
Walking on horizontal smooth surface at 6 km/h	23.9
Walking on horizontal smooth surface at 7 km/h	31.9
Walking on grass at 4 km/h	14.9
Walking in pine forest, smooth surface, at 4 km/h	18–20
WALKING AND CARRYING ON SMOOTH SOLID HORIZONTAL GROUND	
1 kg on the back at 4 km/h	15.1
30 kg on the back at 4 km/h	23.4
50 kg on the back at 4 km/h	31.0
100 kg on the back at 3 km/h	63.0
WALKING DOWNHILL ON SMOOTH SOLID GROUND AT 5 KM/H	
5° decline	8.1
10° decline	9.9
20° decline	13.1
30° decline	17.1
WALKING UPHILL ON SMOOTH SOLID GROUND AT 2.5 KM/H	
10° incline, gaining altitude at 7.2 m/min	
No load	20.6
20 kg on back	25.6
50 kg on back	38.6
16° incline, gaining altitude at 12 m/min	
No load	34.9
20 kg on back	44.1
50 kg on back	67.2

(Continued)

TABLE 3.2   Energy Consumption (to be Added to Basal Metabolism) at Various Activities (*cont.*)

	kJ/min
25° incline, gaining altitude at 19.5 m/min	
No load	55.9
20 kg on back	72.2
50 kg on back	113.8
CLIMBING STAIRS OR LADDER	
Climbing stairs, 35° incline, steps 17.2 cm high	
100 Steps per minute, gaining altitude at 17.2 m/min, no load	57.5
Climbing ladder, 70° incline, rungs 17 cm apart	
66 steps per minute, gaining altitude at 11.2 m/min, no load	33.6

Source: Data from Astrand and Rodahl (1977), Guyton (1979), Rohmert and Rutenfranz (1983), Stegemann (1984).

TABLE 3.3   Total Energy Expenditure, in kcal/day, in Various Professions and Employments

	Energy expenditure (kcal/day)		
	Mean	**Minimum**	**Maximum**
MEN: OCCUPATION			
Coal miners	2970	3660	4560
Elderly industrial workers	2180	2820	3710
Elderly Swiss peasants	2210	3530	5000
Farmers	2450	3550	4670
Forestry workers	2860	3670	4600
Laboratory technicians	2240	2840	3820
Steelworkers	2600	3280	3960
University students	2270	2930	4410
WOMEN: OCCUPATION			
Elderly housewives	1490	1990	2410
Middle-aged housewives	1760	2090	2320
Elderly Swiss peasants	2200	2890	3860
Factory workers	1970	2320	2980
Laboratory technicians	1340	2130	2540
University students	2090	2290	2500

Note: The physical demands are likely to be different now from what they were in the 1960/1970s.
Source: Data from Astrand and Rodahl (1977).

TABLE 3.4 Total Energy Expenditure (per kg Body Weight) at Various Sports

	Energy expenditure (kJ/kg/h)
Badminton	53
Bicycling, 9 km/h	15
Bicycling, 16 km/h	27
Bicycling, 21 km/h	40
Cross-country skiing, 9 km/h	38
Cross-country skiing, 15 km/h	80
Ice skating, 21 km/h	41
Jogging, 9 km/h	40
Running, 12 km/h	45
Running, 16 km/h	68
Swimming, breaststroke, 3 km/h	45
Walking, 4 km/h	13
Walking, 7 km/h	25

Source: Adapted from Kroemer et al. (2010).

Knowing the energy requirements of work allows one to judge whether a job is (energetically) easy or hard. Given the largely linear relationship between heart rate and energy uptake, one can often simply use the heart rate (see above) to label work as "light" or "heavy." Of course, such labels reflect judgments that rely very much on the current socioeconomic concept of what is permissible, acceptable, comfortable, easy, or hard. Depending on the circumstances, one finds a diversity of opinions about how "hard" a given job is.

3.5.3.1 Defining the "Heaviness" of Work

"Light" work is associated with a rather small energy expenditure (about 10 kJ/min or 2.5 kcal/min, including the basal rate) and is accompanied by a heart rate of approximately 90 beats/min. In this type of work, the energy needs of the working muscles are covered by the oxygen that is available in the blood and by glycogen at the muscle. Lactic acid does not build up. During "medium" work, with about 20 kJ (5 kcal/min) and 100 beats/min, the oxygen requirement at the working muscles is still covered, and lactic acid that was initially developed is resynthesized to glycogen during the activity. In "heavy" work, with about 30 kJ (7.5 kcal/min) and 120 beats/min, the oxygen that is required is still supplied if the person is physically capable of doing such work and specifically trained for the job. However, the lactic acid concentration incurred during the initial minutes of the work is not reduced but remains till the end of the work period; it then returns to normal levels after cessation of the work.

With light, medium, and even heavy work, metabolic and other physiological functions can attain a steady state throughout the work period (provided that the person is capable and trained). This is not the case with "very heavy work," when energy expenditures are in the neighborhood of 40 kJ/min and heart rate is around 140 beats/min. Here, the original oxygen deficit increases throughout the duration of work, making intermittent rest periods necessary or even forcing the person to stop working. At even higher energy expenditures, such as 50 kJ/min, associated with heart rates of 160 beats/min or more, the concentration of metabolites in the blood and the oxygen deficit are of such magnitudes that frequent rest periods are needed, and even highly trained and capable persons may be unable to perform the job through a full work shift.

Static and dynamic work

Only "dynamic work" can be suitably assessed by energy demands. "Static" efforts, during which muscles are contracted and kept contracted, hinder, or completely occlude their blood supply by compression of the capillary bed. Thus, the heart makes an effort to overcome the resistance by increasing its rate and the blood pressure; but because the blood flow remains insufficient, relatively little energy is supplied to contracted muscles and consumed there. Hence, although such static effort may be exhausting, it is not well assessed by energy measures.

Ramp, Stair, or Ladder?

The selection of ramps, stairs, or ladders is a good example of the use of assessing "effort" by oxygen consumption, heart rate, and subjective ratings or preferences. Fig. 3.9 indicates resulting recommendations, depending on the angle of ascent. For angles of up to 20 degrees, ramps are preferred; between 20 and 50 degrees, stairs; at steeper angles, stair ladders; and for angles above 75°, ladders. Recommended design details[16] are shown in Figs. 3.10 and 3.11, but different features may be desirable in specific cases: on ships, an "alternating tread" ladder (Fig. 3.12) is occasionally used instead of a regular ladder because it is safer and easier to use.

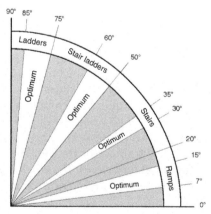

FIGURE 3.9   Selection of ladders, stairs, or ramps according to the angle of ascent. *Source: MIL-STD 1472.*

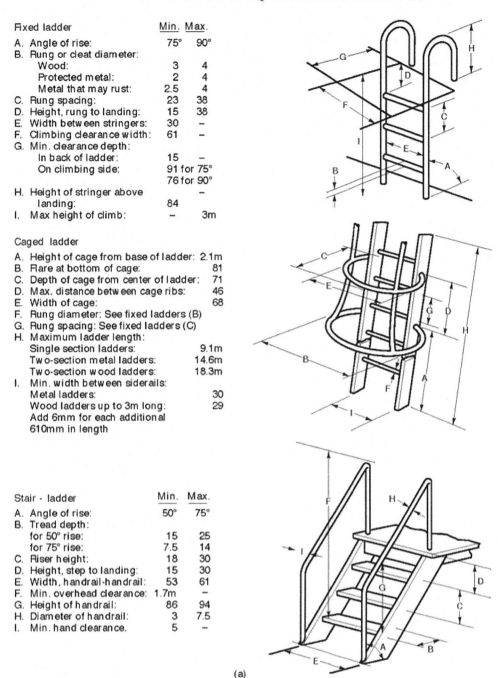

Fixed ladder

	Min.	Max.
A. Angle of rise:	75°	90°
B. Rung or cleat diameter:		
Wood:	3	4
Protected metal:	2	4
Metal that may rust:	2.5	4
C. Rung spacing:	23	38
D. Height, rung to landing:	15	38
E. Width between stringers:	30	–
F. Climbing clearance width:	61	–
G. Min. clearance depth:		
In back of ladder:	15	–
On climbing side:	91 for 75°	
	76 for 90°	
H. Height of stringer above landing:	84	–
I. Max height of climb:	–	3m

Caged ladder

A. Height of cage from base of ladder: 2.1m
B. Flare at bottom of cage: 81
C. Depth of cage from center of ladder: 71
D. Max. distance between cage ribs: 46
E. Width of cage: 68
F. Rung diameter: See fixed ladders (B)
G. Rung spacing: See fixed ladders (C)
H. Maximum ladder length:
  Single section ladders: 9.1m
  Two-section metal ladders: 14.6m
  Two-section wood ladders: 18.3m
I. Min. width between siderails:
  Metal ladders: 30
  Wood ladders up to 3m long: 29
  Add 6mm for each additional 610mm in length

Stair - ladder

	Min.	Max.
A. Angle of rise:	50°	75°
B. Tread depth:		
for 50° rise:	15	25
for 75° rise:	7.5	14
C. Riser height:	18	30
D. Height, step to landing:	15	30
E. Width, handrail-handrail:	53	61
F. Min. overhead clearance:	1.7m	–
G. Height of handrail:	86	94
H. Diameter of handrail:	3	7.5
I. Min. hand clearance.	5	–

(a)

FIGURE 3.10 Recommended design values for ladders (in centimeters unless otherwise stated). *Source: MIL-HDBK 759B.*

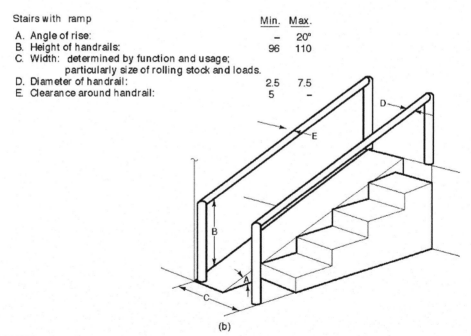

Regular stairs	Min.	Max.
A. Angle of rise:	20°	50°
B. Tread depth:	24	30
C. Riser height:	12.5	20
D. Width, (handrail-handrail)		
One-way stairs:	76	–
Two-way stairs:	1.2m	–
E. Min. overhead clearance:	2m	
F. Height of handrail:	84	94
G. Diameter of handrail:	3	7.5
H. Hand clearance:	5	–

Stairs with ramp	Min.	Max.
A. Angle of rise:	–	20°
B. Height of handrails:	96	110
C. Width: determined by function and usage; particularly size of rolling stock and loads.		
D. Diameter of handrail:	2.5	7.5
E. Clearance around handrail:	5	–

(b)

FIGURE 3.11 Recommended design values for stairs (in centimeters unless otherwise stated). *Source: MIL-HDBK 759B.*

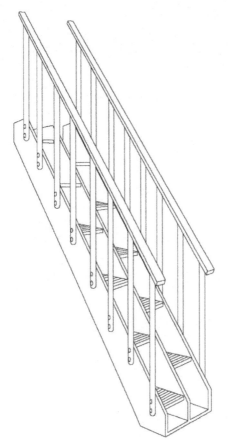

FIGURE 3.12 **Alternating tread stair.**

WHAT YOUR BODY DOES – HOW AND WHY?

Sneezing: The sneeze (*sternutation*) is a very fast semi-autonomous, convulsive expiration of air. It is often caused by an irritation in the nasal passage and serves to clear the cause of irritation. Sneezing can also occur with a sudden change in temperature, a full stomach, an infection, exposure to allergens, or upon sudden exposure to bright light. During a sneeze, the soft palate and uvula depress and the tongue partially closes passage to the mouth, resulting in expulsion through the nose and mouth. The vocal chords come together to plug the windpipe and a sound evolves. Sneezing can occur at any time except during rapid-eye movement (REM) sleep.

Yawning: Yawning (*oscitation*) may be a reaction to either the lack of oxygen, or the accumulation of carbon dioxide, in the lungs or breathing pathways. Yawning may occur when being inactive, such as when sitting and listening to a boring presentation. The frequency of breathing is diminished, and the "air quality" in the lungs and breathing pathways becomes intolerable. A deep inhalation, facilitated by opening the mouth, flushes all the stale air out, taking carbon dioxide with it and introducing oxygen. Another theory about yawning is that yawning and stretching were once part of the same reflex; one could consider yawning as stretching of the face. Another theory is that yawning serves to cool down the temperature of the brain or the body. None of these explains why yawning is an *echophenomenon*, "contagious." Perhaps communal yawning is a social sign of belonging. Children are susceptible to yawning contagion by age two. People even tend to yawn when reading or thinking about yawning.

Shivering: As a muscle contracts to generate energy to do work, heat is produced as a side effect. If no work to the outside is done, a muscle contraction generates heat alone. This occurs in shivering, such as when shivering the jaw makes your teeth chatter or your leg quiver. These quick muscle contractions generate heat to keep the body warm in a cold environment, maintaining homeostasis. The ability to shiver diminishes as we age; this is one reason that the elderly are less resistant to cold temperatures.

Goosebumps: When exposed to cold temperatures or intense emotion, or in response to arousal or other stimulation, we can get goosebumps (also known *piloerection* or *horripilation*). These result from adrenaline release from the adrenal glands, which causes the reflexive contraction of the tiny hair erector muscles. This reaction occurs in all mammals but is more evident in humans as we have less body hair.

Rumbling stomach: Stomach muscles contract and move gas inside the stomach, particularly when it is empty of food. The gases come from air that has been swallowed, and the bubbles make audible noises. Putting food into the stomach takes care of the problem.

Muscle Cramp: Cramps usually occur in response to a contraction signal sent to a muscle that is stretched. Cramps may arise at a sports event, or while asleep or just waking up. A cramp may involve *edema*, the accumulation of fluid inside the muscle but outside its blood vessels. The difference in accumulation generates a pressure, which causes pain. A cramp may also be caused by an inadequate supply of oxygen to muscle tissues of people suffering from low blood pressure. Alternately, local controllers or controllers in the CNS may have run amok.

Cracking Knuckles: The sound created by cracking the knuckles, typically in the metacarpophalangeal joint at the base of each finger, is caused by a dynamic shift in pressure in the synovial fluid. Specifically, the "pop" one hears is caused by the collapse of air bubbles that form in the synovial fluid. Contrary to popular opinion, cracking your knuckles does not result in pain or damage to the joints, either immediately or in the long term. This may be good news for habitual knuckle-crackers but less so for the people around them.

Snoring: One of the main causes of snoring has to do with the *uvula*, a small soft structure hanging from the palate above the root of the tongue; it is at the entrance to the pharynx, which carries air from the nasal cavity to the larynx (the voice box). When the uvula vibrates in the airstream and touches other structures, a snoring sound results. Snoring is most likely to occur when you are lying on your back, so turning over on your side is likely to stop the snoring. Other contributors to snoring are having a low soft palate, being overweight, consuming excessive alcohol, being sleep deprived, and having obstructive sleep apnea—as all of these serve to narrow the airway due to either relaxed throat muscles or tissue abnormalities.

Swelling of Feet: When you sit without moving your legs for a long time, such as in an airplane, swelling (*edema*) of the lower legs (particularly of the feet and ankles) is common. The swelling occurs because lymph (a serum-like fluid, mostly plasma) seeps from the capillaries into the interstitial spaces of the connective tissue, making the tissue expand. The resulting pressure from the expanding tissue hinders the flows of blood and lymph. To prevent swelling, move the lower legs and feet often while sitting, or even better, get up and walk around. After swelling has occurred, it is best to raise the feet and let the accumulated fluid dissipate.

Catching a Cold: Contrary to popular belief, a "cold" is not brought about by a chill, wetness, or draft; none of these predispose a person to infection. Instead, any one of about 200 known cold viruses can infect the upper respiratory tract, including the nose, sinuses, and throat, at any time of the year. The increased occurrence of colds during the colder weather seasons likely comes from people spending more time crowded together indoors; this makes it easy for viruses to spread from person to person, mostly by direct contact such as shaking hands or touching contaminated surfaces. The "flu" (influenza) develops abruptly from fall through spring; it is also a viral respiratory infectious disease but is generally more dangerous than a cold. Frequent handwashing is highly recommended to reduce the risk of "catching" these viruses.

3.6 CHAPTER SUMMARY

The human body converts foodstuffs, in the presence of oxygen, into energy, which then is used to perform work. The respiratory system absorbs oxygen into the bloodstream and extracts from the blood carbon dioxide, water, and heat, which are expelled into the exhaled air. The breathing rate and breathing volume are not convenient for measuring the physical effort of the body.

The circulatory system transports oxygen and energy carriers (primarily glycogen) to various consumer organs (such as muscles) and moves metabolic by-products (CO_2, water, and heat) to the lungs and water and heat to the skin for dispersion. The heart pumps the blood through the system. The number of heartbeats per minute provides a convenient and reasonably accurate measure for the circulatory efforts and for the physical workload of the body in general.

The metabolic system breaks the energy carriers of food into molecules that the body can use. Carbohydrate is broken into polysaccharides (glycogen) and glucose. Fat is broken into glycerol and fatty acids. Protein is broken into amino acids. Glucose and glycogen are the most rapidly used energy carriers, while fat serves to store energy and is used mostly in long-lasting efforts.

The metabolic process of converting polysaccharides and fats into energy used by the muscles occurs in several steps, some of which are anaerobic; yet, overall, oxygen must be provided. Since it is known how much oxygen must be furnished for the conversion of each energy carrier, the measurement of oxygen consumption is an accurate means of assessing the body's energy needs. Oxygen consumption and heart rate are highly correlated.

Another way to assess the workload of the body is to subjectively rate the perceived effort (RPE). This procedure has been well standardized and can be used in combination with, or instead of, the objective measures.

It is one of the tasks of the ergonomist to keep the requirements of a job, in terms of energy demands, within reasonable limits. This can be done if the energy demands of the job are measured and matched with the energy capabilities of the operator.

3.7 CHALLENGES

How would special training through exercise, or the development of deficiencies by disease, of any one of the respiratory, circulatory, and metabolic systems affect a person's ability to perform work: either of the physical, strenuous type, or of the psychological, mental type?

Which events or functions in the respiratory system are easily measurable by the ergonomist and can be used to assess work-related loading of the body?

What are the specific tasks of the blood in supporting the functions of the working body?

By what means can the heart increase blood flow into the aorta? How is the blood supply to a working organ regulated? What are the functions of an artificial pacemaker for the heart?

Why do well-trained athletes often have a very low resting heart rate?

Under what conditions would you use a BMI measurement versus a direct adiposity measurement to obtain body metrics?

What specific features limit the use of the heart rate as an indicator of circulatory loading, particularly of metabolic efforts?

The energy-balance equation does not specify either the duration of body processes or the optimal period required to observe them. How does that affect its validity and use?

How would the consideration of energy storage or loss affect the use of the energy-balance equation?

What are the limitations of the energy-balance equation that is purported to be a measure of human energy efficiency? How can one "become fat" if one eats no fat?

What are the important functions of fat deposits in the body?

What are the two major means by which the body stores energy for use as needed?

Which kind of energy storage in the body is used to support a physical effort?

Why does one have anaerobic processes involved if, indeed, the overall process of metabolism is aerobic?

Is it reasonable to assume that "average persons" perform everyday activities?

Various tables of energy consumption for certain activities are found in the literature and also given in this chapter. What limitations and assumptions apply to these tables?

Are there differences in experimental control that must be applied if either objective or subjective measures are taken to assess the workload associated with certain activities?

NOTES

1 The Greek physician Alcmaeon (6th century BC) was apparently the first to deliberately and carefully dissect human cadavers. He saw the difference between arteries and veins and noticed that the sense organs were connected to the brain by nerves. Praxagoras (4th century BC) distinguished between arteries and veins, but he thought that arteries carried air because they were usually found empty in corpses. (The word "artery" is from Greek words meaning "carrying air.") Around 280 BC, Herophilus noticed that the arteries pulsed and thought that they carried blood. (Asimov, 1989)

2 Physiological measures of human capacities have been reported in great detail in numerous textbooks, such as by Astrand, Rodahl, Dahl, and Stromme (2004), Hall (2016), Rodahl (1989), Kenny, Wilmore, and Costill (2015), summarized for engineers by Kroemer, Kroemer, and Kroemer-Elbert (2010).

3 In 1858, the naturalist Jan Swammerdam (1637–1680) used newly improved microscopes and discovered the red blood corpuscle. In the 1840s, the physician Thomas Addison (1798–1866) studied white blood cells, or leucocytes, from the Greek for "white cell." The third type of objects formed in the blood, called platelets, according to their shape, was studied by the physician William Osler (1849–1919) who reported on them in 1873. They are called thrombocytes, from the Greek for "clot cells." (Asimov, 1989)

4 In 1930, the physician Karl Landsteiner (1868–1943) showed that human blood could be divided into four types. This knowledge made blood transfusions safe. (Asimov, 1989)

5 Based on Galen's (129–199) beliefs, it was thought until the early 1600s that blood was manufactured in the liver and carried to the heart, from which it was pumped outward through arteries and veins alike and consumed in the tissues. The Italian physician Girolamo Fabrici (1537–1619) noticed the valves in the veins that prevent the blood from flowing in the direction that Galen had postulated; yet, Fabrici did not dare to contradict Galen's doctrine. Galen had thought that the heart was a single pump and that there were pores in the thick muscular wall separating the two ventricles. In 1242, the Arabic scholar Ibn an-Nafis wrote that the right and left ventricles were totally separated. He explained how blood pumped by the right ventricle was led by arteries to the lungs where the blood collected air. The enriched blood then was collected into increasingly larger vessels, which brought it back to the left ventricle, from where blood was pumped to the rest of the body. Unfortunately, an-Nafis's book did not become known outside the Arabic world until 1924. In 1553, the Spanish physician Miguel Serveto published a book in which he correctly described some features of the circulation. In 1559, the Italian anatomist Realdo Colombo also described major features of the circulatory system. His work was widely used in the medical profession. In 1628, the English physician William Harvey published his book *De Motu Cortis et Sanguinis*, which describes functions of the heart and blood. This book is generally considered the beginning of modern physiology. (Asimov, 1989)

6 Since Luigi Galvani's time (1737–1798), it had been known that muscular contractions are associated with small electric potentials. The demonstration that this phenomenon also applied to the heart muscle was made possible by the development of a specific galvanometer by the physiologist and Nobel prize winner Willem Einthoven (1860–1927) in 1903. The device allowed the recording of the electrocardiogram, abbreviated as ECG or EKG because, in German, cardio is spelled with a "K". (Asimov, 1989)

7 Astrand and Rodahl (1977).

8 Digestion was long thought to be a physical action, the result of mechanical grinding in the stomach. In 1752, the physicist René Antoine Ferchault de Réaumur (1683–1757) showed that digestion was a chemical process. (Asimov, 1989)

9 See Klesges (1995).

10 In 1827, the chemist William Prout (1785–1850) was the first to chemically classify foodstuffs into groups now called carbohydrates, fats, and proteins. Of course, this grouping is dietetically important but not exhaustive, because other substances, such as vitamins, appear in small yet vital quantities in our food. (Asimov, 1989)

11 For further discussion of weight, metabolism, and activity, see, for example, the 2012 consensus statement from the American Society for Nutrition in Hall, Heymsfield, Kemnitz, Klein, Schoeller, and Speakman (2012).

12 In 1856, the physiologist Claude Bernard (1813–1878) discovered a form of starch in the liver. Because it was easily broken down into glucose, he called it glycogen (Greek for glucose producer). He showed that glycogen could be built up from glucose to act as an energy store and that it could be broken down to glucose again when energy was needed. (Asimov, 1989)

13 In 1913, the physiologist Archibald Vivian Hill (1886–1977) demonstrated that heat was developed, and oxygen consumed, not only during, but also after a muscular contraction, when the muscle was at rest. The biochemist Otto Meyerhof (1884–1951) independently demonstrated the same fact in chemical terms. This indicated that during muscular contraction, lactic acid is developed from glycogen. When the muscle is at rest again, lactic acid is oxidized, thus paying off the oxygen debt that was incurred during the preceding reaction of "anaerobic glycosis." Hill and Meyerhof were awarded shares of the Nobel prize for medicine and physiology in 1922. (Asimov, 1989)

14 Astrand et al. (2004), Chaffin, Andersson, and Martin (2006), Kroemer (2017), Kumar (2004), Kumar (2008), Toomingas, Mathijssenm, and Tornquist (2011), and Winter (2009).

15 Borg (2001, 2005).

16 McVay and Redfern (1994) and Konz and Johnson (2007).

Overview

The traditional psychological model of human perception, cognition, and action postulates a sequential system: input is sensed and then processed, and output follows. While this model has been criticized, no suitable substitute has replaced it yet and, therefore, nearly all currently available information is based on the sequential model.

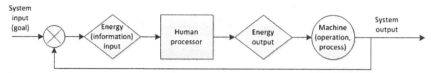

FIGURE 4.1 **The human as energy or information processor.**

The nervous system controls body functions. It receives information from body sensors via its afferent peripheral components. The information is then processed, decisions are made, and control signals are generated in the central nervous system (CNS). Finally, these signals are transmitted to body organs, including muscles, in the efferent peripheral part of the system.

Job and environmental conditions can be very stressful, not only in routine tasks in "normal" environments, but especially in confined and dangerous environments, such as during space exploration. The assessment of the workload is important for subsequent design recommendations. Ergonomic recommendations should also consider the impacts of intrapersonal and cognitive dynamics on individuals and teams.

4.1 INTRODUCTION

System

In the system concept of engineering psychology, the human is considered a receptor, processor, and generator of information or energy. Input, processing, and output follow each other in sequence. The output can be used to run a machine, which may be a simple hand tool or a spacecraft. This basic model is depicted in Fig. 4.1.

Human–technology system

The actual performance of this combined human–machine system, or human–technology system (in the past often called a man–machine system) is monitored and compared with the desired performance. Hence, one or more feedback loops connect the output side (or one of its elements) with the input side of the system. The difference between output and input is registered in a comparator, and corrective actions are taken to minimize any output–input difference. The human in this system makes comparisons, decisions, and corrections.

This "human processor" is the object of research to understand basic human functions or to observe human actions and reactions within the system.[1] Input and output are the sites of application of ergonomics and human-factors engineering. The design of the machine is the classical engineering task, albeit with significant help from the ergonomist.

4.2 THE "TRADITIONAL" AND THE "ECOLOGICAL" CONCEPTS

Linear sequence of stages

In the traditional concept of engineering psychology, our activities can be described as a linear sequence of stages, from perceiving to encoding to deciding to responding. Research is done separately on each of these stages, on their substages, and on their connections. Such independent, stage-related information is then combined into a linear model to provide information for the engineering psychologist.

"Ecological" psychologists believe that this linear model is inadequate; they consider human perception and action to be based on simultaneous rather than sequential inter-actions, and place emphasis on understanding the bidirectional interaction of the human with the environment to understand stages of information processing.[2] The approach has been described as "ask not what's inside your head, but what your head's inside of." An ecological model is useful to the human factors engineer since the interaction of the human and environment is fundamental to the understanding of ergonomics. Two major concepts in the ecological approach are *affordance* and its *perception*.

Affordance is the property of an environment that has certain values to a human. For example, consider a stairway which affords passage for a person who can walk but not for a person confined to a wheelchair.[3] Thus, passage is a property of the stairway, but its affordance value is specific to the user. Accordingly, ergonomics provides affordances.

Perception is how information about affordances is taken in by the individual, directly and simultaneously, by various senses. Thus, the closed-loop coupling between percep-tion and action of a human in a certain environment is not modeled as a simple linear sequence of the stages perceiving–encoding–deciding–responding. Instead, the ecologi-cal model recognizes that information is distributed throughout the closed-loop system.

[margin: Simultaneous stages]

4.3 ORGANIZATION OF THE NERVOUS SYSTEM

4.3.1 Central and Peripheral Nervous System

Anatomically, the nervous system has two major subdivisions. The *central nervous system* (CNS) includes the brain and spinal cord; it has primarily control functions. The *peripheral nervous system* (PNS) includes the cranial and spinal nerves; it transmits sig-nals but usually does not control anything. The essential function of the PNS is to carry information from receptors throughout the body into the CNS and back out to effectors.

[margin: Anatomic division]

Functionally, one divides the nervous system into two major subdivisions: the *auto-nomic nervous system* (ANS), and the *somatic* (from the Greek *soma*, which means body) *nervous system*, which controls mental activities, conscious actions, and skeletal muscle.[4]

[margin: Functional division]

The *ANS* refers to collections of motor neurons (*ganglia*) situated in the head, neck, thorax, abdomen, and pelvis and their axonal connections, and connects the brain to the internal organs. The ANS is responsible for general activation of the body, emergency response, and emotion. The ANS is not generally under conscious control. It consists of the *sympathetic* and the *parasympathetic* subsystems which, together, regulate involun-tary functions, such as those of smooth and cardiac muscle, blood vessels, digestion, and glucose release in the liver.

[margin: Autonomic system]

The sympathetic subsystem is generally responsible for arousal mechanisms, the parasympathetic for relaxation mechanisms. When an organism is startled or experi-ences a "flight or fight" response, the process is mediated by the sympathetic subsystem.

Flight-or-fight states involve a disposition to act (run, hit, bite, kill) and are coupled with strong negative emotions (fear, anger) and an upregulated body state (pounding heart, tense muscles). The part of your brain most responsible for feeling afraid is the *amygdala*. Recent research has suggested another chain of neural connections which links a specific part of the *cerebellum*, called the *pyramid*, to the spinal cord, causing the body to "freeze" when experiencing fear. The freeze response can be thought of as the flight-or-fight response placed on hold and involves the parasympathetic nervous system.

Circulatory processes, digestive processes, heart rate, respiration rate, blood pressure, and other mechanisms may respond to threat via autonomic arousal in a process that has been called a "sympathetic storm." (Excess, inappropriate, or maladaptive sympathetic arousal may play a role in stress responses, chronic anxiety disorders, and certain other medical problems.) In contrast, a parasympathetic activation can be referred to as a "dorsal dive" (i.e., the dorsal aspect of the vagus nerve is activated). In either "flight-or-fight" or "freeze" the individual is focused on responding to significant threat. Higher forebrain areas are involved in early threat responses, including the assignment and control of fear, whereas imminent danger results in fast, likely "hard-wired" defensive reactions, mediated by the midbrain.[5] This has implications for an organism's ability to process, adapt, and react in relation to potential or anticipated stressors compared to sudden or unanticipated stressors.

Somatic system

The *somatic nervous system* refers to afferent (sensory) nerves that relay sensory information from the body to the brain and efferent (motor) nerves that control muscles and glands. This allows control of voluntary actions as well as mental activities.

4.3.2 Brain and Spinal Cord

Brain

The brain is the control center of the body. It creates the human "mind," which generates within its roughly 100 billion neurons and their various connections emotions, moods, beliefs, memories, behaviors, and thoughts. Throughout the ages, philosophers and scientists have debated the relationship between the mind and the brain; this relationship depends on one's orientation, perspective, and purposes. For this book, we will focus on the brain as the primary organ for the mind, much as one can view the lungs as the primary organ for respiration.

The brain continually receives sensory signals regarding body position and movement, and sensations of touch, smelling, hearing, and seeing. It then coordinates and interprets these, developing reactions and consequent actions, often in milliseconds. For this, the brain requires the continuous presence of oxygen. Providing the necessary supply of blood takes about 15–20% of the total cardiac output.

The human brain weighs about 6.6 kg, has a volume of about 1250 cm^3, and is protected by the bony case of the skull. The brain is suspended within the cerebrospinal fluid, which provides both mechanical protection via shock absorption and certain aspects of nutritional support to the brain. The brain is usually divided into the forebrain, midbrain, and hindbrain. Neuromuscular control resides in the *forebrain* with the cerebrum, which consists of the two (left and right) cerebral hemispheres, each divided into four lobes: the frontal, temporal, parietal, and occipital lobes. The lobes together control basic attention

---

THE SEARCH FOR THE SOUL

The search for the location of the human soul is not new. Termed *atman* by ancient Indian philosophers, *psyche* by the Greek and *anima* by the Romans, the soul has been considered resident within but distinct from the human body. When reviewing the rich history of its presumed locales, keep in mind the words of Robert Musil (1890–1942): there is "an abiding miscommunication between the intellect and the soul. We do not have too much intellect and too little soul, but too little intellect in matters of soul."

Pythagoras (570–495 BC) described the soul as consisting of three parts: intelligence and reason (located in the brain) and passion (located in the heart). Leonardo da Vinci (1452–1519) placed the soul above the optic chiasm in the region of the anterior–inferior third ventricle. René Descartes (1596–1650) believed that the soul resided in the pineal gland. Giovanni Maria Lancisi (1654–1720) endorsed the corpus callosum while Albrecht von Haller (1708–1777) placed the soul in the medulla oblongata.

Other historical attempts to define "soul" consisted not only of descriptors of its locale but also of its mass or weight. Perhaps most (in)famously, in 1907, Dr. Duncan MacDougall of Haverhill, Massachusetts, attempted to measure the soul by weighing a human being in the process of dying.[6]

"My first subject was a man dying of tuberculosis. … The patient was under observation for three hours and forty minutes before death … [S]uddenly coincident with death the beam end dropped with an audible stroke hitting against the lower limiting bar and remaining there with no rebound. The loss was ascertained to be three-fourths of an ounce" (21.3 g).

He "measured" the soul in six patients to weigh between 0.5–1.5 ounces (14–43 g). From this experiment derived the popular notion that the soul weighs 21 g (also the title of a 2003 film by Alejandro González Iñárritu). Not surprisingly, MacDougall's findings have long been considered erroneous due to flawed methodology, small sample size, and imprecise measurements.

---

and consciousness, fine and gross motor control, verbal fluency, language functions, various aspects of intelligence (e.g., comprehension, memory functions, abstract thought, planning, and sequencing), and sensory perception. The cerebrum is mostly wrapped by the multifolded cortex, which controls voluntary movements of the skeletal muscle and interprets sensory inputs. The basal ganglia of the *midbrain* are composed of large pools of neurons, which control semivoluntary complex activities such as walking. Part of the *hindbrain* is the cerebellum, which integrates and distributes impulses from the cerebral association centers to the motor neurons in the spinal cord and thus coordinates muscular activity. Fig. 4.2 is a schematic sketch of the major parts of the brain.

The spinal cord is an extension of the brain, and many aspects of human behavior are organized and integrated within it as well. The spinal cord enables communication from

Spinal cord

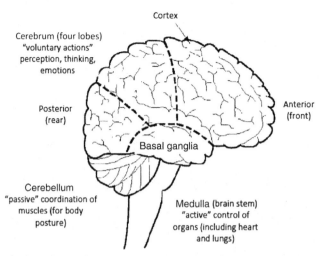

FIGURE 4.2   **Major components of the human brain (as seen from the right).**

the brain outward to the body and relays sensory information inward to the brain. It also organizes certain behaviors, largely without utilizing higher functions of the CNS, principally by means of *reflexes*. In a reflexive behavior, sensory information enters a layer (or layers) of the spinal cord and then directly links to a motor nerve, which then exits to the PNS. The role of the CNS in controlling reflexes is generally limited to determination of speed or ease of triggering a reflex.

Reflexes are divided into three categories: superficial, deep, and special. *Superficial* reflexes (also called *cutaneous* reflexes) are elicited by stimulation of the skin. For example, we blink involuntarily when the cornea is touched. A *deep* reflex is demonstrated by striking the patellar tendon with a soft hammer: the stretching of the tendon results in a reflexive muscle contraction in the quadriceps muscle, which causes the knee to jerk. Similar deep reflexes are also elicited in the jaw, biceps, triceps, internal and external hamstrings, pectoralis, adductor, finger flexors, and several other muscle systems. *Special* reflexes are often used to determine developmentally and clinically significant functions and dysfunctions. For example, the extensor plantar response (*Babinski sign*) in infants consists of dorsiflexion of the great toe and fanning of the remaining toes when the plantar surface of the foot is stimulated.

The uppermost section of the spinal cord contains the 12 pairs of cranial nerves, which serve structures in the head and neck, as well as serving the lungs, heart, pharynx, larynx, and many abdominal organs. The cranial nerves control eye, tongue, facial movements, and the secretion of tears and saliva. Their main inputs are from the eyes, the taste buds in the mouth, the nasal olfactory receptors, and touch, pain, heat, and cold receptors of the head. Thirty-one pairs of spinal nerves extend from the brain to the appropriate vertebrae and pass out between them to serve defined sectors of the rest of the body. Nerves are mixed sensory and motor pathways, carrying both somatic and autonomic signals between the spinal cord and the muscles, articulations, skin, and visceral organs. Fig. 4.3 shows how the spinal nerves emanating from sections of the

Reflexes

Nerves

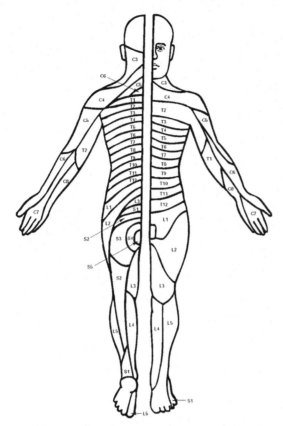

FIGURE 4.3 **Sensory dermatomes with their spinal nerve roots.** The relevant nerve roots are cervical (C2-C8), thoracic (T1-T12), lumbar (L1-L5), and sacral (S1-S5). *Source: Adapted from Sinclair (1973).*

spinal column (called spinal nerve roots) innervate defined areas of the skin (called dermatomes). A spinal nerve root is the initial or proximal segment of one of the 31 pairs of spinal nerves leaving the CNS from the spinal cord. Each spinal nerve is formed by the union of a sensory dorsal root and a motor ventral root, meaning that there are 62 dorsal/ventral root pairs, and therefore 124 nerve roots in total. Fig. 4.4 associates major nerves of the body with their respective areas of cutaneous sensitivity.

4.3.2.1 Neuroplasticity

The ability of the brain to change is called *neuroplasticity*. In youth, *developmental plasticity* occurs as neurons sprout branches, form synapses, and develop neural connections and pathways. In young childhood (ages 2–3), the number of synapses is about 15,000 per neuron, while the adult brain has 7000–8000 synapses per neuron. This massive pruning of synapses normally occurs in adolescence.

Plasticity

Well into the 20th century, it had been commonly accepted that, after growing and developing during youth for about 20 years, the anatomy and functions of the brain became "fixed" like in a wondrous computer with hardwired circuits set to perform

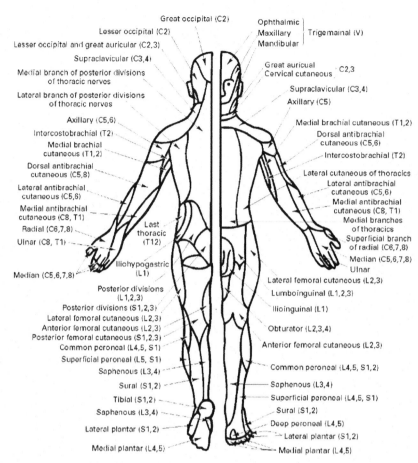

FIGURE 4.4   Territories of the major nerves. *Source: Adapted from House and Pansky (1967).*

certain unchangeable functions. If not injured, the brain would change only in its decline with age. Around the middle of the 20th century, it became evident that, in fact, the adult brain is not immutable. It improves its functioning with every activity performed; at times, the brain can re-organize itself and may alter its structure and develop new ways to function, for example when young brain cells fail to develop properly or when injury damages cells.

Four types of neuroplasticity have been described[7]:

- *Homologous area adaptation*: One brain region takes over the cognitive function of its homologous region in the opposite hemisphere; for example, when one region of the brain is damaged and its normal operations shifted to a mirror-image area of the opposite hemisphere.
- *Cross-modal reassignment*: Structures previously devoted to one sensory input now accept another sensory input; for example, when an adult who has been blind

since birth redirects somatosensory input to the visual cortex in the occipital lobe (a change in the specific functional assignment of an area of the brain).

- *Map expansion*: A specific functional brain region enlarges on the basis of practice and performance; for example, upon the acquisition of musical training.
- *Compensatory masquerade*: A specific cognitive process is assigned to a new location to enable task performance if the previously utilized region is impaired; for example, when someone suffers brain trauma that impairs the spatial sense, a new strategy for navigation can arise, such as memorizing landmarks (a reorganization of pre-existing neuronal networks). Neuroplasticity is the fundamental process underlying real-time visual biofeedback using functional magnetic resonance imaging (fMRI) technology to improve self-regulation. Additionally, cognitive retraining, aerobic exercise, physical training, neuropharmacology, deep brain stimulation, and noninvasive brain stimulation have been used to facilitate adaptive neuroplasticity in stroke and neurocognitive rehabilitation, to reverse symptoms of Parkinson's disease, to reduce chronic pain, to treat mental and addictive disorders, and to address neurodegenerative changes in the aging process.[8] Neuroplasticity is complex, time-sensitive, experience-dependent, and strongly influenced by features of the environment, motivational factors, and aging. However, it clearly can play a significant role in understanding and ameliorating a wide spectrum of brain states and conditions.

4.3.2.2 *Sensors*

The CNS receives information from certain internal receptors, called *interoceptors*, which report on changes within the body—changes in digestion, circulation, excretion, hunger, thirst, sexual arousal, feeling well or sick. *Exteroceptors* respond to sight, sound, touch, temperature, electricity, and chemicals. Since all of these sensations come from various parts of the body, external and internal receptors together are also called *somesthetic* sensors.

Other internal receptors, called *proprioceptors*, include the muscle spindles (nerve filaments wrapped around small muscle fibers), which detect the amount of stretch of the muscle. The *Golgi receptors* are proprioceptors associated with muscle tendons, detecting their tension, which corresponds to the strength of contraction of muscle. *Ruffini corpuscles* are kinesthetic receptors located in the capsules of articulations. They respond to the degree of angulation of the joints (joint position), to change in general, and to the rate of change.

The sensors in the *vestibulum* also are proprioceptors. They detect and report the position of the head in space and respond to sudden changes in its attitude. This is accomplished by sensors in the semicircular canals, of which there are three, located orthogonally to one another. To relate the position of the body to that in the head, proprioceptors in the neck are triggered by displacements between the trunk and head (see also Chapter 5).

Another set of interoceptors, called *visceroceptors*, reports on events within the visceral (internal) structures of the body, such as organs of the abdomen and chest, as well as on events within the head and other deep structures. The usual modalities of visceral sensations are pain, burning sensations, and pressure. Since the same sensations are also

Somesthetic sensors

Proprioceptors

Vestibulum

Visceroceptors

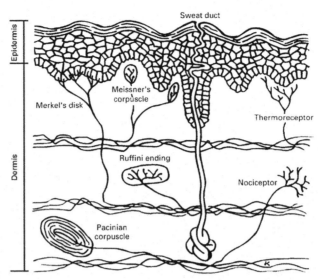

FIGURE 4.5   **Skin receptors.** *Source: Adapted from Griffin (1990).*

provided by external receptors, and because the pathways of visceral and external receptors are closely related, information about the body is often integrated with information about the outside.

External receptors provide information about the interaction between the body and the outside via the senses of sight (vision), sound (audition), taste (gustation), smell (olfaction), temperature, electricity, and touch (taction). Several of these sensations are of particular importance for the control of muscular activities: touch, pressure, and pain, for instance, can be used as feedback mechanism to the body regarding the direction and intensity of muscular activities transmitted to an outside object. Free nerve endings, Meissner's and Pacinian corpuscles, and other receptors are located throughout the skin of the body, although in different densities. They transmit the sensations of touch, pressure, and pain. Since the nerve pathways from the free endings interconnect extensively, the sensations that are reported are not always specific to a modality; for example, very hot or very cold sensations can be associated with pain, which may also be caused by hard pressure on the skin. Fig. 4.5 sketches receptors in the skin.

4.3.2.3 Feelings and Reactions: How and Why

Butterflies in the stomach: The CNS and the gastrointestinal tract are in constant bidirectional communication through neural pathways: thus, the gut microbiome can influence the brain, while the brain can influence microbial composition and function.[9] Various types of psychological stress, including separation, restraint conditions, crowding, heat stress, and anxiety, can affect the composition of the gut microbiota resulting in functional gastro-intestinal distress.[10] A further mechanism involved in the "butterflies" sensation is the "fight-or-flight" sympathetic system which kicks in under stress, causing release of adrenaline, which increases heart rate (to pump more blood and faster), releases huge amounts of glucose from the liver, and shunts blood away from the gut.

(The blood is redirected toward the muscles in the arms and legs which makes them ready to either defend you or run away faster.) This acute shortage of blood to the gut slows digestion as the muscles surrounding the stomach and intestine slow down their mixing of their partially digested contents. While adrenaline contracts most of the gut wall to slow digestion, it relaxes a specific gut muscle called the external anal sphincter, which is why some people report an urgent need to visit a bathroom when they're nervous. This reduction in blood flow through the gut also produces the characteristic "butterflies" feeling in the pit of your stomach: sensing this shortage of blood, and oxygen, the stomach's own sensory nerves let us know that something is amiss.

Blushing: Blushing is a bodily response, mediated by the ANS, wherein capillaries near the surface of the skin fill with blood. This reaction can be caused by sudden and strong emotional triggers, such as embarrassment or stress. It can also be caused by the consumption of alcohol, hot or spicy foods and drinks, fever, strenuous exercise, and certain medical conditions or medications. *Idiopathic craniofacial erythema* is a condition in which an individual blushes strongly with minimal or no apparent provocation. Ironically, having an irrational fear of blushing (*erythrophobia*) can itself trigger an episode of blushing when an individual becomes stressed about the prospect of blushing.

Lump in the throat: The feeling of a lump in the throat is another example of the actions of the ANS. When a person feels threatened, the sympathetic nervous system becomes active and, in this case, makes saliva thick. The thickened saliva is felt as something interfering with swallowing, like a lump in the throat. It can be difficult to distinguish swallowing difficulties due to organic impairment (such as tumors or neurological events) from the perception of a lump in the throat due to autonomic arousal in chronic anxiety conditions.

Hiccups: Having hiccups (*singultus*, or *synchronous diaphragmatic flutter*) consists of breathing going a bit out of control. In normal breathing, impulses regularly progress along the phrenic nerve leading to the diaphragm that separates the abdomen from the chest cavity. The signals cause the diaphragm to contract and reduce the lung volume for expiration. If there is a sudden burst of nervous impulses, the diaphragm contracts abruptly, generating quick expirations. Hiccups can occur with no apparent stimulus but tend to happen at certain times, such as when a person has just eaten a heavy meal, has consumed alcoholic or carbonated beverages, or has experienced sudden excitement. While there are many folk remedies for hiccups, they usually go away by themselves. Rarely, hiccups may persist for months, resulting (not surprisingly) in weight loss and exhaustion.

Gut feelings: We often talk about a sense of intuition coming from the body that is different from an instinct (which is biologically hard-wired). The popular notion of making a decision based on gut feelings may have an actual neurobiological basis related to brain–gut interactions, and to interoceptive memories related to such interactions. Intuitive decision making can be defined as the rapid assessment of the probability of a favorable or unfavorable outcome of a planned behavior in an uncertain situation, and is dependent on previous experiences rather than on inductive and or deductive reasoning. Some studies suggest that such intuitive decision making is based on an interoceptive map of gut responses, acquired during infancy and refined during development.[11]

Goosebumps: Goosebumps (*horripilation*) are the puckering of skin around hair follicles due to the contraction of a muscle at the base of the follicle, which is the pilomotor reflex. The hair stands up and, together with the now enlarged and irregular skin surface, traps air: this insulates the body as a response to cold. Goosebumps are involuntary events generated by the sympathetic nervous system. They can appear when a person is stricken by fear or other strong emotions or sexual arousal: a reaction to either "chill" or "thrill." They are likely a vestigial trait, meaning that they are an archaic reaction, to create a "thicker fur" which might protect better against the bite of an animal or against cold. It is not clear, however, why we do not get goosebumps on our faces.

Head jerk: When you start to doze off, your head drops. Suddenly it jerks back up. Apparently, a sensor in a joint or in a neck muscle received a signal of excessive stretch and, in a reflex, tightened muscles to reduce this stretch. In general, muscles relax as you begin to fall asleep. The brain starts off being very active, but apparently, there is a phase in falling asleep when muscle contraction signals suddenly come through. When this occurs in a meeting or while driving, a sudden head-bob is helpful to quickly wake the dozer and thus avoid social or physical adverse consequences. It is possible that this is a vestigial evolutionary reflex tied to protecting tree-dwelling primates snoozing on branches. When prone and falling asleep, nerve fibers sometimes fire in unison, causing a *hypnic jerk*. The likelihood of a bedtime twitch is increased with sleep deprivation, excessive alcohol or tobacco consumption, and the use of stimulant medications or substances.

NEURO-MYTHS!

"Right brained versus left brained": Common wisdom is that people are either "right-brained" (creative, abstract, artistic thinkers) or "left-brained" (rational, logical, sequential thinkers). However, while some regions seem to be more active in one hemisphere than another, there exists no overall profile of people that dominantly activate one hemisphere over another in their cognitive processes.

"People only use about 10% of their brains": This is an enduring myth perpetuated around the world. Some attribute this quote to Einstein (though if he said this, it was probably in jest, demonstrating his well-known dry sense of humor). As it turns out, the lump of "wetware" between our ears consumes about 20% of our body's energy even as it comprises only about 2% of our average body weight. And, like the heart, it is always "on."[12] A more accurate statement would be that all parts of our brain are in use 100% of the time.

4.3.3 The "Signal Loop"

Input, process, output

Following the traditional psychological concept, the human can be modeled as a processor of signals in some detail, as shown in Fig. 4.6. Information (energy) is received by a sensor and a signal is sent along the afferent (sensory) pathways of the PNS to the CNS, where it is compared with information stored in the brain's short- or long-term

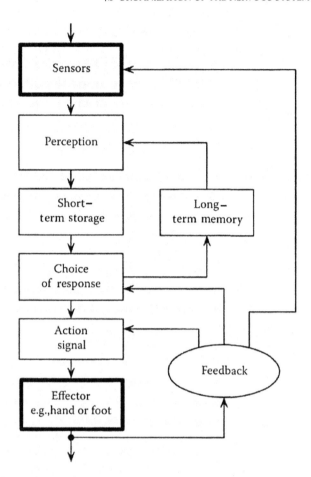

FIGURE 4.6 **The human as signal processor.**

memory. The signal is processed in the CNS, and an action (or no action at all) is chosen. Appropriate feedforward impulses are then generated and transmitted along the efferent (motor) pathways of the PNS to the effectors (the voice, hand, etc.). Of course, many feedback loops exist, although only a few are shown in the model.

Both sides of the processor model can be analyzed further. Fig. 4.7 shows how distal stimuli provide information that may be visual, auditory, or tactile. To be sensed, the stimulus must appear in a form to which human sensors can respond; it must have suitable qualities and quantities of electromagnetic, mechanical, electrical, or chemical energy. If the distal events do not generate proximal stimuli that can be sensed directly, the distal stimuli must be transformed into energies that can trigger human sensations. To accomplish this, the ergonomist designs transducers. For example, a display of some kind (such as a computer screen, dial, or a light) can serve as transducer.

On the output side, the actions of the human effector (such as the hand or foot) may directly control the "machine," or another transducer may be needed. For example, the

Input transducers

Output transducers

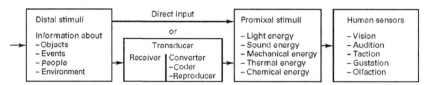

FIGURE 4.7   **Energy input side.**

movement of a steering wheel by the human hand may be amplified by auxiliary power ("power steering"). Fig. 4.8 portrays the model. Of course, recognizing the need for a transducer and providing information for its suitable design is again a primary task of the ergonomist.

4.3.4 Ergonomic Uses of Nervous Signals

Our growing understanding of the human nervous system will surely continue to increase the technologies available to pick up and use nervous signals. Many successes based on empirical knowledge have been achieved in rehabilitation engineering. For example, one may electrically stimulate paralyzed muscles for feedforward control, or stimulate skin sensors for feedback control of prostheses.

EEG

In the 1920s, Hans Berger first attached metal electrodes to the skin of the head and recorded "brain waves," electrical neural activity of the brain via *electro-encephalograms* (EEGs). At about that time, Walter Hess demonstrated that electrical impulses in cat brains affect their emotions and behaviors. Several decades later, José Rodríguez Delgado showed related events in humans.

Afferent impulses are difficult to identify and separate from efferent signals, mostly because of the anatomical intertwining between the sensory and motor pathways of the PNS. Electrical events occurring in the cortex (the *encephalon*, from the Greek for wrapping) that covers the central brain are recorded in the EEG by means of electrodes. The recorded signals are empirically interpreted and are used for a number of purposes, including to classify sleep stages. (See also Chapter 7.) EEG can be used to diagnose

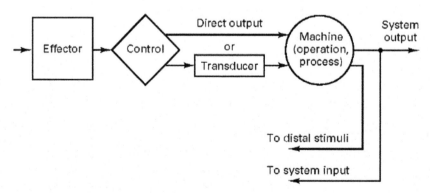

FIGURE 4.8   **Energy output side.**

and differentiate epilepsy and other seizure or seizure-like activity, or to assess depth of anesthesia or coma, encephalopathies, and brain death. EEG research helps us better understand the meaning of brain waves (neurological oscillations) and the electrical response of the brain to a sensory stimuli, be they auditory, visual, or somatosensory.

The effects of motoric signals arriving via alpha-motoneurons at muscle fibers are picked up by electrodes as an *electromyogram* (EMG) at skeletal muscle, as an *electro-oculogram* at eye muscle, and as an *electrocardiogram* (ECG or EKG) at heart muscle. Recording these signals is fairly easily done.[13] The clinical purpose of an EMG is to assess the health of muscles and motor neurons controlling muscles to reveal nerve dysfunction, muscle dysfunction, or problems with nerve-to-muscle signal transmission. EMG signals can be used to analyze the biomechanics of an organism's movement, the amount of muscle fatigue, a muscle's force via its maximal voluntary contraction (MVC; see also Chapter 2). EMG electrodes can be placed on the skin above a muscle or intramuscularly.

EMG

Magnetic resonance imagining (MRI) is a radiologic technique in which magnetic fields, radio waves, and field gradients are used to generate images of the body. Unlike a CT (computerized tomography) scan, it does not utilize ionizing radiation (x-rays). MRI technology is used in medical diagnostics as well as in biomedical research. It can detect contrast between white and gray matter of the brain and can assess functional and structural brain abnormalities. A functional MRI (fMRI) measures the hemodynamic distribution in the brain to determine which area is in use, most commonly using the BOLD (blood-oxygenation-level dependent) imaging technique. BOLD fMRI technology does not depict absolute regional brain activity but instead indicates changes in regional activity over time; it relies on a principle of cognitive subtraction, which assumes that signal differences between two behavioral conditions that are identical in all but one variable are indeed due to this variable. Thus, for research purposes, an appropriate control group is critical in the interpretation of fMRI data. EEG and fMRI technology have been instrumental in brain mapping.[14]

MRI

4.3.5 Responding to Stimuli

The time that passes from the appearance of a proximal stimulus (e.g., a light) to the beginning of an effector action (e.g., movement of the foot) is called the *reaction time*. The additional time required to perform an appropriate movement (e.g., stepping on the brake pedal) is called the *motion*, or *movement*, *time*. Adding the motion time to the reaction time results in the *response time*. (Note that these terms are often not clearly distinguished.)

The experimental analysis of the reaction time goes back to the very roots of experimental psychology: many of the basic results were obtained in the 1930s, with additional experimental work done in the 1950s and 1960s.[15] Innumerable experiments have been performed; hence, many different tables of such times have been published in engineering handbooks. Some of these tables apparently have been consolidated from various sources; however, the origin of those data, the experimental conditions (including the intensity of the stimulus) under which they were measured, the accuracy of the measurements, and the subjects who participated are no longer known.

Reaction times

The following listing of reaction times is typical of generally used, but fairly dubious, information, often applied without much consideration or confidence. Approximate minimal reaction times to stimuli include:

- electric shock: 130 ms
- touch, sound: 140 ms
- sight, temperature: 180 ms
- smell: 300 ms
- taste: 500 ms
- pain: 700 ms

There appears to be little practical time difference in reactions to electrical, tactile, and sound stimuli. The slightly longer reaction times for sight and temperature stimuli may be well within the measuring accuracy or within the variability among persons. However, the time following smell and taste stimuli appears distinctly longer, while it takes by far the longest to react to the infliction of pain.

Time delays   Time passes between appearance of a signal on the input side at a given section of the nervous system and its re-appearance on the output side. Time delays, or lags, occur at the sensor, in afferent signal transmission, in central processing, in efferent signal transmission, and, finally, in muscle activation. A review of the literature suggested that the fastest possible hand reaction times depended on a series of delays that occur between the start and arrival of a signal at different sections of the nervous system.[16]

Estimated time delays include:

- at receptor: 1–38 ms
- along the afferent path: 2–100 ms
- in CNS processing: 70–100 ms
- along the efferent path: 10–20 ms
- muscle latency and contraction: 30–70 ms

Summing the shortest times leads to the theoretically shortest possible delay. There is little reason to assume a situation in which all the delays are shortest (although that may be the condition desired by human factors engineers). The best chances to reduce delays are in the afferent path length and in the CNS processing time.

4.3.5.1 Simple and Choice Reaction Times

If a person knows that a particular stimulus will occur, is prepared for it, and knows how to react to it, the resulting reaction time (RT) is called the *simple reaction time*. Its duration depends on modality and intensity of the stimulus. If one stimulus out of several possible stimuli occurs, or if the person has to choose among several possible reactions, the resulting reaction time is called the *choice reaction time*. Choice RT is a logarithmic function of the number of alternative stimuli and responses; mathematically:

$$RT = a + b \log_2 N \tag{4.1}$$

where a and b are empirical constants and N is the number of choices.

N may be replaced by the probability of any particular alternative. In that case, $p = 1/N$; and we have

$$RT = a + b \log_2 \left(p^{-1} \right) \tag{4.2}$$

which is called the Hick–Hyman law. (To be exact, $RT = a + b\,H$, where H is the transmitted information.)

Under optimal conditions, simple auditory, visual, and tactile reaction times are about 0.2 seconds. If conditions deteriorate, so that, for example, there is uncertainty about the appearance of the signal, the reaction slows. For instance, the simple reaction time to hearing tones near the lower auditory threshold (30–40 dB) may increase to 0.4 seconds. Similarly, the visual reaction time is dependent on the intensity (luminance), duration, and size of the stimulus. For reasonable sizes of the light source (between 0.5 and 1.7 degrees), the reaction time shortens with increasing luminance and by increased duration of the flash. Reactions to visual stimuli in the periphery of the visual field (such as 45 degrees from the fovea) are about 15–30 ms slower than to centrally located stimuli.[17] (More on visual signals in Chapter 5.)

Simple reaction times

Reaction times of different body parts to tactile stimuli vary only slightly, within about 10% for the finger, forearm, and upper arm. When these times are divided into the premotor time (the time from the stimulus to the onset of electromyographic activity in muscles) and the motor time (the time from the onset of EMG activity to the beginning of movement), there are no differences in premotor time.[18] This indicates that it takes longer to move a more massive limb than a lighter one—as one would expect from simple mechanical considerations.

The simple reaction time barely changes with age from about 15–60 years but is substantially slower at younger ages and slows moderately as one grows old. (More on the effects of aging in Chapter 14.)

The choice reaction time expands if it is difficult to distinguish between several stimuli that are quite similar, but only one of them should trigger the response—as expected from the Hick–Hyman law. Measured reaction times are shown in Table 4.1 and Fig. 4.9.

Choice reaction time

4.3.5.2 Motion Time

Motion time follows reaction time. Movements may be simple, such as lifting a finger in response to a stimulus, or complex, such as swinging a tennis racket. Swinging the racket contains not only more complex movement elements, but also larger body and object masses that must be moved, which takes more time. Motion time also depends on the distance covered and on the precision required. Related data are contained in many systems of time and motion analyses, often used by industrial engineers.

In the early 1950s, Paul Fitts performed well-designed and well-controlled studies of motion times.[19] He found that when precision of the target was fixed, motion time increased with the logarithm of distance. If the distance was fixed, motion time increased with the logarithm of the reciprocal of the width of the target. Distance and width almost exactly compensated for each other. These relations have been expressed in a motion time (MT) equation called Fitts's law, expressed as

Motion studies

$$MT = a + b \log_2 \left(\frac{2D}{W} \right) \tag{4.3}$$

TABLE 4.1   Merkel's Data on Reaction Times for Visually Presented Numerals

Number of alternatives	Reaction time (ms)
1	187
2	316
3	364
4	434
5	487
6	532
7	570
8	603
9	619
10	622

Source: Adapted from Keele (1986).

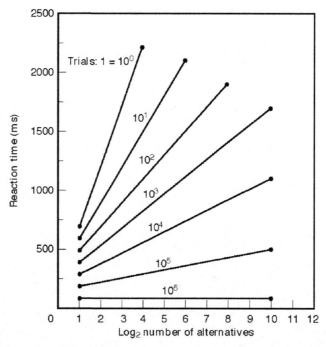

FIGURE 4.9 Schematic relation between reaction time, practice, and number of alternatives. *Source: Adapted from Keele (1986), who used data from Teichner and Kreb (1974). Reprinted by permission of John Wiley & Sons, Inc.*

where D is the distance covered by the movement, and W is the width of the target. The expression $\log_2(2D/W)$ is often called the *index of difficulty*. (The factor 2 is used simply to help avoid negative logarithms.) The constants a and b depend on the situation (such as the body parts involved, the masses moved, and the tools or equipment used), on the number of repetitive movements, and on training.[20] Fitts's law has been found to apply to many movement-related tasks—even to the "capture time" of a moving target.

4.3.5.3 Response Time

The reduction of response time (which is the sum of the reaction and motion times) is a common engineering goal. It can be achieved by optimizing the stimulus and selecting the body member that is best suited to the task.

The best proximal signal is the one that is received quickly (by modality and intensity) and is different from other signals; in the extreme, one may want to bypass receptors and directly act on nerves. Afferent and efferent transmissions depend on the composition, diameter, and length of nerve fibers and on synaptic connections. Practically, the engineer usually selects the shortest distance between sensor location and brain. The best chances for reducing delays are in the processing time needed in the CNS: for perceptual tasks such as the detection, identification, and recognition of the signal, and for cognitive tasks like deciding and planning. Thus, a clear signal leading to an unambiguous choice of action is the most efficient approach to reducing delays.

Choosing the most suitable body member for the fastest response includes first selecting a short afferent distance between sensor and CNS and a short efferent distance to quickly activate muscle; and then making sure that the minimal segment mass must be moved. For example, moving an eye is faster than moving a finger, which is faster than moving a leg.

4.3.6 Mental Workload

The assessment of a workload, whether psychological or physical, commonly relies on the "resource construct," meaning that there is a given (measurable) quantity of capability and attitude available, of which a certain percentage is demanded by the job. If less is required than is available, a reserve exists: see Fig. 4.10. Accordingly, the workload is often defined as the portion of resource (i.e., of the maximal performance capacity) expended in performing a given task.

In accordance with this concept, one should obviously avoid any condition in which more is demanded from the operator than can be given. The performance of the task will not be optimal and the operator is likely to suffer, physically or psychologically, from an overload. However, a task demand that is below the capacity of the operator leaves a residual capacity. Its measurement provides an assessment of the actual workload. Refined models have been proposed: for example, a multiple-resource model in which separate reservoirs are postulated, such as for stages of information processing (afferent, central, and efferent), codes of processing (verbal or spatial), and input/output modalities (visual, auditory or verbal, and manual).[21]

Resource use

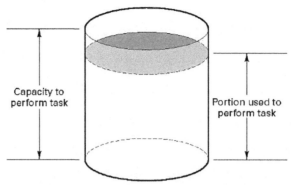

FIGURE 4.10   Traditional resource model.

4.3.6.1 Measuring Workload

Workload is empirically assessed using four different approaches: objective measures of primary-task performance, of secondary-task performance, of physiological events, and subjective assessments. Measures of task performance as well as subjective assessment presume that both "zero" and "full" capacities are known, since they assess the portion of capacity loading. Measuring performance on a secondary concurrent task is intended to assess the spare capacity that remains after allocating resources to the primary task. If the subject allocates some of the resources that are truly needed for primary-task performance to the secondary task, the secondary task intrudes on the primary task; such an invasive secondary task would modify the workload.[22]

4.3.6.2 Measures of Workload

People are different, and individuals differ from each other in their capacities to perform tasks. Thus, the workload imposed by a given task differs from person to person. Further, the workload may depend on the temporal state of an individual—for example, on the person's training, fatigue, and motivation.

One approach to measuring workload is to focus on measuring the primary-task performance by observing how non-critical components of the primary task are performed.[23] The hypothesis is that, as the workload increases, performance changes measurably. Candidates for such unobtrusive measures include the status of a person's speech; the depletion of stock, disorder, or clutter at the workplace; and the length of a line of customers. Such embedded measures of workload do not interfere with the primary task.

Another approach is to focus on measuring secondary tasks. Candidates for secondary tasks to be assessed to measure workload include:

- Simple reaction time: draws on perceptual and response execution resources.
- Choice reaction time: same as for simple reaction time, but with greater demands.
- Tracking: requires central processing and motor resources, depending on the order of control dynamics.
- Monitoring of the occurrence of stimuli: draws heavily on perceptual resources.
- Short-term memory tasks: places heavy demand on central-processing resources.

(Margin notes) Primary task performance

Secondary tasks

- Mathematics: draws most heavily on central-processing resources.
- Shadowing (subject repeats verbal or numerical material as presented): places heaviest demands on perceptual resources.
- Time estimations where subject estimates time passed: draws upon perceptual and central processing resources.
- Time estimations where subject indicates sequence of regular time intervals by motor activity: makes large demands on motor output resources.

Heart rate, eye movements, pupil diameter, and muscle tension can often be measured without intruding on the primary task. However, these measures may be insensitive to the task requirements or may be difficult to interpret. **Physiological measures**

In these tests, we are able to internally integrate the demands of the task, but the subjective assessment of the perceived workload may be unreliable, invalid, or inconsistent with other performance measures. On the one hand, if subjective measures are taken after the task has been completed, they are not real-time evaluations; on the other hand, if performed during the task, they may intrude on the task. **Subjective assessments**

A number of pragmatic measures of the workload have been widely used, in spite of criticism on both theoretical and technical grounds. These measures include: the modified Cooper–Harper scale,[24] the overall workload (OW) scale,[25] the NASA task load index (TLX),[26] and the subjective workload assessment technique (SWAT).[27] The first two are unidimensional rating scales; the last two are subjective techniques using multidimensional scales. While there are good theoretical and statistical reasons for using TLX, SWAT, and OW,[28] they are more complex to administer than the Cooper–Harper scale, which is fairly self-explanatory. That scale is shown in Fig. 4.11, modified to be applicable to systems other than piloted aircraft. **Widely used measures**

The ratings obtained from the Cooper–Harper scale are highly correlated with those in the more detailed TLX and SWAT scales, which might be employed after initial ratings have been obtained from the modified Cooper–Harper scale. On the other hand, TLX and OW were found to have superior sensitivity and operator acceptance.[29] Often, it is advisable to use a combination of workload measurement techniques.[30]

4.4 STRESS

Stress and anxiety are core concepts of psychopathology. A prevalent general model (the diathesis–stress model) assumes that most disorders arise from complex interactions between environmental stressors and (usually biological) predispositions. (For more on job stress within work organizations, see also Chapter 11.)

There is significant evidence that stress and physical illness are related. Early research using the social readjustment rating scale[31] listed life events (ranging from the death of a spouse to being fired from a job to taking a vacation), and subjects were asked to indicate those life events that they have experienced recently. High scores on that scale were found to have been related to heart attacks, other health problems, and depression. Current research is rife with findings linking stress to disease and longevity. Mechanisms involved include the activation of the hypothalamic–pituitary–adrenal (HPA) axis, glucocorticoid metabolism, metabolic hormone production and **Stress and health**

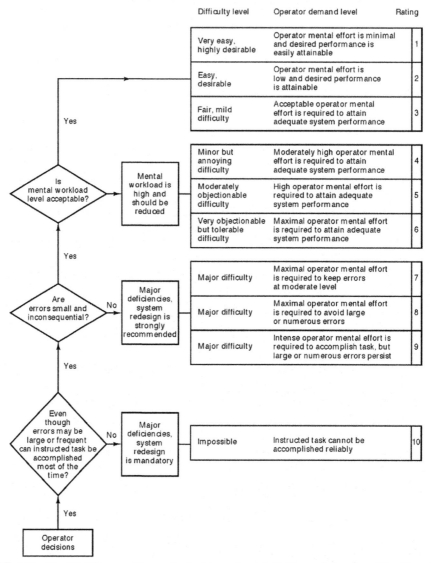

FIGURE 4.11 **Modified Cooper–Harper Scale.** *Source: Reprinted with permission from Wierwille and Casali (1963).*

release, genetic factors, sociodevelopmental factors (including early life experiences) and many others.[32]

It becomes important, then, to consider multivariate and other measures of the association between stress and disease. Many models have been proposed to better understand these complex associations and examine variations in types of stressors and within subgroups of individuals.

In his classic studies in the 1930s, Hans Selye saw stress primarily as a physical trauma to which the human responded. Today, it is generally accepted that stress is not only simply connected with physical events, but also with the appraisal of those events by the individual; hence, stress is also a cognitive phenomenon. Stress involves two-way communication between the brain and the cardiovascular, immune, and other systems via neural and endocrine mechanisms. Beyond the response to acute stress, events of daily life produce a type of chronic stress and may lead, over time, to wear and tear on the body. Fortunately, hormones associated with stress protect the body in the short-term and promote adaptation (referred to as *allostasis*).[33] The hippocampus and the hypothalamus were the first brain regions to be recognized as a target of the glucocorticoid-mediated stress response. Stress hormones produce both adaptive and maladaptive effects on this brain region. Early life events may influence life-long patterns of emotionality and stress responsiveness, but stress and stress-responses throughout the lifespan may alter the rate of functional brain and body aging.[34]

Physical and cognitive

There is some debate in the psychological literature over whether objective assessments or subjective reports of the severity of various stressors provide a better measure of the impact of an event on an individual.[35] Methods of assessing stress include checklists, semistructured interviews, and structured interviews. In each category, several assessment instruments are widely used. Perhaps the most representative checklist questionnaire is Holmes and Rahe's 1967 Social Readjustment Rating Questionnaire (SRRQ)[36] which has been modified many times over the years. Other checklist questionnaires include the Perceived Stress Scale and the Profile of Mood States, which assess the effects of stress, or are designed specifically for job stress, women, children, teenagers, the elderly, Type A behavior (see below), depression, anger, anxiety, etc.

Assessment of stress

Semistructured interviews are used in the Life Event Scale[37] and the Recent Life Events Interview.[38] The Structured Event Probe and Narrative Rating Interview[39] provides a highly organized method of inquiry. A useful assessment of work-related stress is provided by the Workplace Stress Survey,[40] which assesses 10 dimensions of work-related stressors on a 10-point Likert scale. A score between 70 and 100 is indicative of significant work stress that could be problematic to the individual and/or the organization.

Utilizing more advanced measures of stress, we now know that stress can have wide-ranging effects on emotions, mood, and behaviors. (See Table 4.2 for common signifiers of stress.)

The concept of stress on the job is both common and elusive. We all have had the experience of being driven to the edge of our physical and psychological capabilities by strenuous physical exertion, a hot or humid climate, the pressure of deadlines, an oncoming illness, or the feeling that we are expending useless efforts. Some of these stressors are physical, others psychological; some are self-imposed or external; some are short-term or continual. As noted above, assessments of job stress allow the evaluation of an individual's work stress over points in time.

Job stress

Stress isn't necessarily bad. The body's initial response to stress heightens alertness by increasing breathing and heart rate as well as brain oxygenation and activity. However, constant stress can be problematic: long-term exposure to stress and the body's stress response can impact immune, digestive, and cardiovascular functions. Further, what may be stimulating under one condition may become excessive under other circumstances.

"Good" stress or distress?

TABLE 4.2   Common behavioral signs and symptoms of stress

Behavioral signs and symptoms of stress	
1. Frequent headaches, jaw clenching, or pain	26. Insomnia, nightmares, disturbing dreams
2. Gritting, grinding teeth	27. Difficulty concentrating, racing thoughts
3. Stuttering or stammering	28. Trouble learning new information
4. Tremors, trembling of lips, hands	29. Forgetfulness, disorganization, confusion
5. Neck ache, back pain, muscle spasms	30. Difficulty in making decisions
6. Light headedness, faintness, dizziness	31. Feeling overloaded or overwhelmed
7. Ringing, buzzing, or popping sounds	32. Frequent crying spells or suicidal thoughts
8. Frequent blushing, sweating	33. Feelings of loneliness or worthlessness
9. Cold or sweaty hands, feet	34. Little interest in appearance, punctuality
10. Dry mouth, problems swallowing	35. Nervous habits, fidgeting, feet tapping
11. Frequent colds, infections, herpes sores	36. Increased frustration, irritability, edginess
12. Rashes, itching, hives, "goose bumps"	37. Overreaction to petty annoyances
13. Unexplained or frequent "allergy" attacks	38. Increased number of minor accidents
14. Heartburn, stomach pain, nausea	39. Obsessive or compulsive behavior
15. Excess belching, flatulence	40. Reduced work efficiency or productivity
16. Constipation, diarrhea, loss of control	41. Lies or excuses to cover up poor work
17. Difficulty breathing, frequent sighing	42. Rapid or mumbled speech
18. Sudden attacks of life-threatening panic	43. Excessive defensiveness or suspiciousness
19. Chest pain, palpitations, rapid pulse	44. Problems in communication, sharing
20. Frequent urination	45. Social withdrawal and isolation
21. Diminished sexual desire or performance	46. Constant tiredness, weakness, fatigue
22. Excess anxiety, worry, guilt, nervousness	47. Frequent use of over-the-counter drugs
23. Increased anger, frustration, hostility	48. Weight gain or loss without diet
24. Depression, frequent or wild mood swings	49. Increased smoking, alcohol, or drug use
25. Increased or decreased appetite	50. Excessive gambling or impulse buying

Source: Adapted from the American Institute of Stress (https://www.stress.org/stress-effects/, accessed 31 July 2017).

The simple "stress produces strain" sequence (used in engineering and physics, depicting the mechanical relationship between applied force and resulting displacement) may dissolve into the complex relations familiar to psychologists: a stressor may generate a positive "stress" that spurs more activity, or it may result in "dis-stress" which overloads the person and generates ineffectiveness, evasive behavior, anxiety, and even illness.

4.4.1 Demand, Capacity, Performance

The confusing situation regarding stress and its effects may be clarified by the model shown in Fig. 4.12 which reflects three major aspects of ergonomic concern. First, *job demands* depend on the type, quantity, and schedule of tasks; the task environment (in physical or technical terms); and the conditions of the task including the psychosocial relations existing on the job. These (and possibly other related) work attributes are the job stressors that are imposed on the human. Second, a *person's capability* to fulfill the demands of the job, and, third, the *person's attitude* (influenced both by her or his physical and psychological well-being) must be matched with those job demands. If the job demands require only a portion of the person's abilities and attention, the person is likely to feel underloaded and underestimated, and eventually might become bored, inattentive,

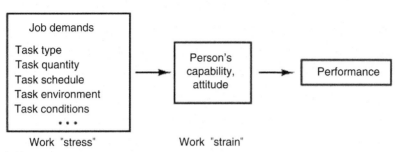

FIGURE 4.12 **Simple model of the relations between job demands ("stress"), human responses ("strain"), and performance.**

and underachieving, or, alternatively, might seek more of a challenge. To the contrary, if the job demands exceed the person's capabilities, then the individual is liable to feel overloaded and might seek either to reduce the workload or to increase his or her capabilities. Of course, other sources of stress besides those at work might require a portion of the person's capabilities or attitude, influencing physical and psychological well-being. In that case, the strain experienced by the individual depends on the sum total of the job and other demands in relation to the person's capabilities and attitude. A buffering factor may be the individuals' hardiness (or resiliency): a constellation of behavioral and attitudinal factors believed to be associated with resistance to the effects of stress.

All of these attributes, conditions, and reactions affect the *person's performance*, which is, of course, an overriding concern privately, in business, and in our everyday life. Importantly, physical health is also greatly affected by stress as key body systems, including the nervous system, the musculoskeletal system, the respiratory system, the cardiovascular system, the endocrine system, and the gastrointestinal system are all vulnerable to the primary and secondary effects of stress.

MATCHING DEMANDS TO CAPACITY

The U-shaped function shown in Fig. 4.13 relates the stress imposed by the work to the resulting strain experienced by the person. According to the U-theory, if the job demands are far below the person's abilities, an *underload* condition exists, and the on-the-job performance is most likely (but not necessarily) diminished. On the other hand, if the work requires more than a person is able and willing to give, an *overload* condition exists, and both work performance and well-being are likely to suffer. If the job demands (work stress) match the person's capabilities and attitude, the proper amount of strain exists and the on-the-job performance is satisfying, both objectively and subjectively.

An optimal experience may occur when the person is in a state of "flow" and the skill level is perfectly balanced to the challenge level of a task that has clear goals and provides immediate feedback. This state of flow is thought to occur when the person is deeply focused, during which time the individual experiences an almost euphoric state along with high alertness and the sense of losing track of time.[41]

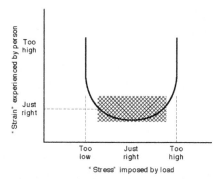

FIGURE 4.13   Postulated "U-function" relating stress and strain.

Monotony

Monotony is the opposite of variety, either of which can be perceived by an individual as stressful. Monotony is produced by an environment in which either there is no change or else changes occur in a repetitive and highly predictable fashion over which the individual has little control. Monotony can be described as qualitative under-stimulation. A varied environment often provokes interest and the human emotion of excitement; in contrast, an unvaried environment produces *boredom*, which also can be considered an emotion. Thus, boredom is an individual's emotional response to an environment that is perceived as monotonous. A bored person often complains of feeling tired or fatigued.

Mental fatigue

The term "fatigue" is commonly used to indicate a physiological status; however, some psychologists prefer that it be used exclusively to define a subjectively experienced disinclination to continue performing the task at hand.[42] Of course, physis and psyche are related, but we must agree on one definition to be able to discuss the topic. We can use the term *mental fatigue* to describe a disinclination to continue a task due to psychological factors.

4.4.2 Stress Experienced by Individuals and Groups

Stress–strain overloads

Understanding people's capabilities and developing job demands and conditions that are matched to those capabilities is the main focus of the ergonomist. Considering the attributes and conditions of work helps us understand proper stress–strain relationships and reduces the risk of particular occupational overload conditions. This occupational overload, together with a person's behavior and mental and physical attributes, constitutes a primary source of variance in explaining individual and organizational distress. (Many of these attributes fall into the domain of psychopathology, a subject beyond the scope of this text.) For example, overworked nurses exhibit behavioral and emotional symptoms of fear, dread, anxiety, irritation, annoyance, anger, sadness, grief, and depression in response to their occupational conditions.[43]

4.4.2.1 Occupational Health Psychology

The extent of strain a person experiences in an occupational setting depends on job experience, age, attitude, self-esteem, and coping ability, among other factors. The growing field of occupational health psychology addresses the convergence of public health and clinical psychology in an industrial or organizational context with an emphasis on preventative management.[44] Occupational health psychologists focus on both environmental modification and individual behavioral factors that are involved in a work environment. The latter category includes such concepts as "workaholism"[45] and cognitive–affective stress propensity.[46] Perhaps one of the most interesting areas of investigation of individual factors has been in the domain of Type A and Type B behavior.

Types A and B behavior patterns are highly correlated with the experience of strain. The type A behavior pattern is exhibited by individuals who are engaged in an (often chronic) struggle to obtain (often poorly defined) things from their environment in the shortest possible period of time—if necessary, against the opposing effects of other things or persons. Type A behavior is characterized by aggressiveness, competitiveness, impatience, and urgency in overcoming obstacles. Individuals exhibiting this behavior may also approach obstacles presented by circumstances or other individuals with an attitude of hostility. Type A individuals are likely to act in ways that make job events more stressful for themselves, and then they find the resulting strain particularly intense. Type A individuals tend to be more successful overall in work accomplishments but may suffer health consequences from the body's response to stress.[47] In contrast, Type B individuals are more relaxed, unhurried, less aggressive, and less competitive, and usually have fewer disorders.[48] In a general population, the Type A/B personality is a normally distributed continuum with relatively few people being extremely driven and competitive or consistently and totally laid back. Most individuals are probably slightly on one side or the other of the A/B divide.

Types A and B

Unlike Type A/B classifications, personality disorders are diagnosable mental disorders and are characterized by enduring maladaptive patterns of behavior, cognition, and intrapsychic experiences exhibited across contexts and time. They are all associated with significant distress or disability in personal, social, and occupational contexts. Official criteria for diagnosing personality disorders are listed the Diagnostic and Statistical Manual of Mental Disorders (currently DSM-5) and the International Classification of Diseases (currently ICD-10).[49] Personality disorders are usually classified based on three clusters of descriptive similarities. Cluster A includes odd or eccentric personality disorders that fall short of schizophrenic criteria (although some may develop into schizophrenia and other psychotic disorders). Cluster B includes dramatic, emotional, or erratic personality disorders. Cluster C includes anxious or fearful personality disorders. It is important to note that there is absolutely no overlap of Type A/B personality "types" with diagnosable personality disorders (Clusters A/B/C). Personality disorders are also distinct from mood disorders (such as anxiety and depressive disorders) but may co-exist with them. Personality disorders are particularly resistant to intervention, and consistent dysfunction observed secondary to a personality disorders in a workplace would also be evident in personal and social settings.

Personality disorders

Traits of a personality disorder observed by a layperson do not confirm the existence or diagnosis of a personality disorder. Anyone suspected of a personality disorder should be referred to a licensed mental health professional for assessment and intervention.

Stress management is a popular term that has been used by psychologists in a variety of ways. The term "management" implies that stress is felt to be an unavoidable component of living with which individuals can learn to cope. Hans Selye stated that to talk about being under stress is as pointless as talking about running a temperature.[50] What is really of concern is an excess of stress, i.e, a distress. Various categories of stressors have been identified: personal ones as well including cataclysmic ones that happen to several people at once (e.g., natural disasters and massive corporate lay-offs).[51]

Various life events have been used to scale personal stressors. The Social Readjustment Rating Scale, previously mentioned, is a 43-unit checklist of different stressful events, ranging from the death of a spouse to taking a holiday, that have been correlated with symptoms of physical illness and psychological distress. Background stressors, so-called daily hassles, have been ordered into a Hassles Scale and have been related to health and well-being.[52] Examples of daily hassles in the workplace include inconsiderate coworkers and noisy work environments. On an intuitive level, it is easy to recognize the cumulative impact that such daily annoyances may have on the physical and emotional well-being of an individual, especially if they are perceived as being uncontrollable. Indeed, research confirms that the strain felt by an individual is greater if the stressor is perceived to be outside one's realm of control. Job strain has been operationalized in several research studies by settings involving a high demand and a low control over decisions. A high level of this kind of job (task) strain has been implicated as a risk factor for coronary disease, as well as for hypertension, in a variety of cross-sectional and prospective epidemiological reports.[53] Sustained uncontrollable stress is also thought to be one factor in the development of "learned helplessness," one model for the development of depression.

Sustained or chronic stress has been defined in varied ways but generally assumes a duration of longer than 4 weeks. Chronic conditions that have been found to increase one's risk for poor health and decreased psychological well-being include poverty, sustained disability, and prolonged marital conflict. In a recent study, subjects enduring severe chronic stressors developed disease upon being inoculated with common cold viruses, while subjects with severe acute stressors did not,[54] and these associations were not attributable to personality variables, health practices, the season of the year, body mass index, demographic variables, or preventive endocrine or immune measures. In addition, individuals are often not adequately aware of the frequency or intensity of background stressors, particularly if they increase gradually over time. An analogy may be found in the animal world: in 1888 (when such studies were not subject to panel review to determine whether they were ethical), William Scripture found that frogs would submit to being boiled alive if the water temperature was increased gradually.[55]

In a growing body of research on professional leadership, the personality traits of affability, resiliency, and optimism have been noted to be associated with high performance.[56] These traits have been conceptualized as components of an *emotional (intelligence) quotient* (EQ) in a somewhat controversial theory which postulates that individuals possessing such traits excel at work more so than those with a high intelligence quotient (IQ). EQ is sometimes referred to as a facet of "pop psychology" but the concept has utility. We may all know someone who is highly intelligent conventionally but has notably poorer emotional awareness or "people skills." Whereas one's IQ is relatively fixed, one's EQ can, to a much larger degree, be built and learned.[57] At the same time, generally EQ and IQ are somewhat correlated. Accordingly, companies can theoretically test an employee's emotional intelligence and teach the employee how to increase it, perhaps especially with individuals with higher IQs. Employers may even be able to incorporate EQ training into more traditional stress management programs. It stands to reason that different emotional skills correlate with success in different professions. For example, success in sales requires the ability to gauge a customer's mood and various other interpersonal skills that are quite different from the more self-directed skills required for a person to be a successful athlete or musician. Individuals with higher EQs may have better emotional and physical health[58] even when accounting for IQ. Leadership performance, in particular, may be enhanced with traits typical of five EQ components called *emotional competencies*[59]:

1. Self-awareness: knowing one's own strengths and weaknesses, values and goals, current emotional states.
2. Self-regulation: controlling or redirecting disruptive emotions and impulses; being adaptable; anticipating consequences.
3. Social skills: managing relationships to ensure desired outcomes and direction, inspiring others.
4. Empathy: understanding others' feelings and perspectives when making changes or decisions.
5. Motivation: being driven to achieve, enjoying the learning process, tenacity.

Individual attitudinal variables (such as beliefs and character predispositions) mitigate the degree of distress elicited in an individual by a given environmental stressor: people have different coping styles. Stress reactions can be compared to allergy attacks[60]: in an allergic reaction, the overmobilization of the immune system—not the allergen itself—causes individuals to feel ill. *Coping* can be defined as the cognitive and behavioral efforts to manage specific external or internal demands that are appraised as taxing or exceeding the resources of the person.[61] The resources of an individual may be material, physical, intrapersonal, interpersonal, informational educational, and cultural. Cognitive components of effective coping strategies include rationality, flexibility, and farsightedness.[62]

| Coping |

Further, effective coping strategies are either problem-oriented or emotion-focused. Problem-oriented coping is directed at controlling an environmental stressor to reduce its impact and is effective if a stressor is objectively controllable. Examples include time management, environmental control or environmental adaptations, assertive communication, and the setting of limits. Emotion-focused coping strategies include structured relaxation exercises, the use of humor, exercise or hobbies, and guided cognitive re-evaluation ("reframing") of a stressor.[63]

The workplace has become a common setting for stress management programs. These often emphasize developing healthier lifestyles (through smoking cessation, weight management, exercise programs, or treatment for drug and alcohol abuse). Such programs are obviously beneficial to the employees who need them, but they also affect the corporate economy if they enhance productivity, reduce absenteeism, and reduce the expense of medical insurance coverage.

Approximately one-half of large companies in the United States provide some sort of stress management training for their workforces. However, stress management programs have two significant weaknesses. First, the effect of stress management training may be short-lived as programs may not be comprehensive or ongoing and as individuals tend to resume well-rehearsed maladaptive patterns upon presentation of new stressors. Second, stress management training programs may ignore organization-specific causes of stress by focusing only on the worker and not on the environment. Thus, *organizational change* is often needed, facilitated by a thorough evaluation to assess and address working conditions incorporating factors such as excessive workload, poorly defined roles and responsibilities, and work schedules that are incompatible with demands outside of the job. To improve working conditions, actions to reduce job stress should give top priority to organizational change. Of course, even the most diligent efforts to improve working conditions at the organizational level will not eliminate stress completely for all workers. Therefore, a combination of organizational change and stress management is the most effective way to prevent stress and distress at work.[64]

US NATIONAL ASSESSMENT OF WORK STRESS

The General Social Survey (GSS) is a biannual sociological survey funded by the National Science Foundation since 1972. The Quality of Working Life (QWL) module is a joint project between the National Institute for Occupational Safety and Health (NIOSH) and the National Science Foundation, and is administered as part of the General Social Survey. The QWL captures how work life and the work experience have changed over the past four decades and helps maintain a baseline for future research. The GSS of 2016 found that 50% of nationwide respondents describe being consistently exhausted because of work, compared with 18% two decades earlier.

Burnout

Burnout is a term often used in recent years to describe a general feeling of being overwhelmed and weary, ineffective, or even traumatized in one's workplace. It was originally applied to health care workers, firefighters, law enforcement officers, paramedics, and others who deal with trauma and human services. The term is now used by workers in many occupations as the workforce in general is more connected, as demands and competition increase for many, and as many societal subcultures have adopted a multi-tasking and hyper-productivity ideal. Common work stressors leading to the phenomenon of burnout include challenges associated with new technologies and processes, unrealistic deadlines, value conflicts, frequent work interruptions and/or over-scheduling, added responsibilities without compensation, and exposure to extreme environmental, medical, and psychosocial situations. The term now even appears in the ICD-10 and is understood as a disease process involving feelings of exhaustion that is similar to (and probably related to) depression rather than feelings of surrender or failure. Common symptoms of burnout are feeling mentally drained and unwell (predicting decreased quality of work and productivity), insomnia, feeling alienated or unappreciated, feeling ineffective, increased feelings of cynicism (predicting turnover), and aggressive outbursts and angry feelings in the workplace.[65] Reducing worker burnout requires addressing both individual and organizational factors.

THE PURSUIT OF HAPPINESS

Typically, 3 in 10 Americans say that they are very happy, 6 in 10 describe themselves as pretty happy, and only 1 in 10 is not happy. Happiness cuts across almost all demographic classifications of gender, age, economic class, ethnic group, and educational level, even wealth. Happy people tend to like themselves, feel personal control over their lives, and are optimistic and extroverted.[66]

Money is important for happiness but only to a fairly specific point.[67] Higher incomes do not always correlate with more life satisfaction due to the *Easterlin paradox*[68] which states that people adapt to higher levels of income over time, and (more importantly) compare their own income to that of their peers. Most research suggests that a sense of purpose or meaning is a fundamental aspect of happiness. Aristotle first discussed the concept of happiness as *eudaimonia* which was an activity rather than an emotion or state of being. Research suggests that many aspects of happiness are under personal control. Such activities can include indulging in small pleasures, engaging in altruistic behaviors, setting and meeting goals, maintaining some close social ties, and maximizing mental health.[69]

WHERE DO THE HAPPIEST PEOPLE LIVE?

Recent studies have investigated and compared rates of happiness internationally. Scandinavian countries seem to lead the way in measures of social happiness, amongst them health and longevity, income and income equality, social supports, employment and employee benefits, and other measures. In 2017, the United Nations' World Happiness Report named Norway the happiest country, followed by Denmark, Iceland, Switzerland, and Finland. The five least happy countries were Rwanda, Syria, Tanzania, Burundi, and the Central African Republic.

4.4.3 Stress Experienced by Confined Groups

Crews in confined spaces such as in spaceships experience a special kind of on-the-job stress. Even during the relatively short missions carried out by US astronauts, severe intra-individual and inter-individual problems appeared, ranging from physiological deficiencies, to anxiety, to difficulties in interaction with other crew members and disagreements with ground control. Problems of this nature have also occurred under other isolated and confining working conditions, such as in the Antarctic.[70]

During one of the US Skylab missions, Commander Gerald Carr and his fellow astronauts not only plotted to hide items from ground control but also went on a strike until some disagreements could be worked out. Similar problems with ground controllers in the USSR, as well as interpersonal conflicts among cosmonauts, developed aboard the Salyut and Mir space stations.[71]

Changes in environment and physiological functions affect a person's psychological and psychosocial well-being. Among the physiological effects of space flight (further discussed in Chapter 6) on astronauts are the following:

- Dimming of vision, peripheral light loss, and, eventually, blackout under increasing plus-Z-direction acceleration.
- Diminished vision, red-out, increased accommodation time, and blurring or doubling of vision in minus-Z-direction acceleration.
- Lengthening of visual reaction time by increased g-levels.
- Degradation of estimates of sizes and distances of objects in space because the objects appear sharper and in higher contrast while foreground and background distance cues are absent.
- More time required for the eyes to adapt to the existing wide ranges of illumination and luminance levels when shifting the gaze.
- Reduction of visual performance by vibration such as during liftoff and landing, when vision is particularly important.
- Shift in perceived colors, reduction in sensitivity to contrast and in near-field acuity but improved visual acuity for distant objects (all of these according to anecdotal and contradictory reports).
- Sensitivity to noise appears to increase, whereas auditory functions in general do not change significantly in space.
- Reduced cabin pressure requires crew members to talk louder to be heard, owing to the reduction in transmission of sound.
- The sensitivity of smell is diminished (probably due to the everpresent nasal congestion in microgravity).
- Unpleasant odors (for instance those associated with medical symptoms, food, or body waste) can become very annoying; yet, pleasant odors are also met with increased responsiveness.
- The sense of taste is degraded, so that food judged to be well seasoned on Earth tastes bland in space. (This effect may be associated with nasal congestion and the upward shift of body fluids.)
- Vestibular effects are evidenced mostly in space sickness and spatial disorientation.
- Larger forces must be applied to generate the same kinesthetic sense stimuli as in a 1-g environment.
- Motor skills are impaired upon entering microgravity, but the deficiency is compensated for after a short period of adaptation.

- Perception of body language and facial expression is changed because crew members appear in different postures and their faces are swollen by the accumulation of fluid in the tissues.
- Anxiety, often present in the initial and final phases of a space mission or during emergency situations, can greatly reduce the effectiveness of individual crew members (particularly of the commander of the space ship) and of the team in general.
- Rigid hierarchical structures (such as those prevalent in the US armed forces) usually become diluted during long missions and develop into a more collegial, democratic sharing of decisions and activities.

While it is known that personal, interpersonal, and psychosocial factors play major roles in the success or failure of long missions, exact information on these factors is still sparse. Compatibility and cohesion of crew members and the leadership style of the commanders of a ship are very important during long periods of confinement, isolation, monotony, and danger, as was experienced in the joint US–Russian Mir space expeditions in the mid-1990s and in more recent missions on the International Space Station.

Three stages in reacting to isolation have been observed. In the first stage, there is heightened anxiety, apparently dependent on the degree of present danger that a person perceives. Heavy workloads appear to reduce the anxiety level during this time. The second stage is associated with feelings of dysphoria or depressed mood, usually during the segment of time while one settles down to routine duties. These feelings may result from the absence of familiar social roles, such as spouse, parent, or club member. The final stage is at the end of the mission and manifests itself in increased expression of affect and anticipatory behavior exhibiting higher levels of anxiety and aggressiveness. In this stage, work performance is commonly degraded, and serious errors of judgment or omission may occur. Table 4.3 summarizes the main issues that affect a crew's behavior and performance. It also lists measures that can be taken to influence behavior and performance positively.

Isolation

Until the early 1990s, crews of men performed all long space missions. Mixed-gender crews generate new facets of intragroup behavior. One idea for stabilizing relations is to use married couples. However, even then, the conditions of confined spaces, working together closely, and social and sexual desires are likely to generate difficult relationships among the crew members, with possibly large changes over time. A related problem is that of pregnancy in space: it is at present unknown whether microgravity, radiation, and other space conditions might affect the development of a fetus.

4.5 ENHANCING PERFORMANCE

There is generally good reason to achieve and perform at one's best. Learning, knowledge, communication, creativity, concentration, and skill, as well as performance under stress are important on the job or in sports; the military is particularly interested in attaining fearlessness, cunning, courage, effectiveness, reversal of fatigue, and nighttime fighting capabilities.

TABLE 4.3   Factors that Influence Crew Behavior and Performance

Psychological, psychosocial, and psychophysiological	Environmental	Space system	Support measures
Limits of performance (perceptual/motor)	Spacecraft habitability —Confinement	Mission duration and complexity	In-flight psychosocial support
Cognitive abilities	—Physical isolation	Organization for command and control	Recreation
Decision making	—Social isolation		Exercise
Motivation	—Lack of privacy	Division of work between human and machine	Work–rest/avoiding excess workloads
Adaptability	—Noise		
Leadership	Weightlessness	Crew performance requirements	Job rotation
Productivity	Artificial life support		Job enrichment
Emotions/moods	Work–rest cycles	Information load	Training
Attitudes	Shift change	Task load/speed	—Preflight environmental adaptation
Fatigue (physical/mental)	Desynchronization of body rhythms	Crew composition	
Crew composition		Spacecrew autonomy	—Social sensitivity
Crew compatibility	Hazards	Physical comfort/quality of life	—For team effort
Psychological stability	Boredom		—For self-control
Personality variables	Stresses	Communications (intracrew and space/earth)	In-flight maintenance of proficiency
Social skills	—Single		
Human reliability (error rate)	—Multiple	Competency requirements	Earth contacts
Space adaptation	—Sequential	Time compression	
Spatial illusions	—Simultaneous		
Time compression			

Source: Adapted from Christensen and Talbot (1986).

4.5.1 Enhancing Individual Performance

Training

Training should be structured to help attain performance ideals. Techniques for enhancing human performance have received a great deal of attention in both the scientific and the popular press. To maximize effectiveness, any training protocol should incorporate an understanding of how new neurological pathways are formed. A fundamental aspect of training is effective practice. (See also Chapter 13 for injury prevention training.)

Practice makes perfect

There are biological mechanisms underlying why practice is a key element of training. The CNS consists of both gray matter and white matter. Gray matter (actually pinkish-gray in color) contains the cell bodies, dendrites, and axon terminals of neurons: all synapses occur in gray matter. White matter consists of fatty tissue and axons, which connect different parts of gray matter to each other. During practice, the myelin sheath covering axons in the white matter becomes more dense, allowing faster and more efficient neural pathways to form. However, practicing optimally is critical to effect the desired changes. Practice (or training) sessions should:

- be distraction-free,
- focus on the aspects of the skill that need improvement,
- consist of repetition and increasing speed,
- incorporate immediate feedback (such as bio-feedback), and
- include frequent and intense activity with regular breaks for consolidation of new pathways.

Both quality and quantity of practice matter. Cognitive rehearsing (using imagination and visualization) of skills improves performance as well. Contrary to popular myth, there is not a specific number of hours of practice, whether physical or cognitive, that will lead to mastery across all skills.

Other methods to improve performance for an individual in the workplace include providing rewards for individual performance incrementally, setting realistic goals, providing meaningful rationales for implementation of new skills or protocols, and utilizing various wellness-oriented strategies. Cognitive focus can be enhanced by incorporating meditative and stress-management techniques.

Pre-performance routines are also commonly used to reduce a sense of pressure or levels of preceding anxiety that interfere with upcoming performance. Patterns of thoughts, actions, and images, consistently engaged in before the performance of a skill, are meant to divert attention from negative and irrelevant information and to establish the appropriate physical and mental state with respect to the action that is to be carried out. Performed at a critical time, these cognitive-behavioral techniques appear to better focus attention and result in improved overall performance. Predictable and well-established routines also, over time, become "cues" for performance and signal the end of one task as well as transition to the next, reducing anticipatory performance anxiety. Thus, the ideal mental performance state includes a clear focus on the requirements of the task, total absorption in the task, high intrinsic motivation, concentration of one's consciousness on the task without experiencing distraction, and a feeling of exercising control by using well-set, rehearsed cognitive and motor routines.

Routines

MENS SANA IN CORPORE SANO ("A HEALTHY MIND IN A HEALTHY BODY")

In Latin text, this axiom is followed by the verb *sit* (so be it), indicating the hope that there be a sane mind in a sane body. The relationship between fitness and psychological well-being has received considerable attention, given how profoundly anxiety and depression disrupt the lives of many people. Exercise may be the single most effective behavioral intervention in the treatment of mild depression and anxiety disorders. Carefully performed moderate levels of exercise can have health benefits (such as weight control and increased muscle mass, cardiovascular benefits, and metabolic activation), which may in turn help to promote a person's overall well-being and fitness. Among cardiac rehabilitation patients, exercise alone improves various measures of psychological functioning. Studies in healthy individuals have shown that exercise improves self-confidence and self-esteem, attenuates cardiovascular and neurohumoral responses to stress, and reduces some Type A behaviors. Exercise promotes neurogenesis in the dentate gyrus of the hippocampus, as well as the release of neurotransmitters associated with mood, mimicking the effects of many antidepressant medications.[72] "Wellness" campaigns in the workplace can be very effective in promoting fitness and camaraderie, especially when incentivizing employees and facilitating healthy competition in cohesive work groups.

4.5.2 Enhancing Team Work

Enhancing team performance is of interest at work, in sports, or in the military; situations in which one does not perform alone, but rather together with other people who do similar, parallel, or complementary tasks. Teams may be highly coordinated (such as the pilot and copilot in an airplane) or only loosely organized (such as academic research teams). When it is desired to improve team work and produce better group outcomes, one should be aware of several aspects of group performance and group behavior in establishing protocols.

Participants and loafers

The performance of a team of individuals is not simply the sum of the individual efforts. Rather, a team should provide greater resources than an individual does, but teamwork also often involves difficult interpersonal coordination and management problems. It is a common experience (although exceptions exist) that a team's performance falls short of initial expectations. The proper selection of members according to their physical, mental, and interpersonal capabilities may enhance the team's performance. If members are not motivated to participate in the team's efforts, some will not contribute their share but will do "social loafing." However, not all suboptimal group performance is due to a lowered input on the part of individual members; instead, it may be the result of faulty interpersonal processes for combining individual capabilities.

Brainstorming

Long believed to be an effective and idea-stimulating team technique, in practice, brainstorming may be no more successful than individual theorization, possibly because of "blocking" (the inability of a team member to produce ideas while others are talking). Effective brainstorming sessions generally involve clear objectives for the session, facilitation to ensure everyone participates and is allowed to be heard, opportunity for individual as well as group ideation, and a preference for quantity rather than quality of ideas (at least, at first).

Group decisions

A consequence of making decisions by consensus is that group decisions tend to be more "risky" than individual decisions. One approach to group decision making is the Delphi technique: individual judgments or opinions are privately elicited and then summarized, and the results are circulated to all team members for further modification until individual positions stabilize. Thus, anonymity of the individual members' inputs is retained. Another approach is the Nominal Group technique: after the initial stages of generating ideas, group members meet face-to-face to discuss and then collectively rank ideas. In spite of the popularity of such techniques, the evidence on team performance available today is not encouraging: neither technique appears to improve on the performance of freely interacting groups. Groups are also susceptible to *Groupthink*[73] which occurs when a group's desire for harmony and conflict-avoidance leads to poor decision-making without critical evaluation of ideas. Groupthink in general involves risks such as illusions of invulnerability, ignoring possible moral problems, rationalization, stereotyping, suppression of minority opinion, illusions of unanimity, and pressure to conform.

Diffusion of responsibility

In group settings, particularly in large groups, an individual's sense of responsibility and willingness to take action is diluted under both pro- and anti-social conditions. In pro-social conditions, an individual tends to believe that someone else will take action to correct a problem or assist an individual and may in fact by less willing to take action because they fear how observers may view them. This is known as the *bystander effect*. However, this effect occurs at all levels of acuity and across work situations as well, as demonstrated by the decreased likelihood of a response to a group email. Not only is someone more likely to respond to an individual email, but the response is likely to be more helpful, specific, and lengthier as compared to responses to a group email.[74] In an anti-social situation, negative behaviors may be more likely to be carried out by an individual because of the de-individuating effects of group membership when the group is in consensus or similarly motivated. Hence, a sense of personal responsibility for an anti-social act is diminished. Diffusion of responsibility in anti-social and risk-taking behavior of groups is a factor in group behaviors ranging from arranging sick-outs in a work setting even to crimes against humanity (see Groupthink, above).

4.6 WHEN OUR MIND PLAYS TRICKS ON US

Perceptions

Our perceptions cannot always be trusted. In both physiological and psychological ways, we may not always make accurate assessments of situations and surroundings. We can make poor decisions and draw faulty conclusions even when completely confident in our beliefs and actions.

Many things can go wrong in the brain, even when the brain does its best to compensate for injuries or disease. Neurological and neuropsychiatric conditions can affect our decision making, perceptions, and actions even though they would seem to be easily demonstrated as illogical or faulty in the face of clear contradictory evidence. A few of the many neurologic and neuropsychiatric perceptual distortions are mentioned here to demonstrate their variety and import. *Visual agnosia* is the inability to recognize familiar objects or faces even while being able to see them. *Cotard's syndrome* is a condition in which an individual believes he/she is dead. The *Capgras delusion* is the belief that loved ones have been replaced by imposters, robots, or aliens. *Phantom limb syndrome* occurs when a limb has been amputated but pain or itching is still perceived in the missing body part, representing a maladaptive form of neuroplasticity and possibly also a synesthetic manifestation. Such syndromes and conditions have diverse origins and can be caused by injury or disease.

We may also fall prey to various psychological distortions that mislead us into thinking that we are making rational and independent decisions even though we are not. Since these cognitive fallacies are not reflective of brain pathology, we can teach ourselves to be aware of them. Examples of common cognitive fallacies include:

- the *priming effect* (not being aware of what has initiated our thought process)
- *confirmation bias* (we tend to ignore information that does not conform with pre-existing beliefs)
- the *availability heuristic* (we are more likely to believe something is common if we can recall having heard about it and are far less likely to believe in something that is rarely reported)
- *apophenia* (the belief that coincidences have meaning)

- *brand loyalty* (we actually prefer the things we own because we rationalize our past choices, not because we necessarily made a rational choice when we bought them)
- the *self-serving bias* (we all tend to think we are "above average")
- the *third person effect* (we believe that our opinions are based on fact and experience, while differing opinions are based on misinformation or propaganda)
- the *gambler's fallacy* (believing that we can predict random events or be "on a streak")
- the *fundamental attribution error* (believing that other people's behavior is a reflection of their personality rather than their circumstances)

4.6.1 Detecting Deception

"White" and other lies

Since time immemorial, humans have both lied and been concerned about detecting lies in others. Deception is an important survival skill for animals; it is also common in humans—for example, in giving the impression that one is more knowledgeable than one actually is, in bending the truth in a "white lie," or in intentionally proposing something known to be false. The ability to detect deception is also an important survival skill, however. Humans need to know when and whom to trust in the interest of their well-being and even their survival.

Folk wisdom has it that a liar can be detected by blushing, facial expressions, shifts in one's gaze, and involuntary body movements ("body language"), as well as by changes in loudness of the voice, rate of speech, and in other speaking patterns. These cues can be suppressed by an experienced liar or may depend on one's attitude and upbringing. Thus, such cues are not reliable indicators of either lying or truthfulness. Studies have shown that even professionals such as customs inspectors, police detectives, law enforcement agents, and judges may not detect deception any more significantly better than by chance. Altogether, there is a large discrepancy between scientific evidence and anecdotal subjective assessments.[75]

Lie detection

In the Middle Ages, primitive and superstitious methodologies were used in the judicial system to determine the veracity of a prisoner's proclamations of innocence. These methods were called "trial of ordeal" or "judgments of God," in which accused individuals' hands would be exposed to extreme conditions (such as fire, or icy or boiling water) with the assumption that a just God would not let a righteous man suffer.[76] The ancient Chinese (ca. 1000 BC) used a dry rice method: the suspect's mouth was filled with a handful of dry rice. If the rice remained dry upon expectoration, the suspect would be deemed guilty based on the assumption that a guilty person's fear and anxiety would be evidenced by a decreased salivation, leaving his/her mouth dry.[77] Unfortunately but predictably, there were very few findings of innocence using these methods.

Contemporary research interest in "scientific lie detection" originated in Italy, Germany, Austria, and Switzerland in the late 1880s. Psychophysiological measurement techniques were introduced at that time and have been under development ever since.[78] The *polygraph* ("lie-detector") is based on the concept that lying provokes specific physiological responses and that the associated emotions of the liar can be qualitatively detected and qualitatively interpreted. Most of the physiological events recorded by the polygraph are associated with breathing, blood pressure, the heartbeat, and galvanic skin responses, particularly in the palms of the hands.

Polygraphs are in wide use in the United States in criminal investigation, security vetting, and selection of personnel for hiring or promotion. Some (but not all) US states find polygraph tests admissible in legal court. Polygraphs are also used in Canada, Japan, Australia, Israel, and the United Kingdom. In Norway, Sweden, Holland, and Germany, polygraphs are of less interest and are generally frowned upon. **Use of polygraphs**

Lie-detector procedures rely on an assumed intimate interconnection between mind and body. They purportedly discover facts about the mind through the observation of physical responses to questions.[79] The polygraph is simply an instrument that records physiological events in connection with questions asked by the tester. The subject's reactions are then interpreted and conclusions are drawn regarding the truthfulness of the examinee's statements. However, key aspects of polygraphic lie detection remain highly controversial: the validity of the theoretical foundations of the practices, the consistency of procedures used, the accuracy and reliability of the results, and the interpretation and biases of the results.

With advances in the neurosciences, techniques in lie detection have begun to focus on measuring brain activities with the use of transcranial magnetic stimulation (TMS), fMRI, positron emission tomography (PET), and EEG.[80] However, problems with these methodologies remain, including sampling errors, brain differences, replicability, technology cost and availability, subject cooperation, and defensive attention strategies.[81] Commonly used deception-generating methods include the Comparison Question Test (CQT) and the Concealed Information Test (CIT), which can be used in polygraphy with EEG and fMRI technologies. However, the CQT is vulnerable to control problems while the CIT suffers from specificity issues. **Brain waves**

Some businesses incorporate problematic techniques of Voice Stress Analysis and Facial Action Coding System protocols in prospective employee selection. These techniques have very little scientific validity or reliability compared to interviews conducted without their use. The US Department of Homeland Security has started using a rapid screening system called the Automated Virtual Agent for Truth Assessments in Real-Time in efforts to flag suspicious behavior. It has been used in simulations and at several borders and in eastern Europe.[82] Time will bear out the efficacy of such protocols. **Voices and faces**

In the discussion of deception assessment, it is vital to remember that the conclusions drawn regarding the person's character and any ensuing legal judgments can have far-reaching implications for his/her future. Therefore, any techniques used should possess the greatest possible degrees of reliability and validity obtainable.

Aldrich Ames, the "master spy" arrested in the CIA in February 1994, dismissed the polygraphs he regularly was given as a CIA employee as "witch-doctory." With regard to passing the lie-detector tests, he said, "Confidence is what does it. Confidence and a friendly relationship with the examiner... [a] rapport, where you smile and you make him think that you like him."[83]

4.7 CHAPTER SUMMARY

In a "person-machine-environment" system, the human perceives information simultaneously through various senses and plans and executes actions. Thus, the traditional concept of a linear sequence from stimulus to sensory input to perception to processing to effector output is probably overly simplistic, and even unrealistic.

Accordingly, the CNS, as processor, receives information concerning the outside world (and the inside of the body) from receptors that respond to light, sound, touch, temperature, electricity, chemicals, and acceleration vectors. The received stimuli are transmitted in a converging manner along the afferent paths of the PNS to the CNS, where they are processed and action signals are generated. These signals are sent along the efferent paths of the PNS to effectors such as the hand or voice box. The selection of appropriate external stimuli and their transformation so that they can be reliably sensed are among the major tasks of the ergonomic engineer. On the output side, the ergonomic design task is to select the proper actions to perform the task.

Job demands and environmental conditions can be highly stressful to the human, both physically and psychologically. Effective methods of coping with stress help to investigate and intervene in the complex interactions of individual coping mechanisms with the environment; these methods can be taught to some degree. Different people use different coping styles, not all of them healthy or effective, which may be ingrained in character or depend on personality dynamics. The most effective way to reduce worker stress is a combination of stress management and organizational change.

Procedures for assessing individual factors, job demands, and for reorganizing and designing the system accordingly are in use. These can help in everyday situations but are of particular importance in unusual situations, such as in space, where tasks, the environment, and the relations among crew members are novel.

Improving the performance of individuals and groups is the goal of learning and training. There are well-established methods of maximizing learning and training, including effective practice parameters.

The performance of groups is vulnerable to various group dynamics. The performance of individuals is also vulnerable to intrapersonal and cognitive dynamics. An individual cannot always make rational decisions or accurate assessments of their own or other's performance or behavior.

4.8 CHALLENGES

What difference would it make if we discussed mental functions simply in terms of the brain instead of the mind?

How would our understanding of human mental processing change if we went from the traditional concept to the ecological approach? Would the concept of "affordance" be modified?

Would the consideration of the various types and sheer number of internal and external sensory receptors make the traditional or ecological concept more plausible?

What transducers are imaginable to make distal stimuli into signals that trigger human sensors?

When driving fast in dense traffic, is a driver more likely to have an accident when he can either brake or switch to another lane, or when braking is the only possible action?

Why should the presentation of an emergency signal by several modalities (such as light and sound) together be advantageous over presenting the signal to only one type of sensor?

What might be a better design of an automobile braking system than the one currently used?

How could Fitts's law be applied to designing a workstation for manually assembling many small parts?

Why is it difficult to maintain a conversation while looking up somebody's telephone number?

What means of measuring the workload of a driver might be employed during rush-hour traffic?

Under what conditions and to what extent should stress management programs be carried out by an employer?

What means can be used to maximize employee health?

What might be suitable training programs for crews that will go into long, confined missions, such as space exploration?

How can one reduce "end-of-mission" stressors?

Discuss the benefits and concerns associated with mixed-gender crews during long isolated missions, such as space travel to Mars.

How could one measure the effectiveness of training in terms of how it affects performance?

Why is it important to receive immediate feedback on performance of a new skill when learning it?

What carefully planned and executed measures might assure the highest possible team performance?

How can we maximize the probability that we make rational decisions? Alternately, should all decisions be rational?

NOTES

1 Information theories, including signal channeling and processing, were major research topics particularly in the 1970s and 1980s; see related publications in the field of cognitive psychology.

2 See, for example, Brunswick (1956), Flach (1989), Gibson (1966), Meister (1989), Vincente and Harwood (1990), Meyer and Kieras (1997), and Mace (1977).

3 See Flach (1989).

4 While at the Museum in Alexandria, Egypt, about 280 BC, Herophilus distinguished sensory and motor nerves. Erasistratus (about 250 BC) distinguished between the cerebrum and the cerebellum of the brain and thought that the many convolutions of the human brain, more numerous than in any other animals, were related to superior intelligence. In 1810, the German physician Franz Joseph Gall (1758–1828) stated that the gray matter on the surface of the brain and in the interior of the spinal cord was the active and essential part of human thinking, and that the white matter was just connecting material. Gall also believed that the shape of the brain had something to do with mental capacity, even trying to relate the shape of the brain and skull with emotional and temperamental qualities, which started the pseudoscience of

phrenology. The French anthropologist Pierre-Paul Broca (1824–1880) demonstrated Gall's theory that the brain controlled different parts and functions of the body. (Asimov 1989)

5　As discussed by Mobbs (2007, 2009).

6　As recorded by MacDougall (1907) but since then discredited.

7　By Grafman (2000).

8　As discussed in a review article by Cramer, Sur, Dobkin, and O'Brien (2011).

9　Mayer (2014).

10　DePalma (2014).

11　Mayer (2011).

12　For more on neuromyths, see Howard-Jones (2014).

13　Basmajian and DeLuca (1985), Kumar and Mital (1996), Perotto (1994), and Soderberg (1992).

14　Appropriate control groups must be in place so as to avoid the fate of the infamous "dead salmon" study, which statistically "proved" that a (dead) salmon subjected to fMRI scanning was responding to photographs of people with various emotions. This study, published by Bennett, Baird, Miller, and Wolford (2010), was honored by the Ignoble Science award in 2012 (accessed 22 July 2017, https://blogs.scientificamerican.com/scicurious-brain/ignobel-prize-in-neuroscience-the-dead-salmon-study/), a tribute most non-ironic researchers would like to avoid.

15　See Wargo (1967).

16　See Wargo (1967).

17　Boff and Lincoln (1988).

18　Anson (1982).

19　See Fitts (1954).

20　See Wing (1983) or Keele (1986) for more details on Fitts's law.

21　See the overviews by Best (1997), Moray (1988), O'Donnell and Eggemeyer (1986), and Wickens and Carswell (1997) for more details.

22　See Tattersall and Foord (1996).

23　Seven (1989) and Doherty (1991).

24　Wierwille, Rahimi, and Casali (1985).

25　Vidulich and Tsang (1985).

26　Hart and Staveland (1988); this has also been adapted to assess workload during surgery as the SURG-TLX—see Chapter 14.

27　Reid and Nygren (1988) and Colle and Reid (1998).

28　Nygren (1991).

29　Hill et al. (1992).

30　Wilson and Eggemeyer (1994) and Wierwille and Eggemeyer (1993).

31　Holmes and Rahe (1967).

32　For more information, see McEwen (2007).

33　Sterling and Eyer (1988).

34　McEwen (2007).

35　Mazure (1998).

36　Holmes & Rahe (1967).

37　Tennant and Andrews (1976).

38　Paykel (1997).

39　Dohrenwend, Raphael, Schwartz, Stueve, and Skodol (1993).

40　Vagg and Spielberger (1999).

41　Csikszentmihalyi (1990).

42　Brown (1994).

43　Motowidlo, Packard, and Manning (1986).

44　Quick (1999).

45　Porter (1996).

46　Wofford and Daly (1997).

47　For reviews of the theories of hostility on cardiovascular and neuroendocrine activity and health consequences, see King (1997).

48　Hackett, Rosenbaum, and Tesar (1988), Kamarck and Jennings (1991).

49 These manuals are periodically reviewed and updated by the American Psychiatric Association (APA) and the World Health Organization (WHO), respectively.

50 Selye (1978).

51 Lazarus and Cohen (1977).

52 Kanner, Coyne, Schaefer, and Lazarus (1981).

53 See, for example, Kamarck et al. (1998), Karasek, Baker, Marxer, Ahlborn, and Theorell (1981), Karasek et al. (1988), Schnall, Schwartz, Landsbergis, Warren, and Pickering (1992).

54 Cohen et al. (1998).

55 As reported by Sedgwick (1888) and often quoted.

56 Cooper and Sawaf (1998).

57 Goleman (1995).

58 Martins (2010).

59 Goleman (2006, 2011).

60 Tache and Selye (1986).

61 Folkman and Lazarus (1988).

62 Antonovsky (1979).

63 Lazarus and Folkman (1984).

64 CDC/NIOSH (1999).

65 Maslach and Jackson (1981), Maslach, Schaufeli, and Leiter (2001).

66 For more on happiness research, see Baumeister, Vohs, Aaker, and Garbinsky (2013), Clark, Fleche, Layard, Powdthavee, and Ward (2017), Kahneman (2005), and Seligman (2002).

67 Myers and Diener (1996).

68 Kahneman and Deaton (2010).

69 Easterlin (1974).

70 Harrison, Clearwater, and McKay (1991).

71 Holland (1991).

72 See, for example, Miller, Balady, and Fletcher (1997).

73 Janis (1972).

74 Barron (2002).

75 Druckman and Bjork (1990).

76 Sullivan (2001).

77 As reported by Ford (2006).

78 See, for example, Ben-Shakhar and Furedy (1990) and Gale (1988).

79 Blinkhorn (1988).

80 Grubin (2010).

81 For more on these concerns, see Vicianova (2015).

82 Elkins (2017).

83 Reported by Walter Pinkus in the Washington Post, May 4, 1994.

Overview

Seeing, hearing, smelling, tasting, and touching are the primary "traditional" human senses. The human is also sensitive to electricity and to pain and has a posture and motion (vestibular) sense. Of all senses, seeing and hearing are most thoroughly understood, and information related to those senses is well suited for ergonomic applications. However, information on the sense of touch, including the perception of temperature and sensitivity to electricity and pain, is much less researched, and engineering applications are rather haphazard. The senses of smell and taste are usually not used in engineering.

Proper ergonomic procedures include measures to correct sensory deficiencies, such as in seeing, and to protect sensory functions from damage, such as hearing. Little is commonly done to enhance or protect the other sensory capabilities.

5.1 INTRODUCTION

The human is able to receive signals through several senses, originally classified by Aristotle (384–322 BC) into five different categories: seeing (*vision*), hearing (*audition*), smelling (*olfaction*), tasting (*gustation*), and touching (*taction*). Taction might include more than one sense, because one feels several mechanical stimuli, such as contact and pressure, often together with pain (*nociception*), electricity (*electroreception*), and temperature (*thermoception*). After more than two millennia, this classification is still commonly used, although it excludes other human sensory abilities for stimulus detection such as the kinesthetic sense (*proprioception*), the sense of balance (*equilibrioception*), and the ability to detect salt and carbon-dioxide concentrations (*chemoreception*), among others. What constitutes a sense perhaps becomes debatable, especially as some senses can be seen as subtypes or combinations of other senses. Thus, we will address human senses in the conventional classification as consisting of the five "traditional" senses and various other "non-traditional" senses.

Only five senses?

Other classification methods exist, delineating sensory systems by anatomy, by the sensory organs that perceive various stimuli, and by the stimuli themselves.

5.2 BODY SENSORS

Every human sensor (sensory organ) can be modeled as a *transducer* that consists of two components. The first part is the *receptor*, which is stimulated by an appropriate proximal stimulus to produce some reaction. The second part is the *converter*, which codes or reproduces the reaction and generates an electric potential (a signal) to be sent along a nervous pathway.

Sensors as transducers

The receptor must be stimulated by the right kind and quantity of the proximal stimulus (see Chapter 4) so that it can react. How this functions in reality is not fully understood, nor is it known either how the sensor acts as the converter of the signal received by the receptor or how it generates the potential that is sent as a sensory signal to the next neuron and on to the central nervous system (CNS).

Stimulation

The problem of how signals converge is quite complex, because numerous sensors are distributed over the receptive field: consider, for example, the many tactile sensors in the tip of the finger. The primary receptor cells (transducers) are linked either directly or in branching ways to a first-order afferent neuron which is the first collecting and switching point that receives the input from all sensors within its receptive field. This neuron may or may not be triggered sufficiently by the incoming signals to send an action potential to the next higher, second-order afferent neuron which also receives inputs from other first-order neurons. Thus, numerous first-order afferent neurons are linked to second-order neurons, which report to the next level in the hierarchy. The signal flow from several low-order neurons to a few higher neurons in the CNS, possibly repeated several times, follows the "principle of convergence," shown in Fig. 5.1.

Convergence of signals

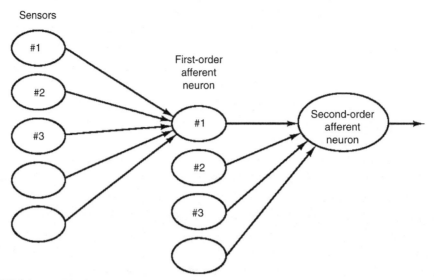

FIGURE 5.1   **Model of converging receptive fields.**

5.2.1 Stimuli

Some stimuli trigger only one type of sensor, but there are also many nonspecific stimulations. For example, electromagnetic and mechanical energies can trigger sensations of vision, hearing, touch, and pain as well as sensations of the vestibulum, and thermal stimuli trigger sensations of cold and warmth, as well as pain sensations. The auditory sense responds to pressure changes at frequencies between 20 and 20,000 Hz. Chemical stimuli affect especially the senses of taste and smell.

5.2.2 Adaption and Inhibition

We are bombarded every minute by a variety of stimuli which would overwhelm us if we could not filter them out according to their importance. Since each of the afferent neurons may or may not transmit an arriving signal further up the chain toward the CNS, this converging network functions as a filtering or coding system, letting only those signals through that are sufficiently strong (i.e., of sufficient importance). The sensory system *adapts* by reducing the nervous discharge even while the strength of the proximal stimulus remains the same. A typical example is the adaptation of taction to the feeling of clothes worn. The system may be *inhibited* also by the reduction of nervous discharge, as when a second stimulus appears in the sensory field. A typical example is vigorous rubbing of a stubbed toe: the rubbing triggers new taction sensations which mask the pain that was originally felt. Except for the auditory and vestibular systems, very little is presently known about the excitation or inhibition in the CNS above the first-order neuron.

There are several explanations of sensory adaptation. The processes may involve the reduction in the firing rate or intensity of the trigger at the sensor, the reduction in transmission along the afferent nervous path, or the masking of signals. Different researchers have attributed various meanings to the term "adaptation," and diverse research approaches have been used, none of which have yet yielded coherent results.

Getting used to a background stimulus level so that we distinguish only important signals is probably a mixture of adaptation and inhibition that results from the presence of several stimuli.[1] However, some receptors adapt slowly, keeping their nervous discharges at the same level for a second or more while others adapt quickly, their discharge rate falling within milliseconds to a low value or zero.

5.2.3 Sensory Thresholds

When a stimulus is of such weak intensity that one cannot sense it, it is below the lower, or absolute, or minimal threshold. The upper or maximal threshold is the limit above which the sensor does not respond any more (such as to sound frequencies above 20 kHz), but at which the stimulus might still irritate or even damage the sensors.

Difference threshold: The proportionality of sensation to a change in the stimulus is often of interest. For example, how do we react to a change of 10 dB in sound pressure at low pressure levels as opposed to high ones? The *difference threshold* is the smallest physical difference in the amount of stimulation that produces a just noticeable difference in the intensity of our sensation. This threshold was investigated in detail by Gustav Fechner and Ernst Weber in the 19th century. Usually, difference thresholds increase with increasing intensity: Weber's law states that the difference threshold is a constant proportion of the intensity of the stimulus. (The law may not hold for extremely small or large intensities of stimuli.) Table 5.1 shows the smallest energies that are detectable and the largest energies that are tolerable or practical. Stimuli near either of these limits may lead to unreliable sensing.

There is an interplay between sensory adaptation and the difference threshold that is strikingly demonstrated when a person immerses one hand in warm water and the other in cold water. The initial sensations of warm and cold, respectively, subside slowly as one adapts to the water temperatures.

When both hands are then immersed in lukewarm water, the warm hand signals the sensation of cold, the cold hand the feeling of warmth. Apparently, the difference in water temperatures makes each hand report the change from the previous state.

TABLE 5.1   Characteristics of The Human Senses[a]

	Vision	Hearing	Touch	Taste, Smell	Vestibular
Stimulus	Light-radiated electromagnetic energy in the visible spectrum	Sound-vibratory energy, usually airborne	Tissue displacement by physical means	Particles of matter in solution (liquid or aerosol)	Accelerative forces
Spectral range	Wavelengths from ~400 to ~700 nm (violet to red), 10^{-6} mL–10^4 L	20–20000 Hz 20 μPa to 200 Pa	Temperature: 3 s exposure of 200 cm^2 of skin, 0–400 pulses/s	Taste: salty, sweet, sour, bitter; Smell: flowery, fruity, spicy, resinous, burnt, foul	Linear and rational accelerations
Spectral resolution	120–160 steps in wavelength (hue), varying from 1 to 20 nm	~3 Hz for 20–1000 Hz; 0.3% above 1000 Hz	~10% change in number of pulses/s	?	?
Dynamic range	~90 dB (useful range); for rods = 3×10^{-5} –0.0127 cd/m^2; for cones = 0.127–31,830 cd/m^2	~140 dB	~30 dB (0.01–10 mm displacement)	Taste: ~50 dB (3×10^{-5}–3% concentration of quinine sulphate); Smell: 100 dB	Absolute threshold is ~0.2 deg/s
Amplitude resolution ($\Delta I/I$)[a]	Contrast = 0.015	0.5 dB	~0.15	Taste: ~0.20; Smell: 0.10–50 dB	~0.10 change in acceleration
Acuity	1 min of visual angle	Temporal acuity (clicks) ~0.001 s	Two-point acuity ranges from 0.1 mm (tongue) to 50 mm	?	?
Response rate for successive stimuli	0.1 s	~0.01 s (tone bursts)	Touches sensed as discrete to 20/s	Taste: ~30 s; Smell: ~20–60 s	~1–2 s; nystagmus may persist to 2 min after rapid changes in rotation
Best operating range	500–600 nm (green yellow) at 34–69 cd/m^2	34–69 Hz at 40–80 dB	?	Taste: 0.1%–10% concentration	~1 g acceleration directed to foot

[a]I = intensity level; ΔI = smallest detectable change in intensity from I.
Source: Adapted from Van Cott and Kinkade (1972), Boff and Lincoln (1988).

5.3 SEEING—THE VISION SENSE

The characteristics of human vision are well researched and described in the literature.[2]

5.3.1 Architecture of the Eye

As sketched in Fig. 5.2, the eyeball is a roughly spherical organ about 2.5 cm in diameter, surrounded by a layer of fibrous sclera.

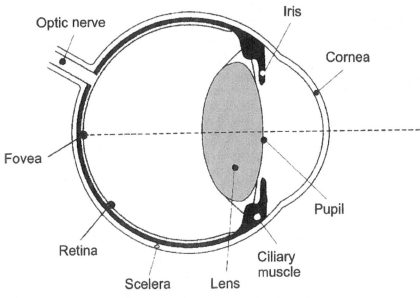

FIGURE 5.2   Horizontal section (top view) through the human eye.

When parallel beams of light from a distant target reach the eye, they first encounter the *cornea*, a translucent, bulging domed section of the sclera at the front of the eyeball, kept moist and nourished by tears. The cornea provides all the refraction needed to focus on an object more than about 6 m (20 ft) away (if the eye is young and healthy).

Behind the cornea is the *iris*, the tissue surrounding a round opening called the *pupil*. The dilator muscle opens—and the sphincter muscle closes—the pupil like the aperture diaphragm of a camera, regulating the amount of light entering the eye.

Having passed through the pupil, light beams enter the *lens*. If a distant object needs to be seen, suspensory ligaments keep the lens thin and flat, so that the light rays are not bent. For close objects (which cannot be focused by the cornea), the ciliary muscle around the lens makes it thicker and rounder, so that the light beams are refracted for suitable focus.

The space behind the lens, the interior of the eyeball, is filled with the *vitreous humor*, a gel-like fluid which has refractory properties similar to those of water.

Light focused by the cornea and lens finally reaches the *retina*, a thin tissue that lines about three-quarters of the inner surface of the eyeball, opposite the pupil. The retina is amply supplied with blood by many arteries and veins and contains about 130 million light sensors.

There are two kinds of light sensors on the retina, named for their shape. The majority, about 120 million, are the *rods*, which respond to even low-intensity light and provide black, gray, and white vision. About 10 million *cones* respond to colored bright light.

Pigments in rods and cones serve to convert light into electrical signals. Each cone contains a pigment that is most sensitive to either blue, green, or red wavelengths. An arriving light beam, if intense enough, triggers chemical reactions in one of the three types of pigmented cones, creating electrical signals that are passed along the *optic nerve* to the brain which can distinguish among about 150 color hues. Rods contain only one pigment, which, when bleached by light, sets off electrical impulses that are sent along the optic nerve to the brain for the perception of white, black, and shades of gray.

On the retina, cones are concentrated in the center, directly behind the pupil where only few rods are found. This area is called the *fovea*. Together with its yellowish sur-rounding area, known as the *macula*, the fovea is mostly used to read fine print.

The optic nerve exits the eye at its rear, about 15 degrees off center toward the inside. Since there are no light sensors in this area, an image at this *blind spot* cannot be seen. However, since the blind spots of both eyes are medially located, they do not overlap in our field of vision and therefore we are unaware of their existence.[3]

5.3.2 Mobility of the Eyes

Each eye is theoretically capable of movements in six degrees of freedom, illustrated in Fig. 5.3. Used most often are rotational movements in *pitch* (around the y-axis), *roll* (around the x-axis) and *yaw* (around the z-axis), but the entire eye also moves linearly— most appreciably forward and backward (along the x-axis). Thus, the center of rotation of the eye is not fixed, but the displacements of the center are fairly small. Therefore, we usually assume the center of rotation remains in place, approximately 13.5 mm behind the cornea.

Six striated muscles are attached to the outside of the eye and control its movements (see Fig. 5.4). Since the muscles do not attach at orthogonal directions to each other, they interact. The superior and inferior recti muscles are responsible primarily for *pitch*:

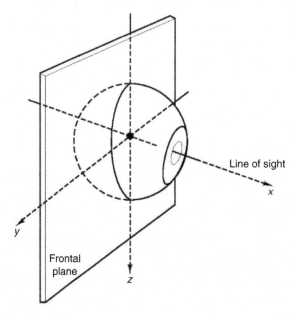

FIGURE 5.3   **Mobility axes of the eye.**

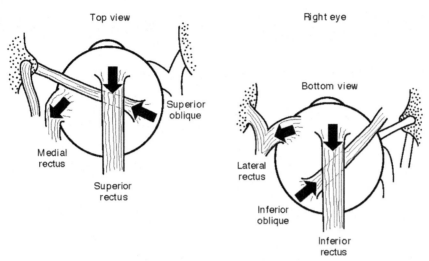

FIGURE 5.4   **Muscles moving the right eyeball.**

the up and down eye rotation. The medial rectus and the lateral rectus muscles provide *yaw*: the left and right movement. The oblique muscles predominantly provide roll movement.

The six muscles work in concert to produce eye movements. When one looks straight ahead, the eyes are said to be in primary position. The primary movements of the eye are in the two degrees of freedom associated with up and down gaze motions (pitch) and left and right movements (yaw). Given the complex attachments of the muscles moving the eye, these primary movements are accompanied by some roll. The further the eye pitches and yaws, the more the eye also rolls. The amount of eyeball roll at various pitch and yaw angles away from the primary position is about 1 degree with 10 degrees each of pitch or yaw; 3 degrees with 20 degrees of pitch or yaw; 8 degrees with 30 degrees of pitch or yaw; and 15 degrees with 40 degrees each of pitch and yaw. We are not aware of this tilting of the visual image as we move the eyes, because the brain compensates for it.

The eye can track continuously left and right (yaw) a visual target which is moving less than 30 degrees per second or cycling at less than 2 Hz. Above these rates, the eye is no longer able to track continuously, but lags behind and must move in jumps (*saccades*) to catch up to the visual target.

5.3.3 Line of Sight

If the eye is fixated on a point target, the *line of sight* (LOS) runs from the object, through the lens (pupil), to the receptive area on the retina—most likely the fovea. Thus, the LOS is clearly established within the eyeball. To describe the LOS direction external to the eye, a suitable reference is needed.

The horizon is often used as reference for the external LOS angle in the medial (*x–z*; see Fig. 1.1) or a sagittal plane. Yet, when looking at an object in front of us, we

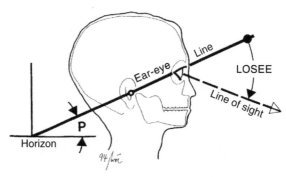

FIGURE 5.5   **Line-of-sight (LOS) angle against the Ear–Eye line (LOSEE) and the horizon.** The horizon is the reference for the LOS angle (LOSEE-P) in the medial plane.

unconsciously adjust two pitch angles: that of the eyes within the head and that of the head itself.

Line-of-sight angle

Various approaches have been taken to define the LOS angle with respect to the head. One is to define it against the *Frankfurt* plane (or auriculo-orbital plane) which defines the anatomic position of the skull via two landmarks; the *tragion* (the notch just above the tragus of the ear) and the lowest point on the rim of the orbit. The orbit is difficult to palpate, whereas it is easy to determine the Ear–Eye plane because its landmark attachments to the head (the right ear canal and the outside juncture of the eyelids of the right eye) can be seen and do not need to be palpated. (For converting angles, the Ear–Eye plane is, on average, 11 degrees more pitched than the horizon or the Frankfurt plane.[4]) For yaw and roll, perpendicular planes meeting at the eye can be defined. The angle between the LOS and the E–E line (LOSEE) in the medial plane is illustrated in Fig. 5.5 (same as Figure 1.3 in Chapter 1).

For ergonomic design of work and the workstation, it is important to know each of two angles: the eye pitch angle relative to the head (LOSEE) and the posture of the head as the pitch angle of the head with respect to the neck, trunk, or horizon.

Head angle

It used to be difficult to establish the position of the head with respect to the trunk. The links between skull and trunk are ill defined mechanically and consist of at least the cervical vertebrae on top of the thoracic column. These vertebrae have various mobilities in their intervertebral joints. The skull also rotates in three degrees of freedom at the atlas of the first cervical vertebra. Additionally, there is no easily established reference system within the trunk by which one could describe the relative displacements of the neck and head. The use of the E-E line and its pitch angle against the horizon partially circumvents this posture problem.

Visual target size

If a visual target is not a point, but can be expressed as the length of a line perpendicular to the LOS, then the target size is usually expressed as the subtended visual angle—the angle formed at the pupil. The magnitude of this angle depends on the distance D of the object and on its size L. The subtended visual angle α is described in Fig. 5.6 and is usually given in degrees of arc (with 1 degree = 60 min = 3600 s of arc). The equation is:

$$\alpha \text{ (in degrees)} = 2 \text{ arc tan} (0.5\,L/D) \tag{5.1}$$

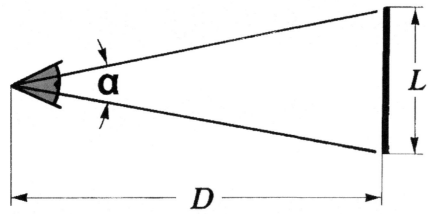

FIGURE 5.6   The subtended visual angle (α) formed at the pupil for an object at a distance (D).

For visual angles not larger than 10 degrees, this can be approximated by:

$$\alpha\,(\text{in degrees})=57.3\,L/D \tag{5.2}$$

$$\alpha\,(\text{in minutes of arc})=60\times57.3\,L/D=3438\,L/D \tag{5.3}$$

Note that the equations do not take into account the distance between the pupil and the lens, because that approximately 7 mm distance has no perceptible effect under most conditions.

The human eye can perceive, at a minimum, a visual angle of approximately 1 minute of arc, or 1 second of vernier acuity. Table 5.2 presents the visual angles of familiar objects. For ease of use, technical products should be designed so that the angle subtends at least 15 minutes of arc, increasing to at least 21 minutes of arc at low light levels.

Visual angle

If the target distance D is measured in meters, the reciprocal ($1/D$) is measured in *diopter*. The diopter indicates the optical refraction needed for best focus. Thus, a target

Diopter

TABLE 5.2   Visual Angles of Familiar Objects

Object	Distance	Visual angle (Arc)
Gage, 5 cm diameter	0.5 m	5.7 degrees
Sun	150 Gm	About 30 min
Moon	384 Mm	About 30 min
Character on CRT screen	0.5 m	17 min
Pica letter at reading distance	0.4 m	13 min
U.S. quarter coin at arm's length	0.7 m	2 degrees
U.S. quarter coin	82 m (90 yards)	1 min
U.S. quarter coin	4.8 km (3 miles)	1 second

TABLE 5.3 Target Distances and Associated Focal Points

Target distance D (m)	Focal point (diopter)
Infinity	0
4	0.25
2	0.25
1	1
0.67	1.5
0.50	2
0.33	3
0.25	4
0.2	5

at infinity has the diopter value zero, while a target at 1 m distance has the diopter value unity (one). Table 5.3 shows values for some typical target distances.

5.3.4 The Visual Field

Visual field

The visual field is the area, measured in degrees, within which the form and color of objects can be seen by both fixated eyes. In its center, the visual field of each eye is occluded by the nose. To the sides, each eye can see a bit over 90 degrees, but only within the inner about 65 degrees can color be perceived. Farther in the periphery, shades of gray dominate, because of the locations of cones and rods on the retina, as discussed earlier. Upward, the visual field extends through about 55 degrees, where it is occluded by the orbital ridges and eyebrows. However, color can be seen only to about 30 degrees upward. The downward vision is limited by the cheek at about 70 degrees; the area in which color can be seen extends down to about 40 degrees. The "blind spot" on the retina occurs at approximately 15 degrees to the outside, for each eye.

Rotating the eyeball increases the visual area to the *field of fixation*; adding about 70 degrees to the outside of the visual field, but nothing in the upward, downward, or inside directions, because the orbital ridges, cheeks, and nose stay in place. The field of fixation is largest in downward gaze and smallest in upward view.[5]

If, in addition to the eyeballs, the head moves, nearly everything in the environment can be seen, as long as it is not occluded by the body or other structures. Mobility of the head is achieved by neck muscles, and may be sharply reduced in persons with a "stiff neck" (often the elderly). These individuals should locate visual targets close to their naturally chosen LOS.

5.3.5 Accommodation

Focusing

The action of focusing on targets at various distances is called *accommodation*. (There is some confusion in the meaning and use of optical and visual terms.[6]) The normal young eye can accommodate from infinity to very close distances, meaning that a diopter range from 0 to about 10 can be achieved.

Normally, with visual objects at distances of 6 m or more, the relaxed lens refracts the incoming parallel rays (i.e., it focuses them) on the retina. To accommodate closer objects, the lens is made thicker by the pull of the enclosing ciliary muscle: the radius of the frontal lens is reduced from 10 to 6 mm and the radius of the rear surface from 6 to 5 mm. Thus, the "optical refracting power of the eye" is changed by adjusting the curvature of the lens so that, by refraction of the incoming light rays, the image of the target falls focused upon the retina at every distance. (One speaks of the retina being "conjugated to the object" at all accommodations.) When not focusing at anything in particular, the lens accommodates at a distance of about 1 m, with much variation among individuals.

The closet point at which the eye can focus called the *near point*. The farthest point at which the eye can focus without conscious accommodation is called the *far point*. The difference between near and far points is called the *amplitude of accommodation*. Most young people can focus at a near point of about 10 cm, but this minimal distance changes to about 20 cm at age 40 and to about 100 cm at age 60, on average.

Amplitude of accommodation

If one aims both eyes at the same point, an angle exists between the two lines of sight connecting each eye with the target. This angle is very small (in fact negligible) at objects more than 6 m away, but increases at shorter viewing distances. While it takes about 200 ms to fixate on one point at reading distance, the time is reduced to about 160 ms when one focuses at a point at 6 m away or farther. If the eyes are not made to converge, an *error of vergence* exists, also called *fixation disparity*. This condition commonly occurs when the quality of the binocular stimulus is low—for example, when the observer is greatly fatigued or under the influence of alcohol or barbituates.[7] A *phoria* exists if the images of one target are not focused on the same spots on the retinas of both eyes, resulting in double images.

Convergence

5.3.6 Visual Fatigue

People doing close visual work, such as who often look at the screen of a computer display, frequently complain of eye discomfort and visual fatigue,[8] or eyestrain (*asthenopia*), all of which are vaguely called "subjective visual symptoms or distress resulting from the use of one's eyes".[9] While the occurrence of such eyestrain (and its intensity) varies much among individuals, it seems often related to the effort of focusing at a distance that is different from the personal minimal "resting distance of accommodation." Many instances of eye fatigue of which computer users often complain are related to the poor placement of monitor, source documents, or other visual targets (see Chapters 9 and 10) or to unsuitable lighting conditions at their workplace (discussed later in this chapter).

For most people whose eyes look ahead, but do not focus on a target, that minimal distance of binocular vergence (also called dark vergence, minimal refractive state, or resting distance of accommodation) is about one meter away from the pupils. This finding is in contrast to the older assumption that the automatic resting position is at *optical infinity*—that is, with parallel visual axes of the eyes. The actual point of vergence is characteristic of the individual's oculomotor resting adjustment and, therefore, is also called a *resting tonus posture*. The automatic selection of the dark vergence (resting) distance apparently is not only due to the biomechanics of the extraocular muscles, but is also dependent on neural processes that integrate information about head and eye positions. As one lowers the angle of gaze or tilts the head down, the resting point gets closer to the eyes, to about 80 cm for a LOS angle of 60 degrees downward. However, the resting distance increases as one elevates the direction of sight, to about 140 cm on average at a 15 degree upward direction. Many elderly people find it more difficult to focus on an elevated target than on a lower one, partly because of reduced mobility of the head due to a "stiff neck."[10]

> It is natural to look down at close visual targets, such as a written text. This is why optometrists place the reading section of (bifocal or trifocal) corrective lenses at the bottom of the lens.

Several aspects are of particular interest for the design of work systems that require exact binocular eye fixation.

- The binocular resting position and the accommodation distance to which the eyes return at rest are quite different between individuals but are constant for a given person. Thus, one should allow and encourage each person to select their preferred personal distance for a visual target.
- Tilting either the eyes or the head has a similar effect on the natural vergence distance. This explains why people may find it more comfortable to lean back in a chair while tilting the head forward and down to look at a close object (at about eye height) rather than to sit upright.
- Targets at or near "reading distance" should be distinctly below eye level, particularly for elderly viewers.
- It is difficult to see a visual target of low optical quality clearly when the observer is fatigued or under the influence of drugs or alcohol.

> Scientists and grandmothers can now agree that visual fatigue is related to near-work because it is easier to look down on it than to look up to it.[11]

5.3.7 Vision Problems

Worse with age

With increasing age, the accommodation capability of the eye decreases, because the lens becomes stiffer by losing water content. This condition is known as *presbyopia*. The result is difficulty in making light rays converge exactly on the retina. If the convergence is in front of the retina, the condition is called *myopia*; if the focus point is behind the retina, one speaks of *hyperopia*.

A nearsighted (myopic) person finds it difficult to focus on far objects, but has little trouble seeing close objects. This condition often improves with age, when the lens's increasing stiffness tends for it to remain flattened. In fact, even then far objects are not exactly in focus, but the rays emanating from them strike the retina (although not exactly in focus) enough so that these distant objects can be sufficiently identified.

Better with age?

In contrast, farsightedness (hyperopia) becomes more pronounced with age, meaning that it becomes more difficult to focus on near objects. Both nearsightedness and farsightedness can be fairly easily corrected with corrective lenses (eye glasses or contact lenses). In many people, the pupil shrinks with age. This means that less light strikes the retina, and therefore, many older people need to have increased illumination on visual objects for sufficient visual acuity.

Worse with age!

Another problem often encountered with increasing age is *yellowing of the vitreous humor*. The more yellow it gets, the more energy it absorbs from the light passing through; consequently, increased illumination of the visual target is needed to maintain good acuity. Also, light rays are refracted within the vitreous humor, bringing about the perception of a light veil (like a mist) in the visual field. If bright lights are in the visual field, the resulting "veiling glare" can strongly reduce one's vision. Yellowing cannot be corrected with artificial lenses.

Floaters are perceived as small flecks in front of the eye. In reality, they consist of small clumps of gel or cells suspended in the vitreous humor. Floaters are visible to the individual only when they are on the LOS, casting a shadow on the retina. Frequently they are not noticed as the eye adjusts to these imperfections. They are more easily perceived when one is looking at a plain background. Occasional floaters are usually harmless, although in rare cases, especially when they occur suddenly, they may be precursors of retinal damage.

Cataracts are pattern of cloudiness inside the normally clear lens. When vision is severely impaired, the cloudy lens may be surgically removed and replaced with a man-made lens.

Glaucoma is the leading cause of blindness in the United States, especially for older people. Glaucoma is a preventable disease of the optic nerve, related to high pressure inside the eye. Regular eye examinations help to detect the beginning of glaucoma and to prevent further damage.

Aging certainly impacts vision, as discussed above and in Chapter 14. Other vision deficiencies are not typically related to aging, such as astigmatism, night-blindness, and color-weakness.

Unrelated to age

Astigmatism occurs if the cornea is not uniformly curved, so that an object is not sharply focused on the retina (depending on its position within the visual field). Often, the astigmatism is a spherical aberration, meaning that light rays from an object located at the side are more strongly refracted than those from an object at the center of the field of view, or vice versa. *Chromatic aberration* is fairly common: an eye may be hyperopic for long waves (red) and myopic for short waves (violet or blue). Placing an artificial lens in front of the eye usually solves problems associated with astigmatism.

Night blindness (nyctalopia) is the condition of a person having less than normal vision in dim light—that is, with low illumination of the visual object. It can be related to, or caused by, other visual problems and thus can often be corrected. Some degree of reduced vision is, of course, normal in dim light (*scotopic* vision). Specifically, impaired color perception, reduced visual acuity, decreased central vision, and seeing moving objects better than stationary ones are normal aspects of night vision.

Color weakness exists if a person can see all colors, but tends to confuse them, particularly in low illumination. Defective color vision is rather common in men, about 8% of whom are color-defective, compared with less than 1% of women. Some people are *colorblind,* meaning that they confuse, for example, red, green, and gray. Only very few people can see only one color or no color at all.[12]

5.3.8 Vision Stimuli

Humans are sensitive to light, meaning that they detect changes in visual stimulation depending on the wavelength, intensity, location, or duration of the stimulus. Usually, discussions of light sensitivity are limited to wavelength and intensity.

The human eye is sensitive and can adapt to increases and decreases in illumination over a wavelength range of about 380–720 nm, that is, from violet to red. The minimal intensity required to trigger the sense of light perception is 10 photons, or an illuminance of the eye of about 0.01 lux. At such low intensity, shorter wavelengths (such as blue-green) are more easily perceived than longer wavelengths; the main perception is of light, not of color.

5.3.9 Viewing Conditions

Cones and rods

Cones perceive colors only if an object's illumination is "bright," above about 0.1 lux. This is called the *photopic* condition. If the illuminance falls between 0.1 and 0.01 lux, both cones and rods respond. This is called the *mesopic* condition and is present at twilight and dawn, when we can see color in the (brighter) sky, but (dimmer) objects appear only in shades of gray. In dim light, below about 0.01 lux, only rods respond. In this *scotopic* condition, only black and white, and gray shades in between, can be perceived.

Fig. 5.7 shows, schematically, the sensibility of rods and cones to visible wavelengths: the spectral sensitivity. Rods respond mostly to shorter wavelengths, while cones cover the whole spectrum. The schematic is drawn by making the maximum of each curve equal to 100%; in reality, the three different pigments of cones have their maximal sensitivities at around 440, 540, and 570 nm, while rods have their single maximal sensitivity at about 510 nm. In terms of luminous intensity, rods are more sensitive because they respond to lower intensities than cones. This means that, as the ambient light level increases, visual detection shifts from domination by the rod system to predominant use of the cones. Under darker conditions, the threshold for detecting light of almost any wavelength is determined by the sensitivity of the rods; under brighter conditions, the cones provide most of the information, independently of the wavelengths of the stimuli. Between these two extremes of illumination, the spectral distribution of

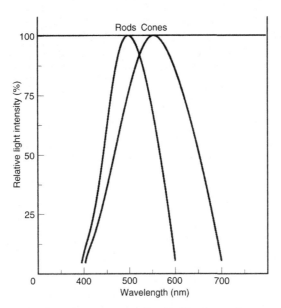

FIGURE 5.7 **Sensitivity of rods and cones to wavelength.** The figure shows conditions at illuminations below 0.01 lux (scotopic) and above 0.1 lux (photopic) but not in the intermediate (mesopic) range. Note that luminous intensity has been set to 100% for the rod and cone curves.

light determines (and other parameters co-determine) which system is more sensitive. The transition from using rods to using cones is due largely to a desensitization of the rods by relatively weak light that leaves the cone sensitivity intact. These conditions are shown schematically in Fig. 5.8. Throughout the wavelength spectrum, cones need higher intensities of light to function.

Some interesting phenomena associated with vision at night:

- In darkness, the color-sensitive cones at the retina are inactive, and what is "seen" is perceived only through the rods. Since there are only few rods directly behind the lens (at the fovea), this area constitutes a blind spot where objects may not be detected at night.
- If one stares at a single light source on a dark background, the light seems to move. This is called the *autokinetic effect*.
- If the nighttime horizon is void of visual cues, the lens relaxes and focuses at a distance of about 1–2 m, making it difficult to notice far objects. Becoming more myopic under dim light is known as *night myopia*.
- Night vision capabilities deteriorate with decreasing oxygen. Thus, at an altitude of 1300 m (4000 ft), vision is reduced by about 5%; at 2000 m, the reduction is about 20% and up to 40% in smokers whose blood has lost some capability to carry oxygen.

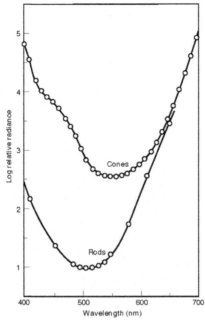

FIGURE 5.8   **Thresholds of cones and rods to light intensity.** *Source: Modified from Snyder (1985).*

5.3.10 Visual Adaptation

Adaptation to light and dark: The eye can change its sensitivity through a large range of illumination conditions. This property is called visual (or ocular) adaptation, and one commonly distinguishes between adaptation to light and to dark conditions. Adaptation is achieved by several measures: adjustments of the pupil, the spatial summation of stimuli, and photochemical functions, including the stimulation of rods and cones. The actual change in response thresholds of the eye during adaptation to the dark or light depends on the luminance and duration of the previous condition to which the eye was adapted, on the wavelength of the illumination, and on the location of the light stimulus on the retina.

Adaptation from darkness to light is very fast, and is fully achieved within a few minutes. Full adaptation from light to darkness, however, takes about 30 minutes; during this period, initially the cones are most sensitive, and the rods thereafter. After adaptation, the sensitivity at the fovea (with a preponderance of cones) is only about one-thousandth of that at the periphery of the retina (with a preponderance of rods). Therefore, weak lights can be noticed in the periphery of the field of view, but not if one looks directly at them (i.e., when they are refracted onto the fovea).

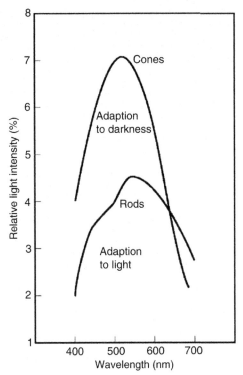

FIGURE 5.9 **Adaptation of cones and rods.**

As Fig. 5.9 shows, adaptation to the dark is governed mostly by the change in threshold of the cones. Maximal sensitivity is at about 510 nm, a shift from adaptation to the light, when the highest sensitivity is at about 560 nm. People who suffer from *dark blindness* have nonfunctional rods and can adapt only via cones. Persons who are colorblind have nonfunctional cones (or only one or two of the three types of cones) and adapt via rods only.

In adaptation to the light, wavelengths of about 560 nm (in the yellow region of the visible spectrum) are most easily perceived at the fovea, with a shift to wavelengths of about 500 nm (yellow-green) in adaptation to the dark. Above 650 nm, where there are large differences in response to the different wavelengths, the spectral sensitivity of the fovea is not very different during adaptation to either the dark or the light: with increasing adaptation to the dark, shortwave stimuli are perceived as having higher luminance. This effect can be an important consideration for the selection of colors in dark rooms. Further, persons who must adapt to the dark after having been in a lighted environment adapt faster to long-wavelength lights (red and yellow) than to short wavelengths (blue).

Human visual response characteristics can be fruitfully considered, for example, in regard to the night illumination of instruments. If the illumination is only by wavelengths longer than about 600 nm (orange), primarily the photopic (cone) system is excited, but the adaptation of the scotopic (rod) system is largely maintained: we can still observe the (mostly black–gray–white) events on the road while driving a car at night. Therefore, instruments are often illuminated with reddish or yellowish light, with the yellow color probably more suitable for the older eye.

The relative sensitivity of the visual system across the visual spectrum can be measured in various ways, but the variable that is most often used is the intensity needed to detect a stimulus of a given wavelength, called the *threshold intensity*.[13]

5.3.11 Visual Acuity

Visual acuity is usually defined as the ability to detect small details and discriminate small objects with the eye. Visual acuity depends on the shape of the object and on the wavelength, illumination, luminance, contrast, and duration of the light stimulus. Acuity is usually measured at viewing distances of 6 m (20 ft) and 0.4 m (1.3 ft), since the factors that determine the resolution of an object can differ in the far and near viewing requirements.

Visual acuity may be limited by either optical or neural causes. Optical problems lead to a degraded retinal image, which can often be improved by corrective lenses. Neural limits stem from the coarseness of the retinal mosaic (distribution of the receptors) or the sensitivity limits of neural pathways.

ACUITY TESTING

To assess acuity, high-contrast patterns are presented to the observer from a fixed distance. The smallest detail detected or identified is taken as the threshold, expressed in minutes of visual arc. Visual acuity is then expressed as the reciprocal of the resolution threshold. It is assumed that a person should normally be able to resolve a detail that subtends about 1 minute of visual arc.

The most common testing procedures use either Landolt rings or Snellen or Sloan letters, all standardized black stimuli against a highly contrasting white background. A measurement of 20/20 on the Snellen chart is regarded as "perfect" vision. A person is defined as "partially sighted" if her or his vision (after correction) is worse than 20/70, but still better than 20/200 (meaning that this person can see an object at 20 ft (6.1 m) distance that others can see at 70 or 200 ft, respectively). By definition, a person is legally blind if the vision in the better eye is 20/200 or less after correction.

Conditions affecting acuity

Since the results of acuity measures are dependent on the testing conditions, such as the test pattern and its distance from the eyes, the US National Academy of Sciences[14] recommended standards for assessing acuity. The following are among the conditions that affect acuity:

- Luminance level: for dark targets on a light background, acuity improves as the background luminance increases (up to about 150 cd/m^2; measurement units are discussed later in this chapter).

- Locus of stimulation on the retina: the highest resolution is obtained under photopic illumination with the target viewed foveally. At scotopic illumination, acuity is highest when the target is offset approximately 4 degrees from the fixation point.
- Pupil size: the highest acuity is observed when the pupil is at an intermediate diameter.
- Viewing distance: for a perfect lens, there should be no effect of distance if the visual angle is used to describe the conditions of viewing. However, acuity does change with the viewing distance, because the lens of the eye changes its shape to fixate at different distances. If accommodation errors occur, the image may be blurred. These errors are more pronounced at low luminance levels.
- Age: spatial vision capabilities change considerably with age, generally leading to a decline in acuity starting in the forties.

Visual acuity depends primarily on the ability to see edge differences between black and white stimuli, measured at relatively high illuminance levels. Such measurement of static edge acuity is simple, but it is neither the only nor the best measure of visual resolution capabilities. For example, people with 20/20 Snellen visual acuity may not do so well in other measures of contrast sensitivity, such as the ability to detect targets from a busy background or to see highway signs at given distances. As our gaze sweeps, the visual details in the field of view generate an image of ever-changing spatial frequencies and contrasts on the retina. Thus, we can consider the visual world as a constantly changing array of textures composed of varying contrasts and spatial frequencies. Accordingly, contrast sensitivity changes with differing viewing fields, and this "field dependency" is operationally defined by the tests that measure it. Many of these tests use a certain geometric shape (e.g., a triangle) which is embedded in a more complex geometric figure.

5.3.12 Visual Contrast

Contrast is defined as the relative change in luminance of a pattern, or, mathematically, $(L_{max}-L_{min})/(L_{max})$ where L_{max} is the highest luminance and L_{min} the lowest luminance of the alternation pattern. To test perceived thresholds of contrast, it is customary to plot the reciprocal of the contrast threshold function, measured as the modulation $M = (L_{max}-L_{min})/(L_{max}+L_{min})$, and to call this quantity *contrast sensitivity*. Complete measures of contrast sensitivity assess visual resolution through ranges of spatial frequencies and contrasts.

5.3.13 Measurement of Light (Photometry)

The measurement of light energies and the perception of those energies by the human observer has been an area of much confusion. The confusion stems from several sources, one being the differences between physically defined and humanly perceived lighting conditions. For example, the subjective descriptors "bright" or "light" and "dim" or "dark" are not solidly related to physical measurements of luminous intensity, illuminance, or luminance, defined shortly. A second source of confusion lies in various competing terminologies, although clarifications are available.[15]

5.3.13.1 Radiometry

Light can be defined as any radiation capable of causing a visual sensation. The natural source of such electromagnetic radiation is the sun. Lamps (also called *luminaires*) are common artificial sources. Measurement of the quantity of radiant energy may be done by determining the rise in temperature of a blackened surface that absorbs radiation.

There are the four fundamental types of energy measurement:

- The total radiant energy emitted from a source, per unit of time: *radiant flux*
- The energy emitted from a point in a given direction: *radiant intensity*
- The energy arriving at (incident on) a surface at some distance from a source: *irradiance*
- The energy emitted from or reflected by a unit area of a surface in a specified direction: *radiance*

Lines radiating from a point p define a cone with a solid angle ω with a spherical surface A at a radius r from p (see Fig. 5.10). The unit of ω is 1 steradian (sr) when $A = r^2$.

The energy emitted from point p is the radiant flux P_e in watts (W). With ω expressed in steradians, the *radiant intensity* per solid angle (in units of W/sr) is:

$$l_e = P_e / \omega \tag{5.4}$$

Most artificial light sources do not radiate uniformly in all directions; hence, the radiant intensity may be different in different directions.

The *Irradiance* E_e (the radiant flux incident on a unit area of surface) of a sphere is $E_e = P_e / \omega / r^2$ (in units of W/m²). Irradiance can also be expressed in terms of the radiant intensity (in units of W/m²) as:

$$E_e = l_e / \omega / r^2 \tag{5.5}$$

The radiant flux emitted by a point source falls on successively greater areas as the distances from the source increases. Since the irradiance E_e changes inversely with the

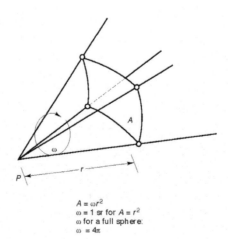

$A = \omega r^2$
$\omega = 1 \text{ sr for } A = r^2$
ω for a full sphere:
$\omega = 4\pi$

FIGURE 5.10   Definition of the steradian (sr).

square of the distance r from the source p, the relation between irradiance and distance is called the "inverse-square law of energy flux." Of course, if the surface is flat (instead of being a section of a sphere around p), not all surface elements are at the same distance from the point source p. Thus, areas of the flat surface are irradiated at varying angles, rather than perpendicularly. If a flat surface section is at an angle α from the normal, the irradiance at any point on the section is:

$$E_e = l_e r^{-2} \cos \alpha \qquad (5.6)$$

The preceding definitions and equations and the inverse-square law are valid only for a true point-source light, such as a distant star. However, for most calculations, the error made with artificial light sources is negligible. Most light sources are not points but have finite dimensions: they are called extended sources. The *radiance L_e* describes the actual density of emitted radiation in a given direction or arriving at a surface, taking into account the projection angle.[16]

Symbols and units used in radiometry and the corresponding terms used in photometry are given in Table 5.4.

5.3.13.2 Photometry Adapted to the Human Eye

The optical conditions of the human eye, the sensory perception of stimuli, and CNS processing modify the physical conditions described so far. To take account of this fact, in 1924 the Commission Internationale de l'Eclairage (CIE) developed the standard luminous efficiency function for photometry. In 1951, the CIE adopted a standard luminous efficiency function for dark-adapted (scotopic) vision. Both standards are still valid.

Accordingly, photometric energy is radiant energy modified by the luminous efficiency function of the "standard observer." Thus, a set of units parallel to those specifying radiant energy defines photometric energy (see Table 5.4). Several other terms also are still in use, a number of them quite unnecessarily, such as candlepower instead of luminous intensity.

Units

With regard to human vision, luminance, the light energy reflected (or emitted) from a surface, is the most important variable, unless we look directly into a light source. Manufacturers of lamps often use the expression "lumens per watt" as a measure of

TABLE 5.4   Radiometric and Corresponding Photometric Terms

	Radiometry		Photometry	
Term, symbol	**Units**		**Term, symbol**	**Units**
Radiant flux, P_e	watts (W)		Luminous flux, F_v	lumen (lm)
Radiant intensity, I_e	Watts per steradian (W/sr)		Luminous intensity, I_v	candela (cd) = lm/sr
Irradiance, E_e	Watts per square meter (W/m^2)		Illuminance (or illumination), E_v	lux (lx) = lm/m^2
Radiance, L_e	watts per steradian per square meter (W/sr/m^2)		Luminance, L_v	candela per square meter (cd/m^2) = lm/sr/m^2

Source: Adapted from Pokorny and Smith (1986).

luminaire efficiency, because measurement in watts describes only the electrical power intake of the light source. Typical values are 1700 lm for a 100-W incandescent lightbulb or 18-W light-emitting diode (LED) bulb, and 3200 lm for a 40-W fluorescent light.

Among the nonmetric units still in use[17] are the following:

- *Illuminance*, the amount of light falling (incident) on a surface, is occasionally measured in footcandles, where 1 footcandle (fc) = 1 lm/ft^2 and 10.76 fc = 1 lx. Also, 1 phot = 1 lm/cm^2; and 105 phot = 1 lx. The Troland is a unit for the illuminance on the retina from a source with 1 cd/m^2 luminance, viewed through an artificial pupil of 1 cm^2.
- *Luminance*, the amount of light energy reflected (or emitted) from a surface, is occasionally measured in lamberts (L) or cd cm^{-2} π^{-1}; 3.183 L = 1 cd/m^2. Also, we have the footlambert (fL) measured in cd ft^{-2} π^{-1}; 3.426 fL = 1 cd/m^2. In addition, we have the stilb, or 10^{-5} cd/cm^2; 105 stilb = 1 cd/m^2. And finally, 1 apostilb = 1 cd/m^2 /π, and 0.3183 apostilb = 1 cd/m^2; and the nit = 1 cd/m^2 = 105 stilb.

5.3.14 Color Perception

Two colors that appear white when combined are called *complementary*. For example, when yellow and blue lights in proper wavelength proportions are projected together onto a screen, the resultant light appears white.

Sunlight contains all visible wavelengths, but objects onto which the sun shines absorb some radiation. Thus, the light that objects transmit or reflect has an energy distribution different from the light they received. A human looking at the (transmitting or reflecting) object does not analyze the spectral composition of the light reaching the eyes; in fact, what appears to be of identical color may have different spectral contents. The brain simply classifies incoming signals from different groups of wavelengths and labels them colors by experience. Human color perception, then, is a psychological experience, not a single, specific property of the electromagnetic energy we see as light.[18]

Depending largely on the distribution of cones (and rods) over the retina, not all its areas are equally sensitive to all colors. We see all colors while looking fairly straight ahead, but cannot perceive any colors at the very periphery of our visual field. Green, red, and yellow all can be perceived within an angle of about 50 degrees to the side from straight ahead, while blue can be seen to about 65 degrees sideways and white even at 90 degrees.

The concept of equivalent-appearing stimuli provides a system for measuring and specifying color.[19] *Colorimetry* is an experimental technique in which one simultaneously views two spectral fields and tries to adjust the spectral content of one to make both appear identical.

Color-matching experiments have shown that the human can perceive the same color if one variously mixes three independent adjustable primary colors: red, green, and blue. (This "additive" combination of spectral radiations is not the same as mixing pigments, discussed below.) Lights that contain dissimilar spectral radiations, but that are nevertheless perceived as the same color by the observer, are called *metameric*. Since it is possible to find a metamere for any color by varying only the three primary colors, human color vision is called *trichromatic*.

MIXING PIGMENTS

When an artist mixes yellow- and blue-pigmented paints to generate green, this is done by "double-subtraction." The yellow pigment absorbs all blue, but reflects yellow as well as red and green. The blue pigment absorbs yellow and red, but reflects blue and green. Both pigments mixed together reflect only green, since the blue pigment absorbs the yellow and the yellow pigment the blue. However, if light of a different wavelength were to fall on the pigments, they would each absorb different "colors" and, combined, reflect a different wavelength that might not appear green.

We see an object as white when it reflects light about equally throughout all wavelengths; but so do gray and black pigments. The differences among gray, black, and white are not of color, but of how much light they reflect. Freshly fallen snow reflects only about 75% of the sunlight falling on it, but it appears bright white to us. Black velvet appears very dark black because it reflects only very little of the light that falls on it. Fig. 5.11 shows, schematically, how pigments reflect light of different wavelengths (i.e., colors).

Pigments reflect different wavelengths

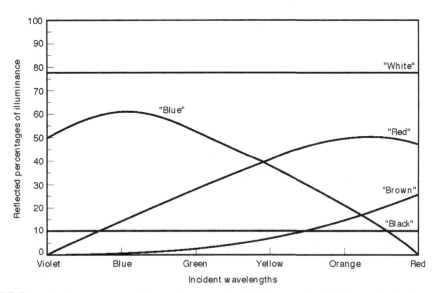

FIGURE 5.11   Schematic representation of how different colors are reflected by different pigments.

Trichromaticity is characterized by two facts:

1. Any color can be produced by combining (adding together) a suitable mixture of three arbitrarily selected spectral radiations (*primaries*).
2. Colors (wavelength components) can be specified precisely, making it possible to compare colors.

The data of a color-matching experiment can be expressed in terms of vectors in a three-dimensional (red, green, and blue) space. The results of the experiment can be shown in a color plane, or chromaticity diagram, which is a two-dimensional representation of what happens when colors are mixed. The diagram is obtained by converting the values of the three stimuli into a form such that the sum of the three equals unity. By convention, X corresponds to the red region of the spectrum, Y to the green region, and Z to the blue region, and $X + Y + Z = 1$.

If the proportions of red and green are plotted along the abscissa X and ordinate Y, respectively, the combined wavelength—the third trichromatic coefficient—falls within a horseshoe-shaped curve called the *spectrum locus*. This curve is shown in Fig. 5.12.

The chromaticity diagram, standardized in 1931 by the CIE, plots colors by the amounts of standard red, blue, and green primaries that, when mixed together, yield the given color. The red, green, and blue lights are mathematically specified and called the three *primaries*. Because the sum of red, blue, and green required to match any color is expressed as unity, any one standard color component can be determined by subtracting the other two from unity.

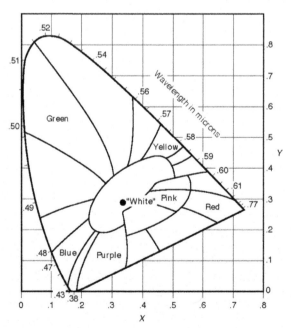

FIGURE 5.12   Trichromaticity diagram.

The most saturated colors are located on the horseshoe curve. Other perceived colors fall inside this boundary. The white located in the center of the diagram, consisting of one-third red, one-third green, and one-third blue, is designated as *illuminant C* by the CIE; it is approximately the color of light from the summer sun in the northern-hemisphere at noon on a clear day.

The numbers along the curve are wavelengths. A color appears greenish in the upper corner, reddish in the lower right corner, and bluish in the lower left corner. The "purity" of each color is determined by its proximity to the center of the diagram, where it would appear white. Although the CIE diagram is many decades old, it is still in common use.[20]

The trichromatic or Young-Helmholtz theory is not the only one attempting to explain human color vision. Alternate approaches include Hering's Opponent Colors theory and Judd's Zone or Stage theories; but other theories are still being developed and tested.[21] The complex nature of color appearance is not yet fully understood, and the mathematical models available are not entirely satisfactory. A human's judgment of the perceived color of a visual stimulus depends on the person's subjective impressions experienced when viewing the stimulus, and the judgment varies with the viewing conditions and the kind of stimuli.

Other theories of color vision

Many systems that order colors according to various variables and criteria are possible; several have been well-developed and are often used. There are three major groups of color systems. The first group is based on principles of additive mixtures of color, such as the Maxwell disk and the Ostwald Color System. The second group uses the principles of color subtraction, such as are applied in mixing pigments; this group is widely used in printing. The third group is based on the appearance and perception of colors. Physical standards are selected to present scales of, for example, hue, chroma, and likeness. Examples from this group are the Munsel Color System and the Optical Society of America (OSA) system.[22]

Color-ordering systems

The following terms are often used, although not in a universal, uniform, or standardized manner:[23]

Terminology

- *Brightness* and *dimness* are individual perceptions of the intensity of a visual stimulus.
- *Chroma* is the attribute of color perception attained by judging to what degree a chromatic color differs from an achromatic color of the same lightness. The 1942 Munsell Book of Color provides many good examples (see also *saturation*).
- *Chromatic* or a*chromatic induction* is a visual process that occurs when two color stimuli are viewed side by side, when each stimulus alters the color perception of the other. The effect of chromatic or achromatic induction is usually called simultaneous contrast or spatial contrast.
- *Hue* is an attribute of color perception that uses color names and combinations thereof, such as yellow or yellowish green. The four unique (unitary) hues are red, green, yellow, and blue, none of which is judged to contain any of the others.
- *Illuminant color* is that color perceived as belonging to an area that emits light as the primary source.
- *Lightness* is the individual perception of how much more or less light a stimulus emits in comparison to a "white" stimulus also contained in the field of view.

- *Object color* is that color perceived as belonging to an object.
- *Related color* is a color seen in direct relation to other colors in the field of view.
- *Saturation* is the attribute of color judgment regarding the degree to which a chromatic color differs from an achromatic color regardless of lightness (see also *chroma*).
- *Surface color* is that color perceived as belonging to a surface from which the light reflected or radiated.
- *Unrelated color* is a color perceived to belong to an area seen in isolation from other colors.

All of the preceding terms rely on perception and judgment. Occasionally, some of them are inappropriately used in place of physically determined concepts. For example, (physical) luminance is not the same as (psychological) brightness. Thus, confusion about these terms is not unusual.

5.3.14.1 Esthetics and Psychology of Color

While the physics of color stimuli arriving at the eye can be well described (although often with considerable effort), perception, interpretation, and reaction to colors are highly individual, nonstandardized, and variable. Thus, people find it very difficult to describe colors verbally, given the many possible combinations of individually perceived hue, lightness, and saturation values.

People believe in and describe nonvisual reactions to color stimuli. Reds, oranges, and yellows, browns, and tans are usually considered "warm" and stimulating. Violets, blues, blueish-greens, and most grays are often felt to be "cool" and to generate sensations of cleanliness and restfulness. Note however that the attraction to certain colors and their combinations vary culturally and regionally as travel, say, between Asia and Europe, readily shows. Still, pale colors often seem cooler than dark colors, cold colors more distant than warm colors, weak colors more distant than intense colors, soft edges of color patches more distant than hard edges. Although experimental evidence regarding these effects is controversial,[24] color schemes are often applied to work and living areas to exploit these stereotypical associations.

Keeping in mind that the impact of color is culture-dependent and is also highly susceptible to individual differences in perception and association, chromology theory as applied to work environments suggests some of the following guidelines:

- Red can be used to activate nervous system activity. Displayed with other colors, it will appear to be the closest to the viewer. It is universally used for traffic lights and stop signs. It is thought to increase appetite as well.
- Blue is thought to have a more calming effect. Some studies have indicated that it improves study capabilities in students and increases worker's concentration.

- Certain shades of green may be most strongly associated with positive emotional states universally, as these are most frequently associated with nature. Darker shades of green may be associated with money and wealth and are frequently used in corporate boardrooms and financial firms.
- White may evoke the perception of sterility and/or blandness.
- Other colors, such as neon shades, bright yellow, and bright pink can be initially attention-getting but may become irritating over time to individuals as bright and vibrant colors reflect more light.

Eye strain and wakefulness can be affected by "screentime." To reduce eye strain when working on displays for long periods of time, it can be helpful to reposition the screen or head subtly in order to change the angle of viewing. One can also adjust brightness and contrast on the screen. Blinking purposefully and more frequently can be helpful, as can closing and rolling the eyes, as both of these increase lubrication. Cleaning the screen can be helpful (and easy), and updating with physical interventions can be as well (e.g., using anti-glare, high-resolution upgrades). It is recommended to take short breaks several times per hour, for 20–30 seconds, and focus on something that is farther away. (The "20-20-20-rule" reflects taking a 20 second break every 20 minute and looking at something 20 ft away.) Adjusting color temperature on the monitor either manually or using software programs can reduce eye strain. It may be helpful to adjust monitor colors based on time of day to compensate for changes in the natural lighting (see also Chapter 10).

Computer screens

5.3.15 Illumination Concepts in Engineering and Design

The characteristics of human vision just discussed provide the bases for engineering procedures to design environments for proper vision. The most important concepts are the following:

1. Proper vision requires sufficient quantity and quality of illumination.
2. Special considerations regarding visibility, especially declining visual abilities in the elderly, require particular care in the arrangement of proper illumination.
3. Illumination of an object is inversely proportional to the distance from the light source.
4. Use of colors, if selected properly, can be helpful, but color vision requires sufficient light.
5. What counts most is the luminance of an object, that is, the energy reflected or emitted from it that meets the eye.
6. Luminance of an object is determined by its incident illuminance, and by its reflectance:

$$\text{luminance} = \textbf{illuminance} \times \text{reflectance} \times \pi^{-1} \qquad (5.7)$$

Reflectance is the ratio of reflected light to received light, in percent. (The numerical value for illuminance is in lux and for luminance in cd/m^2; see Table 5.4.) The factor π^{-1} is omitted when the following nonmetric units are used: luminance in footlamberts (fL) and illuminance in footcandles (fc). Fig. 5.13 shows luminance levels experienced by humans.

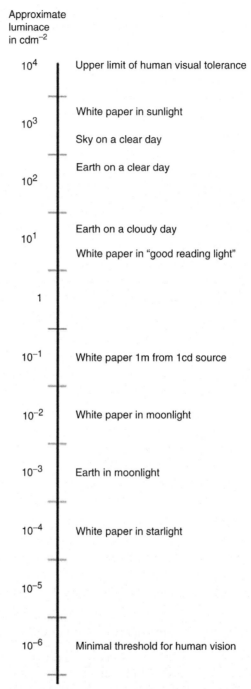

FIGURE 5.13 Luminance levels experienced by humans. *Source: Adapted from Van Cott and Kinkade (1972).*

7. The ability to see an object depends largely on the luminance contrast between the object and its background, including shadows. Contrast is usually defined as the difference in luminances of adjacent surfaces, divided by the larger luminance:

$$\text{Contrast (in percent)} = 100\frac{\left(L_{\text{max}} - L_{\text{min}}\right)}{\left(L_{\text{max}}\right)} \tag{5.8}$$

8. Avoid unwanted or excessive glare. Direct glare meets the eye directly from a light source (such as the headlights of an oncoming car). Indirect glare is reflected from a surface into the eyes (such as the headlights of a car seen in the rearview mirror). Often, kinds of reflected glare are described as either specular (coming from a smooth, polished surface such as a mirror), spread (coming from a brushed, etched, or pebbled surface), diffuse (coming from a matte, nonglossy painted surface), or compound (a mixture of the various types of glare).

Direct glare can be avoided by the following practices:

- Place high-intensity light sources outside the cone of 60 degrees around the LOS.
- Use several low-intensity light sources placed away from the LOS, instead of one intense source.
- Use indirect lighting, where all light is reflected at a suitable surface (within the luminaire or at the ceiling or walls of a room) before it reaches the work area. This generates an even illuminance without shadows. (However, shadows may be desirable to see objects better).
- Use shields or hoods over reflecting surfaces, or visors over a person's eyes, to keep out the rays from light sources (potentially a clumsy way to overcome a condition of bad design).

Indirect glare can be reduced by the use of:

- Diffuse, indirect lighting
- Dull, matte, or other nonpolished surfaces
- Properly distributed light over the work area

Direct lighting (when rays from the source fall directly on the work area) is most efficient in terms of illuminance gain per unit of electrical power; but it can produce high glare, poor contrast, and deep shadows. Indirect lighting (when the rays from the light sources are reflected and diffused at some suitable surface before they reach the work area) helps to provide an even illumination without shadows or glare, but is less efficient in terms of the use of electrical power. A third way uses diffuse lighting while the light source is enclosed by a large translucent bowl, so that the room lighting is emitted from a large surface. This can cause some glare and shadows, but is usually more efficient in using electrical power than indirect lighting is. For more discussion on the ergonomics of lighting, see Chapter 10.

What is most suitable depends on the given conditions, such as the task to be done, the objects to be seen, and the eyes to be accommodated. Thus, general recommendations are difficult to compile. Nevertheless, the IES Lighting Handbook[25] contains recommended illuminance values for certain applications. An excerpt from these recommendations is provided in Table 5.5. Before the recommendations are applied, note

TABLE 5.5 Recommended Illuminance Values and Examples

Illuminance categories and illuminance values for generic types of activities in interiors

Type of activity	Illuminance category	Ranges of illuminances (in Lux)	Reference workplane
Public spaces with dark surroundings	A	20–50	
Simple orientation for short temporary visits	B	50–100	General lighting throughout space
Working spaces where visual tasks are occasionally performed	C	100–200	
Performance of visual tasks of high contrast or large size	D	200–500	Computer workstations
Performance of visual tasks of low contrast or very small size	E	500–1000	Illuminance on task critical
Performance of visual tasks of low contrast and very small size over a prolonged period	F	1000–2000	
Performance of very prolonged and exacting visual tasks	H	5000–10,000	Illuminance directed on task obtained by a combination of general and local (supplementary) lighting
Performance of very special visual tasks of extremely low contrast and small size	I	10,000–20,000	

Examples

Area/activity	Illuminance category
Auditorium, assembly	C
Social activity	B
Bank lobby, general	C
Bank lobby, writing area	D
Bank teller	E
Assembly work, simple	D
Assembly work, moderately difficult	E
Assembly work, difficult	F
Assembly work, very difficult	G
Assembly work, exacting	H
Bakery, mixing or oven room	D
Bakery, decorating/icing by hand	E
Barber shop/Beauty parlor	E
Club room, lounge and reading Conference room	D
Courtroom, seating area	C
Courtroom, court activity area	E
Dancehall/discotheque	B
Brewery, brew house	D
Brewery, boiling and key washing	D
Brewery, filling (bottles, cans, kegs)	D

Source: Adapted from Kaufman and Haynes (1981).

that high reflectances at surfaces (such as that of a ceiling or wall) can allow illumination at a lower level, while surfaces with low reflectance probably require high illumination.

5.3.16 Vision Myths

The following are seven myths about vision and the related facts.[26]

Myth 1: Straining your eyes can damage your eyesight. Examples of strain are working in dim or glaring light, reading fine print, wearing glasses with the wrong prescription, or staring at a computer screen.

Truth 1: Prolonged use of the eyes under any of the stated conditions can cause eyestrain, since the eye muscles struggle to maintain a clear or unwavering focus. In addition, prolonged staring can dry the front of the eye somewhat, because it reduces blinking, which helps lubricate the cornea. But fatigue and minor dryness, no matter how uncomfortable, cannot permanently harm your vision. Minimize the discomfort by using the following techniques:

- *Lighten up*: Age tends to cloud the lens of the eye and shrink the pupil, sharply increasing the need for luminance. Consider installing brighter lights or at least moving the reading lamp closer to the page.
- *Cut the glare*: Position the reading lamp so that light shines from over your shoulder, but make sure it does not reflect into your eyes from the computer monitor. Do not read or do computer work near an unshaded window. And wear sunglasses if you are reading outside.
- *Stop and blink*: When you are working at the computer or reading, pause frequently (several times an hour) to close your eyes, or gaze away from the screen or page, and blink repeatedly. Every hour or so, get up and take a longer break.
- *Find and maintain the right distance*: Keep your eyes at the same distance from the screen as you would from a book. If that is uncomfortable, buy a pair of glasses with a prescription designed for computer work, or use "progressive-addition" bifocals, which have gradually changing power from the top to the bottom of the lens.
- *Lower the screen*: Keep the top of the screen below eye level. Gazing upward can strain muscles in the eye and neck.
- *Use a document holder*: Put a support for reading matter next to the screen, at the same distance from your eyes.
- *Clean your screen and your glasses*: Dust and grime can blur the images.
- *Consider having an eye examination* to confirm whether you need to start wearing glasses or to have your prescription changed.

Myth 2: The more you rely on your eyeglasses or contact lenses, the faster your eyesight deteriorates.

Truth 2: This myth is based on the misconception that artificial lenses do the work of the eyes, which then supposedly grow lazy and weak. However, artificial lenses merely compensate for a structural defect of the eye—an improperly shaped eyeball or an excessively stiff lens—that prevents proper focusing, despite the best efforts of the lens muscles. When you wear glasses or contact lenses, the eye muscles no longer need to be tensed or relaxed (depending on whether you are farsighted or nearsighted) any more often than usual. Instead, they work just as hard as the muscles in a normal eye.

Myth 3: Eye exercises can help many people see better.

Truth 3: Eye exercises can help some children whose eyes have major binocularity problems, such as crossing, misalignment, or inability to converge. But claims that the exercises can help many children read better are unsubstantiated. Erroneous claims that eye exercises not only help one read better, but also sharpen visual acuity, boost athletic performance, and help correct numerous problems in both children and adults, are unsupported and implausible.

Myth 4: The more carrots you eat, the better your eyesight will be.

Truth 4: Carrots are rich in beta-carotene, an orange pigment used by the body to manufacture vitamin A, which is essential for night vision. A reasonably well-balanced diet supplies enough beta-carotene for the eyes. Extra doses of either beta-carotene or vitamin A do not improve vision. However, beta-carotene and other carotenoid nutrients in fruits and vegetables may reduce the risk of cataracts and slow the progression of macular degeneration.

Myth 5: The darker the sunglasses, the better the protection against harmful ultraviolet (UV) light.

Truth 5: In theory, prolonged exposure to the UV rays in sunlight increases the risk of cataracts and possibly macular degeneration, because the UV rays oxidize tissues in the eye. However, the darkness or tint of the sunglasses has no effect on how much UV light they absorb, so check the label (or have an optician test the lenses) to ensure that they provide adequate protection. At minimum, consider sunglasses labeled (in the United States) as meeting ANSI Z80.3 General Purpose UV requirements: these block at least 95% of the high-energy UVB rays and 60% of the lower energy UVA rays. If you spend lots of time in the sun (particularly if you have light-colored eyes) seek stronger protection, indicated by the label "Special-Purpose UV Requirements": these glasses block at least 99% of the UVB rays. Dark lenses do have one potential advantage over paler ones: they block more visible light and reduce glare. However, nearly all sunglasses block enough visible light for safety and comfort under ordinary conditions. Only people exposed to brilliant sunlight (on ski slopes or bright beaches, for example) may need extra-dark, wrap-around sunglasses.

Myth 6: Once you develop a cataract, it must be removed.

Truth 6: Most cataracts are so minor that they cause little loss of vision. Even when objective tests show a substantial drop in visual acuity, surgery may not be required, because subjective factors matter more: if the loss of acuity does not significantly affect your everyday activities or nighttime driving, you may not need surgery, regardless of what any test shows.

Myth 7: If you undergo surgery to correct nearsightedness, you will never need glasses again.

Truth 7: Successful surgery can reduce or eliminate nearsightedness by flattening the cornea, but there is no guarantee that surgery will eliminate the need for glasses, and any surgery has risks. Regardless of the type of eye surgery, some patients will still require glasses for distance vision at least occasionally, and others will eventually need reading glasses as they age.

5.4 HEARING—THE AUDITORY SENSE

Acoustics is the science and technology of sound, including its production, transmission, and effects. The acoustical design goal is to establish an environment that:

- transmits desired sounds reliably and pleasantly to the hearer,
- is satisfactory to the human,
- minimizes sound-related annoyance and stress,
- minimizes disruptions of speech, and
- prevents hearing loss.

Acoustics describes the physical properties of sound, such as frequency or amplitude; *psychoacoustics* establishes relations between the physics of sound and our individual perception thereof, using descriptors such as pitch, timbre, loudness, noise, and speech comprehension.[27]

5.4.1 Sound

Sound is any vibration (passage of zones of compression and rarefaction through the air or any other physical medium) that stimulates an auditory sensation. Unless under water or in space, we are concerned with sound that arrives at the ear by air. A sketch of the ear is given in Fig. 5.14.

5.4.2 Ear Anatomy and Hearing

Airborne sound waves arriving from outside the body are collected by the *outer ear* (auricle or pinna) and funneled into the *auditory canal* (meatus) to the *eardrum* (tympanic membrane), which vibrates according to the frequency and intensity of the arriving

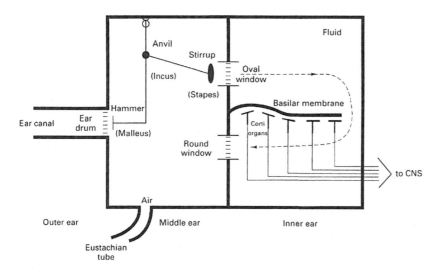

FIGURE 5.14 Schematic representation of the human ear (See also Fig. 5.22).

sound wave. Resonance effects of the auricle and meatus have amplified the intensity of sound by 10–15 decibels (dB, see later for an explanation of the units) by the time it reaches the eardrum.

In the *middle ear*, the sound that arrived via the eardrum is mechanically transmitted by the *ear bones* (ossicles)—the *hammer* (malleus), *anvil* (incus), and *stirrup* (stapes)—to the *oval window*. The ear bones mechanically increase the intensity of the sound from the eardrum to the oval window. The area of the eardrum is about 25 times larger than the surface of the oval window. Both factors together increase the effective sound pressure by a factor of approximately 22.

Air fills the outer and the middle ear, and the *Eustachian tube*, which connects with the *pharynx*, allows the air pressure in the middle ear to remain equal to the external air pressure.

> In an airplane, during rapid descent or ascent, a "clogged" Eustachian tube can delay the equalization of pressure between the inner ear and the environs. If you feel ear pressure or pain, or if you cannot hear well, you may try to open the tube by chewing gum or by willful excessive yawning, but pumping on your outer ears with your hands will not help your middle ears.

The *inner ear* contains the receptors for hearing (and for body position, via the vestibulum, discussed later in this chapter). The inner ear is filled with a watery fluid (called *endolymph* or *perilymph*) that propagates sound waves as fluid shifts from the oval window to the round window through the *cochlea*, an opening shaped like a snail shell with about two and one-half turns. The motion of the fluid deflects the *basilar membrane* that runs along the cochlea and stimulates sensory hair cells (*cilia*) in the *organs of Corti*, located on the basilar membrane. Depending on their structure and location, these organs respond to specific frequencies. Impulses generated in the organs of Corti are transmitted along the *auditory (cochlear) nerve* to the brain for interpretation.

5.4.2.1 The Human Hearing Range

A tone is a single-frequency oscillation, while a sound contains a mixture of frequencies. The frequencies (frequency distributions) of both are measured in hertz (Hz), their intensities (amplitude, sound pressure levels [SPLs]) in logarithmic units known as decibels (dB).

Log scale

One reason for the use of a logarithmic scale is that the human being perceives sound pressure amplitudes in a roughly logarithmic manner. Infants can hear tones of about 16–20 kHz, while elderly people can rarely hear frequencies above 12 kHz. The minimal pressure threshold of hearing is about $20 \times 10^{-6}\,N/m^2$ (or 20 µPa; 1 Pa = 1 N/m^2) in the frequency range of 1000–5000 Hz. The ear experiences pain when the sound pressure exceeds 140 Pa (see Fig. 5.15).

SPL

The sound pressure level (SPL) is the ratio between two sound pressures; P_0 is used as reference and is defined as the minimal pressure threshold of hearing (generally 20 µPa for a young child). The definition of SPL (in units of decibel or dB) is:

$$SPL = 10\log\left(P^2/P_0^2\right) = 20\log_{10}\left(P/P_0\right) \tag{5.9}$$

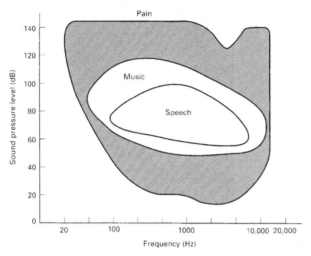

FIGURE 5.15 Ranges of human hearing.

where P is the root-mean-square (rms) sound pressure for the existing sound. Thus, in decibels, the dynamic range of human hearing from 20×10^{-6} to 200 Pa is $20 \log_{10}[200/(20 \times 10^{-6})] = 140$ dB.

Regarding *sound intensity* ("power"), the sound intensity level is similarly defined (in units of dB) as:

Sound intensity

$$\mathrm{SIL} = 10 \log \left(I^2 / I_0^2 \right) \tag{5.10}$$

where I is the rms sound intensity and I_o is used as a reference and is defined as the minimal intensity threshold of hearing (10^{-12} W/m^2 for a young child at 1 kHz). Ranges of sound intensity levels are shown in Fig. 5.16.

ADDING SOUNDS

Because decibel is a logarithmic unit, doubling the SPL causes an increase of 6 dB; a 1.41-fold increase in sound pressure causes an increase of 3 dB, which means doubling the sound energy. If two sounds (of frequencies and temporal characteristics that are both random) occur at the same time, their combined SPL can be calculated from

$$\mathrm{SPL}_{\text{combined}} = 10^{\mathrm{SPL1}/10} + 10^{\mathrm{SPL2}/10} + \cdots + 10^{\mathrm{SPL}n/10} \tag{5.11}$$

Accordingly, one can approximate the combined SPL from the difference d in the intensities as follows: if d is 0 dB, add 3 dB to the louder sound; if d is 1 dB, add 2.5 dB; for $d = 2$ dB, add 2 dB; for $d = 4$ dB, add 1.5 dB; for $d = 6$ dB, add 1 dB; for $d = 8$ dB, add 0.5 dB; for $d = 10$ dB or more, take the louder sound by itself.

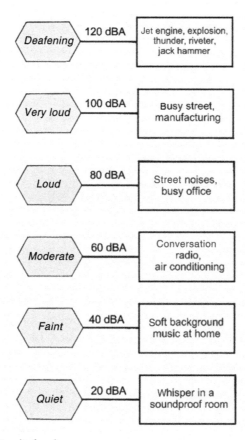

FIGURE 5.16   Sound intensity levels.

5.4.3 Pathways of Sound

In the human body, sound can reach the ear via two different paths. Airborne sound travels through the ear canal and excites the eardrum and the structures behind it, as described earlier. Sound may also be transmitted through bony structures to the head and ear, but this requires 30–50 dB higher intensities (depending on the existing frequencies) to be similarly effective.

The propagation of sound through air or some other medium depends on the intensity and frequency of the sound and on the transmission characteristics of the medium. Moving from terrestrial gravity and atmospheric conditions to microgravity and the vacuum of in space, or from air to immersion in water, changes the transmission characteristics of sound. The velocity V of sound in solids, liquids, and gases is determined by the bulk modulus E of the medium and its density D according to the formula:

$$V = \left(\frac{E}{D}\right)^{1/2} \tag{5.12}$$

with E in Pascals (N/m^2) and D in kg/m^3 resulting in velocity V in units of m/s.

5.4.4 Human Responses to Music

Music is probably one of the oldest human expressions and art forms. Making music has long accompanied, even promoted and accelerated, activities—for example, singing during fieldwork and chanting while marching. A fixture of every human civilization in history, music has been reported as one of the most enjoyable of human activities. We may experience a complex array of sensory activation and emotional responses to music, feeling activated and engaged while listening to some types of music and not to others.

Background music[28] is like "acoustical wallpaper" as found in shops, hotels, and waiting rooms. Such music is meant to create a welcoming atmosphere, to relax customers, to reduce boredom, and (sometimes) to cover other disturbing sounds. Its character is subdued, its tempo intermediate, and vocals are avoided. It may produce a monotonous environment for those continuously exposed to it, while it may appear pleasant to the transient customer.

"Music while you work" is in many respects the opposite of background music. It is not continuous, but is programmed to appear at certain times, it has varying rhythms with vocals, and it may contain popular music. Music played while a person works is meant to break up monotony—to generate mild excitement and an emotional impetus to demanding physical effort or drudgery in a dreary impersonal environment. While improved morale and productivity have been reported in many circumstances, a clear linking of the underlying arousal theory and the specific components of the music is difficult.[29] Thus, the musical content, the rhythm, the loudness, the presentation at certain times, and the selection of particular kinds of music for certain activities, environments, and listener populations are still a matter of art (or preference) rather than science.

An interesting psychophysiological reaction that many of us may have experienced is *frisson*: the aesthetic "chills" and goosebumps (*piloerections*) we may experience when listening to particular passages of music. Some research suggests that people who occasionally experience frission while intensely listening to music have more nerve fibers connecting the auditory cortex (the part of the brain that processes sound) to the anterior insular cortex (a region involved in processing feelings) than those who don't.[30]

Music

With the increased use of open-space design for offices has come a corresponding increase in the use of music over headphones to help the user gain auditory privacy. Wearing of headphones in an open floor-plan office has also become an analog of a "do not disturb" sign.

5.4.4.1 "White" and "Pink" Noise

"White" noise is a sound whose spectral density is the same at all frequencies. This can be demonstrated by using an analyzer with filters that are 1 Hz wide: this results in a flat spectrum over the entire bandwidth being analyzed. For the ear, white noise sounds like electronic static.

"Pink" noise sounds more like a waterfall. Pink noise has a continuous frequency spectrum and constant power within a bandwidth: it has equal energy in each octave (or fractional octave) band. The display of an analyzer using octave (or fractional octave) filters would indicate a flat spectrum. However, if pink noise were analyzed using an analyzer equipped with filters that are 1 Hz-wide, the spectrum would decrease as the frequency increases. The high-frequency filters are wider than the low-frequency filters, and thus, there is less energy per hertz at the higher frequencies.

HUMAN RESPONSE TO NOISE

Noise is psychological and subjective. Single, short tones of low intensity (produced, for instance, by the dripping of water) may under certain conditions be considered noise, just as loud, lasting, complex sounds (such as a neighbor's music) may be under other circumstances. Noise is defined as any unwanted, objectionable, or unacceptable sound. The threshold for noise annoyance varies depending on the conditions, including the sensitivity and mental state of the individual. Noise can:

- create negative emotions and feelings of surprise, frustration, anger, and fear;
- delay the onset of sleep, awaken a person from sleep, or disturb someone's rest;
- make it difficult to hear desirable sounds;
- produce temporary or permanent alterations in body chemistry;
- temporarily or permanently change one's hearing capability; and interfere with some human sensory and perceptual capabilities and thereby degrade the performance of a task.

5.4.5 Physiological Effects of Sound

TTS, PTS

Exposure to intense sounds may result in a *temporary threshold shift* (TTS) from which the hearing eventually returns to normal with time away from the source; or it can cause *a permanent threshold shift* (PTS), which is an irrecoverable loss of hearing. A PTS may be the result of damage to the cochlear cilia, the organs of Corti at the basilar membrane in the inner ear, or the nerves leading to the CNS. Which of these are damaged depends on the frequency and, of course, intensity of the incident noise. The damage is probably due to an overstimulation of cell metabolism, leading to oxygen depletion and destruction, or to exceeding the elastic capacities of the physical structures. The metabolic activity taking place at the moment of the arrival of noise is important: heat, heavy work, infectious disease, and other causes of heightened metabolism increase the vulnerability of the sensory organs. Above about 130 dB, the induced turbulence in the ear may also do mechanical damage.[31] The timing and severity of the loss are dependent upon the duration of exposure, the physical characteristics of the sound (its intensity and frequency), and the nature of the exposure (whether it is continuous or intermittent). Damage may be immediate, such as by an explosion, or may occur over some time, such as with continuous exposure to noise. The effect of continuous exposure is usually insidious and cumulative.

TABLE 5.6 Physiological Effects of Noise

Condition of exposure

SPL (dB)	Frequency spectrum	Duration	Reported disturbances
175	Low frequency	Blast	Rupture of tympanic membrane
167	2000 Hz	5 min	Human lethality
161	2000 Hz	45 min	Human lethality
160	3 Hz		Pain in ears
155	2000 Hz	Continuous	Rupture of tympanic membrane
150	1–100 Hz	2 min	Reduced visual acuity; vibrations of chest wall; gagging sensations; respiratory rhythm changes
120–150	OASPL		Mechanical vibrations of body felt; disturbing sensations
120–150	1.6–4.4 Hz	Continuous	Vertigo and occasionally disorientation; nausea; vomiting
135	20–20,000 Hz		Pain in ears
120	OASPL		Irritability and fatigue
120	300–9600 Hz	2 s	Discomfort in ears
110	20–31.5 Hz		TTS occurs
106	4000 Hz	4 min	TTS of 10 dB
100	4000 Hz	7 min	TTS of 10 dB
100		Sudden onset	Reflex response of tensing, grimacing, covering the ears, and urge to avoid or escape
94	4000 Hz	15 min	TTS of 10 dB
75	8–16 kHz		TTS occurs
65	Broadband	60 days	TTS occurs

Abbreviations: *SPL*, sound pressure level relative to 20 μPa; *TTS*, temporary threshold shift; *OASPL*, overall sound pressure level.

Table 5.6 lists a number of physiological effects of various levels of sound. Intense sound can have other physiological effects.

- The muscles of the middle ear contract, mostly affecting the stapes and therefore reducing the transmission of force to the cochlea. However, it takes about 30 ms after the onset of the sound to activate this protective reflex and about 200 ms for complete contraction to occur, which lasts for less than a second. Thus, the contraction response may be too late or too short to provide sufficient protection to the inner ear.
- The concentration of corticosteroids in the blood and brain is increased and noise also affects the size of the adrenal cortex. Further, continued exposure is correlated with changes in the liver and kidneys and with the production of gastrointestinal ulcers.

- Electrolytes in the body become imbalanced and blood glucose levels change. Sex hormone secretion and thyroid activity may also be affected by noise. Changes in cardiac muscle, fluctuations in blood pressure, and vasoconstriction have been reported with 70 dB SPL and above, becoming progressively worse with higher exposure.
- Abnormal heart rhythms have been associated with exposure to noise.

> High-intensity sound, whether "pleasant but loud" or "noise," can permanently damage hearing.

5.4.5.1 *Effects of Noise on Human Performance*

A number of effects of noise on the performance of tasks have been observed:[32]

- As noise becomes more intense, we become more stimulated and our performance of certain tasks improves.
- Beyond a certain level of intensity, however, task performance degrades.
- Sudden unexpected noise can produce a startle response that interrupts one's concentration and physical performance of a task.
- Continuous periodic or aperiodic noise interferes with the performance of complex tasks, such as visual tracking; performance is diminished with increasing noise levels.
- Psychological effects of noise include anxiety, helplessness, narrowed attention, and other adverse effects that degrade task performance.

Other effects of noise on performance are listed in Table 5.7.

5.4.6 Noise-induced Hearing Loss

Sounds of sufficient intensity and duration regularly result in temporary or permanent hearing loss. Permanent hearing loss may range from mild to profound, but can largely be prevented or, after it has occurred, alleviated by hearing aids. However, if the underlying nervous structure is damaged, the condition is generally not treatable with current technology.

Traumatic exposure

Sound that is of high intensity and brief duration, such as that produced by a cannon or an explosion, can damage any or all of the structures of the ear; in particular the hair cells in the organs of Corti which may be torn apart. This type of trauma results in immediate, severe, and permanent hearing loss with a PTS.

Moderate exposure

We are usually exposed to sound levels of less than 100 dB, often over a period of hours, that may initially cause only short-term hearing loss, measured as a TTS. During quiet periods, hearing returns to its normal level. Yet, a TTS may include subtle mechanical intracellular changes in the sensory hair cells and swelling of the auditory nerve endings. Other potentially irreversible effects include vascular changes, metabolic exhaustion, and chemical changes within the hair cells.

TABLE 5.7 Effects of Noise on Human Performance

SPL (dB)	Conditions of exposure spectrum	Duration	Performance effects
155		8 h; 100 impulses	TTS 2 min after exposure
120	Broadband		Reduced ability to balance on a thin rail
110	Machinery noise	8 h	Chronic fatigue
105	Aircraft engine noise		Visual acuity, stereoscopic acuity, near-point accommodation all reduced
100	Speech		Overloading of hearing due to loud speech
90	Broadband	Continuous	Decrement in vigilance, altered thought processes, interference with mental work
90	Broadband		Performance degradation in multiple-choice, serial-reaction tasks
85	One-third octave at 16 kHz	Continuous	Fatigue, nausea, headache
75	Background noise in spacecraft	10–30 days	Degraded astronaut performance
70	4000 Hz		TTS 2 min after exposure

See Table 5.10 for effects on person-to-person voice communication. *TTS*, temporary threshold shift.
Source: Adapted from NASA (1989)

Repeated exposures to sounds that cause TTSs may gradually bring about a PTS: that is, a *noise-induced hearing loss* (NIHL). Experiments with animals have shown that with each exposure, cochlear blood flow may be impaired, and some hair cells may be damaged. Often, the damage is confined to a special area on the cilia bed on the cochlea, related to the frequency of the sound. With continued exposure to noise more hair cells are damaged, which the body cannot replace; also, nerve fibers to that region in the ear degenerate, a process that is accompanied by corresponding impairment within the CNS.

NIHL

Impairment of hearing ability at special frequencies indicates exposure to noise at these frequencies. In western countries, NIHL usually occurs initially in the range of 3000–6000 Hz (particularly at about 4000 Hz) then through high frequencies, culminating at around 8000 Hz. However, reduced hearing near 8000 Hz is also characteristic of aging, which often makes it difficult to distinguish between environmental and age-related causes. With continued exposure to noise, NIHL increases in magnitude and extends to lower and higher frequencies. NIHL increases most rapidly in the first years of exposure; after many years, it levels off in the high frequencies, but continues to worsen in the low frequencies.[33]

NIHL frequencies

Noise-induced changes start at the pinna (the outer ear) which becomes harder, less flexible, and may change in size and shape. Wax build-up is frequent in the ear canal. Often, the Eustachian tube becomes obstructed, leading to an accumulation of fluid in the middle ear. There may also be arthrosic changes in the joints of the bones (anvil, hammer,

Biologic changes

stirrup) of the middle ear, which, however, do not usually impair sound transmission to the oval window of the inner ear. There, atrophy and degeneration of hair cells in the basilar membrane of the cochlea often occur. Deficiencies in the bioelectric and biomechanical properties of the inner ear fluid and mechanical degeneration of the cochlear partition occur, often together with a loss of auditory neurons. These degenerations cause either frequency-specific or more general deficiencies in hearing capabilities.

Hearing sensitivity

While it is estimated that 70% of all individuals over 50 years of age have some degree of hearing loss, the changes are individually quite different. Typically, in populations that do not suffer from industry or civilization-related noises, the hearing sensitivity in the higher frequencies is less reduced than in people from so-called developed countries. Such changes related to the environment overlap with, and in some cases mask, age-dependent changes.

Audiometry

Hearing ability (especially hearing loss) is assessed by measuring the auditory thresholds (sensitivity) at various frequencies. Such pure-tone audiometry is often combined with measures of an individual's understanding of speech.

Understanding speech

NIHL is associated mostly with problems in differentiating speech sounds in their high frequency ranges. Especially affected is one's ability to make out the relatively high-frequency, low-intensity consonants. With NIHL occurring so often in the higher frequency ranges, important informational content of speech is often unclear, unusable, or inaudible. Also, other sounds, such as background noise, competing voices, or reverberation, may interfere with the listener's ability to receive information and to communicate.

5.4.6.1 Sounds that Can Damage Hearing

Some sounds are physically so weak that they are not heard. Other sounds are audible but have no temporary or permanent aftereffects. Some sounds are strong enough to produce a temporary hearing loss. Sounds that are sufficiently strong or long lasting, and that involve certain frequencies, can damage one's hearing.

The exact distinctions among these sounds cannot be stated simply, because not all people respond to sound in the same manner. Yet, in general, about 85 dBA of sound level is potentially hazardous. (The unit dBA refers to A-weighted decibels, which describes the relative loudness of sounds in air that is perceived by the human ear.) The particular hazard depends on the actual frequency spectrum and duration of the sound. Most environmental sounds include a wide band of frequencies above and below the range from 20 Hz to 20 kHz that humans can hear.

"A" FILTER

The differences in human sensitivity to tones of different frequencies are imitated by filters that are applied to sound-measuring equipment. These filters, today often in the form of software, "correct" the physical readings to what the human perceives. Different filters are identified by the first letters of the alphabet. The "A" filter is most often used because it corresponds best to the human hearing response at 40 dB. A-corrected SPL values are identified by the notation dBA or dB(A).

The level, frequency, and duration of exposure to a sound are critical for determining whether the sounds can damage hearing. It appears that sound levels below 75 dBA do not produce permanent hearing loss, even at about 4000 Hz, a frequency to which people are particularly sensitive. At higher intensities, however, the amount of hearing loss is directly related to the sound level (for comparable durations). In the United States, current OSHA regulations allow 16 hours of exposure to 85 dBA, 8 hours to 90 dBA, 4 hours to 95 dBA, etc. In Europe, 8 hours at 90 dBA are also allowed, but 4 hours at 93 dBA or 16 hours at 87 dBA: the energy trade-off is 3 dBA for 4 hours. If the sound level is about 140 dB, damage does not follow the simple energy concept; apparently, impulse noise above that level generates an acoustic trauma from which the ear cannot recover.

Simple subjective experiences can indicate whether one is exposed to a hazardous sound: for example, a sound that is appreciably louder than conversational level, a sound that makes it difficult to communicate, experiencing that sounds are muffled after leaving a noisy area, or a perception of ringing in the ear (*tinnitus*) after having been exposed to a noisy environment. (Unremitting tinnitus may be due to inner ear damage caused by long term exposure to loud sounds.)

5.4.6.2 Individual Susceptibility to NIHL

In young children, there is little difference in hearing thresholds between girls and boys. Between ages 10 and 20, males begin to show reduced high-frequency auditory sensitivity. Women continue to have better hearing than men into advanced age. These differences may be due to typically greater lifetime noise exposure for males, and not to any inherent sex-linked susceptibility to hearing loss. There is also a broad range of individual differences and sensitivities to a given noise exposure. The biological reasons for this are unknown but, for example, a TTS and/or PTS in response to a given noise may differ as much as 50 dB among individuals. It is suspected that the anatomy and mechanical characteristics of the individual ears play a role, as may the use of ototoxic drugs or previous exposure to noise.[34]

5.4.6.3 Means to Prevent NIHL

Exposure to excessive sound in both level and duration contributes to the risk of NIHL. Common sources of noise implicated in NIHL are guns, power tools, heavy equipment, airplanes, farm vehicles, firecrackers, automobile and motorcycle races, music (whether heard live or through loudspeakers or headphones), and many occupational environments. NIHL can occur whether one likes the sound or not. There are three major approaches to countering the damaging effects of excessive noise.

1. Avoid generation: the first, fundamental, and most successful strategy is to reduce or avoid the generation of sound by properly designing machine parts such as gears or bearings, reducing rotational velocities, changing the flow of air, or replacing

a noisy apparatus with a quieter one. "Active countermeasures" involve using sounds with the same frequency and amplitude but in the direction opposite (180 degrees off phase) to the noise source, physically erasing the source noise. Currently, this works best at frequencies below 1 kHz.

2. Impede transmission: the second strategy is to impede the transmission of sound from the source to the listener. In occupational environments, one might try to put mufflers on the exhaust side of a machine, encapsulate the noise source, put sound-absorbing surfaces in the path of the sound, or physically increase the distance between source and ear.

3. Leave the area: the third strategy is to remove humans from noisy places altogether, at least for parts of the work shift.

5.4.6.4 Hearing Protection Devices and NIHL

HPDs

The last resource for protecting the hearing (in fact, a sub-classification of the second strategy) is to wear a *hearing protection device* (HPD) that reduces the harmful or annoying subjective effects of sounds. HPDs are either worn externally (sound-isolating helmets or muffs, generally called caps) or inserted into the ear canal (plugs). These hearing protectors are variously effective, depending partly on the intensity and the frequency spectrum of the sound arriving at the ear and partly on the fit of the protector to the wearer's head or ear. An inappropriate initial fit, loosening of the device during activity, and outright failure to wear the equipment reduces its effectiveness.

Unfortunately, most commercially available HPDs cannot differentiate and selectively pass speech versus noise energy. Therefore, the devices do not directly improve the signal-to-noise relation. However, they can occasionally improve intelligibility in intense noise by lowering the total energy of both speech and noise that is incident on the ear, thereby reducing distortion due to overload in the cochlea. An HPD has little or no degrading effect on intelligibility in noise above about 80 dBA, though an HPD can cause considerable misunderstanding at lower levels (at which protection usually is not needed anyway). Some of the negative effects of the device may be due to the tendency to lower one's own voice because the bone-conductive voice feedback inside the head is amplified by the presence of the protector, mostly at low frequencies. Therefore, one's own voice is perceived as louder in relation to the noise than is actually the case, often resulting in a compensatory lowering of the voice by 2–4 dB. Thus, one should make a conscientious effort to speak louder when one wears an HPD.[35]

Nonverbal signals, such as warning sounds or sounds of machinery, are also affected by the wearing of an HPD. Signals above 2000 Hz are most likely to be missed, due to the high-frequency properties of conventional HPDs, particularly if the person wearing one has impaired hearing at these or higher frequencies. This indicates that warning signals should be specifically designed to penetrate the device—for example, by using low frequencies (below 500 Hz). Such frequencies diffract easily around barriers, which is a positive side effect. Yet, for a person with normal hearing, wearing an HPD does not usually compromise the detection of signals. Altogether, the wearing of an HPD is highly advisable if the ambient noise arriving at the human ear cannot be lowered otherwise.

TABLE 5.8 Airborne High-Frequency and Ultrasonic Limits

One-third octave band center frequency (kHz)	One-third octave band level (dB)
10	80
12.5	80
16	80
20	105
25	110
31.5	115
40	115

Source: Adapted from NASA (1989)

Audible noise with a constant sound level of 85 dBA or greater is hazardous. If humans must be subjected to such noise, HPDs are necessary. People should not be exposed to continuous noise levels exceeding a defined level in overall SPL under any circumstances. The OSHA limit is 115 dBA rms in the United States; other countries have different regulations.

5.4.6.5 *Infrasound and Ultrasound*

Below and above the regular hearing limits, sound levels, although inaudible, may still have vibrational effects on the human body. To protect the human, the following intervention strategies can be adopted:

- Infrasound pressure levels shall be less than 120 dB in the frequency range of 1–16 Hz, for 24-hour exposure.[36] To achieve this goal, well-fitted ear plugs provide attenuation at frequencies below 20 Hz similar to that in the 125-Hz band; in contrast, earmuffs are not effective.[37]
- Hearing conservation measures shall be initiated when the ultrasonic criteria listed in Table 5.8 are exceeded. Earmuffs and plugs provide protection with attenuation of at least 30 dB at frequencies between 10 and 30 kHz.[38]

5.4.7 Pychophysics of Hearing

While physical measurements can explain acoustical events, people interpret and react to them in very subjective ways—for example, whether certain sounds are interpreted as attractive or noisy. The sensation of a tone or complex sound depends not only on its intensity and frequency, but also on how we feel about it.

The subjective experience of the combined frequency and intensity of a sound is called *loudness.* Compared to the intensity at 1000 Hz, at lower frequencies the SPL must be increased to generate the feeling of "equal loudness." For example, the intensity of a 50-Hz tone must be nearly 100 dB to sound as loud as a 1000-Hz tone with about 60 dB. However, at frequencies in the range of approximately 2000–6000 Hz, the intensity can be lowered and still sound as loud as at 1000 Hz. Above about 8000 Hz, the intensity must be increased again above the level at 1000 Hz to sound equally loud.

Loudness

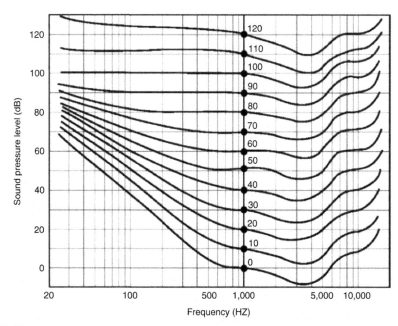

FIGURE 5.17 Curves of equally perceived loudness ("phon curves") are in congruence with the sound pressure level at 1000 Hz.

"Equal-loudness contours" (called *phon curves*) are available[39] (see Fig. 5.17). The actual shapes of the curves are slightly different if the stimulus sounds are presented either by loudspeaker or by earphone.

> To be heard, a low-frequency sound must be louder than a high-frequency sound.

Pitch and timbre

Perceptions of equal loudness indicate that there are nonlinear relationships between *pitch* (the perception of frequency) and *loudness* (the perception of intensity). *Timbre* (tone quality) is even more complex because it depends on changes in frequency and intensity over time.

5.4.8 Voice Communications and Intelligibility

Intelligibility

The ability to understand the meanings of words, phrases, sentences, and entire speeches is called *intelligibility*. This is a psychological process that depends on acoustical conditions. For the satisfactory communication of most voice messages over noise, at least 75% intelligibility is required. Direct face-to-face communication provides visual cues that enhance intelligibility of speech, even in the presence of background noise. Indirect voice communications lack the visual cues. The distance from speaker to listener, background noise level, and voice level are important considerations. The ambient air pressure and gaseous composition of the air affect the efficiency and frequency of the human voice and, consequently, of speech communication.

SPEECH-TO-NOISE "RATIO"

The intensity of a speech signal relative to the level of ambient noise is a fundamental determinant of the intelligibility of speech. The commonly used speech-to-noise ratio S/N is not a fraction but a difference; for a speech of 80 dBA in noise of 70 dBA, the S/N is simply +10 dB. With an S/N of +10 dB or higher, people with normal hearing should understand at least 80% of spoken words in typical broadband noise. As the S/N falls, intelligibility drops to about 70% at 5 dB, to 50% at 0 dB, and to 25% at –5 dB. People with NIHL may experience even larger reductions in intelligibility, while those used to talking in noise do better.

5.4.8.1 Speech Intelligibility

Speech communication is highly dependent on the frequencies and sound energies of the interfering noise. Therefore, for predicting intelligibility. one should use a narrowband instead of a broadband dosimeter to measure the intensities of both the signal and noise sounds. Several techniques predict speech intelligibility on the basis of narrowband measurements.

The articulation index (AI) is an often-used metric for assessing intelligibility. It requires that noise levels in the centers of 15 one-third-octave bands, which range from 200 to 5000 Hz, be measured in the work environment and compared with speech peaks in the same bands. The noise level in each band is subtracted from the speech level. The differences are weighted, with the highest weight given to the most critical voice frequencies. The weighted differences are then summed to produce a single AI value. This technique is described in Table 5.9. Excellent to very good intelligibility can be expected with 1 < AI < 0.7, good intelligibility with 0.7 < AI < 0.5, acceptable with 0.5 < AI < 0.3, marginal or unacceptable with AI < 0.3.[40] Fig. 5.18 facilitates comparison between AI and other measures of intelligibility.

Articulation index

The AI is a relatively complete, but fairly complex means of predicting intelligibility. A more convenient alternative is the speech interference level (SIL) which requires less complex instrumentation and fewer data points. The SIL is simply the average of the decibel values of the existing noise levels of the octave bands 600–1200, 1200–2400, and 2400–4800 Hz.

Speech interference level

The SIL is now often used in a slightly modified form, called the preferred SIL (PSIL). The PSIL is computed as the arithmetic average of the three noise measurements taken in the octave bands centered at 500, 1000, and 2000 Hz. The higher the PSIL, the poorer the communication; for a typical distance between speaker and listener of 1 meter, a PSIL of 80 or above would indicate difficulties in communication. The PSIL should not be used if the noise has powerful low- or high-frequency components.

Preferred speech interference level

TABLE 5.9 Sample Calculation of the Articulation Index (AI)

Band centers	Observed speech peaks minus noise (dB)	Weight factor	Result
200	30	0.0004	0.0120
250	26	0.0010	0.0260
315	27	0.0010	0.0270
400	28	0.0014	0.0392
500	26	0.0014	0.0364
630	22	0.0020	0.0440
800	16	0.0020	0.0320
1000	8	0.0024	0.0192
1250	3	0.0030	0.0090
1600	0	0.0037	0.0000
2000	0	0.0038	0.0000
2500	12	0.0034	0.0408
3150	22	0.0034	0.0748
4000	26	0.0024	0.0624
5000	25	0.00200	0.0500
			AI = 0.473

Source: Adapted from Sanders and McCormick (1987).

Preferred noise criteria

Similar to, or even simpler than, the SIL or PSIL are the preferred noise criteria (PNC) curves, which also use an octave-band analysis of noise. The highest PNC curve penetrated by the noise spectrum is the value indicative of a given situation. Rooms in power plants, for example, have recommended PNCs of 50–60, while offices should be at 30–40.

5.4.8.2 Shouting in Noise

People have a tendency to raise their voices to speak over noise, and to return to normal when the noise subsides. (This is called the *Lombard reflex* although it is more of a conditioned response than a reflex.) In a quiet environment, males normally produce about 58 dBA, when using a loud voice 76 dBA, and, when shouting 89 dBA. Women normally have a voice intensity that is 2 or 3 dBA less at lower efforts and 5–7 dBA less at higher efforts. Thus, people can fairly easily increase the S/N by raising their voices at low noise levels, but the ability to compensate lessens as the noise increases. Above about 70 dBA, raising one's voice becomes inefficient, and it is insufficient at 85 dBA or higher. Furthermore, this forced effort of a shouting voice often decreases intelligibility because articulation becomes distorted at the extremes of voice output. The S/N is a rough estimate for predicting the effectiveness of communication, but the loss of intelligibility by "masking" through noise depends not only on the intensity, but also on the frequency, of both the signal and the noise. In general, at small S/N differences, low-frequency noise causes more speech degradation than high-frequency noise does.

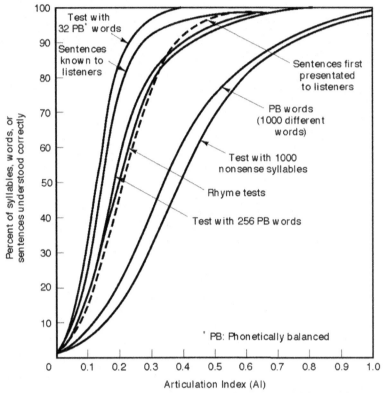

FIGURE 5.18 Articulation index and speech intelligibility. These relations are approximate and depend upon the type of spoken material and the skill of talkers and listeners. *Source: Adapted from NASA (1989).*

5.4.8.3 Masking and Filtering of Speech

The frequencies in voice communications range from about 200 to 8000 Hz, with the range of about 1–3 kHz most important for intelligibility. Men use more low-frequency energy than women do. Intelligibility is barely affected by either filtering or masking frequencies below 600 or above 3000 Hz, but interfering with voice frequencies between 1000 and 3000 Hz drastically reduces intelligibility.

Speech frequencies

In speech (as in written text), consonants are more critical for understanding words than are vowels. Unfortunately, consonants have higher frequencies and, concurrently, generally less speech energy than vowels and therefore are more readily masked by ambient noise; hence, they are more difficult to understand, especially for older people.

Consonants

Noises of a predominantly low-frequency nature, particularly those with single-frequency components, have masking effects that spread upward in frequency, intruding upon speech bandwidths if the noise is between 60 and 100 dBA. Speech is masked when environmental sound inhibits its perception. A given frequency sound can mask signals at neighboring (especially higher) frequencies, possibly rendering them inaudible.

Masking

TABLE 5.10   Speech Interference Level (Noise) Criteria for Voice Communications

Speech interference level, SIL (dB)	Person-to-person communication
30–40	Communication in normal voice satisfactory
40–50	Communication satisfactory in normal voice at 1–2 m; need to raise voice at 2–4 m. Telephone use satisfactory to slightly difficult
50–60	Communication satisfactory in normal voice at 30–60 cm; need to raise voice at 1–2 m. Telephone use slightly difficult
60–70	Communication with raised voice satisfactory at 30–60 cm; slightly difficult at 1–2 m. Telephone use difficult. Earplugs or earmuffs can be worn with no adverse effects on communications
70–80	Communication slightly difficult with raised voice at 30–60 cm; slightly difficult with shouting at 1–2 m. Telephone use very difficult. Earplugs or earmuffs can be worn with no adverse effects on communications
80–85	Communication slightly difficult with shouting at 30–60 cm. Telephone use unsatisfactory. Earplugs or earmuffs can be worn with no adverse effects on communications

Source: Adapted from NASA (1989)

The efficiency of teams is often impaired when noise interferes with voice communication. If masking occurs, the time required to accomplish communication is increased through slower, more deliberate verbal exchanges. Not only is this annoying, but it can result in increased human error due to misunderstandings. Table 5.10 and Fig. 5.19 indicate speech-interference-level criteria for voice communications.

Filtering

For the transmission of speech—for example by telephone—frequency or amplitude filtering is often used, and the following considerations apply:

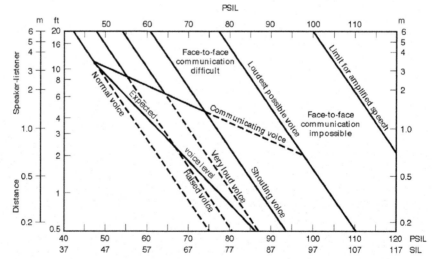

FIGURE 5.19   Face-to-face communications. *Source: Adapted from NASA (1989).*

I. THE ERGONOMIC KNOWLEDGE BASE

- Frequency clipping affects vowels if the clipping occurs below 1000 Hz, but has little effect if it is above 2000 Hz. Peak clipping is usually not critical if it is below 600 Hz or above 4000 Hz. Center clipping is highly detrimental, particularly in the range from 1000 to 3000 Hz.
- In amplitude clipping, cutting the peaks affects primarily vowels and reduces the quality of transmission in general, but is not usually a great problem. Surprisingly, peak clipping and then reamplifying improves the perception of consonants, which carry most of the message. (For example, in English, most written messages can be read and understood even if all vowels are missing.) Center clipping, in contrast, garbles the message because it affects primarily consonants.

5.4.8.4 Components of Speech Communication

Speech communication has five major components: the message itself, the speaker, the means of transmission, the environment, and the listener.

- The message itself becomes clearest if its context is expected, its wording is clear and to the point, and the ensuing actions are familiar to the listener.
- The speaker should speak slowly and use common and simple vocabulary with only a limited number of terms. Redundancy can be helpful (e.g.: "Boeing 747 jet"). Phonetically discriminable words should be used, including using the International Spelling Alphabet (shown in Table 5.11).
- The speech should be transmitted by a high-fidelity system that has little distortion in frequency, amplitude, or time.
- The environment affects communication effectiveness: consider the size of the room, its décor and lighting and temperature, how attentive or distracted or fatigued the audience is.
- The hearer's ability to understand the message is, of course, affected by noise, which can be assessed by the SIL, discussed earlier. Wearing an HPD (a plug, muff, or helmet) produces some filtering and reduces the overall sound level. Yet, it may be advisable to use special ear protection that is penetrable by specific frequencies, or one may have to adjust the signal so that it will not be masked by noise.

TABLE 5.11 International Spelling Alphabet

A	Alpha	J	Juliet	S	Sierra
B	Bravo	K	Kilo	T	Tango
C	Charlie	L	Lima	U	Uniform
D	Delta	M	Mike	V	Victor
E	Echo	N	November	W	Whiskey
F	Foxtrot	O	Oscar	X	Xray
G	Golf	P	Papa	Y	Yankee
H	Hotel	Q	Quebec	Z	Zulu
I	India	R	Romeo		

PRO-WORDS

Procedure words (*pro-words*) are specific words or word-pairs that are used in radio transmissions. They are standardized ways to communicate specific information in a condensed verbal format. Prowords based on the English language include: Acknowledge, All after, All before, All stations, Confirm, Correct, Correction, In figures, In letters, Over, Out, Radio check, Read back, Received, Repeat, Say again, Spell, Standby, Station calling, This is, Wait, Word after, Word before, Wrong.

PHONETIC NUMBERING SYSTEM

Radio communication also uses a phonetic numbering system based on English and Roman languages: Wun, Too, Tree, Fow-er, Fife, Six, Sev-en, Ait, Nin-er, Zero, Decimal; or alternately, unaone, bissotwo, terrathree, kartefour, pantafive, soxisix, setteseven, oktoeight, novenine, nadazero.

5.4.8.5 *Hearing Protection Devices and Speech*

HPDs are used in noisy environments in industry (their use is mandated by OSHA in the United States), the military, at auto races or fireworks displays, and in other circumstances where human hearing would be in danger. Earplugs are inserted into the ear canal, canal caps seal the canal near its rim, earmuffs encircle the outer ear, and helmets seal the ear via built-in muffs.

Passive HPDs

In the past, HPDs were "passive," their attenuation achieved by making sounds pass through material that absorbs, dissipates, or otherwise impedes energy flow. These conventional HPDs can be highly efficient when properly selected and correctly worn; they have little or no degrading effect on the wearer's understanding of speech and other sounds in ambient noise above about 80 dBA, but they do cause misunderstandings during quieter periods.

Conventional HPDs do not selectively pass speech versus noise at given frequencies; thus, they do not improve the S/N ratio. In fact, most passive devices are designed to attenuate high-frequency sound more than low-frequency sound, thereby reducing the power of consonants and distorting speech. People who already have suffered a hearing loss at higher frequencies therefore experience a further elevation of their hearing threshold when they use the conventional HPD.[41]

Active HPDs

New HPD designs incorporate electronics to improve communication and the reception of signals by the wearer. These "active" devices can

- provide diminished attenuation in low-level noise and increased protection during loud periods;
- reduce noise by destructive interference at selected frequency bands;
- let pass or boost desired critical bands, especially those needed for speech; and
- transmit desired signals such as those for speech, warnings, or even music via built-in loudspeakers.

Combining the desired features of active and passive devices may lead to effective HPDs for special applications and general use.

5.4.8.6 Improving Hearing

Many people, especially as they age, experience reduced hearing ability, predominantly in the higher frequency ranges. Devices to aid impaired hearing can be a great improvement toward understanding speech, especially when they utilize electronics to amplify certain bandwidths of sound, and filter out ambient noise.

Many people with impaired hearing are greatly frustrated when they try to use hearing aids, often because the device was not properly selected in the first place or is difficult to use. Nearly three-quarters of American hearing-aid users expressed dissatisfaction due to ergonomic issues[42]:

- Having to remove the hearing aid to use the telephone
- Not being able to maintain the device at the proper volume
- Not being alerted to the battery running low
- Having trouble changing the battery
- Experiencing discomfort while wearing the hearing aid
- Hearing whistling from the device at certain volumes

5.4.8.7 Reverberation

Reflection of sound from surfaces is called *reverberation*. Reverberation time in a room is defined as the time it takes the SPL to decrease by 60 dB after the source of sound is shut off. A certain amount of reverberation is desirable, because it makes speech sound alive and natural. Too much reverberation is undesirable if reflections arrive at the same time a new word is uttered and hence interfere with its perception. A room with little or no reverberation is called acoustically "dead." In such a room, there is little interference between words and intelligibility is near 100%, but because the sound of the word decays before it can propagate through the room, verbal communication may be difficult. If the delay between reflected and original sounds is long, separate sounds (the original and its echo) are heard. Long room reverberation times can produce a bouncing or booming sound. As the reverberation time increases, intelligibility decreases in a nearly linear fashion. If the reverberation time increases beyond 6 seconds, intelligibility is cut in half. A highly reverberant room is called "live" or "hard."

5.4.8.8 Communication at Altitude or Under Water

Communication that take place at a high altitude or under water requires specific technical means. At high altitude (where the ambient pressure is low), the human voice and earphones, as well as loudspeakers, become less efficient generators of sound, and microphones become less sensitive at certain frequencies. This combination of conditions requires amplification to be incorporated into either the source, the transmitter, or the receiver of signals.   **At altitude**

Under water, hearing is limited by the reverberation of sound and noises made by movement and by an increased minimal threshold of hearing, which may be raised about 40 dB by the impedance mismatch between water and air at the ear. The difficulties can be overcome by amplifying the transmitted sound and by using a directional receiver to discriminate against sounds coming from other directions. One can talk under water, but the noise that accompanies the bubbles that are emitted masks speech,   **Under water** so that the listener hears mainly vowel sounds. If a face mask with a built-in micro-phone is not available, the diver should wait for all bubbles to die away before saying the next word and use simple vocabulary, mostly vowels.

5.4.9 Acoustic Phenomena

Directional hearing

Humans are able to tell where a sound is coming from by using the difference in arrival times (phase difference) or intensities (as a result of the inverse-square law of energy flux) to determine the direction of the sound. However, the ability to use ste-reophonic cues varies among individuals and is much reduced when earmuffs are worn.

Distance hearing

The ability to determine the distance of a source of sound is related to the fact that sound energy diminishes with the square of the distance traveled, but the human per-ception of energy depends also on the frequency of the sounds, as just discussed. Thus, a source of sound appears more distant when it is low in intensity and frequency and appears closer when it is high in intensity and frequency.

Difference and summation tones

Two tones that are sufficiently separated in frequency (so that they excite separate areas on the basilar membrane) are perceived as two distant distinct (or independent) tones. When two such tones are very loud, one may hear two supplementary tones. The more distinct tone is at the frequency difference between the two tones; the qui-eter tone is the summation of the original frequencies. For example, two original tones, at 400 and 600 Hz, generate a difference tone at 200 Hz, and a summation tone at 1000 Hz.

Common-difference tone

When several tones are separated by a common frequency interval (of 100 Hz or more), one hears an additional frequency based on the common difference. This effect explains how one may be able to hear a deep bass tone from a sound system that is physically incapable of emitting such a low tone.

Aural harmonics

One may perceive a pure sound as a complex one because the ear can generate har-monics within itself. These "subjective overtones" are more pronounced with low- than with high-frequency tones, especially if these are about 50 dB above the threshold for human hearing.

Intertone beat

If two tones differ only slightly in their frequencies, the ear hears only one fre-quency (the *intertone*) that is halfway between the frequencies of the original tones. The two tones are in phase at one moment and out of phase at the next, causing the intensity to wax and wane; thus, one hears a beat. Beating occurs at a frequency equal to the difference in frequencies of the two tones. If the beat frequency is below six per second (i.e., if the frequencies of the original tones are close together), the beat is very distinct, appearing as a variation in loudness; only the intertone is then heard. When the beat is above eight per second (i.e., with an enlarged frequency interval between the two original tones), the intertone appears to be pulsating or throbbing, and one may hear the original tones as well. When the beats occur more often than about 20 times per second (with an even larger frequency separation between the original tones), the intertone becomes faint, and the two original tones are predominant.

As the distance between the source of sound and the ear decreases, one hears an increasingly higher frequency; as the distance increases, the sound appears lower. The larger the relative velocity, the more pronounced is the shift in frequency. The Doppler effect can be used to measure the velocity at which source and receiver move against each other. `Doppler effect`

When two tones of the same frequency are played in phase, they are heard as a single tone, its loudness being the sum of the two tones. Two identical tones exactly opposite in phase cancel each other completely and cannot be heard. This physical phenomenon (called *destructive interference* or *phase cancellation*) can be used to suppress the propagation of acoustical or mechanical vibrations and, with current technology, is particularly effective at frequencies below 1000 Hz. `Concurrent tones`

5.5 SMELLING—THE OLFACTORY SENSE

Smells may be dear to us, but we have no names for them. Instead, we compare them with other smells, such as of flowers; or we describe how they make us feel, such as intoxicating, sickening, pleasing, delightful, or revolting. Some smells are fabulous when they are diluted, but truly repulsive when they are not. Odors can play major roles in eating, drinking, health, therapy, stress, religion, personal care, and relationships.[43]

5.5.1 Odor Sensors

In the upper rear part of each human nostril, several million smell receptors (bipolar olfactory neurons) are located in a patch of 4–6 cm^2 called the *olfactory epithelium*. Each neuron has a dendrite that ends in an *olfactory knob*, from which hair-like *cilia* protrude into the nasal mucus. The nasal airways are bent, so normally little airflow passes along the olfactory cilia, but the flow can be increased by sniffing. In still unknown ways, certain molecules trigger the sensors, which then send signals directly to the olfactory bulbs of the brain.

Also distributed throughout the mucus of the nasal cavity (and parts of the oral cavity) are free endings of the fifth (trigeminal) cranial nerve, which are connected to a different region of the brain. The trigeminal receptors provide the so-called common chemical sense, and are triggered by, among other odorants, substances that generate irritating, tickling, and burning sensations, which then initiate protective reflexes, such as sneezing, or interrupted breathing.[44] In high enough concentration most, if not all, odorants stimulate both the olfactory and common chemical sensors.

5.5.2 Odorants

Not all odorants are external: the body also generates its own smells. Body odor arises from the apocrine glands, which are small in infants, but develop during puberty. Most of these glands are at the armpits, face, chest, genitals, and anus.

Odorants are chemical substances and can be analyzed by chemical methods. Odors are sensations and must be assessed by measuring human responses to them.

If the physical and chemical determinants of odor were fully understood, it would be possible to predict the sensory properties of odorous materials from their chemical analysis—in practical terms, one could construct an "odor meter" analogous to a decibel meter for sound. Neither such an understanding nor any such device is yet at hand. Nonetheless, various instrumental and sensory methods of measurement have been developed and have been applied to sources of odor and to the ambient atmosphere. Many of the available techniques are costly and time consuming, and not all have been validated.

Many atmospheric contaminants are odorless or very nearly so; carbon monoxide is a notorious example. By contrast, many other substances are readily detectable even in minute concentrations. For example, an organic sulfur compound at a concentration of one molecule per billion molecules of air is likely to be smelled.

Most existing odorants are mixtures of basic components, generating complex odors to which different people, in different environments and at different lengths of exposure, react quite differently. Accordingly, various theories have been proposed to explain the sensitivity of our sense of smell, among them John Amoore's (1970) stereochemical model, which assumes that the dimensions and shapes of certain odor molecules must fit certain receptor sites to generate the sensation of an odor.

While we use the olfactory sense daily, little is known systematically about it, partly because a smell is not easily quantified. The usual test procedure is to present the substance that has an odor in varying concentrations; the threshold concentration is found when it provokes a sensation in 50% of the cases. The sensation (the "smell") is described in relation to other smells; it may be considered pleasant or unpleasant. The assessment of "odor annoyance" has been attempted in a manner similar to that of noise classification.[45]

5.5.3 Describing Qualities of Odor

Several systems to describe and distinguish the qualities of odor have been used in the past. The classic Linacus-Zwaardemarker nine-category system has been largely displaced by either of two other procedures:

1 The Crocker-Henderson approach ranks every smell in four categories along a nine-point scale. The categories are fragrant, acid, burnt, and caprylic ("soapy").

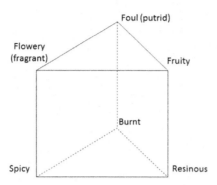

FIGURE 5.20   Hans Henning's 1916 smell prism.

2 Henning's "smell prism" (see Fig. 5.20) employs six primary odors (flowery-fragrant, fruity, foul-putrid-rotten, spicy, resinous, and burnt) arranged in the form of a prism. Stimuli that typically evoke olfactory responses in these categories are violets, oranges, rotten eggs, cloves, balsam, and tar, respectively.

5.5.4 Effects of Odors

The effects of odors can be physiological and independent of the actual perception, or perceived, psychologic, and even psychogenic. *Physiological effects* may stimulate the CNS (especially the hypothalamus and the pituitary gland), eliciting changes in body temperature, appetite, arousal and other physiologic reactions, by modifying olfaction and by triggering potentially harmful reflexes. Unfavorable responses include nausea, vomiting, and headache; shallow breathing and coughing; problems with sleep, stomach, and the appetite; irritation of the eyes, nose, and throat; destruction of the sense of well-being and the enjoyment of food, home, and the external environment; feelings of disturbance, annoyance; or depression. Exposure to some odorous substances may also lead to a decrease in heart rate, constriction of the blood vessels of the skin and muscles, the release of epinephrine, and even alterations in the size and condition of cells in the olfactory bulbs of the brain.[46]

A reduced sense of well-being among multiple coworkers in an office accompanied by consistent symptom resolution when these people are away from the building or office area has been referred to as *sick building syndrome* (SBS). No specific medical tests exist to diagnose SBS, and some of the symptoms (scratchy throats, nausea, respiratory irritation, irritated skin, headaches, fatigue, stuffy or runny noses) are subjective. Odors may contribute to this syndrome, including the presence of mold or mildew, the use of certain cleaning agents, formaldehyde, styrene (a sweet-smelling benzene derivative) or VOCs (volatile organic compounds) in carpeting or fabrics, or even ozone emissions from office equipment, especially in nonventilated or well-insulated work spaces. Indeed, both odorific and nonodorous substances can have effects on the health and well-being of workers.

SBS

5.5.5 Utility of Odors

> The fact that only minute quantities of some stimulating agents are required to bring about a sensation of smell helps the designer who wishes to use smell as means of engineering. For example, the addition of methyl mercaptan in the amount of only 25×10^{-6} gram to each liter of natural gas suffices to render the gas detectable by smell. Similarly, the addition of 3–10 ppm pyridine to argon, another inert gas, endow the argon with an odor.[47] Increasing the concentration of the stimulant to 10 or 15 times the threshold evokes the maximal intensity of smell. Another application is spraying sweetener in the ambient air: if the wearer of a respiratory mask or hood smells the sweetener, the seal on the mask is proven inadequate.

The sense of smell for engineering purposes has some utility: for example, smell can penetrate vast and complex areas, such as underground mines, where a "stench" can signal an emergency evacuation. But the olfactory sense easily adapts and is quite different from person to person.[48] Further, smells change with concentration and time, and many odors can be easily masked by other odors. Interestingly, most children and men cannot smell exaltolide, while many women find it very strong; but sensation varies with the menstrual cycle. Smoking and aging also affect one's smelling capabilities. A blockage of the nasal passages—for example, from a cold—can temporarily eliminate the ability to smell.

The magnitude of the human sensory response to odor (i.e., the perceived intensity of the odor) decreases as the concentration of the odorant gets smaller. (This effect is used to control indoor odors by ventilation or outdoor odors by the use of odor-diluting tall stacks.) However, the relationship between the intensity and concentration of an odorant is not a direct proportionality. When odorous air is diluted with odor-free air, the perceived odor decreases less sharply than the concentration. For example, a ten-fold reduction in the concentration of amyl butyrate in air is needed to reduce its perceived odor intensity by half. Nor do all odorants respond by the same ratios: some, like amyl butyrate, show sluggish changes in odor intensity with changes in concentration, while others change more sharply.[49,50]

> As noted by Diane Ackerman, "How we delight our senses varies greatly from culture to culture .... Masai women, who use excrement [cattle dung] as a hair dressing, would find American women's wishing to scent their breath with peppermint equally bizarre."[51]

5.6 TASTING—THE GUSTATION SENSE

Like the sense of smell, the sense of taste is also not thoroughly understood. While we can detect over 10,000 different odors, we only have a very small number of basic taste categories. A human can readily distinguish primary qualities (such as categories of taste) but cannot distinguish quantitative differences so easily. We are much more sensitive to odors than to taste. The sense of taste interacts strongly with sensations of smell, temperature, and texture, all of which are present in the mouth. Pepper, for example, tastes as it does because it stimulates several types of receptors. The impact of smell on our taste perception is achieved by two routes: *orthonasal* smell is the process of sniffing through the nose, while *retronasal* smell is the aroma released through the back of the mouth to the nose when we chew and swallow. Food becomes almost tasteless if a cold causes temporary loss of the sense of smell largely because of the inhibition of retronasal smell.[52]

5.6.1 Taste Sensors

Nearly ten thousand taste buds, the receptors for taste qualities, are located mostly on the human tongue but are also found at the palate, pharynx, and tonsils. The taste buds (really a budlike collection of cells), which are continuously replaced every two weeks, are arranged in clusters. It appears that some taste buds react only to one stimulus of our four taste qualities, while others respond to several or all of the stimuli. The tip of the tongue is particularly sensitive to sweet, the sides to sour, the back to bitter, and all to salty stimuli. The number of taste buds seems to diminish with aging after the middle forties, when the remaining buds may also atrophy. Taste sensitivity differs from person to person and decreases with age. Some people cannot taste certain substances; for example, 25%–30% of people cannot perceive the bitter chemical 6-*n*-propylthiouracil (PROP for short). These individuals are sometimes referred to as "non-tasters", while "super-tasters" (another 25%–30% of the population) are extremely sensitive to PROP.

"Super-tasters" have many more taste papillae than "tasters" or "non-tasters" of PROP. They are much more sensitive to bitter tastes and somewhat more sensitive to sweet, salty, and unami tastes. They tend to not like hot/spicy foods (due to enhanced pain reception), many vegetables (due to enhanced bitter perception), or alcohol and tobacco. Supertasters also have a reduced preference for sweet, fatty foods, and tend to have lower BMIs and better cardiovascular profiles.[53]

To taste a substance, it must be soluble in the saliva. Taste sensitivity depends on several interactive variables, including the nature of the stimulus, its concentration, its location of activity, and its time of application. Furthermore, the sensitivity depends on the previous state of adaptation to the taste in question, on the chemical condition of the saliva, on temperature, and on other variables.

5.6.2 Taste Stimuli and Qualities

Our sensations of taste qualities are evoked by specific stimuli: sodium chloride (salty), acid hydrogen ions (sour), nitrogenalkaloids (bitter), inorganic carbon (sweet), and glutamate (umami). The first four taste qualities were believed to describe all that which humans could taste until the early 1900s when the specific taste of umami was identified. The Japanese term *umami* (a savory taste) was coined in 1908 by the chemist Kikunae Ikeda. People taste umami through receptors specific to glutamate, which is present in savory foods such as meat broths and fermented products. Since umami has its own receptors rather than arising out of a combination of the traditionally recognized taste receptors, umami is one of the five basic tastes. It can be described as a pleasant "brothy" or "meaty" taste with a long-lasting, mouthwatering and coating sensation over the tongue. Its effect is to balance taste and round out the overall flavor of a dish.

In 2015, Richard Mattes proposed a sixth taste, *oleogustus*, the sensation of fat, which may be provoked by nonesterified fatty acid stimuli.

Taste, smell, and flavor are distinct constructs. The sense of taste is genetic and present soon after birth, while recognizing smells is a learned experience. Flavor is the combination of taste and smell. Our taste sense allows us to detect nonvolatile molecules that we cannot smell but which we need for survival. Sweet tasting foods provide rapid sources of energy and signal both insulin release and eventual satiety. Unami allows us to detect essential amino acids, while the detection of salt in our foods permits the regulation of fluid in the body. Sour and very bitter tastes signal spoiled or toxic foods to be avoided.

One example of the complexity of taste is how wine is described: smelling the aromas in wine is the primary sensation for "tasting" it. Wine served at warmer temperature is more aromatic than when cooler because warmth increases the volatility of aromatic compounds; as does whirling, or aerating, which increase the available surface and with it the rate at which aroma molecules volatilize. When wine is sipped, it is warmed in the mouth and mixes with saliva to vaporize the volatile aroma compounds. These are then inhaled through the back of the mouth, where we have about five million nerve "smell" cells. Some trained humans can distinguish among many smells but even they can usually only name a handful; however, a list of possible choices helps to identify aromas. Individuals have varying sensitivities and their own ways of describing scents. This is why one taster may recount different aromas and flavors than another taster sampling the same wine.[54]

5.7 TOUCHING—THE CUTANEOUS SENSES

5.7.1 Touch Sensors and Stimuli

The sensory capabilities located in the skin are called *cutaneous* (from *cutis*, Latin for skin) or *somesthetic* (from *soma*, Greek for body). They are commonly divided into four groups:

- The *mechanoreceptors*, which sense taction (contact or touch), tickling, pressure, and related commonly understood, but theoretically ill-defined, stimuli.

- The *thermoreceptors*, which sense warmth or cold, relative to each other and to the body's neutral temperature.
- The *electroreceptors*, which respond to electrical stimulation of the skin. (It is disputed whether such specific sensors exist.)
- The *nocireceptors* (from the Latin *nocere*, to damage), which sense pain. We feel what is commonly but imprecisely called sharp or piercing pain, usually associated with events on the surface, and dull or numbing pain, usually felt deeper in the body. (The point is controversial, because some researchers state that there are no specific pain sensors, but that pain is transmitted from other sense organs.)

The study and engineering use of the cutaneous senses is hampered by many uncertainties. First, stimuli are often not well defined, particularly in older research. Second, sensors are located in different densities over the body. Third, the functioning of sensors is not exactly understood. Many sensors react to two or more distinct stimulations simultaneously and produce similar outputs, and it is often not clear whether or which specific sensors respond to a given stimulus. Fourth, the pathways of signal conduction to the CNS are complex (as discussed earlier) and may be joined by other afferent paths from different regions of the body. Finally, arriving signals are interpreted in unknown manners at the CNS. Given all these uncertainties, it is obvious that many of our engineering applications are based mostly on everyday experience and rely on very limited experimental data. Much more research needs to be performed to provide the ergonomist with complete, reliable, and relevant information on the cutaneous senses for the design of systems.

5.7.2 Sensing Taction

The taction sense reacts to touch at the skin. The term *tactile* is often used if the stimulus is received solely through the skin, while the term *haptic* is applied when information is obtained simultaneously through cutaneous and kinesthetic senses—that is, through the skin and through proprioceptors in muscles, tendons, and joints. Most of our everyday touch perception is actually haptic perception.

5.7.2.1 Taction Sensors and Stimuli

Ernst Heinrich Weber demonstrated in 1826 that skin sensors react to the location of the stimulus, and to the stimuli of force or pressure, warmth, and cold. (Pain and other more diffuse feelings were originally not considered part of the tactile sense, but are nowadays included insofar as they respond to stimuli on the skin.) In spite of its everyday use and of much research, the sense of taction is not fully understood. Taction stimuli often are not well defined: what is the relation between pressure, force, and touch? Which sensors respond to each stimulus in what way? Do the sensors respond singly or in groups? If in groups, in what patterns do they respond? Does one sensor respond to only one stimulus or to several stimuli?

Questions exist regarding the association of specific nervous functions with the stimulation of specific sensors and how the two interact to provide a signal to the brain. Often, one postulates that specialized nerve endings are present, one for each sensation, and that nerves are connected to specific centers in the brain. However, pattern theory assumes that the particular experience of a sensation depends on the coaction of several separate and elementary nervous events, but not all need to be present in each pattern.[55]

5.7.2.2 Architecture of the Taction System

All tactile sensors are triggered to discharge signals by a stimulus of the appropriate type and intensity. The signal that is generated varies in frequency and amplitude and travels toward the CNS. Several types of sensors have been identified. The most common is a *free nerve ending*, a proliferation of a nerve that distally dwindles in size and then disappears. Thousands of such tiny fibers extend through the layers of skin. They respond particularly to mechanical displacements and are very sensitive near hair follicles, where *basket endings* surround hair bulbs and respond to displacement of the hair shaft.

In smooth and hairless (*glabrous*) skin, encapsulated receptors are also common. Among these are *Meissner* and similar *Merkle corpuscles*, which are particularly numerous in the ridges of the fingertips. The receptors transmit transient electrochemical surges to the nerves in response to ambient pressures. A single Meissner corpuscle may connect with up to nine separate nerves, which may also branch to other corpuscles. This is an example of simultaneous convergence and diversion of the neural pathways. (How this arrangement can reliably code neural signals is not yet understood).

The *Pacinian corpuscle* is an encapsulated nerve ending of a single, dedicated nerve fiber. These highly responsive tactile receptors are located in profusion in the palmar sides of the hand and fingers and in distal joints. They are also prevalent near blood vessels, at lymph nodes, and in joint capsules. *Krause end bulbs* are particularly sensitive to cold, but probably respond to other stimuli as well.

Nerves from the various receptors are colligated in peripheral cutaneous nerve bundles, which proceed with their neighbors to the dorsal roots of the spinal cord. Fibers originating at the same receptor may pass on to separate dorsal roots (i.e., dermatomal segments). It is not yet understood how such a complex nervous signal pattern is interpreted by the brain.

Nerve fibers have been differentiated according to their conduction velocities. Conduction speeds in the human peripheral nerves range from 1 to 120 m/s. In general, conduction velocity is greater with higher a fiber threshold, a thicker fiber, and the presence of myelin sheathing.

5.7.2.3 Tactile Sensor Stimulation

Classic experiments concerned a subject's ability to perceive the presence of a stimulus on the skin, to locate one stimulus or two simultaneous stimuli, and to distinguish between one stimulus or two stimuli applied at the same time. While earlier research procedures were highly individual, modern research uses mostly three classes of stimulation:

1. Step functions, in which a displacement is produced quickly and held for a period of approximately a second.
2. Impulse functions, in which a transient of some given waveform is produced for a few milliseconds.
3. Periodic functions, which displace the skin at constant or variable frequencies for several milliseconds.

These forms of mechanical energy can be imparted to the skin by different transducers. Most research has used skin displacement, but other experiments rely on units of force or on measures of energy transmitted. (Again, much work remains to be done.) A compilation of "classic" tactile sensitivity of body parts has been published.[56] Others carefully measured the tactile sensibility of the surface of the inner hand and found the thumb region most sensitive, the palm of the hand average, and the finger region least responsive.[57]

5.7.3 Sensing Temperature

5.7.3.1 *Temperature Sensors and Stimuli*

While there is no question that temperature can be sensed, there is no agreement on what its sensors are in the body, including whether the previously mentioned Krause end bulbs are involved. Research is hampered by the conventional experimental procedure of applying pointed metallic cylinders of different temperatures (*thermodes*) to the skin, a technique that confoundingly generates both touch and temperature sensations. Cooling or warming by air convection or radiation may be better ways to stimulate the receptors involved. An additional complication arises from the fact that temperature sensations are relative and adaptive.

Objects at skin temperature are judged as neutral or indifferent, a value of temperature called *physiological zero*. A temperature below this level is called cool, a temperature above, warm. Slowly warming or cooling the skin near physiological zero may not elicit a change in sensation. This range in which no change in sensations occurs even though the temperature varies is called the *zone of neutrality*. For the forearm, the neutrality zone ranges approximately from 31–36 degrees C. Neutrality zones are different for different body parts.

5.7.3.2 *Cold and Warm Sensations*

Apparently, some nerve sensors respond specifically to cold and falling temperatures, while others react to heat and increasing temperatures. The two scales may overlap, which can lead to paradoxical or contradictory information. For example, spots on the skin that consistently register "cold" when stimulated at less than physiological zero, may also report cold when they are stimulated by a warm thermode of about 45C. An opposite paradoxical sensation of warmth has also been observed. Further, the sensation of warmth can be aroused, in some instances, by applying a pattern of alternating warm and cold stimulation. This occasionally generates the sensation of heat even if a cold probe is applied.

Changes toward warm temperatures from physiological neutral are more easily sensed than changes toward cold, at a ratio of about 1.6 to 1. The longer the stimulus is applied and the larger the area of skin to which it is transmitted, the smaller the temperature change that can be discerned. Warm sensations adapt within a short period of time, except at rather high temperature levels. Adaptation to cold is slower and does not seem to occur completely, perhaps because both vasodilation and sweating are responses to a rise in skin temperature, while vasoconstriction is the only countermeasure to cold. Rapid cooling often causes an "overshoot" phenomenon; that is, for a short time one feels colder than one physically is.

Chapter 6 discusses, in detail, interactions of the body with the environments.

5.7.4 Sensing Pain

Touch, pressure, electricity, warmth, and cold can arouse unpleasant, burning, itching, or painful sensations. Cutaneous pain is transmitted through two types of nerve fibers. A-delta fibers relay sharp stabbing-type pains, while C fibers relay duller or burning pains. All pain impulses are relayed to the spinal cord from which specialized neurons transmit signals to the brain, primarily to the thalamus.

> Per researchers Sherrick and Cholewiak, "the full array of devices and bodily loci employed in the study of pain would bring a smile to the lips of the Marquis De Sade and a shudder of anticipation to the Graf von Sacher-Masoch."[58]

Many research results are difficult to interpret because of the various levels and categories that fall under the label "pain." Pain can range from barely felt to unbearable. The threshold for pain is a highly variable quantity, probably because pain is so difficult to separate from other sensory and from emotional components. One can even adapt to pain, at least under certain circumstances and to certain stimuli. Many factors affect how pain is perceived and pain thresholds vary by individual and by occasion. Intensely concentrated activity may diminish pain perception, probably via the activation of serotonin and endorphin release. *Referred pain* indicates the displacement of the location of the pain, usually from its visceral origin to a more cutaneous location; an example is cardiac anginal pain, which may be felt in the left arm.

5.7.4.1 Pain—All in The Brain?

More than three centuries ago, René Descartes taught that pain is a purely physical phenomenon: tissue injury stimulates specific nerves that transmit an impulse to the brain, causing the mind to perceive pain. This was the medical model until 1965, when psychologist Ronald Melzak and physiologist Patrick Wall proposed their *gate-control* theory of pain. According to this theory, sensory signals must go through a gating mechanism in the dorsal horn of the spinal cord, which would either let them pass, or stop them. Melzak and Wall's most startling suggestion was that individual emotions controlled the gate. Indeed, studies have shown different pain thresholds and tolerances in individuals and groups of people: women appear to be more sensitive than men to pain (except during the last few weeks of pregnancy), extroverts have greater pain tolerance than introverts have, and training can diminish one's sensitivity to pain.

Today, it is evident that the brain is actively involved in the experience of pain. Gate-control theory accepts Descartes' view that what you feel as pain is a signal from tissue injury transmitted by nerves to the brain, yet it adds the notion that the brain controls the gateway for such a signal. In 1994, the neurosurgeon Frederick Lenz noticed that areas of the brain governing ordinary sensations could become abnormally sensitized, generating extreme pain in response to perfectly harmless sensations. According to current theory, the brain generates the experience of pain, which (together with other sensations) is conceived as a set of "neuromodules" akin to individual computer programs on a hard drive. The neuromodule is not a discrete anatomical entity, but a network linking components from many regions of the brain. It gathers inputs from sensory nerves, your memory, or your mood—and if the signals reach a certain threshold, they trigger the neuromodule. When you feel pain, it is your brain running a neuromodule that generates your personal pain experience.[59] Nociceptors are activated by tissue damage and send a signal from the peripheral nerve to the dorsal horn of the spinal cord, where neurotransmitters are released that activate other nerves that pass signals to the brain. The brain, via the thalamus, then relays the signals on to the somatosensory cortex (activating sensation), the frontal cortex (activating thinking and response) and the limbic system (activating emotion).

Nociceptive cutaneous pain is the type of pain most frequently addressed in engineering applications. However, several other types of pain can be experienced.

Neuropathic pain is felt as a burning, shooting or even electric sensation that is often described as "deep." Dysesthesia in neuropathic pain involves stimuli that are not normally painful (such as light touch or cold) being quite painful in a phenomenon termed *allodynia*. Neuropathic pain is also characterized by *hyperalgesia* (severe pain from a stimulus that would normally only cause slight discomfort) and *paresthesia* (pain without any stimulus.) Neuropathic pain can be caused by nerve damage or inflammation, increased signal transmission at neuronal synapses, brain or spinal cord compromise, or persistent and excessive bombardment of the spinal cord from C fibers. Neuropathic pain is experienced in such conditions as shingles, diabetes, multiple sclerosis, neuralgias, spinal cord injuries and many other disease processes.

Psychogenic pain is caused or—more typically, worsened—by psychological factors as the degree of pain and disability render a vulnerable individual more dysfunctional.

Pain is also subclassified as *acute* or *chronic* pain. Acute pain is due to a particular injury or disease and serves a useful, biologic purpose (such as telling you to release a hot object). Chronic pain is pain that has persisted for 6 months after the injury or past when usual healing is completed, and "is associated with chronic pathologic processes that cause continuous or intermittent pain for months or years, that may continue in the presence or absence of demonstrable pathologies; may not be amenable to routine pain control methods; and healing may never occur."[60] Chronic pain occurs when connections between neurons are altered, when pain signals are not dampened effectively by the brain's mechanisms to do so, or when nerve damage occurs. *Chronic pain syndrome* can develop as a function of chronic pain with psychogenic contributors regardless of whether the initial injury is nociceptive or neuropathic. It is sometimes colloquially characterized by physiatrists (physicians specializing in physical medicine and rehabilitation) and psychologists as consisting of the "6 Ds" (depression, dependency, drug misuse, dramatization of complaints, dysfunction, and disability) and is a particular area of focus in pain management medicine. People with chronic pain syndrome (as opposed to chronic pain) are very unlikely to do well in a workplace environment and often become unable to work.

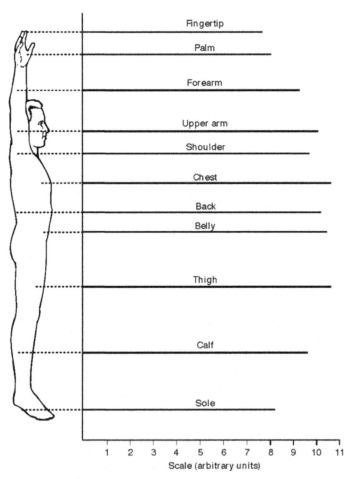

FIGURE 5.21 **Relative sensitivity of body sites to two-joint electrical stimulation.** *Source: Modified from Boff and Lincoln (1988).*

5.7.5 Sensing Electrical Stimulation

While we can feel electrical currents in the skin, it is an open question as to whether we do so through the presence of specific sensors for electricity. Currently, there is no known receptor that is specialized to sense electrical energy or that is particularly receptive to it. In fact, electricity apparently can arouse almost any sensory channel of the peripheral nervous system. Fig. 5.21 schematically shows the sensitivity to two-point electrical stimulation at various body sites.

Research has shown that the threshold for electrical stimulation depends heavily on the configuration and location of the electrodes that are used, on the waveform of the electric stimulus, the rate of stimulus repetition, and on the individual subject. Generally, the threshold is about 0.5–2 mA with a pulse of 1 ms duration. During shorter pulse durations, a temporal summation of single pulses occurs.[61]

5.8 MAINTAINING BALANCE—THE VESTIBULAR SENSE

5.8.1 Vestibular Sensors and Stimuli

Located next to the cochlea in the inner ear, on each side of the head, are three semi-circular canals with two sack-like *otolith* organs: the *utricle* and the *saccule*, shown schematically in Fig. 5.22. These nonauditory organs are called the *vestibulum*.

The arches of the three vestibular canals are at about right angles to each other, with one canal horizontal and two vertical when the head is erect. Each canal functions as a complete and independent fluid (endolymph) circuit, in spite of the fact that all of the canals share a common cavity in the utricle. Thus, the three canals are sensitive to different rotations of the head. Each canal, near its junction with the utricle, has a widening (an *ampulla*) which contains a protruding ridge (the *crista ampullaris*) which carries sensory hair-like structures (*cilia*) that respond to displacements of the endolymph. Cilia also are located in both the utricle and saccule.

5.8.2 Response to Accelerations

The vestibular system responds to the magnitude and direction of accelerations, including the acceleration due to gravity. The sack-shaped utricle and saccule are sensitive to gravity and other linear accelerations of the head. The response latency to linear acceleration is fairly long, about 3 s at 0.1 G, diminishing asymptotically to approximately 0.4 s at accelerations of 1 G or higher. The response time to rotational acceleration

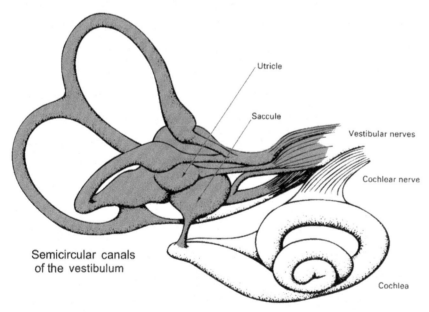

FIGURE 5.22 The cochlea, the three semicircular canals and the otolith organs of the vestibulum, with their nervous connections to the 8th cranial nerve. *Source: Adapted from a 1934 drawing by Max Broedel.*

is about 3 s at accelerations slower than 1 deg/s^2, falling to roughly 0.3 s with accelerations exceeding 5 deg/s^2. However, the system adapts to constant acceleration, and small changes in acceleration may not be perceived. Very little is known about acceleration thresholds in different planes of rotation.[62]

A number of peculiar features are associated with the pea-sized vestibular system. Bringing the head into various postures requires that the brain compare signals not only with a new spatial reference system, but also with new reference inputs from the sensors, because the endolymph now loads the cilia in different ways. Sideways rotation (roll or yaw) induces the two vestibular systems in the head to generate different base signals, which must then be consolidated in the brain.

The complex signals from the vestibulum interact with other sensory inputs arriving simultaneously at the cerebellum and cerebral cortex. Consequently, several "vestibular illusions" can occur,[63] including the following:

- Illusionary tilt: interpretation of linear acceleration as body tilt.
- Unperceived tilt: when the body is aligned with the gravitation vector, a person in an airplane that is performing a bank does not feel this roll.
- Inversion illusion: a person in zero gravity, or lying in a prone position, may feel as if being upside down.
- Elevator illusion: a change in gravitational force produces an apparent rise or lowering of seen objects.
- Coriolis crosscoupling effect: the feeling of falling sideways when the head is tilted forward while the body rotates about a vertical axis.
- Motion or space sickness: this condition is probably due to conflicting inputs from the vestibular and other sensors.

5.9 ENGINEERING USE OF SENSORY CAPABILITIES

In spite of thousands of years of everyday experience with the senses, fairly little has been done to systematically and conscientiously apply our knowledge of human sensory capabilities to design engineering. We all feel with our fingertips whether a surface is smooth or not, a welder brings his hand cautiously close to an object to find out whether it is still hot, and a blind person uses Braille to "read" text that he or she cannot see. But surprisingly little of the existing information about human sensory capabilities has been used purposefully by engineers. For example, round door knobs give no indication, by feel, in which direction they must be turned to open the door, and emergency bars that one must press to open a hinged door usually provide no cues (either for touch or for vision) as to which way the door will swing open.

While we rely heavily on our senses of vision and audition, other input senses, such as taction, olfaction, and gustation, are underused. The cutaneous senses of the hands are commonly utilized in ergonomic design, but other body segments that have the same or similar sensitivities are hardly ever employed.

The CNS receives signals from the various human senses simultaneously, allowing us to get a general picture of the events taking place within and outside the body. For example, the sensation of exerting a force on an object in order to move it (such as is felt in lifting) is presented to the CNS by muscle spindles, which report on muscle stretch (length, and change in length); by Golgi tendon organs, which report on muscle tension (the development of force); by Ruffini joint organs, which report the location of the limbs and angles of the joints; and by the cutaneous senses, because the bending of any joint stretches some regions of skin around the joint and relaxes others. Furthermore, the sense of vision provides information about the movement of object and body segments, and sounds associated with the movement supply additional information.

Redundant information: The body, and various equipment and procedures that we use, often provide redundant information to us in several sensory modalities at the same time. For example, the pilot feels acceleration through body sensors, an instrument displays the attitude of the plane, and a recorded voice tells the crew to pull up if the aircraft noses down.

The most widespread use of human tactual capabilities has been in the coding of controls by shape. In the decade after World War II, much research on controls and displays was performed. That time has affectionately been called the "knobs and dials era" by human factors engineers. Nearly all the shape-coding knowledge that we use today was derived at that time. Knobs on controls were formed like wheels or like airplane wings. Shape and size coding was used to indicate what would happen after activating these controls, with the information conveyed both by vision (if one looked at the control handle, which required sufficient illumination) and by feel (as one touched the control, which might be too late). For specific design recommendations on controls and displays, see Chapter 12.

Incomplete research findings are one explanation for the fact that so little systematic engineering use has been made of human sensory skin capacities. To provide the needed information, research must be based on solid theories regarding the various receptors, their stimulation, and their responses; regarding the screening and propagation of the signals to the CNS; and regarding the interpretation of the signals there. Theories of somesthetic sensitivity require testable models of the absorption and propagation of various forms of energy in the path from the surface of the skin to the receptors. The condition of research regarding both theoretical underpinnings and methodological procedures can still be considered to be in a "primitive state."[64]

"Knobs and dials"

ENHANCING HUMAN PERCEPTION

Just as hair in the human skin amplifies mechanical surface distortion so that the associated sensor may be more easily activated, a number of engineering means exist to make perception more intense. For taction, a thin layer of cloth between the object's surface and the fingertips enhances the perception of unevenness (probably by filtering out tactile noise). For taste and smell, the concentration of an active substance may be increased to ensure and hasten its perception. Emergency signals can simultaneously provide sound, light, color, and smell cues to enhance the speed and accuracy of the accompanying information is received, recognized, and processed.

5.9.1 Changing Sensory Modalities

It is difficult to submit sensory information that is habitually conveyed by one sense through another sense, such as transmitting visual information through taction (as is often done for blind people). The coding of the signal from one sensory system to the other is difficult, particularly if one has never experienced the first kind of sensory input. In one study, it took nine days of extensive practice for a blind person to triple his or her reading performance from an initial rate of about 10 words per minute using an optical-to-tactile converter. However, two sighted persons became highly proficient on the same converter after about 20 hours of experience, at which time they were able to read 70–100 words per minute.[65] It might be better, at least in certain circumstances, not to use a natural code that preserves the actual spatial and temporal relationships between the original character (signal) and the one displayed in the other sensory mode; perhaps an artificial code should be employed that allows better and more exact perception. Thus, in the earlier example of the blind person, instead of using Braille to convey the shape of letters, it might be advantageous to have the text spoken.

5.9.1.1 Using the Taction Sense

Given the lack of reliable experimental information, current guidelines for design applications of the sense of taction still rely much on extrapolations of previous knowledge, common-sense experiences, and guesses. On this basis, the following engineering recommendations are made, though with much caution.

Touch information, transmitted through mechano-receptors, can be differentiated by the human regarding

- the magnitude of mechanical deformation,
- the temporal rate of change, and
- the size and location of the area of the skin that is stimulated (i.e., the number of receptors stimulated).

Vibration at the fingertips shows the following minimal thresholds:

- 200 Hz with a displacement of about 2×10^{-4} mm and 800 Hz with 10^{-3} mm deformation.
- Below 10 Hz and above 1000 Hz only general pressure, not vibration, is sensed.
- The highest sensitivity for vibrations appears to be at about 250 Hz, but "flying by the seat of the pants" may be rather dangerous since sensitivity at the buttocks is very low.

One may prefer certain body areas for tactile input sites. For static two-needle point stimulation, two-point resolution starts at about 2 mm separation at the tip of the finger, at about 4 mm separation at the lips, and at about 40 mm separation at the forearm, but requires about 70 mm separation on the back. Sensitivity to any taction stimulation is highest in the facial area and at the fingertips, and fair at the forearm and lower leg. Some body areas, such as the eyes, though even more sensitive, are unacceptable for tactile input.

Tactile sensitivity is highly dependent on the strength of the stimulus, the rate change of the stimulus, and the temperature. For example, if a stimulus of low intensity appears slowly on cold skin, it may not be noticed.

Measured by step-function inputs, the minimal threshold for force varies from 5×10^{-5} N on the face to 35×10^{-5} N on the big toe. Such stimuli, if they last but 1 ms each, must be separated by at least 5.5 ms to be perceived as two stimuli at the fingertip. If the break between stimuli is too short, they fuse into the sensation of a single stimulus; this phenomenon is called *temporal fusion*.

Under normal conditions of vigilance, auditory signals are better detected than weak mechanical vibratory signals and electrocutaneous signals. In response to electrocutaneous signals, a long response latency, many misses, and false alarms occur. However, in complex environments requiring a heightened level of vigilance, electric signals will provide redundancy. Experiments indicate that the cutaneous system has utility as either a sole or an additional channel for information input.[66]

5.9.1.2 *Using the Temperature Sense*

For several reasons, our sense of temperature is difficult to use for communicative purposes: it has a relatively slow response time, a poor ability to identify location, the capability to adapt to a stimulus over time and it may integrate several stimuli that are distributed over a certain area of skin. Furthermore, interactions exist between mechanical and temperature sensations (e.g., a colder weight feels heavier on the skin than a warm weight) and thermal sensations can be stimulated chemically (for instance, by applying menthol, alcohol, or pepper to the skin) but not mechanically.

The strength of thermal sensation depends on the location and size of the sensing body surface. The temperature sensation is made stronger by increasing:

- the absolute temperature of the stimulus, and its difference from physiological zero,
- the rate of change in temperature, and
- the exposed surface (e.g., immersion of the whole body in a bath as compared to only partial immersion).

Assuming a "neutral" skin temperature of about 33°C, the following general guidelines apply for naked human skin:

- A skin temperature of 10°C appears "painfully cold"; 18°C feels "cold"; and 30°C still feels "cool." The highest sensitivity to changes in coolness exists between 18°C and 30°C.
- Heat sensors respond well throughout the range of about 20–47°C.
- Thermal adaptation—that is "physiological zero"—can be attained in the range of approximately 18–42°C, meaning that changes are not felt when the temperature difference is less than 2°C.
- The ability to distinguish between different temperatures is best in the 18–30°C range for cold sensations and best between 20°C and 47°C for warm sensation. Distinctly cold temperatures (near or below freezing) and hot temperatures above 50°C also provoke sensations of pain. At about 45°C, both cold and heat fibers are stimulated, which may result in the paradoxical sensation of cold when the

stimulus is hot. Another interaction may occur between the sensation of pressure and temperature: a force applied under cold conditions appears to be greater than when it is applied under hot conditions.

• Compared to warmth, cold is sensed more quickly, particularly in the face, chest, and abdominal areas. The body's ability to feel warmth is less distinct, but it is best in hairy parts of the skin, around the kneecaps, and at the fingers and elbows.

5.9.1.3 Using the Smell Sense

Other than for a few, specific applications, olfactory information is seldom used by engineers, because few research results are available, because people react quite differently to olfactory stimuli, because smells can be easily masked, and because olfactory stimuli are difficult to arrange. As mentioned earlier, among the few industrial applications is adding odoriferous methyl mercaptan to natural gas and pyridin to argon to allow people to smell leaking gas.

5.9.1.4 Using the Taste Sense

The sense of gustation is not used in engineering applications at present, but is of much importance to the food and beverage industry.

5.9.1.5 Using the Electrical Sense

Electricity is rarely used as an information carrier, although it has great potential for transmitting signals to the human. Attaching electrodes is convenient. The energies that are transmitted are low, requiring only about 30 μW at the electrode-skin junction, up to a tolerable limit of about 300 mW. Coding can be via placement, intensity, duration, and pulsing. Electrical stimulation can provide a clear, attention-demanding signal that is resistant to masking. Its major drawback, which it shares with mechanical stimulation, is the problem of pain: aching of deep tissues, stinging, or burning can appear if electrodes or energies are improperly applied; and there may be fear of electrical shock. The effect of electric shocks of various intensities is listed in Table 5.12.

5.9.1.6 Using the Pain Sense

Pain does not lend itself to engineering applications, primarily because one is ethically bound not cause pain. Further, damage is already done by the time the sensation

TABLE 5.12 Probable Effect of Electric Shocks of Various Intensities

Effects	Intensity of shock: 60 Hz AC	Intensity of shock: DC
Perception	< 1 mA	< 4 mA
Surprise	1–4 mA	4–15 mA
Reflex action	4–21 mA	15–18 mA
Muscular inhibition	21–40 mA	80–160 mA
Respiratory block	41–100 mA	161–300 mA
Usually fatal	> 100 mA	> 300 mA

Source: Adapted from [US] Department of Defense (2000).

of pain is registered. Pain from muscle microtrauma (caused by muscle-lengthening or *eccentric* as opposed to *concentric* exertions) can take 24 hours to fully manifest. Even pain from acute cutaneous exposure to a hot surface, or from stubbing one's toe, is not registered immediately but may take several microseconds or even seconds—which is too long to effectively abort the action resulting in pain. A-alpha nerve fibers transmit position and spatial awareness signals at 80–120 mm/s and A-beta nerve fibers transmit touch sensations at 35–75 mm/s. However, the transmission of pain is a bit slower: A-delta fibers for sharp pain conduct signals at 5–35 mm/s and C fibers for dull pain or itches conduct signals at a relative "snail's pace" of 0.5–2 mm/s.

5.9.2 Dysfunction in Sensory Processing

What if things go a little haywire? Most of this chapter has discussed the senses and sensory acuity (the physical ability of the sensory organs to receive input) but there is more to say about dysfunction in sensory processing (the ability to interpret the information the brain has received.) Acuity deficits are addressed with devices such as hearing aids and glasses. Processing deficits, on the other hand, are addressed with changes in activity, instruction, instrumentation, environment, and with adaptive practice. There are various categories of sensory malfunctions in which more than one sensory pathway is compromised in an enduring and atypical amalgamation. The visual, auditory, olfactory, gustatory, tactile, and vestibular sensory systems can have implications in the development of sensory-motor, language, perceptual-motor, cognitive, intellectual and attentional functioning. Examples of developmental malfunctions are sensory processing disorders (SPD), synesthesia, dyspraxia, Tourette's, autism-spectrum disorders, some anxiety and mood disorders, and others. Often falling within the purview of psychiatry/psychology and neurology, most, but not all, are formally recognized by the ICD-10 or the DSM-5 (see Chapter 4) if they cause significant distress and disability.

Even subclinical manifestations of a dysfunction in sensory processing can have significant implications for functioning in a social or a workplace environment and are not uncommon. Jean Ayres was the first to scientifically discuss the theory and its assessment,[67] and treatment methods for sensory integration dysfunction, now called *sensory processing disorder* (SPD). While not a formal diagnosis, SPD has a high comorbidity with attention deficit hyperactivity disorder (ADHD), autism-spectrum disorders and psychopathology, but also can exist in isolation.[68]

SPDs

Dysfunction in sensory processing can be understood along two dimensions: neurological thresholds (the way the nervous system responds to sensory input) and self-regulation strategies (the ways the individual manages the input that is available). The latter may be correlated with temperament and personality, and involves the degree of responsiveness of the individual's nervous system. Everyone has times of reacting in an overly responsive fashion (hyper-responsively) to an activity or an environment or, conversely, reacting too little (hypo-responsively). It is when a response style becomes problematic and interferes with effective functioning that it needs to be addressed. There are four basic patterns of potentially problematic responses to sensory input along the two dimensions described above:[69]

- *Sensation seeking* is the combination of high neurological thresholds and an active self-regulation system an individual. These individuals may particularly enjoy generating extra sensory input for meeting their high thresholds of neurological tolerance. They may both seek out and tolerate higher levels of activity and stimuli than others. Problems in the workplace and elsewhere can develop when the need for sensations seeking compromises attention to required activities.
- *Low registration* is the combination of high neurological thresholds and a passive self-regulation system in an individual. These individuals may not notice what is going on around them, and come across as uninterested and affectively flat. They may miss social and physical cues. People with autism or schizophrenia are significantly more likely to experience low registration at the extreme end. Corrective environmental measures include increasing the intensity of sensory input so as to make a signal for action less likely to be overlooked.
- *Sensation avoiding* is the combination of low neurological thresholds and an active self- regulation system in an individual. These individuals tend to be bothered more by sensory input than others, and take action to avoid sensory input. Unfamiliar sensory input may be overwhelming and perceived as very unpleasant. These individuals tend to do best in a highly structured, predictable, unchanging, and rule-bound environment.
- *Sensory sensitivity* is the combination of low neurological thresholds and a passive self-regulation system in an individual. These individuals may detect stimuli that others do not. They are likely to be easily distracted and easily uncomfortable. They differ from sensation avoiders in that they do not remove obstacles or leave situations as readily, and may verbally express their discontent or discomfort instead. People with autism and ADHD are significantly more likely to engage in sensory sensitivity patterns at the extreme end.

It is common for individuals to have varying responses with each sensory system. For example, one might be easily distracted by noise and yet not particularly sensitive to visual input. Alternatively, one could be sensitive to touch but tolerant of loud noise, etc. If a SPD is present, it is typically first observed in childhood and can be addressed with specialized occupational therapies. In an adult, a sensory processing dysfunction can be a residual or a continuation of a developmental dysfunction or can develop upon traumatic exposure to a strong stimulus. Some signs of a worker with SPD in the workplace (depending on the threshold/self-regulation rubric) include difficulty with busy workspaces, loud coworkers, discomfort with uniform or business attire fabric or fit, poor social interactions, exaggerated startle response, processing delays, distractibility, inattention, irritability, sensitivity to smells or food odors, clumsiness, and work avoidance.

Synesthesia
What about if sensory "wires" themselves get crossed? Some individuals have *synesthesia* (from the Greek *syn*, together, and *aisthesis*, perception), where stimulation in one sensory modality leads to an internally generated perceptual experience of another, not stimulated sensory modality. Most commonly, individuals with synesthesia perceive written words or numbers in color (*grapheme-color synesthesia*) while others associate sounds with colors (*chromesthesia*). EEG studies point to early visual processing

differences within the visual cortex of these individuals.[70] Studies have confirmed that the phenomenon is biological, automatic and apparently unlearned, and is distinct from both hallucination and metaphor. The condition runs in families and is more common among women than men. In the workplace, it should not be assumed that all individuals process sensory inputs alike. In fact, synesthesia is now considered an extreme form of multisensory processing that exists within the spectrum of normal perceptual processes.[71]

Research suggests that about one in 2000 people are synesthetes, and some experts suspect that as many as one in 300 people have some variation of the condition. The writer Vladimir Nabokov was reputedly a synesthete, as were the artist Vincent van Gogh, the composer Olivier Messiaen, and the physicist Richard Feynman. Russian mnemonist Solomon Shereshevsky, studied for decades by neuropsychologist Alexander Luria, appears to have used his natural synesthesia to memorize amazing amounts of data. The most common form of synesthesia, researchers believe, is colored hearing: sounds, music, or voices seen as colors. Some synesthetes report experiencing sensory overload, becoming exhausted from so much stimulation. But usually the condition is not a problem—indeed, most synesthetes treasure what they consider a bonus sense. "If you ask synesthetes if they'd wish to be rid of it, they almost always say no," says Simon Baron-Cohen, who studied synesthesia at the University of Cambridge. "For them, it feels like that's what normal experience is like. To have that taken away would make them feel like they were being deprived of one sense."

A final consideration of synesthesia is that most individuals are prone to some degree of crossmodal processing. This is demonstrated in the association of shape with language using the "Takete-Maluma" (also called "Kiki-Bouba") diagram. This diagram with two figures—one spiky and angular, one rounded—shows that, crossculturally, neurotypical individuals do not attach sounds to shapes arbitrarily. When asked to assign the two names to the figures, 88% of neurotypcial individuals refer to the rectilinear object as a "Kiki" (or "Takete") and the curvilinear object as a "Bouba" (or "Maluma"). Individuals with Autism Spectrum Disorder (ASD) show this assignment only 56% of the time (nearly the same as if by chance).[72] When the object pairs were shown to ASD subjects (both and low functioning) as opposed to age-matched neurotypical controls, the degree of severity in ASD corresponded with their degree of multisensory integration abilities.[73]

It is difficult to provide specific ergonomic guidelines to mitigate the impact of varying SPDs, since SPDs and synesthesia present in so many different ways and may only require accommodation for a subset of tasks, if at all. As with any ergonomic design, one needs to address the needs of the user. Consider incorporating variations in activity, individualized instruction, instrumentation accommodations, customizable environment, and specific training—all adapted to the specific needs of the individual.

5.10 CHAPTER SUMMARY

The eye plays a major role in collecting information. Thus, the provision of proper visual signals, through the selection of proper illumination and contrast, and the avoidance of unbecoming circumstances such as glare, are important ergonomic tasks to provide sensory inputs. The eye is seldom used as a means for output, with "eye tracking" being one of the exceptions.

The sense of hearing provides inputs only. Recommendations for the design of communication and sound-signal systems are based on complex, but well-researched, relations between the frequency and intensity of sounds. Protection of the ear from damaging noises is also well understood, and related technical means are documented and ready to be applied.

The sense of smell for engineering purposes has some utility, such as being used as a marker for gases. However, since the olfactory sense easily adapts and varies from person to person and over time, and since odors dissipate and can be masked by other odors, the use of smell for ergonomic design is limited. Similarly, the sense of taste is of limited use for ergonomics as it is not well understood and is less sensitive than the sense of smell. The senses of smell and taste are not currently used for many engineering applications, but may be widely used in industry.

The sense of touch (for pressure, vibration, electricity, and pain) is often employed in human-operated systems. Temperature sensing is rather well understood, although not very reliable as input. The vestibular sense provides information about head and body posture, but is largely dependent on the existence of a well-defined gravity vector (which is missing in space). Furthermore, the human processor is easily confused by conflicting information from the vestibular and other senses—leading, for example, to nausea.

Dysfunctions in sensory processing may require an understanding of the specific needs of the user. Means of addressing processing deficits include changes in activity, instruction, instrumentation, environment, and adaptive practice.

5.11 CHALLENGES

What might be reasonable approaches to classifying the various human senses?

Regarding the classical experiment of immersing one hand in warm water and the other in cold water, how would the sensations change if the overall temperature ranges were decreased? If they were increased? Further, what would happen if one got used to higher or lower temperatures through repeated immersion with stepwise increased temperatures?

People who work in high or low temperatures can "tolerate" these more easily. Is that a form of adaptation? How do the just-noticeable differences change in this case?

How can one avoid interactions between different senses during experiments? Or would it be reasonable to present signals that simultaneously cover two or more modalities?

What might be a practical and easily used, but still scientifically exact, reference plane or system for identifying the direction of the LOS?

Why do many older people need higher illumination to see objects clearly?

Does one person see a given color—say, green—exactly the same as another person? If not, what can be done, in a practical sense, to compare the two perceptions?

Does the definition of the border between dark and light objects have any effect on the perception of contrast?

Do different physical explanations underlie the mixing of three independent primary colors and the artist's mixing of pigments in paint to generate a given color?

How might one determine whether colors have specific effects on a person's mood, and on the performance of a task?

How could one assess the effects of different kinds of "music" on mood and productivity?

How might the generation of specific smells (such as in perfumes) be made a systematic science, instead of remaining an individual art?

What is needed to turn "making things taste good" into a scientific or technological systematic process instead of remaining an art?

Does the replacement of metal objects at different temperatures (thermodes) by the flow of air at different temperatures solve the problem of the simultaneous stimulation of temperature and tactile sensors?

Which sensory modalities should be combined, under certain conditions and for certain signals, to increase the likelihood of perception? To increase the speed of recognition? To enhance CNS processing?

How might a work environment be modified to accommodate individuals with the various types of sensory processing dysfunctions? For example, what are some considerations to redesign an airplane cockpit to better accommodate all users (pilots, maintenance technicians, cleaning crews, and others)? Or a large manufacturing plant with complex electromechanical equipment?

NOTES

1 "Attention is an intentional, unapologetic discriminator…Right now, you are missing the vast majority of what is happening around you. You are missing the events unfolding in your body, in the distance, and right in front of you. By marshaling your attention to these words, helpfully framed in a distinct border of white, you are ignoring an unthinkably large amount of information that continues to bombard all of your senses: the hum of the fluorescent lights, the ambient noise in a large room, the places your chair presses against your legs or back, your tongue touching the roof of your mouth, the tension you are holding in your shoulders or jaw, the map of the cool and warm places on your body, the constant hum of traffic or a distant lawn-mower, the blurred view of your own shoulders and torso in your peripheral vision, a chirp of a bug or whine of a kitchen appliance." from cognitive scientist Alexandra Horowitz (2013), describing "adaptive ignorance."

2 See Boff, Kaufman, and Thomas (1986) who devoted seven chapters to a detailed description of human vision.

3 The Arab physicist Alhazen (965–1039) was the first to discover that vision was made possible by rays of light falling on the eye rather than the result of rays of light sent out by the eyes. Alhazen studied lenses and attributed their magnifying effect to the curvature of their surfaces. His work represents the beginning of the scientific study of optics. At the middle of the 13th century, convex lenses for eyeglasses used by the farsighted (i.e., mostly by the elderly) were well known in both China and Europe. In 1451, the German scholar Nicholas of Cusa used concave lenses for the nearsighted. (Asimov, 1989)

4 Kroemer (1994).

5 von Noorden (1985).

6 See Miller (1990) for a thorough discussion.

7 Heuer and Owens (1989).

8 Chi and Lin (1998).

9 As defined by the National Research Council Committee on Vision (1983) p. 153.

10 For more on visual fatigue, see Heuer et al. (1989, 1991), Jaschinski-Kruza (1991), Miller (1990), Owens and Leibowitz (1983), Rabbitt (1991), Ripple (1952), Tyrrell and Leibowitz (1990), von Noorden (1985).

11 Paraphrasing Tyrrell and Leibowitz (1990), p. 342.

12 Isaac Newton (1642–1727) performed experiments in which he made a beam of light pass through a glass prism. What he saw emerging was a band of colors, with red being the least bent portion of the light, followed by orange, yellow, green, blue, and violet. Making these colorful lights pass through another prism, he showed that they merged to become white light again. In this way, Newton showed that white light was not "pure," but a mixture of different colors. (Asimov, 1989)

13 Hood and Finkelstein (1986) and Snyder (1985a,b) provide more detailed discussions of this topic.

14 As by the National Academy of Sciences (1980).

15 See, for example, Boyd (1982) and Pokorny and Smith (1986).

16 Pokorny and Smith (1986).

17 Throughout the ages, regions developed their own system of measurements, which was functional as long as trade and communications were slow and infrequent. By 1790, a commission was appointed in France to develop a system of reasonable measurements. Members of the commission included such great scientists as Pierre-Simon Laplace, Joseph-Louie Lagrange, and Antoine Lavoisier. The commission generated a system founded on basic natural units, such as the meter (from the Greek, "to measure"), which was equal to 10^{-8} times the distance from the North Pole to the equator. Other units were developed that interconnected with the meter, and larger and smaller units were derived by multiplying or dividing by 10. This metric system was slowly accepted throughout the globe, with the United States one of the last holdouts against it. (Asimov, 1989)

18 Color naming depends on the region and the language. "Primitive languages first develop words for black and white, then add red, then yellow and green; many lump blue and green together, and some don't bother distinguishing between other colors of the spectrum. … [T]he Maori of New Zealand … have many words for red—all the reds that surge and pale as fruits and flowers develop, as blood flows and dries." (Ackerman, 1990, p. 253)

19 Pokorny and Smith (1986).

20 Sayer, Sebok, and Snyder (1990).

21 As discussed by Pokorny and Smith (1986).

22 Adapted from Wyszecki (1986).

23 Wyszecki (1986).

24 Kwallek and Lewis (1990).

25 See Kaufman and Haynes (1981).

26 Adapted from Consumer Reports on Health, April 1997, pp. 42–43.

27 Gaver (1997).

28 See Fox (1983) for a discussion of background and industrial music.

29 Fox (1983), Lesiuk (2005).

30 Sachs, Ellis, Schlaug, and Psyche (2016).

31 Stekelenburg (1982).

32 NASA (1989a,b).

33 National Institutes of Health (1990).

34 National Institutes of Health (1990).

35 Berger and Casali (1992).

36 NASA (1989a,b).

37 Berger and Casali (1992).

38 Berger and Casali (1992).

39 Originally developed by Fletcher and Munson (1933) and revised by Robinson and Dadson (1956).

40 NASA (1989a,b).

41 Casali and Berger (1996).

42 Barnes and Wells (1994).

43 Ackerman (1990), Ballard (1995).

44 Cometto-Muniz and Cain (1994), National Research Council (1979).

45 Hangartner (1987).

46 Since ancient times, it has been hypothesized that pleasant aromas preserve health and that unpleasant odors are injurious. These suspicions formed the basis for the use of aromatic "eau de cologne" and of pomanders stuffed with balsams, and for the attribution of diseases to atmospheric "miasmas." The word "malaria" is derived from the Italian expression for "bad air," *mala aria*. (National Research Council, 1979, p. 3)

47 Cain, Leaderer, Cannon, Tosun, and Ismail (1987).

48 After spending months in Bombay, India, in the 1990s when strong odors abounded, one of the previous authors (KHEK) could no longer smell the "wild" animals in a circus.

49 National Research Council (1979).

50 Only 20% of the perfume industry's income comes from making perfumes to wear; the other 80% comes from perfuming the objects around us. Thus, used-car dealers have a "new-car" spray, real-estate dealers sometimes spray the aroma of a cake baking around the kitchen of the house before showing it to a client, and shopping mall managers add the smell of food to their air-conditioning system to put shoppers in the mood to visit their restaurants. (Ackerman, 1990, p. 39)

51 Ackerman (1990) pp. xvi, 24.

52 "How strange that we acquire taste as we grow. Babies don't like olives, mustard, hot pepper, beer, fruits that make one pucker, or coffee. … No two of us taste the same plum. Heredity allows some people to eat asparagus and pee fragrantly afterward (as Proust describes in "Remembrance of Things Past"), or eat artichokes and then taste any drink, even water, as sweet. Some people are more sensitive to bitter tastes than others and find saccharin appalling, while others guzzle diet sodas. Salt cravers have saltier saliva. Their mouths are accustomed to a higher sodium level, and foods must be saltier before they register as salty." (Ackerman, 1990, p. 141)

53 Crosby (2016). Super-tasters and nontasters: Is it better to be average? Harvard school of public health. https://www.hsph.harvard.edu/nutritionsource/2016/05/31/super-tasters-nontasters-is-it-better-to-be-average/. Accessed 31 July 2017.

54 "The following excerpt is from an article on wine: 'Number one was a youngster: fresh, light, passion-fruity, straightforward, not oakey, fruit-sweet, fine and understated, well-balanced and clean—a modern style with low-key oak and showing cool-fermentation high-tech winemaking, but not very complex, and needing age. Number two was much more complex, showing some maturity: very smoky, lightly leesy, toasted nut aromas; strongly constituted, tightly structured, powerful, spicy with plenty of wood-derived complexity, very long in flavor and excellent drinking now, but in no danger of falling apart. Number three was, for me, even bigger and richer, fatter and more complex than number two. Strong aromas of butter, grilled nuts, toasted bread, smokiness, very full-bodied, ripe, and alcoholic, quite fat and reminding everybody of a good 1978 white Burgundy.' Let's just take this slowly. By my count, and excluding what might be repetitions, there are 28 different qualifiers in that quotation, implying 28 scales of assessment. This writer is asking me to believe that it is possible to discriminate 28 taste scales, let alone assess points along each one? And to do this reliably?" (Applied Ergonomics, 1988, 19(4), 324).

55 An overview of the unsettled state of theories and knowledge is provided by Sherrick and Cholewiak (1986).

56 By Boff and Lincoln (1988).

57 Fransson-Hall and Kilbom (1993).

58 Sherrick and Cholewiak (1986) pp. 12–39.

59 Gawande (1998).

60 Manchikanti, Singh, Datta, Cohen, and Hirsh (2009).

61 Boff and Lincoln (1988).

62 Howard (1986).

63 Boff and Lincoln (1988).

64 Sherrick and Cholewiak (1986).

65 Kantowitz and Sorkin (1983).

66 As discussed by Sherrick and Cholewiak (1986).

67 Ayres (1998).
68 Leekam et al. (2007), Van Hulle et al (2012).
69 Dunn (2001).
70 Sinke et al (2014).
71 Bien, ten Oever, and Sack (2012).
72 Oberman and Ramachandran (2008).
73 Ocecelli et al. (2013).

How the Body Interacts with the Environment

Ergonomics
http://dx.doi.org/10.1016/B978-0-12-813296-8.00006-2

Overview

Most of us work in a "normal" environment: in a moderate climate, on solid ground, at or slightly above sea level. However, experiencing climate changes from summer to winter, working at a high-altitude location, enduring vibrations and impacts like a truck driver, flying like a pilot or astronaut, diving in the deep sea—all these challenge the body in many ways. Work environments may contain dust, toxic fumes, chemicals, vibrations and impacts, and other conditions that can affect performance or even be harmful to the human. Recognizing and avoiding such hazards is the domain of the fields of Industrial Hygiene and Occupational Safety. Fortunately, we have a number of technical and behavioral means to avoid or minimize negative effects on human health, performance, and comfort.

6.1 CLIMATE

The human body generates energy and simultaneously exchanges—gains or loses—energy with (to) the environment. Since we must maintain an essentially constant core temperature, our body needs to shed heat in a hot climate and avoid excessive heat loss in a cold environment.

Thermoregulation of the body

Thermoregulation is the ability of an organism to keep its body temperature within defined parameters, even if the temperature of the environment is markedly different. Human bodies must be efficient thermoregulators in order to function; by contrast, a thermo-conforming organism is able to adopt the temperature of its environment as its body temperature, thereby avoiding the need for thermoregulation completely. We humans achieve thermoregulatory homeostasis when our internal conditions are in a state of dynamic stability.

	Degrees F	Degrees C	K
Water boils →	212	100	373.15
		90	363.15
at about 85 deg C, skin burn damage occurs when touching wood or plastic for 4 seconds		80	353.15
		70	343.15
at about 60 deg C, skin burn damage occurs when touching metal or water for 4 seconds	140	60	333.15
		50	323.15
at about 40 deg C, temperature in the shade of a hot summer day in New York city	104	40	313.15
37 deg C, temperature of the body core		30	203.15
at about 27 deg C, highest comfortable summer office temperature in New York city		20	293.15
at about 18 deg C, lowest comfortable winter office temperature in New York city	50	10	283.15
Water freezes →	32	0	273.15

FIGURE 6.1   Temperature scales in common use.

There are several approaches to modeling the energy production in the body and its energy exchange with the environment: consider an average skin temperature; or, as we've done here, postulate a "core" that must be kept at nearly constant temperature.

The human body has a complex control system to maintain the deep body core temperature very close to 37°C (about 99°F),[1] as measured in the intestines, rectum, ear, or estimated by the temperature in the mouth. While there are minor temperature fluctuations throughout the day due to diurnal changes in body functions (see Chapter 7), the main task is to regulate the energy exchange between metabolic heat generated within the body and external energy in the environment.

Maintain core temperature

The primary task of the human thermoregulatory system is to keep the core temperature close to 37°C, whether we are in cold or hot environments. Changes in core temperature from 37°C even by just 2°, plus or minus, severely affect body functions and task performance; deviations of ±6° are usually lethal. At the skin, the human temperature-regulation system must keep temperatures well above freezing and below the 40s in its outer layers. Although our core temperature remains relatively constant, at the skin, temperatures differ considerably from region to region. For example, the skin at our toes may measure 25°C, legs and upper arms at 31°C, and the forehead at 34°C all at the same point in time: for most of us, this feels comfortable, as noted in Fig. 6.1.

6.1.1 The Energy Balance

Our body derives energy from the food and drink we ingest, and energy expenditure depends on how we move, how much we rest, how we interact with our physical environment, and how our body's cells and tissues perform. Energy expenditure is continuous, but it varies throughout the day. A simple equation describes the energy exchange between inputs to the body and outputs from the body; this is the same Eq. (3.1) as used previously:

$$I = M = H + W + S \tag{6.1}$$

where I is the energy input via food and drink that the body transforms into metabolic energy M. The energy quantity M consists of the heat H that must be dispelled to the outside, the external work W done, and the change in energy storage S in the body. A gain in S is counted as positive, a loss as negative. (See also Chapter 3.)

Heat balance

Rearranging Eq. (6.1), we can calculate the heat energy exchange with the external thermal environment according to

$$H = I - W - S \tag{6.2}$$

The system is in balance with the environment if all heat energy H is dissipated to the environment and stored energy S remains essentially unchanged. However, the case gets complicated by the fact that the body often receives additional heat energy from warm surroundings—or may lose heat in a cold environment.[2]

6.1.2 Energy Exchanges with the Environment

We do not live in isolation; instead, the human thermoregulatory system interacts with the environment. This is thermodynamically easy in a cold environment, but we may feel uncomfortably chilly; dissipating heat energy is more difficult in warm environs where we may feel overly warm. In a comfortable climate, which is what we find in most offices, the regulatory task is to dissipate the heat energy generated by the body's metabolism. Energy (heat) is exchanged with the environment[3] through radiation R; convection C; conduction K; and evaporation E.

6.1.2.1 Heat Exchange by Radiation

Radiation

Heat exchange by radiation, R, is a flow of electromagnetic energy between two opposing surfaces, for example, between a windowpane and a person's skin. Heat always radiates from the warmer to the colder surface. Thus, the human body can either gain or lose heat through radiation. The amount of radiated heat depends primarily on the temperature difference between the two surfaces but not on the temperature of the air between them (the Stefan-Boltzmann Law of Radiative Heat Transfer); see Figs. 6.2 and 6.3.

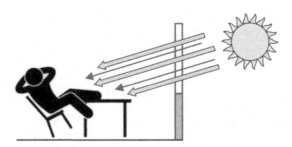

FIGURE 6.2   Gaining heat through radiation. *Source: Adapted from Kroemer and Kroemer (2016).*

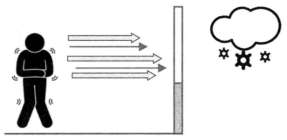

FIGURE 6.3 **Losing body heat through radiation.** *Source: Adapted from Kroemer and Kroemer (2016).*

Fig. 6.2 illustrates how heat from the sun warms our skin, without directly affecting the temperature of the air. Heat may also radiate from a fire, a sun-warmed windowpane, a heating radiator, or any other "warm body" whose surface temperature is higher than that of our skin. Of course, if the skin is warmer than, say, a cold window pane, and if there is no insulator such as clothing between the skin and the glass, then we radiate our heat to the window, as sketched in Fig. 6.3. In this case, our body is losing heat, and we are growing colder.

The amount of radiating energy Q_R lost from or gained by the body through radiation depends essentially on the size A of the participating body surface and on the difference Δ between the temperatures T (in Kelvin degrees) of the surfaces to the fourth power:

$$Q_R = f\left(A, \Delta T^4\right) \tag{6.3}$$

approximated by

$$Q_R \approx A\, h_R\, \Delta t \tag{6.4}$$

where h_R is the heat-transfer coefficient and Δt the temperature difference in degrees Celsius.

6.1.2.2 Heat Exchange through Convection and Conduction

Convection C and conduction K both follow the same thermodynamic rules (Newton's law of cooling). The heat transferred is again proportional to the area of human skin participating in the process, and to the temperature difference between skin and the adjacent layer of the external medium. Accordingly, in general terms, heat exchange by convection or conduction is

$$Q_{C,K} = f(A, \Delta t) \tag{6.5}$$

approximated by

$$Q_{C,K} \approx A\, k\, \Delta t \tag{6.6}$$

with k the coefficient of conduction or convection.

Heat exchange through convection C takes place when the human skin is in contact with air or other gas, and with water or other fluid. Heat energy transfers from the skin

Convection

to a layer of colder gas or fluid next to the skin surface, or transfers to the skin if the surrounding medium is warmer. *Convective heat exchange* is facilitated if the medium moves quickly along the skin surface, thus maintaining a temperature differential. There is always some natural movement of air or fluid as long as a temperature gradient exists: this is called *free convection*. More movement of the medium can be produced by forced action, such as a fan creating a breeze: this is called *induced convection*. These thermodynamic mechanisms are in effect when we feel overly hot and seek the cooling effect of a refreshing breeze or a swim in a cool lake.

Conduction

Conductive heat exchange K exists when the skin touches a solid body. As long as there is a difference in temperature, heat flows, especially if the conductance of the piece is high; less energy flows if the skin touches an insulator having a low conduction value k.

Cork and wood are perceived as feeling warm because their heat-conduction coefficients are below that of human tissue. Even if they are all the same temperature, insulators such as cork or wood feel warmer than metal, which accepts body heat easily and conducts it away; therefore, metal is perceived to feel colder.

6.1.2.3 *Heat Exchange by Evaporation*

Evaporation

Heat exchange by evaporation E works in only one direction: the human loses heat by evaporation. There is never condensation of water on living skin, which would add heat. Some water is evaporated in the respiratory passages but most appears as sweat on the skin. Evaporation of sweat, which is mostly water, requires energy of about 580 cal/cm^3; this energy is primarily taken from the body and therefore reduces the heat content of the body by that amount.

The heat lost by evaporation Q_E from the human body depends mostly on the participating wet body surface A and on air humidity h:

$$Q_E \approx \mathrm{f}(A, h) \tag{6.7}$$

Evaporative heat loss is more difficult in higher humidity than in dryer air conditions. Movement of the air layer at the skin increases the actual heat loss through evaporation since it replaces humid air by drier air. The warmer the air, the more water vapor it can accept. Heat exchange by convection or conduction increases also when the gas, fluid or solid moves along the warmer or colder skin (see Fig. 6.4). This is why using a fan to blow air at us in a hot climate feels so refreshing: the airflow takes away the layer of humid air around our body, which is largely saturated with our evaporated sweat, and replaces it with dryer air. Although we might prefer a cool breeze, even hot air blown toward us will accept our sweat.

Evaporative heat loss occurs even in a cold environment, due to both perspiration and exhalation. There is always some perspiration (and hence sweat evaporation),

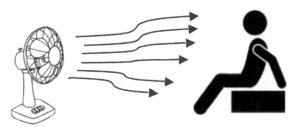

FIGURE 6.4 Air flow enhances heat exchange by convection and body cooling by evaporation. *Source: Adapted from Kroemer and Kroemer (2016).*

called *insensible water loss*. (This explains why our clothes smell when worn too long, even if we are not engaged in strenuous activity.) Further, in the lungs, water evaporates into the exhaled air. As we exert ourselves, ventilation increases, with corresponding increase in volume of exhaled air and evaporated water.

Fig. 6.5 schematically depicts the body's heat loss via radiation, convection (and, similarly, conduction), and evaporation in cool or warm environments. These heat exchanges all depend, directly or indirectly, on the difference in temperature between the participating body surface and the environment.

Temperature difference

6.1.3 Body Heat Balance

Heat balance exists when the heat energy H developed in the body (Eq. 6.2) is in equilibrium with the heat exchanged with the environment by radiation R, convection C, conduction K, and evaporation E: that is, they sum to zero. This relationship can be expressed as:

Heat balance

$$H + R + C + K + E = 0 \qquad (6.8)$$

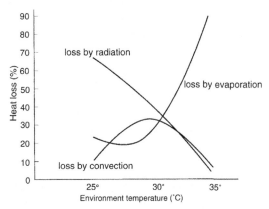

FIGURE 6.5 Effectiveness of convection, radiation, and evaporation in generating body heat loss at different temperatures of the environs.

When the body loses energy to the environment, the quantities R, C, K, and E are net-negative; if the body gains energy from the environment, they are net-positive. Note that E is always negative.

6.1.4 Temperature Regulation and Sensation

Body heat flow

In the human body, heat is generated in "metabolically active" tissues—primarily by skeletal muscles, but also in internal organs, fat, bone, and connective and nerve tissue. The heat energy is circulated throughout the body by our blood. Blood vessel constriction, dilation, and shunting regulate the blood flow within the body. Heat exchange with the environment takes place at the body's respiratory surfaces (primarily in the lungs) but mostly through the skin. The equations shown earlier indicate that the energy exchange with our surroundings depends largely on the area of participating skin surface, especially skin that is naked and exposed to the environment. Clothing helps insulate the skin surface from heat transfer.

In a cold environment, body heat must be conserved. The body does this by reducing blood flow to the skin, making it colder, and we do this deliberately by increasing insulation with warm clothing. In a hot environment, body heat must be dissipated and heat gain from the environment must be minimized. The body does this by increasing the blood flow to the skin (making it warmer) and by increased sweat production and evaporation, and we do this deliberately by dressing more lightly.

Diverse body temperatures

Temperatures in the body are not uniform throughout; there can be large differences between "core" and "shell" temperatures. Under normal conditions, the average gradient between skin and deep body is about 4°C at rest, but in cold environments the difference in temperature may be 20° or more. The temperature-regulation system, located in the hypothalamus, has to maintain various temperatures at numerous locations under varied conditions.

Body heat and muscular work

If the body is about to overheat, it needs to minimize internal heat generation. For this to happen, our bodies reduce muscular activities. Consider warm-climate countries in which inhabitants routinely take a "siesta," a midday break; such breaks allow us to respond to our bodies' signals and rest our muscles. In the opposite case, when more heat must be generated, increasing our work or exercise level and thus augmenting our muscular activities will help us warm up. We also have involuntary reactions that help us heat up, such as shivering. Muscular work generates heat due to its low efficiency (see Chapter 3).

Muscles can generate more heat or less heat but cannot cool the body. Sweat production influences the magnitude of energy loss but cannot bring about a heat gain. Vascular activities affect the heat distribution through the body and thus control heat loss or gain, but they do not generate energy. Efferent motor, sudomotor, and vasomotor actions cooperate in regulating the body heat content, interacting with the external climate, as seen schematically in Fig. 6.6.

Temperature sensors

The human body has various temperature sensors in the core and the shell of the body (see also Chapter 5). Cold sensors are most sensitive from about 15°C to 35°C. Hot sensors generate signals (sent to the hypothalamus) particularly in the range of approximately 38–43°C. There is some overlap in the sensations of "cool" and "warm" in the

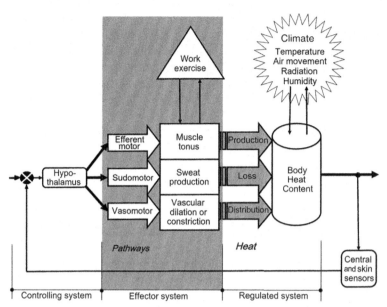

FIGURE 6.6 **Model of the regulation of body heat content.** *Source: Adapted from Kroemer, Kroemer, and Kroemer-Elbert (2010).*

intermediate range. Between about 15 and 45°, perception of either "cold" or "hot" conditions is highly adaptable. Below 15 and above 45°, the human temperature sensors are less discriminating but also less adapting. A paradoxical effect is that, around 45°, sensors again signal "cold" while in fact the temperature is rather hot.[4]

6.1.5 Achieving Thermal Homeostasis

The human thermo-regulatory system must control two temperature gradients: from the core to the skin, and from the skin to the surroundings. The gradient between core and skin is most important because overheating or undercooling of the key tissues in the brain and the chest must be avoided, even at the cost of overheating or undercooling the shell. *(Two temperature gradients)*

Thermal equilibrium, called *homeostasis*, is primarily achieved by regulation of the blood flow from deep tissues and muscles to skin and lungs. By far the most heat is exchanged with the environment through the skin; in the lungs, from 10% to 25% of the total body heat is dissipated. *(First, blood flow)*

Secondary actions to establish thermal homeostasis take place through the muscles. They generate heat through both voluntary or involuntary efforts: work and exercise are voluntary; shivering is involuntary. If the goal is to gain heat, the regulatory system initiates skeletal muscle contractions; if heat gain must be avoided, it reduces or abolishes muscular activities. *(Second, muscle activities)*

Changes in clothing and in the climate of our shelter (office, home) represent our purposeful actions to establish thermal homeostasis. Together with blood-flow regulation and muscle activities, these efforts allow us to achieve the appropriate temperature gradient between the skin and the environment. How we dress and what our *(Third, clothing and shelter)* environment is like also affect radiation, convection, conduction, and evaporation. "Light" or "heavy" clothing differs in terms of permeability and ability to establish stationary insulating layers. Clothing prevents body heat from escaping into the environment by preserving the warm layer of air that exists between skin and clothing. Since air is a relatively bad conductor of heat, it acts as a good insulator. Clothes also affect conductance, that is, energy transmitted per surface unit, time, and temperature gradient. The color of clothing affects how much external radiation energy is absorbed or reflected. Especially in sunlight, light-colored clothes reflect radiated heat, while dark-colored surfaces absorb it.

ASSESSING THE THERMAL ENVIRONMENT

The thermal environment is determined by four physical factors: air (or water) temperature, humidity, air (or water) movement, and temperatures of the surfaces that exchange energy by radiation. The combination of these four factors determines the physical conditions of the climate and our perception of the climate.

Temperature: Measurement of air temperature is performed with thermometers, thermistors, or thermocouples. Whichever technique is used, we must ensure that the ambient temperature is not affected by the other three climate factors, particularly humidity, but also air movement and surface temperatures. To measure the dry temperature of ambient air, the sensor is kept dry and shielded with a surrounding bulb that reflects radiated energy: a dry-bulb (DB) thermometer.

Humidity: Air humidity may be measured with a psychrometer, hygrometer, or other electronic devices. These usually rely on the fact that the cooling effect of evaporation is proportional to the humidity of the air, with higher vapor pressure making evaporative cooling less efficient. Therefore, we can measure humidity using both DB and wet-bulb (WB) thermometers. The highest absolute content of water vapor in the air is reached when any further increase would lead to the development of water droplets. The amount of possible vapor depends on air pressure and on the temperature of the air, with lower pressure and higher temperature allowing more water vapor to be retained than lower temperatures. *Relative humidity* indicates the actual vapor content in relation to the possible maximal content (*absolute humidity*) at the given air temperature and air pressure.

Air Movement: Air movement is measured with various types of anemometers that use mechanical or electrical principles. We can also measure air movement with two thermometers, one dry and one wet, relying on the fact that the wet thermometer shows more increased evaporative cooling with higher air movement than the dry thermometer.

Radiation: Radiant heat exchange depends primarily on the difference in temperatures between the surfaces of the person and the surroundings, on the emission properties of the radiating surface, and on the absorption characteristics of the receiving surface. One easy way to assess the amount of energy acquired through radiation is to place a thermometer inside a black globe (BG) that absorbs practically all arriving radiated energy.

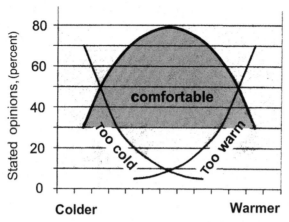

FIGURE 6.7   Indoor summer and winter comfort zones and thermal tolerance for appropriately dressed, seated people doing light work. *Source: Adapted from [United States] Department of Defense (2000).*

A person's feeling of comfort is not determined only by the physics of thermal balance shown in Eq. (6.8). For example, skin wetness (in a warm environment) and skin temperature (in a cold environment) play major roles in how we comfortable we feel.

Thermal comfort

Various theories and proposals have been developed to assess the combined effects of some or all environmental factors and to express these in one number, model, chart, or index. Two scales, the Bedford and the ASHRAE scales, are widely used in North America to assess individual thermal comfort. Other empirical thermal indices establish a *reference* or *effective climate* that "feels the same" as various combinations of the several climate components. A well-known example is the *effective temperature* (ET), which considers the impact of various combinations of temperatures, humidity, air movement, and length of exposure time on perceived comfort and thermal tolerance—see Fig. 6.7.

Thermal indices

6.1.6 Reactions of the Body to Hot Environments

In hot environments, the body produces heat that must be dissipated. This is difficult to do in hot climates; to achieve this, the body directs warm blood to its surfaces in order to raise skin temperatures above the immediate environment. This facilitates heat transfer from the body through convection, conduction, and radiation.

If heat transfer is not sufficient, sweat glands are activated to product sweat, which then evaporates to cool the skin. Recruitment of sweat glands from different areas of the body varies among individuals. The overall amount of sweat developed and evaporated depends on clothing, environment, work requirements, and the individual's acclimation—see below.

Evaporating sweat

If heat transfer by blood distribution and sweat evaporation is still insufficient, muscular activities must be reduced to lower the amount of heat energy generated; this is the final and necessary action of the body, if otherwise the core temperature would exceed tolerable limits.

Stopping work

> If the body has to choose between either unacceptable overheating or continuing to perform physical work, the choice will be to maintain core temperature, which means work activities are reduced or discontinued.

Heat indices

The literature contains numerous proposals for a simple-to-use yet accurate and comprehensive *heat index* which indicates what conditions of hot climates are suitable to perform work, with various levels of activities and individual acclimation, with varying clothing and for assorted periods of time. Currently, the Wet-Bulb Globe Temperature (WBGT) is in general use, in spite of some shortcomings.[5]

WBGT

Sets of instruments can be arranged to simulate the apparent heat stress due to air temperature, humidity, wind, and (solar) radiation. A dry thermometer placed inside a black globe responds both to air temperature and radiation; a wet thermometer responds to both air velocity and air humidity. The WBGT index is generated by three sensors whose readings are automatically weighted and then combined. For outdoors with solar load:

$$WBGT = 0.7\,WB + 0.2\,GT + 0.1\,DB \tag{6.9}$$

For indoor and outdoor conditions without solar load:

$$WBGT = 0.7\,WB + 0.3\,GT \tag{6.10}$$

where WB is the wet-bulb temperature of a sensor in a wet wick exposed to natural air current; GT is the globe temperature at the center of a black sphere of 15 cm diameter; and DB is the dry-bulb temperature measured while shielded from radiation.

Charts, occasionally updated, contain so-called "safe" WBGT numbers which depend, among other factors, on the work load (metabolic rate) of the person, on the person's clothing and state of acclimation, and on air movement. For current permissible heat exposure limit values, please refer to the latest version of ISO 7243 or (in the United States) of OSHA manuals.

HEAT STRAIN

Strenuous physical activities in the presence of high air temperatures, radiant heat sources (including the sun), high humidity, and direct physical contact with hot objects have a high potential for inducing heat strain.

- The most serious sign of heat strain is a *rise in core temperature*, which must be counteracted before the temperature exceeds a sustainable limit (about 38°C).
- Heat strain increases the circulatory activities: cardiac output is increased, which is mostly brought about by a higher *heart rate*. This may be associated with a reduction in systolic blood pressure.

- The most obvious sign is *sweating*. In very strenuous exercises and hot climates, several liters of sweat may be produced in an hour. Sweat begins to drip off the skin when the sweat generation has reached about one-third of the maximal evaporative capacity. Of course, sweat running down the skin contributes very little to heat transfer.
- The water balance within the body provides another sign of heat strain. Severe *dehydration*, as indicated by the loss of 1 or more percent of body weight, can critically affect the ability of the body to control its functions. Maintaining an appropriate fluid level is best accomplished by frequently drinking small amounts of water.[6] Normally, it is not necessary to add salt to drinking water; for most of us, the salt in our food suffices to resupply the salt lost with sweat.

Table 6.1 lists symptoms, causes, and treatment of heat-related disorders.

TABLE 6.1   Symptoms, Causes, and Treatment of Heat-related Disorders

Disorder	Signs and symptoms	Additional information
Heat stroke	Confusion; irrational behavior; loss of consciousness; convulsions; hot, dry skin; usually lack of sweating; and an abnormally high body temperature, such as a rectal temperature of 41°C	Heat stroke is caused by a combination of highly variable factors, and its occurrence is difficult to predict. It occurs when the body's system of temperature regulation fails and body temperature rises to critical levels. Heat stroke is a medical emergency
Heat exhaustion	Headache, nausea, vertigo, weakness, thirst, giddiness, possibly fainting	Heat exhaustion responds readily to prompt treatment. People suffering from heat exhaustion should move out of the hot environment, drink water, and rest
Heat collapse, "fainting"	Collapsing and/or fainting (losing consciousness or syncope) if the brain does not receive enough oxygen because blood pools in the extremities This reaction is similar to that of heat exhaustion but does not affect the body's heat balance	The onset of heat collapse is rapid and unpredictable. To prevent heat collapse, the worker should gradually become acclimatized to the hot environment
Heat cramps	Muscle spasms, caused by both too much and too little salt in the body, resulting from electrolyte imbalance due to sweating and insufficient water replenishment Heat cramps mostly occur when hard physical labor is done in a hot environment	Water must be drunk every 15–20 min in hot environments. Cramps may occur after quickly drinking large amounts of fluid, which dilutes the body fluids Under extreme conditions, such as working for 6–8 h in heavy protective gear, drinking carbohydrate-electrolyte replacement liquids is effective in replacing lost sodium
Heat rash, "prickly heat"	Red pimples (papules), causing a prickly feeling, usually appearing in areas under restrictive clothing where skin remains wet from unevaporated sweat Heat rash is the most common problem in hot work environments	In most cases, heat rashes disappear when the affected person moves to a cool environment In some cases, topical treatments are required to soothe itching and prevent duct blocking
Heat fatigue	Impaired performance of skilled sensorimotor, mental, or vigilance jobs	There is no treatment for heat fatigue except to remove the heat stress before a more serious heat-related condition develops. A program of acclimatization and training for work in hot environments can prevent heat fatigue

Source: Adapted from Spain, Ewing and Clay (1985).

TABLE 6.2   Surface Temperatures (in °C) Which Human Skin Can Tolerate Without Risk of Burn

Material	Maximal surface temperature (°C) for contact times of				
	1 s	4 s	1 min	10 min	8 h
Metal: uncoated smooth surface	65	60	50	48	43
Metal: uncoated rough surface	70	65	50	48	43
Metal: surface coated with varnish 50 μm thick	75	65	50	48	43
Ceramics and concrete	80	70	55	48	43
Ceramics: glazed (tiles)	80	75	55	48	43
Glass and Porcelain	85	75	55	48	43
Plastics: polyamid with glass fibers	85	75	60	48	43
Plastics: duroplast with fibers	95	85	60	48	43
Teflon and plexiglass	–	85	60	48	43
Wood	115[a]	95	60	48	43
Water	65	60	50	48	43

[a]*Higher by 25° or more for very dry and very light wood.*
Source: Adapted from Siekmann (1990).

Skin burns

When we touch hot surfaces, the risk of pain and of skin burn exists. Skin burns until recently were classified as first, second, third, or fourth degree burns based on the thickness of the skin burned but have been reclassified now as superficial, superficial partial-thickness, deep partial-thickness, and full-thickness burns, respectively. The actual critical contact temperature[7] depends on the duration of the contact and on the material of the contacted object or fluid. The shorter the contact, the higher the temperatures that can be tolerated, as listed in Table 6.2.

6.1.7 Reactions of the Body to Cold Environments

The human body has few natural defenses against a cold environment. Most of our counter-actions are behavioral, such as putting on warm clothing, covering facial skin, seeking shelter, or using external sources of warmth.

Conserving body heat

In a cold climate, the body must conserve its heat, which it does automatically by lowering the temperature of the skin. This reduces the temperature difference compared to the outside. The body accomplishes this by keeping the circulating blood closer to the core, away from the skin. For example, the blood flow in the fingers may be reduced to just a few percent of that in a moderate climate. This results in cold fingers and toes, with possible damage to the tissues if temperatures get close to freezing.

Cutaneous vasoconstriction

Activation of cutaneous vasoconstriction is under the control of the sympathetic nervous system; there are also local reflex reactions to direct cold stimuli. An interesting phenomenon is the so-called hunting reaction (a *cold-induced vasodilation*): after initial vasoconstriction has taken place, there is a sudden opening of blood vessels which allows warm blood to return to the skin, such as that of the hands, which rewarms that body section. Then, constriction returns, and this sequence may be repeated several times.

Developing "goosebumps" (*horripilation*) on the skin helps to retain a relatively warm layer of stationary air close to the skin. The stationary envelope acts like an insulator, reducing energy loss through the skin. If we had more or thicker hair (or fur), this response would be more effective.

Goosebumps

The other natural reaction of the body to a cold environment is to increase metabolic heat generation. We can achieve this voluntarily or involuntarily.

Increasing body heat

- Voluntary muscle contractions: we can raise body temperature intentionally by increasing the dynamic muscular work performed, moving body segments, moving arms and shoulders, flexing our hands, stomping with our feet. Such dynamic muscular work may easily increase the metabolic rate to ten or more times above the resting rate.
- Involuntary shivering: such thermogenesis usually begins in the neck, apparently because warmth is needed to supply blood to the brain. The onset of shivering is normally preceded by an increase in overall muscle tone in response to body cooling. With increased firing rates of the motor units (but no actual limb movements), the affected person generally experiences a sense of stiffness. Then suddenly shivering begins, caused by muscle units firing at different frequencies of repetition (*rate coding*) and out of phase with each other (*recruitment coding*). Since no mechanical work is being done, the total activity is transformed into heat production, allowing an increase in the metabolic rate to up to four times the resting rate. If the body does not become warm, shivering may become rather violent: motor-unit firings synchronize enough so that large muscle groups also contract. While such shivering can generate heat that is five or more times the resting metabolic rate, it can only be maintained for a short period of time.

Being in cold water can quickly bring about hypothermia by convection. While we can endure up to 2 hours in water at 15°C, we become helpless in water of 5° after only 20–30 minutes. Wearing clothing that provides insulation increases the survival time in cold water; further, obese people with more insulating adipose tissue have an advantage over leaner people. Floating motionless results in less metabolic energy being generated and spent than when swimming vigorously, while also maintaining the insulating envelope of warmed water around the body.

Cold water immersion

6.2.7.1 *How Cold Does It Feel?*

A person's decision to stay in or escape from the cold depends on subjective assessments of how cold the body surface and/or core are. It is dangerous not to perceive (and hence not to react to) the body's signals of cold since body temperature may continue falling until it is lower than the threshold of perception.

Perceived coldness

The perception of the body's decreasing body temperature depends upon signals from surface thermal receptors and from sensors in the core. As skin temperatures decrease below 35.5°C, the intensity of the cold sensation increases; it is strongest near 20°C, but at lower temperatures of perception the intensity decreases. It is often difficult to separate feelings of cold from pain and discomfort.

COLD STRAIN

The body lowers skin temperature in a cold environment. This reduces the temperature difference versus the outside, and hence reduces heat loss. However, this exposes uncovered fingers, toes, face and ears to cold damage, even while the body core is protected.

- As the skin temperature is lowered to about 15–20°C, manual dexterity begins to decline. Tactile sensitivity is severely reduced when the skin temperature falls below 8°C.
- If tissue temperature approaches freezing, ice crystals develop in the cells and destroy them, a result known as *frostbite*.
- At local tissue temperatures of 5–8°C, peripheral motor nerve velocity is decreased to near zero; this generates a *nervous block*. Severe cooling of skin and adjacent body tissue decreases the ability to perform activities, even if they could save the person ("cannot light a match"), leading to apathy ("let me sleep") and finally hypothermia.
- Severe reductions in skin temperatures are usually accompanied by a fall in core temperature. Lowering the core temperature has very serious consequences because vigilance drops when the body reaches temperatures below 36°C. At core temperatures below 35°, we may not be able to perform even simple activities. When the core temperature drops even lower, the mind becomes confused, with loss of consciousness occurring around 32°C. At core temperatures of about 26°C, heart failure may occur. At very low core temperatures, such as 20°C, vital signs disappear, but the oxygen supply to the brain may still be sufficient to allow revival from hypothermia.

The conditions of cold exposure may greatly influence the perceived coldness. It can make quite a difference whether we are exposed to cold air (with or without movement) or to cold water, whether or not we are wearing protective clothing, and what we are actually doing.

When the temperature plunges, each downward step can generate an "overshoot" sensation of cold receptors, which react very quickly not only to the difference in temperature but also to the rate of change. However, if the temperature stabilizes, the cold sensations become smaller as we adapt to the condition. Exposure to very cold water accentuates the overshoot phenomenon observed in cold air, possibly because the thermal conductivity of water is much greater than that of cold air at the same temperature. Thus, cold water causes a convective heat loss that may be many times that of cold air. People apparently are not able to reliably assess how cold they actually are: neither their core nor surface temperatures correlate well with their cold sensations.[8]

Subjective sensation of cold is a poor, unreliable, and possibly dangerous indicator of core and surface temperature of the body. Objectively measuring ambient temperature, humidity, air movement, and exposure time and taking appropriate actions is a better strategy than relying on subjective sensations.

6.1.8 Acclimation

Continuous or repeated exposure to hot and, to a lesser degree, cold conditions brings about a gradual adjustment of body functions, resulting in a better tolerance of the climate stress.

Successful acclimation to heat is demonstrated by increased sweat production, lowered skin and core temperature, and reduced heart rate, compared with a person's first reactions to heat exposure. The process of *acclimation* (also called *acclimatization*) is very pronounced within about a week, and full acclimation is achieved within about two weeks. Interruption of heat exposure of just a few days reduces the effects of acclimation and is entirely lost after about two weeks after returning to a more moderate climate.

Acclimation to heat

A healthy person can adjust to heat, whether it is dry or humid. Heat acclimation does not depend on the type of work performed, whether heavy and short or moderate and continuous. Somebody who is healthy and fit acclimates more easily than a person in poor physical condition. Training for physical work cannot replace acclimation, but if specific work must be performed in a hot climate, then such work should be included in the acclimation phase. Since the body can adapt to heat but not to dehydration, liberally drinking water is helpful, both during acclimation and then throughout heat exposure, to replace fluid lost by evaporation of sweat.

Working in heat

It is questionable whether the body actually achieves physiological adjustments to moderate cold when appropriate clothing is worn. There are local acclimations, particularly in the flow of blood in the hands and face. However, normally the adjustment to cold conditions is predominantly reliant on choosing proper clothing and work habits with the result that, in cold temperatures, the body has little or no need to change its rate of heat production or, relatedly, of food intake.

Acclimation to cold?

There are no substantial differences between females and males with respect to their ability to adapt to either cold or hot climates, with women possibly at a slightly higher risk for heat exhaustion and collapse and for cold injuries to extremities. Any slight statistical tendencies can be easily counteracted by ergonomic means and may not be obvious at all when observing only a few people of either gender.

Males and females

6.1.9 Working Strenuously in Heat

Hot or cold climates (as well as air pollution and high altitude, discussed later) affect our abilities to perform short or long, moderate or heavy work in various ways. When exposed to whole-body heating, the human body must maintain its core temperature near 37°C. It does so by raising its skin temperature, increasing blood flow to the skin, accelerating heart rate, and enlarging cardiac output. These circulatory demands compete with the supply needs of internal organs and, specifically, of working muscles.[9]

Cardiovascular effects

The pumping capacity of the heart is between about 25 ("average" adults) and 40 (elite athletes) L/min. The blood vessels in skin and internal organs can accept up to 10 L/min, and all muscles together up to 70 L/min. Since the available cardiac output is half or less of these 80 L, the ability of the heart to pump blood is the limiting factor for muscular work in a hot climate.

Effects on muscles

An increase of muscle temperature above normal does not affect the maximal isometric contraction capability of muscle tissue. However, power output of muscles is reduced at higher (and lower) temperatures. Muscle overheating accelerates the metabolic rate, which can make the muscle ineffective if it must work over an extended period of time. Cooling muscle before exercise and keeping its temperature down during exercise can counteract the loss of power and endurance owing to excessive muscle temperature. Cooling reduces the cardiovascular strain and blood lactic acid concentration, and depletes muscle glycogen at a lower rate.

Dehydration

When working in a hot environment, the body loses water by sweating and thus risks dehydration. Even acute water loss incurred in a short time (in a few hours or less), called *hypohydration*, does not reduce isometric muscle strength (or reaction times) if the water loss is less than 5% of body weight. However, fast and large water loss (such as caused by diuretics) generates the risk of heat exhaustion resulting from fluid volume depletion. Dehydration reduces the body's capacity to perform aerobic endurance-type work.

To counteract water loss, we must drink fluid; plain water is best. If strenuous activities last longer than 1 or 2 hours, diluted sugar additives may help postpone the development of fatigue by reducing muscle glycogen utilization and improving fluid-electrolyte absorption in the small intestine. Regular, liberally salted food (common during meals in many countries) is normally sufficient to counteract salt loss. In fact, salt tablets have been shown to generate stomach upset, nausea, or vomiting in up to 20% of all athletes who took them.

Acclimation

As discussed previously, most heat acclimation takes place during the first week of exposure. First, the cardiovascular system adjusts, enlarging blood plasma volume and decreasing heart rate from the conditions at first exposure to heat. Body core temperature returns to normal after 5–8 days in the heat. The chloride concentration in sweat takes up to 10 days to adapt, as does the production of sweat volume and its controlled evaporation on the skin. Within two weeks of staying in a hot climate, the body has acclimated completely.

Effects on mental performance

It is difficult to evaluate the effects of heat (or cold) on mental or intellectual performance because of large subjective variations and a lack of practical yet objective testing methods. However, as a rule, mental performance deteriorates with rising room temperatures, starting at about 25°C for the nonacclimatized person. That threshold increases to 30 or even 35° if the individual has acclimated to heat. Brain functions are particularly vulnerable to heat; keeping the head cool improves tolerance to elevated deep body temperature. A high level of motivation may also counteract some of the detrimental effects of heat. Thus, in laboratory tests, mental performance is usually not significantly affected by heat as high as 40°C WBGT.

6.1.10 Working Strenuously in Cold

In a cold environment, as in a hot climate, the body must maintain its core temperature near 37°C. When exposed to cold, the human body first responds by peripheral vasoconstriction, which reduces blood flow and consequently lowers skin temperature, in order to decrease heat loss through the skin. If control of blood flow away from the periphery is insufficient to prevent excessive heat loss, shivering sets in.

Vasoconstriction and reduces oxygen to the heart, such that the heart must pump harder. Muscular activities of shivering and of physical work require increased oxygen uptake, which brings about increased cardiac output. With mild hypothermia, this results in an increase in both heart rate (*tachycardia*) and blood pressure. If the body cools below 28°C, moderate or severe hypothermia results in a very slow heart rate (*bradycardia*).

Cardiovascular effects

During light work in the cold, core temperature tends to fall after about 1 hour of activity. Cold sensations in the skin regularly initiate reactions leading to lowered skin temperature, yet areas over active muscles can remain warmer due to the heat generated by muscle metabolism. Thus, in the cold, a relatively large amount of heat is lost through convection (and evaporation). Which of the opposing physiological cold responses predominates depends on the specific conditions, particularly on ambient temperature, type of body activity, and insulation (clothing).

Effects on body temperature

Warming of inhaled air in the respiratory passages is sufficient to preclude cold-related injuries to lung tissues under normal conditions, even if we might sense the coldness of the air in the upper respiratory tract or feel some discomfort and constriction of airways when breathing in very cold air through our mouth.

Effects on respiration

During submaximal work in the cold, oxygen consumption may increase as compared to working at normal temperatures. Some of this increased oxygen cost may relate to shivering, and some may relate to the extra effort required to "work against" heavy clothing worn to insulate against heat loss. At medium work intensities, oxygen consumption due to workload in the cold is about the same as at normal temperatures. The ability for maximal work in a cold climate is not affected as long as the exposure does not exceed about 5 hours: for that period of time, the physiological stimuli provoked by exercise appear to override those of cold. However, if core temperature falls, maximal work capacity is reduced, apparently mostly by suppressing heart rate and thus reducing the transport of oxygen to the working muscles in the bloodstream. Endurance work during cold exposure is negatively impacted if there is a decrease in muscle temperature since this reduces contraction capability and induces early onset of fatigue.

Effects on physical work

Water loss occurs surprisingly easily in a cold environment due to suppressed thirst sensation, sweating (which increases in response to increased energy demands of working in the cold), and more frequent urination (*cold diuresis*). While the dryness of cold air may cause respiratory irritation and discomfort, severe dehydration through the lungs does not occur, since exhaled air is cooled on its way out to nearly the temperature of the inhaled air, returning water vapor by condensation onto the surface of the airways. This explains the common experience of a "runny nose" in the cold.[10]

Dehydration

Most of the counteractions we use to guard against cold exposure serve to improve insulation, such as by wearing proper clothing and staying within sheltered areas.

Acclimation?

Consequently, our body usually carries a fairly normal microclimate during its time in the cold, and hence has little or no need to acclimate. Mostly, decreases in shivering have been observed. It is uncertain whether fitness training facilitates the adaptation to cold environments.

Effects on dexterity

While muscle spindles are initially more active as muscle temperature drops, at about 27°C their activity is reduced to 50% and is completely abolished at about 15°. If the core temperature of the body drops to around 35°, vigilance is reduced and nerve coordination suffers; in the low 30s, apathy sets in and loss of consciousness occurs. Manual dexterity is reduced if finger skin temperatures fall below 20°. Tactile sensitivity is reduced at about 8°; near 5°C, skin receptors for pressure and touch cease to function; the skin feels numb. It becomes difficult to perform finely controlled movements in the cold.

Effects on mental performance

Hypothermia reduces cognitive performance and mood corresponding to declining core body temperature and increasing task complexity. The effects of non-hypothermic cold on mental performance are less consistent, but often also show a deterioration with decreasing temperature. When combined with sleep loss and poor nutrition, such as during disasters or military operations, acute cold stress further degrades vigilance and mood.[11]

6.1.11 Designing the Thermal Environment

There are many ways to generate a thermal environment that suits the physiological functions of the human body. The primary approach is to adjust the physical conditions of the climate (namely temperatures, air humidity, and air or water movement) to influence heating or cooling of the body via radiation, convection, conduction, and evaporation, as listed in Table 6.3.

TABLE 6.3   Designing the Environment to Change the Heat Content of the Human Body

Environment	Radiation	Convection	Conduction	Evaporation
Dry air	Unrelated	Unrelated	Unrelated	Increase
Humid air	Unrelated	Unrelated	Unrelated	Reduce
Stagnant air or water[a]	Unrelated	Little change	Unrelated	Little change
Fast moving air or water[a]	Unrelated	Much change	Unrelated	Much change
Hotter[b] air or water[a]	No direct effect	Increase	Unrelated	No direct effect
Colder[b] air or water[a]	No direct effect	Reduce	Unrelated	No direct effect
Hotter[b] solid	No direct effect	Unrelated	Increase	Unrelated
Colder[b] solid	No direct effect	Unrelated	Reduce	Unrelated
Hotter[b] opposing surface	Increase	Unrelated	Unrelated	Unrelated
Colder[b] opposing surface	Reduce	Unrelated	Unrelated	Unrelated

[a]Water in convection only.
[b]Compared to skin temperature.

MICROCLIMATE

What is important to the individual is not the climate in general (*macroclimate*) but the conditions within which the person interacts directly. We each prefer an individual *microclimate* that feels "comfortable" under our individual conditions of adaptation, clothing, and work.

The optimal microclimate is not only highly individual but also variable. It depends on age and gender: for example, aging people tend to become less active, to lose muscle strength, to reduce their caloric intake, and to initiate sweating at higher skin temperatures. Microclimate also depends on the surface-to-volume ratio, which in children is much larger than in adults, and on the fat-to-lean body mass ratio, generally larger in women than in men.

Variable micro-climate

Thermal comfort depends strongly on the type and intensity of work performed. Physical work in the cold may lead to increased heat production and hence to decreased sensitivity to the cold environment, while hard physical work in a hot climate can become intolerable if an energy balance cannot be achieved.

Type and intensity of work

What we wear also affects the microclimate. Air bubbles contained in the clothing material or between layers provide insulation, both against hot and cold environments. Permeability to fluid (sweat) and air plays a role in heat and cold. The color of the outside of clothes is important in a heat-radiating environment, such as in sunshine, with darker colors absorbing radiated heat and light colors reflecting incident energy.

Clothing

The insulating value of clothing is measured in units of clo, with 1 clo = $0.155°C\ m^2/W$. The clo is not the SI unit of thermal resistance but is analogous to the R-value of house insulation used in the United States. One clo is the amount of clothing needed by an inactive person to feel comfortable in a room at 21°C and 50% relative humidity: such as a shirt, slacks, and a sweater.

Clothing worn determines the surface area of exposed skin (note that the mean surface area of the human body is approximately $1.8\ m^2$). More exposed surface area allows better dissipation of heat in a hot environment but can lead to excessive cooling in the cold. Fingers and toes need special protection in the cold because they have large surfaces of small volumes, away from the warm larger body masses. Head and neck have warm surfaces which release a great deal of heat: this is desirable in a hot environment but not in the cold.

Convection heat loss increases as air moves more swiftly along exposed skin surfaces: this is an important consideration for those who will be spending time outside in cold weather because a still-air temperature measurement does not reflect the impact of wind

Wind-chill

Wind speed (km/h)	Actual air temperature (degrees C)											
	5	0	-5	-10	-15	-20	-25	-30	-35	-40	-45	-50
5	4	-2	-7	-13	-19	-24	-30	-36	-41	-47	-53	-58
10	3	-3	-9	-15	-21	-27	-33	-39	-45	-51	-57	-63
15	2	-4	-11	-17	-23	-29	-35	-41	-48	-54	-60	-66
20	1	-5	-12	-18	-24	-30	-37	-43	-49	-56	-62	-68
25	1	-6	-12	-19	-25	-32	-38	-44	-51	-57	-64	-70
30	0	-6	-13	-20	-26	-33	-39	-46	-52	-59	-65	-72
35	0	-7	-14	-20	-27	-33	-40	-47	-53	-60	-66	-73
40	-1	-7	-14	-21	-27	-34	-41	-48	-54	-61	-68	-74
45	-1	-8	-15	-21	-28	-35	-42	-48	-55	-62	-69	-75
50	-1	-8	-15	-22	-28	-35	-42	-49	-56	-63	-69	-76
55	-2	-8	-15	-22	-29	-36	-43	-50	-57	-63	-70	-77
60	-2	-9	-16	-23	-29	-36	-43	-50	-57	-64	-71	-78
65	-2	-9	-16	-23	-30	-37	-44	-51	-58	-65	-72	-79
70	-2	-9	-16	-23	-30	-37	-44	-51	-58	-65	-72	-80
75	-3	-10	-17	-24	-31	-38	-45	-52	-59	-66	-73	-80
80	-3	-10	-17	-24	-31	-38	-45	-52	-60	-67	-74	-81

FIGURE 6.8 **Wind-chill equivalent temperatures as a function of wind speed and air temperature:** WCT = 13.12 + 0.6215 T − 11.37 $V^{0.16}$ + 0.3965 T $V^{0.16}$. Shading indicates temperatures with high risk of frostbite. *Source: Adapted from the Canadian and US National Weather Services, and Osczevski and Bluestein (2005).*

on body cooling. An appropriate index of the *wind-chill* considers not only how we feel outside but also how long we can tolerate the cold. One of the first wind-chill measures, called Siple–Passel (named after the two explorers who derived it in the early 1940s), was based on how quickly water froze in the Antarctic wind.[12] In 2001, the Canadian and US National Weather Services developed a new chart that reflected the cooling of exposed human skin more realistically than the older data.[13] The wind-chill chart of equivalent temperatures (Fig. 6.8) depicts the impact of wind speed on bare human skin compared to air temperature. For example, a wind of 30 km/h at a DB air temperature of −5°C generates an equivalent calm-air temperature of −13°C. With stronger wind and colder air, the cooling of exposed skin (and risk of frostbite) increases. Covering the skin, for example, with gloves or a face mask, increases insulation against heat loss by wind chill.

Frostbite

Frostbite can occur to exposed warm human skin if the wind chill temperature is below −25°C. Frostbite sets in quickly; it can happen within 10 minutes to warm skin that is suddenly bared to wind chill below −35° and might take less than 2 minutes below −60°. These times are further reduced if the skin is already cool at the start of exposure.

Acclimation

Thermal comfort is also affected by acclimation, the status of the body (and mind) of having adjusted to changed environmental conditions. A climate that was perceived as rather uncomfortable and reducing one's ability to perform physical work during the first day of exposure may be agreeable after two weeks. Seasonal changes in climate, unusual work, different clothing, and attitude have major effects on what we are willing to accept or even to consider comfortable. In the summer most people find warm, windy, and rather humid conditions comfortable, while during the winter we feel that cool and dry weather is normal.

Standards, recommendations

Various combinations of climate factors including temperature, humidity, and air movement can subjectively appear similar. The WBGT (discussed previously) is currently most often used to assess the effects of warm or hot climates on humans; for a cold climate, various similar approaches have been proposed but are not yet universally accepted.[14] Information specific to the built environment in North America, such as in offices, is contained in ANSI-ASHRAE Standard 55. For outdoor activities, recommendations in military and International Standards Organization (ISO) standards are applicable.[15]

6.1.12 Climate: Summary

When working in hot or cold climates, skin temperatures in the range of 32–36°C, associated with core temperatures between 36.7 and 37.1°, are considered comfortable.

With appropriate clothing and light work, comfortable temperature ranges are about 21–27°C WBGT. In the summer, or in a generally warm climate, we are willing to work in the upper end of this range, whereas in the winter or in a generally cool climate, we easily stay in the lower portion. The preferred range of relative humidity is between 30% and 70%. Deviations from these zones are uncomfortable and can make performing physical work difficult or even intolerable. Mental performance is usually neither affected by rather cold temperatures nor by heat as high as 40° WBGT.

Indoors, air temperatures at floor level and at head level should differ by less than about 6°C. Differences in temperatures between body surfaces and side walls should not exceed approximately 10°. Air velocity should not exceed 0.5 m/s, preferably remaining below 0.1 m/s.

For outdoor activities, proper clothing can generate a suitable microclimate for our bodies. Breaks are important to maintain a work-rest regimen.

In a hot environment, short-term maximal exertion of muscle strength is not affected by heat or water loss. However, the ability to perform high-intensity endurance-type physical work is reduced during acclimation to heat, which normally takes up to two weeks. Even after we have achieved acclimation, our cardiovascular system's demands for heat dissipation and our muscles' demands for blood supply continue to compete. The body prefers heat dissipation, resulting in a proportional reduction of performance capability. Dehydration further reduces the ability of the body to work; hypohydration poses acute health risks. Consequently, heat often limits the intensity and duration of work because of the difficulty to dissipate body heat to the hot environment.

In a cold environment, strong isometric muscle exertions are impaired only if the muscles are cold. Overall, cold by itself does not affect dynamic physical work although the ability to do light exercise is a bit reduced. Endurance activities are impaired only if core or muscle temperatures are lowered and if dehydration occurs. Clothing worn for insulation may hinder motions. In extreme cold, dexterity and mental performance suffer.

6.2 POLLUTED AIR

Natural events such as forest fires, dust storms, and volcanic eruptions can fill the air with contaminants, mostly smoke, soot, and dust. Air pollution is often a man-made problem, well known from the smog conditions too often found in large cities such as Los Angeles, Mexico City, London, Beijing, and others. Primary pollutants are particulate matter (PM): solid or liquid particles that are airborne and dispersed and include carbon monoxide and oxides of sulfur and nitrogen. PMs directly affect the body's respiratory system and hence indirectly impact circulation and metabolism (see Chapter 3).

Outdoor and indoor

Outdoor air pollution can be carcinogenic to humans, with the smallest particulates in the air posting the greatest health risk. The risk is further increased with indoor air pollution such as smoke (from tobacco and from fumes that result from cooking and heating with biomass fuels and coal). Exposure to pollution is linked to heart and lung disease, including heart attacks, cancers, asthma, and other respiratory issues.[16]

PM

Particulate matter is a complex mixture of solid and liquid particles of organic and inorganic substances that are suspended in the air. The major components are sulfate, nitrates, ammonia, sodium chloride, black carbon, mineral dust, and water. The most health-damaging particles have a diameter of 10 μm or less ($\leq PM_{10}$), which allows them to lodge deep inside the lungs. Chronic exposure to particles contributes to the risk of developing cardiovascular and respiratory diseases as well as of lung cancer, even at very low concentrations.

Carbon monoxide

Carbon monoxide (CO) is the pollutant that lowers physical work performance most significantly. Hemoglobin in the human blood has an affinity for CO that is 230 times greater than that for oxygen. As a consequence, CO easily attaches to hemoglobin and takes the place of oxygen, thus reducing the ability of blood to provide cells with oxygen. Furthermore, CO attached to hemoglobin causes the remaining binding sites on the hemoglobin molecule to develop a high affinity for oxygen, thereby making it more difficult to release oxygen to the cells that need it.

Sulfur dioxide and nitrogen dioxide

Sulfur dioxide (SO_2) can affect the respiratory system and lung function and causes eye irritation. Inflammation of the respiratory tract causes coughing, mucus secretion, aggravation of asthma and chronic bronchitis; it also makes people more prone to infections of the respiratory tract. Nitrogen dioxide (NO_2) is harmful to children and potentially harmful to adults but does not seem to affect submaximal exercise capabilities, although it can be an irritant in the upper respiratory tract.

Ozone

Ozone (O_3) is formed by the interaction of oxygen, nitrogen dioxide, hydrocarbons, and ultraviolet light; thus, ozone formation is closely tied to sunlight. At ground level, ozone is a potent airway irritant; it can cause breathing problems, trigger asthma, reduce lung function and cause lung diseases. Nonetheless, although we suspect it has a negative impact on performance, no clear physiological impairment of submaximal or maximal performance capability has been definitively demonstrated. Ozone is a secondary pollutant, since it evolves from primary pollutants interacting with each other and with water, salt, and ultraviolet light.

Exhausts and aerosols

Secondary pollutants besides ozone include peroxyacetyl nitrate (PAN) and other aerosols. The primary source of atmospheric PAN is exhaust from combustion engines. While blurred vision and eye irritation are known symptoms of exposure, effects on submaximal or maximal work efforts have not been sufficiently studied. Aerosols, which are formed by the interactions of various acids and salts, can cause discomfort, but have not been definitively linked to decrements in work-performance capabilities.

Effects on physical work

Only carbon monoxide shows a clear detrimental effect on maximal aerobic performance capabilities. Other compounds can cause irritations, but currently there is no evidence that they decrease work capabilities. However, the lack of definitive studies makes providing recommendations difficult.

6.3 HIGH ALTITUDE

The ability to perform strenuous work depends greatly on the supply of oxygen to the working muscles (see Chapter 3). While the oxygen content in the ambient air remains constant at about 21% up to an altitude of at least 100 km above the Earth surface, the barometric pressure reduces significantly with increasing height. Because the air molecules are more dispersed, each breath delivers less oxygen to the body. At an altitude of 3.7 km (12,000 ft), we're breathing in 40% less oxygen than at sea level. At 5.5 km (18,000 ft), we take in 50% less oxygen.

Multiplying the percentage of oxygen by the barometric pressure yields the partial pressure of oxygen. At sea level, where the barometric pressure is 760 Torr (mmHg), the partial pressure of oxygen in the ambient air is 160 Torr. At 3000 m altitude, the barometric pressure is about 530 Torr; hence the partial pressure of oxygen is about 110 Torr, a reduction of 30% from sea level.

Cognitive functions in people who climb high mountains (4200 m or higher) decline: in comparison to performance at sea level, at altitude they show diminished success in standardized perceptual, cognitive and sensory-motor tasks as well as reduced learning and retention of new perceptual and cognitive skills.[17]

Effects on cognition

Lack of appetite is very common at altitudes of 5000 m or higher, as is not feeling thirst. Thus, weight loss and dehydration impair physical function and energy expenditure after just a few days.[18]

Effects on body mass

Oxygen, like any other gas, naturally moves from higher to lower concentrations. In accordance to this general rule, the partial pressure of oxygen in higher altitudes is reduced, so our bodies' ability to supply cells with oxygen is also reduced. This is critically important for the mitochondria in our muscles, where most of the energy for physical work is generated. A series of processes determines the body's ability to bring oxygen to the mitochondria.

Effects on gas exchange

- Breathing: air is moved in and out of the lungs.
- Blood oxygenation: in the lungs, oxygen transfers from the air across lung tissue and into the bloodstream, where it combines with hemoglobin.
- Circulation: oxygen-carrying hemoglobin is then transported in the bloodstream to the muscle cells.
- Muscle supply: oxygen diffuses out of capillaries into the cell and finally to the mitochondrion.

For physical and physiological reasons, ventilation, the first process of oxygenating mitochondria, is actually facilitated at altitudes above 3000 m. Physical work also increases ventilation. Consequently, effects on breathing at increasing altitude do not limit our capacity to work.

Effects on breathing

The second process is the diffusion of oxygen from the lungs to the blood. With increasing altitude, partial oxygen pressure in the lungs is reduced and hence diminishes the

Effects on blood oxygenation

difference between lung air pressure and blood pressure. This in turn slows diffusion. Since blood flow velocity is increased during physical work because of increased cardiac output, the time available for oxygen diffusion in the lungs to each passing hemoglobin cell is reduced. Hence, blood oxidation falls with increasing altitude.

Effects on circulation

The next step in the process is to make oxygen-rich blood available in the arterial (systemic) branch of circulation. The oxygen content of arterial blood depends on the hemoglobin concentration in the blood, and on the ability of the hemoglobin to attract oxygen. Within the first few hours of exposure to altitude, the ability to carry oxygen remains the same as at sea level, but the actual oxygen content in the blood is reduced due to reduced diffusion in the lungs. However, after a few hours of altitude exposure above 3000 m, there is a shift in fluid distribution in the body: blood plasma volume in the circulatory blood stream decreases, because up to 30% of volume moves into cells. But with acclimation, blood volume may increase slightly and red-cell production is stimulated, resulting in an increase in hemoglobin concentration in the blood. As a result, in spite of the volume shift and because of the increase of hemoglobin concentration in the flowing blood, the ability of arterial blood to carry oxygen remains at approximately sea-level values.

Effects on oxygen supply to muscles

The next step in the oxygen-transport process is providing oxidized blood to the working muscles, specifically the mitochondria. During the first few days of exposure to altitude, their oxygen supply is diminished because of the reduced oxygen content in the blood (which is the result of reduced diffusion in the lungs). As the exposure to altitude continues and altitude acclimation is achieved (see below), the oxygen content in arterial blood returns to sea-level values.

The final process in the oxygen transport chain is the oxygenation of tissue. At altitude, hemoglobin releases oxygen to the tissues more easily than at sea level. With acclimation to altitude, the capillaries in muscle tissues are enlarged, and as a result, the diffusion of oxygen from blood to the cell becomes easier.

Cardiovascular effects

A person's ability to perform stressful aerobic work depends strongly on the heart's ability to move blood through the body, so that oxygen can be provided to metabolizing muscles and that metabolic byproducts can be removed from them. The cardiac output (minute volume) is essentially the product of stroke volume and heart rate. Cardiac output is less affected at lower altitudes but shows marked changes in heights above about 1500 m. The pumped blood volume actually increases at rest and during submaximal efforts, but after about two days, exposure to altitude causes the volume to become progressively reduced. After about two weeks of staying at altitude, cardiac output is lowered—at all levels of effort—to a minute volume below that at sea level, and stays at that low volume for the duration of the stay at altitude. This reduction is primarily due to reduced stroke volume, which, in turn, mostly follows from the reduced blood-plasma volume.

Physiological adjustments

The body's initial response to high altitude is to increase the number of breaths taken per minute and the depth of each breath (tidal volume). This increased ventilation increases the pressure of oxygen within the lungs, and facilitates the release of carbon dioxide to the air.

Another adjustment is the redistribution of fluid in the body, as discussed earlier. The reduction in blood-plasma volume occurs within hours of arriving at altitudes of more than 3000 m. With long altitude stays, some of the blood is redistributed, but it does not return to sea-level conditions.

During the first few hours of exposure to altitude, reduced oxygen content in the arterial blood is caused by the difficulty of diffusing sufficient oxygen from the air in the lungs into the blood. Adaptation to height begins thereafter, mostly via relative and absolute increases of hemoglobin in the circulating blood and by a slight increase in blood volume. This restores the blood's ability to bring oxygen to the working cells. However, cardiac output capability remains suppressed throughout the stay at higher altitudes. Therefore, human capability for highly demanding physical activity remains reduced at altitudes above 1500 m, even after altitude adaptation.

Rapid change in fluid distribution is associated with several well-known altitude-related medical problems. Acute mountain sickness (AMS) commonly occurs at heights above 3000 m and is directly related to the rate of ascent and to the final altitude. It brings about headaches, lassitude, nausea, insomnia, irritability, and depression along with impaired emotional control. Symptoms become apparent after several hours of exposure and become most severe within one or two days, after which they recede. AMS can be reduced or eliminated by ascending gradually and through medication. While AMS is often debilitating, it is also self-limiting. In some people who rapidly gain altitude, excessive accumulation of fluid in spaces between cells or in the cells (*edema*) can occur in the brain or lungs. Edema is potentially life threatening but can be counteracted by immediate evacuation to lower elevations and through medical intervention and assistance.[19]

"Altitude sickness"

6.3.1 High Altitudes: Summary

Short-time high-intensity activities ("explosive efforts") do not suffer with increasing altitude because they are anaerobic and hence do not depend on oxygen transport. Likewise, short exertions of muscle strength are not reduced during acute exposure to high altitude.

The capacity for both submaximal and maximal aerobic work remains at sea-level ability up to about 1500 m altitude. Above 1500 m, maximal work capacity decreases at a rate of approximately 10% per 1000 m (with significant variation among individuals). This reduction persists for the entire stay at that elevated height.

Submaximal work capacity is not affected up to about 3000 m, but any given task requires a larger percentage of the available (reduced) maximal capacity than at sea level. As a result, the ability to endure such submaximal efforts is also reduced at altitude: the longer the effort, the greater the decrement.

Cognitive performance is reduced at high altitudes above 4200 m.

6.4 VIBRATIONS AND IMPACTS

Vibration is an oscillatory motion around a fixed point, called *periodic* when the oscillation repeats itself. People commonly experience vibration when it is transmitted to them: from a seat, such as when in a vehicle; from the floor on which they stand, such as when in a ship near its engine; or from a handheld tool that oscillates, such as when operating a jackhammer.

Vibrations, especially when they last over a considerable amount of time, and *impacts* (short, often violent shocks) are of ergonomic concern particularly in human-operated systems. Operators who experience and must tolerate such effects include those in earthmoving equipment, tanks, airplanes, spacecraft, and water vessels. Models of vibration generally consider that the body continues to vibrate, at the same frequency, over a considerable period of time; the simplest way of describing this is by a sinusoidal equation. In contrast, non-periodic vibrations are generally considered as one-time shocks (impacts).

Parts vibrate differently

The human body reacts to different kinds of vibration in various manners. When the imposed vibration is translational (linear), and if the body is considered rigid, then all its parts undergo the same motion. When the imposed vibration is rotational, even for a rigid body, not all of its parts follow the same path of motion. Since the human body is not actually rigid, different body parts vibrate dissimilarly even if subjected to the same linear and/or rotational vibration.

Observed health effects

Epidemiological investigations of disorders possibly caused by vibration[20] are complicated because many adverse health effects could also be related to other conditions, including hazardous working conditions, awkward body postures, repetitive activities, and poor diet.

Interest in the effects of impacts and vibration on the spinal column has been spurred in particular by complaints from heavy equipment and vehicle operators of frequent and persistent back problems. Issues resulting from head vibrations are also widely reported, many with respect to the ability to see visual targets and to avoid motion sickness. Vibrations of the hand (such as from jackhammer use) remain problematic. Other well-established concerns that arise as a result of vibrations include gastro-intestinal events, cardiac problems, and hearing loss.

In 1862, the physician Maurice Raynaud described a disease of jackhammer operators which was clearly vibration-induced: their fingers became pale and cold because the vibrations caused the muscles around blood vessels to contract, reducing blood supply to the area. The condition was aptly (if not charmingly) called "white" or "dead" finger; it now carries the name *Raynaud's disease*.

Measurements

To understand the impact of vibrations on the human body, we must first measure them; this is not always easy. Measurements of spinal vibrations are difficult to perform: x-rays are no longer used due to concerns about radiation danger; chemically inert nails inserted into the vertebrae are unappealing (to say the least) to volunteers. Measurement of vibrational motion of the head is much simpler: record the movement of a bar extended from a dental mold which is clenched between the teeth. Hand vibration responses are easily observed, but vibrations transmitted at the

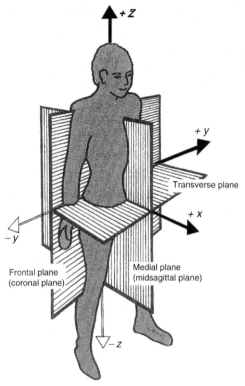

FIGURE 6.9   **Convention on directions of vibrations and impacts. Note the directions of *x*, *y*, and *z*—see text.** *Source: Adapted from Kroemer, Kroemer, and Kroemer (2010).*

interface between the tool and the hand of the operator can be difficult to measure without affecting the handling of the tool. One indirect approach is to attach accelerometers to the tool and/or the finger/hand/forearm. In spite of these difficulties, extensive information is available from experimental measurements and epidemiological investigations.[21]

6.4.1 Describing Vibration

In this book, the coordinate system we use to describe the direction of mechanical vibration and impact on the human features is described in Fig. 6.9 (same as Fig. 1.1 in Chapter 1):

- the $+x$ axis pointing in forward direction from the body,
- the $+y$ axis goes to its left side of the body, and
- the $+z$ axis points head-ward up.

Note that vibration/impact and aerospace research occasionally employs different notations.

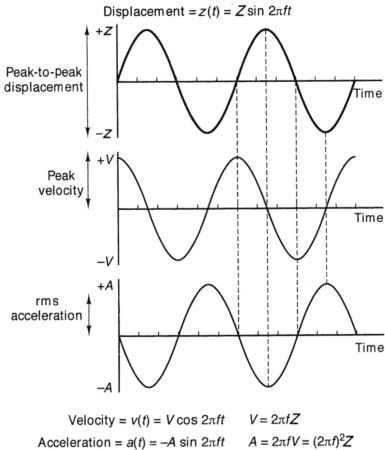

FIGURE 6.10   Displacement, velocity and acceleration, over time, are descriptors of sinusoidal vibration.

Forcing functions

Vibrations or impacts are generated by energy external to the equipment and are often transmitted to the human through a seat. Typically, the exciting function stems from a vehicle moving along an uneven road; these irregular events can be recorded and simulated in the laboratory. Such simulations can be realistic but experimental control is difficult. Therefore, the *forcing function* is often generated in the form of a harmonic excitation wave (or as one-time triangular or square excitation for impact research—see below). Although a mathematically harmonic forcing function (also called *potential function*) is not likely to be encountered in the real world, understanding the behavior of a system undergoing sinusoidal harmonic excitation helps to understand how this system would respond to more realistic inputs.

Sinusoidal vibration

Fig. 6.10 shows the plot of a basic sinusoidal vibration. Essential responses of the system include: the amplitude of the displacement, time derivatives of displacement (velocity and acceleration), and the involved forces.

The rms value is the square root of the mean of the squared values. The rms value is often more useful than the statistical mean of either displacement, velocity or acceleration of a periodic vibration: the average numerical value over each complete period is zero.

If the oscillation of a mass follows a sine wave, its displacement has a peak amplitude A above the stationary location and the opposite peak $-A$ below. Its peak-to-peak displacement is $2A$; its rms magnitude is $0.707A$. It's important to accurately denote the descriptor's peak, peak-to-peak, or rms.

Displacement

Describing the magnitude of oscillation by velocity is a perfectly acceptable but seldom used because measuring acceleration is, at present, more convenient.

Velocity

The common measure of acceleration is by its rms value. Acceleration can be stated in g units, with $1\,g = 9.80665\,m/s^2$: the acceleration on Earth due to gravity. The logarithmic decibel (dB) scale is also used—see Tables 6.4 and 6.5.

Acceleration

The duration of the vibration is important to ergonomics; unfortunately this is not at all reflected in the above measures of displacement, velocity and acceleration. For example, there are many possible rms values for the same vibration depending on the time period during which it is calculated. The effects of the duration of exposure are usually explained in terms of "dose" values.

Effects of time

In a simple harmonic motion, there is sinusoidal oscillation at a single frequency, but more realistic forcing functions contain vibrations of several frequencies. Fourier analysis is the analysis of a complex waveform expressed as a series of sinusoidal functions, the frequencies of which form a harmonic series; this analysis can decompose such complex vibrations into single super-positioned sinusoidal events. In some cases

Mixes of frequencies

TABLE 6.4 Conversions between Decibels and Units of Acceleration and Velocity

Decibel (dB)	Acceleration (m/s²)	Velocity (m/s)
−20	10^{-7}	10^{-10}
0	10^{-6}	10^{-9}
20	10^{-5}	10^{-8}
40	10^{-4}	10^{-7}
60	10^{-3}	10^{-6}
80	10^{-2}	10^{-5}
100	10^{-1}	10^{-4}
120	1	10^{-3}
140	10	10^{-2}
160	10^{2}	10^{-1}
180	10^{3}	1
200	10^{4}	10

TABLE 6.5 Reference Quantities for Decibel Levels

Description	Definition (dB)	Reference quantity
Vibration acceleration level	20 log (a/a_0)	10^{-6} m/s^2
Vibration velocity level	20 log (v/v_0)	10^{-9} m/s
Vibration displacement level	20 log (d/d_0)	10^{-11} m
Vibration force level	20 log (F/F_0)	10^{-6} N
Power level	10 log (P/P_0)	10^{-12} W
Intensity level	10 log (I/I_0)	10^{-2} W/m^2
Energy density level	10 log (w/w_0)	10^{-12} J/m^3
Energy level	10 log (E/E_0)	10^{-12} J
Sound pressure level in air	20 log (p/p_0)	2×10^{-5} Pa
Other-than-air sound pressure level	20 log (p/p_0)	10^{-6} Pa

Note: log = log$_{10}$.
Source: Adapted from ISO 1683 (1983b and 2015).

these frequencies are *harmonics*, integer multiples of the lowest frequency, also called the *fundamental frequency*. A *spectrum* describes how the vibration magnitude varies over a range of frequencies.

Octave bands

Oscillation is most often measured in either octave or part-octave bands. An octave is the interval between two frequencies when one frequency is twice that of the other. If f$_1$ and f$_2$ are the lower and upper frequencies of the bands, then

1/1 octave: $f_2 = 2f_1$
1/2 octave: $f_2 = 2^{1/2}f_1$
1/3 octave: $f_2 = 2^{1/3}f_1$
1/6 octave: $f_2 = 2^{1/6}f_1$

For example: when centered on 1 Hz, the octave band is from 0.707 to 1.414 Hz, while the third-octave band extends from 0.891 to 1.122 Hz. In these cases, the bandwidth increases in proportion to frequency. Another method is to determine the frequency content using a constant bandwidth (such as 0.1 or 1.0 Hz) at all frequencies.

Period, frequency

For a sinusoidal waveform of the vibratory motion, the reciprocal of the *period T* is the *frequency f*, which describes the number of cycles of motion per second, expressed in Hertz (Hz). Also often used is the *angular frequency ω*, expressed in radians per second: since a complete cycle (360 degrees) corresponds to 2π radians, $\omega = 2\pi f$, in units of rad/s^2.

Instantaneous values

The phases of velocity and acceleration are ahead of displacement by 90 and 180 degrees, respectively, as shown in Fig. 6.10. At the maximum displacement, the velocity is zero and the acceleration is at a minimum; when the displacement is zero, the velocity is maximal and the acceleration is zero.

At a given time *t*, the *instantaneous displacement z* is described as a function of *t* by:

$$z(t) = Z \sin (2\pi ft + \varphi) \tag{6.11}$$

where Z is the peak displacement and φ is the phase angle (time delay).

TABLE 6.6A Conversion of Parameters for Sinusoidal Vibration in z Direction

	Displacement Z	Velocity V	Acceleration A	Jerk J
Displacement Z	Z	$Z = V/(2\pi f)$	$Z = A/(2\pi f)^2$	$Z = J/(2\pi f)^3$
Velocity V	$V = Z(2\pi f)$	V	$V = A/(2\pi f)$	$V = J/(2\pi f)^2$
Acceleration A	$A = Z(2\pi f)^2$	$A = V(2\pi f)$	A	$AV = J/(2\pi f)$
Jerk J	$J = Z(2\pi f)^3$	$J = V(2\pi f)^2$	$J = A(2\pi f)$	J

The *instantaneous velocity v* of the motion is the first time-derivative of displacement:

$$v(t) = 2\pi f Z \cos(2\pi ft) = V \cos(2\pi ft) \tag{6.12}$$

where $V = 2\pi f Z$ is the peak velocity.

The *instantaneous acceleration a* of this motion is the time-derivative of velocity:

$$a(t) = -(2\pi f)^2 Z \sin(2\pi ft) = -A \sin(2\pi ft) \tag{6.13}$$

where $A = (2\pi f)^2 Z = 2\pi f V$ is the peak acceleration.

Higher order derivatives are also sometimes used. For example, the *jerk* is the time-derivative of acceleration, where peak jerk equals $(2\pi f)^3$, or $(2\pi f)^2 V$, or $2\pi f A$.

In a simple system, there are fixed relations between acceleration, velocity, and displacement in sinusoidal vibration. Changing from velocity to acceleration adds a 90-degree phase lag, and the values change according to $(2\pi f)$, as listed in Tables 6.6A and 6.6B. The table also shows how to convert the measures of displacement, velocity, acceleration, and jerk; it also contains conversions between peak, peak-to-peak, and rms values.

Phase lags

6.4.2 Whole-body Vibration

Local vibration exists when one single body segment, such as the hand, is subject to vibration. Whole-body vibration occurs when a person stands, lies, or sits on a vibrating surface. Vibration is then transmitted unevenly throughout the body—consider a man operating a fork lift while sitting on a vibrating seat; his upper body experiences displacement while his feet remain still. Furthermore, people sitting in moving vehicles are often exposed to simultaneous local vibrations—consider the driver of a passenger vehicle, her head vibrating from a headrest, her back from a backrest, her hands from

TABLE 6.6B Conversion of Parameters for Sinusoidal Vibration in z Direction

	Peak	Peak-to-peak	rms
Peak	Peak	0.5 (peak-to-peak)	1.414 rms
Peak-to-peak	2 (peak)	Peak-to-peak	2.828 rms
rms	0.707 (peak)	0.3535 (peak-to-peak)	rms

a steering wheel, and her feet from the floor, possibly all in various and varying directions. Thus, the distinction between whole-body and local vibration is not always clear and simulation of real-world conditions may be extremely complicated.

For simplification and standardization of measurements, the exciting vibration is commonly sinusoidal (unless specifically defined otherwise); in one of the three basic orientation axes as shown in Fig. 6.9 (usually in the z direction); and centered in and pointed along a main anatomic axis of the body.[22]

Newton's second law states that a force F applied to an object produces an acceleration a which is proportional to and in phase with the force. The constant of proportionality is the mass m: $m = F/a$. If force F is applied to a (massless, ideal) damper, it produces a velocity v which is proportional to and in phase with the force. This constant of proportionality is called *damping* c: $c = F/v$. If a force F is applied to a (massless, ideal) spring, it produces a displacement x which is proportional to and in phase with the force. The constant of proportionality is called *spring stiffness* k: $k = F/x$.

Different parts of the body react to the same external vibration in disparate manners. Therefore, individual body components need to be represented by their specific masses, springs, and dampers. An undamped mass-spring component is shown in Fig. 6.11. Its mass m oscillates freely in form of a sine wave:

$$z = Z \sin(\omega_n t + \phi) \tag{6.14}$$

where ω_n denotes its natural frequency in rad/s, ϕ is the phase angle, k describes the spring. The acceleration is:

$$a = \omega^2 Z \sin(\omega_n t + \phi) \tag{6.15}$$

If the system includes damping (coefficient c), as in Fig. 6.12, the damper absorbs energy and over time decays the motion of the system. The equation $ma + cv + kz = 0$ describes the involved forces when the system is in balance.

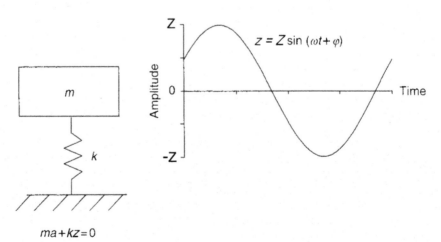

$$ma + kz = 0$$

FIGURE 6.11   Vibration of a mass-spring system.

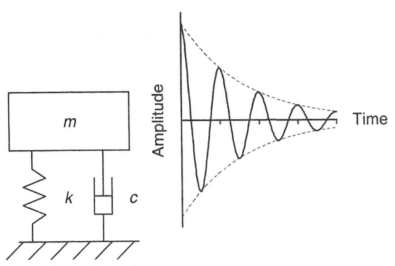

FIGURE 6.12 **Vibration of a damped mass-spring system.**

These idealized elements are arranged to be either parallel or in series with each other, such as shown in Fig. 6.13, which depicts a model used for assessing the effects of single-axis (z) vibration on a standing human.

6.4.3 Modeling the Elastic Human Body

Understanding and modeling the responses of the human body to vibrations (and to impacts) is complex. The motions of the whole system and of its elements are commonly approximated by sinusoidal displacements and forces. However, such simulation is subject to many assumptions, both about the causing vibrations and about the body components' characteristics. The input signal may not truly represent what, for example, a given road imposes on a vehicle seat in terms of various intensities and types of vibrations (and impacts) in varied directions. Furthermore, there is great inter- and intra-individual variability in body responses, with differences depending on such aspects as body size and posture, and on support and restraint systems. The response of the body cannot be simply explained in terms of resonances: body elements, if taken in isolation, do show specific natural frequencies, but these are highly damped. Further, the interactions between differently-vibrating body segments may generate a complex network of causal and temporal sequences.

In most studies, the dynamic response of the body is assumed to be *linear*, that is, proportional to the excitation. Dynamic responses of the body or its parts are usually described by *transfer functions* which apply at certain frequencies.

Impedance transfer function relates two measures obtained at different points of the body. If we are interested only in the magnitude, for example, of head excursion in relation to seat displacement, we are referring to *transmissibility*. *Comfort* curves are assumed to be the inverse of transmissibility. Figs. 6.14 and 6.15 depict examples of observed transmissibility.

Transfer functions

Impedance, transmissibility

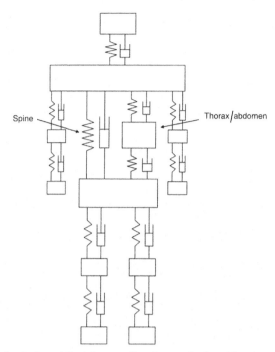

FIGURE 6.13   A mechanical model of the standing human body with masses, springs, and dampers. *Source: Adapted from Village and Morrison (2008).*

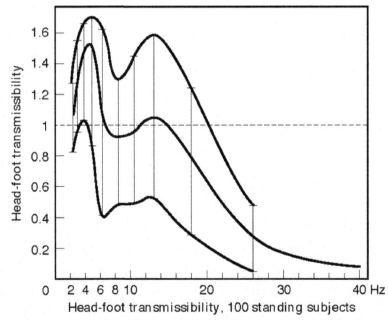

FIGURE 6.14   Typical head-foot transmissibility: the curves represent the mean and 67% boundaries observed on standing subjects. Note the resonances at about 4 and 14 Hz. *Source: Adapted from Oborne (1983).*

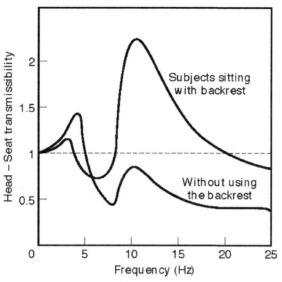

FIGURE 6.15 **Example of head-seat transmissibility when either leaning against a backrest or not using the backrest.** *Source: Modified from Griffin (1990).*

Mechanical transfer function relates two different measures obtained at the same point of the body. It considers the ratios between the force that drives a system at a particular frequency and the resulting movements, such as displacement, velocity, and acceleration.

> Mechanical impedance

In a rigid body, force and acceleration are always in phase and, at any frequency, the ratio of the rms magnitudes reflect the mass of the object. At most frequencies, however, the human body does not behave as if rigid, and force and acceleration are out of phase depending on the stiffness and the damping at each frequency. Of course, we can still calculate the ratio of force to acceleration, but it no longer equals the static mass of the object. Consequently, we reference *apparent* or *effective mass*. Similarly, the properties of dampers and springs have different effects on movement of masses at changing frequencies; thus the term *impedance* is used instead of *damping*; *dynamic stiffness* or *dynamic modulus* replaces *stiffness*. The inverse ratios also have distinct names: acceleration divided by force is called *accelerance* (or *inertiance*), velocity divided by force is called *mobility*, and displacement divided by force is called *dynamic compliance*.

> The human body is not rigid

The basic example in Fig. 6.16 points out how to model a complex system as a set of simple subsystems with discrete components (of mass, damping, and stiffness) which, together, should have the same mechanical impedance as the complex body. A typical model of the sitting body involves at least two or three masses, as sketched in Fig. 6.16. There, m_1 is the mass of the upper trunk that moves relative to the platform supporting the seated body; m_2 is the partial mass of the lower trunk and of the thighs that follows the platform in phase. If the feet and lower legs do not move with the seat, their stationary mass is identified by m_3. Properly determining these masses and their stiffness and damping parameters is essential to making the model realistic.

> Sets of subsystems

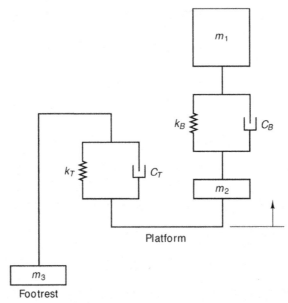

FIGURE 6.16   Two-degree-of-freedom model of a subject sitting on a vibrating platform. *Source: Modified from Griffin (1990).*

A more complex model is needed to describe the reactions of the different body parts of a person standing on a vibrating surface. Fig. 6.17 shows a solution according to 1987 ISO Standard 7962, meant to be applicable up to a frequency of 31.5 Hz. This solution lists parameters for masses, springs and dampers, all of which need to be adjusted carefully for different subjects and conditions.

USE OF STANDARDS

A *standard* is a document that provides requirements, specifications, guidelines or characteristics that can be used consistently to ensure that materials, products, processes and services are fit for their purpose. All standards reflect consensus, often compromise, among various parties such as scientists, engineers, industry insiders, business representatives and regulators. For businesses, standards can be strategic tools.

Currently available norms on vibrations and impacts, from the ISO or from others, must be used with great caution. It remains difficult to correctly model specific individuals' responses to vibration, even of the human body in general. In some cases, the data base is too meager to allow to establish, with certainty, forces on and movements of the body; and insufficient to predict the effects on comfort, health, or performance. Michael Griffin's statement, demonstrating candor and humor, is still true: "Many models have achieved complexity without representing the known behavior of the body."[23]

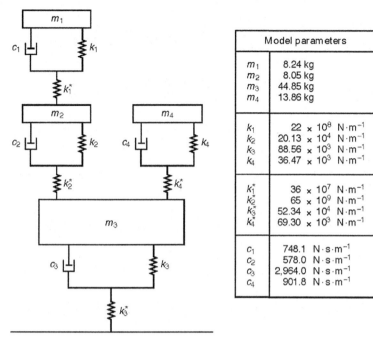

Model parameters

m_1	8.24 kg
m_2	8.05 kg
m_3	44.85 kg
m_4	13.86 kg
k_1	22×10^8 N·m^{-1}
k_2	20.13×10^4 N·m^{-1}
k_3	88.56×10^3 N·m^{-1}
k_4	36.47×10^3 N·m^{-1}
k_1^*	36×10^7 N·m^{-1}
k_2^*	65×10^9 N·m^{-1}
k_3^*	52.34×10^4 N·m^{-1}
k_4^*	69.30×10^3 N·m^{-1}
c_1	748.1 N·s·m^{-1}
c_2	578.0 N·s·m^{-1}
c_3	2,964.0 N·s·m^{-1}
c_4	901.8 N·s·m^{-1}

FIGURE 6.17 Four-degree-of-freedom model for calculating the vertical transmissibility of a 75 kg subject standing on a vibrating surface. *Source: Adapted from ISO 7962 (1987).*

6.4.4 Vibration Effects on Comfort, Performance, Health

Vibration can produce a wide range of effects, depending on frequency, intensity, and direction of pulsation, and on the body parts to which it is transmitted. The cause-and-effect relationships between vibrations and human responses are complex and often difficult to research. In the laboratory, vibration equipment may not be able to generate pure sinusoidal motions; or it may not be able to generate concurrent high accelerations, large displacements and large forces.

The reactions of the human body are quite different from person to person; they depend on muscle tension and posture of the subject as well as on restraining devices used. In the real world, vibration effects on the human are often not the sole stressors; for example, it may be quite difficult to determine specifically how vibrations or impacts experienced by a truck driver may affect that person's performance and well-being because the person was driving the whole day in varying conditions of traffic, weather, and noise.

> While oscillation is often unwanted, there are also good vibrations, such as those associated with the pleasant feelings of shaking hands, sitting on a rocking chair, or laughing. Pulsations can be a source of excitement; imagine thrilling rides at an amusement park, skiing on bumpy slopes, sailing over a wave-etched lake. Vibration has even been advocated for improving joint mobility of athletes or of patients suffering from arthritis.[24]

Vertical vibration

Most parts of the human body move together under vertical oscillation at frequencies below 2 Hz. The associated sensation is that of alternately being pushed up and then floating down. The eyes are usually able to follow objects that either move with the body or are stationary. However, free movements of the hand may be disturbed, which can cause problems in activities that require exact positioning of the hands. If the vertical oscillation is below 0.5 Hz, it may cause motion sickness.

Vertical oscillation at frequencies above 2 Hz brings about amplification of the vibration within the body; consequently, body parts no longer move uniformly. The frequencies with greatest amplification, the resonance frequencies, are different for different parts of the body, for different individuals, and for different body postures. At frequencies between 4 and 5 Hz, resonances occur, for example, in the head and hands; discomfort is strongly felt.

At frequencies above 5 Hz, the force required to generate a given vertical acceleration falls rapidly with increasing frequency; thus, the vibration reaching the head, and its associated discomfort, decrease. But at frequencies between 10 and 20 Hz, the voice may warble, and vision may be affected, particularly at frequencies between 15 and 60 Hz because of resonances of the eyeballs within the head. Table 6.7 lists approximate resonance frequencies for the body and its parts.

Standing or sitting

When standing, keeping the knees straight or bent can greatly influence the effects of frequencies above 2 Hz in vertical vibration. When sitting, the design of the seat has little effect at frequencies below 2 Hz, but "soft" seats, such as in many automobiles, can greatly amplify vertical vibrations, even doubling the experienced frequency. The design of seats to be used in vibrating environments can greatly influence the vibration effects experienced by the seated person.

Horizontal vibration

Sideways or fore-and-aft vibration of the seated body below 1 Hz sways the body, even if resisted by muscle action. Between 1 and 3 Hz, it is difficult to stabilize the upper parts of the body, a situation that is associated with great discomfort. With increasing frequency, horizontal vibration is less readily transmitted to the upper body, so that at frequencies above 10 Hz the vibration is mostly felt at the seat surface. A backrest can greatly influence the effect of horizontal vibration: at low frequencies it can help to stabilize the upper body, but at high frequencies it strongly transmits vibration to the upper body, primarily in the anterior–posterior direction (see also Fig. 6.15 for head-seat transmissibility in the vertical direction).

XYZ vibration

Often, horizontal and vertical impulses are present at the same time. Body parts may move in several planes even if stimulated in only one direction: for example, the head may perform pitch motions in the medial plane even if the vibration applied to the body is strictly vertical. The main head movements induced by vibrations in x, y, or z directions, transmitted through a rigid seat without backrest, are listed in Table 6.8. However, not only are there large variations in vibration responses among individuals, but their posture and muscle tension can greatly affect the reactions as well.

Subjective assessment

Stanley Smith Stevens's power law proposes that the magnitude of a subjective sensation increases in proportion to a power of the stimulus intensity. It relates the perceived sensation P to the magnitude (intensity) of the physical stimulus I by $P = k\,I^n$, where k is a constant and exponent n has mostly been in the range of 0.9 and 1.2.[25] This allows us to establish subjective ratings of the sensations associated with vibrations.

TABLE 6.7 Body Resonances Caused by Vibrations in z Direction

Body part	Resonances (Hz)	Symptoms
Whole body	1–2 4–5, 10–14	Motion sickness, sleepiness General discomfort
Upper body	6–10	
Head	5–20	
Brain	Below 0.5	
Eyeballs	1–100; mostly above 8, strongly 60–90	Difficulty seeing
Skull, jaw	100–200	
Larynx	5–20	Changes in pitch of voice
Shoulder	2–10	
Lower arm	16–30	
Hand	4–5	
Trunk	3–7	
Chest	5–7	Chest pain
Spine	Around 5	
Pelvis	around 5 and 9	
Chest wall	60	
Heart	4–6	
Stomach	3–6	
Abdomen	4–8	Abdominal pain
Lower intestine	Around 8 when seated	
Bladder	10–18	Urge to urinate
Cardiovascular and respiratory systems	2–20	Similar to responses to moderate work

Note: 1. The frequency interacts with displacement and acceleration; certain combinations can generate specific effects on sensation, performance, and health. 2. Excitation in z direction can produce body reactions in x and y directions as well. 3. Body posture, muscle tension, and body restraints may affect responses strongly.

TABLE 6.8 Large Head Movements Caused by Vibrations in x–y–z Directions Transmitted through a Rigid Seat without Backrest

	Driving vibration in z direction	Driving vibration in x direction	Driving vibration in y direction
Causes linear motion in z direction	Up to 20 Hz	2–6 Hz	below 3 Hz
Causes linear motion in x direction	5–10 Hz	Below 4 Hz	below 4 Hz
Causes linear motion in y direction	–	–	below 3 Hz
Causes yaw rotation about z	–	–	–
Causes roll rotation about x	–	–	Around 2 Hz
Causes pitch rotation about y	About 5 Hz	2–6 Hz	–

Source: Adapted from data from Griffin (1990).

I. THE ERGONOMIC KNOWLEDGE BASE

Semantic scale assessment of vehicle rides was first issued in 1973 by the Society of Automotive Engineers (SAE). These and similar approaches have been used to establish "comfort" contours similar to those developed to assess noise. In general, people of differing body sizes, ages and genders report the same sensations associated with vibration, although larger subjects tend to be less sensitive to frequency below 6 Hz. Various standards (including ISO 2631 and BS 6841)[26] delineate time-dependent comfort boundaries in whole-body vibrations for combinations of frequencies and accelerations, indicating that vibrations might be tolerable over a defined time period.

Vibration and noise

In many cases, noise accompanies vibrations. Excessive noise can damage hearing and impair performance. Further, subjects' responses indicate similarity between the sensations of noise and vibration, as expressed by Stevens's power law, which may allow formulating conditions of equivalence. As a result, when both mechanical and acoustical vibrations are present, reduction in one may not only improve the overall perception of the environment, but might in fact reduce the perception of the other, unchanged stimulus.

Performance

There are general regions of interest where vibrations impact performance—see Fig. 6.18. However, research on the effects of mechanical vibrations on mental activities, state of arousal, ability to make decisions, or attitudes is incomplete. Vibration does not seem to affect performance on simple perceptual tasks involving auditory or visual detection of signals; but complex cognitive tasks may suffer.[27] However, known biomechanical effects of body vibration concern the eyes as principal information input ports, and mouth and hands as the principal output means. Performance can be similarly affected when either the body oscillates while the outside remains stationary, or when the visual or manual target vibrates while the body is still.

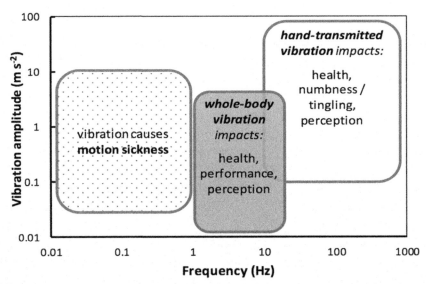

FIGURE 6.18 General ranges and frequencies of interest for motion sickness and for vibration of the whole body or transmitted through the hand. *Source: Adapted from Mansfield (2005).*

If a visual target oscillates slowly, our eyes pursue the movement and maintain a stable image on the retina as part of the reflex response. The human eyes are able to perform pursuit reflexes for display oscillations of up to about 1 Hz. At higher frequencies, the saccadic movements are too slow and the image becomes blurred. When the observer is vibrating, the head and eyes experience both translational and rotational movements. The body's motoric compensatory activities become increasingly insufficient as the frequencies exceed 8 Hz; vision problems occur when the apparent displacement of the visual object gets larger than 1 min of arc.[28] In general, the effects of translational vibration decrease with large viewing distance, while the effects of rotational oscillation are independent of distance.

Visual task performance

Although the mechanical characteristics of many machines and of their controls fortunately attenuate vibration at frequencies above 1 Hz, manual operation of highly sensitive systems, or tasks that require small or precise movements, may nonetheless be strongly affected by vibratory environments (see Table 6.8). Manual activity may be discrete, such as pressing a button, or the effort may be continuous, such as driving an automobile. In either case, vibrations between 6 and 10 Hz induced in the upper body, and 2 and 10 Hz in the shoulder region, cause performance reductions. Tactile perception is affected in many ways:[29] the hands have resonances at 4–5 Hz, depending on the direction of vibration, on seating condition, on posture, etc. There are also strong acceleration effects, with larger accelerations impacting control operations more than small accelerations. The effects of (induced or resulting) vibration that occurs on several axes simultaneously are complex and difficult to model.

Manual task performance

The airflow through the larynx as well as breathing irregularities, and related changes in general body tension, may change the pitch of the voice, particularly in vertical oscillations at frequencies between 5 and 20 Hz.

Voice performance

Whole-body vibration with a peak magnitude below $0.01\ m/s^2$ is hardly felt, while accelerations of $10\ m/s^2$ rms or higher can be hazardous. The effects of intermediate accelerations depend on the actual frequency, direction, and duration of the vibration.

Disorders and injuries

Whole-body vibration has acute physiological effects on different body systems, either caused by movement of organs and tissues or by general stress response likely caused by stimulation of the sympathetic nervous system. Initial exposure to whole-body vibration often brings about short-lived changes in physiological functions: vertical vibration in the range of 2–20 Hz can produce increases in heart rate, cardiac output, respiration rate, and oxygen consumption, similar to responses to light or moderate exercise. At the onset of vibration exposure, increased muscle tension and initial hyperventilation have been observed. The musculoskeletal system is, by mechanical reasoning, often strongly affected by the motions and energies that it must resist or counteract, and reflex responses may be inhibited. Muscular contractions do not necessarily protect the body but may in fact enhance the strain beyond that of a passive system because of untimely contraction due to phase lags. Reflex responses can be inhibited. Since vibration and noise often occur together, negative effects on hearing have been reported, in some cases apparently even in the absence of noise, probably caused by vasoconstriction and ischemia in the cochlea.[30]

Acute responses

The assertion of chronic effects is more difficult because only few case-control studies with cohorts have been conducted; consequently, we must largely rely on retrospective

Chronic responses

or cross-sectional comparisons between vibration-exposed and non-exposed populations. Observed effects are usually not specific to vibration environments but also occur in, or may be aggravated by, other daily or ordinary work conditions. The most common lasting health problems associated with whole body vibrations are back pain and spinal disorders and, to lesser extents, gastrointestinal disorders, as well as problems in cardiac, renal, urinal, and genital systems. Raynaud's "white-finger" disease in workers using vibrating tools such as jack hammers has been well established for over 150 years. Conversely, vibration may be beneficial for the skeletal system for tissue maintenance or (re)growth, per Wolff's law for bone adaptation.

> Since experimental studies suggest that low-amplitude high-frequency vibration encourages bone growth and strength, researchers have been interested to see if it helps patients with osteoporotic or otherwise weakened bone or joints. In one study,[31] transmitting vertical sinusoidal vibration to the human body over a range of clinically appropriate amplitudes (from 0.05 to 3 mm) and frequencies (from 10 to 90 Hz) showed that substantial amplification of peak acceleration could occur between 10 and 40 Hz for the ankle, 10 and 25 Hz for the knee, 10 and 20 Hz for the hip, and at 10 Hz for the spine. As might be expected, nonlinearities in the human musculoskeletal system mean that transmission of vibration to the body is a complicated phenomenon. Care must be taken to balance safe attenuation of the higher frequencies with intensified accelerations at larger amplitudes. Clinical studies have shown mixed results with no definitive benefit.

Motion sickness   Vibration can induce "motion sickness" (*kinetosis*), often accompanied by feelings of fatigue or tiredness and by nausea and vomiting. Describing motion sickness as being "caused by mismatched perceptions about motions" seems correct, but does not identify the factors that cause or aggravate it, nor their interactions: direction, frequency, amplitude, duration of motion, and absence or presence of visual information. The prevailing theory is that the problem arises from a conflict among information received from the vestibular and the visual sensory systems. The involvement of the vestibular system is critical: motion sickness can be caused just by simple head movements during body oscillation. Motion sickness is particularly prevalent on ships, in aircraft and in automobiles, but also exists in flight simulators. (Curiously, a person prone to kinetosis may not suffer from it when in control of a vehicle.) Many astronauts suffer from motion sickness, which may or may not subside within hours or days after launch into space.

The following advice is often given to prevent motion sickness:[32] avoid instances of wide fields of view, time lags in visual perception, head movements during acceleration and oscillation, and head movements while wearing distorting optics including magnification lenses.

There is no convincing experimental evidence that the relationship between motion exposure and meals eaten has any effect on the occurrence of motion sickness, except that there may be truth in the saying that motion sickness thrives on an empty stomach. Consumption of fluid is advisable—even if little is retained. Mental activity may be beneficial for minimizing sickness (possibly as a distraction from mild nausea), but head movements should not accompany such mental activity. Quoting Michael Griffin, "Fighting at sea and singing have both been said to suppress symptoms, although only the latter can be recommended here!"[33]

6.4.5 Describing Impacts

While we might intuitively understand the transitions from impact to shock to bump to vibration, their actual delineations are arbitrary. Therefore, different descriptions of these events and their effects on the body have been proposed. Impacts with serious damage often come from falls, both of our own body falling or from being hit by a falling object. Impact injuries are frequent in vehicle crashes and in sports, where impacts—especially to the head—are prevalent among boxers and soccer or football players.

Kinds of impacts

Impacts are usually defined[34] as events with sudden onset, of less than 1 second duration, and with rapid change in velocity of the impacting or impacted object. The time-profiles of actual events are often difficult to describe, measure, and recreate in the laboratory, so they are commonly (and perhaps unrealistically) generated in terms of triangular or trapezoidal accelerations or decelerations over time. For convenience, accelerations and decelerations are commonly expressed in multiples of g (or G),[35] with $1\text{ g} = 9.80665\text{ m/s}^2$.

Human tolerance to an impact depends, among other factors, on the direction of impact, the magnitude of acceleration or deceleration, and the total time of exposure. G-tolerances in aerospace environments are discussed further below.

Impact tolerance

In whole-body impacts, such as falls, linear impacts occurring at right angles to the spinal axis are better tolerated than those parallel to the spine. Among the most severe life-threatening skeletal fractures are head injury and damage to components of the vertebral column. Of course, a wide variety of conditions affect impact survival. For example, a 2-m head-first fall of a child onto a flat solid surface can result in skull fracture; the same can happen to an adult who falls backward on the head from standing upright when there is not enough time for the body to assume a protective posture. Human skull tolerance limits appear to be about 150–200 g for 3 m/s^2 average acceleration and 200–250 g for peak accelerations.[36] Fig. 6.19 shows various conditions under which humans did survive impacts; generally, of course, people do not live through such impacts.[34]

Human tolerance to multiple-G conditions sustained for a few seconds depends on the direction of the resulting force relative to the body. Experience and experimentation have shown that such impacts are best tolerated in the ±x direction (backward or forward), when the body is supported either on its belly (prone) or on its back (supine), with the support perpendicular to the direction of impact. Fairly little is known about sideways actions, that is, in ±y direction.

Direction affects impact tolerance

6.4.6 Falling through the Atmosphere

A mass (person or object) falling vertically through the atmosphere experiences two opposite external forces: weight (due to gravitational attraction) and drag (due to air resistance).

Long falls

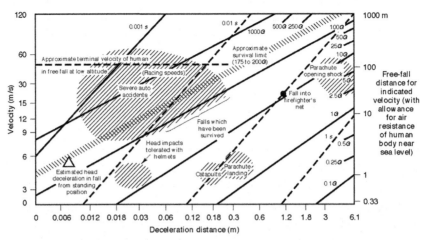

FIGURE 6.19 **Approximate conditions under which humans have survived impacts.** *Source: Adapted from Snyder (1975).*

Newton's second law states that force equals mass m times acceleration a. With the acceleration due to gravity of about 9.81 m/s², we calculate that near the Earth's surface, weight $W = m\,g$.

The drag force D depends on the object's drag coefficient d (often labeled c_d), the area A of the object that produces drag, the atmospheric density r (often labeled with the Greek symbol ρ, "rho"), and the square of the velocity v:

$$D = d\,A\,r\,v^2 / 2 \tag{6.16}$$

Weight and drag are vector quantities that oppose each other. Therefore, the force that actually determines the motion of a falling body is the difference between its weight W and the drag D. So, the net external force F is

$$F = W - D \tag{6.17}$$

The acceleration of an object with mass m falling with drag becomes

$$a = (W - D)/m \tag{6.18}$$

Terminal velocity

Fall velocity v increases until drag D becomes equal to weight W. When that is case, no net external force F remains to act on the object; so, its vertical acceleration goes to zero. Without further acceleration, by Newton's first law of motion, the object falls at constant velocity, called the terminal velocity. At terminal velocity V_t, drag D equals weight W:

$$D = W = d\,A\,r\,V_t^2 / 2 \tag{6.19}$$

This allows us to solve for the terminal vertical velocity V_t:

$$V_t = \sqrt{(2W)/(d\,A\,r)} \tag{6.20}$$

Accordingly, all objects falling through the atmosphere reach their own specific terminal speed if they fall far enough. That terminal velocity will be large if the object is heavy (W), if it is "streamlined" and therefore produces little drag (d), and if the path is through thin air (far above ground): terminal velocity is higher for a ball of lead than for a feather.

Regardless of the altitude at which they begin their jumps, two people of similar shape and weight end up with about the same falling speed after plummeting a long distance: without parachute or wingsuit, their terminal velocity is about 55 m/s (nearly 200 km/h), which is about ten times faster than a raindrop falls.

LONG FALLS SURVIVED[37]

1. In January of 1943, Alan Magee was a turret gunner in a B-17 bomber flying over the Atlantic coast of German-occupied France. His plane took enemy fire and broke up. Magee jumped from the bomber without a parachute. He lost consciousness during the fall before smashing through the glass roof of the St. Nazaire train station. He awoke to find German doctors putting him back together. His injuries included a broken right leg and ankle, a nearly severed right arm, and 28 wounds from shards of glass.

2. In 1998, adventurer Steve Fossett attempted to circumnavigate the globe solo in a helium balloon. A storm above the Coral Sea off Australia produced a barrage of hail which shredded the balloon. During the ensuing fall, he lay across a bench in the capsule to prepare for impact. He plummeted about 8 km; after the remnants of his vessel had splashed down, he found himself unhurt. As the capsule filled with water, he scrambled into a life raft and was eventually rescued after more than 10 h at sea.

3. In August 2004, Christine McKenzie was an experienced skydiver, but during jump number 112 neither her main parachute nor the emergency chute opened. After plummeting more than 3000 m, she landed across a string of power lines; the give in the cables absorbed most of the impact energy. She fell to the ground with only a broken pelvis.

4. Late in January 2007, after too may drinks, a reveler ran down a hotel corridor in Minneapolis, Wisconsin, and smashed headfirst through a glass window on the 17th floor. The 125 kg man fell almost 50 m before landing on his feet atop an asphalt-covered first-floor overhang. He sustained just a broken leg and a few scratches.

5. On July 30, 2016, skydiver and stuntman Luke Aikins jumped from an airplane at 7575 m altitude without a parachute or wingsuit. During a 2-min fall, belly-down, arms extended, he soared toward a special two-layer 30m × 30m net set up near Simi, California. Just before impact, he turned his body and landed on his back in the net's center. Without injury, he climbed to the ground.

6.4.7 Accelerations in Aerospace

The pilots of a high-performance aircraft, or a space crew departing from or returning to Earth, are commonly subjected to linear or rotational accelerations. These may remain sustained for some time, but often they vary; as discussed previously, accelerations of quick onset and cessation are called impacts.

Describing directions

In aerospace vernacular, the direction of acceleration is often graphically described by the direction in which the eye (or another body part) is displaced by the acceleration. "Eyeballs in" means acceleration in x direction (back-to-chest), "eyeballs right" describes acceleration in y direction (right-to-left) and "eyeballs down" denotes acceleration in z direction (toe-to-head). Similarly descriptive words apply to angular motions: "cartwheel" is a roll around the x axis, "somersault" is a pitch around y, and "pirouette" is a yaw around z.

Magnitude of impacts

Fairly high impact accelerations may be encountered as a result of thruster firing, ejection, seat or capsule firing, escape-device deployment, flight instability, air turbulence, and crash landings. NASA Standard 3000 (1989) included guidelines for designing aerospace equipment so that 1-second impacts are limited as shown in Table 6.9. Sample accelerations include:

- Violent maneuvers: up to 6 G, omnidirectional.
- Parachute opening shock: approximately 10 G_z.
- Ejection firings: up to 17 G_z.
- Crash landings: from 10 to more than 100 G, omnidirectional.

G levels

On Earth, we experience a (nearly) constant 1 G level, pointing toward the center of the Earth. Inside a spacecraft (with 2017 technology), the crew experiences up to +2 G during stage separation; during launch, entry, and abort operations, they can face up to +6 G. In trans-orbital flight, sustained accelerations are very low, such as 10^{-6} G, omnidirectionally. During orbital maneuvers, angular accelerations up to ± 1.5 deg/s^2 are possible in all directions.

Effects of G

Depending on the magnitude and direction of acceleration, the effects may be imperceptible and normal, they may generate discomfort and impairment, or they may be dangerous, even life-threatening. During acceleration in z direction, we experience increased weight (mass times acceleration), drooping skin and body tissue at +2 G_z; inability to raise body parts and dimming of vision at +3 G_z; blackout and loss of consciousness at 5–6 G_z. Downward acceleration effects include facial congestion

TABLE 6.9   Limits of Accelerations Lasting up to 1 second

Direction	Impact limit (G)	Rate of impact (G/s)
$\pm x$	20	10,000
$\pm y$	20	1000
$\pm z$	15	500
45 degrees off x or y or z	20	1000

Source: Adapted from NASA Standard 3000 (1989).

and reddening of vision at $-3\ G_z$. Few can tolerate $-5\ G_z$ for over more than 5 seconds.

Strong vertical accelerations, such as from a very fast airplane or spacecraft, can cause gravity-induced loss of consciousness (G-LOC). Even a few Gs can impede the heart from generating enough pressure to supplying sufficient blood (and oxygen) to the brain. G-LOC occurs when the necessary blood pressure at the brain falls under 22 mmHg. Since there is some residual oxygen stored in the brain, a pilot can sustain high G levels only for a very short time (less than about 5 seconds). Signs of impending G-LOC include tunnel vision and *grey-out* (loss of vision). Then *black-out* (loss of consciousness) sets in: it usually lasts about 15 seconds, followed by amnesia for another 10 seconds. Of course, during G-LOC the pilot is not able to perform any tasks at all. After regaining consciousness, physiological impairments persist and performance usually remains severely degraded. G-LOC onset, duration, and recovery depend on magnitude and onset rate of G_z, time at G_z, and the number of earlier exposures.

G-LOC

Of course, the principal counteraction to G-LOC is to avoid severe flight maneuvers, but that may not be possible, especially for fighter planes. Temporary use of an automated flight control (auto-pilot) is possible but does not solve the problem of G-LOC. Selecting pilots who have higher than average $+G_z$ tolerance, and training them to expect, recognize and act upon early symptoms may be helpful. Alternately, pilots can use anti-G straining maneuvers or an antigravity body suit to increase their arterial blood pressure. Another option is a design solution: change the direction of the $+G_z$ vector relative to the pilot by placing the person, at least during the severe flight path, in a different body position.

Countering G-LOC

> To avoid G-LOC, some military aircraft have a pilot seat that automatically reclines during high-G flight maneuvers so that the body experiences the acceleration vector in the backward $(-x)$ direction. For that same reason, astronauts have been placed onto reclined body-contoured seats during rocket blast-off from Earth.

6.4.8 Vibrations and Impacts: Summary

Plenty of empirical and anecdotal experience exists regarding the reactions of the human to vibrations and impacts with respect to comfort, performance or health. In experimental research, vibrations are usually modeled as single or combined sinusoidal displacements of the body, while impacts are modeled in terms of short-term, high-onset single events.

The body's responses to vibrations are often expressed in terms of resonances at certain frequencies. Modeling the dynamic responses of the body through mechanical functions of mass properties, dampers, and springs, is an often complex undertaking; moreover, unfortunately, the resulting models often do not encompass all possible physiological and psychological reactions.

Actual vibration responses show large interindividual variations and depend on factors including body posture, restraint systems, and the actual directions and kinds of impulses that exist in the real world; imagine, for example, a vehicle driving over rough terrain. Consequently, we must use currently available standards and design recommendations with great caution. Nevertheless, the information we do have helps us provide guidelines with respect to those vibrations and impacts. These guidelines have profound effects on human comfort, health, and performance, and provide information on technical means to avoid those conditions that do adversely affect the human.

6.5 SPACE

Traveling through extraterrestrial space is an ancient human desire. While flying by airplane became feasible around 1900, well into the 1950s humans could only dream of seeing Mother-Earth from far away, of circling the Moon or landing on another celestial body. With advances in technology and increasing knowledge about how humans can live and perform "out there" (see Table 6.10), visits into space became reality in the 1960s.

Early space travel

In the early years of preparation for space travel, researchers tried to extrapolate knowledge from experiences gained in other long, isolated, dangerous missions, such as in submarines, polar expeditions, mental hospital wards, prison stays, bomber crew missions, prisoners of war, professional athletic teams, and shipwrecks and other disasters.[38] From the 1960s on, space travel advanced, initially in particular by Soviet space crews with several international guests. Cosmonauts stayed aboard the Salyut and then the Mir space stations for many months, some a year or longer: Valeri Polyakov stayed in the Mir space station in 1994–1995 for almost 438 days.

American astronauts did not perform long-term space missions for about two decades after their last Skylab activity in early 1974, but joined their Russian and international colleagues on Mir in the 1990s and then on the International Space Station (ISS), whose first sections were launched into Earth orbit in 1998. With the experiences of several hundred humans who actually lived and worked in space (even though still relatively near to Earth), we have rich data to draw upon. However, their overall number is still a relatively small sample size for drawing population-wide inferences with any statistical certainty.

The Russian term is *cosmonaut*, the American term is *astronaut*, the Chinese term is *taikonaut*. *Naut* (from the Greek word for sailor) has become a suffix meaning traveler, voyager, explorer and, by analogous use, a specialist in a particular field which is identified in the prefix. So, cosmonauts, astronauts and taikonauts are all "spacenauts."

6.5.1 Humans in Space

Live and perform

In the 1960s, the decision was made to send humans into space, instead of just shipping remote- or self-controlled machines. There were two main reasons: one, the ability

TABLE 6.10 Spaceflight Developments

1926	Robert Goddard launches the first successful fluid-powered rocket
1931	August Piccard and Paul Kipfer ascend by balloon to 16 km altitude
1934	Wernher von Braun and coworkers start developing overall rockets
1942	A V2 rocket attains 90 km altitude
1944	V2 missiles bombard allied cities
1947	Chuck Yeager flies faster than sound in the X-1 rocket-engine powered plane
1957	Sputnik 1 carries a sphere into orbit around the Earth. Sputnik 2 follows with dog Laika aboard
1958	NASA shoots his first satellite into space
1959	Lunik 3 spacecraft transmits pictures of the back side of the Moon
1961	Yuri Gagarin is a first human being to circle the Earth
1963	Valentina Tereshkova is the first woman in earth orbit
1968	Apollo 8 spacecraft circles the Moon
1969	Neil Armstrong and Edwin "Buzz" Aldrin land on the Moon in the Apollo 11 spacecraft
1970	The crew of Apollo 13 survives an explosion on board the spacecraft
1971	Salyut-1, the first space station, launched, however three cosmonauts perished in their capsule
1973	First Skylab space station launched
1975	Soyuz-19 and Apollo 18 dock
1975	Venera 9 spacecraft lands softly on Venus and transmits pictures of its surface
1976	Viking spacecraft lands on Mars and transmits pictures of its surface
1977	Voyager spacecraft flies toward Jupiter, Saturn, Uranus, and Neptune
1980	Indian Rohini-1 satellite launched
1981	First reusable space shuttle lifts off
1984	Bruce McCandless and Robert Stewart make first untethered spacewalks
1986	Challenger shuttle explodes shortly after start, seven astronauts are killed
1986	Mir space station is launched and used by international crews until 2001
1990	Chinese Asiasat-1 satellite launched
1995	Shuttle Discovery docks with the Mir space station
1997	Mars Pathfinder with its Sojourner rover land on Mars
1998	First modules of the international space Station ISS launched into earth orbit
1999	Chinese Shenzhou 1 launched
2001	Dennis Tito is the first paying space tourist aloft
2003	Space Shuttle Columbia disintegrates as it reenters the Earth's atmosphere, killing the crew
2003	Yang Liwei becomes China's first taikonaut aboard Shenzhou 5
2005	The SpaceShipOne rocket-powered plane flies twice at altitudes above 100 km
2011	Chinese Tiangong-1 space station launched
2011	NASA's Dawn spacecraft enters orbit around asteroid Vesta and Juno is launched to orbit Jupiter
2012	NASA's Curiosity rover lands on Mars
2012	North Korea places its first satellite into orbit
2016	Shenzhou-11 spacecraft docks to the Tiangong-2 space station

of the human to respond to unexpected situations, the other, sheer curiosity and love of adventure. To send humans so far aloft was a risky idea: it wasn't even known whether people could survive weightlessness and, if so, for how long and whether they could function well enough to survive.

> In space, humans must survive, perform tasks within the spacecraft, and do extravehicular activities (EVAs) for construction and maintenance of the space station and for exploration. A spacenaut must overcome major difficulties—physiological, psychosociological, and medical. The magnitudes of most problems increase the longer the space mission lasts. The principal technical challenges are to construct a shell and its propulsion: they must function reliably, safely, and suitably for space travelers.

Inside the space station, we must keep people healthy and functioning by providing a breathable atmosphere and a suitable climate along with managing food, drink, and waste, as listed in Table 6.11. Specific requirements include generating a pressurized envelope, protecting the humans from radiation, and possibly generating artificial gravity (discussed in more detail below). The tasks of designing human-centered systems for use in a new and dangerous environment were challenging in the early years of space travel when there was no past experience—and there will continue to be new tasks as unknown and unexpected challenges arise.

TABLE 6.11   Some Examples of Technology Needs for Long-duration Space Travel

Issue	Technology need
Atmosphere pressure and composition control	Reliable hardware for long-duration missions for total and partial pressure and composition monitoring and control, fire detection and suppression, etc.
Temperature and humidity control	Substantial improvement in thermal control technology
Atmospheric revitalization	CO_2 control, removal, and reduction; O_2 and H_2 balance; trace gas monitoring and control
Food supply: storage, processing, and growing	Closed-loop bioregenerative food production systems; automated systems for harvesting and processing edible biomass and space crops
Water management	Waste water collection and processing; water quality monitoring; storage and distribution of recovered water
Waste management	Physical, chemical, but regenerative systems to automatically collect and process urine, collect and store fecal matter and other recyclable waste
Portable life-support systems	Minimize size, weight, pressure, power demands
Health maintenance	New procedures in occupational medicine including personal hygiene, exercise, diagnostics and therapeutics, surgery, and medical aid

The internal space of the celestial vehicle must be of sufficient volume; free from unacceptable pollution and contamination such as intrusive odors, noises, and lights; and provide some privacy to each individual. In addition to the psychosocial aspects of living in severely confined quarters under new and possibly dangerous conditions ("stress" is discussed in Chapter 4), special problems can arise if international crews of males and females are put together for long-duration space activities and if medical problems arise.

Habitability

A space crew builds its own microsociety, necessary in particular on long missions such as to Mars. Space travelers are separated from their loved ones and from Earth. They are confined to a small crowded space. In space, there is some sensory deprivation, especially in the vestibulum, and senses may change, especially in their intensity: sounds and smells, for example, appear altered and more intrusive. The body's natural internal clock, its circadian rhythm, is generally disrupted, so sleep and relaxation patterns are changed. Crews of different cultural backgrounds and mixed gender must get along with each other, which may be quite difficult, especially during long missions. Some examples of interpersonal challenges that were originally listed by NASA in the mid-1990s include:

Microsociety

- Assertiveness, politeness, trust
- Patience and tolerance
- Language, nonverbal communications
- Respect for other cultures and habits
- Command structure, decision-making, conflict management and resolution
- Gender roles and stereotypes
- Interpersonal interests
- Personal hygiene
- Scheduling and time management

6.5.2 Radiation

Radiation can be life-threatening, is unpredictable, and increases over time and as the spacenaut moves further away from Earth.

The three primary natural sources of radiation are the Earth's magnetosphere, solar flares and wind, and cosmic radiation. The magnetosphere's radiation is strongest between about 2400 and 19,000 km above ground. Standard spacecraft shielding can protect space crews from exposure, particularly since they usually pass quickly through the radiation zone. More substantial shielding is necessary to protect humans against solar winds, which are composed of high-energy particles. The most serious threats are solar flares and cosmic radiation.

Radiation threats

Solar flares run in about 11-year cycles, corresponding to sunspot activity. During each cycle, between 20 and 30 solar flares expose the spacenaut to up to 100 rem of radiation, while up to five flares may generate up to 1000 rem, and two flares may be even more powerful than 5000 rem. (For comparison: a standard X-ray exposes a patient to about 0.01 rem; residents in Denver, Colorado [USA], are exposed to approximately 0.2 rem per year. Exposure to 100 rem causes acute radiation sickness, 300 rem can be

Solar flares

lethal.) Another major problem is that solar flares are difficult to predict; particles that precede a flare do so by only about 1 hour. This gives space crews little time to seek protection.

Space ships and permanent space stations must provide radiation shielding against even the most intense bursts of solar radiation, which usually last for less than a day. Solar storms must be monitored at all times, and people on the moon be kept within 1 h of travel time to a radiation shelter.

Cosmic radiation

It is very difficult to protect spacenauts against cosmic radiation, which has very high energy particles. Very dense shielding is required. This may be fairly easily accomplished on a permanent station, for example, on Mars, because one can use natural topographic features, such as valleys, which provide some shelter, or burrow into rock, or heap solid material on the roof of a built shelter.

On-board radiation

While in the space vehicle, the crew must be protected not only from external solar and cosmic high-energy particles, but also from any radiation generated on-board. For example, a nuclear-powered rocket engine (currently the most promising technique for generating propulsion for space exploration) might be used and require shielding. This adds additional mass to the space ship; using current technology, the ship's mass would be at least 500 tons.

Radiation effects

Radiation is abundant in space, and there was and is currently no effective protection available to people in space. This is of great concern because radiation (depending on wavelengths, intensities, and exposure) can seriously damage or destroy spacenauts' cells, permanently affecting them. Moreover, it can also harm genetic material, affecting the offspring of space travelers. Exactly what harm specific doses of space radiation do is still a matter of educated guesses because, so far, observation of radiation effects is hampered by the fairly small number of exposed humans, most of whom have been in space only for days, weeks, or months.

6.5.3 Pollution and Contamination

On-board pollution is a serious problem. Besides radiation, as discussed above, pollution may also consist of chemical releases, microbes and particulate debris, dust, and noise. Toxic chemicals or disease-causing organisms can endanger space crews. Humans can be breeding grounds for bacteria and viruses. Recycled air, water and wastes can carry contamination continuously and throughout. These are major challenges similar to those experienced by hygienists and infectious disease specialists in hospitals or on board ships.

6.5.4 Protective Spacesuits

During extravehicular activities, humans must wear "space suits": individual enclosures that protect against the near-vacuum conditions in space or against greatly reduced atmospheric pressure such as on Mars. This requires bulk and matter, both critically reducing mobility and manipulation for the person inside. While not shielding against radiation, current space suits provide suitable pressure and a breathable internal atmosphere as well as cooling or heating: in orbits around Earth, temperatures can range from −130°C (in the shade) to 120°C (in sunlight). Failure of the suit, or even its puncture, for example, by a micrometeoroid, is likely to cause catastrophic results.

6.5.5 Impact with Objects in Space

Man-made debris is in orbit around Earth, and meteoroids and asteroids are everywhere in space. Impact with an object of sufficient size can have disastrous consequences for spacecraft and crew, either by direct physical injury or via losing pressure and air and consequently causing decompression sickness (see below). The risk near Earth is relatively small since larger objects can be tracked from the ground and thus the crew may be able to evade damage, and perhaps help could be sent if needed. In interplanetary space, there are relatively few objects, and even if the space ship is large, the likelihood of a collision is small, but if it occurs, no help or rescue is to be expected coming from Earth.

6.5.6 Microgravity

Protection of humans from the immediate and delayed effects of the lack of gravity is still a largely unsolved technical challenge.

Per the laws of physics, some weak gravity pull exists everywhere in space either because adjacent celestial masses exert attraction forces or because the spacecraft is being accelerated, linearly or angularly. Since there is no true "zero gravity" (except in the center of a spacecraft in a stable orbit), we call any acceleration level below 10^{-4} G *microgravity*, or popularly, *weightlessness*.

"Weightlessness"

Since our bodies (and minds) have always been accustomed to Earthly gravity pull, microgravity has detrimental physiological effects which—especially in spaceflights lasting 60 days or longer—can become serious and need to be counteracted by technical, behavioral and medical means.[39] Consequences for the musculoskeletal system, for fluids and blood, and nervous control are discussed below.

Space effects

6.5.6.1 Musculoskeletal System

A major problem associated with long duration spaceflight is the deconditioning of the human's musculoskeletal system. Because of reduced use in low gravity, muscles lose some of their volume. Muscle atrophy, if untreated, creates a condition similar to that experienced by paraplegics, whose muscles diminish from lack of use. Such loss of muscle mass, together with cardiovascular deconditioning (see below), can cause significant problems. For example, after a 211-day mission, Soviet cosmonauts could not maintain a standing position on Earth for more than a few minutes. Such weakness could cause severe problems for spacenauts who want to step onto Mars, for example. To counter the effects of muscle disuse in low gravity, the muscles must be exercised extensively; cosmonauts Kornienko and Kelly did so for about 2 hours daily during their 340 days in 2015–2016 on the ISS.

Muscle atrophy

In space, there is also a change in body posture, and in some body dimensions. If relaxed, the body assumes a semicrouched position: that is, knee and hip angles are at about 130 degrees, the curvature of the lumbar and thoracic spine section flattens, and

Body posture and size

the pelvic angle changes, resulting in an extension of body stature of up to 10 cm. Head and cervical column bend forward, and the upper arms "float up" against the trunk to about 45 degrees. This posture, reported by NASA in 1978 and shown in Fig. 6.20, is quite similar to one found in relaxed underwater postures[40] and shown in Fig. 6.21. This posture seems to result mostly from a new balance of muscular and other tissue forces

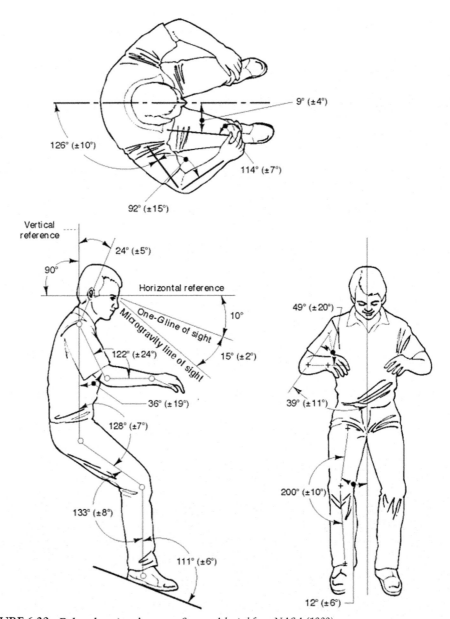

FIGURE 6.20   Relaxed posture in space. *Source: Adapted from NASA (1989).*

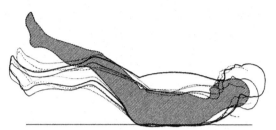

FIGURE 6.21   **Relaxed posture under water.** *Source: Used with permission from Lehmann (1962).*

acting around the various body joints. While the relaxed space posture by itself has no ill effects, it needs to be considered when workstations are designed. For example, it is difficult to work at waist level, since the spacenaut must force his or her arms continually down to this level. Bending forward requires abdominal muscles' actions, and it is difficult to sit or stand erect. Obviously, design of space clothing must also take into account the "neutral" space posture.

Physical exercise also helps combat another skeletal problem in space: demineralization of bones. Calcium is lost through urine and feces. This increases the possibility of a bone fracture, with those in the legs, the hips, and the lower lumbar vertebrae most susceptible. These bones support much of the human body weight when in a gravity environment. Previous long-duration spaceflights have shown about 10% mass loss of certain bones. Altogether, the effects on bones of long-term space travelers appear similar to osteoporosis in aged people.

Demineralization of bones

6.5.6.2 *Blood and Fluid Distribution*

Another major problem that occurs in low gravity is the redistribution of body fluids. The body's circulatory system operates well in a 1 G environment: for example, valves in the veins prevent blood from pooling in the feet and legs and instead force it toward the right ventricle of the heart. The anti-pooling system still operates in space, although there is little gravity to work against. This results in an accumulation of fluids in head, neck, and chest and a reduction of volume in the legs.

Fluid redistribution

Body fluid redistribution has important ramifications. The body's internal sensors indicate that there is too much blood in the upper body, for which the body compensates by reducing blood flow. This aids the upper body, but worsens the problem in the lower body, where the fluid level is already low. Using a reduced-pressure suit around the legs helps fluid flow to the legs.

Fluid imbalance

A side effect of fluid redistribution is decreased thirst. Insufficient water intake reduces the overall fluid level in the body. A decrease in fluid volume adds to the demineralization of bones, and increases the retention of sodium.

Changes in fluid distribution within the body also have a negative effect on the heart. The increased volume of fluid in the upper body increases blood pressure in the head and neck veins. Sensing this heightened pressure, the heart actually reduces its output by lowering its contracting rate and shortening the length of the diastole. In spacenauts who stayed in microgravity for months, actual decreases in heart size were noted.

Reduced heart size

I. THE ERGONOMIC KNOWLEDGE BASE

Changes in blood composition

Together with the changes in the cardiovascular system, blood composition also alters: plasma volume and the mass of erythrocytes decrease, cholesterol level increases, and the concentration of phosphorus decreases.

If there seems to be an excess of blood, even if just in the upper body, the body reduces the production of new red blood cells, which diminishes the amount of oxygen in the blood. This may result in anemia after many days in microgravity.

Reduced ATP level

A long stay in microgravity decreases the level of ATP and the intensity of glycolysis (see Chapter 3). This is of little concern in microgravity, where less energy is needed to move the body than on Earth. However, the reduction in the ability to generate energy within the body through metabolic processes (for which ATP glycolysis and available oxygen are important) can reduce a person's overall work capacity level, which is needed upon reentry into a gravity environment. Such decreased capacity could be quite significant after a long trip to Mars, where there is only one-third Earth gravity.

Altered vision

Shifts in body fluids may also cause swelling in the back of the eye, which leads to poorer eyesight. These changes seem temporary after short trips to space but appear much more persistent after months in microgravity. In fact, returning spacenauts have been reporting vision issues with increasing and disturbing frequency. For years, researchers regarded vision problems as minor and largely reversible. But now that space excursions can last for many months at a time, vision disturbances are becoming more of a concern. Several theories have been put forth: bloodstream toxins that may put pressure on the eye, radiation exposure, oversights during exercise sessions. The prevailing hypothesis is that weightlessness moves fluid surrounding the brain, and some of the excess fluid may be traveling down the optic nerve sheath until it presses into the back of the eye.

Kidney stones

While astronauts seemed to have few issues with kidney stones after two-week Space Shuttle deployments, the incidence of kidney stones after a 6-month ISS mission is much higher. Reasons may include high concentrations of calcium oxalate in the bloodstream, which cause stones to form in the kidneys; bone loss; or most likely the high level of carbon dioxide in the ISS air supply, which is 10–20 times higher than in Earth air.

Dizziness and fainting when back in gravity

After some time of weightlessness, the body establishes a new circulatory pattern that serves the body while in space. However, upon returning to Earth, space travelers usually experience episodes of dizziness and even fainting. This is probably due to reduced heart rate and reduced total blood volume developed in space. Back in gravity, the antipooling mechanisms of the body's circulation again move the blood from the upper to the lower body. This causes the brain to lack sufficient blood supply, resulting in dizziness and fainting. The problem is usually resolved within a week or so on Earth, since more blood and red blood cells are produced, thus supplying enough blood throughout, but the recovery may be much slower in a low-gravity environment such as on the Moon or Mars. During this period of recovery, astronauts are quite sensitive to rapid changes in posture, such as in standing up rapidly.

> Upon return to Earth after the Skylab 3 flight in 1973, the astronaut Jerry Carr said, "I didn't faint, but I felt pretty clumsy. My head felt like a big watermelon and I had to work hard to support it. I'd been a butterfly for 84 days and suddenly weighed something again."[41]

6.5.6.3 Nervous Control

During the first few days of spaceflight, spacenauts commonly face disturbances in their vestibular system, possibly mostly related to the otolith receptors. Conflicting stimuli from the visual and proprioceptive systems often cause disorientation, vertigo, dizziness, and postural and movement illusions, often accompanied by nausea. Roughly every other space crew member suffers profound discomfort and motion sickness over some time, which varies by individual from hours to days, until the body adapts and reorients itself.

Similar symptoms occur when space travelers reenter a gravity environment, to which body (and mind) must adjust. This usually causes reduced work capacity and generates errors in interpreting the visual environment, (such as movement illusions), affects sensory-motor coordination, and brings about internal discomfort. Such symptoms can be expected for the first few days after landing on extraterrestrial bodies.

Depending on the work-rest cycle and external time clues, a person's circadian rhythm (see Chapter 7) may be altered. So far, spacenauts near the Earth have been kept on a 24-hour cycle, with 8–10 hours of work and 8 hours allocated to sleep. This practice has avoided problems with conflicting internal rhythms. It might be advisable to maintain this 24-hour cycle even in space ships traveling to Mars.

Microgravity affects sleeping and eating. Sleep disturbances have been common during spaceflight, but conditions other than low gravity may play a role. With improved food technology, eating and drinking practices have been developed that are not substantially different from those on Earth, although they can become monotonous and boring.

A puzzling phenomenon that may be more psychological than physiological in origin is an increased sensitivity to vibrational and acoustic stimuli. After about 30 days in low gravity, the tolerance level to sounds and vibrations has been found greatly reduced, which may affect the sleep patterns, interpersonal relations, and communications of spacenauts, and which can generate general annoyance.

Another unexplained problem is the suppression of the body's immune system in space. Reasons for the suppression are not well understood, but cosmonauts returning to Earth after more than 200 days in space have displayed severe allergic reactions which they had not had prior to their mission. Currently no countermeasures are known, which is unfortunate; even if there is nothing on Mars that could harm humans, upon return to Earth the travelers could be susceptible to many diseases here.

6.5.6.4 Performance

When entering microgravity, many astronauts initially experience nausea, and a few suffer from it throughout the space mission. This can hinder, possibly severely, the execution of tasks. Most crew members become accustomed to working and living in the new environment within just a few hours, and are able to perform well in about 3 days. Complete adjustment so that one feels "at ease" takes about three weeks—and then, upon returning to Earth, about as long again to readjust to gravity.[42]

Long-term space flight without countermeasures would produce major negative changes in the cardiovascular, respiratory, muscular–skeletal, and neuromuscular systems of the space crew. Careful selection of physically and psychological fit candidates,

Margin notes:
Disturbed vestibular sense

Reduced sensory-motor coordination

Circadian rhythms

Sleeping, eating

Vibrations and sounds

Immune system

suitable training and, when aloft, physical exercises are key countermeasures to many of these unwanted adaptations.

Large movements

Absence of gravity is in general a bonus for locomotion in space. Once a spacenaut is accustomed to microgravity, he or she accomplishes motion with minimal effort and performs acrobatic-appearing maneuvers with ease. However, for force exertions, the body must be braced so that counter-forces can be developed to actively exert force vectors. For example, spacenauts doing repair or assembly work must be tethered or otherwise anchored to the machinery on which they are working so that forces or movements, reactive to those that they are applying, don't hurl them into space.

Motor performance

A problem encountered usually only in the early stages of spaceflight is a decrease in motor performance. Newly arrived crew find it difficult to accurately estimate the amount of physical work required to perform certain tasks, and it also takes them longer to perform activities due to the lack of gravity-created references and resistances. After a few days in low gravity, the movements and activities become more precise, and the perceived level of difficulty decreases.

> When back on Earth, spacenauts tend to drop things for the first few days. They seem to be carrying on the habit of simply letting an item go—which would float in low gravity.

Emotion, cognition

A recent study supports that short (13−15 days) real and analogue spaceflights do not impact positive emotion (mood) and performance in male and female astronauts, however long-duration analogue spaceflight decreases it significantly. A spacenaut's cognition and mood may be impaired over time by specific and nonspecific effects of microgravity; the latter includes cumulative insomnia, interpersonal relationships, isolation, illness, and discomfort.[43]

Stability and stamina

Upon returning to Earth gravity after a prolonged stay in space, spacenauts' vertical stability was found to have declined, together with reductions in muscle strength and endurance. The stability problems disappeared in about a week upon return to full gravity, while regaining strength and stamina took longer. It is unknown, however, how long it would take to recover in a 1/3 G environment, as on Mars.

6.5.7 Artificial Gravity

If artificial gravity that mimics the one on Earth could be provided in space ships, nearly all the problems associated with microgravity could be avoided or alleviated. For this, two techniques come to mind.

Rotation

The first is to rotate the space structure, possibly by using the metabolic energy generated by the crew's exercise regimen, thus generating a centrifugal force from the center of rotation. This force is the product of mass of the astronaut, of the radial distance away from the center of centrifugation, and of the squared angular velocity. The larger the velocity, and the farther away from the center of rotation, the larger the centrifugal force experienced by a person.

There is a complication with rotating acceleration: according to the law of physics, an object moving within the rotating environment experiences Coriolis force. The Coriolis force is an artifact of the Earth's rotation; specifically, it is an effect whereby a mass moving in a rotating system experiences this Coriolis force acting perpendicular to the direction of motion and to the axis of rotation. On the Earth, the effect tends to deflect moving objects to the right in the northern hemisphere and to the left in the southern and is important in the formation of cyclonic weather systems. In our space scenario, the Coriolis force would confuse the spacenaut's vestibular system and is likely to cause severe motion-sickness problems which, unfortunately, would probably last throughout the entire duration of being in this environment.

Another solution, one that would likely work out better for spacenauts, is to generate a linear acceleration environment. For example, on a trip to Mars, the spaceship could be linearly accelerated in the desired direction for half the distance. Then, it would be turned around and, through use of the same accelerating engines, be decelerated until it arrived at Mars with zero speed. The creation of constant linear acceleration would keep space travelers in familiar gravity environment.

Linear acceleration

Scott Kelly, along with cosmonaut Mikhail Kornienko, flew to the International Space Station in March of 2015, orbiting 230 miles above Earth, and remaining aboard for a full year. The thrill of space travel comes at a personal price: backaches, poor sleep, balance disturbances, headaches, weakened muscles, rashes, radiation exposure, nausea, bone loss, kidney stones, skin soreness, vision problems.

Upon returning from nearly a full year in space, Kelly said the muscle soreness and fatigue were a lot worse. In space, he spent about 2 hours a day exercising to keep his muscles functional for his return to Earth. When asked which muscle groups in particular were bothering him most upon his return, he replied "Most of 'em." He continued, "I think coming back to gravity is harder than leaving gravity. So, I don't know, maybe the aliens got it a lot easier than we do."

After a year of living in microgravity, he had become an inch and a half taller due to the temporary expansion of his spine, but he quickly shrunk back to his former height after touching down in Kazakhstan. As Kelly put it, "Gravity pushes you back down to size."

6.5.8 Space: Summary

The technology to construct and launch space vehicles exists, but protection of the crew from solar and cosmic radiation is still an unsolved problem. There are also slow, lingering and dangerous effects of the lack of gravity in space: the body's musculoskeletal, vascular, cardiac, and sensory systems suffer in substance and performance. This, in turn, reduces the ability to perform tasks both in space and on the ground, either upon return to Earth or upon landing on another celestial body. Generating artificial gravity in the space vehicle would alleviate these problems, but doing this has still not been perfected.

6.6 UNDER WATER

Functioning in the underwater world presents thrilling excitement along with daunting challenges to humans, physically, physiologically and emotionally. Plenty of people occupy jobs that require underwater work—such as HazMat team diver, aquarium staff, fish farm worker, underwater photographer or welder or archaeologist. Of course, underwater diving is also a popular recreational sports activity.

Regardless of the reason that a human is spending time underwater, the human factors issues associated with that fall into two main groups: challenges to the senses, and effects of water pressure.

6.6.1 Sensory Inputs and Perception

Vision

Water absorbs energy predominantly in the red-orange end of the visible spectrum, so things under water appear mostly in shades of blue/green. Further, light is dispersed by suspended particulates. The combination of reduced radiant energy and of color contrast leads to decreased visual acuity.

Human eye tissue does not tolerate high external water pressure well at all. This mandates that some type of lenses be worn underwater; the frame of this eye wear generally severely limits peripheral vision. Additionally, the headgear creates an air–water interface, causing light rays passing through this interface to change their trajectory. This refraction creates distortions in the perception of size and distance of underwater objects.

Hearing

Compared to transmission in air, the speed of sound spread and the distance over which it is conveyed are markedly increased in water due to its higher density than air. Our ability to perceive the direction of sound depends on one ear (the more distant from the origin) receiving the sound waves fractions of a second later than the other (closer) ear. The human auditory system cannot, however, easily make this distinction underwater because of the increased speed of the waves. The combination of relatively amplified and seemingly directionless sound can be quite disconcerting: a novice diver may suddenly begin fearing death by blunt trauma from a boat propeller when said boat is actually far away.

Taste, smell

Except for the taste of aging rubber from a less-than-pristine breathing mouthpiece and, perhaps, the "smell of fear," our senses of taste and smell are effectively functionless in the underwater world.

Touch

The relative cold of the underwater environment tends to blunt the senses of fine touch and, to a lesser extent, pain. Protective and/or warming gloves amplify the loss of touch significantly. Moreover, the combination of cold and glove decrease fine motor coordination.

Spatial orientation

We orient ourselves in space using internal cues, for example, intracranial and intravascular pressures; we sense the pull of gravity, and we interpret external cues such as the sky and the normal position of objects in our environment. In deep water, significant pressure is applied to the whole body of a diver, the effects of gravity are minimized, the sky may not be visible, surrounding objects may not be present or noticeable.

Fortunately, the diver quickly learns to use one cue that is present underwater but not on land: bubbles of exhaled air, which always travel up.

6.6.2 Effects of Water Pressure

Most of us can stay underwater by holding our breath for less than a minute, but people who dive regularly for sponges and pearls are able to stay underneath a couple of minutes longer. This time limit is set by the need to breathe. A snorkel allows us to access air if we remain very close to the water surface but it does not work at greater water depth because of the increased water pressure—and the need, of course, for the tip of the snorkel to be above water.

Breathing

To understand basic physics and physiology of the human under water, we can model the body (for the moment and for purposes of this explanation, we suspend disbelief and we disregard solid bone structures) as compressible tissues permeated with cavities that are filled with watery fluids or, as the middle ear and lungs, with air. Water is not compressible, but the gas is.

Body under water

At sea level, the human body is under "one atmosphere" of air pressure (1 atm = 10.13 N /cm^2 = 101.3 kPa = 760 Torr = 14.7 psi). With each 10 m (33 ft) of depth in salt water (about 34 ft in fresh water) the body is subject to an additional atmosphere of external pressure. This compresses its few gas-filled spaces, such as the lungs. Breathing from an air supply becomes impossible with depth unless the breathed air is also under pressure.

Pressure

After having stayed at depth, and with the body's air volumes compressed accordingly, upon ascent the gas expands. Excess gas volume in the human gastro-intestinal tract can be expelled by belching or flatulence but air in the lungs must be constantly exhaled on ascent or lung rupture with air embolism can occur; this is a critical and potentially fatal event.

Gas expansion with ascent

Increased pressure can have other serious effects on human physiology. Oxygen and nitrogen are dissolved in body fluids in proportion to the atmospheric pressure and to the composition of air. At extreme pressure (i.e., depth), air can actually be toxic to tissues, by virtue of the high oxygen levels dissolved therein. Deep divers therefore must breathe a modified gas mixture to compensate.

Nitrogen comprises about 78% of air. Increased ambient pressures increase dissolution of nitrogen in human tissues. High brain concentrations of nitrogen lead to impaired cognition and clouded sensorium, a condition known as *nitrogen narcosis*. Because nitrogen does not dissolve easily in body fluids, this condition takes time to occur and is, fortunately, unlikely to occur with depth/time combinations achieved by recreational divers.

Because of its low affinity for dissolution in body fluids, nitrogen rapidly exits tissues as pressure decreases with ascent from depth. But if the amount of dissolved nitrogen or the speed of ascent exceeds the ability of the body to exhale it, nitrogen bubbles form in any organ, causing tissue damage; in joints, causing pain and joint destruction; in the blood, causing brain ischemia (insufficient blood supply to the brain to meet metabolic demand).

Decompression illness

The general term *decompression illness* applies to two ailments loosely related by the development of gas bubbles: arterial gas embolism and decompression sickness.

In the early 1800s, geotechnical engineering workers who needed to stay and work under water—to build quays and bridges and tunnels, for example, or to work on submerged portions of ships—managed to stay under water for extended periods of time through use of an air pump that filled their helmet with air, or through a caisson. A caisson is a multiperson chamber; laborers enter a caisson through an air lock to work in compressed air that allows breathing and prevents flooding. However, when improperly decompressed to one atmosphere, they often developed joint pain and more serious problems including numbness and paralysis, even death in some cases.

Caisson disease received its name because it appeared in construction workers when they left the compressed atmosphere of the caisson and rapidly reentered uncompressed atmospheric conditions. For example, construction of New York's Brooklyn Bridge, which was built with the help of caissons, resulted in numerous workers being either killed or permanently injured by caisson disease during its construction.

Arterial gas embolism

When a diver holds her breath during rapid ascent, water pressure quickly decreases, while her obstructed airway prevents the escape of gas that is expanding within her lungs and bronchial airways. The expanding air can rupture organs and enter her blood. Air bubbles are carried with the blood flow into the arterial circulation and often to the brain. A possible result is sudden loss of consciousness, convulsions, and paralysis.

Decompression sickness

For deep dives, inert gases (usually nitrogen or helium) are part of the mixture used to fill air tanks. When the diver is deep, water pressure is intense; the gas dissolves from his lungs into the blood from where it diffuses into tissues, especially in the spinal cord and the brain, which are well supplied with blood. Given enough time, the gas taken up by the tissues during the dive is washed out by the blood, taken to the lungs, and exhaled. However, if the diver surfaces too quickly, the pressure of the dissolved gas in the tissues exceeds ambient pressure, and bubbles can form. The various combinations of intra- and extravascular gas bubbles can lead to limb pain, coughing and shortness of breath (called "chokes"), numbness or paralysis. Nitrogen bubbles in the respiratory system can cause excessive coughing and difficulty in breathing. Other symptoms include chest pain, dizziness, paralysis, unconsciousness or blindness. In almost half of the cases, symptoms do not appear immediately but rather an hour after completion of a dive, so observation is indicated for up to 24 hours after ascent (especially if air travel follows a dive). In extreme cases, decompression sickness (DCI, also called *caisson disease* or "the Bends") can cause death.

The name "The Bends" was applied to decompression sickness in the early 1870s: caisson workers building the St. Louis Bridge walked with a stoop which fashionable women then adopted as the "Grecian Bend." Of the roughly 600 caisson employees in St. Louis, 119 were seriously afflicted and 14 died. Currently, close to 1000 cases of decompression illness are reported annually among recreational divers. Decompression illness occurs in 0.02%–0.04% of dives, occurs more frequently in men than in women, and occurs more frequently in cold waters than in warm waters.

Knowledge of the underlying physics and physiologic mechanisms allows human-engineered equipment to supply appropriate breathing air and proper diving procedures, especially for descent and ascent. Self-contained underwater breathing apparatus (SCUBA) gear with compressed air is widely used by recreational divers to depths less than 40 m: they can stay underwater for up to 10 min without special control of descent or ascent. Longer underway dives generally require a decompression (or "safety") stop prior to full ascent. To go deeper, much more complicated equipment and strictly controlled procedures are necessary: a dive 90 m deep requires, with today's technology, several separate tanks with their own breathing rigs (called *regulators*) for air, oxygen, helium, nitrogen; and a carefully observed regimen for descent, stay, and ascent. A dozen or so decompression stops, each lasting several minutes, can make for a long time under water: diving down to 70 m takes, with current equipment, at least 2.5 hours, of which only half an hour is spent at depth.

SCUBA

Design of proper diving equipment itself is a challenging ergonomic task. Of course, the apparatus itself must function reliably even under great water pressure, but it must also be safely and quickly usable under conditions of limited visibility and reduced mobility due to under-water protective clothing; in coldness; in confined space when working inside a shipwreck or cave; and by a diver who is possibly scared or in danger. The procedures employed in training novice drivers are critical and must be carefully designed, taught, and executed.

6.6.3 Under Water: Summary

Diving and working under water brings with it many exciting scenarios, but also a host of dangers. Anyone interested in spending time under water, whether for recreation or for employment, should understand the physics and physiology that underlie underwater activities, as well as the physiological mechanisms that make it possible to spend extended time there.

The effects of water pressure on the human body can be serious, even life-threatening. Breathing from an air supply becomes impossible during deep dives unless the breathed air is also under pressure. Moreover, even with appropriate breathing air, without proper diving procedures, humans can suffer diseases including arterial gas embolisms and decompression sickness.

6.7 CHAPTER SUMMARY

The human body does not exist in a vacuum; it functions in interaction with its environment.

Our climate is characterized by three aspects: temperature, humidity, and air movement. The exchanges of heat energy between the environment and the body follow the physical processes of radiation, convection, conduction and evaporation. Their effectiveness depends on several conditions including the clothing worn, the energetic content of the work conducted, and the exposure time. The temperature difference between exposed skin and the environment is very important, but humidity and airflow also play major roles. Within reasonable limits, the human can function in both hot and cold environments, given appropriate job demands and clothing.

Air pollution is a detriment not only to health, but also to the ability to perform physically demanding work, particularly if carbon monoxide is present. Further studies may also show negative effects of other pollutants.

Arrival at altitude, after having lived near sea level, influences the ability to perform physical work in various ways. Up to about 1500-m elevation, the ability to work aerobically is largely not affected; but at higher altitudes, both the short-term maximal capacity and the long-term submaximal ability are diminished, and they remain reduced for the entire stay at altitude.

The effects of vibration on the human depend very much on the amplitude, frequency, direction, and point of application. Vibration in one direction, for instance from foot-to-head, can bring about response movements in other directions, such as head nodding. Whole-body vibration can have consequences quite different from vibration of only body parts, such as the hands. Body posture and body restraint systems can greatly affect the results of vibrations. Some information is available about appropriate ergonomic measures, depending on circumstances. However, more research and better modeling are needed.

Since the middle 1900s, flying airplanes at high speed has been possible. One factor in making high-speed flight possible is that we can now control—through behavioral means and through human engineering—the deleterious effects of sharp accelerations upon the supply of oxygen to the brain that would otherwise lead to unconsciousness.

Space is an exciting and challenging new environment for humans. So far, the exposure effects of space and resultant gravity changes have been experienced by just a few hundred people, most over periods of days, some over months. The general finding is that microgravity can diminish the function of the body in many ways, particularly regarding the musculoskeletal and the circulatory systems. Countermeasures to avoid health and performance problems, especially upon reentry into gravity, continue to be developed. The design of equipment, tasks, and work environments suitable for long-term space missions is still a great challenge.

Diving and working deeply under water can result in great danger to the human, but is made possible by proper use of specialized and complex equipment and adherence to a carefully-monitored ascent regimen.

6.8 CHALLENGES

What are the effects, in theory and practice, of the concept of either "constant core temperature" or "average skin temperature."

Why is it difficult to control heat transfer through the head?

The energy-balance equation, given in this chapter, does not consider the time domain. What are the consequences?

While it is true that the temperature of air between a radiating surface and the body does not affect the energy transferred by radiation, it should have an effect on energy transfer by convection. How?

Why is there no energy transfer between the human body and the environment through condensation?

Why is it more important to avoid overheating and undercooling of the body's core than of the shell?

Is it conceivable that procedures other than the current subjective ones could be used to establish indices of "effective climate"?

Which engineering means exist to control the environmental climate in (a) offices, (b) workshops, such as a machine shop or a foundry, and (c) outdoors?

Are sensations of feeling hot, or cold, reliable indicators of climate strain?

Is exercising a practical means of acclimatizing oneself to working in heat, in cold, or at altitude?

Through which body parts is vibration most likely transmitted?

Is it reasonable to assume that vibration is transmitted to the body either only horizontally, or only vertically, as done in most research?

How would the descriptors of vibration change if the acting vibration were not sinusoidal?

Can you think of other professionals besides truck drivers who are particularly exposed to vibrations, or impacts?

What might be the effects of several sources of vibration arriving simultaneously at different body parts?

Could you imagine that vibrations on the job might, under certain circumstances, be helpful in job performance?

What explanations other than "conflicting CNS information" might be applied to the problem of motion sickness?

What factors are likely to affect the likelihood of withstanding impacts?

Which body postures, and physical behaviors, might be appropriate to combat the effects of sustained strong accelerations?

Why is it difficult to make crash-dummies anthropomorphic?

What are reasons for, or against, sending humans into space?

Discuss the aspects of "systems engineering" associated with space engineering.

Consider details that make a confined living (or working) space "habitable."

What means exist to physically exercise astronauts in space?

NOTES

1 The freezing temperature of water is 32°F in the Fahrenheit scale and 0°C in the metric Celsius scale. The boiling temperature of water is 212°F and 100°C. The Kelvin scale has its lowest temperature (absolute zero) at −273.15°C; its temperatures are stated in Kelvin units (K) where one K has the same magnitude as 1°C.
To convert temperature degree values between Celsius and Fahrenheit, scale the degrees to boil water ((212−32)/100, or 9/5) and align the water freezing temperatures (32°F and 0°C):
Celsius to Fahrenheit: (9/5)°C + 32 = °F
Fahrenheit to Celsius: (°F − 32) (5/9) = °C

2 The physicist Daniel Gabriel Fahrenheit (1686-1736) called the lowest temperature he could achieve (in a mixture of ice, water, and ammonium chloride) "zero." He defined the temperature of a mixture of ice and water as 32 degrees, and the temperature of boiling water as 212 degrees: the eponymous Fahrenheit scale. In 1742, the astronomer Anders Celsius (1701-1744) proposed a centigrade scale (from the Latin for "hundred steps") with 100 degrees for the freezing temperature of water and 0 degrees for the boiling temperature. This scale was reversed a year later, and was renamed the Celsius scale by international agreement in 1948. It is used in all countries except the USA.

3 For equations for heat exchange by convection or conductance, radiation and evaporation, see Bernard (2002); Bernard and Dukes-Dobos (2002); Havenith (2005); Youle (2005).

4 Boff and Lincoln (1988).

5 d'Ambrosia Alfano, Malchaire, Palella (2014); Bernard and Iheanacho (2015); Parsons (2005, 2014); Youle (2005).

6 Carroll (2015).

7 Siekmann (1990); Ungar and Stroud (2010); ISO 13732.

8 Hoffman and Pozos (1989).

9 For more information on the ergonomics of working strenuously in heat, see Kroemer (1991).

10 Glove manufacturers provide an ergonomic intervention for the cold-temperature-induced runny nose by adding specialized absorbent padding on the index finger of ski gloves.

11 Palinkas (2001); Maekinen et al. (2006); Lieberman et al. (2009).

12 The earlier Wind Chill Index was based on experiments conducted by Paul Siple and Charles Passel, who measured the time to freeze water in bottles suspended from the top of the expedition's building in the Antarctic. The current Wind Chill Equivalent Temperatures were formulated by Randall Osczevski and Maurice Bluestein, who redefined the experimental measures by considering the steady-state of exposed facial skin of a 150 cm tall adult walking into the wind.

13 The latest version of the Wind Chill Index is available at www.weather.gov. This table was developed in the United States jointly with the Canadian weather service and is used in both countries with no regionalization. Other countries have their own versions, available from their meteorological agencies or the World Meteorological Organization. The research was conducted by Randal Osczevski and Maurice Bluestein and is described in their 2005 publication.

14 See Youle (1990); d'Ambrosia Alfano et al. (2014); ANSI-ASHRAE Standard 55; Bernard and Iheanacho (2015); Parsons (2014).

15 Note that ANSI-ASHRAE, ISO, and other standards are often updated, so one should check for the most recent version.
ANSI-ASHRAE: https://www.ashrae.org/resources–publications/bookstore/standards
ISO International Organization for Standardization, Chemin de Blandonnet 8, 1214 Vernier, Geneva, Switzerland: www.iso.org

16 For further reading, see the WHO air quality guidelines (updated September 2016) by the International Agency for Research on Cancer (IARC) of the World Health Organization (WHI); from www.who.int and https://www.epa.gov/pm-pollution. Accessed November 30, 2016.

17 Kramer et al. (1993).

18 Cymerman (1996) and Tschöp and Morrison (2001).

19 Some people function better at high altitude than others. Nepalese Sherpas are well-known for their ability to climb very high mountain peaks while carrying heavy loads. There seems to be a genetic component to their metabolic adaptations that contribute to enhanced performance: allowing more efficient oxygen utilization, better muscle energetics, and less oxidative stress. (Horscroft et al. 2017).

20 Griffin (1990, 1997, 2004); Guignard (1985); Putz-Anderson (1988).

21 See, for example, Griffin (1990, 1997), Guignard (1985), Putz-Anderson (1988), Wasserman et al. (1997).

22 ISO 2631, 8727.

23 Griffin (1990, p. 181).

24 Griffin (1990).

25 Griffin (1990).

26 For a comparison of standards, see Griffin (1998).

27 Nakashima and Cheung (2006).

28 See Griffin (1990) and Oborne (1983).

29 Griffin (1990).

30 For more information on acute physiological effects, see Griffin (1990), Guignard (1985); Dupuis and Zerlett (1986); Martin et al. (1986); Village and Morrison (2008).

31 Kiiski et al. (2008).

32 CSERIAG GATEWAY Vol. 4, No.1, 1993.

33 From MJ Griffin (1990), pages 327–328.

34 On 11 September 2001, an estimated 200 people jumped from the higher floors of the burning towers of the World Trade Center's tower in New York City. Their falls lasted about 10 seconds. Tragically, not one person survived.
Impacts are often defined: see, for example, ISO 2631-5:2004.

35 Strictly, the standard acceleration of free fall (due to gravity) is denoted by g_0 or g_n, the nominal gravitational acceleration of an object in a vacuum near the surface of the Earth. Its value is defined as 9.80665 m/s^2 and may be denoted either g or G.

36 Personal communication between original author KHE Kroemer and RG Snyder, February 18, 1991.

37 Based in part of an article by Joe Hasler (2010), http://www.popularmechanics.com/outdoors/survival/stories/4344037?src=social-email. Accessed August 16, 2016.

38 Holland (1991).

39 See Jenkins (1991).

40 As reported by Lehman in 1962, see Kroemer et al. (2003), pp. 282, 283.

41 Carr, J. (1989) Final Frontier, 2(3), p. 29.

42 Personal communication between original author KHE Kroemer and astronauts Jerry Carr and Jack Lousma, January 11, 1990.

43 Liu et al. (2016).

Overview

Because of circadian rhythms, the human body undergoes changes in its physiological functions throughout the day and night. Our attitudes and behaviors also fluctuate rhythmically over a 24-hour period. For most of us, our body is prepared for physical and mental work during daily (daylight) waking hours, while during the night, sleep is normal. We should arrange our work schedules to align with our physiological, psychological, and behavioral rhythms in order to maintain our performance standards and avoid negative health and social consequences. Substance use (alcohol and drug use) usually affects our physiological and behavioral states and as a result impact our work performance, generally in a negative way.

7.1 BIOLOGICAL RHYTHMS

Introduction

The human body follows a set of daily fluctuations, called *circadian rhythms* (from the Latin *circa*, about, and *dies*, the day). Most of us are diurnal organisms (from the Latin *diurnus*, of the day) and are active during the day, rather than being nocturnal. Circadian rhythms are regular physiological events that are observable, including body temperature, heart rate, and hormone excretion. Each is thought to be controlled in the body by a self-sustained pacemaker or internal clock which runs on a daily cycle. Some rhythms, such as core temperature, blood pressure, and sleepiness, are coupled to each other, entrained and synchronized by time markers called *zeitgeber* (from the German *Zeit*, time, and *geber*, giver). Daylight and darkness are examples of zeitgeber: nature's clocks, they establish temporally based activities such as work tasks, office hours, mealtimes, etc. Human social behavior (the inclination to do certain activities, as well as to rest and sleep) both follows and reinforces biological rhythms.[1]

Female menstrual cycle

One well-documented chrono-biological rhythm is the female menstrual cycle, which is regulated through synchronization of the activities of the hypothalamus, pituitary, and ovary. The 28-day period is usually divided into five phases: preovulatory or follicular, ovulatory, postovulatory or luteal, premenstrual, and menstrual. The main hormonal changes occur in the release of estrogen and progesterone around the 21st day of the cycle; estrogen shows a second peak at ovulation. Hormonal release is low during the premenstrual phase. While there is neurophysiological evidence that estrogen and progesterone affect brain function, these two hormones have antagonistic effects on the central nervous system (CNS): estrogen is stimulatory and progesterone is inhibitory. Variations in hormone production during the menstrual cycle may moderately affect the capacity to perform certain tasks, but this is often offset by increased effort. (See also Chapter 14.) The occurrence of mental and physical complaints in some women during the premenstruum is well established, but its impact varies. When premenstrual symptoms are pronounced, have been regularly present for at least a year, and are accompanied by disturbances in mood, premenstrual dysphoric disorder (PMDD) may be diagnosed. Less severe symptoms are sometimes classified as premenstrual syndrome (PMS). PMDD seems to occur in around 2% of menstruating women; estimates of rates of PMS vary widely but it is a mistake to assume that all, or even most, menstruating women experience it. Both PMDD and PMS do not occur after menopause.

Another well-documented chrono-biological rhythm is the relationship between mood disorders and the seasons. These have been noted for centuries and are now termed *seasonal affective disorder* (SAD), a type of depression that is linked to the changes in seasons. SAD symptoms usually start in the fall and intensify in the winter months, easing as we move into spring and summer. Symptoms include feelings of depression and hopelessness, sluggishness, lack of interest in activities that used to be appealing, changes in appetite and weight, and changes in sleep patterns. While the exact causes of SAD are still unknown, the disorder may be triggered by dysregulations in the neurotransmitter serotonin, and/or in the production of the hormone melatonin, and/or in the production of vitamin D. Risk factors for SAD include being female, being young, having a family history, living far from the equator, and having depression or bipolar disorder.

Seasonal mood disorders

7.1.1 Circadian Rhythms

Human health is maintained through a state of balanced control called *homeostasis*, where physiological variables are balanced in spite of external disturbances. However, upon close examination, we can see that underneath this "steady state" of homeostasis, many physiological functions occur in a pattern of rhythmic variations. Rhythms with a cycle length of 24 hour are called *circadian* (or *diurnal*) rhythms; those which oscillate faster than once every 24 hour are called *ultradian*; those which repeat less frequently are known as *infradian*.

24-hour cycle

Among the circadian rhythms, the best known physiological variables are body temperature, heart rate, blood pressure, and potassium excretion. (See Fig. 7.1.) Most of these variables show higher values during the day and lower values during the night, but hormones in the blood tend to be more concentrated at night, particularly in the early morning hours. The amount by which the variables change over a day and the times at which extremes occur are quite different among individuals and even within the same person.

One way to observe the diurnal rhythms and assess their effects on performance is simply to follow the person's activities. During the daytime, we normally expect the person to be awake, active, and eating, while at night the person sleeps and fasts. Physiological events do not completely follow that general pattern. For example, body core temperature falls even after the person has been sleeping for several hours and is usually lowest between 3:00 and 5:00 a.m. The temperature then begins to rise, more quickly upon awakening, and continues to increase, with some variations, until late in the afternoon. This suggests that body temperature is not a passive response to our regular daily behavior (getting up, eating meals, and performing work) but rather is self-governed.

Diurnal rhythm

In the human being, the circadian pacemaker is located in the *suprachiasmatic nucleus* of the *hypothalamus* in the brain. The hypothalamus is located below the thalamus and right above the brain stem. In humans, it is roughly the size of an almond. Under

Circadian pacemaker

constant living conditions, the underlying physiological rhythms of the body are solid, self-regulated, and remain in existence even if one's daily activities change.

Variations   Variations in observed rhythmic events (due to exogenous influences) may mask internal regular fluctuations. For example, skin temperature (particularly at the extremities) increases with the onset of sleep, regardless of what time it occurs, and turning the lights on increases the activity level at any time. Therefore, a person's observed skin temperature or activity level does not necessarily indicate the individual's internal rhythm, but may in fact mask it.

Still, under regular circumstances, signs of external activity and internal events do tend to coincide. During the night, the low values of physiological functions—for example, core temperature and heart rate—are due primarily to the diurnal rhythm of the body; however, they are further aided by nighttime inactivity and fasting. During the day, peak activity usually coincides with high values of the internal functions. Typically, the observed diurnal rhythm is the result of internal (endogenous) and external (exogenous) events that concur. If this balance of concurrent events is disturbed, consequences in health or performance may appear.

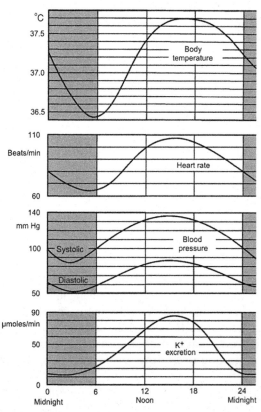

FIGURE 7.1 **Typical variations in body functions over the day.** *Source: Adapted from Colligan and Tepas (1986).*

The number of deaths due to cardiac arrest is highest during the morning hours. Health experts have known for more than 30 years that the erratic heartbeat responsible for *sudden cardiac death* (SCD) strikes most often in the morning, with the peak risk hours ranging from 6:00 a.m. to 10:00 a.m., and scientists have long suspected a link between SCD and the body's circadian rhythm. Apparently, a protein (KLF15) that helps regulate the heart's electrical activity occurs in rhythmically fluctuating levels that change like clockwork throughout the day, which then helps maintain a normal, steady heartbeat. Research suggests that persons with low KLF15 levels are more susceptible to SCD.[2]

When a person is completely isolated from external factors that could act as zeitgebers, including regular activities, the internal body rhythms "run free." This means that the circadian rhythms are free from external time cues and are only internally controlled. Experiments have consistently shown that circadian rhythms persist when running free, but their cycles are slightly different from the regular 24-hour duration. Most rhythms run freely at about 25 hours.

24-hour cycle

If rhythms run at different phases, they are *desynchronized*. However, if a person is returned to regular 24-hour time markers and activities, internal rhythms resume their coordinated steady cycles. The strongest *entraining* (determining or modifying the phase or period of) factor appears to be the light-dark sequence[3]: very bright lights and sunshine are effective in advancing or slowing the diurnal rhythm.[4] Activity is also a strong enforcer.[5]

Desynchronization

Providing new time markers and enforcing related new activity times can shift the body's internal rhythm. This happens, for example, when we travel to and stay for an extended time in a new time zone. It takes most people from 3 days to a full week to adjust their circadian rhythms by the requisite 5 or 6 hours when they fly from Europe to the eastern United States, for example (jet lag). The amount of time needed to entrain a new diurnal rhythm depends on the individual, the magnitude of the time shift, and the intensity of the new time markers. In laboratory simulations, researchers can manipulate zeitgebers to analyze the effects of jet lag or shift work. Entraining or synchronizing the internal rhythms so that they follow periodic cues has been demonstrated at cycle durations between 23 and 27 hours.

Entrainers

It appears to be a bit easier to set our internal clock "forward," such as occurs in the spring, when daylight saving time is introduced, or during a west-to-east flight, such as from North America to Europe.

Some people have consistently shorter (or longer) free-running periods than others. Females have, on the average, a free-running period that is about 30 min shorter than that of males.

Circadian rhythms change with a person's age; rhythm amplitudes especially diminish with increasing age. This phenomenon is particularly obvious in the case of body temperature. In the elderly, there is also a pronounced shift toward morning activity. The oscillatory controls seem to lose some of their power with aging, so we are more susceptible to rhythm disturbances as we age. Disruption in the circadian rhythm may also influence the "sundowning" phenomenon experienced by many individuals with dementia, wherein increased confusion, tremors, mood swings, and agitation occur at day's end when the natural light begins to fade.

7.1.2 Daily Performance Rhythms

Given the systematic changes in physiological functions during the day, we could reasonably expect corresponding changes in mood and performance. Of course, attitudes and work habits are also affected—and often strongly—by the daily organization of getting up, working, eating, socializing, relaxing, and going to bed. Experimentally, we could separate the effects of internal circadian rhythms and of external daily organization. But for practical purposes, we want to look at the combined results of the internal and external factors as they affect performance.

Daily fluctuations

On average, after the day begins, we take a few hours to reach our peak levels of alertness and energy—and that peak does not last long. Not long after a midday meal, those levels begin to decline, hitting a low at around 3:00 p.m. Lunch is often blamed—activities are interrupted for a noon meal and digestion lowers our psychological arousal level, bringing about increased lassitude, together with increased blood glucose and changed blood distribution in the body—but it is also just a natural part of the circadian process. After the 3:00 p.m. dip, alertness tends to increase again until hitting a second peak at around 6:00 p.m. After this, our level of alertness declines for the rest of the evening and throughout the early morning hours until hitting the very lowest point at approximately 3:30 a.m. Then, levels of alertness tend to increase for the rest of the morning until hitting the first peak shortly after noon the next day.

There are of course individual differences in circadian rhythms, and these can be quite dramatic. The typical pattern is indeed very common, and the general shape of the curve describes almost everyone. However, some people have a circadian rhythm that shifts noticeably in one or another direction. Morning people—often referred to as "larks"—have peaks and troughs in alertness that are earlier than the average person, while "night owls" prefer to rise late and stay up late; their alertness levels are shifted in the opposite direction.

Many different activities, physical and mental, can be performed during the day. They may follow a circadian rhythm or may differ because of external requirements. Personal interest, fatigue or boredom, and rewards or urgency usually have stronger effects on performance than diurnal rhythms have during daytime hours. For example, information processing in the brain (including immediate or short-term memory demands), mental arithmetic activities, and visual searches are strongly affected by personality, by the length of the activity, and by motivation.

7.2 SLEEP

Most of us sleep between 6 and 9 hours a night, which means that we spend about a third of our lives asleep. This may seem like a long time, but we actually sleep the least among all the primates.

While most Western societies now sleep, in specific bedrooms, at night and for 6–9 hours at a time, this was not always so. Before the 1800s, communal sleeping was common. Beds and rooms were shared due to space limitations and more communion within the nuclear family in general. Before indoor lighting, people often retired at sunset and rose at sunrise, but such long periods of sleep could not be sustained. Thus, sleep was biphasic: segmented into two phases with 1–3 hours of purposeful wakefulness in-between. This allowed for middle-of-the-night chores such as adding wood for heat, for meals, and for socializing. In the mid-1800s, dedicated bedrooms became more common in North America and Europe. Communal sleeping faded as Victorian aesthetics and high-mindedness introduced concerns about hygiene and propriety. By the late 1800s, sleep was condensed into one block with the advent of artificial light. As people became more active in the evenings thanks to artificial lighting, their bedtimes were pushed back until segmented sleep was largely impractical. In the early 1900s, scientists studied sleep cycles, and sleep research was promoted (and sleep monitored) as a way to promote both health as well as corporate and military efficiency.[6]

7.2.1 Early Theories of Sleep

Two millennia ago, Aristotle postulated that during wakefulness, a substance he termed "warm vapors" built up in the brain and that these vapors needed the sleep cycle to dissipate. In the 19th century, there were two opposing schools of thought: that sleep was caused by congestion of blood in the brain and, contrarily, that sleep was caused by blood drawn away from the brain. In the 1800s, behavioral theories were common to explain why we sleep. According to some of those theories, sleep was the result of an absence of external stimulation, or sleep was not a passive response but an activity to avoid fatigue from occurring. Early in the 20th century, it was thought that various sleep-inducing substances accumulated in the brain, an idea taken up again in the 1960s. During the 1930s and 1940s, various neural inhibition theories were discussed, which included the assumption that the brain possessed sleep-inducing centers.

7.2.2 Current Theories of Sleep

There is still a great deal of unknown about why we sleep. Restorative theories about the function of sleep focus on types of recovery from the wear and tear of wakefulness. Alternative theories claim that sleep is not restorative, but simply a form of instinct or diminution of behavior to occupy the unproductive hours of darkness; the relative immobility of the body during sleep can be a means of conserving energy. It appears that these three aspects of sleep function—restoration, occupying time, and energy conservation—all explain certain characteristics of sleep, but none of them does so completely or sufficiently.

Even though there is still a great deal of mystery surrounding sleep, scientists understand some of sleep's critical functions and reasons we need it for our health and wellbeing. One of the vital roles of sleep is to help us solidify and consolidate memories. As we go about our day, our brains absorb an immense amount of information. These experiences and bits of information first need to be processed and stored before our brains can

Critical functions

"log" and record them, and it appears that many of these steps happen while we sleep. Overnight, these pieces of information are transferred from short-term memory, which is more fleeting, to long-term memory, which is stronger and more lasting. This process is called *consolidation*. Research shows that after people sleep, they tend to retain information and perform better on memory tasks. Our bodies all require long periods of sleep in order to restore and rejuvenate, to grow muscle, repair tissue, and synthesize hormones.

Two central clocks

One approach to model the regulation of alertness, wakefulness, sleepiness, sleep, and many other physiological functions is to consider the body as being under the control of two "central clocks."[7] One clock controls sleep and wakefulness, and the other controls physiological functions such as body temperature. Under normal conditions the internal clocks are linked together, so that body temperature and other physiological activities increase during wakefulness and decline during sleep. However, this congruence of the two rhythms may be disturbed, such as by night-shift work, in which a person must be active during nighttime and asleep during the day (more on shift work later in this chapter). As such patterns continue, the physiological clocks adjust to the external requirements of the new sleep/wakefulness regimen. This means that the formerly well-established physiological rhythm flattens out and, within a period of about 2 weeks, reestablishes itself to the new schedule.

7.2.3 Sleep Stages

As outlined above, sleep is an active period in which a lot of important processing, restoration, and strengthening occurs. Sleep is not homogeneous, but has several stages that repeat, more or less regularly, during the night. These are commonly observed and labeled according to brain and muscle activities. The brain and muscles show large changes from wakefulness to sleep, which can be observed by electrical means (*polysomnography*).

EEG

Electrodes attached to the surface of the scalp pick up electrical activities of the brain-specifically, the cortex, which is also called the *encephalon* (because it wraps around the inner brain). The technique of *electro-encephalography* (EEG) records signals that can be described in terms of amplitude and frequency. Certain constituents of EEG waves that appear to be regularly associated with sleep characteristics are labeled vertices, spindles, and complexes. The EEG amplitude is measured in microvolts, and it rises as consciousness falls from alert wakefulness through drowsiness to deep sleep. EEG frequency is measured in cycles per second, called Hertz (Hz). Sleep researchers call frequencies above 15 Hz fast waves and those under 3.5 Hz slow waves. Frequency falls as sleep deepens; *slow-wave sleep* (SWS) is of particular interest to sleep researchers.

Certain frequency bands have been given Greek letters. The main divisions are typically described as follows:

1. Beta, above 14 Hz: these fast waves of low amplitude (under 10 microvolts) occur when the cerebrum is alert or even anxious.
2. Alpha, between 8 and 14 Hz: these frequencies appear during relaxed wakefulness, when little information is taken in through the eyes (particularly when the eyes are closed).

3. Theta, between 3.5 and 8 Hz: these frequencies are associated with drowsiness and light sleep.
4. Delta, under 3.5 Hz: these are slow waves of large amplitude, often over 100 microvolts, and occur more often as sleep becomes deeper.

As discussed in Chapter 2, muscle activities can be recorded via surface *electro-myography* (EMG). In observing sleep, the electrical activity of the muscles that move the eyes (monitored via *electro-oculography*, EOG) and those in the chin and neck regions are of particular interest. After a person falls asleep, the sleep stages become progressively deep; this is indicated by more synchronous and less frequent brain activity. In deep sleep, heart rate and respiration are slow, and muscles retain their tonus. However, in its deepest sleep phase, the brain becomes nearly as active as during wakefulness, heart rate and breathing vary, dreams are frequent, and eyes move rapidly under closed lids. Accordingly, this phase is called rapid eye movement (REM) sleep. The REM phase becomes longer and the light-sleep phases get shorter as the sleep-stage sequence repeats itself, which occurs approximately every 90–100 minutes throughout the night.

> EMG, REM

Currently, EOG outputs of the eye muscles are most often used as the main measure identifying REM and non-REM sleep stages. Non-REM sleep is further subdivided into four stages according to their associated EEG characteristics. (See Table 7.1.)

The human brain assumes a physiological state during sleep that is unique to sleep and cannot be attained during wakefulness. While muscles can rest during relaxed wakefulness, the cerebrum remains in a condition of "quiet readiness," prepared to act on sensory input, without any diminution in responsiveness. Only during sleep do cerebral functions show marked increases in their thresholds of responsiveness to sensory input. In the deep-sleep stages associated with slow-wave non-REM sleep, the cerebrum apparently is functionally disconnected from subcortical mechanisms. The brain needs sleep to restitute, a process that cannot take place sufficiently during waking relaxation.[8]

It seems that, on average, the first 5–6 hours of regular sleep (which contain most of the slow-wave non-REM sleep and at least half of the REM sleep) are required for a

TABLE 7.1   Stages of Sleep

Condition	Muscle EMG	Brain EEG	Sleep stage	Average percent of total time asleep
Awake	Active	Active, alpha and beta	0	-
Drowsy, transitional "light sleep"	Eyelids open and close, eyes roll	Theta, loss of alpha, vertex shape waves	1, non-REM	5
"True" sleep		Theta, few delta, sleep spindles K-complexes	2, non-REM	45
Transitional "true" sleep		More delta SWS (<3.5 Hz)	3, non-REM	7
Deep "true" sleep		Predominant delta SWS (<3.5 Hz)	4, non-REM	13
Sleeping	Rapid eye movements, other muscles relaxed	Alert, much dreaming, alpha and beta	REM	30

Source: Adapted from Horne (1988).

I. THE ERGONOMIC KNOWLEDGE BASE

person to keep performing at normal levels. Further sleep may be called facultative or optional, because it serves mostly to "occupy unproductive hours of darkness," with dreams being the "cinema of the mind."[9]

7.2.4 Normal Sleep Requirements

While there are, as usual, variations among individuals, certain age groups in the Western world show regular sleeping habits. Newborns sleep 16–18 hours a day, mostly in sets of a few hours' duration. Young adults sleep, on the average, 7.5 hours, with a standard deviation of about 1 hour. Some adults are well rested after 6 or 7 hours of sleep, or even less, while others habitually sleep 8 hours or more. The amount of SWS in both short and long sleepers is about the same, but the amounts of REM and non-REM sleep periods differ considerably.

Many people who sleep for just a few hours per day are able to keep up their performance levels even if their total sleep time is shorter than normal. A common lower limit seems to be around 5 hours of sleep per day, with even shorter periods still having at least some benefit.

7.2.5 Sleep Loss and Tiredness

If a person does not get the usual amount of sleep, he or she gets tired, and the obvious remedy is to get more sleep. Fig. 7.2 shows the effects of sleep loss on body temperature-the temperature is elevated, but keeps its phase.

However, simply getting more sleep (or better sleep) is not always a straightforward proposition. Sleep loss and sleep disorders are particularly common in various disease and disability states, while some sleep disorders are themselves a disability. Disabled and ill people sleep poorly for a number of reasons, including pain, emotional stress, or

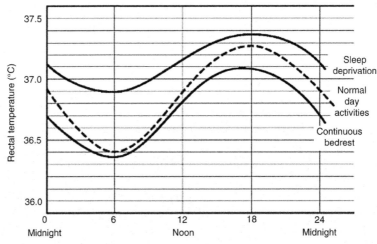

FIGURE 7.2 **Changes in body temperature associated with normal activities, bed rest, and sleep deprivation.** *Source: Adapted from Colligan and Tepas (1986).*

the challenging sleeping environments in which many people reside. (Anyone who has spent even a few nights in a hospital can attest to the latter.) Additionally, familial and institutional care-takers also experience disrupted sleep. Thus, disordered sleep may represent a primary or a secondary condition. With the advent of sleep research technology, an entire medical subspecialty of Sleep Medicine exists and a number of primary sleep–wake disorders (*parasomnias*) have been identified.

Sleep disorders involve problems with the quality, timing, and amount of sleep, which cause problems with functioning during the daytime. Sleep difficulties are linked to both physical and emotional problems. Sleep problems can both contribute to or exacerbate mental and physical health conditions can and be a symptom of various mental health and physical health conditions. There are many different types of sleep disorders, of which insomnia is the most common. In primary care, 10–20% of people complain of significant sleep problems. About one-third of adults report insomnia symptoms and 6–10% meet the criteria for insomnia disorder. Insomnia is not just defined as having difficulties falling asleep (initial insomnia). Individuals can experience mid-night awakenings or terminal insomnia (waking up too early and not being able to fall back asleep)—or even a combination of all three. Sleep–wake disorders also include hypersomnolence disorders, narcolepsy, breathing-related sleep disorders, circadian rhythm sleep–wake disorders, non-REM sleep arousal disorders, nightmare disorder, REM sleep behavior disorder, restless legs syndrome, and substance/medication-induced sleep disorder. Individuals with these disorders typically present with sleep–wake complaints of dissatisfaction regarding the quality, timing, and amount of sleep. Resulting daytime distress and impairment are core features shared by all of these sleep–wake disorders.

The restorative benefits of sleep to the brain are fairly well researched. Two or more nights of sleep deprivation diminish both a person's actual performance and the motivation to perform (but apparently, not the inherent cognitive capacity to perform). Irritability, suspiciousness, and slurred speech are the most common side effects. After up to two days of sleep deprivation, even though a person "feels tired," his or her mental performance is still rather normal on stimulating and motivating tasks; boring tasks, however, show a reduction in performance. All task performance is reduced after more than two nights of sleep deprivation.[10]

When people must work continuously for long periods, such as 24 hours or more, they are working without interruption and without sleep. Accordingly, any deleterious effects they experience are partly a function of the long working hours and partly a function of sleepiness. The nature of these effects depends on the types of tasks performed, on the motivation of the worker, and even on timing, because wakefulness and sleepiness appear in cycles during the day. As a general rule, the performance of a task is influenced by three factors:

- the internal diurnal rhythm of the body,
- the external daily organization of work activities, and
- the individual's motivation and interest in the work.

Each factor can govern, influence, or mask the effects of the others on task performance.

The performance of different types of work is affected differently when the person performing the work task has endured a long period of work with no sleep.[11]

Sleep disorders

Sleep deprivation

Long work periods

- A shorter task is affected less by sleep loss than a task that must be performed uninterruptedly for half an hour or longer.
- Performance of repeated tasks is likely to become worse with each successive repetition.
- Performance of monotonous tasks is highly diminished by sleep deprivation, whereas the performance of a task that is new to the operator is less affected.
- Performance of complex tasks is affected more than simple tasks.
- Tasks that are paced by the work itself deteriorate more with sleepiness than do operator-paced tasks.
- A person's accuracy in performing tasks may still be quite good even after losing sleep, but it takes longer to perform the tasks.
- A task which is interesting and appealing, even if it involves complex decision-making, can be performed rather well even over long periods of time. But if the task is disliked and unappealing, decision-making is prolonged.

In general, loss of sleep (particularly if it is associated with long periods of work, such as a full day) diminishes a person's performance of a task, prolongs the individual's reaction time, causes the person to fail to respond or produce a false response, slows cognition, and disturbs short- and long-term memory capabilities. Most of the deterioration occurs when the circadian rhythms are set for a night's sleep. Performance is further diminished following two or three nights of sleep deprivation. After missing four nights, very few people are able to stay awake and perform, even if their motivation is very high.

The performance of all mental tasks (except brief, interesting ones) diminishes with long hours of work coupled with sleep loss. The longer the work period and the more monotonous, repetitive, uninteresting, and disliked the task, the more performance degrades.

7.2.5.1 *Using Caffeine to Stimulate Wakefulness*

Caffeine

Caffeine is a stimulant that is quickly absorbed into the bloodstream. For about half an hour after drinking a strong cup of coffee, most people feel more awake and better able to pay attention; heart rate is increased by 2–10 beats per minute over a period of 5–15 minutes. Drinking 5–10 cups of strong coffee is likely to have an "overdosing" effect, generating a condition called *caffeinism*: symptoms include light-headedness, tremor, headache, palpitation, and difficulty falling asleep. Caffeine is found in coffee, teas, cocoa, and most chocolate products. It is added to many soft drinks and is a component of some medications for headache, cold, and allergy. Cocoa and chocolate products often contain *theobromine*, which has effects similar to those of caffeine on behavior and body functions. *Theophylline*, with analogous effects, is contained in tea. The amounts of caffeine and related substances are as follows:

- In coffee, 60–150 mg per cup (240 cm^3), with instant coffee usually in the low range.
- In tea, 8–50 mg per cup, with instant tea usually in the middle range.
- In cocoa, about 15 mg per cup.
- In soft drinks, between 40 and 70 mg per 355 cm^3 (12 fluid ounces).

- In chocolate, about 12 mg caffeine and 120 mg theobromine per 50 g (2 oz), while most baked chocolate goods have about half these amounts.

7.2.6 Deprivation and Recovery

During long working times after sleep deprivation of at least one night, short periods of reduced arousal or even of light sleep, known as lapses or gaps, occur. With increased time at work, so-called microsleeps happen increasingly often: the person falls asleep for a few seconds, but these short periods—even if frequent—have little if any recuperative value, because the individual still feels sleepy and performance still degrades.

Gaps, microsleeps

Naps, short sleeps lasting 1–2 hours, improve a person's subsequent performance during long working spells after a sleepless night. (A caveat: if a person is awakened from napping during a deep-sleep phase, sleep inertia with low performance can appear and may last up to 30 min.) The temporal placement of a nap may have differing effects; naps of at least 2 hours taken in the late evening or at night, when the diurnal rhythm is low, have beneficial effects on subsequent performance lasting several hours, while a daytime nap may be of less value.[12]

Naps

If long periods of mental work are unavoidable, consider interspersing the work with physical activities or exercises. Further, "white noise" may improve performance slightly. Hot snacks are often welcome, as are hot or cold (usually caffeinated) beverages. Music may also have an invigorating effect. Bright illumination of 2000 lux or more helps to suppress the production of the hormone melatonin, which causes drowsiness. Certain drugs, particularly amphetamines, can restore one's performance to nearly normal level, even when given after three nights without sleep.[13] (Note that while we report scientific findings, we do not recommend the use of amphetamine or other drugs to overcome the effects of challenging work schedules.)

Other

Recovery from sleep deprivation is quite fast. A full night of undisturbed sleep lasting several hours longer than usual will restore someone's efficiency almost fully.

Recovery

7.3 SHIFT WORK

Shift work—defined here as any work shift that veers significantly from the traditional "9-to-5" daytime schedule—is a reality for many employees. Shift workers help meet the demands of 24-hour work cycles. From an organization's competitive standpoint, shift work is an excellent way to increase production and customer service without major increases in infrastructure. Shift workers include physicians and nurses, pilots, road construction crews, police officers, shipping service workers, call center employees, manufacturing plant operators, and commercial drivers.

The 24-hour period is usually covered with either two 12-hour works shifts or three 8-hour work shifts. Since the industrial revolution, when 12-hour shifts on 6 or 7 days of the week were common, drastic changes in work schedules have occurred. In the first part of the 20th century, the then-common 6-day work-weeks with 10-hour shifts were shortened. Since the 1960s, many work schedules standardized to nominal 8 hour per day, 5 days a week, leaving 2 days per week free of (paid) work.

7.3.1 Advantages and Disadvantages

Shift work has both advantages and disadvantages, and it fits some individuals better than others. Often, working the second (evening) or third (night) shift means more pay than working the day shift; and sometimes better jobs are available at night because there is less competition for these positions. This can also mean that second and third shift workers get promoted more quickly than their daytime colleagues might. Some third shift workers also describe a looser, friendlier work environment at night, with more camaraderie and a shared focus on completing the work. Night shift may also be less busy than a regular day shift, with less traffic and fewer distractions, greater autonomy, and fewer meetings (management is sleeping).

How a person copes with shift work, however, depends on the individual. There are other factors that influence the suitability of shift work. Consider parents with young children: one parent may opt for permanent evening or night shift work so that, if the other parent works days, someone can always be home with the kids. According to one study from California, more than a fourth of night shift workers choose that schedule because of childcare needs.[14]

The drawbacks of shift work are clear: shift work inherently disrupts the natural circadian rhythms (discussed above) as well as social patterns. Adjusting to shift work requires physical and mental adaptations, and may result in consequences to health.

7.3.2 Effect on Health

It is also no surprise that shift work can impact health. Shift work has been linked to increased risk of sleep disorders, physical impairments, mental and emotional impacts, and motor-vehicle and work-related accidents and errors.

SWSD

It is no surprise that many shift workers suffer from sleep disorders. Night-time shift workers may get on the order of 1–4 hours less sleep than daytime workers.[15] When they do sleep, it is often split into two different sessions of few hours when they get home from work and then a shorter nap before going back to work. Because of natural circadian rhythms and the natural light of daytime, it is quite simply more difficult to sleep. Some shift workers suffer from a more serious daytime sleepiness, called shift work sleep disorder (SWSD). SWSD consists of recurring or constant sleep issues in shift workers, marked either by difficulty sleeping or by excessive sleepiness. Other symptoms include lack of concentration, headaches, and drained energy. Consequences of this disorder include increased accidents and work-related errors, increased sick leave, and irritability or mood problems. Indeed, patients with sleep disorders spend twice as many days in bed per year due to sickness and miss 3–4 more days at work.[16]

Some general guidelines can alleviate impacts of shift work. These include minimizing exposure to light after work to keep sunlight from activating the circadian clock, following bedtime rituals and maintaining a regular sleep schedule, and keeping a quiet, dark, and peaceful environment during sleep times. Shift workers may opt to maintain the sleep schedule on days off work. In some cases, prescription medication to promote wakefulness and sleep aids to fall asleep can be carefully used.

A link to obesity, digestive (gastrointestinal, GI) issues, and cardiovascular (CV) disease can develop with insufficient sleep and disruptions to the circadian rhythm. The circadian rhythm guides bodily functions such as heart rate, blood pressure, body temperature, digestion, and brain activity, so ignoring—even defying—this natural rhythm causes disruptions. Further, insufficient sleep adversely affects metabolism and appetite, and studies have shown that shift workers have higher levels of triglycerides than day workers. Shift workers suffer significantly more upset stomachs, ulcers, and bouts of constipation and indigestion than day workers do.[17]

GI and CV effects

Shift work disrupts social interactions, which can have implications for mood and mental health. If the family and friends of a shift worker are on opposing schedules, the shift worker may begin to feel isolated and uninvolved (which may be mitigated as new social relationships with others on the same schedule are formed). The extent to which this affects the worker is highly individualized (as well as culturally influenced) but may be severe. Further, disrupting the circadian rhythm affects the ebb and flow of brain hormones, impacting mood, and along with sleep deprivation, increases the risk of developing psychiatric conditions.

Mood disorders

Tolerance for shift work differs from person to person and varies over time. Three out of 10 shift workers have been reported to leave shift work within the first 3 years due to health problems.[18] The tolerance of those remaining on shift work depends on various personal factors (including age, personality, stressors, and diseases, as well as the ability to be flexible in one's sleeping habits, in order to overcome drowsiness), on social-environmental conditions (including family composition, housing conditions, and social status), and, of course, on the work itself (the workload, shift schedule), and the worker's income and other compensation. These factors interact, and their importance differs widely from person to person and changes over one's work life.

Errors at work and even accident probability can also increase for shift workers. During the initial period of adjusting to night-time shift work, performance is likely to be impaired between midnight and the morning hours, with the lowest performance around 4:00 a.m. Such impairment, which may be absent or minimal in the performance of cognitive tasks, varies in level, but is similar to that induced by legal blood alcohol limits.[19] However, as the worker continues on night shifts, her or his internal clock realigns itself with the new activity rhythms, and a daily routine of social interactions, sleeping, and going to work is established.

Work errors

As mentioned earlier, shift workers tend to get less sleep every daytime cycle than nonshift workers. Since they are less likely to sleep the full amount their bodies require, they can amass a large "sleep debt" over time, and that can mean slowed reaction times, delayed thinking, and lowered ability to solve problems. It might also be harder to stay focused and on-task. These issues are exacerbated by the disruptions to circadian rhythms, as described earlier. For those individuals working in professions where mistakes come with potentially immediate and dire consequences—physicians and nurses, pilots, drivers—the potential for danger is very real. Research shows increased safety risks for shift workers[20]: those who work nights are more likely to report nodding off at work and "drowsy" driving to and from work. Note also that major accidents in recent history—the Exxon Valdez oil spill, Three Mile Island, and Chernobyl melt-down—all took place during overnight shifts.

7.3.3 Disruptiveness of Shift Work

The main features of the many shift systems are shown in Fig. 7.3. For convenience, they can be classified into several basic patterns, but any given shift system may contain aspects of several patterns. Four questions are particularly important regarding the features of shift systems[21]:

1. Does a shift extend into hours that would normally be spent asleep?

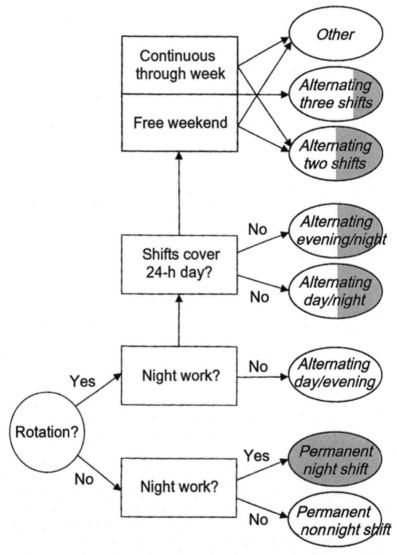

FIGURE 7.3   **Key features of typical shift systems (other shift attributes are also possible).** *Source: Adapted from Kogi (1985) with permission of John Wiley and Sons, Limited.*

2. Is the shift worked throughout the entire 7-day week, or does it alternate with days of rest, such as a free weekend?
3. Into how many shifts are the daily work hours divided? Are there two, three, or more shifts per day?
4. Do the shift crews rotate, or do they work the same shifts permanently?

Other important features are as follows:

- The starting and ending time of a shift
- The number of workdays in each week
- The hours of work in each week
- The number of shift teams
- The number of free days per week or per rotation cycle
- The number of consecutive days on the same shift, which may be a fixed or variable number

All of these various aspects may affect the welfare of the shift worker, the performance of her or his work, and the work schedule of the organization.

The most common schedules contain either a permanent shift assignment or a weekly rotation schedule. Several examples of such schedules are shown in Table 7.2. In most systems used today, the same shift is worked for 5 days and is followed by 2 free weekend days. This regimen does not, however, evenly cover all the 168 hours of the week; additional crews are needed to work on weekends or other "odd" periods.

If a company runs three shifts a day with four teams, the shift system (for one team) is 1–1–2–2–3–3–0–0 with a 6:2 work-day-to-free-day ratio and a cycle length of 8 days; this is known as the metropolitan rotation. The continental rotation, which also has three shifts per day and four crews, has the sequence 1–1–2–2–3–3–3, 0–0–1–1–2–2–2, 3–3–0–0–1–1–1, 2–2–3–3–0–0–0; its work-day-to-free-day ratio is 21:7, with a cycle length of exactly 4 weeks. Other shift schedules may utilize 12-hour shifts (two shifts a day) to accommodate the needed work coverage with available teams.

TABLE 7.2 Examples of Shift Schedules with 5 Workdays per Week

System	Workdays:free days	Shift sequence[a]
Permanent day shift	5:2	1–1–1–1–1–0–0
Permanent evening shift	5:2	2–2–2–2–2–0–0
Permanent night shift	5:2	3–3–3–3–3–0–0
Alternating day-evening	10:4	1–1–1–1–1–0–0, 2–2–2–2–2–0–0
Alternating day-night	10:4	1–1–1–1–1–0–0, 3–3–3–3–3–0–0
Alternating day-evening-night	15:6	(forward rotation) 1–1–1–1–1–0–0, 2–2–2–2–2–0–0, 3–3–3–3–3–0–0 or (backward rotation) 1–1–1–1–1–0–0, 3–3–3–3–3–0–0, 2–2–2–2–2–0–0

[a]1 represents day shift, 2 evening shift, 3 night shift, 0 a free day, that is, without a scheduled shift.
Source: Adapted from Kogi (1985). Reproduced by permission of John Wiley & Sons, Limited.

7.4 COMPRESSED WORK-WEEKS, EXTENDED WORK-DAYS

We refer to a compressed work-week when the required weekly work hours (such as 40 h) are performed in only 4 or even 3 days, allowing the worker to have 3 or 4 free days each week. A compressed work-week can reduce the number of commuting trips and the number of times the worker sets up and closes down work activities, as well as offering longer continuous days away from work. Disadvantages include potential for increased fatigue and reduced performance and safety levels, because compressing the same weekly working hours into fewer days means extending the duration of each work shift.

7.4.1 Appropriateness

Whether or not an extended work-day is appropriate for a given job is not always easy to determine. One approach is to have a trial period in the work place, survey the employees, and monitor their health and safety in addition to evaluating productivity and quality of work. Factors other than the actual work contribute to whether or not an extended workday is acceptable or advisable, including environmental conditions like temperature and job characteristics like boredom or repetition.

Some general guidelines can help assess the appropriateness of an extended work-day. Compressed work-weeks may be suitable for

- Jobs that do not require a high degree of physical exertion (such as a university lecturer who has downtime between classes and lectures).
- Jobs with natural resting periods (such as a firefighter who is on "stand by" for much of the shift, or a machinist who has cycle time between set-ups that permit periods of reduced attention while the machine is running).
- Jobs with creative activities (such as writers or artists who often work intensively on projects with high engagement).

Jobs that require repetitive tasks in stationary positions and with high degree of concentration are largely not suited for extended work-days. For example, a data entry job that requires continual keyboarding while seated would likely not be a candidate for an extended work shift.

Working very long shifts, such as 12 hours, is likely to cause drowsiness, with concomitant reductions in cognitive abilities and motor skills. Consequently, performance may diminish during the course of each shift and as the work-week progresses. A fatigued worker may become careless and work practices may grow less safe.

7.4.2 Advantages and Disadvantages

For employees and employers, potential advantages of a compressed work-week include[22]:

- Extra day(s) off work while retaining base pay
- Increased options for nonwork activities with two or more consecutive days away from the job
- Reduced commuting costs and stresses

- Additional time during the work-day for scheduling meetings or training sessions
- Fewer and therefore reduced start-up and warm-up expenses
- Increased productivity and quantity/quality of services
- Enhanced opportunities to hire skilled workers in tight labor markets

Potential disadvantages of extended workdays include:

- Decreased job performance due to long work hours or to moonlighting on free days
- Difficulty in scheduling child care and family life during the work-week
- Increased impact of tardiness or early departure from work or absenteeism
- Increased on-the-job and off-the-job accidents
- Decreased productivity
- Scheduling problems if the operations of the organization are longer than the work-week
- Increased energy and maintenance costs
- Increased daily exposure of workers to physical and chemical hazards (acceptable doses are usually calculated for an 8-hour exposure)

7.5 FLEXTIME

While the daytime, continuous workday is still very common, work–life balance issues have become an important concern for employees and companies. Employees with good work–life balance are happier and healthier, take fewer sick days and incur fewer health-care costs; they're also more productive and help maintain a positive office morale. One way some companies help employees achieve their optimal work–life balance is through *flextime*. Flextime is a work schedule that requires employees to be present for predetermined core hours but allows them the flexibility to vary the rest of their schedule based on their needs and preferences. Using flextime, workers can distribute the prescribed number of working hours per shift (e.g., 8 hours) over a longer block of time, such as 10 or 12 hours.

Work–life balance

For employees and employers, potential advantages of flextime include[23]:

- Extra day-to-day free time while retaining base pay
- Increased options for scheduling nonwork activities
- Reduced commuting costs and stresses
- Less fatiguing for workers
- Reduced job dissatisfaction and increased job satisfaction
- Means to recognize and facilitate employees' individual differences
- Reduced tardiness, absenteeism, and employee turnover
- Increased productivity
- Ability to adjust workforce to reflect short-term fluctuations in demand
- Enhanced opportunities to hire skilled workers in tight labor markets

Potential disadvantages of flextime include:

- Scheduling problems to cover necessary jobs at all times
- Increased difficulty in scheduling meetings or training sessions
- Poorer communication within and between organizations

- Increased energy and maintenance costs and health/food service hours
- Irregular work hours to reflect short-term changes in demand
- Need for more sophisticated planning, organization, and control
- Need for additional supervisory personnel and means to record time worked
- Reduced quantity/quality of services

In addition to condensed work-weeks and flextime, other work arrangements that vary from the regular 8 hour a day, 5 days a week routine have become popular. Work schedules may include staggered work hours, seasonally adapted hours of work or lengths of shifts, part-time work (often combined with job sharing), and telecommuting. All of these arrangements decouple the individual from strict job requirements in time or location.

Flexibility requires that the employer change from traditional organizational patterns regarding supervision, communication, job requirements, work equipment, and scheduling; accordingly, the employee must increase her or his self-reliance, independence, responsibility, and skills. These major changes require a careful assessment of the nature of the job, as well as extensive cooperation between management and employees.

SCHEDULE ARRANGEMENTS FOR AIRPLANE FLIGHT CREWS

Crew expenses are the second largest operating expense airlines have (jet fuel is the largest). The airline industry is heavily unionized, and there are strict rules and limitations on work schedules—how many hours a day a crew is in the air, how long crew members can be away from their home base before they lay-over in a hotel, etc. Airplane flight crews that cross time zones during long-distance flights and need to sleep between flights experience a number of problems. For one, the quality and length of sleep at the stopover location is frequently far less satisfactory than at home. The resulting tiredness is often masked or counteracted by the use of caffeine, tobacco, and alcohol. The second problem is the extended time on duty, which includes preflight preparations, the flight period itself, and the wrap-up after arrival. Detrimental effects are substantially worse after an eastward flight than a westward one; further, crew members over the age of 50 appear to be more affected than their younger colleagues.[24]

Recommendations for the shift arrangement for flight crews center around trying to maintain their regular diurnal rhythms. Particularly when flying eastward, flight crews should adhere to well-planned timing that duplicates, as far as possible, the sleep–wakefulness periods at home, meaning that crew members should try to go to sleep and get up at their regular home times, and their next flights should be during their regular time of wakefulness.

7.6 BODY RHYTHMS AND ALTERNATE WORK SCHEDULES: SUMMARY

Human bodily functions and social behavior follow internal rhythms. Aside from the female menstrual cycle, the best-known rhythms are a set of daily fluctuations called circadian or diurnal rhythms, which appear in such functions as body temperature, heart rate, blood pressure, and hormonal excretions. Under regular living conditions, these temporal programs are clearly established and persistent.

The well-synchronized rhythms and the associated behavior of sleeping (during the night) and being active (during the day) can be desynchronized and put out of order if the time markers during the 24-hour day are changed and if activities are required at unusual times. The resulting sleep loss and tiredness influence human performance in various ways. Performance of mental tasks, attention and alertness usually are diminished, but most physical activities may still be carried out without any significant degradation in performance. Some researchers are concerned that disturbing the internal rhythm through certain types of shift work can be detrimental to a person's health in a number of ways. Moreover, the sociopsychological effect of shift work is that it prevents those who are on it from participating in some family and other social activities.

Different aspects of shift work, including the worker's health and well-being, performance and accidents, and psychological and social adjustments, interact with each other, but not always in the same direction. The following recommendations, based on physiological, psychological, and social behaviors associated with performance, present considerations for working hours and shift work:

- Job activities should follow entrained body rhythms.
- Work during the daylight hours is preferable.
- Evening shifts are preferred to night shifts, and appropriate lighting and environmental stimuli should be provided.
- If shifts are necessary, two opposing rules apply: (1) Either work only one evening or night shift per cycle, then return to day work and keep weekends free or (2) stay permanently on the same shift.
- A duration of 8 hours of daily work per shift is usually adequate, but shorter times for highly (mentally or physically) demanding jobs should be considered, and longer times (such as 9, 10, or even 12 hours) may be acceptable for some types of routine work-or especially creative work.
- Compressed work-weeks are often acceptable (even preferable, depending on the individual) for routine jobs.
- Together with changing work tools and habits, new and much less rigid time schedules are becoming common, requiring different attitudes for workers and managers; flexibility is becoming more and more important.

7.7 EFFECTS OF ALCOHOL, MARIJUANA, AND OTHER DRUGS ON PERFORMANCE

Alcohol and marijuana are the most widely used and abused substances in the United States, with alcohol at the top of the list, and marijuana second. The cause of alcohol and marijuana's popularity (and increased likelihood of abuse) is two-fold:

- Both are relatively easy to access, despite the age limit for purchasing alcohol and the limited legalities in acquiring marijuana.

- Both are relatively cheap in comparison to other drugs; even in an economic recession, alcohol and drugs are still widely purchased.

The breakdown of the number of Americans over 12 who used, misused or abused the most commonly used drugs is as follows (2015 data, with a base of approximately 268 million Americans over the age of 12)[25]:

- Alcohol: 52% (138.8 million) had any use in the last month, 25% (66.7 million) were binge-users, 6% (17.3 million) were heavy users
- Marijuana: 8% (22.2 million)
- Misuse of prescription pain relievers: 1% (3.8 million)

Other drugs were noted with fewer than 2 million users (listed in order): cocaine, misuse of prescription tranquilizers, misuse of prescription stimulants, hallucinogens, methamphetamine, inhalants, misuse of prescription sedatives, heroin.

We cover the effects of the two most commonly misused substances—alcohol and marijuana—in this chapter to show how they impact human performance. As an aside, with decriminalization of marijuana increasingly occurring on a state-by-state level in the United States, and globally, it is likely that the impact of marijuana use (and misuse) will rise.

7.7.1 Alcohol and Human Performance

Alcoholic beverages contain many chemical substances; approximately 200 *congeners* (products of fermentation) have been identified in wine. It is still not clear which of these (in addition to ethanol) are responsible for the feelings of temporal euphoria and freedom from inhibition associated with drinking alcohol—or for the less desirable physiological and psychological effects. While there is some evidence that alcoholic beverages in limited doses may be beneficial for the cardiovascular system, taken excessively, alcohol harms human beings.

Alcohol in the human bloodstream has neurological effects:

1. First, it impairs the functioning of the cerebral cortex, which houses intelligence.
2. Then the limbic system is affected, which, among other functions, controls mood.
3. Finally, alcohol impairs the brain stem, where the "fight-or-flight" response is generated and where such functions as heart rate, blood pressure, and respiration are controlled.

Effects on CNS

Consequently, alcohol impairs the central nervous system. The greater the blood alcohol level, the more the impairment in

- judgment and reasoning
- language, both expressive and receptive
- insight
- memory and discernment of alternative scenarios
- attention and concentration
- fine and gross motor control and body posture
- delay in reaction time
- affective stability versus emotional volubility.

Exceptions to reliable dose dependent reactions occur in individuals with genetic or underlying organic conditions (such as renal or hepatic insufficiency) in whom small doses of a substance may produce a disproportionately severe intoxicating effect. Disinhibition and pronounced reactivity due to social context should also be taken into account, as intoxicated individuals tend to "absorb" the mood and mileau of their social surroundings. (This is particularly concerning in violent or aggressive escalations.)

Long-term excessive users of alcohol (sometimes called *alcoholics*) are likely to show pathological effects. Toxic changes occur in organs such as the brain and in muscles. In the intestinal system, metabolic processes suffer, since alcohol interferes with absorption, digestion, and with the metabolism and utilization of nutrients and vitamins.

Alcoholism

There is some confusion about the terminology one should use in describing patterns of alcohol use. The concept of inveterate drunkenness as a disease appears to be rooted in antiquity. The Roman philosopher Seneca classified it as a form of insanity. The term *alcoholism* appeared first in the classical essay "Alcoholismus Chronicus" (1849) by the Swedish physician Magnus Huss. The phrase *chronic alcoholism* rapidly became a medical term for the condition of habitual inebriety, and the bearer of the "disease" was called an *alcoholic* or *alcoholist*. While the term "alcoholic" remains in favor in the lay community (and also unfortunately in professional circles), it has just about as much utility as older terms with a derogatory connotation, such a "lush," "barfly," or "boozer."

One of the difficulties with the word *alcoholism* is that its meaning varies depending on who is using it. Members of the group Alcoholics Anonymous will use it to refer to a disease, but in common speech *alcoholism* can be used to describe anyone who has a problem with alcohol. Even those individuals who are not dependent or do not show negative consequences of alcohol abuse may be referred to as alcoholic if they drink "too much" by someone else's standards. Medical professionals tend not to use the term because it is too vague and lacks parameters or criteria for its definition. A preferred term might be *"problem drinker"* as this captures the concern that the quantity, manner, or frequency of drinking is causing problems in the person's functioning and well-being.

Alcoholism can be defined in a number of ways.

Definition

- A purely *pharmacological-physiological* definition classifies it as a drug addiction that requires imbibing increasing doses to produce desired effects and that causes a withdrawl syndrome when drinking is stopped.
- A *behavioral* definition notes that alcohol assumes marked importance in the individual's life and in which the individual experiences a loss of control over its desired use. This may or may not involve physiological dependence, but invariably it is characterized by alcohol consumption that is sufficiently great to cause regret and repeated physical, mental, social, economic, or legal difficulties.
- A *sociological* definition posits that the medicalization of alcoholism is an error and that its diagnosis often lies in the eyes and value system of the beholder. Unlike

most disease symptoms, the loss of control over drinking does not hold true at all times or in all situations. The "alcoholic" is not always under internal pressure to drink and can sometimes resist the impulse to drink or can drink in a controlled way. Further, symptoms of alcoholism (and societal acceptance of differing levels of intoxication) vary from culture to culture. In the general population, variation in daily alcohol consumption is distributed along a smooth continuum. (This characteristic is inconsistent with the *medical* model, which implies that alcoholism is either present or absent, as a pregnancy or a tumor would be.)

- An *epidemiological* definition to identify alcoholics within a population would rely on quantity and frequency measurements of reported community drinking and alcohol-related hospitalizations, perhaps utilizing a formula based on the frequency of deaths from cirrhosis within the population, or on arrests for alcohol-related misbehavior.

Alcohol-related disorders are defined by the World Health Organization (WHO) in the International Statistical Classification of Diseases and Related Health Problems (ICD-10, issued in 1992) and in the United States by the American Psychiatric Association in the Diagnostic and Statistical Manual of Mental Disorders (DSM-5). (Note that what applies to alcohol applies to all substances of use/abuse.) The DSM-5, released in 2013, introduced a paradigm shift in the understanding of alcohol (and other substance-related) disorders. Importantly, terminology such as substance abuse and substance dependence has been eliminated. Instead, all substance use (including alcohol) recognizes degrees of illness and impairment. The DSM-5 identifies substance-related disorders resulting from the use of ten separate classes of drugs: alcohol, caffeine, cannabis, hallucinogens, inhalants, opioids, sedatives, stimulants (including amphetamines and cocaine), tobacco, and other or unknown substances. The activation of the brain's reward system is central to problems arising from drug use—the rewarding feeling that people experience as a result of taking drugs may be so profound that they neglect other normal activities in favor of taking the drug. While the pharmacological mechanisms for each class of drug is different, the activation of the reward system is similar across substances in producing feelings of pleasure or euphoria (often referred to as a "high"). Clinicians can specify how severe or how much of a problem the substance use disorder is, depending on how many symptoms are identified. Two or three symptoms indicate a mild substance use disorder; four or five symptoms indicate a moderate substance use disorder; and six or more symptoms indicate a severe substance use disorder. Clinicians can also add "in early remission," "in sustained remission," "on maintenance therapy," and "in a controlled environment."

According to the DSM-5, there are two groups of substance-related disorders: *substance-use disorders* and *substance-induced disorders*. Substance-use disorders are patterns of symptoms resulting from the use of a substance that the person continues to take, despite experiencing problems as a result. Substance-induced disorders, including intoxication, withdrawal, other substance/medication-induced mental disorders, are detailed alongside substance-use disorders.

The DSM-5 recognized eleven criteria for substance-use disorders:

1. Taking the substance in larger amounts or for longer than intended.
2. Wanting to cut down or stop using the substance but not managing to.

3. Spending a lot of time getting, using, or recovering from use of the substance.
4. Cravings and urges to use the substance.
5. Not managing to function at work, home, or school because of substance use.
6. Continuing to use, even when it causes problems in relationships.
7. Giving up important social, occupational, or recreational activities because of substance use.
8. Using substances again and again, even when it puts the user in danger.
9. Continuing to use, even when the user knows he/she has a physical or psychological problem that could have been caused or made worse by the substance.
10. Needing more of the substance to get the desired effect (tolerance).
11. Development of withdrawal symptoms, which can be relieved by taking more of the substance.

Substance/medication-induced mental disorders are mental problems that develop in people who did not have mental health problems before using substances. These include substance-induced psychotic disorder, bipolar and related disorders, depressive disorders, anxiety disorders, obsessive-compulsive and related disorders, sleep disorders, sexual dysfunctions, delirium, and neurocognitive disorders.

AMERICANS AND ALCOHOL ABUSE[26]

Alcohol-Use Disorder (AUD) in the United States: 15.1 million adults ages 18 and older (6.2% of this age group) had AUD: 9.8 million men and 5.3 million women. About 1.3 million adults received treatment for AUD at a specialized facility in 2015 (representing less than 10% of adults who needed treatment).

Alcohol-related deaths: An estimated 88,000 people (approximately 62,000 men and 26,000 women) die from alcohol-related causes annually, making alcohol the fourth leading preventable cause of death in the United States.

7.7.1.1 Blood Alcohol Content

The effects of alcohol are usually stated in relation to the blood alcohol content (BAC, in percent). BAC is best measured in a blood sample, but is often approximated from a sampling of exhaled air (such as with a Breathalyzer). In 1932, the Swedish scientist Eric Widmark developed a basic formula for assessing BAC:

$$\% \text{BAC} = (5.14 \, A W^{-1} r^{-1}) - 0.015 \, H \tag{7.1}$$

where A is the liquid ounces of alcohol consumed; W is the person's weight in pounds; r is a gender constant of alcohol distribution (0.73 for men and 0.66 for women); and H is the hours elapsed since drinking commenced. To find A in the Widmark formula above, multiply the number of liquid ounces of alcoholic beverages consumed by the percentage of alcohol in the beverage. For example, a typical bottle of regular beer contains

12 oz of liquid and is 5% alcohol, so it contains 0.60 liquid ounces of alcohol. The 5.14 appearing in Eq. (7.1) is a conversion factor of 0.823 × 100/16, wherein 0.823 is used to convert liquid ounces to ounces of weight, 100 is used to convert the final figure to a percentage, and 16 is used to convert pounds to ounces. The 0.015 figure appearing in the formula is the average alcohol elimination rate. The alcohol content of a given beverage can be determined using the *Alcohol Content of Common Beverages* list; and assume that all beverages were consumed on any empty stomach. The BAC formula assumes that the body begins to eliminate alcohol immediately upon absorption and continues to eliminate it until it has been completely removed from the body.

HOW TO CALCULATE % BAC

A 170-pound man consumed seven 12-oz bottles of regular beer between 8:00 p.m. and midnight. To calculate his % BAC at 12:30 a.m., we determine the variables as follows:

A = 7 beers × 12 oz × 5% = 4.2 oz.

W = 170 lbs

r = 0.73 for a man

H = 4.5 h

Then, using the % BAC formula:

$$\%\,BAC = (5.14\,AW^{-1}r^{-1}) - 0.015\,H \tag{7.1}$$

Therefore, the calculated % BAC is

% BAC = (5.14 × 4.2/170/0.73) − 0.015 × 4.5 = 0.174 − 0.068 = 0.106%

Note that this % BAC is above the level of legal impairment (0.08%) for driving in the United States.

Absorption

Consumed orally, alcohol is absorbed into the bloodstream by simple diffusion through cell boundaries in the linings of the stomach and the gastrointestinal tract, with large intra- and interindividual differences in the resulting BAC. Once absorbed, the alcohol is distributed by the blood. Since ethanol mixes freely with water, it quickly reaches all of the body's cells. (About half the blood volume is water, and cells are bathed in it.) Hence, the alcohol content of organs with a good blood supply (such as the brain, lungs, liver, and kidneys) quickly becomes the same as that of the blood. The highest alcohol content in the blood occurs about half an hour after ingestion. Full absorption may take up to 6 hours with a heavy meal.

Distilled alcohol ("hard liquor") is absorbed faster than the alcohol in beer. The more alcohol consumed over a given time, the slower its absorption, because alcohol diminishes the stomach's ability to empty itself, slowing the arrival of alcohol at the lining of the intestinal tract. This effect may explain the higher tolerance (that is, slower absorption rate) of drinkers compared to abstainers. Delayed gastric emptying also explains why absorption is also slowed by a "full stomach." Absorption is slowest after a meal high in carbohydrates; moderate after a meal high in fats; and fastest after a meal high in proteins.

Alcohol is freely diffused in the body. Consequently, people who have less body water (women, obese people, and the elderly) generally will have a higher BAC than those with more body water (men, slim persons, and younger people) after drinking the same dosage of alcohol. Women taking oral contraceptives metabolize ethanol more slowly than other women or men, meaning that they could stay intoxicated longer. The strength of the effects of the alcohol also depends on the circadian rhythm (discussed earlier in this chapter). For example, the effects of alcohol are stronger in the early afternoon than in the evening.

7.7.1.2 Aftereffects of Alcohol

Alcohol is eliminated from the body at a uniform rate until its concentration is very low. The rate per hour is about 0.015% for men and 0.019% for women. Alcohol is oxidized, and the by-products—carbon dioxide and water—are excreted by breathing; the water that is produced increases the need to urinate. Engaging in physical exercise to speed up the process of elimination is largely ineffective, as is consuming coffee.

Elimination

DOES COFFEE HELP US SOBER UP?

The short answer to this widely believed myth is a resounding *no*. Alcohol leaves our body through oxidization, and this process is not affected by caffeine. In reality, consuming coffee after alcohol may actually do more harm than good, since coffee can trick us into thinking that we are less inebriated than we actually are. Imagine having four glasses of wine at a restaurant meal, then ordering coffee after dinner in the hopes of perking and sobering up. Even though alcohol continues to make our brain and body more sluggish, the caffeine may cause us to believe we are more energized and less inebriated. This in turn might lead to the extremely poor decision of driving a car or engaging in other risky behavior.

Too much alcoholic "good cheer" can result in a hangover. Headache, nausea, and stomach irritation are caused by undigested by-products of alcohol, particularly acetaldehyde and lactic acid, that build up in the bloodstream as the liver falls behind in digesting alcohol. Since the speed of alcohol conversion cannot be changed, neither by drinking coffee nor by breathing fresh air, the simple (and sole) cure is time: wait long enough to give the liver the time it needs to eliminate the alcohol from the body. Symptoms of a hangover may be relieved by over-the-counter medications, including antacids, aspirin, and other pain relievers; however, these remedies can cause further stomach irritation.

Hangover

7.7.1.3 Effects of Alcohol on the Nervous System

The nervous system is made up of the CNS and the peripheral nervous system (PNS). (See Chapter 4.) The CNS includes the brain and the spinal cord. The most important parts of the CNS are protected by bones: the skull protects the brain and the spine protects the spinal cord. Duties of the CNS include taking in information through the

CNS, PNS

senses, controlling motor function, thinking, understanding, and reasoning. It also controls emotion. The PNS includes the neurons, or nerves, that form a network designed to carry information to and from the neck and arms, trunk, legs, skeletal muscles, and internal organs. Neurons can send nerve signals to and from the brain at up to 200 miles per hour. The PNS is remarkably efficient and smooth when it is healthy and unimpaired.

Not surprisingly, alcohol affects both the CNS and the PNS quite dramatically. Alcohol acts as a depressant on the CNS, slowing it down. Specific effects on the CNS, including the brain, are generally as follows:

- At BAC levels below 0.05%, judgment is impaired and inhibitions are reduced (resulting in the illusion of stimulation).
- At 0.1% BAC, sensory and motor functions are depressed.
- At 0.2% BAC, control of emotion is lost.
- At 0.3% BAC, cognition is affected and stupor occurs.
- At 0.4-0.5% BAC, the drinker becomes comatose.
- At 0.6% BAC, breathing and heartbeat are depressed, and death can occur.

Effects on the PNS include a reduction in the transmission of neural impulses at the synaptic junction by the depressant effects of alcohol. Further, a low BAC increases nerve excitation and a high BAC inhibits nerve excitation.

Becoming "tolerant" to alcohol does not increase the lethal threshold: The LD-50 (meaning a lethal dose in 50% of those who drink) remains at 500 mg/dl.

7.7.1.4 Effects of Alcohol on Human Senses

Vision

Human vision is affected in various ways by alcohol. Visual acuity is relatively insensitive to the BAC, as is adaptation to light and dark. But alcohol increases the sensitivity to dim lights and decreases the ability to discriminate among various bright lights. Resistance to glare is also reduced, and color sensitivity is affected: light red, green, and yellow are less easily discerned from white, but blue and violet hues are more easily discriminated.

Alcohol causes the eyes to converge at long viewing distances and diverge at short ones. Depth perception is impaired at a relatively high BAC (around 0.1%) and the ability to judge distances is also reduced. Visual accommodation is impaired and eye-movement latency increased. The critical fusion frequency, which is the highest rate at which light is perceived as flashing, is decreased by large dosages of alcohol (0.1% BAC), but not by smaller doses. Peripheral vision is somewhat reduced by alcohol, but only under a heavy general information load. Together, these findings lead to the conclusion that alcohol impairs the ability to see rapidly changing events.

Other senses

Like vision, hearing is variably affected by alcohol. Auditory acuity appears relatively insensitive to alcohol, but the ability to glean information from auditory stimuli is impaired. Both smell and taste sensitivities are diminished with as little as 0.01% BAC. The sensitivity of touch is reduced, particularly with respect to two-point discrimination, and sensitivity to pain is diminished by alcohol.

7.7.1.5 Effects of Alcohol on Motor Control

Motor control

Motor control is greatly reduced by alcohol. For example, standing without swaying, touching the index fingers together, and other measures of hand steadiness and gait control show significant diminution, even at 0.1% BAC. This explains the effectiveness and validity of the *standard field sobriety test* (SFST), used by US law enforcement during a traffic stop to determine if a driver is impaired. Two of the three tests that make up the SFST test motor control skills (and the ability to follow directions): the walk-and-turn test, and the one-leg stand test. When used correctly, the SFST can be used to correctly identify alcohol-impaired drivers over 90% of the time.

Simple reaction time is increased by alcohol, but only at BAC levels of 0.07% or more; a BAC of 0.08%-0.1% increases reaction time by about 10%. The time it takes to make a choice is even more affected by alcohol, and the incidence of wrong choices is increased as well. Response to an auditory signal seems to be more prolonged than to a visual signal.

Reaction time

7.7.1.6 Effects of Alcohol on Cognition

Regarding verbal performance, alcohol increases superficial, egocentric, and inappropriate associations. Alcohol also reduces fluency and mastery of words. The ability to make arithmetic calculations is impaired, and time seems to pass more quickly to a person who is inebriated. While retrieving information from memory seems unaffected by alcohol, drinking does lead to memory deficiencies regarding events that took place when the person became intoxicated, especially if a large amount of information had to be stored.

Cognition

Simple auditory or visual vigilance tasks are not affected by alcohol, but complex ones are impaired. The impairment of judgment under alcohol is well documented, both regarding judging one's own performance as well as somebody else's. Alcohol increases the willingness to take risks.

7.7.1.7 Epidemiology of Alcohol Use

To assess consumption patterns in a population, we need to be able to measure dosages, or "serving sizes." For the purposes of alcohol research, one drink refers to 0.5 oz of alcohol. This corresponds to approximately 12 oz (350 ml) of 5% alcohol content beer, 8 oz (240 ml) of 7% alcohol content ale or malt liquor, 5 oz (150 ml) glass of 12% alcohol content wine, or 1.5 oz (44 ml) of most distilled spirits (with 40% alcohol content, or 80 proof).[27] "Binge" drinking is defined as drinking within a short period time to reach a 0.08% BAC or higher and is usually accomplished with 5 drinks for men and 4 drinks for women. Binge drinking is often associated with engaging in other unhealthful or risky behaviors. "Risky" drinking has been defined as reaching a peak BAC between 0.05% and 0.08%. A "bender" is typically defined as 2 or more days of sustained heavy drinking. Most heavy drinkers (defined as 8 or more drinks a week for women and 15 or more drinks a week for men), however, are not considered "excessive" or "problem" drinkers nor alcohol dependent.[28] Women display lower rates of lifetime alcohol use than men. The largest gender difference is noted at 55 years of age and older, while the smallest gender difference is found among young adults ages 18–24 years old. Approximately two-thirds of men and one half of women who are arrested in the United States are frequent alcohol users.[29]

In 2014, excessive alcohol use was deemed the fourth leading preventable cause of death in the United States at a societal cost of $223.5 billion, or about $1.90 per drink.

Alcohol is implicated in 1 of 10 deaths among working-age adults in the United States.[30] Alcohol is a contributing factor to deaths in 30–50% of motor vehicle accidents, 40% of falls, 26% of fires, 49–70% of homicides, and 25–37% of suicides. Short-term risks of alcohol are injuries, accidents, violence, sexual assault, alcohol poisoning, and risky sexual behaviors. Long-term risks include physical, mental health, and socioeconomic problems. Long term medical complications of excessive alcohol consumption include dementia, anemia, pancreatitis, cirrhosis, gastritis, insomnia, impotence, hepatitis, peripheral neuropathy, myopathies (including cardiomyopathy), *delirium tremens* (in withdrawal), alcoholic hallucinosis, and irreversible brain damage. Excessive alcohol also increases cancer risk, particularly for esophageal, breast, head/neck, colorectal, and liver cancers. The most vulnerable brains are those of unborn children: fetal alcohol syndrome (when brain development is disrupted in gestation due to the mother's consumption of alcohol) manifests with smaller infant brain size, a reduced number of brain cells, an array of facial characteristics, and chronic learning and behavior problems.[31]

Contrary to the popular image of the down-and-out unemployed drug or alcohol abuser, most drug users, binge drinkers, and heavy drinkers are employed; this implies that substance use and abuse is a significant concern for employers. 2007 National Household Survey on Drug Abuse data in the United States showed that of the illicit drug users 18 and over, 75% were employed; among adult binge drinkers, 79% were employed; of the people reporting heavy alcohol use, 80% were employed; and of the adults classified with other substance dependence or abuse, 60% were employed full-time. These workers increase employer health care costs and decrease overall corporate productivity. Programs such as mandatory drug testing and employee assistance programs have been found to reduce medical claims and absenteeism related to alcohol and substance abuse.

7.7.1.8 Screening for Problematic Alcohol Use

CAGE

The acronym "CAGE" has long been, and continues to be, an effective and quick screening method that is useful in clinical and employment settings for identifying problematic drinking patterns. To apply the technique, just ask a person, "Have you ever thought you should *C*ut back on your consumption of alcohol? felt *A*nnoyed by those who criticize your drinking? felt *G*uilty or bad about your drinking? had a morning "*E*ye-opener" (alcoholic drink) to relieve a hangover?" Answering yes to two or three of these questions invokes high suspicion of problematic alcohol use while answering yes to all four indicates a serious condition.[32] Of course, there are many other measures, including structured interviews, but these are generally not appropriate or adaptable to a workplace setting.

7.7.1.9 Effects of Alcohol on Performing Industrial Tasks

Task performance

Trends for performance deficits at various alcohol dosages indicate[33] that psychomotor performance is least impaired and perceptual-sensory performance most impaired. Cognitive performance is moderately impaired. Alcohol reduces the ability to perform industrial work tasks. As BAC rises, errors increase and output decreases; in assembly tasks, at 0.09% BAC, productivity was reduced up to 50%. Deleterious effects on the quantity and quality of work performed were also seen in machine-tool (punch or drill presses, for example) operators, and in welders. While different operators react differently to different alcohol dosages over different application times and with different work tasks, the fall-off in performance as BAC increases is clear and definitive.[34]

7.7.1.10 Effects of Alcohol on Automobile Driving

Consuming alcohol prior to driving greatly increases the risk of car accidents, highway injuries, and vehicular deaths. The greater the amount of alcohol consumed, the more likely a person is to be involved in an accident.

Driving

As outlined earlier, alcohol is a depressant that affects both the CNS and the PNS. With the CNS slowed, normal brain function is delayed and a person cannot perform properly. Alcohol affects a person's information-processing skills (cognitive skills) and hand-eye coordination (psychomotor skills). When alcohol is consumed, many of the skills that safe driving requires—such as judgment, concentration, comprehension, coordination, visual acuity, and reaction time—become impaired. A driver who is under the influence of alcohol appears to have a shrunken visual field; this means that information is collected at shorter viewing distances and less frequently. Response latency and response errors are increased. Exact steering, which we require to stay in lane or to park, is impaired, as is the ability to judge driving speed. Willingness to take risks is increased.

Here are some signs that a driver is impaired by alcohol:

- approaches a red traffic light quickly and then makes a sudden stop
- changes lanes often ("weaves")
- straddles the centerline
- changes speed often and/or erratically
- drives too fast or too slowly
- drives in darkness without headlights.

SOBERING STATISTICS[35]

- An estimated 32% of fatal car crashes involve an intoxicated driver or pedestrian.
- Over 1.2 million drivers were arrested in 2011 for driving under the influence of alcohol or narcotics.
- Car crashes are the leading cause of death for teens, and about a quarter of those crashes involve an underage drinking driver.
- On average, two in three people will be involved in a drunk driving crash in their lifetime.

7.7.1.11 Effects of Alcohol on Pilots

Every country makes its own rules on BCA and pilots; in the United States, the FAA blood alcohol limit for airline pilots is 0.04%, and pilots are banned from consuming alcohol within 8 h of reporting for duty (the *bottle-to-throttle* rule). Professional pilots must also comply with their employer's policies, which tend to be more restrictive; for example, most airlines impose a 12-h consumption restriction.

Even as airplanes become increasingly autonomous (see Chapter 12), flying an aircraft remains a profoundly demanding job. Pilots are trained to perform the tasks necessary to get the plane into the sky, navigate safely to the destination location, and safely land the aircraft—all complex tasks. In order to perform these procedures, the pilot needs full control of his or her cognitive abilities, even while adhering to a host of national and international regulations. Additionally, pilots must constantly be prepared for an unexpected event—sudden cabin depressurization, an engine malfunction, or even a medical emergency on board—and it is obvious why pilots need their faculties intact and unaffected by alcohol.

In an experiment using a flight simulator, both younger (mean age 25 years) and older (mean age 42 years) pilots showed reduced flying performance, including communication, for at least 2 hours after having reached 0.10% BAC. Their overall performance remained impaired for up to 8 hours.[36]

Twelve male pilots, all with relatively few flying hours (50–315) and without instrument rating, performed simulated flight activities either under placebo conditions or with alcohol dosages that brought about a BAC of about 0.04%. Although these pilots' performances of the main flying tasks were relatively unaffected, the performances of many other tasks were reduced. Pilots who consumed alcohol were often inattentive to important secondary tasks and violated safety procedures. The researchers concluded that, even under low alcohol levels, pilots' performance would reduce the margin of safety in routine flying conditions and that, in circumstances of increased demands on the pilot, the probability of an accident would be significantly increased.[37]

7.7.2 Marijuana and Human Performance

After alcohol, marijuana is the most widely used substance. In 2015, almost half of the people in the United States had tried marijuana, 12% had used it in the past year, and 7.3% had used it in the past month.[38] In 2014, daily marijuana use amongst US college students had surpassed daily cigarette use. Young adults are trying marijuana at increasing younger ages. In many of the US states as well as in countries around the world, marijuana is increasingly being decriminalized for medicinal or even recreational use. Medical cannabis use also currently takes place in Canada, Belgium, Australia, the Netherlands, Germany, Spain, Columbia and other countries. Consequently, the effect of marijuana on humans and human performance is relevant in the workplace.

Marijuana comes from the hemp plant *Cannabis*. The major strains are *Cannabis sativa* and *Cannabis indica*, each possessing distinctive psychoactive qualities. In both cases, then main active ingredient is *tetrahydrocannabinol* (THC); however, this is only one of 483 known compounds in the plant, including at least 65 other *cannabinoids*. To get the drug into their systems, most users smoke the plant's dried leaves, flowers, stems, and seeds. Marijuana can also be mixed into food (such as baked goods), brewed as a tea, or inhaled with a vaporizer.

7.7.2.1 *Effects of Marijuana on Physiology*

No matter how it gets into the system, marijuana affects almost every organ in the body as well as the nervous system and the immune system. Smoking marijuana has almost instantaneous results because the body absorbs THC right away; when marijuana is eaten, as in a baked good laced with the drug, it takes much longer for the body to absorb THC because it must be broken down in the stomach before it enters the bloodstream. The effects of inhaled or ingested marijuana usually end after 3 or 4 hours.

Marijuana also contains *cannabidiol* (CBD); this chemical, unlike THC, does nothing to get a user "high." CBD is thought to be responsible for many of marijuana's potential therapeutic effects, including:

- Pain relief: this is by far the most common reason people request medical marijuana; studies have shown that the compound may help reduce inflammation, which is a component of rheumatoid arthritis and inflammatory bowel diseases like Crohn's.
- Controlling epileptic seizures: CBD may help in the treatment of rare forms of childhood epilepsy.
- Weight gain: some compounds in marijuana may help stimulate appetite in those suffering from weight loss due to treatments like chemotherapy.

Smoking pot can increase the heart rate by as much as two times for up to 3 hours: within just a few minutes of inhaling marijuana, heart rate can increase by between 20 and 50 beats a minute. This can last anywhere from 20 minutes to 3 hours, according to the [US] National Institute on Drug Abuse. A January 2017 report[39] found insufficient evidence to confirm the effect of cannabis on overall risk of a heart attack, but reported some limited evidence that smoking it increases the risk. There is yet little formal evidence of the effect of marijuana on lung cancer risk, but we certainly do know that its smoke irritates lungs—which is why regular marijuana smokers are more likely to have an ongoing cough and to have lung-related health problems like chest colds and lung infections.

Heart, lungs

Other physical effects of marijuana include

- Dizziness (marijuana affects activity in the cerebellum and basal ganglia, two brain areas that help regulate balance, coordination, reaction time, and posture)
- Shallow breathing
- Red eyes and dilated pupils
- Increased appetite
- Slowed reaction time (using marijuana doubles the risk of being in a car accident)
- Lowered testosterone levels in men (heavy marijuana use may lower testosterone levels and sperm count and quality, reducing libido and fertility)
- Prenatal effects in pregnant women (substantial evidence links prenatal cannabis exposure and lower birth weight; and marijuana use may increase other pregnancy complications)
- Physical withdrawal symptoms (cravings, irritability, sleeplessness, and reduced appetite) are likely when long-time users stop using the drug

7.7.2.2 *Effects of Marijuana on Mind and Mood*

Most people use marijuana because the resulting high makes them feel happy, relaxed, creative, or detached from reality. THC interacts with the brain's reward system, the part that has been primed to respond to things that make us feel good, like eating and sex.

Smoking marijuana can also have less-pleasant effects on mind and mood. Potential negative side effects include

- A distorted sense of time (feeling as if time is sped up or slowed down is one of the most commonly reported effects of using marijuana)
- Random thinking
- Paranoia
- Anxiety and anxiety attacks (some conflicting evidence suggests that marijuana may ease some anxiety disorders, but it may also be associated with social anxiety)
- Depression (it is not clear if marijuana increases the risk of depression or whether depressed people are more likely to smoke)
- Short-term forgetfulness (marijuana can interfere with memory by changing the way the brain processes information)

Overuse

Although people have theorized that marijuana does not cause physical dependence, the fact is that marijuana can be highly habit-forming: nearly 10% of people who use it become dependent on it. It still is not clear whether marijuana is uniquely a "gateway" drug that makes people more likely to try drugs like cocaine and heroin. The majority of people who use marijuana do not go on to use other substances.[40] However, cross-sensitization is not unique to marijuana. Alcohol and nicotine also prime the brain for a heightened response to other drugs and are, like marijuana, also typically used before a person progresses to other, more harmful substances.

The amount of THC in marijuana has risen in recent years; most leaves contain 7% as opposed to the 1–4% common many years ago, and some worry this might make it easier to become dependent on or addicted to marijuana. It also strengthens many of the drug's mind-altering effects.

Unknowns

There are still many questions about how marijuana affects the body and brain; not surprisingly, scientists say far more research is needed to asses both harmful and potentially healing aspects of marijuana.

7.7.2.3 *Effects of Marijuana on Automobile Driving*

Driving

Like alcohol, marijuana impairs several driving-related skills and does so in proportion to how much cannabis the driver consumed. The effects of marijuana are more variable from person to person, however, because of an individual's tolerance, how the cannabis was smoked/ingested, and differences in absorption of THC. Marijuana significantly impairs judgment, motor coordination, and reaction time, and studies have found a direct relationship between blood THC concentration and impaired driving ability. However, the role played by marijuana in crashes is often unclear because it can be detected in body fluids for days or even weeks after intoxication (since THC is fat-soluble), and because people frequently combine it with alcohol. The risk from driving under the influence of both alcohol and cannabis can be greater than the risk of driving under the influence of either alone. Marijuana and alcohol are often socially used together. This complicates the effects of both substances on an individual: smoking marijuana while drinking a little alcohol increases THC's absorption, making the high from the drug more intense. Similarly, THC delays the peak of alcohol impairment, meaning that it tends to take longer for someone using both substances to feel drunk. A person who is both drinking and smoking marijuana might very well not realize just how impaired he or she is.

There are still many questions regarding how to define legal levels of impairment. There is as yet no THC analog to alcohol's BAC, which makes the proper regulation of usage and impairment challenging.

Unknowns

7.7.3 Other Drugs that Affect Performance and Behavior

A nonexhaustive summary of various other drugs that might be encountered in performance and behavior problems is provided below.

Opioids: These drugs are often prescribed for pain but can be abused for their mood-elevating or calming effects. Semisynthetic opioids (oxycodone and others) and synthetic opioids (fentanyl and others) historically have been available rather readily by prescription or illicitly. Whether prescribed by their own doctor, obtained from a family member or an acquaintance, or purchased illegally, the availability of opioids and their misuse have resulted in a crosscultural problem sometimes referred to as an epidemic. A subset of prescription opioid addicts will turn to heroin, a cheaper "street" opioid, when they can no longer obtain or afford the prescription medication. When misused, CNS depression, coma, cardiac arrhythmias, and death may occur, depending on a number of factors including the method of administration and the nature of the opioid.

Cocaine (stimulants): Stimulants like cocaine are in a class of drugs that can elevate mood, increase feelings of well-being, and increase energy and alertness, but they also have dangerous effects like soaring heart rate and elevated blood pressure.

Prescription tranquilizers (CNS depressants & sedatives): Depressants, generally prescribed to promote sleep or to reduce anxiety, are usually categorized as either sedatives or tranquilizers. Sedatives primarily refer to barbiturates (e.g., phenobarbitol) but also include sleep medications (such as the brand-name prescription drugs Ambien and Lunesta, which are also classified as sedative-hypnotics). Also included in this category are benzodiazepines (which are very effective for anxiety but can produce dependence and withdrawal problems), first-generation antihistamines, anesthetics, and others. Effects of misuse include drowsiness, oversedation, weakness, disinhibited and dangerous behaviors, confusion, slurred speech, and coma.

Prescription stimulants: This classification of drugs temporarily increase alertness and energy and include amphetamines (such as Adderall) and methylphenidate (e.g., Ritalin and Concerta) which are often, paradoxically, used to treat attention deficit disorders. Historically, they were used to treat asthma, obesity, and a variety of neurological disorders. Other stimulants include biphetamine and dexedrine. Abuse of prescription stimulants can result in addiction and dangerously high body temperature and irregular heartbeat.

Hallucinogens: This classification of drugs includes mind or perception-altering drugs that users take to achieve a state of euphoria. LSD and Ecstasy (methylenedioxymethamphetamine, MDMA) are the most abused drugs in this category. These drugs cause users to see vivid colors and images, hear sounds, and feel heightened sensations that seem real but may not exist. Abusers also may have traumatic experiences and emotions as a result of taking the drugs that can last for many hours. Some short-term physical effects can include increased body temperature, heart rate, and blood pressure; sweating, loss of appetite, sleeplessness, dry mouth, nausea, and tremors. Rapid emotional shifts, life-threatening behaviors, and psychosis may result.

Methamphetamine and crystal methamphetamine: These are a uniquely dangerous and toxic subset of stimulants. Effects include euphoria, sleeplessness, autonomic arousal, heart failure, confusion, slurred speech, mania, dangerous decision making, violent and aggressive behavior, and psychosis amongst many others.

Inhalants: Inhalants are volatile substances found in many household products, such as oven cleaners, gasoline, spray paints, and other aerosols; they induce mind-altering effects. Inhalants are extremely toxic and can cause major damage to the heart, kidneys, lungs, and brain. Commonly abused inhalants include solvents with strong fumes (paint thinner, degreaser, dry-cleaning fluid, cement glue), aerosol sprays (hair spray, air fresheners, spray paint), and gases (gasoline, kerosene, anesthesia, nitrous oxide).

Club drugs: These recreational drugs from varying categories are often used in nightclubs or at concerts, and include GHB, Rohypnol, Ketamine, LSD, and MDMA (Ecstasy a.k.a. "E", "XTC", "Molly"), amongst others. Particulary problematic with the latter "club drug" is that it may consist of a "cocktail" of various (harmful) bulking agents with uncertain percentages of MDMA. Depending on the dosage and the combination of substances, side effects include euphoria, confusion, sedation, amnesia, combativeness, hallucinations, seizures, coma, and worse.

Anabolic-androgenic steroids: These are often abused for their muscle building aspects but may cause mood swings, agitation, sleeplessness, hypertension, mania, stroke, liver and heart problems, and paranoid ideation with or without violent outbursts.

Tobacco/nicotine: The effects of cigarette smoking on health are well-documented and dire. Tobacco smoke and nicotine also have effects on cognition and the brain, and as more workplaces forbid smoking altogether, nicotine withdrawal effects also need to be reckoned with in a workplace.

7.7.4 Human Factors Perspectives on Addressing Alcohol and Drug (Ab)Use

Internationally, mortality and morbidity rates from drug overdoses and drug-related behavior are increasing; the multitude of factors contributing to this unfortunate statistic is beyond the scope of this book. The reality of drug abuse is a factor in all aspects of functioning, including in the workplace. Approaches to stemming the enormity of human casualty include educational and preventive programs, mandatory and voluntary treatment, and punishment (such as fines and incarcerations). The underlying human factors element in addressing alcohol and drug (ab)use is in the approach: from an *abstinence* or a *harm reduction* model. Stringent abstinence ("just say no") approaches may fail to acknowledge people are imperfect and thus may not sufficiently consider how to reduce the impact (harm) that comes from imperfect decisions. Harm reduction is a set of practical strategies and ideas aimed at reducing negative consequences associated with drug use. Abstinence may be required in some settings and with some substances; it may be a long-term goal of harm reduction; it may be a medical necessity in some circumstances. Harm reduction encourages abstinence but recognizes that an abstinence goal is not always possible or even appropriate, both on group and individual levels. While recovery from inappropriate use of alcohols and other drugs ideally achieves the removal of a harmful substance from the life of an individual, getting there may require taking steps to reduce the risks to the person and the social, work, and familial environments in which they live.

7.7.5 Effects of Alcohol, Marijuana, and Other Drugs: Summary

In most people, even at relatively small blood alcohol levels, motor performance is diminished. When people decide to drive (or perform other complex tasks) after drinking, their cognitive performance is reduced while sensory perception, decision making, and response time are most severely impaired. Unfortunately, the affected person usually is not aware of these impairments, because alcohol also reduces the ability to make sound judgments about one's own performance.

Like alcohol, marijuana impairs several driving-related skills and does so in proportion to how much cannabis the driver consumed. Marijuana affects individuals differently, depending on a person's tolerance, how the cannabis was consumed, how the THC is absorbed, etc. The effects of marijuana are more variable from person to person, however, because of an individual's tolerance, how the cannabis was ingested/smoked, and difference in absorption of THC. Nevertheless, marijuana significantly impairs judgment, motor coordination, and reaction time, and studies have found a direct relationship between blood THC concentration and impaired driving ability.

Using other illegal drugs—or misusing prescription drugs—can make driving unsafe, just like driving after drinking alcohol. The effects of specific drugs differ depending on how they act in the brain; for example, someone who has used cocaine might be especially aggressive, even reckless, while driving. A sedated driver may be sleepy or dizzy behind the wheel. Any and all of these impairments can lead to accidents.

Mixing substances, like drugs and alcohol, exacerbates the effects and makes driving (or performing other complex tasks) even more treacherous because even small amounts of some drugs can have a measurable effect. As a result, some countries and certain US states have zero-tolerance laws for drugged driving, meaning an impaired driver can face driving-under-the-influence charges if there is any amount of drugs in the driver's system. It's important to note that many regions are waiting for research to better define blood levels that indicate impairment, such as those they use with alcohol.

7.8 CHAPTER SUMMARY

The human body functions in patterns that, in essence, reflect a pattern of resting at night and being active during the day. This human clock, which is governed by circadian rhythms of the body, ideally should not be interrupted by work requirements; yet, on occasion, we must work over long periods or during the night. In these instances, we are likely to experience detriments in certain kinds of performance, and we may suffer health consequences as a result.

For those workers who do shift work, it is best to keep the body on the same natural rhythm and intersperse only one evening or night shift with regular day shifts. If that is not possible, and longer durations of shift work are required, it is best to fully commit to the alternate schedule and adjust the internal rhythms accordingly.

Alcohol, marijuana, and other drugs, many even in small doses, impair nervous system functions, dull the senses, weaken motor control, slow cognitive processes, and may even result in a loss of contact with reality. Consequently, the performance of work tasks is hampered in proportion to the amount of psychoactive substances.

The use of alcohol, marijuana, and other drugs followed by driving (or performing other complex tasks) is very concerning in the United States and throughout the world. Stakeholders including law enforcement, traffic safety officials, physicians, attorneys, and the public at large: all must worry about the issue of impaired driving, flying, and any other risky behavior that impacts the public. Various legal requirements apply for operating vehicles—driving automobiles or piloting aircraft—vis-à-vis permissible amounts of drugs or alcohol in the bloodstream. What becomes complex is identifying the impaired operator, assessing and documenting the level of impairment, conducting tests to measure the level, and interpreting the results of the test. Further complicating matters is the fact that individuals vary, for example, in terms of their dosage, acute and residual effects, how the drug was administered, even their metabolism. Moreover, drugs are often combined with other drugs or with alcohol.

There are various approaches to reducing the likelihood of a compromised worker (due to sleep impairment, shift work, or the effects of alcohol and other drugs) and the identification of an impaired worker is the first step in addressing remediation approaches.

7.9 CHALLENGES

Discuss the interactions between activities of the body according to internal rhythms and the effects of time markers.

What may explain the large individual differences in daily rhythms, sleep, and activities?

Should one try to design different work schedules for "morning" or "evening" persons?

What would you do if you were forced to take a noon break or forced not to take such a break?

Evaluate the theory that it is only the brain, and not the body, that needs sleep.

What are the interactions between having to work extremely long periods and missing sleep in terms of the impact to performance?

How would your work be affected if you were forced to get along with, say, 5 hours of sleep per night?

Given certain types of jobs involving mental and physical work and combinations thereof, which means might be appropriate to help one's performance during long work periods? Accordingly, which kinds of activities should, and which should not, be required during a night shift?

Suppose you divide long periods, such as a year, into divisions other than 7-day weeks, 24-hour days, and weeks with weekends. Might people consent to work on a basis different from the "5 days on, 2 days off" arrangement now common?

How might these arrangements affect one's social interactions?

Is the absorption of alcohol the only factor that explains the higher tolerance of habitual drinkers, as opposed to that of abstainers?

Does drinking beverages with caffeine have a beneficial effect on a person's behavior, even if caffeine does not influence the oxidation of alcohol?

How important is it to be aware of one's impaired judgment after imbibing alcohol?

What can be done to counter the effects of alcohol on one's work performance and behavior?

To what extent should employers be concerned about their employees use of impairing substances when not at work?

NOTES

1 So-called biorhythms were popular in the 1970s and 1980s; they were said to be regular waves of physiological and psychological events, starting at birth, but running in different phases and wavelengths. According to biorhythm proponents, these rhythms were known in Ancient China more than 3000 years ago, then re-"discovered" in Berlin in the 19th century. This pseudoscience postulates that internal human "master clocks" regulate a 23-day physical cycle, a 28-day emotional cycle, and a 33-day intellectual cycle. Adherents use biorhythms to "chart" out lucky days for physical, emotional, and mental activities. No hard evidence exists to support the existence of biorhythms.

2 According to https://www.acs.org/content/acs/en/pressroom/newsreleases/2013/september/explaining-why-so-many-cases-of-cardiac-arrest-strike-in-the-morning.html, accessed July 31 2017.

3 Aschoff (1981).

4 Czeisler, Kronauer, and Allan (1989), Czeisler, Dumont, and Richards (1990a), and Czeisler, Johnson, and Duffy (1990b).

5 Turek (1989).

6 Reiss (2017).

7 Horne (1988).

8 Horne (1985, 1988).

9 Quoted from Horne (1988), pp. 54 and 313.

10 Nighttime performance levels during long-term sleep deprivation are much lower than levels that are normal during the day, when the body and brain are usually awake. Harrowing tales are told by US hospital interns and residents, many of whom were routinely expected to work 120-hour weeks, including 36 hours at a stretch. Some admit that mistakes are frighteningly common. A California medical resident fell asleep while sewing up a woman's uterus – and then toppled onto the patient. In another California case, a sleepy resident forgot to order a diabetic patient's nightly insulin shot and instead prescribed another medication. The man went into a coma. Compassion can also be a casualty: one young doctor admitted to abruptly cutting off the questions of a man who had just been told he had AIDS, "All I could think of was going home to bed." As reported in the magazine Time, December 17, 1990, p. 80.

11 Froeberg (1985).

12 Naps are not just for young children. Many people have taken up the habit of lying down after lunch for 15 or 20 min ("Nur ein Viertelstündchen" as the Germans say), often falling asleep briefly. In laboratory tests, this kind of nap has been shown to have little effect on the performance of subsequent work, yet many people say they simply need it. Some companies now incorporate nap rooms or nap pods into the work place.

13 Reported by Froeberg (1985) but not recommended to duplicate.

14 According to a 2010 study from the University of California's Center for WorkLife Law and the Center for American Progress, cited by BL Foster (2014), The Night Shift, https://www.psychologytoday.com/articles/201405/the-night-shift, accessed July 31, 2017.

15 As reported by UCLA Health (2015) http://sleepcenter.ucla.edu/coping-with-shift-work, accessed July 31, 2017, and Åkerstedt and Wright (2009).

16 Bajraktarov et al. (2011).

17 Nojkov, Rubenstein, Chey, and Hoogerwerf (2010).

18 Bohle and Tilley (1989).

19 Monk (1989).

20 As summarized in Kroemer (2016).

21 Kogi (1985).

22 Adapted from Kogi (1991) and Tepas (1985).

23 Adapted from Kogi (1991) and Tepas (1985).

24 Graeber (1988).

25 2015 Data reported by SAMHSA (2016), https://www.samhsa.gov/data/sites/default/files/NSDUH-DetTabs-2015/NSDUH-DetTabs-2015/NSDUH-DetTabs-2015.pdf, accessed July 31, 2017.

26 As reported in https://niaaa.nih.gov/alcohol-health/overview-alcohol-consumption/alcohol-facts-and-statistics, accessed July 31, 2017.

27 Turner (1990).

28 Esser et al. (2014).

29 Friis and Sellers (1999).

30 Stahre, Roeber, Kanny, Brewer, and Xingyou (2014).

31 AHA (2015), CDC (2015), National Cancer Institute (2013) *Alcohol and Cancer Risk* https://www.cancer.gov/about-cancer/causes-prevention/risk/alcohol/alcohol-fact-sheet, accessed July 31, 2017; National Health Service (NHS) (2015), *Alcohol-Related Liver Disease,* https://www.nhs.uk/conditions/Liver_disease_(alcoholic)/Pages/Introduction.aspx, accessed July 31, 2017.

32 Ewing (1984).

33 Price (1988).

34 Hahn and Price (1994).

35 Data from NHTSA, NHTSA 2010, FBI, SAMHSA, NHTSA, respectively.

36 Morrow and Jerome (1990).

37 Ross and Mundt (1988).

38 S. Motel (2015). *Pew Research Center.* 2015. http://www.pewresearch.org/fact-tank/2015/04/14/6-facts-about-marijuana/, accessed July 31, 2017.

39 [US] National Academies of Sciences, Engineering and Medicine report on the health effects of cannabis and cannabinoids, http://nationalacademies.org/hmd/reports/2017/health-effects-of-cannabis-and-cannabinoids.aspx, accessed July 31, 2017.

40 Goldenberg, Ishak, and Danovitch (2017).

SECTION II

Design Applications

8

Ergonomic Models, Methods, Measurements

Overview

We humans use *models* to understand our own physiology and psychology, and our roles while cooperating with other people, performing tasks and working with equipment. Models characterize us individually and as representatives of humankind, for example in our biomechanical features. They describe how we drive our cars along a grid of roads; and how we do, or should do, our jobs. The designer of human-operated machinery (such as spacecraft, airplanes, or automobiles) uses computerized models of the human (as operator or passenger) to design proper shells and interfaces so that the human is safe, comfortable, and capable of using the equipment. Other, less complex models are useful for design and evaluation of everyday workplaces.

Models

The model by which we try to understand our functioning provides the underpinnings for the *methods* that we use to assess our performance as individuals or as part of social or human–technology systems. In the description of body strength and

Methods

Ergonomics
http://dx.doi.org/10.1016/B978-0-12-813296-8.00008-6

endurance, for example, we use biomechanical techniques to evaluate the effects of body posture, physiological procedures to ascertain muscular effort, and psychological methods to judge the effects of motivation and stress. In the same manner, models also determine the methods and procedures by which we design human–technology systems, complex or simple, for usability, efficacy, and safety.

Measurements

Models and methods lead to *measurements*, the specialized ways in which we assess specific parameters of our functioning and performance. For example, strength exerted by our hand onto a tool can be measured in terms of acceleration, force or torque, in their magnitudes and directions, all varied over time; by electromyographic signals, oxygen consumption or heart rate; and by the effects of exhortation, by rating of perceived exertion, or by comparison with other stressful situations.

8.1 INTRODUCTION

We all have ideas, images, concepts, constructs, and patterns that help us to understand our roles, or other people's roles, in the private or work environment, while performing tasks and operating equipment. Ergonomists/human-factors engineers prefer well organized and objectively describable patterns, often in mathematical form. These are called *models* (or *paradigms*). They describe or imitate, in systematic manner, the appearance and the behavior of the human, often in some stressful situation.

Humans or machines

Regarding the human's role in the modern work world (in particular for dangerous or inhospitable environments and/or with predictable or repetitive tasks), we often distinguish between what people can and should do, and what "machines" do better. A general distinction is that people can think and feel, are intuitive and vulnerable; machines are strong and logical but without personality, and they may be discarded when they have served their purpose. A more detailed listing of respective roles is presented in Table 8.1.

8.2 MODELS

Definition

In ergonomics and human-factors engineering, the term *model* is usually defined as a *mathematical/physical system, obeying specific rules and conditions, whose behavior is used to understand a real (physical, biological, human–technical, etc.) system to which it is analogous in certain respects.*

Two aspects of that definition deserve special attention.

- The model "obeys specific rules and conditions." This means that the model is itself restricted: for example, the model may involve a simple design template that is only two-dimensional, displaying the static outline of one specific single-percentile size.
- The model is "analogous in certain respects to the real system." This means that the model is limited in its validity (or fidelity), with its boundaries often so tight that they barely overlap the actual conditions. To continue with the example: a two-dimensional, static, average design template does not represent the bodies of all office employees.

TABLE 8.1   People or Machines?

Capability	Machines	People
Speed	Much superior to humans	Lag about 1 s
Power	Consistent and as large as designed	1.5 kW for about 10 s, 0.4 kW for a few minutes, 0.1 kW for continuous work over a day
Manipulative abilities	Specific	Great versatility
Consistency	Ideal for routine, repetitive, precision work	Not reliable
Complex activities	Multichannel, as designed	Single (or few) channel(s)
Memory	Best for literal reproduction and short-term storage	Large store, long term, multiple access; Best for principles and strategies
Reasoning	Good at deductive reasoning	Good at inductive reasoning
Computation	Fast and accurate, but poor at error correction	Slow and subject to error, but good at error correction
Input sensitivity	Can be outside of human senses, as designed	Wide range and variety of stimuli perceived by one unit (e.g., the eye deals with relative location, movement, color)
	Insensitive to extraneous stimuli	Affected by heat, cold, noise, vibration
	Relatively poor for pattern detection	Good at pattern detection; Can detect signals within high noise levels
Overload reliability	Sudden breakdown	Can function selectively, may degrade
Intelligence	None (?)	Can deal with expected and with unpredicted events; Can anticipate; Can learn
Decision-making	Dependent on program and sufficient inputs, as designed	Can decide even on the basis of incomplete and unreliable information
Flexibility, improvisation	None	Have
Creativity, emotion	None	Have
Expendable	Yes	No

Source: Adapted from Woodson and Conover (1964), pp. 1–23.

Thus, the model's internal limitations and its limits of applicability need to be kept in mind while employing it.

In some cases, the model is relatively simple, for example, showing the outline of the human body. Yet, even basic anthropometric models must represent that humans come in many sizes with different proportions, that humans move and don't maintain frozen postures, that they are individually of differing strength capabilities which depend on skill and training and posture and motivation (see Chapter 1). One should be wary of any anthropometric computer-aided design (CAD) tools that do not reflect human variability.

> "We should pause here to ponder that drawing distinctions, making classifications and developing models, usually imposes artificial divisions of our own choosing upon a universe that is, in many ways, all in one piece. We do so because it helps us in our attempted understanding of the universe. It breaks down a set of objects and phenomena too complex to be grasped in their entireties into smaller bits that can be dealt with one by one. There is nothing objectively 'true' about such classifications, however, and the only proper criterion of their value is their usefulness." Isaac Asimov (1963)

8.2.1 Types of Models

Every model represents a (proven) theory or a (tentative) hypothesis. It incorporates the current state of knowledge and can be verified by consulting available data or by conducting new experiments.

Models

The first stage in the formulation of a model is the identification of relevant subsystems or variables, independent and dependent. The next step, the modeling stage, is the formulation of the relations among the subsystems or variables. The final stage is that of validation, or confirmation through objective evidence that the model adequately represents reality for its intended application.

Submodels

Commonly, one submodel is constructed of the human operator, and another submodel of the equipment or processes with which he or she is working. Then these two component-models are linked together to show the interface between them, and how they interact. Thus, a general model of the behavior of the human/equipment/process system is generated. Occasionally, one even models the "user" of the operator–machine system: the office manager, air traffic controller, military officer. In this case, there are three submodels: the operator, the machine, and their supervisor.

Open or closed models

Models may be "open" or "closed." An *open model* is affected by circumstances outside it, such as by climate conditions, vibrations and impacts (see Chapter 6), or changes in workload. A *closed model* excludes these external effects; it functions within its own cocoon.

Loops

Open-loop models do not consider the effects of the activities of the model on itself. An example is a person firing a gun in the dark: after the bullet has left, the shooter does not know whether it hit the target or not, and hence firing a second shot is not affected by the first because there is no feedback about the outcome. A *closed-loop* system, in contrast, utilizes feedback about previous actions to modify the next activities.

Normative or descriptive

Another major distinction is between "normative" and "descriptive" models. A *normative model* assumes there is necessarily some appearance or behavior that is normal, perfect, ideal, in a standardized and nonvarying way; often a singular appearance or behavior which the model represents. Thus, a normative model is often deterministically constructed, presuming that the effects of variables within the system, or acting upon the system from the outside, can be clearly predicted and hence modeled: to reflect what ought to happen. The opposite is a *descriptive model* that reflects actual, variable behavior. Such changes in behavior are due to variations (often assumed to be stochastic) in internal or external variables.

Descriptive models are often used for *simulation*, defined as running the model's variables throughout their given ranges in order to exercise the model and generate a range of results.

Simulation

While in the past most models were physical (such as templates or mannequins), they are now often mathematical and computerized. A *mathematical model* has the advantage of being precise, formal, and often general: the variables can be manipulated easily, and parameters in the equations assumed freely (or changed to assess their relative influence on the model's outcome). But their rigid mathematical structure can be a disadvantage, especially when the form and nature of equations needs to be presumed without a validity test and cannot be changed without changing the model itself. Thus, some mathematical or statistical models fit reality poorly, often being either too general or too specific—see the discussion of inadequate models below. Furthermore, if the boundary conditions are not explicitly and carefully stated and understood, especially for computerized mathematical models, these models can be extrapolated inappropriately beyond their range of applicability.

Mathematical

8.2.1.1 *Good and Bad Models*

The value of a model is judged against a set of criteria:

- *Validity* is the agreement of the outputs of the model with the performance of the actual system that it should represent (this is also called *fidelity*, or *realism*).
- *Utility* is the model's ability to accomplish the objectives for which it was developed.
- *Reliability* is the repeatability of the model, in the sense that the same or similar results are obtained when exercising it repeatedly. Reliability may also be considered the ability to apply the model to similar, but not exactly the same, systems.
- *Comprehensiveness* is the applicability of the model to various kinds of systems.
- *Ease of use* (*usability*) is, obviously, a very important criterion. If highly trained and skillful capabilities are required from the user, a model is not likely to be used often and by many. On the other hand, oversimplification of a model to achieve ease of use is not desirable either.[1]

8.2.1.2 *Ergonomic Models*

Starting in the mid-1990s, the G-13 committee of SAE[2] has undertaken a major effort to model the human, geometrically and behaviorally, in interaction with equipment and environment. This "ergonomic model" has been characterized[3] as

- Computer-based simulation of individual humans,
- Work-related characteristics and performance,
- Required for equipment operation, production, and maintenance, and
- Having the purpose of influencing equipment design.

Such modeling requires collaboration between researchers, developers and users because

- Programers may not understand the science of ergonomics
- Ergonomists may not understand possibilities and limitations of software

- Users are often not expert programers or ergonomists
  - Users want valid tools that are easy to learn and use
  - Users want a simple solution for a complex task

The need for ergonomic models has been apparent since the mid-1980s. Then, as today, two possible solutions are conceivable: the *integrated model* (then called super-model) which has a complete set of analytical functions, shared by all its components; and the *shell model* which has a wide modular architecture allowing various modelers and users to insert their own submodels. The advantages and drawbacks of these two types of models[4] include:

- Integrated models:
  - Pros: High fidelity, compatible integrated functions, shared structure makes for efficient processing, total control of input and processing prevents misuse.
  - Cons: Difficult to modify, for practical reasons currently limited applicability (primarily because needed detail information is only available for specific groups of people and tasks).
- Shell models:
  - Pros: User inserts own models and data, great flexibility, third parties can develop their own models for insertion.
  - Cons: Dissimilar definitions of variables, incompatible functions, amalgamation of noncompatible models and data, flexibility allows misuse of model.

8.2.1.3 Inadequate Models

Models may be false,[5] inaccurate, misleading, at least inadequate, because of misunderstood human (and system) behavior, or because of misuse of modeling procedures, or both.

Some models seem to have been developed by people who know a lot about how to program a computer, but too little about the human and how the human functions with and within the system. Their models of the human are likely to be inaccurate, unrealistic, and overly simplified—but probably "work well" in terms of model mathematics and computerization.

> "A person with a lot of common sense and no technology may not arrive at the optimum solution; however, a person with a lot of technology and no common sense is really dangerous." Steven Johnson (1998)

Animation

Some models incorporate motions of the human based on *animation*: the creation of movement patterns observed under certain conditions and then applied inadequately to other conditions. A typical case is taking a sequence of static positions observed in strength testing and "morphing" these into an apparently dynamic motion pattern allegedly depicting actual lifting of an object. (For example, as noted in Chapter 2, a connecting curve as in Fig. 2.25 implies motion when in fact strength was exerted and measured in separate static positions, independently, one after the other, with a break after each exertion; such data should be used with caution.) As smooth as many of these animations appear, they are often simplified or exaggerated as in cartoons, with a strong potential to mislead the model user.

A related fallacy is the assumption, born from the desire to keep the model simple, of linear behavior: meaning that if one variable (say, workload) increases, the associated dependent variable (say, speed of human activities) will change linearly as well. But many human behavior traits do not simply respond proportionally to changes in the task. Therefore, if a system is based on linear algorithms, then the system behavior is unlikely to be truly descriptive (let alone predictive), increasingly so the more extreme (nonlinear) the conditions are.[6]

Linear behavior

8.2.1.4 *Misuse of Modeling*

Early ergonomic models were less sophisticated than contemporary models. The earliest models of the human body were purely geometric, reducing the body to ellipsoids and spheres with estimated masses and centers of gravity. More sophistication was added in the 1960s by subdividing body segments into additional geometric shapes, with segment-density data and calculated mass moments of inertia. With increasing use of computerization, ever more refined models with more detailed linkages and ranges of motion have been created. However, it remains critically important to understand the underlying assumptions for each model. For example, the following simplifications may still be built into the models to trade off ease of implementation with model validity:

Simplistic assumptions

- dimensions of the mythical 50th-percentile male
- body movement at constant velocity (after initial acceleration and before final deceleration)
- body segments formed as rigid cylinders with unchanging dimensions
- joint locations taken from erect standing posture
- independent segmental movement or muscle activation (that is, no linkage between movement of contralateral body segments or coactivation of muscles)

Because a more simplified model is relatively easier to implement, it is tempting to use it to make predictions—while overlooking or disregarding some of the model's basic assumptions (and the limitations which they impose). Although enticing, it is ultimately inappropriate to expand the use of the model to simulate working conditions that fall outside of the model's applicability.

Overextended application

Another misuse of models is feeding incorrect data to the model. Since the model itself may not be able to evaluate the correctness of the input data, the model spits out results: garbage-in–garbage-out (also known as GIGO). It is bad enough if such output is clearly nonsensical, but it is worse yet if the output is incorrect (but not obviously so) and hence misleading. A related problem is hidden under the euphemism "fitting input data to the model." This may simply mean that the data needs to be formatted to meet input requirements, which is acceptable. However, if fitting data really means modifying their content, their meaning, or their "behavior," then such fitting is really data falsification and is not acceptable.

Incorrect inputs

False use of output

Finally, the user may misinterpret and misapply the model outputs, such as by transferring static strength calculations to dynamic activities.

Three Main Misuses of Models:

1. Model itself is inappropriate: *Whatever in, garbage out.*
2. Inputs to the model are false: *Garbage in, garbage out.*
3. Outputs are misapplied: *Garbage use.*

User beware!

To avoid these problems, the model must clearly specify key underlying assumptions. The user of the model must be knowledgeable, that is, able to judge the appropriateness of the model for the situation. The user must be able to assess the validity of the input data. And the user must refrain from applying model outputs to conditions outside the model constraints. As models have become easier to use, so has the ability to misuse models in ways that may not be apparent to the user.[7]

Validation

Validation is one way to check whether the model (re)presents reality. Validation of the model means, in the simplest case, feeding "true" data into the model and comparing the model output (prediction) with the behavior of the "true" system. If the model does not describe reality correctly, then internal algorithms and/or the basic structure of the model are insufficient. Although this may not be sufficient to validate the model, it is an important and necessary part of the validation. The more sophisticated the model is, or the more critical the model is, the more complex the validation may become.

Neglecting to assess the validity of a model is like buying a car without first driving it.

8.3 METHODS

The model that we have of the human and her or his performance as part of a technical or social system determines the *methods* by which we describe human–system performance, or how we design equipment for ease of use and performance. If the model is static, the model algorithms are static, and consequently the measures that we use as inputs to the model and that we receive as model outputs are static as well. Yet, reality is dynamic, humans move and change; so do their systems, and so does system performance.[8]

Depending on the model, on the given goal and conditions, and on the available time and effort, various types of methods exist to determine possible links between treatments (changes in independent variables, such as positioning the driver in a vehicle) and associated or caused outcomes (changes in dependent variables, such as seeing the instruments). The proper use of these methods is the every-day challenge for the human factors engineer/designer, and the bread-and-butter of the ergonomic researcher.

The easiest task is to measure a given condition: for example the interior space of automobiles, or the body sizes of the user population. This is called a *cross-sectional* survey, because all specimens (or at least a representative sample of them) are measured at the same time, without regard to age. The point-in-time measure does not allow any direct conclusion about ongoing effects, such as aging. Cross-section

There are a great variety of methods to determine relationships between treatments and their effects. The main divisions (which partly overlap) are between observations and experiments.

Observational studies (such as recording the incidences of repetitive trauma in keyboarders) are common in medical research; they are often called *epidemiological* or descriptive studies. They can identify powerful associations (like between noise levels and hearing loss) but are retrospective and often require long duration and large numbers of observations. The two main types are case–control and cohort studies. For a *case–control* study, the researcher compares factors (such as exposure time) found in one group of individuals with an affliction (say, repetitive trauma among keyboarders) to factors in a comparable group without that affliction (keyboarders without repetitive trauma). In a *cohort* study, one simply follows large groups of people over long periods of time with the intent to identify factors (possible causes and preventives) associated with certain outcomes, such as loss of hearing or cumulative trauma. Epidemiological

In planned experiments with treatments (called *clinical trials*, especially in medicine) one assigns people to groups with and without experimental treatments, and observes the effects of the treatments in terms of outcomes in the dependent variables. This is the current "gold standard" of experimental research if the subjects are consenting volunteers, fully informed, and safeguarded; if participant recruitment is unbiased and assignment to treatment groups is random; if neither subjects nor researchers know who is treated or not (part of a double-blinded study); and so forth. Clinical trials

Laboratory experiments allow the tightest control, and often outright elimination, of extraneous variables that could confound or even falsify the results. The laboratory environment permits careful manipulation of the specific independent variables. Lab experiments

However, the sterile conditions in the laboratory are generally far different than the real world of ergonomics, so *field studies* are the realistic approach to assess the effects that changes of the independent variable (say, noise) have on the dependent variable (say, understanding speech)—but extraneous conditions may interfere. Field studies

8.3.1 Turning a Question into a Testable Hypothesis

A lack of rigor in many ergonomic studies has been reported and addressed by compiling a checklist[9] that is useful both for planning experiments and for assessing reported studies. Studies and experiments on human and human–system performance must be carefully planned to test a theory or, more often, a hypothesis; then just as carefully executed, evaluated, and reported. The usual approach is to state a hypothesis (such as of two repeated measurements of hand strength, the second exertion is the Experimental design strongest) and then to determine whether that proposition is true or false based on the experimental results.

Null hypothesis

Testing is commonly done in terms of the Null Hypothesis: "there is no difference between the outcome of the tests" (such as: exertions of the dominant and nondominant hands yield the same results). Whether the null hypothesis is rejected or not is determined by statistical evaluation of the experimental data. Proper development and execution of a research project is important so that the hypothesis can be tested.[10]

Thorough guides for design of experiments and the analysis of their results have been compiled,[11] including guidance for complex studies with multiple variables and how to control for individual differences among the subjects who participate in experiments.[12]

8.3.2 Assessing Performance

Performance assessment of an existing (including prototype) human-operated system is often a complicated and experimentally expansive task, especially if the system is complex and interacts with many other variables (i.e., is an open system). On the other end of the spectrum, overall usability of a new system can often be tested with just a few users.[13]

8.4 MEASUREMENTS

> In measuring, we to aim for the scientifically desirable while employing the art of the feasible.

To measure requires that one knows the relationships between the *dependent variables* (such as strength exertion, or task performance, or heart rate) that respond to the selected *independent variables* (such as pilot work load or flight path) to be experimentally manipulated. Hence, the experimenter must have an ergonomic model of cause–effect relations and accordingly select measuring method and measuring instruments in order to obtain suitable outcome parameters.

Objective or subjective?

The measurement may be in objective or subjective units. Performance measures such as force, torque, work or energy, power, travel (and its time derivatives, speed and acceleration), heart rate, oxygen consumption, incidents, accidents, errors and the like are *objective* parameters. These are preferred by most researchers because of their seeming independence from human subjectivity.

Subjective assessments are self-reports (judgments) of states within the subject or observer, and appear prone to many sources of errors. Self-reports are abhorred by advocates of objective measurements. However, there is an argument that every measurement in science is dependent on the human[14]: by selection of the measure, by data collection, analysis and interpretation. In fact, several scientific disciplines rely mostly on human observation, prominently so psychology and sociology, both significant aspects of ergonomics. Psychologists in particular have developed procedures of training and execution that seek to make subjective assessment reliable, precise and valid. Researchers often successfully use such tools as interview, questionnaire, rating, ranking, paired comparisons, observation, or verbal protocol analysis. It has been argued that objectivity and subjectivity are not useful ways of distinguishing among measures of human performance. The distinction becomes even more blurred as we recognize that many outcome quantities that we measure "objectively" (e.g., keying performance in words per minute, hand strength in Newtons) are codetermined by inherent subjective processes (e.g., motivation) that are difficult to control and quantify.

Of course, there may be as many measurement classes, and units therein, as there are models, methods, and independent variables; past, present, and future. This is especially true for performance measure of humans and human–equipment systems. Yet, for assessment of relatively simple outcomes, certain groups of measurement have become well established, and using these has the advantage of having comparison values in the literature. A review of the first seven chapters of this book reveals a fairly limited number of measures that serve ergonomists; most quantity units rely on physics in terms of length, time, and temperature as well as on basic physiologic and psychologic assessment tools, with some biochemical indicators used as well.[15] If one measurement has the most widespread use, then this must be *heart rate*, which appears to be a truly transdisciplinary tool of psychophysical significance.[16] Further and more in-depth information on the selection of measurements and instrumentation is available.[17]

8.5 EXAMPLES OF MODELS, METHODS, MEASUREMENTS

This book contains a large number of models and related methods for design and assessment.

In Chapters 1 and 2, for example, the human body is modeled, according to Borelli's 17th-century concept, as a skeleton of links that are articulated in the joints, and powered by muscles as engines that move the links about the joints. The functioning of muscles to produce a strength output is modeled in form of a flow system with feedforward and feedback. Common measures referred to are distance, angle, circumference, diameter, force, torque (moment).

In Chapter 3, the body's energy production is compared to the processes in a combustion engine, and models are given of the interactions among the circulatory, respiratory, and metabolic subsystems. Common measures are oxygen uptake, heart rate, blood pressure, endurance time.

In the fourth chapter, control of the human body is modeled in terms of neural networks. The nervous system is constructed of the central and peripheral subsystems. Inputs of information to the body occur in the various sensors. A triggered signal is transmitted along the efferent pathways to the brain, where the information content is processed and decisions made regarding actions. These are initiated by signals sent along the efferent pathways to output effectors such as mouth or hands. Within this complex model, components can be modeled, such as the eye or the ear. Typical measurements are in bits, bytes, volts, amperes, time. In the assessment of performance, discussed at the end of Chapter 4, judgments of various kinds are used including forced decision, true–false determinations, speech patterns; also, force exertion, time measurements, body movements, heart rate, breathing patterns, sweat rate, skin conductivity.

In Chapter 5, paradigms for the human sensory facilities are presented, including the scheme of converging perceptive fields. Common measures include are length of time, angle, distance, acceleration, force, energy, power, frequency, wavelength, amplitude, flux, illuminance, temperature.

In Chapter 6, the interaction of the body with the environment is specified in measurements of energy, temperature, humidity, motion, endurance; distance, velocity, acceleration, pressure; effects of chemicals; and in a variety of physiological measures mentioned above.

Chapter 7 describes models of human body rhythms, with measurement units taken from biochemistry, physiology and physics as well as medicine and psychology; performance in its widest sense is assessed using common engineering and managerial tools and metrics.

Certainly, the more complete and realistic the model of the human operating a technical system is, the better the designer can "fit" equipment and task to the human. Given the complexities of the human body and mind, one should expect that paper-based information and physical models (such as two-dimensional templates and three-dimensional dummies) will be superseded by computer-based models except in the simplest cases. Computer-based models and simulations can incorporate complex information about the human and allow fast and variable use of that knowledge. The key is to understand the limitations and assumptions built into the model, and to not extend the results of the model beyond its validated range. Appropriate modeling facilitates the ergonomic design of equipment (tool, work station, or vehicle) for ease of use and for safe and efficient operation.

8.6 CHAPTER SUMMARY

The more that becomes known systematically about the human body and mind, and its functioning within systems, the better one is able to express that knowledge in formal models. Simple body size analogues are mostly descriptive, while more complex models are based on proven theories and make allowances for the effects of the environment and for adaptability and learning through feedback. Models are useful if they are valid (realistic) and reliable. It is also essential that they do not require excessive specific knowledge and experience from the user.

In efforts to achieve simplicity of model and use, too often the criteria of validity and reliability are neglected, such as by using false ideas of proportions of the human body, or unrealistic animation for body movements. While these problems can be fairly easily recognized and corrected, there are more serious ones that incorporate complex though unproven hypotheses: this is not uncommon in behavioral models, but can often be spotted by such simplistic details as assumed linear relations between variables.

Realistic modeling has enabled many improvements in the design of simple and complex human–machine systems. For example, accessibility of machine parts that need maintenance in equipment can be designed into the product; restraint systems for people in automobiles have become very effective; workplaces can be designed to avoid bending and twisting body movements of the worker.

The intent to develop a model also guides research. To fill the knowledge need of a model concept, information is gathered in a systematic manner to meet exactly that objective. This focuses the research approach. Significant steps have been made in that direction, but many more need to follow. Progress in advanced modeling requires systematic knowledge that, in many instances, is still under investigation.

8.7 CHALLENGES

Discuss the exceptions and boundary conditions made in physical models versus behavioral models of the human.

What are the consequences of either limiting or enlarging the numbers of "specific rules and conditions" which a model obeys?

How can one test a "concept model"?

What are the practical advantages of using a "closed" versus an "open" model?

Consider the models employed in Chapters 5 and 6: are they "normative" or "descriptive"? Also, assess these models in terms of validity, reliability, and comprehensiveness.

What is more important to the modeler: mastery of the computer aspects, or knowledge of the functioning of the human?

Can every model be validated?

When validating a model, is it enough to have an "expert" say the model is valid, or do you need evidence for various populations or applications?

How can you determine what level of accuracy is required of a model for a project?

For a particular model, when is it acceptable to go beyond its assumptions, and what can you do to be sure the results are still valid?

Why does modeling give guidance to future research?

NOTES

1   A CADAM (computer-aided design and manufacturing) model was described in 1991 (as reported by Davids in CSERIAC Gateway 2(3)) in which the body contours of human submodels, called ADAM and EVE, are based on hypothetical 95th-percentile homunculi, which then are multiplied by 0.93 to allegedly represent "average persons" or multiplied by 0.8725 to purportedly depict 5th-percentile phantoms. Ease was achieved by sacrificing almost any promise of validity because linear scaling does not exist in the human body.

2   The Human Modeling Technology Standards committee of the SAE International (originally founded as the Society of Automotive Engineers).

3   McDaniel (1998).

4   As listed by McDaniel (1998).

5   Some models are simply false: the proverbial "average person", for example, does not exist except in the minds of some journalists and politicians but, nevertheless, was used in early biomechanical modeling.

6   Current models of the human have come a long way from simplistic assumptions, such as represented by static two-dimensional contour templates. Yet, realistic behavior of the human, such as passively reacting to external forces or actively performing tasks, or both combined, is still only incompletely understood and modeled.

7   See, for example, the discussion by Ulrich Raschke in "Digital Human Modeling Research and Development User Needs Panel" SAE Technical Paper Series 2005-01-2745.

8   In the 1990s, the automotive industry relied on two-dimensional design templates (static models) of the occupants' bodies to lay out cab interiors. The templates were placed with their heels at the "package design origin" point: the location of the gas pedal. Different drivers' sizes were approximated with rigid body templates described in SAE standards. One of them displayed the so-called eye ellipse, which itself had been determined by placing visitors to automobile exhibitions into a stylized seat and marking the location of their eyes in reference to the seat. However, observation of actual car driving a decade earlier had already shown that the drivers' eyes were often outside the standardized eye ellipses: the model and method had become outdated.

9   By Heacock, Koehoorn, and Tan (1997).

10 Weimer (1995) discussed, in a down-to-earth manner, how to develop and execute a research project: this is also the topic of many more theoretical treatises on methodological and statistical aspects in human factors research—see the listing in Williges (1995) for relevant literature.

11 For example, by Williges (1995). However, as with all studies, do not give in to the temptation to "p-hack" the data to try glean a nominally statistically valid result, also known as a false-positive (as described by Simmons et al., 2011), by excessive and tortuous analysis of subsets of the data outside of the original hypothesis being evaluated.

12 As in Han et al. (1997).

13 As discussed by Lewis (1994), Scerbo (1995), and Virzi (1992). For an example of a more complex simulation model, see, for example, Deutsch et al. (2016).

14 As argued by Muckler and Seven (1992).

15 As noted by Vivoli et al. (1993).

16 Bedny and Zeglin (1997), Jorna (1993), and Roscoe (1993).

17 Radwin (1997), Radwin et al. (1996), and Rodgers (1997).

Ergonomics
http://dx.doi.org/10.1016/B978-0-12-813296-8.00009-8

Overview

Most of us work while walking or standing, or while sitting on a chair or stool, but we often kneel and reach and bend and twist our bodies. Some of us work while lying supine or prone, for example in low-seam mining and in car engine repair work. In many non-Western countries, sitting and kneeling on the ground is very common during work. The sitting posture is particularly useful when a relatively small workspace must be covered with the hands, and finely controlled activities are performed. For this, the workstation and work object must be designed appropriately. Fewer restrictions apply for the ergonomic design of the workspace where we are on our feet, walking and standing.

The workspace of the hands depends on trunk posture and work requirements. Consequently, several workspace design configurations may be described. Additionally, vision requirements at work often codetermine the best workspace design.

Controls are generally operated either with the hands or the feet. Foot operation is stronger but slower; if it is frequent, it should be required from seated operators only. Hand operation of controls is faster and weaker but is more versatile. The operative side of tools and equipment should be designed to fit the hand properly. This requires not only suitable sizing of a handle, but also suitable alignment so that the wrist or arm is not placed into straining position.

Repeated and forceful operations, especially when accompanied by improper postures, may lead to overuse disorders (ODs). These are often associated with the repetitive use of hand tools, particularly if they vibrate. Keyboarding is another common source of repetitive injuries. Ergonomic recommendations for proper design of workstations are available to make work easy and efficient.

9.1 MOVING, NOT STAYING STILL

Design for motions

We want to move around at work and not keep a static posture. Therefore, ergonomic design of work station should support body movement, not maintained body positions. A convenient way to do this is to select the extreme body positions that we expect to occur and to design for motion between these.

> Our bodies are built to move about, not to hold still. It is uncomfortable and tiresome to maintain any single body position without change over extended time periods. We experience this while driving a motor vehicle when the locations of the seat, of the head (in order to see), and of hands and feet (on controls) constrain us to a nearly immobile posture.

Lie, sit, stand

One traditionally distinguishes three categories of body postures: lying, sitting, and standing. Of course, there are many other positions, not just transient ones between the three major positions, but postures that are independently important—for example, kneeling on one or both knees, or squatting, or stooping (often employed during work in confined spaces such as loading cargo into aircraft), in agriculture, and in many daily activities. Reaching, bending, and twisting of body members are usually short-term efforts.

Work is seldom done when lying down but does occur, for example, in repair jobs, or in low-seam underground mining. In high-performance aircraft, prone or supine positions are purposely used to better tolerate the acceleration forces experienced in aerial maneuvers—see Chapter 6. For example, during World War II, experiments were performed using a pilot who was lying down, which reduced the profile of the aircraft; in some current fighter airplanes and tanks, the pilot or driver is in a semireclining posture.

Sitting and standing are usually presumed to be associated with more or less erect posture of the trunk, and of fairly straight legs while standing. Sitting at work was thought to be properly done when the lower legs were essentially vertical, the thighs horizontal, and the trunk upright. This convenient model of all major body joints at zero, 90, or 180 degrees is suitable for standardization of body measurements, but the *0-90-180 posture* is not commonly employed, not subjectively preferred, and not especially "healthy"—see the discussion on posture in Chapter 10.

0-90-180 Postures

International standards (such as ISO 6385) provide ergonomic principles for the design of work systems (including permanent and flexible work places), with balanced attention paid to the human, social, and technical requirements to optimize well-being and overall system performance.

9.1.1 "Suitable" Body Motions and Positions at Work

By the rules of *experimental design*, body movements and postures and their changes can be considered independent variables. If all other conditions and variables are controlled (such as work task, environment, and so forth)—which often requires a laboratory setting—the effects of posture changes on dependent variables can be observed, measured, and evaluated. Various dependent variables have been selected in different disciplines, such as:

- In physiology: oxygen consumption, heart rate, blood pressure, electromyogram, fluid collected in lower extremities.
- In medicine: acute or chronic disorders including cumulative trauma injuries.
- In anatomy/biomechanics: X-rays, CAT scans, changes in stature, disc and intraabdominal pressure, model calculations.
- In engineering: observations and recordings of posture; forces/pressures on seat, backrest, or floor; amplitudes of body displacements; productivity metrics.
- In psychophysics: structured or unstructured interviews, subjective ratings by either the experimental subject or the experimenter.

These techniques have become standard procedures, but they are not always appropriately used and interpreted.

Electromyography (EMG) has been employed extensively to assess the strain generated in the trunk muscles of seated individuals. Most EMG techniques inherently

EMG

assume static (isometric) tension of the observed muscles, with the maximal voluntary contraction used as reference amplitude.[1] Most sitting is not static, however, and the EMG amplitudes are small. The importance of small changes in electromyographic events is questionable; it is potentially misleading to use EMG information alone for design and evaluation of working postures if the EMG signals suggest that less than 15% of maximal voluntary muscle strength is required.[2]

Spinal compression

Results of measurements and model calculations of the pressure within the spinal discs of sitting and standing individuals have been put in doubt.[3] For example, the contributions of the facet joints (*apophyseal* joints) of the spinal column to load bearing have been largely neglected[4]; and the pressure within a bent spine has been calculated to be lower than in a straight column.[5]

IAP

The relationships between intra-abdominal pressure (IAP) and back support have been investigated,[6] as well as the relationship between IAP and spinal compression when lifting.[7] Studies have concluded that abdominal pressure may not relieve the spinal column,[8] as had been assumed.

Fidgeting

Observations of body motions (voluntary and unconscious) while sitting are difficult to interpret. Often, movement may be the result of discomfort or because a chair facilitates changes; sitting still may be enforced by a confining chair design or may indicate that a comfortable posture is maintained. Similarly, seat pressure distribution is not well related with perceived seat comfort in cars.[9]

Subjective assessments supreme?

Table 9.1 lists observational and recording techniques of body motions and postures. Some techniques are well established and easy to use, others are not.[10] For nearly all techniques, however, the threshold values that would separate suitable from unsuitable conditions are unknown or variable. Thus, the interpretation of the results obtained by many of the listed techniques is difficult, to say the least: most currently useful techniques are based on subjective assessments by the seated individual. Their judgments presumably encompass all phenomena addressed in physiological, biomechanical, and engineering measurements, and these appear to be most easily scaled and interpreted.

Discomfort/ pain questionnaires

Many studies use subjective judgments (made by the subject or the experimenter) that assess the suitability of an existing situation. While the procedures vary widely among researchers, most rely on the initial work from the 1960s and 1970s.[11] "Discomfort" or "pain" questionnaires have been developed in different countries.[12] By administrating a questionnaire, employees may experience heightened awareness of problems; this should be in everybody's interest, but management sometimes sees this with apprehension. The manner of administration may affect the outcome.[13] Reliability of questionnaire information can be quite high.[14]

Nordic Questionnaire

An often-used, well-standardized inquiry tool is the "Nordic Questionnaire."[15] It is well structured and requires binary or multiple-choice answers. It consists of two parts: one asking for general information, the other specifically focusing on low back, neck, and shoulder regions. The general section uses a sketch of the human body, seen from behind, divided into nine regions. The interviewee indicates if there are any musculo-skeletal symptoms in these areas. Fig. 9.1 shows the body area sketch (same as used in Fig. 10.1). If needed, further body sketches can be used, showing the body in side or frontal views, or providing further details.[16]

TABLE 9.1 Methods for assessing posture

Observation/Technique	Measurement procedures[a]	Assessment criteria	Threshold values	Relevant?
Oxygen consumption	E	E	V	Probably
Heart rate	E	E	V	Yes
Blood pressure	E	E	V	Yes
Blood flow	E–D	E–D	V	Yes
Innervation	D–V	D–V	U	Yes
Leg and foot volume	E	V	U	Perhaps
Temperature, skin or internal	E–D	U	?	Yes
Muscle tension	E–D	V	U	Yes
Electromyography	D	D	V	Yes
Joint diseases	E–D	D	V	Yes
Musculoskeletal disorders	D	D	V	Yes
Cumulative trauma disorders	D	D	V	Yes
Spinal disc pressure	E–V	D	V	Yes
Spinal disc disorders	D–V	D	V	Yes
Spinal disc shrinkage	D	D	U	Yes
Spinal facet disorders	D–V	V	U	Yes
Spinal alignment of vertebrae	D–V	V	V	Yes
Spine curvature	D–V	V	U	Yes
Spinal mechanical stresses, including model calculations	D–V	V–U	V	Yes
Intra-abdominal pressure	E–D	V	U	Perhaps
Surface (skin) pressure at buttocks	E–D	V	U	Yes
Surface (skin) pressure at back	E–D	V	U	Yes
Surface (skin) pressure at thighs	E–D	V	U	Yes
Upper extremity posture	E–D	V	U	Yes
Posture of head, neck, trunk and legs	E–D	V	?	Yes
Changes in posture	E	?	?	Yes
Change in stature	E	E	V	Perhaps
Sensations (ratings) of ailments	E–D	V	D	Yes
Sensations (ratings) of pain	E–D	V	D	Yes
Sensations (ratings) of discomfort	E–D	V	D	Yes
Sensations (ratings) of comfort/ pleasure	E–D	D	D	Yes

[a]E: well established; D: being developed; V: variable; U: unknown; ?: questionable.
Source: Modified from Kroemer (1991a).

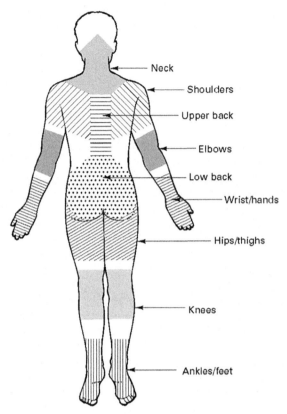

FIGURE 9.1 **Body sketch used in the Nordic Questionnaire.** *Source: Redrawn from Kuorinka et al. (1987).*

A specific section of the Nordic Questionnaire concentrates on body areas in which musculoskeletal symptoms are most common, such as the neck and low back areas. The questions probe deeply with respect to the nature of complaints, their duration, and their prevalence. The test has been modified for use in the United States.[17] The Nordic Questionnaire provides internationally standardized information.

Lying down is the least strenuous posture in terms of physical effort, as measured by oxygen consumption or heart rate. It is not generally well suited to performing physical work with the arms and hands, because for most activities they must be elevated, which causes strain.

Being upright, like walking or standing, is much more energy-consuming, but it generates large work spheres (ranges of acceptable motion) for the arms and hands, and (if walking) a great deal of space can be covered. Furthermore, it facilitates dynamic use of the whole body and its limbs and thus is ideal for large energies and impact forces, such as when splitting wood with an axe.

Sitting, in many respects, falls between the two postures above. Since body weight is partially supported by a seat, energy consumption and circulatory strain are higher than when lying down, but lower than when standing. Arms and hands can be used freely, but their workspace is limited if the worker remains seated. Less energy can be summoned than when standing, but given the better stability of the trunk supported on the seat (and possibly by using arm rests), it is easier to perform finely controlled finger activities. It is also easy to operate pedals or controls with the feet when sitting; the feet are barely needed to stabilize posture or support body weight and thus are fairly mobile. The two most important working postures are walking/standing and sitting. In either condition, the most easily sustained posture of the trunk and neck is one in which the spinal column is straight in the frontal view, but follows the natural *S*-curve in the side view, that is, with *lordoses* (forward bends) in the cervical and lumbar regions, and a *kyphosis* (backward bend) in the thoracic area. However, maintaining this trunk posture over long time-periods becomes very uncomfortable, mostly because of the muscle tension that must be maintained to sustain the body's position. Further, when standing still, unable to move the legs and feet, the feet and lower legs may swell as a result of the accumulation of body fluids—a problem to which many women are particularly prone. Thus, either standing still or sitting still is "unphysiologic"; instead, the posture should be changed often. This includes interludes of walking and moving the head, trunk, arms, and legs.

The posture and movements of the spinal column have been of great concern to physiologists and orthopedists, since many people suffer from discomfort, pain, and disorders in the spinal column, particularly in the low back and in the neck areas. Researchers have postulated that the human body is "not built" for prolonged sitting or standing, or not fit due to diet and/or exercise, or experienced degeneration (particularly of the intervertebral discs), which could be counteracted through physical activities and special exercises. To change the posture of sitting workers, various devices have been proposed, including pulsating cushions in the seat or backrest, or frequent readjustments of the chair configuration, particularly of the angles of seat pan and backrest. Many suggestions have been made regarding the shape and angle of the seat pan and the angle of the backrest. Alternately, the backrest and thus back support can be omitted entirely, so that trunk muscles must be employed to stabilize the body: however, this contradicts the experience that a backrest reduces the load on the spinal column. The muscles that stabilize the spinal column run essentially between the pelvic and shoulder areas. Since they can only contract, not push, intensive use increases the compression force on the spinal column, as sketched in Fig. 9.2. To summarize these considerations:

Sitting and back pain

1. Changes in posture help avoid continued compression of the spinal column as well as muscular fatigue.
2. A seat should be so designed that the sitting posture can be changed frequently. If prolonged sitting is required, then a tall backrest that reclines can help support the back and head while allowing a relaxing break.

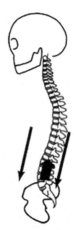

FIGURE 9.2   Activation of longitudinal trunk muscles generates compression of the spine.

Additional information on sitting and seats for the office workspace is available in Chapter 10.

Comfort

The concept of comfort, as related to sitting, was elusive as long as it was defined (simply, conveniently, and falsely) as the absence of discomfort: comfort and discomfort are not extremes of a single judgment scale.[18] Instead, there are two scales, each for comfort and discomfort, which are not even parallel but do partly overlap. Discomfort is mostly related to biomechanically incorrect seat design features, to circulatory problems, and to fatigue that increase with long periods of sitting; comfort is associated with the feeling of well-being, with support, plushness, softness, and even pleasing esthetics. Fig. 9.3 sketches these features, with more detail provided in Chapter 10.

9.2 RECORDING AND EVALUATING POSTURES AT WORK

Postulated postures

One approach to evaluate posture is to identify given postures and observe how often they actually occur, or how much they deviate. Often, certain postures are preferred for some reason and then are labeled as good or healthy. For example, in the late 1800s, the "upright" or "erect" back was thought to be healthy and therefore was promoted; a hundred years later, the so-called "neutral" posture became an ideal, often with little explanation or scientific foundation. (More on this in Chapter 10.)

"NEUTRAL" POSTURE?

In recent years, the term "neutral posture" has become popular, usually suggesting a healthy or desirable body position. Unfortunately, it is often not clear what "neutral" means: Is that the middle of the total motion range in a joint? This would make some sense for the wrist, indicating the hand is straight and in line with the forearm. But there is no obvious significance to the "middle" joint position in elbow or knee, shoulder or hip, or

---

the spinal column. Does the term "neutral" suggest that all tissue tensions about a joint are balanced, so that the position is stable? Does the term imply a minimal sum of tissue tensions (torques) around a body joint? Or does this apply to tensions around several joints, or all body joints? Does "neutral" imply minimal joint discomfort, or a relaxed posture? Or a posture instinctively assumed for a task to generate high body strength or to avoid fatigue? Definition is needed to ensure clarity and usefulness of this term.

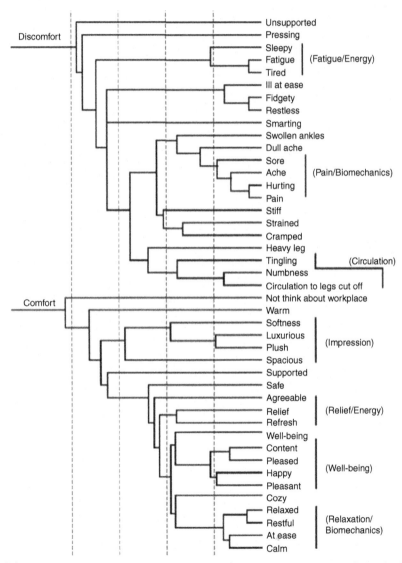

FIGURE 9.3 **Attributes of comfort and discomfort while sitting.** *Source: Courtesy of M. Helander.*

Actual postures

The other approach is to observe and record, in detail, the actual positions of body members. This procedure is facilitated by concentrating on particularly important body parts and recording their positions, using standardized terminology either in descriptive terms[19] or in angles of deviation against a reference.[20] Some techniques provide a set of predrawn body-segment positions from which the observer selects those that best represent the actual conditions. The observations may be documented contemporaneously at the workplace, or transcribed from photographs or movie recordings. Continuous recordings allow the description of motions (instead of one-time postures); some recording techniques utilize markers attached to the observed person to exactly describe body movements.[21]

Many procedures

Various methods and techniques have been developed to describe body postures and to make judgments about their suitability. A *posturegram* records the locations of limbs and joints as well as the direction and magnitude of their movements.[22] This system relies on defined postures of the body, with deviations recorded. The OWAS (Ovaco Working Posture Analysis System) approach is similar in that it postulates basic body segment positions to which one compares to actually observed working postures. This system also includes an evaluation of the suitability of the working postures and interventions.[23] The OWAS approach has been computerized[24] and checked for reliability.[25] Based on OWAS, the RULA (Rapid Upper Limb Assessment) procedure was developed to investigate work places where work-related upper limb disorders are reported.[26] A related method was used in the TRAM system[27] and in PATH[28] where, like with OWAS, postures at the work place are recorded at regular intervals on prepared recording forms. TRAM also includes estimates of forces and of the suitability of body positions. The ARBAN system[29] stressed the ergonomic analysis of the work conditions. It uses videotaping, coding the posture and work load, and includes computerization of the results and their evaluation. The computer routine and evaluation is based on heuristic rules. "Posture targeting"[30] uses prepared sketches of the human body, with targets associated with each major body joint. On these targets, deviations from a standard posture are recorded so that they show the direction and magnitude of displacement of body segments. These posture codes can be combined with assessments of the postural loading.[31] Another approach uses a computer-aided procedure which has a menu of standard postures for trunk and shoulders.[32] Gross deviations from these are read from videotapes and recorded in their frequency of occurrence. One taxonomy uses diagrammatic depictions to record posture,[33] while another uses photographs.[34] The PEO technique[35] uses a hand-held computer for in situ observation at the work place; the program calculates the number and direction of the body postures. More descriptions are available.[36]

Awkward postures?

Some of these techniques include judgments of the suitability of observed postures. Unfortunately, the foundations of these judgments are generally ill-defined and ill-supported; not surprisingly so, given the large number of possible criteria and circumstances that lead to the assessment. While certain body movements and awkward postures are clearly undesirable (such as excessively twisting the trunk), the suitability of others depends very much on the given circumstances.[37] Thus, incorporating an evaluation of observed postures (and motions) in procedures of posture recordings is a rather difficult and (so far) unresolved task.

Many of these techniques have been employed with some success, though with various degrees of fidelity, repeatability, and time consumption.[38] Yet, a truly satisfactory technique still needs to be developed that combines validity and usability in the recording, and possibly judging, of static postures and motions at work.

Improved techniques needed

ERGONOMICS OF TRENCHING

Brennan (1987) described the ergonomic design challenges associated with the posture of the driver of a trenching vehicle. The operator sits looking forward, the direction in which the machine moves, but the trenching tool is attached to the rear of the machine (see Fig. 9.4). To observe the trenching operation, the driver must rotate trunk and neck nearly 180 degrees (see Fig. 9.5). While all the controls to move the vehicle are located, as is common, in front of the operator, the controls for operating the trenching attachment are located to the side (see Fig. 9.6).This arrangement is ergonomically faulty in several aspects: it enforces an overly twisted posture on the operator during the trenching operation and it is likely to result in mistakes in operation of the machine. Unfortunately, similar situations are often found in underground mining equipment, earthmoving machinery, and motorized lift trucks: contorted body postures imposed on the operator; controls improperly located; vision blocked; and noise, jolts, and impacts from the ground transmitted to the operator.

Direction of trenching operation ⟶

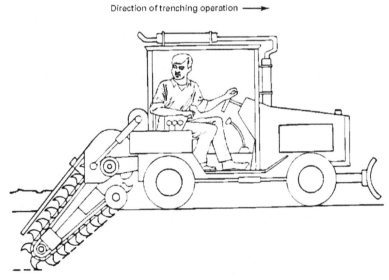

FIGURE 9.4 **Trencher.** *Source: With permission from Brennan (1987).*

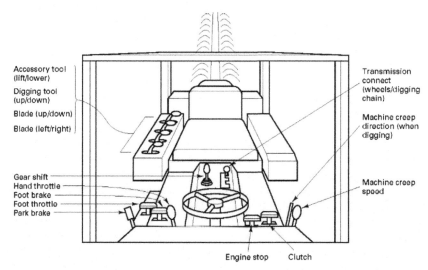

FIGURE 9.5   **Frontal view of the trencher cab.** *Source: With permission from Brennan (1987).*

9.3 DESIGNING FOR THE STANDING OPERATOR

Move, don't stand still

Standing is used as a working posture if sitting is not suitable, either because very large forces must be exerted with the hands and/or because the operator has to cover a fairly large work area. In the second case, walking is much preferable to standing still; standing in place should be imposed only for a limited period. Forcing the worker to stand simply because the work object is customarily put high above the floor or outside sitting reach is not a sufficient justification. For example, in automobile assembly, car bodies have been turned or tilted and parts have been redesigned so that the worker did not have to "stand and bend" and "stand and reach" in order to perform the task. Other examples for workstations designed for standing operators are shown in Fig. 9.7, depending on the need to exert large forces over large spaces, to make expansive strong exertions with visual control, or to work with large objects.

Work near elbow height

The height of the workstation depends largely on the activities to be performed with the hands, and the size of the object. Thus, the main reference point is the elbow height of the worker. Such anthropometric information is presented in Chapter 1; however, the tabulated values must be adjusted to account for a slumped instead of upright body. As a general rule, the strongest hand forces and most useful mobility are between elbow and hip heights. Thus, the support surface (e.g., a bench or table) is determined by the working height of the hands and the size of the object on which a person works.

Shoes and walking surface

Sufficient room for the feet of the operator must be provided, which includes toe and knee space to move up close to the working surface. Of course, the floor should be flat and free of obstacles. Elevated platforms should be avoided, if possible, so nobody trips over an edge. Elastic floor mats and soft shoe soles can reduce foot, leg, and back discomfort.[39] Appropriate friction between soles and the walkway surface helps to avoid slips and falls.[40]

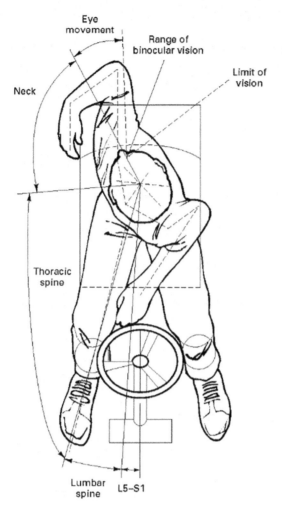

FIGURE 9.6 Contorted body posture of the trencher operator (as viewed from the top) looking at the trenching equipment. *Source: Modified from Brennan (1987).*

FIGURE 9.7 Workspaces designed for the standing operator: useful when large forces over a large area, or forceful hand exertions, or working on large objects are required.

No twists or bends

Body movements associated with work while walking/standing are desirable in physiological respects, but they should not involve excessive bends and reaches, and especially should not include twisting motions of the trunk. Workers should never be forced to stand (or even worse, to stand still) at a workstation just because the equipment was originally poorly designed, or badly placed, as unfortunately often is the case with punch or drill presses used in continuous work. Other machine tools, especially lathes, have long been constructed so that the operator must stand and lean forward to observe the cutting action, and at the same time extend the arms to reach controls on the machine—a difficult position for the operator to safely maintain.

Semi-sitting

So-called stand-seats may allow the operator to assume a somewhat supported posture somewhere between sitting and standing. Examples of these are shown in Figs. 9.8 and 9.9. Occasionally, high stools can be used to allow (rather uncomfortable) sitting at workstations otherwise requiring the operator to stand. Such semiseats usually do not have full backrests and do not support the body fully and therefore, and for

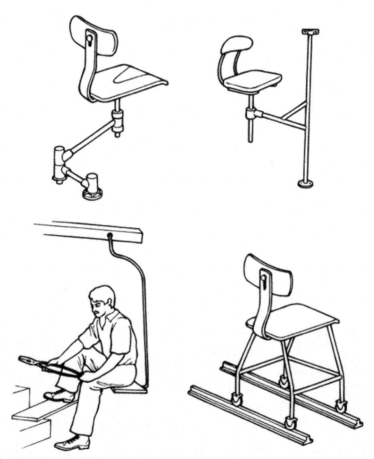

FIGURE 9.8 **Examples of stand-seats.**

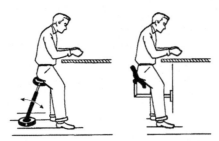

FIGURE 9.9   Stools used to temporarily allow (rather uncomfortable) sitting at workstations designed for a standing operator.

reasons of stability, the feet must still carry much of the bodyweight. Thus, semi-sitting is not a truly satisfactory solution, albeit a better one than enforced and long-duration standing-in-place.

9.4 DESIGNING FOR THE SITTING OPERATOR

Sitting is a much less straining posture than standing, mostly because fewer muscles need to contract to stabilize the body. When seated, the body enjoys plenty of support at mid-section through seat pan and seat back. Sitting allows better controlled hand movements, but coverage is more limited and smaller forces are exerted with the hands. A sitting person can operate controls with the feet and do so, if properly seated, with great force. When designing a workstation for a seated operator, one must particularly consider the required free space for the legs and feet. If this space is severely limited, very uncomfortable and fatiguing body postures result, as shown in Fig. 9.10 (same as

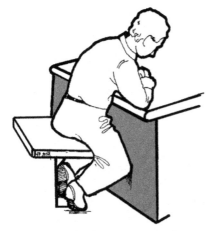

FIGURE 9.10   Leg space for the seated operator must be provided.

FIGURE 9.11   **Working postures when sitting on a flat surface.**

Fig. 1.17). The disadvantages of sitting include low back pain and foot swelling, usually associated with maintaining the same posture for too long.

The preferred working area is in front of the body, at about elbow height with the upper arm hanging (as for a standing operator), where both exact and fast manipulations are most easily done. Many activities of seated operators require close visual observation, which co-determines the proper height of the manipulation area, depending on the operator's preferred visual distance and the preferred direction of gaze (see Chapter 5).

In Western countries, it has become customary to provide chairs that are at about popliteal height of the sitting person (see the anthropometric tables in Chapter 1); thus seat heights range from about 35 to 50 cm. The comfort of a seat is largely determined by design properties of the seat pan and the backrest (for more detail on office chair design, see Chapter 10). Seats for shop use are usually sturdier and simpler, but the general ergonomic design principles are the same: suitable height, width, breadth, and shape for pan and backrest.

Sitting may not require a chair at all. Doing work while sitting crosslegged on the ground, somewhat similar to the "lotus" position, is quite common in some regions of the world. In this position, feet and legs are crossed in front of the body. The body weight is transmitted mostly through the buttocks while the legs and feet serve to stabilize the posture, although some people even sit on their feet. Another common work position is the "squat" posture, in which the soles of the feet are on the ground, the knees severely bent, the thighs close to the trunk, and the person sits nearly on the heels (see Fig. 9.11). One leg may be extended away from the body while the other is kept close. Alternately, the kneeling position may be used, occasionally with the feet rotated outward. This posture is often assumed if work requires that the upper body be bent forward. These "low" postures reduce the heights of eyes and elbows with respect to the ground. For workplace and equipment design, this means that working-height dimensions of sitting Westerners often do not suit Asian operators.[41]

While in India, one of the original authors (KHEK) was invited to visit a modernized small manufacturing plant. In it were many hand-operated machines, mostly drill presses and punch presses, elevated on pedestals; the operators sat on stools. However, the overall impression was that of a "staged" situation. The visitor inquired steadfastly and was finally told that all of these machines originally had been placed directly on the floor, without pedestals; and the operators sat, squatted, and kneeled on the floor as well, as they were accustomed to doing. Then, management decided to "improve" the working conditions according to Western images, and put the machines on pedestals and the operators on stools. When left alone, the operators would assume their regular traditional postures on these stools, with their feet at seat height. For visitors, however, the operators were exhorted to put their feet down.

9.5 DESIGNING FOR WORKING POSITIONS OTHER THAN SITTING OR STANDING

Semi-sitting is one example of a working posture that is neither sitting nor standing. In many cases other postures must be assumed at work, although often only briefly, such as when reaching for a barely accessible object, stooping in a low-ceiling compartment, or straining to perform repair work inside a narrow opening. Little can be done in terms of systematically designing body supports related to such unusual and awkward postures, except not to design equipment that requires them.

A visitor to a high-tech manufacturing facility was impressed by the manager's explanations of how highly automated the production was, and how few workers were needed to run it. "Why," the visitor asked, "are there so many cars parked outside?" The answer was that these belonged to the repair technicians who did the old-fashioned, bloody-knuckles repair work on the automated manufacturing machinery.

Some jobs, though, habitually require bending, stooping and twisting for working: for example, when loading and unloading aircraft passengers' luggage, both behind the airport check-in counter and in aircraft cargo holds. Repair, maintenance, and cleaning jobs often require awkward body postures. Low-seam mining is notorious for requiring bent, stooped, kneeling, even crawling and lying working postures from the miners.[42] In the building and construction trades, bent and twisted body postures are frequent.[43] Other examples are in agriculture,[44] manufacturing and assembly work,[45] and grocery check-outs.[46] Also notoriously bad is entering and exiting rear seats in automobiles.[47]

Awkward postures?

Avoid awkward
postures
These and other examples indicate a need for a systematic ergonomic design approach. First, it must be established whether or not such postures are indeed necessary. If not, better solutions for the work can be found that no longer include them. If they cannot be avoided, special body supports must be designed: for example, military standards and specifications describe the body supports of tank crews, or of aircraft pilots. There are also specific recommendations for the construction industry.[48] A semireclining chair for overhead tasks has been proposed,[49] and space requirements of operator compartments in low-seam mining equipment have been described[50] (see Fig. 9.12).

9.5.1 Work in Restricted Spaces

There are times when work must be performed in restricted spaces, such as in access tunnels, tanks, and mines. The primary restriction usually lies in the lowered ceiling of the workspace. Work becomes more difficult and stressful as the ceiling height forces workers to bend neck and back, or requires squatting, or even lying down. Thus, if restricted spaces are unavoidable, equipment and mechanical aids should be developed that alleviate the human's task. For example, in aircraft baggage handling, it is advantageous to first collect the luggage in containers and then put these containers into the cargo hold, rather than loading individual pieces into the cargo hold.

Other examples of space-restricted spaces are passageways, walkways, hallways, and corridors. Minimal dimensions for these are given in Fig. 9.13. For tight places, where one may have to squat, kneel, or lie on the back or belly, dimensions are given in Fig. 9.14 and Table 9.2. Dimensions for escape hatches, shown in Fig. 9.15, need to accommodate even the largest workers wearing their work clothes and possibly equipment. These openings can be made somewhat smaller for maintenance workers who need to get through access openings in enclosures of machinery; recommended dimensions are shown in Fig. 9.16. The size for openings through which one hand must pass, holding and operating a tool, depends on the given circumstance; some recommended dimensions are shown in Fig. 9.17. These dimensions need to be modified if the operator also must see the object through the opening, and if special tools must be used and movements performed with one hand. In some cases, both hands and arms must fit through the opening, which then needs to be about shoulder-wide. For further information, see the standards issued by ISO, NASA, US Military, and various design handbooks.[51]

9.6 DESIGNING FOR FOOT OPERATION

In comparison to hand movements over the same distance, foot motions consume more energy, are less accurate, and are slower; but they are more powerful, as one would expect from biomechanical considerations.

Standing or
sitting?
If a person stands at work, fairly little force and fairly infrequent operations of foot controls should be required since, during these exertions, the operator has to stand on the other leg alone. Operation of foot controls is much easier for a seated operator since the body is largely supported by the seat. Thus, the feet of a seated operator can move more freely and, given suitable conditions, can exert large forces and energies.

Body position	Height *H* (cm)		Depth *D* (cm)	
	Minimum	Preferred	Minimum	Preferred
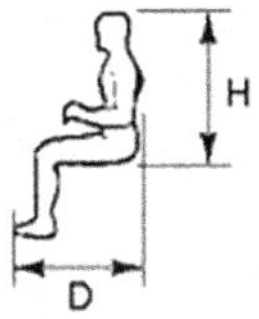	103	110	94	100
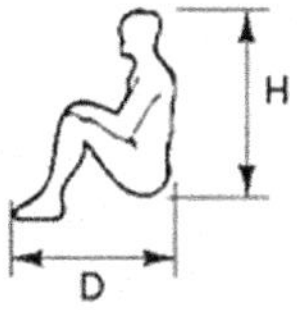	98	110	94	100
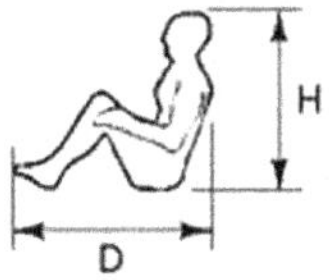	98	110	88	90
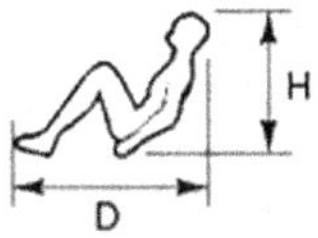	64	65	125	140
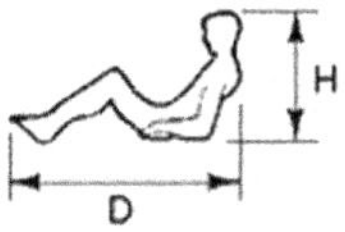	60	62	150	160
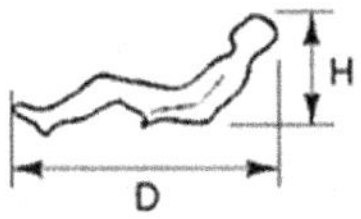	50	54	170	180
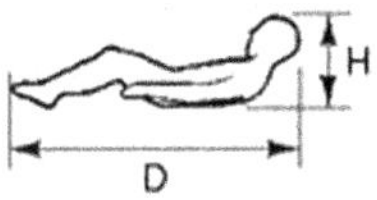	48	52	190	195
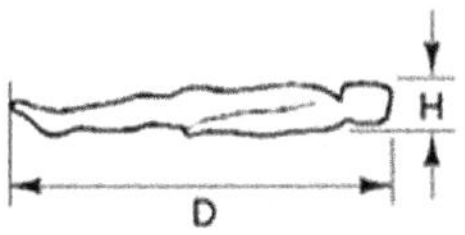	38	46	200	210

FIGURE 9.12 **Spaces required to accommodate US coal miners.** *Source: Adapted from Conway and Unger (1991).*

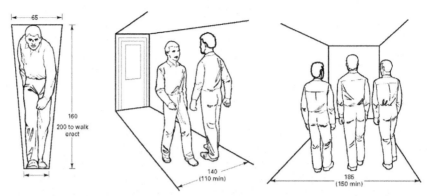

FIGURE 9.13   Minimal dimensions (in cm) for passageways and hallways. *Source: Adapted from Van Cott and Kinkade (1972).*

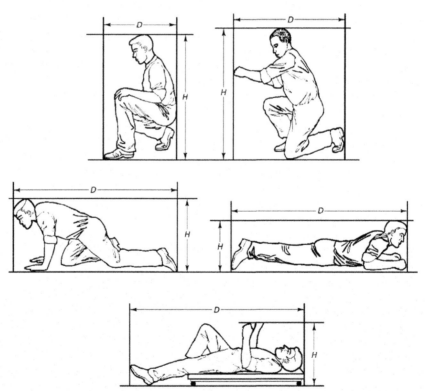

FIGURE 9.14   Minimum height and depth dimensions for "tight" work spaces. *Source: Adapted from MIL-STD 759.*

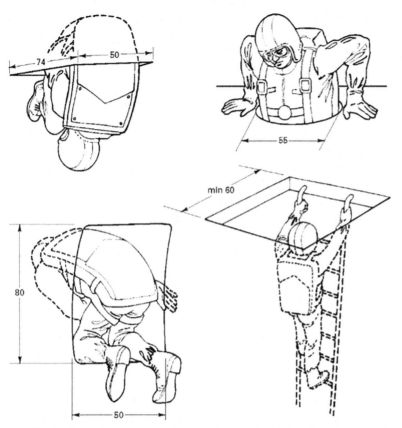

FIGURE 9.15 **Minimal openings for escape hatches.** *Source: Adapted from Van Cott and Kinkade (1963).*

TABLE 9.2 Dimensions (in cm) for "Tight" workspaces

	Height H[a]			Depth D[a]		
	Minimal	Preferred	Arctic Clothing	Minimal	Preferred	Arctic Clothing
Stooped or squatting	66	–	130	61	90	–
Kneeling	140	–	150	106	122	127
Crawling	79	91	97	150	–	176
Prone	43	51	61	285	–	–
Supine	51	61	66	186	191	198

[a]*Dimensions in cm.*
Source: Adapted from MIL-STD 759.

II. DESIGN APPLICATIONS

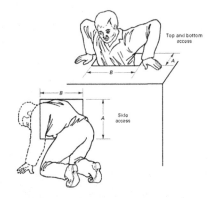

Dimensions	A, depth		B, width	
Clothing	Light	Bulky	Light	Bulky
Top and bottom access	33 cm	41 cm	58 cm	69 cm
Side access	66 cm	74 cm	76 cm	86 cm

FIGURE 9.16 **Access openings for enclosures.** *Source: Adapted from MIL-HDBK 759.*

Bicycling

A good example for such an exertion is pedaling a bicycle: all energy is transmitted from the leg muscles through the feet to the pedals. These should be located directly underneath the body, so that the body weight above them provides the reactive force to the force transmitted to the pedal. The crank radius should be about 15 cm for short-legged individuals, and up to 20 cm for those with long legs. Suitable pedal rotation is usually between 0.5 and 1 Hz, but depends on such factors as the gear ratio of the bicycle (often variable), the ground surface, and the purpose of bicycling (such as for leisure, exercise, or competition). In some cases it is best to lower the center of mass of the combined person-bicycle system. In this case, the cranks may be moved forward and upward, to nearly the height of the seat. Placing the pedals forward makes body weight less effective for generation of reaction force to the pedal effort, hence some sort of backrest should be provided, against which the buttocks and back press while the feet push forward on the pedal. Instead of using the bicycle principle to propel the body, one can use this approach to generate energy, for instance when "pedaling" an electricity generator. (This may also be done with the hands, but the arms are less powerful than the legs.)

> The traditional arrangement of foot controls in the automobile is, by all human factors rules, atrocious: the gas pedal requires that the foot be kept in the same position over long periods of time; the brake pedal must be reached by a complex and time-consuming motion of the foot from the gas pedal toward the body, to the left, and then again forward. It makes no sense that pushing forward on the gas accelerates the vehicle, but also pushing forward on the brake decelerates. The current arrangement encourages use of the right foot alone, while the left foot is usually idle.

	Approximate dimensions (cm)		Task
	A	B	
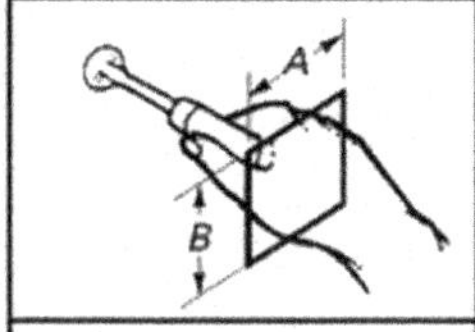	11	12	Using common screwdriver, with freedom to turn hand through 180°
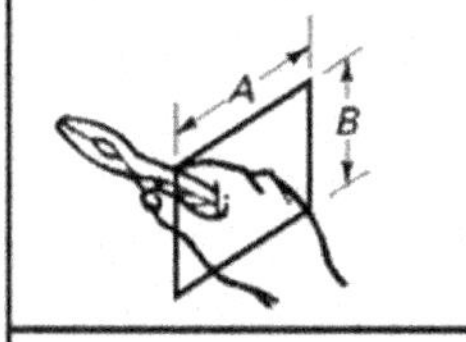	13	12	Using pliers and similar tools
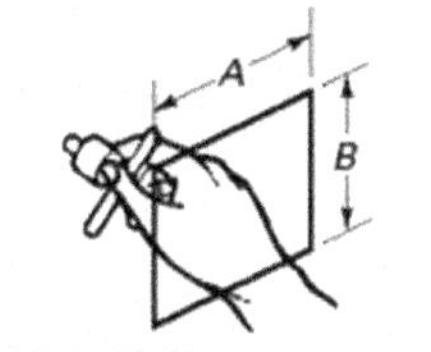	14	16	Using "T" handle wrench, with freedom to turn hand through 180°
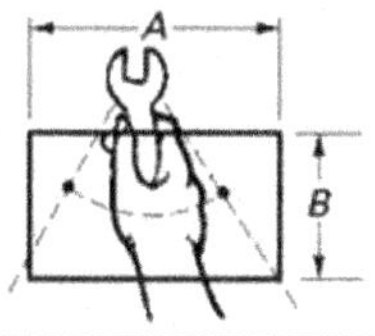	27	20	Using open-end wrench, with freedom to turn wrench through 60°
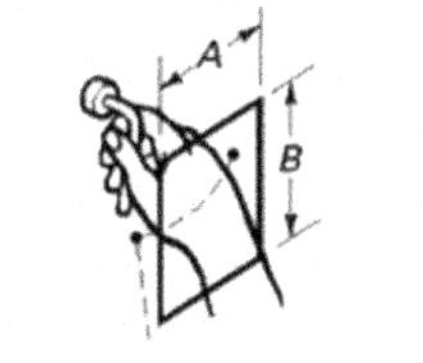	12	16	Using Allen-type wrench with freedom to turn wrench through 60°
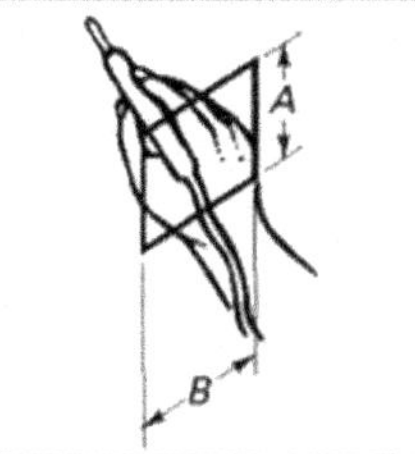	9	9	Using test probe, etc.

FIGURE 9.17 Minimal opening sizes (in cm) to allow one hand holding a tool to pass. *Source: Adapted from MIL-HDBK 759.*

Small forces, such as for the operation of switches, can be generated in nearly all directions with the feet, with the downward or down-and-forward directions preferred. The largest forces can be generated with extended or nearly extended legs, in the downward direction limited by body inertia, in the more forward direction limited both by inertia and the provision of buttock and back support surfaces. These principles are typically applied in automobiles. For example, foot operation of a clutch or brake pedal can normally be performed easily with the knee at about a right angle. But if the power-assist system fails, very large forces must be exerted with the feet: in this case, one must thrust one's back against a strong backrest and extend the legs to generate the needed pedal force.

Figs. 9.18–9.23 provide information regarding the forces that can be generated by the feet with the legs, depending on body support and hip and knee angles. The largest forward thrust can be exerted with the nearly extended legs, which leaves very little room to move the foot control farther away from the hip.[52] However, caution is necessary when applying these data, because they were measured on different populations under varying conditions.

Similar to the hand workspace discussed earlier, a preferred workspace for the feet results from the foregoing discussions—see Fig. 9.24.

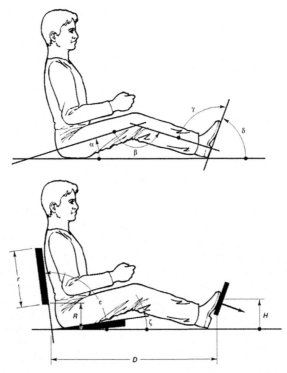

FIGURE 9.18 Conditions affecting the force that can be exerted on a pedal: body angles (*upper part*), workspace dimensions (*lower part*).

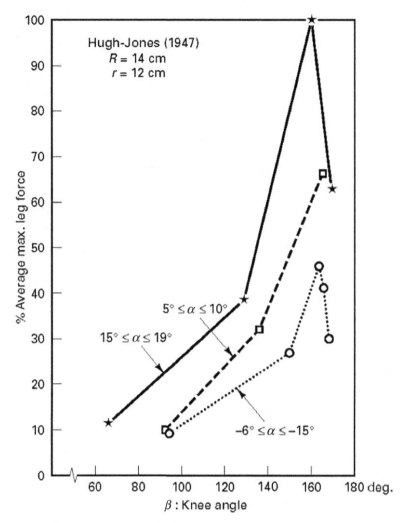

FIGURE 9.19 **Effects of thigh angle and knee angle on pedal push force—maximum force is at least 2100 N.** *Source: From three studies reported by Kroemer (1971).*

In automobiles, power-assisted brakes and steering systems generate a difficult design problem. As long as auxiliary power is available, brakes can be operated easily, in almost any conceivable leg posture. However, if the auxiliary power system fails, suddenly forces are required from the operator which are three to ten times as high for braking (or steering). The driver must not only recognize that much more effort is required, but also must develop this effort quickly and often in a body posture that is not favorable for such large exertion, such as with a strongly bent knee.

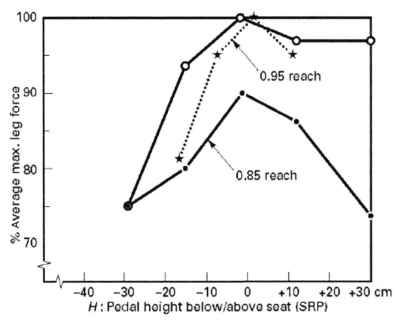

FIGURE 9.20   Effects of pedal height H on pedal push force—maximal force is about 2600 N. *Source: From two studies reported by Kroemer (1972).*

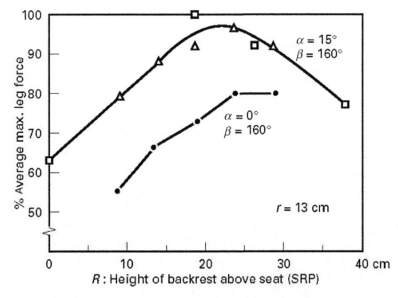

FIGURE 9.21   Effects of backrest height R on pedal push force—maximal force is about 1700 N. *Source: Adapted from Kroemer (1972).*

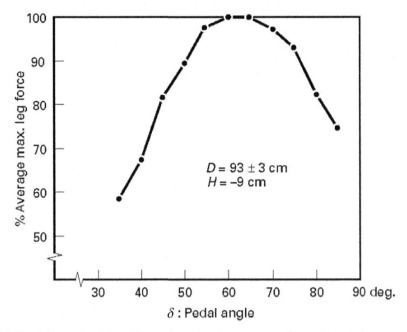

FIGURE 9.22 **Effects of pedal (ankle) angle on foot force generated by ankle rotation—maximal force is about 600 N.** *Source: Adapted from Kroemer (1971).*

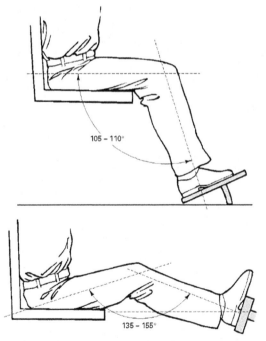

FIGURE 9.23 **Light downward forces can be exerted at knee angles of about 105–110 degrees, while strong forward forces require knee extension at 135–155 degrees.** *Source: Adapted from Van Cott and Kinkade (1972).*

II. DESIGN APPLICATIONS

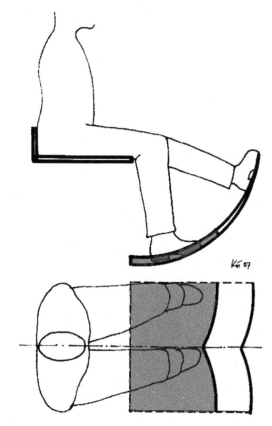

FIGURE 9.24   Preferred (*shaded*) and regular work spaces for the feet, assuming a seated operator.

9.6.1 Design Rules for Foot Controls

1. Repeated operation of a foot control should only be required from a seated operator.
2. Design for push roughly in the direction of the lower leg.
3. Small forces can be exerted by flexing the foot about the ankle.
4. Large forces can be applied by pushing with the whole leg, preferably with a solid back support from the seat.
5. Do not use foot controls for fine control, continuous operation, or quick movements. (Note that this rule is routinely violated by design of foot pedals for speed control of automobiles.)

9.7 DESIGNING FOR HAND USE

The human hand is able to perform a large variety of activities, ranging from those that require fine control to others that demand large forces. (The feet and legs are capable of more forceful exertions than the hand; see above.) Hand tasks may be divided by activity:

- Fine manipulation of objects, with little displacement and force. Examples include writing by hand, assembly of small parts, adjustment of controls.
- Fast movements to an object, requiring moderate accuracy to reach the target but fairly small force exertion there; an example is the movement to a switch and its subsequent operation.
- Frequent movements between targets, usually with some accuracy but little force; an example is an assembly task, where parts must be taken from bins and assembled.
- Forceful activities with little or moderate displacement; examples include many assembly or repair activities, such as when turning a hand tool against resistance.
- Forceful activities with large displacements; such as when hammering.

Accordingly, there are three major types of requirements: for accuracy, for strength exertion, and for displacement. For each of these, certain characteristics of hand-arm movements can be described, if one starts from a "reference position" of the upper extremity: *the upper arm hangs down; the elbow is at right angle, hence the forearm is horizontal, and extended forward; and the wrist is straight. In this case, the hand and forearm are in a horizontal plane at approximately umbilicus height.*

Accurate and fast movements

Fitts's law provides guidance for accurate and fast movements (see Chapter 4): the smaller the distance traveled and the larger the target, the more accurate a fast movement can be. Thus, finger movements are the fastest and most accurate, followed by movements of the forearm only. Among forearm movements (when the upper arm remains still), a horizontal forearm sweep—seemingly rotating about the elbow but, in fact, about the shoulder joint in which the upper arm twists—is faster, more accurate and less tiring than when the forearm flexes/extends in the elbow. The least accurate and the most time-consuming and fatiguing movements are those in which the upper arm is pivoted out of its vertical reference location. This establishes the "preferred manipulation space" mentioned in Chapter 1; its location is sketched in Fig. 9.25.

Forceful exertions

Exertion of force with the hands is a more complex matter. The thumb is the strongest digit and the little finger the weakest. The gripping and grasping strengths of the whole hand are larger than that of the fingers, but depend on the coupling used between the hand and the handle—see Fig. 9.26. The forearm can develop large twisting torques. Torque about the elbow depends on the elbow angle, as depicted in Fig. 9.27. Large force and torque vectors can be exerted when the elbow is bent at approximately a right angle; the strongest pulling/pushing forces toward (or away from) the shoulder can be exerted with the extended arm, provided that the trunk can be braced against a solid structure. The forces exerted with the arm and shoulder muscles are largely determined by body posture and body support, as shown in Fig. 9.28. Likewise, finger forces depend on the finger-joint angles, as listed in Tables 9.3 and 9.4.

Table 9.5 provides detailed information about manual strength capabilities measured in male students and machinists. Female students developed between 50% and 60% the digit strengths of their male peers, but 80%–90% in "pinches." However, the data presented here or found elsewhere must be applied with caution, because they are likely to have been determined on different subject groups, with different techniques, and under different physical and psychological conditions than in the specific application

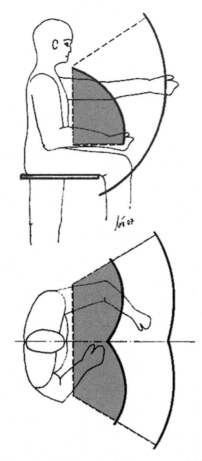

FIGURE 9.25   Preferred (*shaded*) and regular manipulation spaces within the overall reach envelope of the hands.

case. Furthermore, the users may be fatigued or may be especially trained or motivated, with possibly major effects on strength (see also the discussion of strength measurement in Chapter 1).

9.7.1 Designing Hand Tools

Hand tools need to fit the contours of the hand; they need to be held securely with (preferably straight) wrist and suitable arm postures; they must utilize strength and energy capabilities without overloading the body. Hence, the design of hand tools is a complex ergonomic task.

Use of hand tools is as old as mankind. It developed from simple beginnings—using a stone, bone, or piece of wood that fitted the hand and served the purpose—to the purposeful design of modern implements (such as the screwdriver, cutting pliers, or

1. Digit touch:
   One digit touches an object without holding it.

2. Palm touch:
   Some part of the inner surface of the hand touches the object without holding it.

3. Finger palmar grip (hook grip):
   One finger or several fingers hook(s) onto a ridge or handle. This type of finger action is used where thumb counterforce is not needed.

4. Thumb- fingertip grip (tip pinch):
   The thumb tip opposes one fingertip.

5. Thumb- finger Palmar grip (pad pinch):
   Thumb pad opposes the palmar pad of one finger or the pads of several fingers near the tips. This grip evolves easily from coupling #4.

6. Thumb- forefinger Side grip (lateral grip or side pinch):
   Thumb opposes the radial side of the forefinger at its middle phalanx.

7. Thumb–two- finger grip (writing grip):
   Thumb and two fingers (often forefinger and index finger) oppose each other at or near the tips.

8. Thumb- fingertips enclosure (disk grip):
   Thumb pad and the pads of three or four fingers oppose each other near the tips (object grasped does not touch the palm). This grip evolves easily from coupling #7.

9. Finger-palm enclosure (enclosure):
   Most, or all, of the inner surface of the hand is in contact with the object while enclosing it.

10. Grasp (power grasp):
    The total inner hand surface is grasping the (often cylindrical) handle, which runs parallel to the knuckles and generally protrudes from one side or both sides of the hand.

FIGURE 9.26 **Couplings between hand and handle.** *Source: Adapted with permission from Kroemer (1986). Copyright 1986 by the Human Factors and Ergonomics Society, Inc. All rights reserved.*

II. DESIGN APPLICATIONS

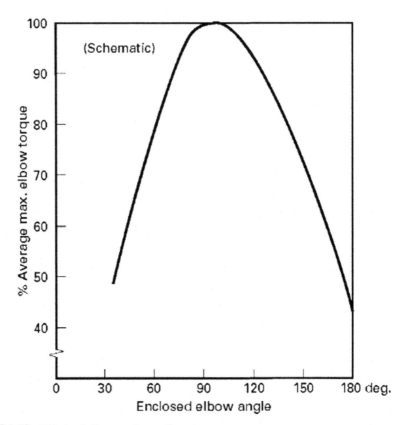

FIGURE 9.27   **Effects of elbow angle on elbow torque.**

power saws) and of controls in airplanes and power stations (see Chapter 12 on controls). Thus, a vast literature is available on tool design.[53]

Fit the tool

The tool must fit the dimensions of the hand and utilize the strength and motion capabilities of the hand/arm/shoulder system. Some dimensions of the human hand were given in Chapter 1, and additional information is available.[54]

Many ways to grip

Some hand tools require a fairly small force but precise handling, such as surgical instruments, screwdrivers used by optometrists, or writing instruments. Commonly, the manner of holding these tools has been called *precision* or *writing grip*. Other instruments must be held firmly between large surfaces of the fingers, thumb, and palm. Such holding of the hand tool allows the exertion of large forces and torques, hence has commonly been called *power grasp*. However, there are many transitions from merely touching an object with a finger (such as pushing a button) to pulling on a hook-like handle, from holding small objects between the fingertips to transmitting large amounts of energy from the hand to the handle. One attempt to systematically arrange the various couplings of hand with handle is shown in Fig. 9.26.

Shaping the grip

For the touch-type couplings (#1–#6 in Fig. 9.26), relatively little attention must be paid to fitting the surface of the handle to the touching surface of the hand. However,

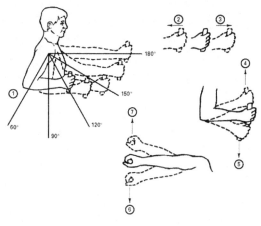

FIGURE 9.28 **Fifth-percentile arm strengths exerted by sitting men.** *Source: Adapted from MIL-HDBK 759.*

	Fifth-percentile arm strength (N) exerted by sitting men												
(1)	(2)		(3)		(4)		(5)		(6)		(7)	
Elbow flexation (deg)	Pull		Push		Up		Down		In		Out	
	Left	Right	Left	Right	Left	Right	Left	Right	Left	Right	Left	Right
180	222	231	187	222	40	62	58	76	58	89	36	62
150	187	249	133	187	67	80	80	89	67	89	36	67
120	151	187	116	160	76	107	93	116	89	98	45	67
90	142	165	98	160	76	89	93	116	71	80	45	71
60	116	107	96	151	67	89	80	89	76	89	53	71

one may want to put a slight cavity into the surface of a pushbutton so that the fingertip does not slide off; to hollow out the handle of a scalpel slightly so that the fingertips can hold on securely; to roughen the surface of a dentist's tool and not make it round to reduce the risk of it turning in the hand. Thus, design details that facilitate holding onto the tool, moving it accurately, and generating force or torque play important roles even for small hand tools.

These "secure tool handling" considerations become even more important for the more powerful enclosure couplings (#8–#10 in Fig. 9.26). These are typically used when large energies must be transmitted between the hand and the tool. The design purpose is to hold the handle securely (without fatiguing muscles unnecessarily, and avoiding pressure points) while exerting linear force or rotating torque at the working end of the tool. — **Strong couplings**

It is important to distinguish between the energy transmitted to the work object and the energy transmitted between the hand and the handle. In many cases, the energy transmitted to the external object is not the same, in type, amount, or time, as energy generated between hand and handle. Consider, for example, the impulse energy transmitted by the head of a mallet compared to the way energy is transmitted between hand and handle, or the torque applied to a screw-head with the tip of a screwdriver compared to the combination of thrust and torque generated by the hand. Thus, the ergonomist must consider both the interface between tool and object, and the interface between tool and hand. — **Transmitted energy**

TABLE 9.3   Average forces (and standard deviations), in newtons, exerted by nine subjects in fore, aft, and down directions with the fingertips, depending on the angle of the proximal interphalangeal (PIP) joint

	PIP joint at 30 degrees			PIP joint at 60 degrees		
Direction	Fore[a]	Aft	Down	Fore	Aft	Down
DIGIT						
2 Index finger	5.4 (2.0)	5.5 (2.2)	27.4 (13.0)	5.2 (2.4)	6.8 (2.8)	24.4 (13.6)
2 Index finger, nonpreferred hand	4.8 (2.2)	6.1 (2.2)	21.7 (11.7)	5.6 (2.9)	5.3 (2.1)	25.1 (13.7)
3 Middle finger	4.8 (2.5)	5.4 (2.4)	24.0 (12.6)	4.2 (1.9)	6.5 (2.2)	21.3 (10.9)
4 Ring finger	4.3 (2.4)	5.2 (2.0)	19.1 (10.4)	3.7 (1.7)	5.2 (1.9)	19.5 (10.9)
5 Little finger	4.8 (1.9)	4.1 (1.6)	15.1 (8.0)	3.5 (1.6)	3.5 (2.2)	15.5 (8.5)

[a]*Average force in N (standard deviation of force in N).*
Source: KHE Kroemer (unpublished data).

Manually driven tools may be classified as follows[55]:

- Percussive tools (ax, hammer) require the human task: swing and hold handle.
- Scraping tools (saw, file, chisel, plane) require the human task: push/pull and hold handle.
- Rotating, boring tools (borer, drill, screwdriver, wrench, awl) require the human task: push/pull, turn, and hold handle.
- Squeezing tools (pliers, tongs) require the human task: press and hold handle.
- Cutting tools (scissors, shears) require the human task: pull/press and hold handle.
- Cutting tools (knife) require the human task: pull/push and hold handle.

Note that in each case the operator must "hold" the tool.

Power-driven tools may use an electric power source (saw, drill, screwdriver, sander, grinder), compressed air (saw, drill, wrench), smoothed internal combustion (chainsaw), or explosive power (bolter, cutter, riveter).

Impacts from power tools

When using manual tools, the operator generates all the energy and therefore is always in full control of the energy exerted (with the exception of percussive tools, such

TABLE 9.4   Digit poke forces exerted by 30 subjects in direction of the straight digits

Digit Poke Forces[a] Exerted by 30 Subjects in Direction of the Straight Digits (see also Table 9.5)			
Digit	10 Mechanics	10 Male students	10 Female students
1 Thumb	83.8 (25.2) A	46.7 (29.2) C	32.4 (15.4) C
2 Index Finger	60.4 (25.8) B	45.0 (30.0) C	25.4 (9.6) DE
3 Middle Finger	55.9 (31.9) B	41.3 (21.6) C	21.5 (6.5) E

[a]*Average force in N (standard deviation of force in N), entries with different letters are significantly different from each other ($P \leq 0.05$).*
Source: Adapted from Williams (1988).

TABLE 9.5   Digit, grip and grasp forces exerted by male machinists and male students

Digit Forces and Grip/Grasp Forces[a] Exerted by 12 Male Machinists and 21 Male Students (see also Table 9.4)

Couplings (see Fig. 9.28)	Digit 1 (Thumb)	Digit 2 (Index finger)	Digit 3 (Middle finger)	Digit 4 (Ring finger)	Digit 5 (Little finger)	
Push with digit tip in direction of the extended digit ("poke")	138 (41) 91 (39)[b]	84 (35) 52 (16)[b]	86 (28) 51 (20)[b]	66 (22) 35 (12)[b]	52 (14) 30 (10)[b]	[b]See also Table 9.4
Digit touch (coupling #1) perpendicular to extended digit	131 (42) 84 (33)[b]	70 (17) 43 (14)[b]	76 (20) 36 (13)[b]	57 (17) 30 (13)[b]	55 (16) 25 (10)[b]	–
Same, but all fingers press on one bar	–	Digits 2, 3, 4, 5 combined: 162 (33)				–
Tip force (as in typing; angle between distal and proximal phalanges about 135°)	–	65 (12) 30 (12)[b]	69 (22) 29 (11)[b]	50 (11) 23 (9)[b]	46 (14) 19 (7)[b]	–
Palm touch (coupling #2) perpendicular to palm (arm, hand, and digits extended and horizontal)	–	–	–	–	–	233 (65)
Hook force exerted with digit tip pad (coupling #3, "scratch")	118 (24) 61 (21)[b]	89 (29) 49 (17)[b]	104 (26) 48 (19)[b]	77 (21) 38 (13)[b]	66 (17) 34 (10)[b]	All digits combined: 252 (63) 108 (39)[b]
Thumb–fingertip grip (coupling #4, "tip pinch")	–	1 on 2: 59 (15) 50 (14)[b]	1 on 3: 63 (16) 53 (14)[b]	1 on 4: 44 (12) 38 (7)[b]	1 on 5: 30 (6) 28 (7)[b]	–
Thumb–finger palmar grip (coupling #5, "pad pinch")	1 on 2 and 3: 95 (19) 85 (16)[b]	1 on 2: 34 (7) 63 (12)[b]	1 on 3: 70 (15) 61 (16)[b]	1 on 4: 54 (15) 41 (12)[b]	1 on 5: 34 (7) 31 (9)[b]	–
Thumb–forefinger side grip (coupling #6, "side pinch")	–	1 on 2: 112 (16) 98 (13)[b]	–	–	–	–
Power grasp (coupling #10, "grip strength")	–	–	–	–	–	366 (53) 318 (61)[b]

[a]Average force in N (standard deviation of force in N), with data taken from 12 male machinists (unless noted).
[b]Data from 21 male students, see also Table 9.4.
Source: Adapted from unpublished data from Higginbotham and Kroemer.

as hammers or axes); when using power-driven tools whose energy is mostly generated and applied to the outside by the auxiliary power source, the human usually just holds or moves the tool. Yet, if that tool suddenly experiences resistance, the reaction force may directly affect the operator, often in terms of a jerk or impact which can exceed the person's abilities and lead to injuries. This occurs particularly often with chainsaws, powered screwdrivers, wrenches and augers. Major problems with many powered hand tools are impacts and vibrations transmitted to the operator, such as by jack hammers, riveters, power wrenches, and sanders. Power-driven tools may transmit vibrations and repeated impacts to the human (see also Chapter 6), particularly if

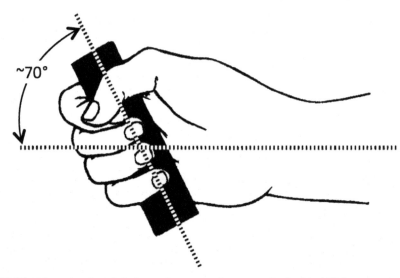

FIGURE 9.29   The natural angle between forearm and grasp center is about 70 degrees.

associated with improper postures, often leading to various kinds of overuse disorders (ODs) (discussed below).

"Handy rule"

Proper posture of the hand/arm system while using hand tools is very important. As a rule, the wrist should not be bent but kept straight to avoid overexertion of such tissues as tendons, tendon sheaths, and compression of nerves and blood vessels (see below).

Oblique thrust line

Normally, the grasp centerline or "thrust line" of a straight handle is at about 70 degrees to the forearm axis (see Fig. 9.29). For example, using common straight-nose pliers often requires a strong bend in the wrist, and neither the direction of thrust nor the axis of rotation correspond with those of the hand and arm. This often results in wrist bending that reduces the force that can be applied.[56] The design rule of "bend the tool, not the wrist" improves that situation, as shown in Fig. 9.30.

Form-fitting

Another technique to avoid unsuitable postures and unnecessary muscle exertions in keeping the tool in the hand is to "form-fit" the handle to the human hand. Instead of using a straight, uniform surface (see top tool surface shown in Fig. 9.31), the handle may be formed to fit the contacting hand parts, for example, the thenar pad (group of muscles at the base of the thumb) and the rest of the palm (as shown in Figs. 9.30 and 9.32). Bulges and restrictions along the handle generate some form-fit, which (along with friction) prevent the handle from sliding out. However, severe notches or serrations, or other extreme form-fits, can make the handle very uncomfortable if it is held differently than the tool designer anticipated (see bottom tool surface in Fig. 9.31).

Friction between hand and handle

Form and surface treatment of the handle can be very important if dirt, dust, oil, or sweat changes the coefficient of friction between the handle and the hand. In such a case, special shapes and surface treatments can be used, either to keep these contaminants away or alleviate their effects.[57]

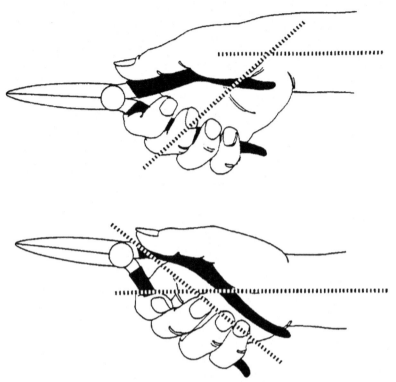

FIGURE 9.30 Common use of straight-nose pliers is often accompanied by a strong bend in the wrist. Rather than bending the wrist (shown at *top*), one should bend the tool (shown on *bottom*). *Source: Modified from Tichauer (1973).*

The thrust force T that can be developed on a handle is a function of the coefficient of friction μ between the handle and the hand (or the glove) and the grasping force G (which is perpendicular to T):

$$T = \mu G \qquad (9.1)$$

Thrust force

The friction between hand and handle is largely determined by the surface texture of the handle and by its cross-sectional shape. Smooth surfaces provide little friction, while it is difficult to slide on rough surfaces. Grooves, ridges, and serrations hinder sliding perpendicularly to them, the more so with increased protrusion and sharper edges. Of course, care must be taken not to exert too much pressure that might damage hand tissues or even cut into the skin, a problem that can be alleviated by wearing gloves (which by themselves might increase the coefficient of friction). Thus, a proper balance must be found in handle shape between easy sliding with a low coefficient of friction and mechanical interlocking with a coefficient of friction at unity value.

The larger the grasping force and/or the friction, the larger the thrust force. The grasp force depends on a person's hand strength which in turn depends on the relation between the sizes of handle and hand. The friction depends on the texture and shape of

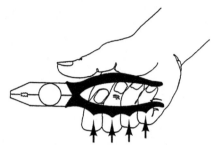

FIGURE 9.31   **Form-fitting a handle can be helpful or painful.** *Source: Modified from Tichauer (1973).*

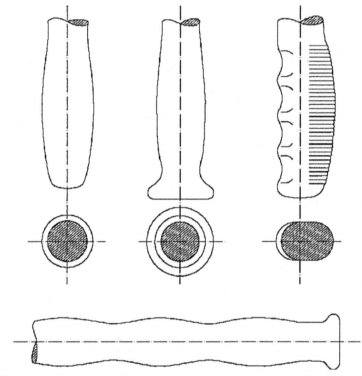

FIGURE 9.32   **Suitable bulges and constrictions along the handle allow many hand positions without severe tissue compression. A flange at the end presents the hand from sliding off the handle.**

the handle, on the skin conditions of the hand (soft or calloused), on the glove or mitten (if worn), and on intermediate materials: sweat and grease increase slipperiness; dust, sand, and other abrasive materials impede gliding.

Wearing gloves   Often, gloves or mittens are worn, either because of temperatures (hot or cold) that must be kept away from the skin, or because mechanical injuries must be avoided, or to alleviate the transmission of vibrations and shocks from the tool to the hand. Usually, wearing suitable gloves increases the friction between hand and handle. However, gloves must fit the hand well; unfortunately, too often, employees are provided working gloves (one-size-fits-all) that are too large (one-size-fits-none). The mismatch can make it difficult to grasp the tool securely and excessive glove material can get caught, for instance by a grinding wheel, and lead to injury.

Design and materials of gloves or mittens affect the amount of force that can be exerted. For example, gloves of space suits have been shown to strongly reduce not only mobility and tactile sensitivity of the hand, but also exertion endurance and the amount of force available to do work. This is because the hand inside the glove must spend a strong effort to bend and move the (stiff) glove itself before energy can be exerted to the object in space.

Space suit gloves

In terms of hand strength, the biomechanical details already discussed in Chapters 1 and 2 apply; in particular, whether intrinsic or extrinsic muscles of the hand are used, how well these muscles are developed, and the skill (experience and/or training) of the user. It is also important to consider the duration of each grasp. Endurance is enhanced and fatigue is reduced if the muscular effort is short and if it requires only a small percentage of the available strength.

Hand strength and endurance

The relationship between handle size and hand size is important in two ways. If the handle is too small, not much force can be exerted, and large local tissue pressures might be generated (such as when writing with a very thin pencil). Conversely, if the handle is too large for the hand, hand muscles must work at disadvantageous lever arms (such as when trying to squeeze the caulking gun common in the United States). Numerous studies on grip strength have been conducted[58]; in most cases, more-or-less cylindrical handles have been used for tests. With these cylindrical handles, diameters between 3 and 4 cm have been found to allow the largest grasp or compression force, with up to 6 cm diameters being suitable for individuals with large hands. However, assessment of the grasp force is not as simple as one might wish, since the contribution of each digit, or of sections of the palm, should be measured separately: each contributes its own "force times coefficient of friction." Few studies shed light on the contributions of hand sections,[59] while mostly an overall, somehow averaged grip strength is measured.[60]

Many tools can be used either with the left or right hand, but about one of ten men and women prefers to use the left hand and has better skills and more strength available with the left hand. Thus, it is advisable to provide hand tools designed for use with either right or left hand, or provide separate right-handed and left-handed tools if required.

"Lefties"

9.7.2 Design Rules for Hand Tools

1. Push or pull in the direction of the forearm, with the handle directly in front of it.
2. Provide good coupling between hand and handle through shape and friction.
3. Avoid pressure spots and "pinch points."
4. Round edges, pad surfaces.
5. Avoid tools that transmit vibrations to your hand.
6. Do not operate tools "frequently and forcefully" by hand: a robot or other machine is better suited for such activities.

9.8 DESIGNING FOR HUMAN STRENGTH

There are many sources for information on forces and torques that operators can apply—for example, see the relevant tables and figures in this chapter. While these data indicate orders of magnitude, the exact numbers should be viewed with great caution, because they were measured on various subject groups under widely varying circumstances (as in Tables 9.3–9.5 and Fig. 9.33). In many cases, it is best to take strength measurements on a sample of the intended user population in actual use conditions to verify that a new design is operable, instead of relying on published data.

Exertable body force or torque depends on individual strength, training, experience and skill, and on body posture. For example, note how the finger forces listed in Table 9.3 depend decidedly on finger posture. Hand forces depend on wrist position[61] and on arm posture, as shown in Fig. 9.28. Exertions with arm, leg, and shoulder or back depend much on the posture of the body and on the support provided to the body (i.e., on the "reaction force" in the sense of Newton's third law) in terms of friction or bracing against solid structures. Fig. 9.33 and Table 9.6 illustrate this: both were derived from the same set of empirical data but extrapolated to show the effects of friction at the feet, body posture, location of the point where force is exerted, and the use of body parts on horizontal push (and pull) forces applied by male soldiers.

Force depends on body posture

EXOSKELETON TO AUGMENT HUMAN CAPABILITY

An exoskeleton is a structure, manufactured from strong material and powered, that moves in synchrony with the body. It may be distant from the human body but mimick the human's motions: this is called a teleoperator. Or it may surround the body, allowing both to move as a unit: here the powered structure can provide superhuman strength.[62]

9.8.1 Design Rules for Operator Strength

The engineer or designer considering operator strength has to make a series of decisions. These decisions include the following:

Is the use static or dynamic? If static, information about isometric capabilities can be used. If dynamic, additional considerations apply, concerning for example physical (circulatory, respiratory, metabolic) endurance capabilities of the operator, or prevailing environmental conditions. Chapters 1, 2, 3, and 6 of this book provide such information.

Most body segment strength data are available for static (isometric) exertions. They provide reasonable guidance for slow motions, but they are probably too high for concentric motions and a bit too low for eccentric motions. Of the little information available for dynamic strength exertions, much is limited to isokinematic (constant velocity) cases. As a general rule, strength exerted in motion is less than that measured in the static positions located on the path of motion.

	Force-plate[1] height	Distance[2]	Force, N	
			Mean	SD
	Percent of shoulder height		With both hands	
	50	80	664	177
	50	100	772	216
	50	120	780	165
	70	80	716	162
	70	100	731	233
	70	120	820	138
	90	80	625	147
	90	100	678	195
	90	120	863	141
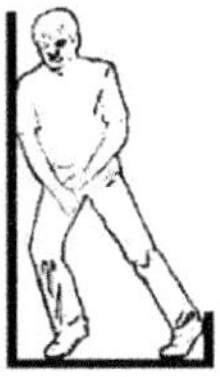	Percent of shoulder height		With one shoulder	
	60	70	761	172
	60	80	854	177
	60	90	792	141
	70	60	580	110
	70	70	698	124
	70	80	729	140
	80	60	521	130
	80	70	620	129
	80	80	636	133
	Percent of shoulder height		With both hands	
	70	70	623	147
	70	80	688	154
	70	90	586	132
	80	70	545	127
	80	80	543	123
	80	90	533	81
	90	70	433	95
	90	80	448	93
	90	90	485	80

	Force-plate[1] height	Distance[2]	Force, N	
			Mean	SD
		Percent of thumb-tip reach*	With both hands	
Force plate	100 percent of shoulder height	50	581	143
		60	667	160
		70	981	271
		80	1285	398
		90	980	302
		100	646	254
			With the preferred hand	
		50	262	67
		60	298	71
		70	360	98
		80	520	142
		90	494	169
		100	427	173
		Percent of span**	With either hand	
	100 percent of shoulder height	50	367	136
		60	346	125
		70	519	164
		80	707	190
		90	325	132

[1]Height of the center of the force plate – 20 cm high by 25 cm wide – upon which force is applied.
[2]Horizontal distance between the vertical surface of the force plate and the opposing vertical surface (wall or footrest, respectively) against which the subjects brace themselves.

*Thumb-tip reach – distance from backrest to tip of subject's thumb as arm and hand are extended forward.
**Span – the maximal distance between a person's fingertips when arms and hands are extended to each side.

FIGURE 9.33 **Maximal static horizontal push forces of males.** *Source: Adapted from NASA (1989).*

TABLE 9.6 Horizontal push and pull forces capable of being exerted intermittently or for short periods by male soldiers

Horizontal force[a]	Applied with[b]	Condition (m: coefficient of friction at floor)
100 N, push or pull	Both hands, one shoulder, or the back	With low traction, $0.2 < \mu < 0.3$
200 N, push or pull	Both hands, one shoulder, or the back	With medium traction, $\mu \sim 0.6$
250 N, push	One hand	If braced against a vertical wall 51–150 cm from, and parallel to, the push panel
300 N, push or pull	Both hands, one shoulder, or the back	With high traction, $\mu > 0.9$
500 N, push or pull	Both hands, one shoulder, or the back	If braced against a vertical wall 51–180 cm from, and or if anchoring the feet on a perfectly nonslip ground (like a footrest)
750 N, push	The back	If braced against a vertical wall 60–110 cm from, and parallel to, the push panel or if anchoring the feet on a perfectly nonslip ground (like a footrest)

[a]*These are minimal forces that may be nearly doubled for two, and less than tripled for three, operators pushing simultaneously. For the fourth and each additional operator, not more than 75% of their push capability should be added.*
[b]*See Fig. 9.36 for examples.*
Source: Adapted from MIL-STD 1472.

Is the exertion by hand or by foot, or in concert with other body segments? For each, specific design information is available (as in this chapter and in Chapter 13). If it is possible to choose the body segment, the selection should aim to achieve the safest, least strenuous, and most efficient performance.

Is a maximal or a minimal exertion the critical design factor? The expected strength exertions range, obviously, from the measured minimum to the maximum. The infamous "average" user strength is just statistical phantom and usually has no design value. The structural strength of the object must be so far above the highest perceivable strength application that even the strongest operator may not break a handle or a pedal. The design value of the object's structural strength is to be set, with a safety margin, above the highest perceivable applied strength. However, the minimum operational strength requirement is determined by the weakest operator who can still achieve the desired result, so that a door handle or brake pedal can be successfully operated or a heavy object be moved.

Measured strength data are often treated, statistically, as if they were normally distributed and reported in terms of averages (means) and standard deviations. This allows the use of common statistical techniques to determine data points of special interest to the designer (see Chapter 2). Yet in reality, body segment strength data are often in a skewed rather than in a bell-shaped distribution. This is not of great concern, however, because usually the data points of special interest are the extremes: maximal or minimal. Maximal exertions are near (often above) the 100th percentile. Minimal strength is at a given percentile value at the low end of the distribution: often one selects the 5th percentile. If these values cannot be calculated, they can often be estimated.

9.9 DESIGNING FOR VISION

For many work tasks, the eyes must focus on the work object, or the tool, or must at least provide general guidance for the manipulation. This is often a problem in repair work or in some assembly tasks, when either only a small opening is available for manipulation and vision or where other objects may interfere with vision.

A particularly difficult ergonomic problem is often found at microscope workstations. Most traditional microscopes are designed so that the eye must be kept close to the ocular. This forces the operator to maintain the same posture, often over extended time periods. In addition, the microscope may be designed or placed so that the operator must bend head and neck to obtain proper eye location in relation to the eyepiece. Large forward bends of the head and cervical column, exceeding 25 degrees from the vertical, are particularly stressful, because neck extensor muscles must be tensed to prevent the unbalanced head from pitching forward even more. The neck bend also affects the posture of the trunk; both neck and trunk muscles must be kept tensed over long periods. Consequently, complaints of microscope operators about pains and aches in the neck and back areas are frequent. Furthermore, some microscopes have hand-operated controls located high in front of the shoulders. This requires that the hands be lifted to that position and kept there, also requiring tension in muscles controlling arm and hand posture. Selecting microscopes that allow a variation of the eye position with respect to the eyepiece, locating it so that the operator need not bend forward, and having proper locations of hand controls can alleviate many of these problems.

Microscopes

In general, visual targets that require close viewing should be placed in front of the operator in or near the medial plane, in "reading distance" (40–80 cm) from the eyes. The angle of the line of sight (from pupil to target) is preferably between 20 and 60 degrees below the Ear–Eye Line plane (i.e., 0–40 degrees below the horizon if the head is held erect); see the discussions in Chapters 1, 5, and 10.

Line of sight

If the vision requirements are less stringent, especially if the person must look at the target only occasionally, the visual targets may be placed on a partial sphere surrounding the operator (see Fig. 9.34).

9.10 DESIGNING TO AVOID OVERUSE DISORDERS IN SHOP AND OFFICE

A review of the literature yields many reports on injuries related to repetitive activities or occupations since 1713[63]: writers' cramp, washer women's sprain, carpenters' arm, bricklayers' hand, shovelers' hip, telegraphists' cramp, typists' tenosynovitis, tennis or golf elbow. Scrivener's palsies or writer's cramps became a well-discussed issue in the 1800s[64] when pianists and violinists were also identified as prone to such "occupation neuroses" together with (among others) harpists, zither players, painters, watchmakers, knitters, engravers, masons, shoemakers … and money counters.[65]

More recently, in Australia during the early 1980s, an epidemic of so-called repetition strain injuries occurred among keyboard operators. Fig. 9.35 shows the frequency of injuries reported in one large company. Similar events associated with

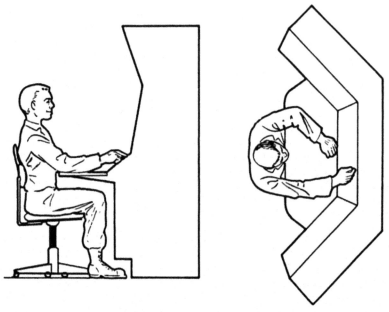

FIGURE 9.34   Console space suitable for placement of displays and controls. *Source: Adapted from MIL-STD 1472.*

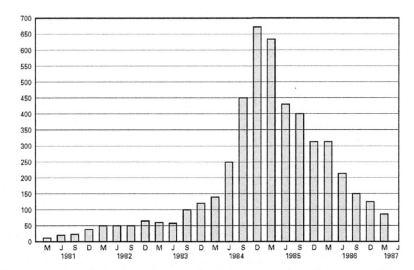

FIGURE 9.35   Overuse disorders at Telecom Australia. *Source: Adapted from Hocking (1987).*

keyboarding, though not on such large scale, have since occurred in Japan and the United States.

In all of these occupations, the provocative actions involved often-repeated movements of small muscles and their tendons; thus, their origins were apparently "peripheral" and could be understood as local physical overexertions (such as in the carpal tunnel syndrome). However, recognizing that these neuroses usually included spasms and other abnormal movements (which got worse the more the victim fought against the inhibition), the disorders were thought to involve the motor cortex of the brain. Disordered brain functions can explain why cramps often accompany biomechanical overloading of local motion systems. *Focal dystonia* includes disorganization in the basal ganglia as well as in the motor and in the sensory cortex; this disturbs the coordination of agonist and antagonist muscles which is necessary for highly repetitive activities. Understanding the underlying injury mechanisms helps us design ergonomic measures to reduce the risk or severity of occurrence.

An *overuse disorder* (OD) is the result of excessive use of a body element, often a joint, muscle, or tendon. In contrast to a single-event injury (acute or traumatic), the OD stems from repeated actions which don't cause injury when occurring only rarely, but whose time-related cumulative effects eventually result in injury. As discussed later, these effects are usually related to body motion or posture, energy or force exerted, and duration or repetitiveness.

> Excessive repetition

Different terms have been used to describe these phenomena, such as occupational OD (or injury, or syndrome); regional or work-related musculoskeletal disorder (MSD); repetitive motion or stress or strain injury (RSI); osteoarthrosis; rheumatic disease; or cumulative trauma disorder (CTD). In this text, the term "overuse disorder," OD, will be used.[66]

> Many names for ODs

In the medical profession, ODs have long been recognized and diagnosed. For example, in the 19th century Raynaud's Disease ("white finger" or jackhammer disease) was known to be caused by insufficient blood supply associated with repetitive impacts; in 1893, Gray described inflammations of the extensor tendons of the thumb in their sheaths after excessive exercise; in 1934 Hammer stated that human tendons cannot tolerate more than 1500–2000 exertions per hour. Inflammations of tendons and tendon sheaths were often reported in typists during the 1930s–1950s. Comprehensive reports on the carpal tunnel syndrome appeared in the 1950s and 1960s.

> Causes of ODs

In the 1940s, engineers and physicians knew and described how ODs were related to design and operation of work equipment and work schedules. From the 1960s on, specific physiological and biomechanical strains of human tissue, especially tendons and their sheaths, were linked to repetitive activities. This resulted in recommendations for design, arrangement and use of tools and equipment (such as keyboards) to alleviate or avoid ODs, especially in the hand/arm region.

> Avoiding ODs

While diagnoses and medical treatments of ODs are well established, their specific relations to work equipment and occupational activities have been hotly debated: for example, in massive and largely unsuccessful lawsuits against US keyboard manufacturers in the 1990s. One point of view is that ODs should not occur in healthy individuals and that, therefore, a connection between work and pathology is inadvertently introduced by a physician (or an attorney). Some people who claim to have an OD are accused of malingering, to have compensation neuroses, to be victims of mass hysteria. The fear of contracting an OD could lead to lowering normally acceptable discomfort thresholds, which may have contributed to the RSI "epidemic" of the 1980s in Australia (Fig. 9.35). In contrast, the prevalent position taken in the current literature is that repetitive activities at work, daily living, and recreation may be causative, precipitating, or aggravating.[67]

> Why do ODs occur?

"Mountain Peaking Through Fog" Analogy

Mountain
analogy

The appearance of health complaints related to cumulative trauma may be compared to a mountain. Its wide base is an accumulation of common everyday cases of tiredness, fatigue, uneasiness, and discomfort during or after a long day's work. The next higher layer consists of instances of occasional movement or postural problems beyond just weariness, often accompanied by small aches and pains that, however, disappear after a good night's rest. The narrower levels above are composed of cases of soreness, pain, and related persistent symptoms; they are present throughout most of the day and do not go away completely during the night or over a weekend. Above this layer, smaller again, is a layer of symptoms that make it difficult to continue related activities and that may lead to seeking advice from friends and co-workers as to how to alleviate the problems. The very pronounced symptoms and health complaints in the next higher level prompt discussions with nurses or physicians, who may recommend managerial and engineering changes at work. On top of this are disorders, injuries, and diseases that need medical attention and often cause short-term disability. At the peak are injuries and diseases that require acute medical treatment such as surgery. The very tip consists of a fortunately small number of disabilities which medical interventions cannot alleviate.

The top of this "mountain of problems," with its broad base and tapering center, is visible above the "fog of psychosocial perception" that usually shrouds its base and lower sections. What becomes visible depends on the prevalent sensitivity to discomfort and pain, the existing willingness to disclose problems to supervisors and health-care givers, and the actual awareness level of society for recognizing existing problems. If the fog reaches high enough, only the peak of the mountain with the severe cases is visible. The lower the fog, the more problems become evident. In clear conditions, even the basic and most widespread layers of the "mountain of cumulative trauma" are in sight.

Fig. 9.36 depicts this analogy. The less serious concerns for health or work appear at the low levels. With increasing height, the risk to the person's health becomes more pronounced, and task performance is affected. Changes in engineering or managerial aspects of work, including work-rest arrangements, often can still alleviate the conditions. If persistent symptoms or acute aches and pains appear, medical advice is usually sought. At the high levels, pathological conditions exist that require medical intervention. If these are not completely successful, disability results.

9.10.1 Biomechanical Strains of the Body

The human body can tolerate low-level mechanical strains (of motion, posture, force, or vibration) from single or repeated mechanical stresses. The risk of injury to tissue increases with augmented extrinsic physical demands of work and life style, and with reductions in a person's physical capacity (intrinsic factors such as age, anatomy, pain response). All soft tissues will fail under excessive loading: muscles, tendons, synovial sheaths, ligaments, fascia, cartilage, intervertebral discs, nerves. But even at strain levels clearly below failure range, injuries to soft tissue include a complex cascade of responses with (muscle) fatigue, inflammation, and structural degeneration that usually can heal but only when the strain is removed. Given the biomechanics of the human

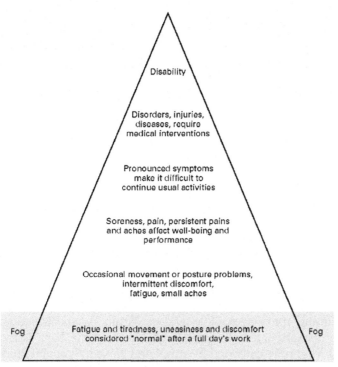

FIGURE 9.36 Analogy of the "mountain of cumulative trauma partly obscured by fog," the level of which indicates perception and awareness of symptoms.

body, internal strains are generally several times higher than the external loads that generate them.[68]

ODs occur particularly in muscles, their tendons and sheaths, and are often associated with impeded blood flow, swelling, and pressure build-up inside tissue spaces, such as inside the carpal tunnel. ODs are frequent in the hand–wrist–forearm area, in shoulder and neck, and in the low back. Repetitive loadings may even damage bone, such as the vertebrae of the spinal column.

Biomechanically, one can describe the human body as consisting of a bony skeleton whose segment links connect in joints and are powered by muscles that bridge the joints—the Borelli model (Chapters 1 and 2). The muscle actions are controlled by the nervous and hormonal systems and supplied through a network of blood vessels, which also serves to remove metabolic byproducts such as carbon dioxide, lactic acid, heat, and water.[69]

Borelli's model

Bones provide the stable internal framework for the body. Bones can be shattered or broken through sudden impacts, and they can be damaged through continual stresses, such as in vibration. Bones are connected to each other and with other elements of the body through the connective tissues of cartilage, ligaments and tendon-muscle units. Tendons are often encapsulated by a fibrous tissue, the sheath, which allows gliding motion of the tendon, facilitated by synovia, a viscous fluid that reduces friction with the inner lining of the sheath.

Structure

Supply

Blood vessels provide oxygen and nutrients through arteries and their ever more branching capillary network to the working muscles and organs, and remove metabolic waste products from them through the venous system (see Chapter 3).

Control

Action signals are generated in the central nervous system and sent through nerves of the feedforward (efferent) pathways of the peripheral nervous system to the muscles. Sensors collect information about the actions and relay this information through nerves of the feedback (afferent) pathways of the peripheral nervous system to the spinal cord and brain for appropriate actions and reactions (see Chapter 4).

9.10.2 Body Components at Risk from Overuse Disorders

While bones (except vertebrae) usually do not suffer from ODs, the joints, muscles, and tendons and their related structures, as well as blood vessels and nerves, are at risk.

Strain

A *strain* is an injury to a muscle or tendon. Muscles can be stretched excessively, which is associated with aching and swelling. A more serious injury occurs when a group of fibers is torn apart. If blood or nerve supply is interrupted for an extended time, the muscle atrophies. Tendons contain collagen fibers, which neither stretch nor contract; if overly strained, they can be torn. Scar tissue may form, which creates chronic tension and is easily reinjured. Also, tendon surfaces can become rough, impeding their motion. The gliding movement of a tendon within its sheath, caused by muscle contraction and relaxation, may be quite large—for example, 5 cm in the hand when a finger is moved from fully extended to completely flexed. Synovial fluid in the tendon sheath, acting as a lubricant to allow easy gliding, may be diminished, which causes friction between tendon and its sheath. Initial signs of injury include feelings of tenderness, warmth, and pain, which may indicate inflammation.

Inflammation

Inflammation of a tendon or its sheath is a protective response of the body: it limits bacterial invasion. The feeling of warmth and swelling stems from the influx of blood. The resulting compression of tissue produces pain. Movement of the tendon within its swollen surroundings is impeded. Repeated forced movement may cause the inflammation of additional fiber tissue, which, in turn, can establish a permanent (chronic) condition.

A *bursa* is a fluid-filled sac lined with synovial membrane: a slippery cushion which prevents rubbing between bone and muscle or tendon. An often-used tendon, particularly if it has become roughened, may irritate its adjacent bursa, setting up an inflammatory reaction, called bursitis, (similar to the inflammation in tendon sheaths) which inhibits the free movement of the tendon and hence reduces joint mobility.

Sprain

A *sprain* occurs when a joint is displaced beyond its regular range, and fibers of a ligament become strongly stretched, torn apart, or pulled from the bone. This can result from a single trauma or from repetitive actions. Injured ligaments may take weeks or months to heal, because their blood supply is poor. A ligament sprain can bring about a lasting joint instability and increase the risk of further injury.

Pressure

Nerve compression may stem from pressure by bones, ligaments, tendons and muscles within the body, or from hard surfaces and sharp edges of work places, tools, and equipment. Increased pressure within the body can occur if the position of a body segment reduces the passage opening through which a nerve runs. Another source of compression, or an additional one, is the swelling of other structures within this opening, often of irritated tendons and their sheaths. The carpal tunnel syndrome (see later) is a typical case of nerve compression.

Impairment of a motor nerve reduces its ability to transmit signals to enervated muscle motor units. Thus, motor-nerve impairment impedes the controlled activity of muscles, and hence reduces the ability to generate force or torque for application to tools, equipment, and work objects. Sensory-nerve impairment reduces the information that can be brought back from sensors to the central nervous system. Sensory feedback is very important for many activities, because it contains information about force and pressure applied, position assumed, and motion experienced. Sensory-nerve impairment usually brings about sensations of numbness, tingling, or pain. The ability to distinguish hot from cold may be reduced. Impairment of an autonomic nerve reduces the ability to control such functions as sweat production in the skin, so a common sign of autonomic-nerve impairment is dryness and shininess of skin areas controlled by that nerve.

Nerve impairment

Blood-vessel compression, often of an artery, reduces blood flow through the supplied area. This means reduced supply of oxygen and nutrients to muscles, tendons, and ligaments; it also means diminished removal of metabolic byproducts, such as lactic acid. Vascular compression reduces blood flow (*ischemia*), which in particular limits the possible duration of muscular actions and impairs recovery of a "fatigued" muscle after activity. Such neurovascular compression is often found in the neck, shoulder, arm, and hand.

Reduced blood flow

Vibrations of body members, particularly of hand and fingers, may result in *vasospasms* that reduce the diameter of arteries. This impedes blood flow to the body areas supplied by the vessels, evidenced by blanching of the area, known particularly as the white-finger (Raynaud's) phenomenon. Exposure to cold may aggravate the problem, because it can also trigger vasospasms, particularly in the fingers. Associated symptoms include intermittent or continued numbness and tingling, with the skin turning pale and cold, and eventually loss of sensation and control. In the fingers, this condition is often associated with the use of powered machinery. Frequent operation of keys on keyboards might also represent a source of vibration strain to the hand-wrist area.

Vasospasm

9.10.3 Carpal Tunnel Syndrome

In 1959, Tanzer described 10 cases of carpal tunnel syndrome. Two of his patients had recently started to milk cows on a dairy farm, three worked in a shop in which objects were handled on a conveyor belt, two had gardened with considerable hand weeding, one had been using a spray gun with a finger trigger. Two patients had been working in large kitchens, stirring and then ladling soup twice daily for about 600 students.

Among the best known ODs is *carpal tunnel syndrome* (CTS), first described 125 years ago. The occurrence of ODs in typists and their possible relation to key force and displacement, key operation frequency, and posture of arms and hands was described in 1964.[70] The typical gradual onset of numbness in the thumb and the first two and a half fingers of the hand supplied by the median nerve was described in 1966 and 1972.[71] The results of a survey of the work and hobby activities of 658 patients who suffered from CTS were described in 1975[72]: four out of five patients were employed in work requiring light, highly repetitive movements of the wrists and fingers. Frequent impairments in hands and arms of operators of accounting machines were discussed in 1980.[73] In 1983, the American Industrial Hygiene Association acknowledged the prevalence and importance of CTS by publishing Armstrong's ergonomic guide to avoid the CTS, termed an occupational illness[74] in the guide.

Thus, in the 1970s and early 1980s, CTS was well recognized as an often-occurring, disabling condition of the hand that can be caused, precipitated or aggravated by light activities in the office or heavy work in shop or construction. Of course, leisure activities may be involved as well. Critical activities include a flexed or hyperextended wrist, especially in combination with forceful exertions, in highly repetitive activities, and often with vibrations. Of course, case studies and epidemiological surveys have inherent attributes (such as the lack of a control group and existence of confounding variables) that make it difficult to definitively prove a causal connection between activities and disorders, but apparent relationships between the two strongly suggest a cause–effect relationship.[75]

Carpal tunnel

The anatomical conditions that explain CTS were described in 1963.[76] On the palmar side of the wrist, near the base of the thumb, the carpal bones form a concave floor. It is bridged by three ligaments (the *radial carpal*, *intercarpal*, and *carpometacarpal*), which in turn are covered by the *transverse carpal ligament*, firmly fused to the carpal bones. Thus, a canal is formed by the carpal bones, constituting the floor and sides, with ligaments covering this canal like a roof—see Fig. 9.37. This structure is called the *carpal tunnel*. Its cross-sectional shape is roughly oval. Flexor tendons of the digits, the median nerve, and the radial artery all pass through the carpal tunnel. This crowded space is further reduced when the wrist is flexed, extended, or laterally pivoted, and exacerbated further if the tendons and/or tendon sheaths are swollen or if outside pressure is applied.

CTS

The median nerve, which passes through the carpal tunnel, innervates the thumb, much of the palm, and the index and middle fingers as well as the radial side of the ring finger. Reduction in the crowded space of the carpal tunnel leads to pressure on the median nerve and/or blood vessels. For example, "dropped" or "elevated" wrists, shown in Fig. 9.38, reduce the available cross-sectional space of the carpal tunnel, and hence generate a condition that may cause CTS, particularly in individuals with small wrists.[77] CTS is medically asserted by reduced conduction velocity of electric impulses along the associated section of the medial nerve.

9.10.4 Occupational Activities and Related Disorders

Table 9.7 lists those conditions that are most often associated with ODs: of course, this list is neither complete nor exclusive. New occupational activities arise, such as computer mouse use in the late 1990s, and several activities may be required to perform the same

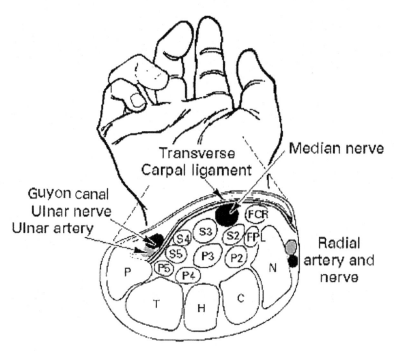

FIGURE 9.37 Schematic view of the carpal tunnel with the tendons of the superficial (S) and profound (P) finger flexor muscles, flexors of the thumb (FCR, FPL), nerves and arteries, carpal bones (P, T, H, C, N) and ligaments. *Source: Adapted from Kroemer (1989b).*

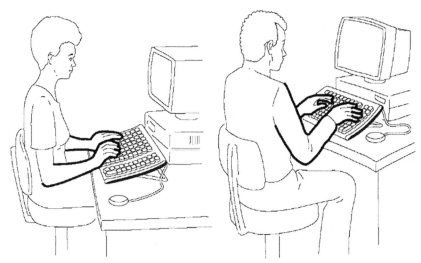

FIGURE 9.38 Dropped and elevated wrists. Note also the pressure at the edge of the table. *Source: Courtesy of Herman Miller Inc.*

TABLE 9.7   Common overuse disorders

Disorder[a]	Description	Typical job activities
Carpal tunnel syndrome (writer's cramp, neuritis, median neuritis) (N)	The result of compression of the median nerve in the carpal tunnel of the wrist. This tunnel is an opening under the carpal ligament on the palmar side of the carpal bones. Through this tunnel pass the median nerve, the digit flexor tendons, and blood vessels. Swelling of the tendon sheaths reduces the size of the opening of the tunnel, pinching the median nerve, and blood vessels. The tunnel opening is also reduced if the wrist is flexed or extended or ulnarly or radially pivoted. Tingling, numbness, or pain in all digits but the little finger	Buffing, grinding, polishing, sanding, assembly work, typing, keying, operating a cash register, playing musical instruments, surgery, packing, housekeeping, cooking, butchering, hand washing, scrubbing, hammering
Cubital tunnel syndrome (N)	Compression of the ulnar nerve below the notch of the elbow. Tingling, numbness, or pain radiating into ring or little finger	Resting forearm near elbow on a hard surface or sharp edge; also when reaching over an obstruction
DeQuervain's syndrome (or disease) (T)	A special case of tendosynovitis that occurs in the abductor and extensor tendons of the thumb, where they share a common sheath. The condition often results from combined forceful gripping and hand twisting, as in screw driving	Buffing, grinding, polishing, sanding, pushing, pressing, sawing, cutting, surgery, butchering, using pliers, operating a turning control such as that on a motorcycle, inserting screws in holes, forceful hand-wringing
Epicondylitis ("tennis elbow") (T)	Tendons attaching to the epicondyle (the lateral protrusion at the distal end of the humerus bone) become irritated. This condition is often the result of the impact of jerky throwing motions, repeated supination and pronation of the forearm, and forceful wrist extension movements. The condition is well known among tennis players, pitchers, bowlers, and people hammering. A similar irritation of the tendon attachments on the inside of the elbow is called medical epicondylitis, also known as "golfer's elbow"	Turning screws, small-parts assembly, hammering, cutting meat, playing musical instruments, playing tennis, pitching, bowling, golfing
Ganglion (T)	A tendon sheath swelling that is filled with synovial fluid, or a cystic tumor at the tendon sheath or a joint membrane. The affected area swells and causes a bump under the skin, often on the dorsal or radial side of the wrist. (Since, in the past, the condition was occasionally smashed by striking the swelling with a Bible or heavy book, it was also called a "Bible bump.")	Buffing, grinding, polishing, sanding, pushing, pressing, sawing, cutting, surgery, butchering, using pliers, operating a turning control such as that on a motorcycle, inserting screws in holes, forceful hand-wringing
Neck tension syndrome (M)	An irritation of the levator scapulae and trapezius group of muscles of the neck, commonly occurring after repeated or sustained overhead work	Belt conveyor assembly, typing, keying, small parts assembly, packing, carrying a load in the hand or on the shoulder
Pronator (teres) syndrome (N)	Result of compression of the median nerve in the distal third of the forearm, often where it passes through the two heads of the pronator teres muscle in the forearm; common with strenuous flexion of elbow and wrist	Soldering, buffing, grinding, polishing, sanding

TABLE 9.7 Common overuse disorders (*cont.*)

Disorder[a]	Description	Typical job activities
Shoulder tendonitis (rotator cuff syndrome or tendonitis, supraspinatus tendinitis, subacromial bursitis, subdeltoid bursitis, partial tear of the rotator cuff) (T)	This is a shoulder disorder, located at the rotator cuff, which consists of four tendons that fuse over the shoulder joint where they pronate and supinate the arm and help to abduct it. The rotator cuff tendons must pass through a small bony passage between the humerus and the acromion, with a bursa as cushion. Irritation and swelling of the tendon or the bursa are often caused by continuous muscle and tendon effort to keep the arm elevated	Punch press operations, overhead work, assembly, packing, storing, construction work, postal "letter carrying," reaching, lifting, carrying load on shoulder
Tendonitis (tendinitis) (T)	An inflammation of a tendon. Often associated with repeated tension, motion, bending, being in contact with a hard surface, or vibration. The tendon becomes thickened, bumpy, and irregular in its surface. Tendon fibers may be frayed or torn apart. In tendons without sheaths, such as within the elbow and shoulder, the injured area may calcify	Punch press operation, assembly work, wiring, packaging, core making, using pliers
Tendosynovitis (tenosynovitis, tendovaginitis) (T)	This is a disorder of tendons that are inside synovial sheaths. The sheath swells. Consequently, movement of the tendon within the sheath is impeded and painful. The tendon surfaces can become irritated, rough, and bumpy. If the inflamed sheath presses progressively onto the tendon, the condition is called stenosing tendosynovitis. DeQuervain's Syndrome is a special case occurring in the thumb, while the trigger finger condition occurs in flexors of the fingers	Buffing, grinding, polishing, sanding, punch press operation, sawing, cutting, surgery, butchering, using pliers, operating a turning control such as that on a motorcycle, inserting screws in holes, forceful hand-wringing, keyboarding
Thoracic outlet syndrome (neurovascular compression syndrome, cervicobrachial disorder, brachial plexus neuritis, costocalvicular syndrome, hyperabduction syndrome) (V, N)	A disorder resulting from compression of nerves and blood vessels between the clavicle and the first and second ribs, at the brachial plexus. If this neurovascular bundle is compressed by the pectoralis minor muscle, blood flow to and from the arm is reduced. This ischemic condition makes the arm numb and limits muscular activities	Buffing, grinding, polishing, sanding, overhead work, keying, operating a cash register, wiring, playing musical instruments, surgery, truck driving, stacking, material handling, postal "letter carrying," carrying heavy loads with extended arms
Trigger finger (or thumb) (T)	A special case of tendosynovitis wherein the tendon becomes nearly locked, so that its forced movement is not smooth, but occurs in a snapping or jerking manner. This is a special case of stenosing tendosynovitis crepitans, a condition usually found with digit flexors	Operating a finger trigger, using hand tools that have sharp edges pressing into the tissue or whose handles are too far apart for the user's hand, so that the end segments of the fingers are flexed while the middle segments are straight
Ulnar artery aneurysm (V, N)	Weakening of a section of the wall of the ulnar artery as it passes through the Guyon tunnel in the wrist; often caused by pounding or pushing with the heel of the hand. The resulting "bubble" presses on the ulnar nerve in the Guyon tunnel	Assembly work

(Continued)

TABLE 9.7   Common overuse disorders (*cont.*)

Disorder[a]	Description	Typical job activities
Ulnar nerve entrapment (Guyon tunnel syndrome) (N)	Results from the entrapment of the ulnar nerve as it passes through the Guyon tunnel in the wrist. The condition can occur from prolonged flexion and extension of the wrist and repeated pressure on the hypothenar eminence of the palm	Playing musical instruments, carpentering, brick-laying, use of pliers, soldering, hammering
White finger ("dead finger," Raynaud's syndrome, vibration syndrome) (V)	Stems from insufficient blood supply and brings about noticeable blanching. Finger turns cold, gets numb, and tingles, and sensation and control of finger movement may be lost. The condition is due to closure of the digit's arteries caused by vasospasms triggered by vibrations. A common cause is continued forceful gripping of vibrating tools, particularly in a cold environment	Chainsawing, jackhammering, use of vibrating tool, sanding, scraping paint, using a vibrating tool too small for the hand, often in a cold environment

[a]*Type of disorder: N, nerve; T, tendon; M, muscle, V, vessel.*

job, such as concurrent mouse and keyboard use. The major activity-related factors in repetitive strain injuries are rapid and often-repeated movements, forceful exertions, static muscle loading (often to maintain posture), vibrations, and cooling of the body. Their negative effects are, or may be, aggravated by inappropriate organizational and social factors,[78] discussed later.

Repetitiveness   *Repetitiveness* is a matter of definition ("more than once per time unit"); what is low or high depends on the specific activities or body part involved. For industrial work, high repetitiveness may be defined as a cycle time of less than 30 seconds, or as more than 50% of the cycle time spent performing the same fundamental motion.[79]

Forcefulness   *Forcefulness* is also a matter of definition; what is low or high depends on the specific activities or body part involved. A high or strong force exerted with the hand, for example more than 45 N, may be a causative factor for OD.[80] The force applied to a computer key is only about 1 N; however, this translates to a multiple thereof in tension force in the finger tendon.[81]

Tension related to posture   When high enough, *static muscle tension* (often generated to maintain body posture) is stressful. If muscles must remain contracted at more than about 15% of their maximal capability, circulation is impaired. This can result in tissue ischemia and delayed dissipation of metabolites (see Chapter 2). Body position can also affect passage space for blood, nerves and tendons, for example, in CTS (see Fig. 9.38). In fact, any strong deviation of the wrist from the neutral position, as well as inward or outward twisting of the forearm especially with a bent wrist, combined with the pinch grip, is stressful.

"Seven sins"   There are seven conditions that specifically need to be avoided:

1. Activities with many repetitions.
2. Activities that require prolonged or repetitive exertion of more than about one-third of the operator's static muscular strength available for that activity.
3. Body segments put in extreme positions.
4. Being forced to maintain the same body posture for long periods of time.

5. Pressure from tools or work equipment on tissues (skin, muscles, tendons), nerves, or blood vessels.
6. A tool vibrating the body, or part of the body.
7. Exposure of working body segments to cold, including air flow from pneumatic tools.[82]

9.10.5 Stages of Overuse Disorders and Their Treatment

The clinical features of ODs are varied, variable, and often confusing. The onset of these symptoms can be gradual or sudden. Three stages are commonly distinguished:

- Stage 1 is characterized by aches and "tiredness" during the working hours, which usually ease overnight and on days off. There is no reduction in work performance. This condition may persist for weeks or months, and is reversible.
- Stage 2 has symptoms of tenderness, swelling, numbness, weakness, and pain that start early in the work shift and do not settle overnight. Sleep may be disturbed, and the capacity to perform the repetitive work is often reduced. This condition usually persists over months.
- Stage 3 is characterized by symptoms that persist at rest and during the night. Pain occurs even with nonrepetitive movements, and sleep is disturbed. The patient is unable to perform even light duties and experiences difficulties in daily tasks. This condition may last for months or years.

The early stage is often reversible through work modification and rest breaks. Exercise as a precautionary measure or for rehabilitation must be applied with great caution and is often of questionable value.[83] In the later stages, the most important factor is abstaining from contra-indicated activities. This may mean major changes in working habits and lifestyle. Further treatments include physiotherapy, appropriate medications, and other medical treatments (including surgery). Medically, it is important to identify an OD early, at a stage that allows effective treatment. Ergonomically, it is even more important to recognize potentially injurious activities and conditions early on and alleviating them through work (re)organization and work (re)design.

9.10.6 Nonbiomechanical Factors in Overuse Disorders

Individual, organizational, and social factors can affect the physiological pathways leading from tissue loading to overload, impairment and possibly disability.[84] Yet, how and whether they do so for an individual is difficult to assess. For personal factors as age or medical status, there are strong relationships to the biological mechanisms. This is also true for body mass and gender, but the mechanism is less clear. Factors such as genetics and general conditioning are less well-established. Even when found statistically significant in (retrospective) studies, these relationships rarely show high predictive value.[85]

Psychosocial factors include social support and workplace stress; job content, variety, demand, control, satisfaction, enjoyment. Social support appears to have a mediating effect on stress. Among the organizational variables, poor job content (e.g., in terms of task identity and integration) and high demand have been found to be related to higher rates of MSDs; to a smaller extent, this also seems to be true for job control. Individual and biomechanical variables likely have greater impact on ODs than psychological stress. Altogether, social and organizational variables, even when found statistically significant for the generation of MSDs, are not large contributors.[86]

9.10.7 Ergonomic Interventions

A variety of management actions can reduce or eliminate the risk of ODs. These include engineering designs or re-designs, work methods including tools and personal protective equipment, administrative controls, and organized exercise. Of these, engineering and method designs are most reasonable and successful because they address the origin of the problem and can mitigate the harm. Administrative controls and training, if appropriately done, may have positive effects by reducing the exposure and hazard.[87] Physical exercise can be useful, but must be carefully selected so as not to inadvertently increase the biomechanical stress already experienced.[88]

Ergonomic Means to Counter ODs

Of course, it is best to avoid conditions that may lead to an OD. Work object, equipment, and tools used should be appropriate designed; instruction on and training in proper postures and work habits is important; managerial interventions such as work diversification (the opposite of job simplification and specialization), relief workers, and rest pauses are often helpful. The basic, most important principle is to "not to do, not to repeat" possibly injurious motions and force exertions and to avoid unsuitable postures.

As a general rule, tools and tasks should be designed to be handled without causing wrist deviations. The wrist should not be severely flexed, extended, or laterally pivoted but in general remain aligned with the forearm. The forearm should not be twisted (not be pronated or supinated). The elbow angle should be varied about a median 90 degrees. The upper arms usually should hang down along the sides of the body. The head should be held fairly erect. The trunk should be mostly upright when standing; when sitting for long periods, a tall, well-shaped backrest is desirable. Severe trunk twisting should not be required at work. There should be enough room for the legs and feet to allow standing or sitting comfortably. It is important that the postures of the body segments, and of the whole body, can and will be varied often during the working time.

Jobs should be analyzed for their movement, force/torque and posture requirements, preferably by using the well-established industrial engineering procedure of "motion and time study."[89] Each element of the work should be screened for factors that can contribute to ODs, and these elements be eliminated by human factors engineering. Table 9.8 provides an overview of generic ergonomic measures to fit the job to the person.

Avoidance of OD, whether by appropriate planning of new work or by redesign of an existing workstation, follows one simple generic rule: let the operator perform "natural" activities, those for which the human body is suited. Avoid highly repetitive activities and those in which straining forces or torques must be exerted or in which awkward posture must be maintained over prolonged time. The opposite approaches of selecting workers who seem to be especially able to perform work that most people cannot do, or letting several individuals work at the same workstation alternately to distribute the

TABLE 9.8 Ergonomic measures to avoid common overuse disorders

Disorder	Avoid in general	Avoid in particular	Do ...	Design ...
Carpal tunnel syndrome	Rapid, often repeated, finger movements, wrist deviation	Dorsal and palmar flexion, pinch grip, vibrations between 10 and 60 Hz	... use large muscles, but infrequently and for short durations	... the work object properly
Cubital tunnel syndrome	Resting forearm on sharp edge or hard surface			... the job task properly
DeQuervain's syndrome	Combined forceful gripping and hard twisting		... let wrists be in the line with the forearm	... hand tools properly ("bend tool, not the wrist")
Epicondylitis	"Bad tennis backhand"	Dorsiflexion, pronation		
Pronator syndrome	Forearm pronation	Rapid and forceful pronation, strong elbow and wrist flexion	... let shoulder and upper arm be relaxed	... round corners, pad
Shoulder tendonitis, rotator cuff syndrome	Arm elevation	Arm abduction, elbow elevation	... let forearms be horizontal or more declined	... placement of work object properly
Tendonitis	Often repeated movements, particularly while exerting a force; hard surface in contact with skin; vibrations	Frequent motions of digits, wrists, forearm, shoulder	... alternate head and neck postures	
Tendosynovitis, DeQuervain's syndrome, ganglion	Finger flexion, wrist deviation	Ulnar deviation, dorsal and palmar flexion, radial deviation with firm grip		
Thoracic outlet syndrome	Arm elevation, carrying	Shoulder flexion, arm hyperextension		
Trigger finger or thumb	Digit flexion	Flexion of distal phalanx alone		
Ulnar artery aneurism	Pounding and pushing with the heel of the hand			
Ulnar nerve entrapment	Wrist flexion and extension	Wrist flexion and extension, pressure on hypothenar eminence		
White finger, vibration syndrome	Vibrations, tight grip, cold	Vibrations between 40 and 125 Hz		
Neck tension syndrome	Static head/neck posture	Prolonged static head/neck posture		

Source: Adapted from Kroemer (1989b).

overload over several workers, are poor choices that should be applied only as a stop-gap measure until a good solution can be found.

> Fit the job to the person; do not attempt to fit persons to jobs.

In addition to keeping the number of repetitions and the amounts of energy (force or torque) small, the following posture-related measures should be considered:

1. Provide a chair with a headrest, so that one can relax neck and shoulder muscles at least temporarily.
2. Provide an armrest, possibly cushioned, so that the weight of the arms must not be carried by muscles crossing the shoulders and elbow joints.
3. Provide flat, possibly cushioned surfaces on which forearms may rest while the fingers work.
4. Provide a wrist rest for individuals operating traditional keyboards, so that the wrist cannot drop below the key level.
5. Round, curve, and pad all edges that otherwise might be point-pressure sources.
6. Select jigs and fixtures to hold workpieces in place, so that the operator does not have to hold the workpiece.
7. Select and place jigs and fixtures so that the operator can easily access the workpiece without contorting hand, arm, neck, or back.
8. Select and place bins and containers so that the operator can reach into them with least possible bending and twisting of hand, wrist, arm, neck, and trunk.
9. Select tools whose handles distribute pressure evenly over large surfaces of the operator's digits and palm.
10. Select hand tools and working procedures that do not require pinching grips.
11. Select the lightest possible hand tools.
12. Select hand tools that are properly angled so that the wrist must not be bent.
13. Select hand tools which do not require the operator to apply a twisting torque.
14. Select hand tools whose handles are shaped so that the operator does not have to apply much grasping force to keep it in place or to press it against a workpiece.
15. Avoid tools that have sharp edges, fluted surfaces, or other prominences that press into tissues of the operator's hand.
16. Suspend or otherwise hold tools in place so that the operator does not have to do so for extended periods of time.
17. Select tools that do not transmit vibrations to the operator's hand.
18. If the hand tool must vibrate, have energy absorbing/dampening material between the handle and the hand (however the resulting handle diameter should not become too large).
19. Make sure that the operator's hand does not become undercooled, which may be a problem particularly with pneumatic equipment.
20. Select gloves, if appropriate, to be of proper size, texture, and thickness.

9.10.8 Research Needs

Our current knowledge about the biomechanical relationships between activities (on the job, in daily living and during leisure) and ODs is largely limited to exertions of fairly large forces, high exertion frequencies, and certain maintained body postures. However, even for those gross muscular activities, the exact causal relationships to ODs are not totally clear.[90] Some causal factors are not well defined, and for most the critical threshold values are not known.

A major problem, both in concept and in application, is how to assess the frequency of activities related with ODs. When numerically describing the frequency of activity, one presumes that the actions occur at regular intervals during the recording time—however, this is commonly not the case over a day's working time. Activities may bunch together in some time periods but occur seldom during others. It is unknown how an uneven distribution of activities over working time may be related to the occurrence of ODs.

Repetitiveness

The force or torque requirement of a task may be measured statically (isometrically) or in dynamic terms. Some definitions of forcefulness are applicable only to static exertions, while many activities are in fact dynamic in nature. Some ODs are explicitly related to motion (dynamics)—as indicated by the term "repetitive motion injury." Exertion of energy in either static or in dynamic conditions establishes very different body strains.[91] Unfortunately, our current knowledge base seems to be largely dependent on the assumption of isometric muscle efforts, that is, in a static condition.

Forcefulness

Shop activities, usually with large muscle-strength exertions, have been most commonly linked to ODs. In the office, operation of keys (formerly on typewriters, now primarily on computer keyboards and with mice) requires, per activation, rather small energies to be exerted by the fingers; but the number of such actions per hour is often high, in some cases up to 20,000. This brings up the under-researched problem of interactions between energy (force and displacement) and repetitiveness with respect to ODs. It is likely that those interrelations are rather complex, and certainly they include factors beyond (static) force, frequency and body posture.[92]

Clearly, much research must be conducted to identify the components of activities that may lead to ODs, and to understand why and how these physical events, singly or combined, overload body structures and tissues. When these relationships are understood, it should be possible to establish exact thresholds or doses—for factors such as posture, force or torque, displacement, their rates of occurrence and duration— which separate acceptable from harmful conditions.

9.11 CHAPTER SUMMARY

The human body is the traditional measure for sizing hand tools, equipment, and workstations—see Table 9.9. We may assume different postures at work: sitting is generally preferred and we can more easily apply force with the feet while seated; but walking and standing allow us to cover more space and exert larger hand forces. There are specific considerations when designing for a standing or sitting operator, or for other work positions, such as in restricted spaces.

The hands are the body's primary work tools; they operate with finest control directly in front of the body, at belly to chest height. Hand tools are often needed when finger manipulation alone is insufficient, but they should be designed so that they are helpful for the intended purpose, not stressful or even potentially damaging for the body. Foot operation of controls is more forceful, but frequent foot operation should be required only if the operator is seated.

ODs are caused by oft-repeated activities, particularly if these require extensive body energy, force and torque; and by "unnatural" postures, especially in wrists, arms,

TABLE 9.9   Folk norms of measurement

Inch	Breadth of thumb; length of distal phalanx of little finger
Phalanx	Length of distal phalanx of thumb; length of middle phalanx of middle finger (2 in.)
Hand	Width of palm across knuckles, length of index finger (two phalanges)
Span (of hand)	Distance between tips of spread thumb and index finger (two hands, four phalanges)
Foot	Length of foot (three hands, six phalanges)
Ell	Length from elbow to tip of extended middle finger (three spans, six hands)
Step	Distance covered by one step (four spans, 16 phalanges)
Fathom	Distance between tips of fingers of the hands with arms extended laterally ("span akimbo") (three steps)

Source: Reprinted with permission from Drillis (1963).

shoulders, neck and back; and by exposure to vibrations, impacts, and cold. Although in many cases exact injury mechanisms for combined stresses are not yet well understood, rules and recommendations for design of proper equipment and its use are available.

9.12 CHALLENGES

Which are the structures that keep the spinal column in balance?

Consider the problems associated with drawing a person's attention to perceived working conditions in the course of an interview.

What changes would be needed in conventional lathe design to allow the operator to sit?

What makes work in tight, confined spaces so difficult?

Why is it difficult to exert large forces with the foot when one is standing?

Consider the effects of trunk muscle use in sitting without a backrest.

How could the attributes of "sitting comfort" change with sitting time?

Do you agree with the argument that sitting may be more conducive to falling asleep on the job than standing?

Can one sit too comfortably?

Consider alternative design solutions for the foot controls in automobiles.

Consider alternative design solutions for the hand-control functions of flight direction and engine speed in an airplane.

What are the disadvantages of contouring hand tools to fit the hand closely?

Propose tests to assess the "usability" of gloves.

Which might be means to measure pressure distribution of surfaces that form-fit the human body, such as a hand tool, or a seat surface?

Often force is applied to an object, not in a continuous way, but in steps, such as in the breakaway force to set an object into motion, and then the force to keep it in motion. How could such force exertions be measured?

Defend or refute the statement: overexertion injuries should not occur in normal use of the body.

What are the advantages of being aware of signs of body strain related to repetitive work? Are there disadvantages as well?

Is "social awareness" for problems at work detrimental or advantageous to productivity?

Which biomechanical models could be applied to explain cumulative trauma injuries to the hand-arm system as a result of manipulations?

Describe specialized job analyses to check for potential ODs.

Under what conditions might it be admissible to select people to do difficult jobs instead of fitting the job better to human capabilities?

Discuss the possible interactions of forcefulness, repetitiveness, and posture in the causation of CTDs.

NOTES

1 Basmajian and DeLuca (1985); Soderberg (1992); Kumar and Mital (1996).
2 Wiker et al. (1989).
3 Jaeger (1987).
4 Adams and Hutton (1986).
5 Aspden (1988).
6 Boudrifa and Davies (1984) and Woodhouse et al. (1995).
7 McGill and Norman (1987) as well as Marras and Reilly (1988).
8 McGill (1999a,b).
9 Gyi and Porter (1999).
10 KHE Kroemer (1991a) judged their status and usefulness.
11 Shackel et al. (1969) and Corlett and Bishop (1976).
12 For example by Occhipinti et al. (1985) in Italy; in Scandinavia by Kuorinka et al. (1987), in the United States by Chaffin and Andersson (1991).
13 Andersson, Karlehagen, and Jonsson (1987).
14 Booth-Jones et al. (1998).
15 Developed by Kuorinka et al. (1987) and modified by Dickinson et al. (1992).
16 van der Grinten and Smitt (1992).
17 Chaffin and Andersson (1991).
18 As shown by Helander and Zhang (1997).
19 Occhipinti et al. (1991).
20 Priel (1974), Corlett et al. (1979), and Gil and Tunes (1989).
21 See, for example, Genaidy et al. (1994).
22 Priel (1974).
23 Karhu et al. (1981), and examples of OWAS being used by, for example, White and Lee Kirby (2003), Nwe et al. (2012), and many others.
24 By Mattila et al. (1993).
25 By de Bruijn et al. (1998).
26 McAtamney and Corlett (1993).
27 Berns and Milner (1980).
28 Buchholz et al. (1996).
29 Holzmann (1981).
30 Refined by Corlett et al. (1979).
31 Wilson and Corlett (1990).
32 Keyserling (1986a,b, 1990).
33 Malone (1991).
34 Paul and Douwes (1993).
35 Fransson-Hall et al. (1995).

36 Several of these techniques are described by their originators in Section II/Section II of Karwowski and Marras (1999) Occupational Ergonomics Handbook.

37 Genaidy et al. (1995) and Keyserling (1990).

38 Burt and Punnett (1999), Fisher and Tarbutt (1988), Malone (1991), and Ziobro (1991).

39 Hansen et al. (1998), Krumwiede, Konz, and Hinnen (1998), and Stuart-Buttle et al. (1993).

40 Chiou et al. (1996), Groenquist and Hirvonen (1994); Jones et al. (1995), Lin and Cohen (1995), Leclercq et al. (1995), and Redfern and Bidanda (1994).

41 Chinese cultural relics from the Shang through the Han dynasties (1600 BC to 220 AC) show that people sat and slept on mats in small rooms with low ceilings. Sitting with crossed legs and kneeling were common postures, whereas squatting or sitting with extended legs and feet was considered impolite or immoral. The opening of the Silk Road allowed envoys of the Chinese Han dynasty to visit western Asia, after which stools and other chair-like furniture were introduced to China. By the fourth century, traditional rituals and formalities had changed. Paintings from that era show people seated on hourglass-shaped stools, on four-legged stools, or on beds raised on legs. During the seventh to the tenth centuries, the traditional life on mats gradually gave way to the folding stool. Around the year 1200, sets of raised furniture, including stools, chairs, tables, screens, dressing tables, racks, etc., were used in China. Drawings of the Ming dynasty (1368-1644) show a variety of furniture styles, often elevated on legs, with forms of classical simplicity. However, in noble households, it was still regarded improper for women to sit on chairs. (Xing, 1988)

42 Gallagher (1999).

43 As described by Buchholz et al. (1996), Helander (1981), and Schneider and Susi (1994).

44 Meyers et al. (1997).

45 Duquette et al. (1997), Helander and Nagamachi (1992), and Kragt (1992).

46 Johansson et al. (1998) and Estill and Kroemer (1998).

47 Giacomin and Quattrocolo (1997).

48 Helander (1981).

49 By Lee et al. (1991).

50 By Conway and Unger (1991).

51 Such as by Eastman-Kodak (1986) or Woodson et al. (1991).

52 Actual force data are compiled, e.g., in NASA and US Military Standards; and by Eastman-Kodak (1986) and Woodson et al. (1991). As with any data set, use these data carefully since they were measured on different populations under varying conditions.

53 For example, as summarized by Bullinger et al. (1997) and by Freivalds (1999).

54 For example, in the publications by Gordon et al. (1989), NASA/Webb (1978), Wagner (1988) and particularly by Greiner (1991).

55 Using in part the listing by Fraser (1980).

56 Zellers and Hallbeck (1995).

57 Bobjer et al. (1993).

58 Reith (1982) wrote a review.

59 See, for example, Bishu et al. (1992), Bjoering et al. (1998), Lowe and Freivalds (1998), and Yun and Freivalds (1995).

60 Imrhan (1998).

61 Zellers and Hallbeck (1995) and Imrhan (1998).

62 Crowell (1995).

63 As compiled by Kroemer (1998a).

64 Bell (1833), as cited by Sacks (2007).

65 Gowers (1888), as cited by Sacks (2007). The incidence of overuse disorder in musicians is further discussed by Fry (1986) and Lockwood (1989).

66 In his famous 1713 book *De Morbis Artificum* (Diseases of Workers, translated by Wright 1993), Bernadino Ramazzini reported on diseases associated with distinct occupations and trades. He described ODs that appeared in workers who did "violent and irregular motions" and assumed "unnatural postures of the body" (Wright 1993; page 43). He also reported ODs among secretaries and office clerks; he said that their diseases were caused by "incessant movement of the hand and always in the same direction", by constant sitting posture, and by prevailing "strain on the mind" (Wright 1993; page 254). Ramazzini's treatise is a basis of the fields of Occupational Medicine and Industrial Hygiene.

II. DESIGN APPLICATIONS

67 Bernard (1997) and National Research and Council (1998, 1999).

68 National Research and Council (1999) and Wasserman (1998).

69 Kroemer et al. (1997).

70 Kroemer (1964).

71 Phalen (1966, 1972).

72 Birkbeck and Beer (1975).

73 Huenting et al. (1980).

74 Kroemer (1998a).

75 Bernard (1997).

76 Robbins (1963).

77 Morgan (1991).

78 See, for example, Bernard (1997), Kuorinka and Forcier (1995), National Research and Council (1998, 1999), and Nordin et al. (1997).

79 Silverstein (1985).

80 Silverstein (1985).

81 Rempel et al. (1997) and Rempel et al. (1999).

82 Over 50 years ago, NJV Peres wrote: "It has been fairly well established, by experimental research overseas and our own experience in local industry, that the continuous use of the same body movement and sets of muscles responsible for that movement during the normal working shift (notwithstanding the presence of rest breaks), can lead to the onset initially of fatigue, and ultimately of immediate or cumulative muscular strain in the local body area" (Peres 1961, p. 1). "It is sometimes difficult to see why experienced people, after working satisfactorily for, say, 15 years at a given job, suddenly develop pains and strains. In some cases these are due to degenerative arthritic changes and/or traumatic injury of the bones of the wrist or other joints involved. In other cases, the cause seems to be compression of a nerve in the particular vicinity, as for example, compression of the median nerve in carpal tunnel syndrome. However, it may well be that many more are due to cumulative muscle strain arising from wrong methods of working ... " (Peres 1961, p. 11)

83 Silverstein et al. (1988) and Lee et al. (1992).

84 Moon and Sauter (1996).

85 National Research and Council (1999).

86 National Research and Council (1999).

87 National Research and Council (1999).

88 Lee et al. (1992).

89 Konz (1990).

90 Bernard (1997).

91 Kroemer et al. (1990, 1997).

92 Latko et al. (1997), Marras and Schoenmarklin (1991), Moore et al. (1991), and Occhipinti and Colombini (1998).

The Computer Workplace

Overview

How an organization or an individual designs a workspace depends on several factors, including the type of work the organization does—for example, whether the work requires privacy or if it encourages collaboration. Other considerations center around cost concerns and technology. Consider the computer: when it first started to appear in offices, it was massive and needed its own room. Then computers shrank to fit on individuals' desks, with employees seemingly tied to their computers. Computers subsequently grew even smaller, becoming mobile devices that fit into their users' pockets. Today, the mobile device seems tied to the person. With computers now portable and wireless and employees no longer tethered to office spaces, ergonomic principles must consider that virtually any space can become a work space—the traditional office or a nontraditional office (be it at home, in a neighborhood café, on a train, at the beach, or at an airport).

One constant in office design and work space evolution is change. What's more, even with all of the design knowledge we have today, another constant in the workplace over the past centuries has been employee discomfort. Office workers have been struggling with some degree of pain and discomfort for as long as offices have existed. Differing work tasks, prevalent use of diverse computers, and changed attitudes means rethinking old recommendations for workspace design. New designs should accommodate a wide range of body sizes, postures, and activities; facilitate visual tasks, support task performance, and allow convivial interaction with co-workers; improve well-being and make people feel comfortable and at ease in their work environment.

10.1 INTRODUCTION

Life for an office worker in the late 1800s and the early 1900s was less physically exhausting than that of a factory worker, but challenging in its tedium; in Herman Melville's description, men were "pent up in lath and plaster—tied to counters, nailed to benches, clinched to desks."[1] Then, as today, technology dictated office (and furniture) design; it shifted shapes as discoveries and technological developments occurred. Gaslight replaced candlelight, typewriters replaced quills, calculators and computers replaced the abacus. Typewriters and their companions, vertical file cabinets, became commercially successful around 1900. Also at that time, standing while working changed to sitting.

Even as office work evolved and office furniture changed with this evolution, office workers throughout the centuries experienced pain: low-back pain and musculoskeletal overexertion, often together with eyestrain and wrist pain, have been common complaints of office personnel since office work debuted. Ergonomically appropriate systems are a critical component of a safe and healthy work environment, but this was not generally considered in early office design, and even now our understanding continues to grow.

Every day we experience the discomfort of maintaining our body posture over extended periods of time. Even when we are resting in bed, we need to reposition ourselves after a while and move around. This need to "shake a leg" becomes urgent after we have forced our bodies into a still position for a period of time—during an extended seated meeting, or standing still "at attention" in a military formation. Our bodies are quite literally built to move around, not to stand or to sit still. Consequently, we should design our workstations for movement, but instead, we have been designing them for rigid sitting for over a century now.

10.2 THEORIES OF POSTURE AND COMFORT

In the 1880s, body posture became a serious concern to physiologists and doctors. Orthopedist Franz Staffel reported that German farmers and laborers often had abnormal back curvatures. He noted that their spinal columns were either too flat or overly bent: *lordotic* (backward), *kyphotic* (forward), and *scoliotic* (sideways). Staffel and his contemporary orthopedic researchers declared that these back curvatures were unhealthy, hypothesized that they were due to malnutrition and bad postural habits, and concluded that they had to be avoided, especially in children. In their opinion, the upright (straight, erect) standing posture was balanced and healthy, while curved and bent backs were unhealthy and therefore had to be avoided. Staffel published papers on seating for school children that recommended a straight torso. Consequently, "straight back and neck, with the head erect" became the recommended posture for standing and, in what was deemed logical at that time, office furniture was designed for sitting up straight. Staffel decried chairs that "were constructed more for the eye than for the back" and further suggested a roll of horsehair be placed in the spine's lumbar curve to avoid scoliosis (an early ergonomic intervention).[2]

10.2.1 Sitting Upright

Staffel's recommendation for an ideal back posture when sitting was similar to what he advocated for standing: straight up. Staffel and his contemporaries were particularly concerned about the postural health of children; they felt that school seats and desks should be designed (and the children urged to maintain) that "normal" erect posture of back, neck, and head. (Staffel is often considered the "father of the modern school chair.") Starting in the late 1880s, many often elaborate and adjustable "hygienic and healthy" designs for school furniture were proposed: seats, desks, and seat–desk combinations were all laid out to promote the upright posture.

Designing for this postural ideal completely disregards the need to change among various postures while sitting rather than to stay in any one position. What is more, unfortunately, is that the typical design of the conventional computer workstation also enforces one long-maintained posture: the prevailing idea of "hands on the keyboard, eyes fixated on computer screen" is still widespread. The ensuing positioning of hands and arms, eyes, and head determines the position of the upper body and allows very little variation in posture.

Imagine the physical freedom we have when a keyboard is not needed or if it is moveable and if the display is large and also moveable; it becomes intuitively clear that no one and only healthy posture exits for which the workstation should be designed.

10.2.2 Assessing "Suitable" Postures

For many decades, experts believed that standing or sitting with a straight back was not only physiologically desirable but also socially proper. Even in ancient Egypt, tomb reliefs illustrate that the dignified Egyptian man was expected to sit with a straight back, horizontal thighs, and vertical lower legs. Although there is nothing wrong with sitting or standing upright, it should be the operator's choice and not be forced because of workstation design. Sitting still over extended periods of time leads to tissue compression, reduced metabolism, poorer blood circulation, and accumulation of extracellular fluid in the lower legs.

Quantitative measures

The effects of managing a posture or of postural changes can be measured quantitatively and evaluated by observing variations in dependent variables.[3] Several disciplines provide their specific techniques:

- *Physiology*: Oxygen consumption, heart rate, blood pressure, electromyograms, fluid collection in the lower extremities.
- *Medicine*: Acute or chronic disorders including cumulative trauma injuries (these are all "after injury" assessments).
- *Anatomy and biomechanics*: X-rays; CAT scans; changes in stature, disk, and intra-abdominal pressure; model calculations.
- *Engineering*: Observations and recordings of posture; forces or pressures at seat, backrest, or floor; amplitudes and frequencies of body displacements; productivity metrics.
- *Psychophysics*: Structured or unstructured interviews and subjective ratings by either the experimental subject or the experimenter.

While most of these techniques are well established, some are not easy to use or are not sensitive to postural variations. Most require a laboratory setting, few are suited for field studies. For nearly all outcomes, the threshold values that separate "good" from "bad" conditions are unknown or variable, so interpreting results of those measurements is difficult.

Qualitative measures

Subjective judgments presumably encompass all the phenomena addressed in the various measurements listed earlier, and they can be reliably scaled and interpreted.[4] A large number of questionnaires have been developed to use qualitative measures to assess people's feelings about uncomfortable or painful working conditions.[5] An often-used inquiry tool is the Nordic Questionnaire.[6] It consists of two parts, one asking for general information and the other focusing specifically on low back, neck, and shoulder areas. It uses a sketch of the human body divided into nine regions—see Fig. 10.1 (same as Fig. 9.1). The interviewee indicates any symptoms that may exist in these areas. If needed, more detailed body sketches can be used.[7] These questionnaires provide information about the nature of problems and complaints—pains and discomfort, their duration, and their prevalence.

10.2.3 Comfort Versus Annoyance

Comfort or discomfort

Comfort (as related to sitting) has long been defined, simply, conveniently, and misleadingly, as the absence of discomfort. In reality, these two aspects are not opposite extremes on a single judgment scale.[8] Instead, there are two scales, one for the desirable feelings of *comfort* and the other for such unpleasant experiences as being ill at ease, fatiguing, straining, aching, and hurting: all terms that indicate some degrees of discomfort or *annoyance*. These two scales partly overlap but are not parallel (as also discussed in Chapter 9).

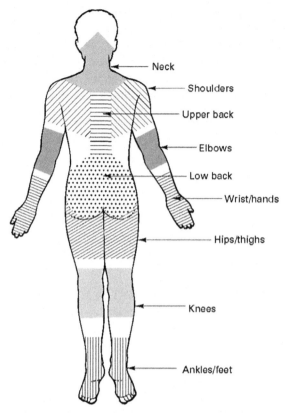

FIGURE 10.1 **Body sketch used in the Nordic Questionnaire.** *Source: Redrawn from Kuorinka et al. (1987).*

Feelings of comfort when sitting are associated with such descriptive words as warm, soft, plush, spacious, supported, safe, pleased, relaxed, and restful. However, and importantly, exactly what feels comfortable depends very much on the individual and his or her habits, on the environment and task at hand, and on the passage of time. Additionally, aesthetics play a role: if we like the appearance, the color, and the ambience, we are inclined to feel comfortable. Appealing upholstery, for example, can strongly contribute to the feeling of comfort especially when it is neither too soft nor too stiff but distributes body pressure along the contact area and if it "breathes" by letting heat and humidity escape as it supports the body.

Comfort scale

To avoid confusion, instead of using the term *discomfort* in this context, we will use the term *annoyance* as the descriptive label for the scale containing the unwelcome statements. Feelings of annoyance are expressed by such words as stiff, strained, cramped, tingling, and numbness or not supported, fatiguing, restless, soreness, hurting, and pain. Some of these attributes can be explained in terms of circulatory, metabolic, or mechanical events in the body; others go beyond such physiological and biomechanical phenomena.

Annoyance

Users can rather easily describe design features that result in feelings of annoyance such as seats in wrong sizes, too high or too low, with hard surfaces or sharp edges; again, we note here that avoiding these mistakes does not, per se, make a seat comfortable.

Ranking

One way to assess relative levels of comfort or annoyance for seating utilizes seven specific statements that are used for each judgement. These are then each assigned a ranking from "not at all" to "extremely." The following statements characterize *comfort*:

1. I feel relaxed.
2. I feel refreshed.
3. The chair feels soft.
4. The chair is spacious.
5. The chair looks nice.
6. I like the chair.
7. I feel comfortable.

The following statements characterize *annoyance*:

1. I have sore muscles.
2. I have heavy legs.
3. I feel uneven pressure.
4. I feel stiff.
5. I feel restless.
6. I feel tired.
7. I feel annoyed.

The researchers found that it is more difficult to rank seats by characteristics of annoyance rather than comfort because the body is surprisingly adaptive, except when the subject has a sore or injured back. In contrast, comfort descriptors proved to be useful and discriminating for ranking seats in terms of preference. The researchers' subjects said it was easy to make an overall statement about the seats in terms of overall comfort or annoyance (responses to #7 in the lists) after having responded to the six more detailed descriptors.[9]

Preference rankings of seats could be established early during the sitting trials; they did not change much over the duration. However, it is not clear whether a few minutes of sitting on seats are sufficient to assess them or whether it takes longer trial periods.

10.2.4 Free-flowing Motion

Slumping

Many of us have asked ourselves why slumping in our chairs should be considered bad for us when it often feels so good. Slumping and other body movements are the instinctive attempts to take strain and tension away from muscles that are working to maintain prolonged postures. Studies have continued to show that individuals in offices often leaned back in their seats, even though the seats were not designed for this.[10] Similarly, individuals who were reading often assumed a kyphotic (forward) lumbar curve even when sitting on a seat that had a lumbar pad that should have produced a lordosis (backward curve).[11] Evidently, our bodies will lead us to sit in any manner that we find comfortable, regardless of how experts think we should sit.

FIGURE 10.2 **Relaxed underwater postures, superimposed on a relaxed posture in weightlessness.** *Source: Adapted with permission from Lehmann (1962) and NASA/Webb (1978).*

In 1962, Lehmann demonstrated the contours of five people "resting" under water where the water fully supports the body. Sixteen years later, NASA astronauts were observed as they relaxed in space. The similarity between the postures under water and in space is remarkable, as shown in Fig. 10.2. It appears that in both cases, the sum of all tissue torques around body joints has been nullified. It was not by accident that the shape of so-called easy chairs is quite similar to the contours shown in both postures.

This suggests that current office seat design should recognize and reinforce free-flowing motions and allow for dynamic rather than static, unmovingly maintained posture. People move around and should be able to take on any posture they wish, as sketched in Fig. 10.3.

The "free-flowing motion" or "floating support" design idea has these basic tenets:

• Allow the user to move freely in—and with—the seat and opt for a variety of sitting postures at will, each of which is supported by the seat, and to get up easily and move around when desired.

FIGURE 10.3 **Free-flowing motion.**

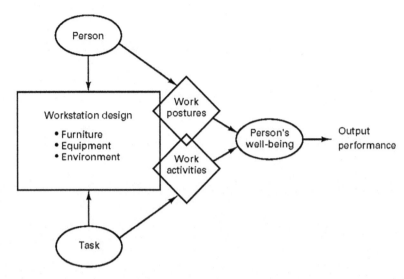

FIGURE 10.4   **The holistic workstation.**

- Make it easy for the user to adjust the seat and other furniture, especially keyboard and display, to the changing motions and postures.
- Design for a variety of user sizes and user preferences.
- Consider that new technologies develop quickly and should be usable at the workstation. For example, radically new keyboards and input devices, including voice recognition, are available and will continue to evolve and improve; display technologies are undergoing rapid changes; wireless transmission no longer limits the placement of display and input devices.

10.2.5 Ergonomic Design of the Workstation

Successful ergonomic design of the workstation in the office is a holistic task, because all the interrelated parts of our workstation environment need to be considered (schematically depicted in Fig. 10.4) as working together to form the station. Work tasks, work movements, and work activities interact with each other. They affect and are influenced by the workstation components, furniture, and other equipment and by the environment. All of these in turn must "fit" the individual to support his or her well-being and contribute to work output. Of course, job content and demands, control over one's job, and many other social and organizational factors also influence feelings, attitudes, and performance (see Chapter 11).

10.3 DESIGNING FOR VISION, MANIPULATION, AND BODY SUPPORT

When designing the layout of a work task and workstation, it helps to consider three main links between a person and the task:

1. The first link is the *visual interface*: We look at the keyboard, the computer screen, or source documents. We cover considerations concerning visual interface below.
2. The second link is *manipulation*: Our hands operate keys, a mouse, or other utensils; they manipulate pen, paper, telephone, and drawing tools. Occasionally, feet operate controls; starting and stopping a dictation machine is an example. The types and intensities of visual and motor requirements can differ substantially among jobs. We cover considerations of manipulation in later in this chapter and in Chapter 12.
3. The third link is *body support*: the seat pan supports our body at the under-sides of buttocks and thighs, and the backrest supports our back. Armrests or a wristrest may serve as additional support links. Body support considerations are covered in later in this chapter.

10.3.1 Designing for Visual Interfaces

The location of the visual targets greatly affects the body position of the computer operator. Objects that require our eyesight should be located directly in front of us, at a convenient distance and height from the eyes. If we are forced to tilt or turn our heads up, or sideways, to view our computer screen or document, we are likely to experience neck, shoulder, and back pain, possibly accompanied by eyestrain.

As a rule, the screen or source document should be about half a meter from the eyes, the proper viewing distance for the operator. A convenient measure is to place the screen and source document at arm's length, or slightly less than arm's length.

Typewriters have fewer keys (less than 50) than current keyboards (more than 100 keys), so most contemporary computer keyboards have become large, nearly horizontal visual target areas. Even those who practice touch-typing (i.e., don't need to look at the keys to type text) may need to revert to scanning keys to complete keyboarding tasks. Thus, many operators have their keyboard as their first visual target. The second visual target is the display area of the computer monitor: the viewscreen is correctly placed in front of the operator at about right angles to the line of sight. The third visual target is often some sort of source document, which may in some cases be large such as a drawing used in computer-aided design.

Visual targets

The problem of placing all these visual targets is mostly one of available space within the center of the person's field of view. Most people prefer to look down to close visual targets at angles between 20 and 60 degrees below the Ear–Eye plane. (This is up to 45 degrees below the horizon when one sits upright, as shown in Figure 1.3 in Chapter 1.) The natural way of focusing on a close-by target with the smallest effort is to incline the head slightly forward and to rotate the eyeballs downward (not holding trunk and neck erect and looking straight ahead at the screen). Mistakenly putting the monitor up high behind the keyboard is uncomfortable for most viewers. Instead, the display should be placed immediately behind the keyboard, with the lower edge of the screen as near as possible to the rear section of the keyboard. The source document should be placed next to it, on either the left or right side depending on user preference.

If the work-tasks require frequent reading from a source document, a document holder can be useful to hold the document close and parallel to the monitor screen,

Document holder

Corrective
eyewear

about perpendicular to the line of sight. A document placed far to one side causes a twisted body posture and lateral eye, head, and neck movements.

Artificial lenses (eyeglasses or contact lenses) are useful for correcting visual deficiencies, but present an additional problem for display users. Generally, lenses are ground for a viewing distance of about 40 cm and a downward tilt of the line of sight; however, many visual targets in the computer area are placed further away (including the screen which is often well behind and above the keyboard). Users may compensate by squinting the eyes while trying to focus or by moving the head forward to bring it closer to the correct focusing distance. Further, an almost unavoidable consequence of middle age is far-sightedness (*presbyopia*), when natural lenses regularly lose their ability to focus on close visual targets and make it difficult to discern characters on the computer screen or in a source document. The use of reading glasses (such as bifocals or trifocals) makes it easier to read a document placed on the work surface. However, when the worker is reading text on the computer screen, bifocals require the worker to throw his/ her head back and tilt the neck at a severe angle to decipher the text (as depicted in Fig. 10.5). Not surprisingly, such postures, including kyphosis of the neck, can cause muscle tension and head- and neck-aches.

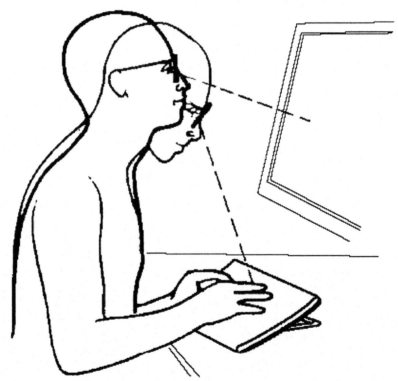

FIGURE 10.5   **Excessive head and neck postures when looking through "reading glasses" at monitor or keyboard.**

10.4.1.1 *Proper Office Lighting*

For the designer, the most important design factors when it comes to office lighting are the distribution of illumination, luminance, and luminous contrast. (See also Chapter 5.)

Office lighting

- *Illumination* is the amount of lighting falling on a surface. The light may come from the sun or from luminaires (lamps).
- *Luminance* is the amount of light reflected or emitted from a surface. Light may be reflected from a ceiling, wall, or tabletop, or it may be emitted from a video display terminal (VDT) screen surface.
- *Luminous contrast ratio* describes the difference between the luminance values of two adjacent areas, assuming a defined boundary between them.

Of the three, *illumination* is the best known (and the least useful) phenomenon. The sun, the sky, or a lamp send visible energy (light) into the surroundings. However, what is emitted is of little direct consequence, because what "meets the eyes" is luminance (unless one stares directly at the light source): examples include looking at a piece of paper, seeing the surroundings about us, or viewing a computer screen.

The quantity that determines human vision is the luminance experienced on the retina. This results from the light energy sent (emitted or reflected) from the visual target and its visual surroundings in our field of view. In rooms, most light energy reaching the eyes is reflected from surfaces, mostly walls and ceilings. Excessive luminance is called *glare*.

Unless we look directly into a light source (sun or lamp), we experience glare indirectly as light reflected from a shiny object. Polished surfaces reflect incoming light at the same angle at which it was received. Mirrors have a background coating that absorbs little light but reflects nearly all of the incoming energy. The *reflectance* value (generally in percent) indicates the portion of incident light that is reflected by a surface. Specular ("directed") glare can be minimized by making surfaces matte with a slightly roughed finish to reflect incoming light in various directions.

Glare

The larger the luminous contrast between two (reflecting or emitting) areas, and the better defined their common boundary, the easier it is for the eye to distinguish them.

Contrast

The distribution of light within a room depends on the location of light sources and the direction of light flow from them, and on the reflectances of ceiling, walls, and other surfaces. As a rule, the reflectances and colors of the room surfaces should be chosen so that there is a continuous decrease in reflectance from ceiling to the floor. This translates into white paint on the ceiling (reflectance about 80%–90%), bright colors like beige, yellow or green on the wall (reflectance at 40%–60%), floors at medium or darker colors like blue-green or brown-beige (reflectance at 20%–40%).

Room surfaces

The eye "sees" what arrives at the retina—the incoming light's energies and wavelengths are distributed over the area of stimulated rods and cones. Ideally, the arriving image solely represents the electronically generated display, but of course, it can contain the reflections of ambient light sources that are mirrored at the surface of the display. If this glare is strong enough, it diminishes the apparent contrast, or color, and hence washes out the original image so that it can be difficult, if not impossible, to be discerned.

Reflecting light

When a light source is reflected from a surface into the eyes, this is called *indirect* (or *reflected*) *glare*. It can occur when the headlights of a trailing car are reflected in the

Indirect glare

FIGURE 10.6 Indirect glare on a display from lamp and sun.

rear-view mirror, striking the driver's eyes and reducing the ability to see the road ahead. In the office, indirect glare from a window, lamp, or even a white shirt can impair the view of text or other images in the computer display. Indirect glare from a lamp and the sun is shown in Fig. 10.6. An imperfect way to reduce glare is to apply some sort of treatment to the front of the reflecting surface of your monitor. A filter can absorb some energy of the incoming light, often in particular wavelengths (colors), but it also absorbs energy from the emitted image, which—unfortunately—reduces the luminance contrast that arrives at the eye. A related approach is to use micromesh or micro-louvers that limit the directions from which stray light may fall onto the monitor surface. The disadvantage of the latter technique is that the observer must position his or her eyes in exactly the proportionate position, forcing head and upper body into a proscribed and immutable posture. Alternately, coatings or rough texture on the surface can be used to alter the reflections of incoming light. The basic problem with all of these measures is that they treat the symptoms (the reflections) instead of eliminating the source of the disturbance in the first place.

Indirect glare from a computer screen can be completely resolved if no sources of high luminance are visible to the human operator. This includes:

- not placing any bright objects within the operator's field of view (within his or her cone of vision), such as redirecting or turning off a lamp, not having a white or reflective surface near the operator (see Fig. 10.7);

FIGURE 10.7   **Indirect glare from a shiny surface.**

- putting a shield between the luminaire and the monitor;
- positioning or angling the screen so that any existing bright objects are outside the operator's cone of vision.

When a light source shines into the eyes, this is called *direct glare*. It can occur when the headlights of an approaching car "blind" the driver for a moment, as the light energy shining directly on the retina overpowers the subtle image of the road that was visible at low-level luminance. In the office, direct glare from a task light pointed at the user or light flooding in from a window (as in Fig. 10.8) can optically overpower the image presented on the monitor. Direct glare is controlled by measures similar to those used to eliminate indirect glare: reducing the intensity of the incoming light (turning off the lamp, drawing a curtain across the window); using shields; repositioning the operator, the monitor, and/or the entire workstation by about 90 degrees.

Direct glare

The best ways to provide glare-free lighting from windows and luminaires are to position them strategically and use indirect rather than direct lighting when appropriate. Locating light sources overhead or to the left and right sides of the operator reduces indirect glare, as long as there are no reflecting (shiny) surfaces at the workstation.

10.3.2 Designing for Motor Interfaces

With current technology, most of the data we need to transfer to our computer is done by hand. The common interface is the conventional flat keyboard, often accompanied by

FIGURE 10.8 **Direct glare through a window.**

other input means such as mouse, trackball, joystick, or light pen. The design of these interfaces affects the workload and the motor activities of the operator and dictates the layout of the computer workstation.

Keyboard

The keyboard that is still most typically used in computing today is the one derived from a long-ago typewriter. We have a number of reservations with the conventional QWERTY keyboard because of its ergonomically contraindicated features, as discussed extensively in Chapter 12. A consequence of the widespread use of this keyboard is overuse disorders, which are common in keyboard operators. Causal or contributing factors are the frequency of key operation combined with awkward forearm and wrist postures—see Chapter 9.

Keyboard modifications

Many proposed improvements to the QWERTY keyboard have been suggested, and they can be grouped into several broad categories. Some relocate letters on the keyboard and change the geometries of the keyboard, such as arranging keys in curved rows and columns. Others divide the keyboard into halves, one for each hand, arranged so that the center sections are higher than the outsides, so hands do not need to pronate like they do on the flat keyboard. Alternately, the keyboard may be fully separated into two sections, one for each hand, arranged like an accordian where the user types with the hands facing each other. Chord keyboards involve activating two or more keys simultaneously (called chording) to generate one character or whole words, or chunks of words, such as those used by stenographers, or keys that have three or more different contact positions. For more detail on keyboard design (past and present), see Chapter 12.

We know that people vary in size, shape, ability, and preference, so there is no one keyboard that is best for every person. The conventional large QWERTY keyboard is likely to remain in common office use because it is mass produced at low cost and it is entrenched. People are familiar with it; however, it is not the best design for all individuals and all situations. Varied ergonomic designs have been developed to address specific needs under specific work conditions, and some are viable alternatives for QWERTY, including voice recognition devices. With new technologies, sensors, and activation devices, we expect that there will be many additional effective ways to transfer input signals to computers other than keyboards.

The design of the keyboard and any other manipulated input device, such as a mouse or puck, determines hand and arm posture. The essentially horizontal surface of a conventional keyboard and mouse requires that the palms also be kept approximately horizontal. This uncomfortable posture requires strong pronation in the forearm, close to the anatomically possible extreme. Hands and wrists are not the only body parts affected by the designs of keys and other input devices. The location of hand-operated input devices of any kind naturally determines the position of the hand, forearm, and upper arm and, consequently, the posture of the trunk. As a rule, the best position for the hand is in front of the body, at elbow height. If mouse, puck, stick, or other input instruments are used jointly with a keyboard, they should all be placed closely together; in fact, many keyboard housings already contain a trackball or trackpad.

Posture

Supporting arms and hands appropriately relieves shoulder and back muscles from their holding tasks. Various kinds of arm, wrist, and hand rests can be employed during work, or at least during breaks in extended work. Such rest periods are helpful but by themselves are not sufficient to overcome ill-fitting equipment and excessive workloads.

Supports

In the United States, ANSI/HFS standards and HFES provide guidelines for specific design features such as key spacing, key size, key actuation force, and key displacement. They also provide useful design dimensions for mouse and puck input devices, trackballs, joysticks, styli and light pens, tablets and overlays, and touch sensitive panels.

10.3.3 Designing the Sit-down Workstation

One of the first steps in designing the office workstation for seated work is to establish the main clearance and external dimensions. The size of the furniture should essentially derive from the users' body dimensions and from their work tasks.

The main anthropometric data that determine the design, especially the height, of the furniture that fits sitting users are lower leg (popliteal and knee) heights, thigh thickness, and the heights of elbow, shoulder, and eye.

Fig. 10.9 (same as the seated body in Fig. 1.7 and numbered as in Table 1.3, Chapter 1) depicts body measurements of particular importance for furniture design: #8 height sitting, #9 eye height, #10 shoulder (acromion) height, #11 elbow height, #12 thigh thickness (clearance) height—all measured above the horizontal flat seat pan. Measured

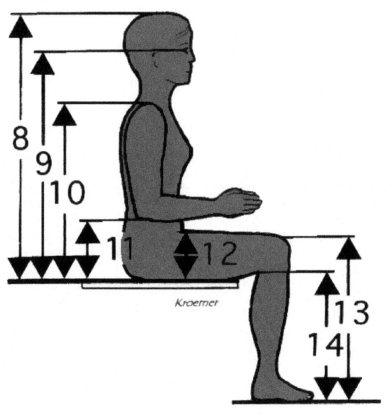

FIGURE 10.9   Body measures commonly taken on sitting individuals.

from the floor up are #13 knee height and #14 popliteal height. Not shown in this illustration is #28 hip breadth (sitting). The numbering follows standard anthropometric practice.[12]

Desks

Conventional offices, often inside cubicles, generally have fixed-height desks and tables set relatively high so that people with comparatively long legs will fit comfortably underneath. Users have to adjust their postures and seats to those desk and table elevations, which can become awkward, especially if trays or drawers in or on the desk reduce the height of available legroom. Newer offices may be equipped with work surfaces, tables, and desks that can be adjusted in height to fit individuals' bodies and work habits.

No desk?

Presently, desk-free work is growing ever more popular. Laptop computers, tablets, virtual keyboards, using voice recognition and handwriting for computer inputs, and software developments together with "wearable" hardware can make tables and desks unnecessary. Much of our office work can be done on our laps, perhaps with the occasional use of a laptop cart. Industrial designer Niels Diffrient (1928–2013) anticipated that trend in his "executive workstation"—see Fig. 10.10.

FIGURE 10.10   Niels Diffrient's proposed "executive" workstation with a reclining body posture.

There are at least four principal master plans, each with a number of variations:

Furniture heights

1. *Fixed seat height*: The classic design plan, used less frequently of late, starts with a fixed seat height, like we might find in an old-fashioned classroom or in public waiting areas. From there the designer derives the heights of the work surface and, if used, of footrests.
2. *Fixed work surface height*: This traditional furniture design strategy assumes a fixed height of the major work surface, usually the top of table or desk in conventional offices. Next, the designer selects the adjustment range of seat heights that fit the elevation of the work area and from there determines whether footrests are needed for individuals with shorter legs.
3. *Fully adjustable furniture*: This modern design procedure assumes that all the furnishings are adjustable in height. The design starts at floor level and from there determines the height adjustments of the seats above the floor. The elevations of seat pans serve to establish the heights of table and of desk surfaces, including the height of supports for keyboard, mouse pad, and other work utensils.
4. *Personal selection*: An individual user chooses and combines particular pieces of furniture to create a custom workplace that suits personal tasks and preferences. For example, if the user does "lap work," conventional tables or desks may not be present, possibly replaced with a wheeled tray or laptop cart and mobile supply and storage units. Often a reclining upholstered "easy-chair," possibly with a foot stool, is used instead of a mainstream office chair.

The clearance depths and widths of the furniture must fit critical horizontal body dimensions of users, especially popliteal and knee depths and hip breadths. The outside

Furniture depth and width

depths and widths of the furniture are largely determined by what amount of space the tasks will require and what the users' reach capabilities are.

The common computer workstation has a seat, a support surface for keyboard and display, tablet, or mouse and (mostly for papers) an additional working surface. It is best, and generally most expensive, to have all of these independently adjustable. Proper workstation dimensions make the office furniture fit almost everybody, tall or short. Suitable approximate height adjustment ranges for office furniture in Europe, North America, and other areas with users of similar body sizes are as follows:

- Seat pan above the floor: 37–51 cm, up to 58 cm
- Support surface for keyboard, mouse: 53–70 cm
- Surface of work surfaces, tables, and desks: 53–72 cm
- Support for the display: 53–90+ cm

Work surface

For work surfaces (and generally the location for the keyboard), there are two main reference points for ergonomic design: the elbow height of the person and the location of the eyes. Both depend on how the workstation user sits or stands, upright or slouched, and how he or she alternates among postures and positions. The table or other work surface of a sit-down workplace should be adjustable in height between about 53 and 70 cm, perhaps even a bit higher for very tall individuals, to permit proper hand/arm and eye locations. Often, a keyboard or other input device is placed on the work surface or connected to it by a tray. A keyboard tray can be useful, especially if the table is a bit high for the user, but it also reduces the clearance height for knees and upper thighs. The tray should be large enough for keyboard and trackball or mouse pad, unless these are built into the keyboard, but its front edge should not cut into palm or forearm.

Monitor support

Using a separate support for the computer monitor is best; this way, the display height can be adjusted independently from the table or work surface. The position of the screen greatly affects the body position of the computer operator. As a general rule, the screen should be about 50 cm from the eyes, low, and behind the keyboard so that the user looks down at it. Fortunately for users of current laptop computers, this position naturally occurs.

> Keep in mind that specific design recommendations in this book may not apply to user populations with widely different body sizes and work practices, who might also prefer different working postures. They would need their own ranges in furniture dimensions and adjustments.

10.4.3.1 Office Chair

Chair

We have already mentioned in this chapter that the long-held ideal for "proper sitting" at work—upright trunk, with thighs (and forearms) essentially horizontal and lower legs (and upper arms) vertical, all major body joints at zero, 90, or 180 degrees—makes for a convenient but misguiding design template. Such a stance is neither commonly adopted nor subjectively preferred and it is not even especially "healthy." Instead, when designing or selecting an office chair, we must consider a full range of motions and postures.

When you sit down on a hard, flat surface, not leaning against a backrest, you feel the seat-bones, the *ischial tuberosities* (the inferior protuberances of the pelvic bones) pressing on the seat surface. The gluteus maximus muscle covers those bones when standing, but they are exposed in the seated position. Our upper body's weight is transmitted predominantly by the ischial tuberosities to the seat of the chair, which is felt especially keenly when the seat pan is hard and unforgiving. The bones act as fulcra around which the pelvic bones rotate under the weight of the upper body. The bones of the pelvic girdle are tightly linked to the lower spine by connective tissue. Rotation of the pelvis, therefore, affects the posture of the lower spinal column, particularly in the lumbar region. If the pelvis rotation is rearward, the normal lumbar spine lordosis is flattened (see Fig. 10.11).

Seat pan

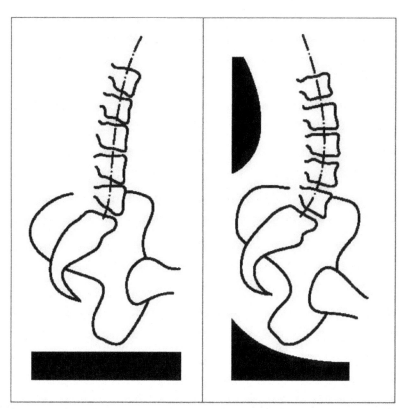

FIGURE 10.11   Positions of the pelvic and lumbar spine on a flat and on an inclined seat surface, showing flattening of the lumbar lordosis. *Source: Adapted from Kroemer and Robinetter (1968).*

SITTING COMFORT AND PRESSURE

Pressure and seat pan size affect your sitting comfort in a seated position: when you sit on a hard surface, the underside of your sitting bones transmits all the upper body weight to the seat pan.

Let's assume an upper body *mass* of 50 kg (about 110 lb) and therefore a weight (force) of nearly 500 N (50 kg × 9.81 m/s^2), and a contacting seat pan area of 50 cm^2 (about 20 in^2). *Pressure* is defined as *force per area*: so 500 N/50 cm^2 = 10 N/cm^2. This is equivalent to 100 kPa (kilopascal) (about 15 psi) which is then the (averaged) pressure under your ischial tuberosities.

If you use a larger seat pan, you reduce the pressure: increasing the area by the factor 5 reduces the pressure to 1/5th. Consequently, if you temper the seat pan surface with a cushion, upholstery, or elastic mesh, your seated pose just became more comfortable pressure-wise.

Strong leg muscles (including hamstrings, quadriceps, sartorius, tensor fasciae latae, psoas major, adductors, gluteus) run from the area of the pelvis across the hip and knee joints to the lower legs. Therefore, the angles at hip and knee affect the location of the pelvis and hence the curvature of the lumbar spine. With a wide-open hip angle, a forward rotation of the pelvis on the ischial tuberosities is likely, accompanied by lumbar lordosis. These actions on the lumbar spine take place even as associated muscles are relaxed; muscle activities or changes in trunk tilt can counter the effects.

Seat pan tilt

A seat pan surface should be able to be tilted throughout the full range from declined forward, kept flat, to inclined backward, as specified by ANSI/HFES 100 (2007) in the United States. This allows the user to assume various postures with differing curvatures of the lower spinal column, from kyphosis (forward bend) to lordosis (backward bend). The surface of the seat pan must support the weight of the upper body comfortably and securely. Hard surfaces generate pressure points that can be avoided or lessened by cushioning upholstery, pillows, mesh, or other materials that elastically or plastically adjust to body contours.

Seat pan depth

The only inherent limitation to the size of the seat pan is that it should be sufficiently short that the front edge does not press into the legs' sensitive popliteal tissues, which are located behind the knees. Usually, the seat pan is between 38 and 42 cm-deep and at least 45 cm-wide. A well-rounded front edge (called a "waterfall") is mandatory. The side and rear borders of the seat pan may be slightly higher than its central part.

Seat pan height

The height of the seat pan must be widely adjustable, preferably down to about 37 cm and up to 58 cm to accommodate individuals with either short or long lower legs. It is very important that the person, while seated on the chair, can easily do all adjustments, especially in height and tilt angle. If making adjustments is cumbersome or difficult, they will probably not be made. Fig. 10.12 illustrates major dimensions of seat pan and backrest; Fig. 10.13 shows the preferred type of chair backrest.

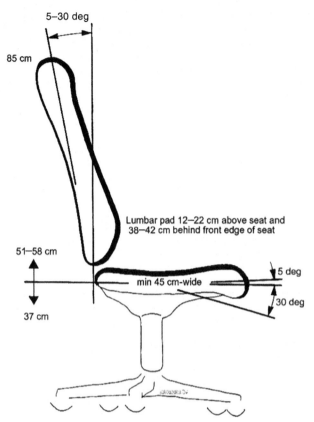

FIGURE 10.12 Major dimensions of seat pan and backrest.

The question regarding backrests is whether one is even needed. Backrest-free proponents embrace the idea of "active sitting," where trunk and leg muscles must remain continually active to keep the upper body in balance—for example, some workstation users like to sit on a flexible ball. However, most want to use a chair with a backrest. Leaning against a backrest allows it to carry some of the weight of the upper body and hence reduces the load that the spinal column must otherwise transmit to the seat pan. In addition, its lumbar pad, protruding slightly in the lumbar area, helps to maintain lumbar lordosis. Finally, most of us find that leaning against a properly formed backrest is relaxing.

Backrest design considerations

The backrest should be as large as can be accommodated at the workspace: this means up to 85 cm-high and up 30 cm-wide. To provide support from the head and neck on down to the lumbar region, it should be shaped to follow the contours of our back, specifically in the lumbar and the neck regions. Many users appreciate an adjustable pad or an inflatable cushion for supporting the lumbar lordosis and a similar but smaller pad at the cervical bend. The lumbar pad should be adjustable from about 15 to 23 cm and the cervical pad from 50 to 70 cm above the seat surface.

Backrest size

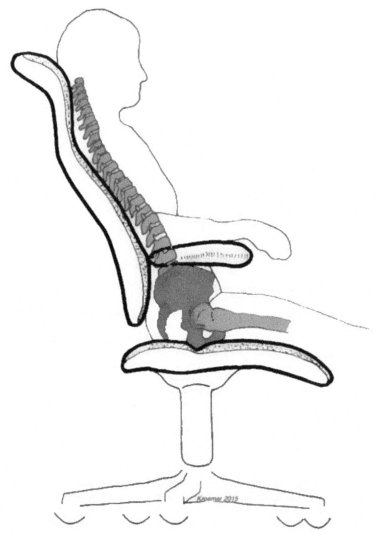

FIGURE 10.13   **Preferred backrest shape.**

Backrest angle

The angle of the backrest must be easily adjustable while seated. It should range from slightly behind upright (95 degree from horizontal) to about 30 degree behind vertical (120 degree), with further declination for rest and relaxation desirable. Whether or not the seat back angle should be mechanically linked to the seat pan angle appears to be a matter of personal preference.

Armrest

Armrests can provide support for hands, arms, and even portions of the upper trunk; they can help us sit down and get up. Even when only used for short periods of time, armrests can be valuable if they have an appropriate load-bearing surface, ideally padded. They should be adjustable in height, width, and possibly direction. Conversely, sometimes armrests prevent or hinder movement of the arm, pulling the seat toward the workstation, or getting in and out of the seat. In these cases, either having only short armrests, ones that can be flipped to the side, or no armrests is appropriate.

Footstools, hassocks, and ottomans have long been popular to prop up feet, but footrests in the office usually indicate deficient workplace design—for example, when a seat pan is too high for the individual, perhaps because the seat must be set excessively elevated at a high work surface, then the person uses a footrest. If a footrest is used, it should be sufficiently high that the sitting person's thighs are nearly horizontal when the foot is supported. A footrest should not consist of a single bar or other small surface because this restricts the ability to change the posture of the legs. Instead, the footrest should provide a support surface that is about as large as the total legroom available in the normal work position.

Footrest

10.3.4 Designing the Stand-up Workstation

Standing up during computer work harkens back to the 1900s office. Moving around and standing can be a welcome change from sitting, provided that the person does it as his or her own choice. We might choose to stand for reading, writing, talking with somebody face to face, or telephoning. Stand-up workstations can feature a dedicated standing-height work surface to hold a handheld or laptop computer. Other stand-up configurations are adjustable from a sitting height to a standing one, smoothly elevating the work surface so the computer or input device simply shift upward as the person stands up and vice versa when it is time to sit back down. For those who prefer moving rather than standing or sitting, some workstations are designed to allow the worker to pedal on a stationary bicycle, climb a stair climber, or walk a treadmill.

Stand-up workstations should be adjustable, with the area used for writing or computer inputs at approximately the user's elbow height when standing (more or less a meter above the floor). As at a sit-down workstation, the display should be located close to the other visual targets and directly behind the keyboard. If the work surface is used for reading or writing, it may slope down slightly toward the person. A foot bar at about two-thirds knee height (approximately 0.3 m) allows the person to temporarily prop up a foot, which allows for welcome changes in pelvis rotation and spine curvature.

Non-resilient floors, such as those made of concrete, can be hard on people's feet, legs, and backs. Carpets, elastic floor mats, and soft-soled shoes can reduce strain. Appropriate friction between soles and the walkway surface helps to avoid slips and falls.

Flooring

The screen of the display unit, particularly its optical quality, is of major ergonomic importance. In general, the screen should provide a stable image (no flickering), showing characters of good contrast against the background. Dark characters on light background are often preferred. The displayed characters and symbols should use clean fonts with clear edge definition, of appropriate size, and properly spaced. Luminance level ("brightness") and contrast should be easily adjustable by the user. Using more than two colors should be done cautiously, since some colors pose visual problems, and because a large variety of colors displayed on the same screen may be more irritating than useful.

Display

10.4 THE OFFICE ENVIRONMENT

Offices inherently expose people to environmental conditions that may affect their health, comfort, willingness, and ability to perform. The primary goal of office design is to ensure the best possible conditions for employees, with safety and comfort top of mind. Lighting, sound, and climate are the most important environmental conditions influencing how office inhabitants feel and perform; naturally, workers' safety is also paramount.[13]

The workforce is currently in the process of "greying." Japan has the world's fastest aging population, as a quarter of its citizens are 65 and over. In only one generation, Japan has shifted from a country with a young population to the one with the highest proportion of older people. The retirement age (*teinen*) has shifted from age 50 to 60 to 65, and some are now proposing a retirement age of 75. Similarly, Britain's pensioners now make up a fifth of the population and soon there will be as many people over 40 as there are under 40. In the United States, social security benefits have been postponed to ages beyond the once-customary 65.

Older workers are encouraged to stay in the workforce longer, as longevity increases and as fewer young people are available than in previous generations to enter the workforce. Thus, on an international basis, it is not unusual to have 3 or even 4 generations represented in a given office environment. It becomes increasingly important, therefore, to consider the needs of all age groups and populations in designing the office environment and workstations (see also Chapter 14).

Lighting

Lighting (discussed earlier in this chapter) strongly affects users in the work environments. Designers should set up lighting in the office so that visual work tasks are easy to do. Appropriate workplace lighting allows users to see what they need to see clearly and vividly but without glare or annoying bright spots in the visual field, and provides an appealing environment in terms of contrast and colors. Task lighting, such as a lamp or light source attached to and controllable at the workstation, is also widely adopted: previously, high levels of overhead lighting were popular, now focused spot-lit "islands" are becoming common. Task lighting saves energy, is aesthetically appealing, and gives us control over our own light source, leading to additional individual comfort and satisfaction. Many companies also recognize the value of natural light and provide skylights and windows to add light and a more open, spacious feel to the workplace.

Sound and noise

Sound and its evil twin, noise, pervade our space in a myriad of forms—such as ringing, beeping, and chiming phones; clattering keyboards, churning printers and copiers; chatty colleagues; a coworker's blaring headphones; external commotion from raucous traffic. We are accustomed to a certain background sound level, often generated by the climate control system or other equipment, which helps to mask some disruptive noises. However, lack of sound can also be disruptive; it can actually be too quiet.

Research and personal experiences suggest that noise usually is the most intrusive of the environmental factors affecting people in the office. Accordingly, acoustics play a prominent role when planning office spaces. Ambient noise can be reduced using items such as sound-absorbent screens, carpets, draperies, and fabric-covered outer shells of filing cabinets or other furnishings in the offices.

Air ventilation, temperature, humidity, and, relatedly, smells and other contaminants also influence comfort and, in their extremes, health at work. Consequently, climate has an impact on our productivity, performance, and willingness to spend long hours in the office. At work, the (physical) climate should be neither too hot nor too cold; neither too damp nor too dry. Fresh air is preferred over stale air, but the airflow should not generate a strong draft. The latest office spaces offer unprecedented levels of employee comfort; some even offer individual climate control at each workstation.

Climate

Employee morale is another function of office space design: to attract and retain good employees and to increase their job satisfaction. Indeed, current initiatives in office design are centered on retaining employees and encouraging them to expand their office hours beyond the usual working hours. With people spending long hours at work, one of the ways to keep them productive is to make them comfortable and happy with their office space. Job satisfaction, after all, affects productivity and performance, and work conditions have been shown to influence the level of an employee's satisfaction on the job (see also Chapter 11).

Morale

Expressing the corporate character is also a function of office design. The design of an office, and how it is run, express the personality of an organization. In a sense, the office serves as a company's "visible motif." A workspace should reflect what the organization cares about, all while facilitating the working styles and preferences of the team.

Character

An office designed to take into account employees' opinions and needs increases creativity, productivity, collaboration, engagement, health, and happiness; conversely, when an office fails to reflect a company's values or personality, that can have a negative impact on employees.[14] Put another way, we shape offices and office designs, but they shape us, too. This is illustrated in an extreme example: imagine a prison, where there is a complete absence of control for the inmate, a total lack of stimuli, and an environment of misery and darkness. Then, consider how that affects mental health, physical well-being, and even the inmate's personality. Clearly, the aesthetics of the immediate work environment impact workers deeply.

The range of office designs is huge; offices can vary from large to small; from "open plans" to single-office spaces for one person. The appropriateness of an office layout depends on the type of work performed, including careful evaluation of the degree of interaction required and the amount of privacy needed. Advantages of open spaces include that they are generally less expensive and more flexible than separate closed offices in terms of building cost and utility usage for lighting and climate control. Open-plan spaces often facilitate communication and collaboration among workers. Finally, open spaces reduce the culture of hierarchy that individual offices may encourage: without the corner office suites often found in more traditional office designs, there is less of a division between the executives and the staff. Disadvantages of open designs include disruptive noise—such as speech and equipment sounds—which may reduce performance and job satisfaction. There is some evidence that noise generated in open

Office design

offices distracts a third of the workforce and also is associated with higher amounts of sick leave taken by employees.[15]

As individuals, we each have our own work preferences. Some are better able to concentrate on the task at hand even with disruptions, while others can concentrate only when they have complete privacy. We may prefer to share a large table with others, where nobody can "hide" behind a closed door; or we may be more comfortable with the security and privacy offered by office walls. Many offices, of course, contain a mixture of office designs—some wide-open spaces, then other sections with shoulder-high cubicle dividers, and maybe closed-off individual offices along the perimeter of the room. In many businesses, tasks that require concentration are interspersed with those that require interaction. Flexible office design can accommodate preferences in how workers do their tasks during the typical workday.

10.4.1 The Home Office

Nearly everything outlined earlier is relevant when discussing a home office: ergonomic principals apply to the home office just as they apply to a traditional office. If one does "office work" only for short periods of time in the home, then any table and chair may suffice. But if the worker is occupied with any task for hours at a time, the working conditions should be addressed more carefully: equipping the home office with an appropriately furnished workstation that fits the tasks and fits the worker.

The worker should feel free to select an easy-chair, a comfortable office chair, a semi-sit perch, or a kneeling chair—whatever feels good to the worker and what supports his/her body well over long periods of time. Furniture should facilitate the worker's choice to stand upright or move about. Items should be comfortable and suitable for the individual, and not just be labeled "ergonomic."

This same guidance applies to keyboards: the selection should be comfortable for the user, and should include voice inputs or similar if preferred. Alternate input devices should also be available to provide relief from overuse of a single keyboard design, such as a docking station with a detached keyboard for home use as an adjunct to portable laptop computers or handheld devices.

10.4.2 On-the-go Offices

Working away from a formal office is increasingly common, and with the rise of mobile devices it is becoming dramatically easier—as seen by the popularity of cafés with patrons poring over their laptops or phones as prodigiously as the baristas are pouring coffee drinks behind the counter. A café (sometimes called "coffice") can be a convenient workspace: offering wireless connectivity, climate control, and refreshments, making meeting with other colleagues easy as well. However, cafés, bistros, lobbies, waiting areas, airport departure lounges, and hotel rooms are not usually designed for workplace ergonomics even as we attempt to use them as our "instant" workstation. With a limited ability to customize an on-the-go office, the user will need to adapt to the environment.

Depending on what is available, select a chair and work surface (table) that is comfortable for you. A café that provides high tables with tall stools makes it easy to alternate between standing and sitting. When standing, one can ease the lumbar spine by using the rung of the stool as a foot prop. This configuration can be transformed into a semi-sitting position by half-leaning, half-sitting on the stool. As at any workstation, the user should ensure the devices used are positioned accordingly.

Sitting

Consider using devices and add-ons that make the laptop or input device more mobile-friendly. Some options include portable laptop computer stands that raise the screen to a higher eye level, portable mouse track pads or pens, touch-screens, and wireless keyboards. A detached keyboard is especially useful to relieve wrist or hand pain because of the large variety of keyboard designs, including split keyboard designs.

Accessorize

Increasing use of smartphones (really a hand-held minicomputer) has led to a renewed awareness of high-tech-induced repetitive motion injuries. Ergonomic interventions apply here as well: use the table surface to support the hands, wrists, and/or elbows; hold the phone near eye level without straining the neck; use digits from both hands rather than just the dominant thumb to type. Add-on devices (such as headsets) can encourage hands-free usage and voice dictation can reduce typing. (This is not always feasible in a busy location without irritating fellow customers or compromising privacy.)

Smartphone smarts

10.4.3 Co-working Spaces

Co-working spaces, also called shared offices, are a viable alternative or supplement to corporate and home offices. The quality of shared spaces varies widely, but there are many options. Most co-working spaces offer benefits beyond the workstation: conference room space, scanners and printers, and, in some cases, other unusual amenities like refreshments, climbing walls, or nap pods. They also often provide opportunities for networking, mentoring, collaborating, and learning. They can reduce feelings of isolation resulting from working from home, and the distraction and unpredictability of on-the-go offices. Shared workstations can also offer the community and collaboration that many people seek in their professions.

Many co-working organizations target their appeal to certain business sectors or even niche groups; for example, they may cater to information technology start-ups, or to writers and artists, or to professionals in the legal industry or international business. To that end, many offer special services geared toward the needs of that particular group, including translation services, photo studios, design labs, incubators, and accelerators.

There is some irony here: for many decades, the collective office-based workforces have decried their plights—the annoying commutes, the pointless office politics, the drudgery of shared spaces, and the irritations of forced togetherness—and yet now, many seek out and even pay for the privilege of sharing workspaces.

10.4.4 Workspace of the Future

Time will tell what tomorrow's office will look like, but it will be shaped in large part by the prevailing advances in technology. As more employees work remotely, workspace options will continue to evolve. If there are enough "road warriors" and digital nomads, then suppliers of potential "landing" places will want to figure out what they are looking for and what they are willing to pay extra for. We expect that consumer spaces for on-the-go office work will adapt and offer more ergonomic workstations, whatever the workstations will be.

As for furniture, ergonomists have emphasized the need for easy adjustments of chairs, of support surfaces for the keyboard, and of the display in height, distance, and angle. At the same time, many formal studies supported our personal observations that current adjustment features are seldom used, possibly because existing means of adjustment are still too complicated. Adjustments should be intuitive and not require any conscious interaction of the individual. For example, an ideal seat might simply follow all movements of the sitting person, interpreting the body, and supporting it throughout. We are hopeful that manufacturers will soon incorporate technological progress so that furniture will automatically adjust itself to our proportions as we shift and move.

10.4.5 Fitting It All Together

An uncomfortable office is not instantly converted into a fantastic one by changing out a computer or installing a better chair. Certainly, improving any one component is helpful, but all of the components, equipment, and furniture (as in Fig. 10.14) plus lighting and climate must fit each other, and the person in the office must be willing and able to take advantage of all the offered possibilities. This reflects the holistic approach mentioned at the beginning of this chapter: each component of the office interacts with the other elements. A more ergonomic office encourages movement, gets people to leave their chairs to communicate and collaborate, and creates an environment where people are stimulated, where they thrive, where they want to succeed and remain. The workplace has a profound effect on individuals' well-being, and the office is the manifestation or representation of the company itself. A company's commitment to its employees through ergonomic office workspaces results in less illness, better performance, higher retention, and, at the end, better profitability.

10.5 CHAPTER SUMMARY

Neither theories nor practical experiences endorse the idea of one single proper, healthy, and comfortable sitting, standing, or working position. Instead, many motions and postures are subjectively comfortable for one particular person, at that particular time, depending on body, preferences, and work activities.

The traditional postulate that everybody should sit upright is outdated, and furniture should be designed for free-flowing motion. Changing from one posture to another one,

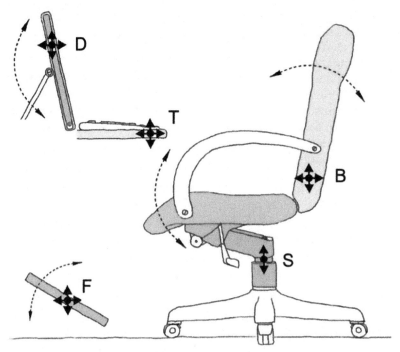

FIGURE 10.14 All components working together: adjustments of the components of a computer workstation.

moving freely among all the comfortable poses, is ideal. Motion, change, variation, and adjustment at an individual's will—this is central to well-being.

Consequently, work spaces should allow for body movements among various postures. For this to happen, furniture should adjust automatically or at minimum must be easily adjustable in all of its primary dimensions, especially seat height, seat pan angle, and backrest position. The entire computer workstation should facilitate easy variations as well, especially location and height of the input devices and height and distance of the display.

These tenets apply not just to a traditional office, but to home offices, on-the-go workspaces, and the workspaces of the future as well.

10.6 CHALLENGES

What is so bad about standing, or walking around, while doing office work?

What are some reasons for recommending an "upright" trunk while sitting?

What are some appealing factors, and unappealing conditions, associated with office work?

Which are the advantages of having uniform heights for (a) seats and (b) work surfaces?

Which control functions, and data input functions, could conceivably be shifted from the fingers to other body parts?

What prevents us from having the computer "talking back" at the operator, instead of simply displaying information on the screen?

Why is changing one's posture desirable, as opposed to maintaining a comfortable posture for a long period?

Under what circumstances may the following links between the operator and the computer be the main determiners of posture: (a) vision, (b) keyboarding, and (c) other finger-control interactions?

Which are the main arguments for (or against) individual offices versus multi-person offices?

Should one strive to have windows in the office, in spite of the difficulties of controlling illumination?

Should eye examinations, and provision of suitable corrective lenses, be of more concern to the individual or to the employer?

Do participatory ergonomics work better than imposed interventions?

Free-flowing motion design ideas are useful and even necessary in the design of comfortable chairs and seats. In what other settings are these design concepts helpful?

Noise pervades our space in many often-distracting ways. However, many of us are accustomed to having music or broadcasts piped into our ears through headphones. What is the impact of listening to music at work?

What technology do you think will become prominent in the near future? How will this development shape office design and work space layout?

NOTES

1 From Herman Melville's *Moby-Dick*, published in 1851.

2 See Staffel, 1884 and 1889.

3 See Grandjean (1997) and Kroemer and Kroemer (2001).

4 See Booth-Jones et al. (1998).

5 Based on initial work by Shackel et al. (1969) and Corlett and Bishop (1976).

6 By Kuorinka et al. (1987), modified by Dickinson et al. (1992).

7 See Van der Grinten and Smitt (1992).

8 Helander and Zhang (1997).

9 As reported by Helander and Zhang (1997).

10 Grandjean, Huenting, & Nishiyama (1984).

11 Bendix et al. (1996).

12 See, for example, Gordon et al. (2014) and Kroemer et al. (2010).

13 For more information on ergonomics in the office environment, refer to Kroemer & Kroemer (2016).

14 Heels, L. (2015). Does your workspace reflect your culture? [Blog]. http://chicagocreative-space.com/does-your-workspace-reflect-your-culture/. Accessed March 10, 2016.

15 Harrison, K. (2015), Best practices for setting up your office to maximize employee productivity. Forbes Magazine. http://www.forbes.com/sites/kateharrison/2015/06/25/a-new-infographic-offers-best-practices-for-setting-up-your-office-to-maximize-employee-productivity/. Accessed March 10, 2016.

Ergonomics
http://dx.doi.org/10.1016/B978-0-12-813296-8.00011-6

Overview

This book was written to address what we know about the human body and mind at work, and to then design work tasks, tools, interfaces, and procedures so that the human can perform safely, efficiently, and even with enjoyment. Given these themes, the importance of understanding the individual within an organization becomes clear. Therefore, we devote this chapter to organizational behavior, the study of the individual inside the organization.

The office is an example of a work system that depends entirely on humans: without them, no work is accomplished. Consequently, ergonomics focuses on the human as the most important component of the workplace and adapts the office to the people involved. This human-centered design requires knowledge of the characteristics of the people in the office—their physical dimensions, capabilities and limitations, and preferences. This chapter focuses on the psychographics of the humans in an office, concentrating on qualitative data pertaining to the individuals rather than quantitative, measurement-based information. Studying organizational behavior helps us understand and deal with the interpersonal and organizational challenges in the workplace.

Understanding an individual in her or his job involves considering not just the person but the environment—the organization—in which the person operates. Understanding an organization involves considering all of its components; organizations are networks of related parts, and all of the elements work together to enable the organization to function as a whole. Macro- and micro-views of the organization start with the elements that define organizations and end with the people that work there. The field of organizational behavior addresses what motivates employees and determines their behaviors, job design and job satisfaction, communications among workers and how to work well with others, job stress, and work–life balance.

11.1 INTRODUCTION

In today's competitive work environment, companies try to keep employees happy, that is, satisfied and motivated. Companies recognize that improving employee satisfaction almost always improves bottom-line profit: happy employees are productive, they treat customers better, work harder, even take fewer sick days than disaffected or disengaged employees. Moreover, they tend to stay employed with the organization, which reduces one of the most significant company expenses—those related to employee turnover. Turnover translates into the cost of recruiting, hiring, and training a new employee; this new hire will likely accomplish less initially (in some studies, more than one-third less in the first 3 months) than an experienced worker. The outright expense of replacing a valuable employee can range from half to several times a year's pay.

11.2 ELEMENTS OF AN ORGANIZATION

The basic organizational model, shown in Fig. 11.1, depicts the internal and external elements that define an organization. Six structural elements interact to define a company: strategy, climate, culture, policies and procedures, structures, and systems.

Strategy refers to the plan—stated or implicit—that the company has for success against its competitors. It guides the company's operations and determines specific tactics that the company will use to meet its overall strategy. Examples include the flooring company that offers next-day carpet installation services, the internet news service that promises hard news without fluff-pieces, the manufacturing plant that promises next-day turn-around for prototype or low-volume 3-D printed and machined parts.

Climate and culture are closely related. *Climate* refers to the emotional state of the people in an organization; how they individually feel about the company, their coworkers, and their jobs. (Note that engineers use the term "climate" in its physical sense, referring to temperature, humidity, and the like, as covered in Chapters 6 and 10.) *Culture* is rather a group phenomenon; it relates to the behaviors, beliefs, values, customs, and ideas that an organization encompasses.

Climate and culture have become more important than ever, with management increasingly realizing that these elements are major determinants of employees' overall happiness within an organization. With societal changes in the approach to work, companies' cultures vary more widely than ever before—and are more important to people than ever before.

Climate and culture can be valuable bargaining chips to attract and retain employees, and prospective recruits should understand the organization's culture before signing on. Both the potential employee and the potential employer are well advised to carefully scrutinize the perceived "fit" of an individual within an organization: if a person does not appear to match the corporate culture, and vice versa, then an employee/employer relationship is often not effective and not advisable.

Strategy

Climate and culture

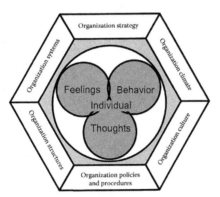

FIGURE 11.1   Basic organizational model.

Policies and procedures are the rules and guidelines that govern a firm's conduct. Policies are official rules and, in larger companies, are often formally written up in detail in an employee handbook. Examples of policies include the amount of medical benefits a company provides, or the number of paid vacation days an employee receives. Procedures (or norms) may not be formally documented but are nevertheless widely understood by employees and are generally applicable to routine tasks. An example of a procedure might be how quickly one is expected to respond to e-mails, or how to request office equipment.

Structures outline the hierarchies within an organization. In larger companies, they are usually depicted in organizational charts, which depict the hierarchy ("chain of command"). An organization's structure determines accountability and authority within its ranks; essentially, it defines the official relationships that exist between employees. Each level in a structured organization has its own degree of authority and responsibility; moving up in the hierarchy means increasing levels of authority and responsibility. In general, under a "unity of command" principle, each employee should only be accountable to one boss.

Structures traditionally fall into seven main categories:

- *Function* means that work is divided by specific task, such as finance or product development, and the various departments carved out by the structure generally report to an executive vice president.
- *Product* means that the staff is divided into all functions necessary to produce and sell a given product or brand; this structure usually features product or brand managers who are entrepreneurs in running their division almost like a separate company. Many consumer packaged-goods companies are organized in this manner.
- *Customer* structure divides all functions by customer need; this is the type of structure often found in service industries.
- *Geographic* structures organize work by location, and often, regional offices are set up to manage the business in a specific geographic area.
- *Division* structure means that each business unit operates like an independent organization under the overall umbrella of a parent corporation. Although a division may run itself almost autonomously, financing is often handled by the parent company.
- *Matrix* structure is unique in that an employee may have more than one boss because two or more lines of authority exist—and consequently, this structure does not adhere to the unity of command mentioned earlier. The matrix structure is generally reserved for organizations that feature complex and time-consuming projects that call for specialized skills. A management consulting firm might have a matrix structure; another example might be found in a law firm, when a paralegal reports to several lawyers.
- *Amorphous* means that the organization has no formal structure; instead, employees forge and dismantle reporting relationships as needed.

Most larger companies use a mix of operational structures, resulting in a hybrid structure.

Systems are developed by companies so that money, people, and things (machines, equipment, supplies) are properly allocated, controlled, and tracked. Systems fall into several categories of distribution and management, including money (accounting, investment, and budgeting), object (inventory and production), people (human resources, employee appraisals), and future (strategic planning, business development, marketing planning).

Systems

11.3 THE INDIVIDUAL

The inner portion and most important feature of the basic organizational model (Fig. 11.1) is the individual. When a company's stock market value is determined, "hard" assets like property, plant, and equipment generally make up one-third to one-half of its value. The remainder is made up of "soft" attributes that are more difficult to quantify: patents, customer base, and employee satisfaction. As a result, there is more incentive than ever to examine the individual feelings, behaviors, and thoughts (the center of the diagram in Fig. 11.1) to assess employees' state-of-mind and find out what influences it. Most companies consider their "human resources" to be their most valuable assets.

Individuals are unique: our upbringing, environment, experiences, and personalities all make us different. People don't always get along even in the best of circumstances, so expecting everyone to get along in a work environment (with all of its inherent tensions) is unrealistic at best.[1] To gain insight into why people act the way they do on the job, there are theories of behavior, motivation, and job satisfaction. This review is by no means comprehensive; instead, it provides an overview of some of the best-known views.

Unique

11.3.1 Behavior: The APCFB Model

One way to understand how an external event can cause an employee's behavior is the *APCFB* model, which stands for *assumptions, perceptions, conclusions, feelings, behaviors*. The model postulates that the assumptions a person holds are an intricate part of the person's overall makeup; these *assumptions* are closely held beliefs that we have about the way the world and the people in it is, or should be. Many of the assumptions we hold dear were created or at least strongly influenced early in our lives by parents and peers. These assumptions make up our value system; these values are so strongly felt that they are difficult for us to change.

When an external event occurs, we all see it through our own "filters" that influence how we view or perceive the event: assumptions affect our *perceptions* of the event. Our filters include internal defense mechanisms that act to protect us from psychological damage. When an event occurs, our existing value system (and the associated filters we have subconsciously created) shapes our view of the event into a given perception, which may differ significantly from reality. Yet our perception leads us to draw *conclusions*, which in turn lead to *feelings*, and these feelings then cause our *behavior*. Behavior includes doing and/or saying something, and this behavior may at times seem wildly out of context or proportion to the actual event, depending on the assumptions (and filters) that we held to begin with.

An example of the APCFB model: A newly-promoted manager with a limited human factors background is asked to supervise the layout of new workstations for company employees. When a respected ergonomist from an outside consulting agency presents the new workstation design, the newly promoted manager feels threatened by the consultant's knowledge while feeling that she herself ought to be more skilled in ergonomics. To her staff's surprise, she rejects the proposed design in spite of its merits. Finding fault with the proposed workstation layout allows her to avoid confronting her own perceived ineptitude in human factors (her filters helped protect her self-esteem). In the meantime, the new manager's staff is left with severe concerns about its future workstations.

Understanding and utilizing the value systems of employees can be extremely effective: this may be called *empowerment*. Returning to the example of the newly-promoted and insecure manager, if company directors had assured the new manager that her lack of human factors knowledge was not a problem and would be addressed with appropriate training, she may have accepted the ergonomist's work. This is called *goal congruence*: the goals of the employee and management, in this instance, become equivalent or congruent. Goal congruence among the employees of a department or an organization enhances the group's productivity.

11.3.2 Motivation: Individual and Environment Models

Motivation incites, directs, and maintains behavior toward goals. Motivation and ability together result in performance: a motivated person with the right skills is willing to expend effort that leads to performance. Performance is moderated by situational constraints at work, such as climate (both psychological and physical) and available equipment.

Understanding what motivates people can help us make them effective and happy at work. A number of theories that explain motivation have been developed and can be roughly divided into two categories. The first group focuses on the *individual* and his or her inherent traits; the second group places the *environment* at the forefront of motivation. Theories that focus on the individual include need-based theories and the expectancy theory. Environmentally-based approaches, which assume that motivation is driven by external factors, include reinforcement, equity, and goal-setting theories. We cover each in turn below.

11.3.2.1 Individual-focused Models of Motivation

Behavior appears to be motivated by the urge to satisfy needs. These needs characterize and drive us, and we act in a continuous quest to satisfy them. *Needs* are defined here as requirements for survival and well-being and are not optional to an

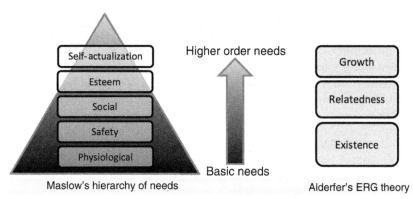

FIGURE 11.2 Motivational theories: Maslow's hierarchy of needs (*left*) and Alderfers' ERG theory (*right*).

individual. Needs fall into two related categories: physical, those that are necessary to physiological survival and comfort, and psychological, required for a fully functioning consciousness.

In 1943, Abraham Maslow published his *Hierarchy of Needs theory*[2] (shown in Fig. 11.2 on the left); later researchers then applied his work to organizational behavioral uses. Maslow and his disciples viewed motivation as a function of meeting an employee's needs, ranging from the basic physiological necessities (air, food, shelter, clothing, and so forth) to the higher-order wants (self-esteem, self-actualization). He believed that individuals act in a never-ending mission to satisfy these needs, covering first the basic needs and then working their way up the needs hierarchy; the lower order needs take precedence over the higher order needs. The hierarchy of needs he proposed has five levels: physiological needs like the need for food and water are at the basis, followed by safety needs that focus on economic and physical security. The third level revolves around social needs for belongingness and love, while the next level focuses on esteem needs, including self-confidence, recognition, appreciation, and respect. The highest-order needs are self-actualization needs; it is at this level that an individual achieves full potential and capability.

Needs Hierarchy

In 1972, Clayton Alderfer proposed another needs-based approach he called the *ERG theory*: *existence*, *relatedness*, and *growth*.[3] *Existence* needs are material needs and include food, water, compensation, and working conditions; *relatedness* needs involve relationships and interactions with others including family, friends, and colleagues, and *growth* needs revolve around the desire for personal development and advancement. Alderfer identified similar needs as Maslow did (in effect compressing Maslow's five needs into three, see Fig. 11.2), but saw these needs arranged in a continuum as opposed to a hierarchy, which allows for movement back and forth among the needs. Further, more than one need may be operational at any given time: the proverbial "starving artist" can concurrently seek to satisfy the needs for both food and for creative growth. He also hypothesized that if a person became frustrated in a fruitless and aggravating attempt to satisfy a higher need, he or she would regress toward fulfilling lower needs instead— he called this syndrome "frustration-regression."

ERG

Another individual-focused approach to motivation is offered by the *expectancy theory*: originated in the 1930s, was not applied to work motivation until the late 1950s, and became prominent in the field of motivation research in the 1960s. Expectancy theory assumes that a person's motivation (and resulting satisfaction) depends on the difference between what his or her environment offers and what he or she expects. The theory can be expressed in form of an equation that outlines the factors underlying motivation:

$$\text{Motivation} = \{\text{expectation that work will lead to performance}\} \times \{\text{expectation that performance will lead to reward}\} \times \{\text{value of reward}\} \tag{11.1}$$

In essence, it is a cognitive approach in which each person is assumed to be rational and knowledgeable about his or her desired rewards. The *expectancy* is the worker's belief that his effort E will result in achieving performance P: $E \rightarrow P$. The *instrumentality* is the worker's belief that performance will result in an outcome O: $P \rightarrow O$. This outcome is a reward provided by the organization, which may be positive (recognition, pay, promotion) or negative (loss of employment). The *valence* reflects the value of the reward as perceived by the worker, which is based on the worker's values, needs and goals. The greater the valance of the reward, the more motivated the worker is.

An example of Expectancy Theory: A professional with a limited human factors background is asked to supervise the layout of new workstations for company employees. She recognizes how important well-designed ergonomic workstations are to employee health and productivity, and that she may get promoted if she performs well on this task. Therefore, she gets additional training in human factors engineering and hires a respected ergonomist from an outside consulting agency to propose the new workstation design. She works closely with the ergonomist to ensure a good final design that meets the employees' and the company's needs, and is rewarded with a bonus in pay.

Note that if her expectation had been the bonus in pay, her motivation for future tasks would be strengthened. Since her expectation had been a promotion, her motivation for future tasks is reduced. Companies must clearly demonstrate the link between performance and outcomes with the rewards that are important to the worker to ensure high levels of motivation.

The "expectancy" component is crucial: the worker must connect how hard he tries with how well he performs, which then results in a reward on which the worker places value. Each of the equation's components can help explain some aspect of motivation. The expectancy theory has received prevailing support in the realm of organizational behavior and is widely applied today.

11.3.2.2 Environmentally Focused Models of Motivation

An environment-focused approach to motivation is offered by the *reinforcement theory* (also called *operant conditioning*), originated by B. F. Skinner and his work with animals in the 1960s. His findings were later applied to organizational behavior and used to describe motivation. The basic reinforcement model involves three key components: stimulus, response, and reward. The *stimulus* elicits a *response*; the *reward* is what is given to reinforce a desired response. Applying reinforcement theory to the workplace can be effective in motivating employees, but picking the appropriate rewards can be challenging. Individuals respond in different ways to varying incentives, as also seen through the expectancy theory, so any "universal" reward systems are inherently imperfect. This is one of the primary limitations of the theory: although in principle it models motivation well, it tends to ignore individual differences in how rewards are valued.

Reinforcement

In the 1960s, J. Stacy Adams[4] proposed that workers seek equity between their work effort or performance and the outcomes they receive, as compared to other workers. As its name suggests, the *equity theory* describes how the employee's motivation is influenced by his perception of how equitable (or inequitable) his treatment by the company will be. Workers consider equity across the company in terms of a ratio of the contributions (such as effort or time) to the benefits (such as recognition or pay) received. Benefits do not need to be equal, as long as this ratio of contributions to benefits is perceived to be similar and fair.

Equity

The *goal-setting theory* posits that people set targets and then purposefully pursue them; it is based on the premise that conscious ideas underlie our actions and that what motivates us are the goals we have set.[5] These goals both motivate us and direct our behavior; they help us decide how much effort to put into our work. To positively influence motivation and behavior on the job, however, employees must be aware of each goal, must know what actions are necessary to achieve it, and must recognize it as something desirable and attainable. Difficult goals may foster higher levels of commitment on the part of the employee; and the more specific the goal, the more focused the efforts of the individual to attain it.

Goal-setting

It is important to note that the workforce is constantly changing and developing. What once motivated people has shifted and continues to do so. Pay and stability ("one job for life") used to be considered the important motivators in a job; now, with an increasingly diverse, life–work balance-conscious workforce, there are many other nonfinancial ways to reward employees and keep them motivated and happy. Childcare, on-site health clubs, flexible work hours, virtual work shifts, and vacation time—these are just a few examples of the rewards that employees might seek.

Consider that, in the United States alone, an estimated 80 million young Americans (born between 1981 and 1997, the so-called Millenials) are entering the workforce. By 2020, it is projected that one-third of all adults and half of all workers will be Millenials.[6] As a group, Millenials tend to be educated (34% hold at least a bachelor's degree), culturally diverse, digitally fluent, and creative; they also tend to dislike bureaucracy and traditional hierarchies and will change jobs frequently. To motivate them, employers should tap into what these workers find important: for example, to satisfy an employee's desire to help people in need, employers could offer paid time-off to volunteer.

11.4 JOB SATISFACTION

Job characteristics influence job satisfaction, which affects productivity and performance at work. Job satisfaction is important to us as a society, as managers, and as employees for several reasons. First, as a society, we generally value high-quality work, and seek fulfilling, rewarding jobs. Additionally, employers are increasingly interested in keeping employees happy because job satisfaction is associated with critical (and revenue-impacting) variables such as turnover, absenteeism, and job performance. Finally, employees who are happy on the job might well enjoy a higher quality of life overall, with fewer stress-related disorders and illnesses, than those who are not happy. There is a correlation between job stress and physical disorders; stress is covered in further detail later in this chapter and in Chapter 4.

Job satisfaction is defined as the extent to which a person derives pleasure from a job. This is distinct from *morale*, which refers to the collective spirit and over-all goodwill of the larger group in which the person functions. Job satisfaction has been extensively researched and, accordingly, there are several theories that strive to explain what causes or prevents it. Note that job satisfaction theories overlap with theories about employee motivation. These approaches can be broadly categorized into need-based or value-based theories, social comparison theories, and job content and context theories.

Need-based

According to *need-* or *value-based theories*, job satisfaction is an attitude that is determined within the individual. The theories in this general category postulate that every person has physical and physiological needs that he or she strives to fill in order to obtain satisfaction. A job defined as "satisfying" meets physical needs (like food and shelter) through rewards such as appropriate income, and meets psychological needs (like self-esteem and intellectual stimulation) through growth opportunities and professional recognition. Needs not only motivate us (as discussed previously), but they provide satisfaction when they are fulfilled.

Value-based theories recognize that although needs may be universal, individuals assign different weights and priorities to them. One person might value personal growth and recognition at work, for example, but care little about monetary rewards; this person would be satisfied even in a low-paying job as long as he/she was recognized for his/her performance and given sufficient professional challenges. According to these theories, job satisfaction is present when a person's needs/values are met; if a person's needs/values change, however, satisfaction would decrease.

Social comparison

While the need-based theories explain much about happiness at work, they do not consider how we do work in a social context and that human nature compels us to compare ourselves with others. This brings us to the second group of theories, which center on *social comparison*. Here, researchers posit that people assess their own feelings of job satisfaction by observing others in similar positions, inferring their feelings about their jobs, and comparing themselves to the other people who work in similar capacities. Intuitively, it makes sense that social comparisons influence job satisfaction: our perceptions of others factors into our lives in many ways, and satisfaction at work is no exception.

Job content and context

Work conditions are the focus of the third group of theories that model job satisfaction; Frederick Herzberg's *two-factor theory* is the best known and best researched of these. Here, intrinsic job *content* factors, when present and positive, are associated with satisfaction, while extrinsic job *context* factors, when negative, are associated with dissatisfaction. In the 1950s, Herzberg and his colleagues' isolated five *content* factors (*motivators*) that they felt were "satisfiers": achievement, recognition, the work itself, responsibility, and advancement. He proposed that a job featuring plenty of these factors would create job satisfaction, and that their absence would result in a neutral or indifferent employee. Further, he speculated that a different set of (*hygiene*) factors influenced dissatisfaction; these are *context* factors such as company policies and administration, salary, supervision and management, interpersonal relationships, job security, and physical conditions at work. He proposed that a job without good context factors would make an employee dissatisfied, while the absence of context factors would lead to a neutral or indifferent employee. A job should ideally be designed with plenty of rewarding content factors (to ensure satisfaction) and positive context factors (to avoid dissatisfaction). Note that the opposite of "job dissatisfaction" is not "job satisfaction" but rather "no job dissatisfaction," because content factors and context factors are not the same. Due to employee expectations, having content factors positively influences motivation but their absence does not necessarily lead to dissatisfaction. Similarly, the presence of context factors may not increase satisfaction but their absence may lead to dissatisfaction.

Ultimately, all the theories described here contribute to an overall understanding of what keeps employees happy on the job. Each theory supplies some degree of explanation for what contributes to job satisfaction. An integration of existing theories may provide the best approach to understanding satisfaction. Clearly, we all strive to fill basic needs; shelter, food, and adequate clothing are mandatory for us and take precedence over higher order needs. Once we have ensured the basics, we want to be intellectually challenged, recognized and praised, promoted, and rewarded. However, we all have different definitions of intellectual stimulation or of what constitutes a reward. Also, we place different weights on what is important. Moreover, we are doubtlessly influenced by what we see around us—are our colleagues earning more than we are for doing the same work? Does our friend who works from home seem happier, or is it the one who works in a corner office? Do we like our peers at work? How supportive is our boss?

11.5 JOB DESIGN

Another way to understand job satisfaction and employee motivation is by considering the way a job is designed. In the early 1900s, Frederick Taylor radically changed workplace organization by introducing specialization. He suggested that jobs be broken down into small tasks that could then be standardized and divided among workers. This usually meant that a given worker did the same task repeatedly, resulting in extreme specialization and, unfortunately, often tedium. Although Taylor's job designs generally improved productivity, workers rebelled against the boredom, depersonalization, and routine of their repetitive jobs and began showing up late (or not at all) and exhibiting symptoms of stress and repetitive work injuries. In the mid-1900s, "job enlargement" (increasing tasks and variety) and "job enrichment" (increasing workers' participation and control) grew popular. Motivation and job satisfaction theories described previously can help to explain how job design influences behavior. The job characteristics model synthesized much of the existing research on job design and became one of the most accepted and examined explanations of job enrichment.

Job characteristics

The *job-characteristics model* proposes that any job can be described via five core dimensions[7]:

1. *Skill variety*, the number of different talents and activities that the job requires.
2. *Task identity*, the extent of work from beginning to end that the job involves.
3. *Task significance*, the job's impact on others.
4. *Autonomy*, the degree of independence in planning, controlling, and determining work procedures.
5. *Feedback*, information about one's effectiveness, how the performance is evaluated and perceived.

All of these core job dimensions influence "critical psychological states" of the jobholder through the experience of meaningfulness of work (the first three dimensions), responsibility for work outcomes (autonomy), and knowledge of results (feedback). According to the theory, high levels of these three critical psychological states lead to favorable work outcomes, including high motivation, higher quality work, higher satisfaction, and lower absenteeism and turnover. Put more simply, the basic tenet of good job design is that employee happiness leads to quality of work life, one in which employees can fulfill their personal potentials.

11.6 POWER AT WORK

In the intricate political landscape of an organization, power affects everyone. Power is a motivator. People tend to crave, use, and occasionally abuse their power, so it is helpful to understand where it originates. Generally, there are considered to be five types of power in business:

- *Coercive* power is based on fear; people who have this power are in the position to inflict some sort of punishment on others. At a factory, the shift manager might reassign late-night shifts to a mechanic whom she does not like to inconvenience the mechanic; an organization's coercive power includes its capacity to fire employees or reduce pay.
- *Reward* power is just the opposite of coercive; it is based on the expectation of receiving praise. A scientist might write an especially detailed technical report to receive praise from the laboratory manager; the manager holds the reward power. A company's reward power lies in its ability to provide incentives such as raises, promotions, and paid vacations.
- *Referent* power reflects someone's ability to attract and inspire others regardless of his/her formal job title or status. An administrative assistant might be widely known and respected by company employees as the person who single-handedly seems to keep the office running, so others will gladly help him when he asks.

- *Legitimate* power is due to the formal status held within the organization's hierarchy. A corporate chief executive makes the final decisions in corporate meetings because she has the authority and even the duty to do so.
- *Expert* power comes from one's own skill or knowledge, regardless of formal position or job status. A low-level equipment operator might have extraordinary power in a company if he is the only one who knows how to keep the machine producing at peak efficiency.

One can have more than one type of power within in an organization—or even all five. Consider the example of a well-known and respected plant manager who hires a machine repair technician. The plant manager holds coercive, reward and legitimate power over the technician: she can fire him at any point, praise him and give pay raises as she is formally the technician's supervisor. Additionally, her expertise in the field of plant operations gives her expert power over the technician, and her much-admired professionalism and knowledge lends her referent power as well.

11.7 COMMUNICATING AT WORK

One activity that pervades and influences (directly or indirectly) all of the topics discussed previously is communication. It is the exchange of information between two (or more) individuals and the inference of its meaning. Individuals understand and establish their roles in an organization through many modes of communication: how they provide and receive feedback, how they make decisions, and how they state and pursue goals. Functions of communication include establishing control, disseminating information, enhancing motivation, and expressing feelings. Communication is the lifeblood of an organization; failing to communicate is its Achilles' heel. Communication occurs throughout the organization and the environment in which the company operates—inter-personally (among employees), intra-organizationally (among the groups and departments of the company), and inter-organizationally (between organizations). The important of effective communication cannot be overstated.

11.7.1 Interacting with Others

Surviving (and thriving) within the intricate political structure of an organization virtually mandates getting along with people. The people with whom you most need to interact effectively are those in your management chain (subordinates and supervisor) and your team members. Managing the relationship upward is as important as managing the relationships downward and laterally. Effectively managing a professional relationship includes understanding the context, goals, job pressures, strengths, weaknesses, and preferred work styles of both the other person and yourself. Only then can you map out, develop, and maintain a relationship that fits both your needs and styles and meets both of your expectations. Ideally, this relationship should be based on trust and honesty.

The steps mentioned earlier may seem unrealistic—how can we truly understand the stated and unstated goals of our bosses—yet even just trying to understand them will bring us closer to effectively dealing with colleagues. Adjust your communication style to the other person involved in the conversation; be flexible in scheduling (to consider, for example, time zone differences, or other work or personal commitments) and medium (for example, augment the spoken aspect of a teleconference with visually sharing your concurrently typed notes, to accommodate those who can better read than follow a spoken conversation in your preferred language, and send out written minutes to document any verbal commitments and ensure alignment on decisions). These basic principles can be similarly applied to dealing with any professional relationships: understand them, understand yourself, and base your relationships with them on these assessments.

One tool that will facilitate communication with supervisors, subordinates, and colleagues alike *active listening.* Too often when we suppose that we are listening, we truly are not; instead, we might be silently disagreeing with the speaker (*he doesn't know what he's talking about*), formulating our response to the person talking (*well, you could have issued a report*) or thinking of something else and tuning out entirely (*I wonder what I should have for lunch*). Active listening helps circumvent this problem by forcing us to really hear what the other person is saying. This lets us get a valid perception of what the speaker is communicating, and gives the speaker the satisfaction of truly being heard. Practice active listening by paying close attention to what the person is saying; responding to the information the speaker provides without leading and without giving advice; and identifying the speaker's feelings along with the content of what is said by absorbing the information and repeating portions back to the speaker to verify the information. Importantly, an active listener gives control of the conversation to the other party.

11.7.2 Performance Appraisals

Performance appraisals are formal means of communicating between managers and subordinates. When done properly and at the right time, performance appraisals can be very useful in achieving organizational/administrative goals, for providing feedback and evaluation, and for coaching and development. *Organizational* or *administrative goals* pertain to personnel actions like placement, promotions, pay, and (when needed) performance correction and improvement. *Feedback and evaluation* refer to the employee's performance, including strengths and weaknesses of the employee's work. *Coaching and development* relate to the critical final goal of the appraisal: how to improve (not punish) the employee's work. Working together, the manager and the subordinate should agree on specific goals and timetables for improvement; this provides a valuable opportunity for encouragement and career coaching.

While they are very valuable in the theoretical sense, in practice, performance appraisals are often misused, mismanaged, even feared. People resist formal performance appraisals for a number of reasons. Managers may dislike giving evaluations because they are uncomfortable criticizing subordinates and may feel that they lack the skills needed to evaluate employees. Subordinates may become defensive and anxious during their appraisals. Accordingly, the task is often delayed until the appraisal is too late to be useful. Management plays a vital role in making the appraisal process succeed by tying the process into the organization's overall goals. Appraisals must have tangible results; for example, if promotions and raises are given based on performance, the appraisal process should be a tool for reviewing performance. Put differently, there should be a high correlation between the thoroughness of the appraisal process and how extensively the resulting information is used.

Performance-appraisal systems often include other types of evaluations in addition to the downward (supervisor-to-subordinate) evaluation. These include self-assessments (where people are asked to evaluate themselves and outline their strengths and weaknesses), peer assessments (where people are asked to evaluate their colleagues), and "upward" or management evaluations (where employees evaluate their manager's supervisory abilities); all provide important data for use in the evaluation.

Once the information is gathered and evaluated, the appraisal conversation takes place; this is the critical concluding link of the appraisal process in which supervisor and employee meet to review and discuss the evaluation. Not surprisingly, few people look forward to the performance appraisal if they fear it will be confrontational. To have a more effective appraisal, consider the method of delivery (a supportive supervisor is more effective than one who is threatening), the existing relationship between the supervisor and the employee, and the nonverbal cues that are projected during the conversation. In general, an interactive mode of communication seems to improve the tenor and outcome of the interview: giving employees substantial participation in appraisal interviews and conducting two-way discussions increases satisfaction with the interview and sparks motivation to improve subsequent job performance. If feedback is frequently given between formal performance appraisals, both supervisor and employee should know that there will be no unpleasant surprises. In addition to a review of past performance and discussion of strengths and weaknesses, the appraisal interview should involve setting job-related plans and goals for the future (an *action plan* for the employee). This action plan should consider the views of both parties and be realistic; otherwise, it may become meaningless or even detrimental. The action plan defines the goals and is the blueprint against which to evaluate performance during the next appraisal.

Setting goals helps clarify objectives, focus efforts, use time and resources efficiently and productively, and increase the chance of success. One approach is to define SMART goals:

- Specific: simple, sensible, significant
- Measurable: meaningful, motivating
- Achievable: attainable, agreed-upon
- Results-focused: relevant, reasonable, realistic, and resourced
- Time-bound: timely, time limited

Sometimes this is expanded to include feedback and thus becomes SMARTER: Evaluated and Reviewed.

11.7.3 Workplace Bullying

Bullies are not just a bane of schoolchildren; they appear at the work place as well. Workplace bullying can be defined as repeated, health-harming mistreatment of one or more persons (the targets) by one or more perpetrators.[8] It is abusive conduct that may include the following: threatening, humiliating, or intimidating; work interference (sabotage) that prevents work from getting done; verbal abuse and, in rare instances, physical abuse.

Workplace bullying can be categorized two ways: overt bullying wherein a person or people engage in oafish (at best) or outright hostile or violent (at worst) behavior and more insidious bullying where small, repetitive acts accumulate over time. These have been called the "hammers" and the "needles"; the hammers must of course be addressed, but letting the less obvious needles continue can swiftly turn a workplace toxic.[9] The effects on the employee being bullied can be devastating, and a British study found that one in three people leave their job after being bullied.[10]

There are subtle but critical distinctions between bullying and managing or co-working. Comments that are objective and are intended to provide constructive feedback are not usually considered bullying but rather are intended to assist employees with their work, even if they are not positive or pleasant. The routine interactions of a workplace, such as expressing differences of opinion; offering constructive feedback, guidance, or advice about work-related issues; taking reasonable disciplinary actions; and managing a staff member's performance are not workplace bullying behaviors.

Bullying is usually considered to be a pattern of behavior where one or more incidents indicate that bullying is taking place. If you are not sure an action or statement could be considered bullying, apply the "reasonable person" test: would most people consider the action unacceptable? Another way to determine if someone is a target of workplace bullying is to consider if there are resulting symptoms including nausea, heightened anxiety, obsessing about work while not there, feeling vulnerable at work, and losing confidence, motivation, enthusiasm and/or enjoyment at work.

Workplace bullying is unacceptable for compassionate reasons, and (secondarily) for its impacts on associates: increased absenteeism, employee turnover, accident/incident occurrence, reduced customer service, decreased morale and productivity. Many companies have formal workplace policies against harassment, conveying management's commitment to preventing workplace harassment and including training on how to deal with complaints and how to arrive at solutions.

Workplace bullying is not routine, acceptable, or deserved. If you see or experience bullying and you feel comfortable dealing directly with the harasser, firmly tell the person that the behavior is not acceptable and ask him/her to stop. If you are uncomfortable approaching the individual on your own, ask a supervisor, union representative, or human resources professional to accompany you. Keep a record of bullying events, detailing the date, time, what happened, and any witnesses. Both the character of the incidents and their frequency or pattern are important. Retain copies or images of any written documentation, texts, e-mails, or other messages that depict the harassing behaviors. Report the harassment as specified in your workplace policy, or to your supervisor or manager.[11]

11.8 STRESS WITHIN ORGANIZATIONS

Stress is a psychological state caused by environmental conditions that lead to a person's specific psychological, behavioral, and physical reactions. Stress does not discriminate; almost every worker in any job at any level faces stress. Because it is initially a psychological phenomenon, stress is closely tied to motivation, behavior, and job satisfaction. For more on the psychopathology of job stress for individuals and crews, see Chapter 4.

Stress at work is real,[12] as shown in surveys:

- One-fourth of employees view their jobs as the number one stressor in their lives.
- Three-fourths of employees believe the worker has more on-the-job stress today than a generation ago.
- Problems at work are more strongly associated with health complaints than are any other life stressor—more so than even financial problems or family problems.

Further, the impact of stress is not declining. A 2017 US survey found that 60% of office workers stated that their work-related pressure has increased in the last 5 years, and just over half of workers reported being stressed at work every day. Their concerns are recognized by senior management: half of the CFOs surveyed acknowledged their teams are stressed and that worker anxiety is on the rise.[13]

Stress as a psychological phenomenon involves interactions between an individual and the demands of the environment. The concept of stress has its foundations in research by Walter Bradford Cannon, early in the last century, and Hans Selye in the 1970s. They described what we now simply call "stress" as *general adaptation syndrome* (GAS), a pattern of physiological reactions to the environment, or the body's reaction to adverse environmental stimulation.[14] GAS is characterized as a mobilization of energy resources that proceeds along three stages: alarm, adaptation (resistance), and exhaustion. In the *alarm* stage, the body mobilizes its resources to meet the assault of a stressor; these bodily resources include increased blood pressure and heart rate, elevated muscle tension, higher levels of hormone production, and the release of energy. During the *adaptation* stage, the body works hard to maintain homeostasis (physiological balance). This effort to normalize its systems strains the body. In the third and last phase of *exhaustion*, the person's biological integrity is endangered if the body's systems are overworked in their efforts to adapt—the "normal" stress has become an overwhelming "distress."

`GAS`

At some point, if overload is continued, the primary biological systems begin to fail, and serious physiological problems can occur. In short, Selye believed that response to a stressor led to heightened use of resources to either resist or adapt to the demand.[15] This applies in a similar fashion to psychological loads and the person's subsequent responses. When an individual is experiencing psychological, behavioral, and physiological effects of stress at work, actual changes in body chemistry occur, and these changes (heightened blood pressure, increased muscle tension, decreased immune system response, and hyperventilation) correlate with heightened risk of depression, headaches, ulcers, cardiovascular concerns, and work-related musculoskeletal disorders.

Selye hypothesized that stress followed a stimulus-response model, in which certain environmental demands (stressors) would invoke a related response. Individual differences are important: different individuals react differently to the same stressors, and the same individual will react differently at different times.

Stressors

Given that stressors vary in terms of the severity of their impact on people, we simply generalize here and state that there are certain environmental demands that tend to exert the most significant effect on employees. Some of the biggest stressors are the following:

- Demanding jobs with high workloads, pressure to perform, a too-fast work pace.
- Lack of control over the process and work and lack of autonomy.
- Task difficulty, with duties perceived as overly complex, or with conflicting demands.
- Overbearing responsibilities for others, lack of social support, isolation.
- Monotony and underload on the job, with overly routine, repetitive, and boring duties and little content variety.
- Poor supervisory practices, including unsupportive superiors or incompetent management.
- Technological problems, like frequent computer malfunctions or equipment breakdowns.

Specific to the effects of computerization on job stress, one review[16] concluded that psychobiological mechanisms do exist that link psychological stress to heightened susceptibility to work-related musculoskeletal disorders (e.g., in keyboarders) because stress can affect our hormonal, circulatory, and respiratory systems. This in turn exacerbates the impact of physical risk factors, such as improper workstations and excessively repetitive movements (see also Chapters 9, 10 and 12). Proper ergonomic design of work tools and equipment is particularly consequential in high-pressure jobs.

Coping

Reducing distress becomes an important goal for all levels of the organization, from the individual employee through supervisors to the company as a whole. The fundamental approach is to reduce factors that are unnecessarily stressful. Lower stress levels may enhance job satisfaction and maximize personal health, help supervisors keep their staff happy and healthy, and help companies achieve and maintain lower healthcare costs and better productivity. Approaches for managing stress can be grouped into six major coping mechanisms[17]:

- Setting priorities and deciding what is really important and what is not.
- Utilizing personal resources, like talking about feelings with family members or friends.
- Recovering and dealing with the problem by moving on to a new activity.
- Distracting oneself with an unrelated activity or by taking the day off.
- Passively tolerating by accepting the situation.
- Releasing emotions by taking feelings out on others; turning to smoking, drinking, drugs, or withdrawing from others. (Clearly, this coping mechanism is neither recommended nor desirable.)

In recognition of the many negative effects stress can exert on employees, some companies have created and implemented formal stress-reduction initiatives and programs. These can include time management seminars, instruction of relaxation techniques, on-site wellness facilities that include exercise equipment and yoga classes, psychologists or counselors available to employees, biofeedback seminars, and flexible schedules.

11.9 WORK–LIFE BALANCE

Work–life balance is critical but also highly individual. Thoughts and concerns about how much to work and how the work schedule is affecting you and those around you are a side effect of having a job. With changes in technology and the pressure to do more at work, the concept of work–life balance becomes more pressing today than ever. Much research supports that taking time off away from work is critical for overall health and well-being; this time off can be short breaks, single days, and more lengthy vacations. Time off is not a luxury but an important component of any healthy lifestyle. Moreover, just as our minds need a respite from the work routine, our backs need a break from the office chair, our eyes need respite from the computer screen, and our fingers and wrists crave some time away from the keyboard.

11.9.1 The Standard Work Week

During the Industrial Revolution, workdays were long—between 10 and 16 hours a day—because factories had to run around the clock to meet production goals (and labor laws protecting workers were lax or nonexistent). In the 1920s, Henry Ford, founder of Ford Motor Company, established the 5-day, 40-hour work week, and he did so for a surprising reason: he wanted to reduce staff work hours so employees would have the time and desire to shop. In fact, he still paid them the same wages he had paid during the longer work week; he believed that potential consumers needed not just the free time but also the available funds to shop. The standard American 40-hour work week was codified into United States law in 1940.[18] Researchers have continued to assess the relationship between work hours and performance to try to determine an optimum. Beyond 40–50 hours per week, the marginal returns from additional work decrease rapidly and become negative.

11.9.2 Human Nature and Work

Adam Smith, the father of industrial capitalism, believed that we as people, at our core, are essentially lazy. As he wrote in 1776, "It is the interest of every man to live as much at his ease as he can."[19] This notion and its reverberations have set the stage for much of how work was framed from that time forward into modern day. Taylorism, which debuted a century after Adam Smith's book was published, was based on the premise that efficiency was the key to profitability and that an individual worker's job satisfaction was not even remotely a factor for consideration. To incite a worker into showing up daily for menial and/or unrewarding work, compensation was necessary (and the only reason any free man would appear). Consequently, for centuries, employers set up systems of work where compensation was the proverbial carrot and institutions (factories, offices, and shops) provided never-ending and often unfulfilling toil.

In 1949, career counselor William J. Reilly deftly captured the essence of this human desire to find our purpose and do what we love in his book *"How to Avoid Work."* It is a remarkable text given the era in which it was written; the book encouraged both men and women of all ages to pursue their passions rather than the safety of conventional professions.[20]

"Most [people] have the ridiculous notion that anything they do which produces an income is work—and that anything they do outside "working" hours is play. There is no logic to that. [...] Your life is too short and too valuable to fritter away in work. If you don't get out now, you may end up like the frog that is placed in a pot of fresh water on the stove. As the temperature is gradually increased, the frog feels restless and uncomfortable, but not uncomfortable enough to jump out. Without being aware that a chance is taking place, he is gradually lulled into unconsciousness. Much the same thing happens when you take a person and put him in a job which he does not like. He gets irritable in his groove. His duties soon become a monotonous routine that slowly dulls his senses. [...] They are people whose minds are stunned and slowly dying. [...] Actually, there is only one way in this world to achieve true happiness, and that is to express yourself with all your skill and enthusiasm in a career that appeals to you more than any other. In such a career, you feel a sense of purpose, a sense of achievement. You feel you are making a contribution. It is not work."

A Gallup poll in 2013 queried 230,000 full-time and part-time workers in 142 countries and found that only 13% of these people felt engaged and fulfilled by their jobs.[21] This suggests that nearly 9 in 10 people surveyed were to some degree unhappy in their jobs—a staggering statistic. Keeping in mind William Reilly's 1949 proposition, we posit that this profound dissatisfaction with work is one of our own making: the product of how we have designed our institutions, with income as the primary motivating factor, and the actual work itself (whether or not it is fulfilling and enjoyable) as less important.

As described previously in this chapter, there are many motivating factors at work and a number of nonmonetary means of reward or compensation. In the context of work–life balance, a fulfilling job at ±40 hours a week is far preferable—and balanced—than one we hate or find boring.

11.9.3 Work–Life Integration

Work–life balance might better be described as work–life *integration*. With today's constant connectivity, margins between being "at work" and being "off the clock" are blurry and ever shifting. Nonetheless, most agree that it is important to disconnect from work. In the first century, Roman philosopher Seneca the Younger uttered a warning that is as valid today as it ever was: "it is inevitable that life will be not just very short but very miserable for those who acquire by great toil what they must keep by greater toil." We have set a formidable trap for ourselves with an aggressive work culture, where it is common not only to accept but actually to praise and admire extreme work schedules.

Regarding the number of hours spent working, Americans are among the hardest (or longest) working people on the planet, working about 1788 hours per year. Japanese work "only" about 1735 hours a year and Europeans average 1400 hours per year.[22] A 2014 Gallup poll found that Americans work on average 47 hours per week: in fact, one in five clocks more than 60 hours per week.[23] With technology ever advancing, we do much of that work in locations other than the traditional office (see Chapter 10), which allows work to encroach on our "free" time even more. And our constant connectivity yields more distractions at (from) work as well—shopping online, updating social media, and checking personal e-mails.

The sheer number of hours worked is a real problem because the propensity to work around the clock contributes to illness. A review of both published and unpublished data for over 600,000 individuals showed that employees who work long hours have a higher risk of stroke than those working standard hours. The biggest overachievers, working 55 hours a week or more, were 33% more likely to suffer a stroke. The researchers speculated that increased tension could lead to other unhealthy behaviors—drinking heavily, overeating, and selecting nutrient-poor and calorie-rich "comfort" food. These behaviors in turn are associated with poorer physical outcomes.[24]

Companies wanting to attract and retain the most promising talent should consider offering a guarantee of work–life balance and a work environment that appeals to the most entrepreneurial among them. The desire for work–life balance may be not only about time off or the length of the workweek, but also about flexibility and the desire for a creative, dynamic working environment.

One more argument in favor of work–life balance: consider how many times you have had a professional insight or innovative solution to a problem pop into your mind during a routine activity, such as while taking a shower or on a brisk walk. This is no coincidence: one of the ingredients considered to fuel creativity is the neurotransmitter dopamine,[25] which is facilitated by a warm shower, exercise, and similar activities that make us feel relaxed and content. Another crucial factor is distraction: something to disengage our brain from fixating on certain topics or issues or moving into a rut. When our minds are distracted and not focused on a given work problem, we tend to look inward, which stimulates our subconscious. This may allow us to make connections that we could not make before, leading to more insightful and creative ideas.

11.9.4 When the Balance Shifts

Appropriate work–life balance is defined by the worker. Work is what we spend the majority of our days doing, how we provide financially for ourselves and our families, and consequently what we do for a living, so it becomes a core part of who we are as individuals. For those lucky enough to make a living doing what they enjoy, work can be especially consuming. Add mobile technology into the mix, and work can easily take over personal lives.

For some, work–life balance is shifted in the direction of work: they are simply that driven, or may work in professions where projects demand round-the-clock attention. Take the examples of successful founders of some of our most innovative international technology companies. Facebook Chief Executive Mark Zuckerberg reportedly spent

50 or 60 hours a week in his office, expending very little energy on anything but work. His practice of wearing the same "uniform" of jeans and a T-shirt every day resulted in very little time spent purchasing or selecting clothing. He said, "If you count all the time I am focused on our mission, that is basically my whole life."[26] Similarly, under the direction of Bill Gates, Microsoft grew into an enormous success, and Gates was known for massive coding sessions where he would fall asleep on his desk, only to awaken and start typing again right away.

This intense passion that some people possess may be understandable and possibly essential in some instances and under certain circumstances. This is true not just today but also in earlier times; Michelangelo didn't have much work–life balance in the early 1500s when painting the ceiling of the Sistine Chapel. The same might be said for Leonardo da Vinci or Albert Einstein or any number of inventors, musicians, artists, and writers. Further, there are plenty of companies where associates are expected to work long, late hours and stay plugged in via technology even when not in the office. Employees may willingly accept the working conditions, provided the work itself is sufficiently engaging and satisfying (or the rewards sufficient) for them to remain in their jobs.

11.9.5 How to Achieve Work–Life Balance

Achieving work–life balance is inherently subjective: what works well for one individual at one moment in time might not work well for someone else. Even if it is understood that some circumstances call for extreme bouts of work, over time, few people can or should keep up this pace over extended periods of time. Generally speaking, most of us would like to experience a comfortable, happy lifestyle, with a satisfying and fulfilling work life rounded out by an equally satisfying and fulfilling non-work life. Some suggestions for this follow.[27]

11.9.5.1 *Define Success for Yourself*

Success

Define what success means in both the professional and the personal realms. Importantly, these definitions change over time and with evolving circumstances. Then, communicate the resulting needs that go along with these definitions by explaining them to colleagues, supervisors, family, and friends. Finally, respect the boundaries of the goals you have set: this develops a routine over time and establishes your commitment to the goals.

11.9.5.2 *Manage Technology*

Disconnect

Always being connected to work can actually undermine performance; as discussed previously, there are neuroscientific reasons why creativity is sparked by non-work routines. Albert Einstein himself noted, "I think 99 times and find nothing. I stop thinking, swim in silence, and the truth comes to me." While mobile technology (smartphones, etc.) has made life easier in many ways, it has also complicated our ability to truly be away from work.

- *Embrace the off button at home*: mobile technology encroaches on our personal time when we cede control to it. Technology comes with an "off" button; use it.

- *Embrace the off button at work*: most of us blend our personal lives with work. A recent poll confirms that nearly 9 of 10 of us admit to wasting time at work every day, in particular on personal use of technology (non-work-related e-mails, texts, and internet usage).[28] If we return our attention to work and control the distractions, our productivity will increase, which translates into fewer hours spent at work.

11.9.5.3 Build Support Networks

Support networks can be formal (such as paid child-care provider at home) or informal, and they should exist both at home and at work. For the latter, trusted work peers serve as advisors or sounding boards, and a supportive boss or helpful colleagues step in when a crisis at home requires time away from work. Similarly, emotional support at home is extremely valuable, with friends or family helping when an individual needs perspective or assistance.

Support

11.9.5.4 Collaborate with Partner/Family

Employees with strong family relationships generally understand the need for shared vision of success for everyone at home.

Collaborate

- If *travel or relocation* is part of the job, these types of decisions are best made as a family unit with involvement from every member of that personal team.
- *Pace yourself*: You and your loved ones presumably plan to have a long, healthy, productive, and happy career and life, so understanding the value of pace is critical. Alternate times when you need to invest more time/effort at work with times when you can slow down. With some insight and self-awareness, pacing yourself helps you to enjoy the journey as much as the destination.

11.10 CHAPTER SUMMARY

To understand why a person behaves the way he or she does at work requires an understanding of the individual, of course, but also of the environment in which the person operates. In this chapter, we reviewed the basic organizational structure and studied how each element interacts to function as a whole. Various theories help explain behavior, motivation, and job satisfaction. Job design is critical; the basic tenet of good job design is that employee happiness leads to quality of work life in which employees can fulfill their personal potentials.

One activity that pervades and influences—directly or indirectly—all aspects of organizational behavior is communication. Individuals understand and establish their roles in an organization through many modes of communication: how they provide and receive feedback, how they make decisions, and how they state and pursue goals. Functions of communication include establishing control, disseminating information, enhancing motivation, and expressing feelings. Communication is the lifeblood of an organization -- and failure to communicate is its Achilles' heel. In particular, performance appraisals are formal means of communicating employees' contributions to their workplace. In addition to a review of past performance and discussion of strengths and weaknesses, the appraisal should involve setting job-related plans and realistic goals for the future.

Workplace bullying is a destructive form of communication among employees; it is defined as repeated, health-harming mistreatment of one or more individuals by one or more perpetrators. This behavior is not routine, acceptable, or deserved, and it should not be tolerated. Workplace bullying leads to increased absenteeism, employee turnover, accident/incident occurrence, reduced customer service, decreased morale and productivity.

Workplace stress influences worker behavior, motivation, and job satisfaction. The psychological, behavioral, and physiological effects of stress at work can compromise the individual's health.

Companies globally are beginning to recognize the negative consequences of overworked and overwhelmed workers. Employees need time off the job to maintain a balance between work and private life. Increasing use of (especially mobile) technology can both simplify work and make it more complex. Constant and ubiquitous connectivity makes balancing work and non-work time far more difficult than ever. In most any career, there are times of intense and long work hours and other times where workloads ebb. Work–life balance (or work–life integration) is a self-defined, self-determined state of well-being, helping effectively manage responsibilities at work, at home, and in the community.

11.11 CHALLENGES

The outright expense of replacing a valuable employee can range from half to several times a year's pay. List some of the factors that would be involved in this steep cost.

Of all the organizational models, what challenges are unique to matrix structures or might be especially difficult to manage?

A company's climate and culture can be valuable bargaining chips in attracting and retaining employees. What kind of corporate climate and culture would appeal to you?

How would you describe the concept of empowerment? Give examples of how to empower an employee.

Using Maslow's hierarchy of needs, provide specific examples of each of the levels of needs.

What motivates people at work has changed a great deal from generation to generation. What do you think are some major differences in what motivated staffers in previous decades compared to today? What changes do you foresee for workers in the generation to follow yours?

Considering the types of power described in this chapter, select five people in your purview (famous or not) and describe the types of power they have and why.

What do you think about performance appraisals? Can you think of a better system for assessing performance?

Have you witnessed or experienced workplace bullying? If so, describe what was involved and how (if at all) the issue was resolved.

What is your opinion about work–life balance? How would you approach achieving it, if you believe in it? If not, why not?

NOTES

1 Stressful or monotonous duties, long hours, and imposed reporting relationships with people you may or may not like make job-related problems are virtually inevitable. The popular media reflects this reality; there are many books and articles, even comic strips, that provide instructions or commentary on dealing with employees and bosses, and many movies and television shows explore the treacherous terrain of interpersonal relationships on the job with a dramatic and/or comedic focus.

2 Published by Maslow (1943).

3 Alderfer (1972).

4 See Adams (1963).

5 Locke and Latham (1990).

6 Willer (2015). Millenials will dominate the workforce. *Forbes.* http://www.forbes.com/sites/sap/2015/06/22/millennials-will-dominate-the-workforce-is-your-business-ready/. Accessed March 10, 2016.

7 Hackman and Oldham (1980).

8 According to Namie (2014). Workplace Bullying Institute (WBI) United States Workplace Bullying Survey, http://workplacebullying.org/multi/pdf/WBI-2014-US-Survey.pdf. Accessed March 10, 2016.

9 Huppke (2015) The impact of workplace bullying. *The Chicago Tribune*, October 16, 2015 http://www.chicagotribune.com/business/careers/ct-huppke-work-advice-1019-biz-20151016-column.html. Accessed July 31, 2017.

10 Trades Union Congress (2015). "Nearly a third of people are bullied at work, says TUC." Issued November 16, 2015. https://www.tuc.org.uk/news/nearly-third-people-are-bullied-work-says-tuc. Accessed March 10, 2016.

11 Adapted from "Violence in the Workplace Prevention Guide," CCOHS (2015). http://www.ccohs.ca/oshanswers/psychosocial/violence.html. Accessed March 15, 2016.

12 As discussed by the National Institute of Occupational Safety and Health (NIOSH) (2014). Stress...at work. http://www.cdc.gov/niosh/docs/99-101/. Accessed March 10, 2016.

13 News release from Accountemps, a Robert Half company (2017). "The Heat Is On: Six In 10 Employees Report Increased Work Stress." http://rh-us.mediaroom.com/2017-02-02-The-Heat-Is-On-Six-In-10-Employees-Report-Increased-Work-Stress. Accessed July 31, 2017.

14 Conway and Smith (1997).

15 Selye (1978).

16 See review by Conway and Smith (1997).

17 See, for example, Muchinsky (2016).

18 A 40-hour work week is the norm in many countries, such as the United States, Korea, and Japan. France has adopted a 35 hour work week, Australia 38 hours, and Brazil 44 hours. Other countries, such as India, may still expect a six-day, 48-hour (or more) workweek. Of course, this does not necessarily reflect the actual number of hours worked; additional hours may be worked for overtime pay or because an employee (or business owner) has more work than can be completed within the allocated time.

19 From Adam Smith's *The Wealth of Nations*.

20 As quoted by Popova (2012) How to avoid work: A 1949 guide to doing what you love. *Brainpickings*.https://www.brainpickings.org/2012/12/14/how-to-avoid-work/. Accessed March 15, 2016.

21 As quoted by S. Crabtree (2013), Worldwide, 13% of employees engaged at work. *Gallup*.http://www.gallup.com/poll/165269/worldwide-employees-engaged-work.aspx. Accessed March 15, 2016.

22 As reported by the Organisation for Economic Cooperation and Development (OECD) (2014).www.oecd.org/gender/data/balancingpaidworkunpaidworkandleisure.html. Accessed March 10, 2016.

23 As reported by N. McCarthy (2014) A 40 hour work week in the United States actually lasts 47 hours. *Forbes.* http://www.forbes.com/sites/niallmccarthy/2014/09/01/a-40-hour-work-week-in-the-united-states-actually-lasts-47-hours/. Accessed March 15, 2016.

24 Kivimäki et al. (2015).

25 For more on dopamine and creativity, see for example Flaherty (2005).

26 As quoted Shandrow (2015). Surprise! Mark Zuckerberg isn't a workaholic. Well, not exactly. *Entrepreneur*. http://www.entrepreneur.com/article/245139. Accessed March 15, 2016.

27 Harvard Business School (HBS) drew on 5 years' worth of interviews with close to 4000 executives world-wide to compile research in 2014 regarding facets of work–life integration; see Groysberg and Abrahams (2014). Manage your work, manage your life. *Harvard Business Review 92* (3), 58–66. https://hbr.org/2014/03/manage-your-work-manage-your-life. Accessed July 31, 2017.

28 As reported by Conner (2015). Wasting time at work: The epidemic continues. *Forbes.* http://www.forbes.com/sites/cherylsnappconner/2015/07/31/wasting-time-at-work-the-epidemic-continues/. Accessed March 15, 2016.

Ergonomics
http://dx.doi.org/10.1016/B978-0-12-813296-8.00012-8

Overview

Controls, also called activators in International Standardization Organization (ISO) standards, transmit inputs to equipment. They are usually operated by hand or foot. When an operator activates a control, he or she sees the results on the equipment's display or indicator or in the equipment's ensuing actions.

The 1940s and 50s are known as the "knobs and dials era" in human factors engineering because at that time, a great deal of experimental work dealt with the use of controls and displays. Consequently, this topic is well researched, and well-proven design recommendations are available for traditional controls and displays.[1] Standards and handbooks for industry and government agencies[2] contain newer detailed design guidelines for established and for recently developed equipment. Different countries, trades, industries, and agencies follow their own regulations, standards, and practices; as a result, their design rules and recommendations are likely to differ from those presented in this chapter.

Most recommendations for selection, design, and arrangement of controls and displays are purely empirical and apply to existing devices and predominately western (European and North American) stereotypes. Hardly any overriding general "laws" based on human motion or energy principles, or on perception and sensory processes, are known or stated which would guide ergonomic control and display design. Therefore, rules for selection and design are likely to change with new kinds of machinery and tasks, and in particular, as equipment becomes more autonomous.

12.1 CONTROLS

Controls exist for any one of these reasons:

- To activate or shut down equipment (such as an on–off key)
- To make a discrete selection (such as choosing a number on a keypad)
- To make a quantitative selection (such as selecting a temperature on a thermostat)
- To apply continuous control (such as steering an automobile).

12.1.1 Selecting the Control

In selecting controls, the underlying principle of ergonomics—designing around the human—applies. Accordingly, controls should be selected for their functional usefulness with the following guidelines:

1. The control type should be compatible with common expectation: like using a toggle switch to turn on a light as opposed to a rotary knob.
2. The control size and motion characteristics should be compatible with past experience: like choosing a relatively large steering wheel for two-handed operation in a vehicle rather than a small rotary control.
3. The direction of operation should be compatible with common expectations: like pushing or pulling an "ON" button rather than turning it to the left.

4. Operations that require fine motor control and minor force should be done with hands, while gross adjustments and larger forces should generally be exerted by feet.
5. Controls should always be secure and safe in that they cannot be inadvertently used or operated in excessive or inappropriate ways.

> Consider the pedal arrangements in modern automobiles: vital and finely controlled operations, usually requiring fairly little force, are performed with feet to operate the gas and brake pedals. Furthermore, the movement of the foot from the accelerator to the brake pedals is complex and potentially disastrous if performed incorrectly, and even more complex if the other foot has to operate a clutch petal. The petal arrangement violates guidelines for good selection of controls, but is commonly used and widely accepted arrangement (and has been studied in many course projects and graduate theses).

12.1.1.1 Compatibility of Control-Machine Movement

Controls should be selected so that the direction of the control movement is compatible with the response movement of the controlled machine, be it a vehicle, equipment, component, or accessory. Table 12.1 lists examples of compatible movements.

TABLE 12.1 Control Movements and Their Expected Results: 1 Means Strongly Preferred, 2 Means Less Preferred

Direction of control movement	Effect[a]											
	On	Off	Right	Left	Raise	Lower	Retract	Extend	Increase	Decrease	Open valve	Close valve
Up	1	–	–	–	1	–	2	–	2	–	–	–
Right	1	–	1	–	–	–	–	–	2	–	–	–
Forward	1	–	–	–	–	2	–	1	1	–	–	–
Clockwise	1	–	2	–	–	–	–	–	2	–	–	1
Press[b], squeeze	2	–	–	–	–	–	–	–	–	–	–	–
Down	–	1	–	–	–	1	–	2	–	2	–	–
Left	–	2	–	1	–	–	–	–	–	2	–	–
Rearward	–	2	–	–	2	–	1	–	–	1	–	–
Back	–	–	–	2	–	–	–	–	–	–	–	–
Counterclockwise	–	1	–	–	–	–	–	–	–	2	1	–
Pull[b]	1	–	–	–	–	–	2	–	–	–	–	–
Push[c]	–	2	–	–	–	–	–	2	–	–	–	–

[a]For Effect, 1 means strongly preferred, 2 means less preferred.
[b]With trigger-type control.
[c]With push–pull switch.
Source: Modified from Kroemer (1988b).

Compatibility   The relationship between control and display, *compatibility*, develops from the context or situation where an association appears manifest or intrinsic; an example is locating a control next to its related display. Importantly, many (if not all) relationships depend on what we have learned and what is habitually used among our peers. In the Western world, red is perceived to mean danger or stop, and green to mean safe and go. Such a relationship is called a *population stereotype*, probably learned during early childhood, which may differ from one user group to another. For example, in Europe a switch is often toggled downward to turn on a light, whereas, in the United States the switch usually is pushed up. Outside the Western world one may encounter quite different stereotypical expectations and uses; for example, in China some durable conventions for direction of motion are different from those in the United States.[3]

12.1.1.2 Control Actuation Force or Torque

The force or torque that the operator applies to activate the control should be kept as low as possible, especially if the control must be operated frequently. If there are jerks and vibrations, it is usually better to stabilize the operator than to increase the control resistance; this helps prevent uncontrolled or inadvertent activation.

Note that many design recommendations for rotational controls list values for tangential force at the point of application instead of torque; this is usually the more practical description.

The force or torque that the operator must apply depends on the control's built-in mechanical resistance; mechanical resistance can be one of five types, listed below:

1. *Intertial*: A function of mass and acceleration (according to Newton's second law); it resists change in any state of motion that it might be in, whether that be at rest or in motion.
2. *Stiction*: Static (resting) friction that must be overcome to initiate motion; it helps avoid accidental movement but provides little feedback.
3. *Coulomb friction*: A dynamic friction that must be overcome to maintain movement; it provides feedback about both presence and direction of motion of the control.
4. *Elastic*: A spring-type linear or non-linear resistance against displacement; it provides feedback about direction and amount of control motion, often employed in "dead person" control because it re-sets toward initial position.
5. *Viscous damping*: The magnitude of resistance depends on activation velocity; it helps to stabilize control movement.

12.1.1.3 Control-Effect Relationships

The relationships between control action and the resulting effect should be obvious and utterly unmistakable. Means to achieve this are common sense and habitual use; appropriate techniques include similarity, proximity, grouping, coding, labeling, and other design procedures, discussed later in this section. Certain control types are preferred for specific applications. Tables 12.1–12.3 assist in the selection of appropriate hand controls.

TABLE 12.2 Control–Effect Relationships of Common Hand Controls: 1 Means Strongly Preferred, 3 Means Least Preferred

Direction of control movement	Effect[a]									
	On–off	On/ standby/ off	Off/mode 1/ mode 2	One of several functions	One of 3 or more alternatives	Select operating condition	One of mutually exclusive conditions	Set a value on scale	Select a value in discrete steps	Engage/ disengage
Key lock	1	–	–	–	–	–	–	–	–	–
Toggle switch	1	2	3	2	–	1	–	–	2	–
Push button	1	1	2	1	–	1	1	–	1	–
Bar knob	3	1	1	–	1	2	–	–	1	–
Round knob	–	–	–	–	–	–	–	1	–	–
Discrete thumbwheel	–	–	–	–	–	–	–	–	1	–
Continuous thumbwheel	–	–	–	–	–	–	–	2	–	–
Crank	–	–	–	–	–	–	–	3	–	–
Rocker switch	1	–	–	2	–	1	–	–	–	–
Lever	–	1	1	–	–	1	1	3	1	–
Joystick or ball	–	–	–	–	–	–	–	3	–	–
Legend Switch	1	1	1	–	–	1	1	–	–	–
Slide[b]	1	1	1	3	1	2	–	1	1	–

[a]For effect, 1 means strongly preferred, 2 means less preferred, 3 means least preferred.
[b]Estimated with no known experimental data.
Source: Modified from Kroemer (1988b).

TABLE 12.3   Selection of Hand- and Foot-operated Controls

Small operating force	Suitable controls
Two discrete positions	Hand operated: key lock, toggle switch, push button, rocker switch, legend switch, bar knob, push–pull switch, slide
Three discrete positions	Hand operated: toggle switch, legend switch, bar knob, slide
Four to 24 discrete positions or continuous operation	Hand operated: key lock, round knob, joystick, continuous thumbwheel, crank, lever, slide, trackball, mouse, light pen
Continuous slewing, fine adjustments	Hand operated: round knob, trackball

Large operating force	Suitable controls
Two discrete portions	Hand operated: push button, detent lever Foot operated: push button
Three to 24 discrete positions	Hand operated: detent lever, bar knob
Continuous operation	Hand operated: hand wheel, lever, joystick Foot operated: pedal

12.1.2 Arranging and Grouping Controls

Several operational rules govern how controls should be arranged and grouped; we describe these below.

Locate for ease of operation: Controls must be oriented with respect to the operator. If the operator moves through different positions, such as when a driver operates a backhoe, the controls and control panels should move with the operator, so that their arrangement in each position remains the same for the driver.

Primary controls come first: The most important and most frequently used controls should have the most prominent positions with respect to ease of operation and reach.

Group related controls together: Controls that have sequential relations, that are associated with a particular function, or that are operated together should be arranged in functional groups (along with their associated displays). Within each functional group, controls and displays should be arranged according to operational importance and sequence.

Arrange for sequential operation: If operating the controls follows a set pattern, controls should be arranged to facilitate that sequence. Left-to-right (preferred) or top-to-bottom are the most common arrangements; think of print in the Western world.

Be consistent: The arrangement of functionally identical or similar controls should be the same from panel to panel.

Include a dead-man control: If the operator becomes incapacitated and consequently releases the control, or continues to hold on to it, the control should be designed so it returns the system to a noncritical operation state or shuts it down.

Guard against accidental activation: There are numerous ways to guard controls against inadvertent activation; examples include placing mechanical shields around them or requiring critical forces or torques to activate. Note that most guarding techniques also reduce the speed of operation.

TABLE 12.4 Minimal Separations (in mm) Between Controls, Edge-to-edge, When the Controls are in Their Closest Positions and Operated With Bare Hands

	Keylock	Bar knob	Detent thumbwheel	Push button	Legend switch	Toggle switch	Rocker switch	Knob	Slide
Keylock	25	19	13	13	25	19	19	19	19
Bar knob	19	25	19	13	50	19	13	25	13
Detent thumbwheel	13	19	13	13	38	13	13	19	13
Push button	13	13	13	13	50	13	13	13	13
Legend switch	25	50	38	50	50	38	38	50	38
Toggle switch	19	19	13	13	38	19	19	19	19
Rocker switch	19	13	13	13	38	19	13	13	13
Knob	19	25	19	13	50	19	13	25	13
Slide	19	13	13	13	38	19	13	13	13

The given values are measured edge-to-edge with the controls in their closest positions. For arrays of controls and operations with gloves, larger distances may be recommended; see Boff and Lincoln (1988).

Pack tightly but do not crowd: If it is absolutely necessary to put a large number of controls into a limited space, the minimal edge-to-edge distances listed in Table 12.4 should be used.[4]

A *dead-man control* or *dead person's switch* is activated automatically if the human operator loses control of the switch (such as being physically removed from control, losing consciousness, or even dying). These switches are usually used as a form of fail-safe to stop or idle a machine. These are commonly used for locomotives, lawn mowers, personal watercraft, chainsaws, treadmills, and other machines.

12.1.3 Designing Controls

If the controlled operation must take place in discrete steps, these should be marked and secured by *detents* or stops in which the control comes to rest. However, continuous controls should be selected if the controlled operation can be located anywhere within the adjustment range of the control with no need to be set in any given position.

Detent or continuous

12.1.3.1 *Detent Controls*

The following seven descriptions present design guidance for controls that use detents to keep them in distinct settings.

Key-operated switches called key locks are used to prevent unauthorized use of the machine or equipment. Key locks are usually set into distinct ON and OFF positions.

Key lock

- The ON and OFF positions should be labeled.
- Keys should be removable from the lock when turned OFF.

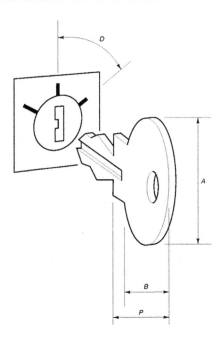

FIGURE 12.1   **Key lock.**

- Keys with teeth or serrations on both edges (preferred) should fit the lock with either side up.
- Keys with a single row of teeth should be inserted the with the teeth pointing up.

  Design recommendations are given in Fig. 12.1 and Table 12.5.

Bar knob

Rotary selectors called bar knobs should be used for discrete functions when two or more detented positions are required.

- Knobs should be bar-shaped, with parallel sides, and the index end should be tapered to a point.

TABLE 12.5   Recommended Dimensions of a Key Lock

	Minimum	Preferred	Maximum
A Height (mm)	13	–	75
B Width (mm)	13	–	38
P Protusion (mm)	20	–	–
D Displacement (degrees)	30	–	90
S[a] Separation (mm)	25	–	25
R[b] Resistance (N-m)	0.1	–	0.7

Note: Letters correspond to measurements in Fig. 12.1.
[a]Edge-to-edge between keys is closest positions.
[b]Key should snap into detent position and not be able to stop between positions.
Source: Modified from MIL-HDBK 759, MIL-STD-1472F.

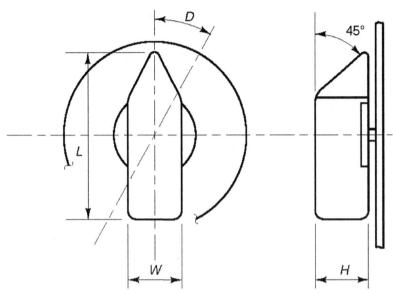

FIGURE 12.2 Bar knob.

Design recommendations are given in Fig. 12.2 and Table 12.6.

Detent thumbwheels for discrete settings may be used if the function requires a compact input device for discrete steps. Design recommendations are given in Fig. 12.3 and Table 12.7.

Push buttons should be used for single switching between two conditions, for entry of a discrete control order, or for release of a locking system. Push buttons can be used for momentary contact or for sustained contact, with or without detent.

Detent
thumbwheel

Push button

TABLE 12.6 Recommended Dimensions of a Bar Knob

	Minimum	Preferred	Maximum
L Length (mm)	25 (38[a])	–	100
W Width (mm)	13	–	25
H Height (mm)	16	–	75
D Displacement (degrees)	15 (30[b])	–	40 (90[b])
Separation (mm): one hand random operation	25 (38[a])	50	63[a]
Separation (mm): two hands simultaneous operation	75 (100[a])	125	150 (175[b])
Resistance[c] (N-m)	0.1	–	0.7

Note: Letters correspond to measurements in Fig. 12.2.
[a]If operator wears gloves.
[b]For blind positioning.
[c]High resistance with large knob only. Knob should snap into detent position and not be able to stop between positions.
Source: Modified from MIL-HDBK 759, MIL-STD-1472F.

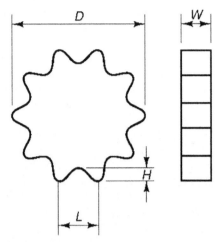

FIGURE 12.3   Detent thumbwheel.

TABLE 12.7   Recommended Dimensions of a Detent Thumbwheel

	Minimum	Preferred	Maximum
D Diameter (mm)	29	–	75
W Width (mm)	3	–	–
L Trough distance (mm)	11	–	19
H Trough depth (mm)	3	–	6
S Separation side-by-side (mm)	10	–	–
R^a Resistance (N)	2	–	6

Note: Letters correspond to measurements in Fig. 12.3.
[a]*Thumbwheel should snap into detent position and should not be able to stop between detents.*
Source: Modified from MIL-HDBK 759, MIL-STD-1472F.

- The push button surface should generally be concave (indented) to fit the finger, or convex for operation with the palm of the hand.
- When either shape is impractical, the surface should provide high frictional resistance to prevent slipping.
- A positive indication of control activation should be included; this could be accomplished by an audible click or visible integral light, for example.

Design recommendations are given in Fig. 12.4 and Table 12.8.

Legend switch   Detent legend switches are particularly well suited to display qualitative information regarding equipment data that requires the operator's attention and action.

- Since they must be easily seen, legend switches should be located within a 30 degree cone along the operator's line of sight.

Design recommendations are given in Fig. 12.5 and Table 12.9.

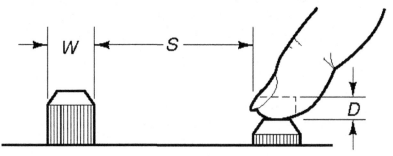

FIGURE 12.4 Push button.

TABLE 12.8 Recommended Dimensions of a Push Button

		Minimum	Preferred	Maximum
W Width of square	Fingertip used	10 (19[a])	–	25
or diameter (mm)	Thumb used	19 (25[a])	–	25
	Palm of hand used	40 (50[a])	–	70
D Displacement	Fingertip used	2	–	6
(mm)[b]	Thumb or Palm used	3	–	40
S Separation (mm)	Single operation by fingertip	13 (25[a])	50	–
	Sequential operation by same finger	6	13	–
	Different fingertips used	6	13	–
	Thumb or Palm used	25	150	–
R Resistance (N)	Single finger used	3	–	11
	Several fingers used at same time	1	–	6
	Thumb used	3	–	23
	Palm of hand used	3	–	23

Note: Letters correspond to measurements in Fig. 12.4.
[a]*If operator wears gloves.*
[b]*Depressed button shall stick out at least 2.5 mm.*
Source: Modified from MIL-HDBK 759, MIL-STD-1472F.

Detent toggle switches may be used when two discrete positions are required. Toggle switches with three positions should be used only when the use of a bar knob, array of push buttons, legend switch, etc. is not feasible.

> Toggle switch

- Toggle switches should be oriented so that the handle moves in a vertical plane, with OFF in the down position.
- Horizontal actuation should be used if and only if compatibility with the controlled function or equipment location makes that desirable.

Design recommendations are given in Fig. 12.6 and Table 12.10.

Detent rocker switches may be used when two discrete positions are required. Rocker switches protrude less from the panel than toggle switches do.

> Rocker switch

- Rocker switches should be oriented so that the handle moves in a vertical plane, with OFF in the down position.

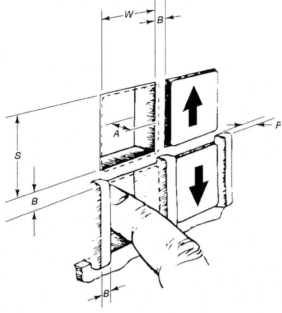

FIGURE 12.5 **Legend switch.**

TABLE 12.9 Recommended Dimensions of a Legend Switch

	Minimum	Preferred	Maximum
W Width of square or diameter (mm)	25[a]	–	38[a]
A Displacement (mm)	3	–	6
B Barrier width (mm)	3	–	6
P Barrier depth (mm)	5	–	6
R Resistance (N)	3	–	17

Note: letters correspond to measurements in Fig. 12.5.
[a]*If operator wears gloves.*
Source: Modified from MIL-HDBK 759, MIL-STD-1472F.

- Horizontal actuation should be used if and only if compatibility with the controlled function or equipment location makes that desirable.
- Narrow switches, as shown at the bottom of Fig. 12.7, are suitable for tactile definition of gloves are worn.

Design recommendations are given in Fig. 12.7 and Table 12.11.

12.1.3.2 *Continuous Controls*

The following six descriptions present design guidance for controls used to achieve any setting within a continuous range of adjustments.

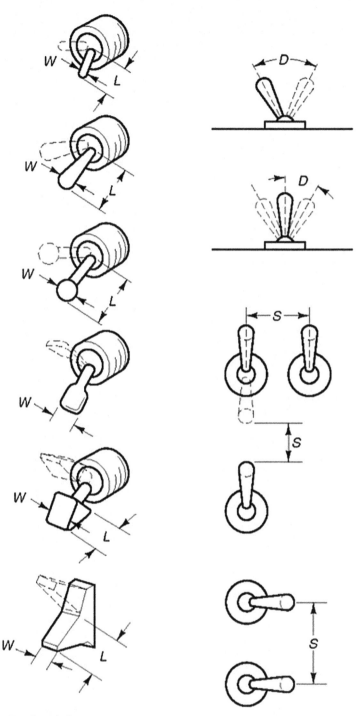

FIGURE 12.6   Toggle switch.

TABLE 12.10   Recommended Dimensions of a Toggle Switch

		Minimum	Preferred	Maximum
L Arm Length (mm)		13 (38[a])	–	50
W Tip width or diameter (mm)		3	–	25
D Displacement (degrees):	2 positions	30	–25	80
	3 Positions	17		40
S Separation (mm):				
Horizontal array, vertical operation, single finger				
Random operation		19	50	–
Sequential operation		13	25	–
Horizontal array, vertical operation, several fingers				
Simultaneous Operation		16 (32[a])	19	–
Vertical array, horizontal operation		25 (38[a])	–	–
Vertical array, operation toward each other, tip-to-tip		25	–	–
R Resistance (N)[b]		3	–	11

Note: Letters correspond to measurements in Fig. 12.6.
[a]*If operator wears gloves.*
[b]*Switch should snap into detent position and not be able to stop between detents.*
Source: Modified from MIL-HDBK 759, MIL-STD-1472F.

Knob

Continuous knobs—also called round knobs or rotary controls—should be used when little force is required and when precise adjustments of a continuous variable are required.

- If positions must be distinguished, an index line on the knob should point to markers on the panel.
- When panel space is extremely limited, knobs should be small and the resistances should be as low as possible without permitting the setting to be changed by vibrations or by inadvertent touching.

Design recommendations are given in Fig. 12.8 and Table 12.12.

Crank

Continuous cranks should be used if the control must be rotated frequently. For tasks involving large slewing (rotating an object around an axis) movements or small, fine adjustments, a crank handle may be mounted on a knob or handwheel.

- The crank handle should be designed so that it rotates freely around its shaft, especially if the whole hand grasps the handle.

Design recommendations are given in Fig. 12.9 and Table 12.13.

Handwheel

Handwheels for continuous (normally two-handed) operation should be used when the breakout or rotation forces are too large to overcome with one-handed control—provided, of course, that two hands are available for this task.

- Knurling (a pattern of straight, crossed, or angled lines is cut into the material) or indentations should be built into the rim to facilitate operator grasp.

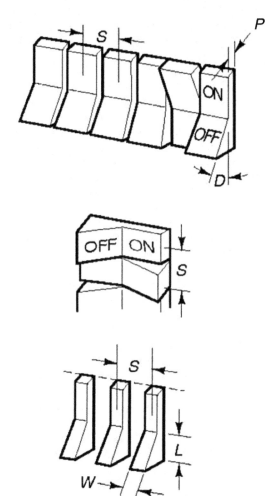

FIGURE 12.7 Rocker switch.

TABLE 12.11 Recommended Dimensions of a Rocker Switch

	Minimum	Preferred	Maximum
W Width (mm)	6	–	–
L Length (mm)	13	–	–
D Displacement (degrees)	30	–	–
P Protusion, depressed (mm)	2.5	–	–
S Separation, center-to-center (mm)	19 (32[a])	–	–
R Resistance (N)	3	–	11

Note: Letters correspond to measurements in Fig. 12.7.
[a]If operator wears gloves.
Source: Modified from MIL-HDBK 759, MIL-STD-1472F.

II. DESIGN APPLICATIONS

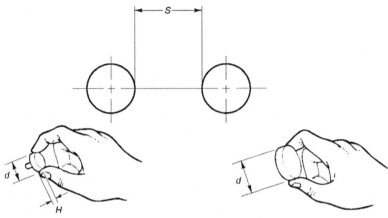

FIGURE 12.8 Knob.

TABLE 12.12 Recommended Dimensions of a Knob

		Minimum	Preferred	Maximum
H Height (mm)[a]		13	–	25
d Diameter (mm)[a]	Fingertip grasp	10	–	100
	Thumb-fingertip grasp	25	–	75
S Separation (mm)	One-hand operation	25	50	–
	Two-hand operation at same time	50	125	–
T Resistance torque (N-m)	Knob up to 25 mm diameter	–	–	0.03
	Knob over 25 mm diameter	–	–	0.04

Note: Letters correspond to measurements in Fig. 12.8.
[a]*Within the ranges, knob size is relatively unimportant if the resistance is low and the knob can be easily grasped and manipulated.*
Source: Modified from MIL-HDBK 759.

- When large displacements must be rapidly made, a spinning handle (called spinner crank) may be attached to the hand wheel unless this is overruled by safety considerations.

Design recommendations are given in Fig. 12.10 and Table 12.14.

Lever

Continuous levers—often called joysticks or simply sticks—are sometimes used when large force or displacement is required at the control, and/or when multidimensional movements are required. There are two kinds of levers:

1. The force joystick does not move (and is isometric) but transmits control inputs according to the force that is applied to it. The force lever is particularly appropriate when "return-to-center" of the system is wanted after each control input.
2. The displacement joystick (sometimes misleadingly called isotonic) moves and transmits control inputs according to its spatial position, movement direction, or movement speed. To improve "stick feel," elastic resistance that increases with displacement may be used. This lever is appropriate when control in two or three dimensions is required, especially when accuracy is more important than speed.

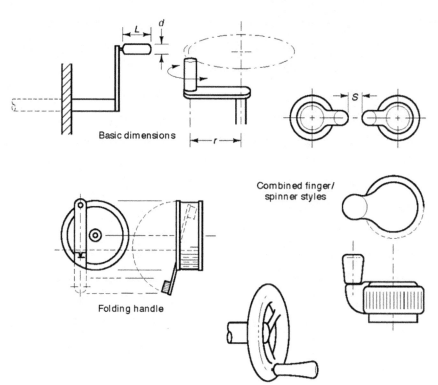

Basic dimensions

Combined finger/
spinner styles

Folding handle

FIGURE 12.9 Crank.

TABLE 12.13 Recommended Dimensions of a Crank

		Minimum	Preferred	Maximum
Operated by Finger and Wrist Movement, Resistance below 22 N:				
L Length of the handle (mm)		25	38	75
d Diameter of the handle (mm)		10	13	16
r Turning Radius (mm)	below 100 rpm	38	75	125
	above 100 rpm	13	58	115
S Separation when handles are closest (mm)		75	–	–
Operated by Arm Movement, Resistance above 22 N:				
L Length of the handle (mm)		75	95	–
d Diameter of the handle (mm)		25	25	38
r Turning radius (mm)	below 100 rpm	190	–	510
	above 100 rpm	125	–	230
S Separation when handles are closest (mm)		75	–	–

Note: Letters correspond to measurements in Fig. 12.9.
Source: Modified from MIL-HDBK 759, MIL-STD-1472F.

II. DESIGN APPLICATIONS

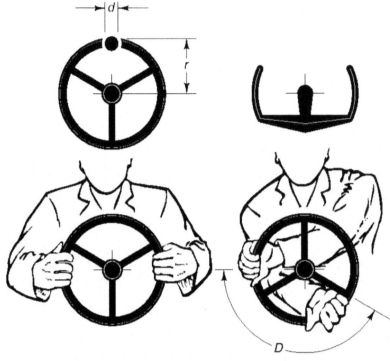

FIGURE 12.10   **Handwheel.** *Source:* MIL-HDBK 759.

TABLE 12.14   Recommended Dimensions of a Handwheel

		Minimum	Preferred	Maximum
r Radius of the wheel (mm)	With power assist	175	–	200
	No power assist	200	–	255
d Diameter of the rim (mm)		19	–	32
T Tilt from vertical (degrees)		30 light vehicle	–	45 heavy vehicle
D Displacement (degrees)	both hands on wheel	–	–	120
R Resistance (N)		20	–	220

Note: Letters correspond to measurements in Fig. 12.10.
Source: Modified from MIL-HDBK 759.

When levers are used to make fine or continuous adjustments, the limb most involved in the work should be supported. For example, for large hand movements, provide elbow support; for small hand movements, provide forearm support; for finger movements, provide wrist support.

When several levers are grouped in proximity to each other, the lever handles should be coded (see below). When at all practicable, levers should be labeled (see below) with regard to function and direction of motion.

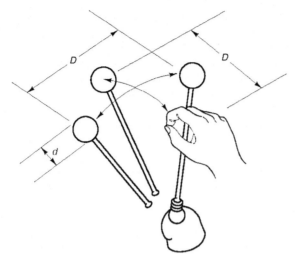

FIGURE 12.11 Lever.

High-force levers may be appropriate for occasional or emergency use. They should be designed so that the operator either pulls the lever up or pulls it back toward the shoulder, with an elbow angle of 150±30 degrees. The force required to operate the lever should not exceed 190 N. The handle diameter should be between 25 and 38 mm, and its length should be at least 100 mm. Displacement should not exceed 125 mm, and clearance behind the handle and along the sides of the handle's path should be at least 65 mm. The lever may have thumb button or a clip-type release.

Design recommendations are given in Fig. 12.11 and Table 12.15.

TABLE 12.15 Recommended Dimensions of a Lever

		Minimum	Preferred	Maximum
d Diameter (mm)	Finger grip	13	–	38
	Hand grip	38	–	75
D Displacement (mm)	Fore-aft	–	–	360
	Left-right	–	–	970
S Separation (mm)	One hand used	50[a]	100	–
	Two hands used	75	125	–
R Resistance (N)	Fore-aft	One hand used	9	–	135
		Two hands used	9	–	220
	Left–right	One hand used	9	–	90
		Two hands used	9	–	135

Note: Letters correspond to measurements in Fig. 12.11.
[a]About 25 mm if one hand operates two adjacent levers at the same time.
Source: Modified from MIL-HDBK 759.

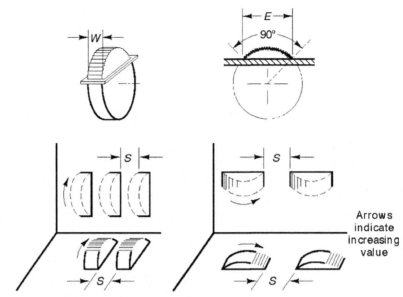

FIGURE 12.12   **Continuous thumbwheel.** *Source: Adapted from MIL-HDBK 759.*

Continuous
thumbwheel

Slide

Thumbwheels for continuous adjustments may be used as an alternative to round knobs when compactness is needed. Design recommendations are given in Fig. 12.12 and Table 12.16.

Slides are used for continuous settings; think of music mix-and-control stations. Design recommendations are given in Fig. 12.13 and Table 12.17.

12.1.3.3 *Remote Control Units*

Remote control units (*remotes*) usually are small hand-held control panels which are manipulated at some distance from a device, machine, or robot with which they communicate, either by cable or wirelessly. In the robotics industry, remote control units

TABLE 12.16   Recommended Dimensions of a Continuous Thumbwheel

		Minimum	Preferred	Maximum
E Rim exposure (mm)		–	25	100
W Width (mm)		–	3	23
S Separation (mm)	Side-by-side	25 (38[a])	–	–
	Head-to-foot	50 (75[a])	–	–
R Resistance (N)		–	–	3[b]

Note: Letters correspond to measurements in Fig. 12.12.
[a]*If operator wears gloves.*
[b]*To avoid inadvertent activation.*
Source: Modified from MIL-HDBK 759, MIL-STD-1472F.

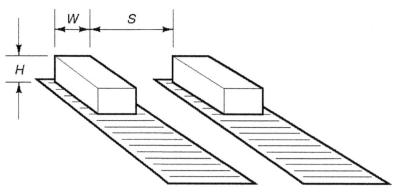

FIGURE 12.13 **Slide.** *Source: Adapted from MIL-HDBK 759.*

TABLE 12.17 Recommended Dimensions of a Slide

		Minimum	Maximum
W Width (mm)		6	25
H Height (mm)		6	–
S Separation (mm)	Single finger use	19	50
	Single finger sequential use	13	25
	Several fingers used at same time	16	19
R Resistance (N)		3	11

Note: Letters correspond to measurements in Fig. 12.13.
Source: Modified from MIL-STD-1472D.

(*teach pendants*) are often used by a technician to specify the point in space to which the robot effector (a gripper perhaps, or other tool) must be moved for its next task or operation. In other industries, remote-operated drones may be used to view or inspect items, especially when altitude or contamination are involved.

Two ergonomic issues are particularly pertinent here: one is danger to the human operator, who must, for example, step into the operating area of the equipment to see the exact location of the robot effector in relation to the workpiece. The second ergonomic concern is the proper design and operation of the controls. In principle, joysticks (levers) are preferable because they allow easy commands for movement in two or three dimensions. Pushbuttons are the more common solution used, but they permit control only in linear or angular direction. However, they also are easier to protect from inadvertent operation and from damage when the unit is dropped. The appropriate arrangements of sticks or buttons in arrays on the manipulation surface of the remote is similar to the design aspects discussed earlier in this chapter. Device designers should pay particular attention to the ability to immediately stop the robot in emergencies, and to quickly and safely more it away from a given position, even if the operator is inexperienced.

TABLE 12.18   Recommended Dimensions of a Pedal for Continuous Adjustments

		Minimum	Maximum
Height or depth (mm)		25	–
Width (mm)		75	–
Displacement (mm):	Operation with regular shoe	13	65
	Operation with heavy boot	25	65
	Operation by ankle flexion	25	65
	Operation by whole-leg motion	25	180
Separation, edge-to-edge (mm)	Random operation with one foot	100	150
	Sequential operation with one foot	50	100
Resistance (N)	Foot does not rest on pedal	18	90
	Foot rests on pedal	45	90
	Operation by ankle flexion	–	45
	Operation by whole-leg motion	45	800

Source: Modified from MIL-STD-1472F.

12.1.3.4 Foot-operated Controls

Foot-operated switches may be used if just two discrete conditions—such as ON and OFF—need only occasionally be set. Long-duration foot control operation should be required only from an operator who is seated. Foot operation can be more forceful than operation by hand, but is less precise and less appropriate for continuous adjustments. That said, pedals are of course widely used in vehicles for continuous adjustments, such as in controlling automobile speed (acceleration and braking).

Design recommendations for pedals are given in Table 12.18.

12.1.4 Coding Controls

There are many ways to code hand-operating controls, however not so many for pedals. Coding serves to identify controls, explain how to operate them, indicate the effects of their operation, and show their status. Examples of *informal coding* includes the use of stick-on notes and creative reminders such as shown in Fig. 12.14 by an aircraft crew.

The designer of the control panel should always employ uniform coding principles. The major coding principles are:

Location: Controls associated with similar functions should be placed in the same relative location from panel to panel.

Shape: Distinguishing controls by their shape can appeal to both vision and action senses. Sharp edges should be avoided, of course! Various shapes and surface textures have been investigated; Figs. 12.15–12.20 show examples.

Size: To discriminate controls by size, we can use up to three different sizes; keep in mind that controls that have the same function on different items or pieces of equipment should have the same size (and shape).

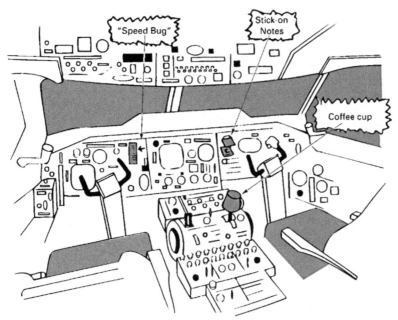

FIGURE 12.14 "Informal coding" done by pilots in their aircraft cockpit. *Source: Adapted from Norman (1991).*

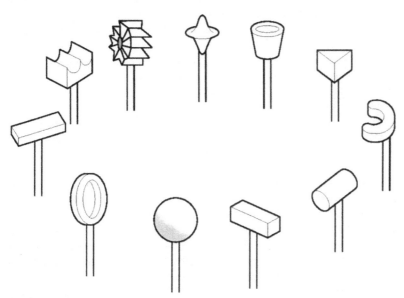

FIGURE 12.15 Examples of shape-coded aircraft controls. *Source: Adapted from Jenkins (1953).*

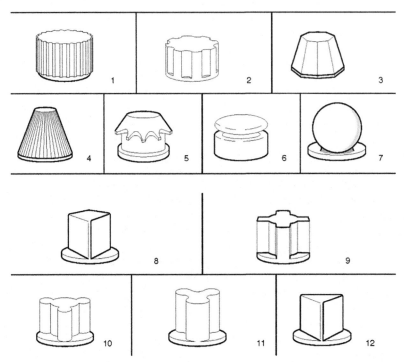

FIGURE 12.16 **Examples of shape-coded knobs of approximately 2.5 cm diameter, 2 cm height.** Numbers 1 through 7 are suitable for full rotation (but do not combine 1 with 2, 3 with 4, and 6 with 7). Numbers 8 through 12 are suitable for partial rotation. Recommended combinations are 8 with 9, 10 with 11, and 9 with 12 (but do not combine 8 with 12, 9 with 10 or 11, 11 with 3 or 4). *Source: Adapted from Hunt and Craig (1954).*

Mode of operation: Controls can be distinguished by their manner of operation, such as push, turn, or slide. If the operator is unfamiliar with the control, he or she might try the wrong manner of operation first; this is likely to increase operation time and risks.

Labeling: While proper labeling is a secure means to identify controls, it only works if the labels are actually read and understood by the operator. Importantly, it takes time to read labels. The label must be placed so that it can be easily read, is well illuminated, and is not obscured. Trans-illuminated (back-lit) labels, potentially incorporated in the control, can be helpful, as long as the content is readily comprehended. See the discussion on labels later in this chapter.

Color: Most controls are either black or grey. For other colors, red, orange-yellow, and white are generally recommended; if an additional color is absolutely necessary, blue can be used. Note that blue is not ideal, because dark blue is too close to black in appearance, and lighter blue is too close to green, particularly for those people (especially men) with green/blue color deficiency. The use of color requires sufficient luminance of the surface area.

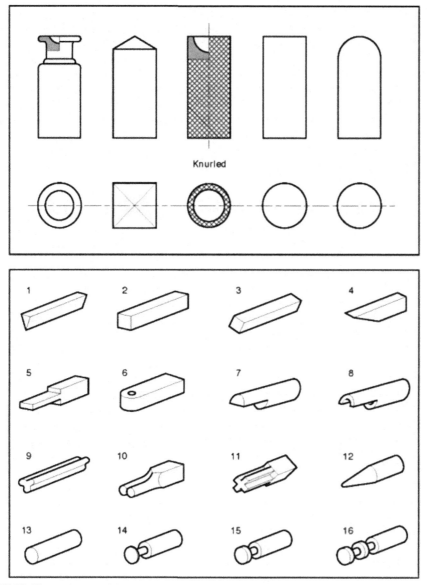

FIGURE 12.17 Examples of shape-coded toggle switches of approximately 1 cm diameter, 2.2 cm length.
Source: Adapted from Stockbridge (1957) top; Green and Anderson (1955), bottom.

Redundancy: Often, coding methods can be combined: location, size, shape, and color with labeling. This offers several advantages:

1. The combination of codes can generate a new set of coding.
2. The combination of codes provides multiple ways to achieve the same kind of feedback, called *redundancy*. For example, if the operator is not able to glance at a

FIGURE 12.18 Examples of push buttons shape-coded for tactile discrimination. Shapes 1, 4, 21, 22, 23, and 24 are best discriminable by bare-handed touch alone, but all shapes were occasionally confused. *Source: Adapted from Moore (1974).*

control, he or she might still be able to feel it. Another example is an ambulance warning signal which can be both a flashing light and a wailing sound.

Advantages and disadvantages of the types of coding are listed in Table 12.19. Table 12.20 shows the largest numbers of coding stimuli that can be used together.

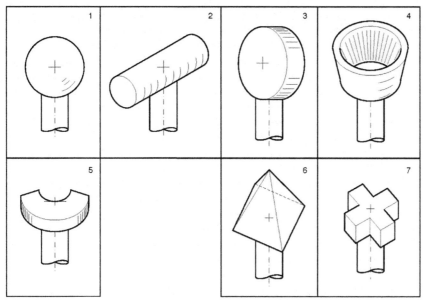

FIGURE 12.19 Shapes proposed for use on finger-operated controls of mining equipment. Recommended for concurrent use: Two handles: 1 and 2, 1 and 5, 1 and 6, 1 and 7, 2 and 3, 2 and 4, 2 and 5, 2 and 6, 2 and 7, 3 and 5, 4 and 6, 3 and 7, 4 and 5, 4 and 6, 4 and 7. Three handles: 1, 2, and 6; 1, 2, and 7; 2, 3, and 6; 2, 3, and 7; 3, 5, and 6. Four handles: 1, 2, 3, and 6; 1,2, 3, and 7; 2, 3, 4, and 6; 2, 3, 4, and 7. Five handles: 1, 2, 4, 5, and 6. Avoid combinations: 1 and 3, 3 and 4, 5 and 7. *Source: Adapted from Kroemer (1980).*

12.1.5 Preventing Accidental Activation of Controls

Accidental activation of controls is something that generally needs to be avoided; it absolutely must be avoided if accidental activation might cause injury to individuals, damage to the system, or degradation of important system functions.

There are various ways to prevent accidental activation, some of which may be combined:

- Locate and orient the control so that the operator is unlikely to strike it or move it accidentally in the normal sequence of control operations.
- Recess, shield, or surround the control through physical barriers.
- Cover or guard the control.
- Interlock several controls so that either the prior operation of a related control is required, or an extra movement is needed to operate the control.
- Add resistance—viscous or coulomb friction, spring loading, or inertia—so that an unusual effort is required for actuation.
- Lock (indent) a path so that the control cannot pass through a critical position without delay.

Note that these techniques usually slow down the operation, which may be detrimental in an emergency.

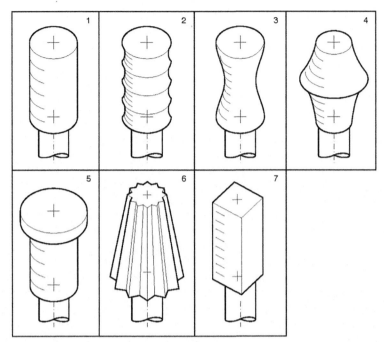

FIGURE 12.20 Shapes proposed for use on lever handles on mining equipment. Recommended for concurrent use: Two handles: 1 and 5, 1 and 6, 1 and 7, 2 and 6, 2 and 7, 3 and 4, 3 and 6, 3 and 7, 4 and 5, 4 and 6, 4 and 7, 5 and 6, 5 and 7. Three handles: 1, 2, and 6; 1, 2, and 7. Four handles: 1, 2, 4, and 5; 1, 2, 4, and 6; 1, 2, 4, and 7. Five handles: 1, 2, 3, 4, and 6; 1, 2, 3, 4, and 7. Avoid combinations: 2 and 5, 3 and 5, 6 and 7. *Source: Adapted from Kroemer (1980).*

12.2 KEYBOARDS AND COMPUTER INPUT DEVICES

Within just a few decades, the computer has changed the ways in which we receive, send and store information; how we communicate; how "office work" is done; and how work systems are designed and run. With the computer being such a dynamic tool, it is surprising that most of our computer inputs still occur through old-fashioned binary (ON–OFF) pushbuttons.

12.2.1 Numerical Keypads

Two different kinds of numerical keysets are widely used. (Both are likely on a contemporary mobile phone.) There is the so-called telephone keyset:

1	2	3
4	5	6
7	8	9
	0	

And the so-called calculator keyset:

7	8	9
4	5	6
1	2	3
	0	

TABLE 12.19 Advantages and Disadvantages of Coding Techniques

Advantages	Location	Shape	Size	Mode of operation	Label	Color
Improves visual identification	Yes	Yes	Yes		Yes	Yes
Improves tactile and kinesthetic identification	Yes	Yes	Yes	Yes		
Helps standardization	Yes	Yes	Yes	Yes	Yes	Yes
Aids identification in low illumination and in colored lighting	Yes	Yes	Yes	Yes	Yes[a]	Yes[a]
Can help identifying current control setting		Yes		Yes	Yes	
Requires no/little training, is not subject to forgetting					Yes	
Disadvantages						
May reduce ease of use	X	X	X	X		
May require additional space	X	X	X	X	X	
Limited number of codings available	X	X	X	X		X
May be less effective if gloves are worn		X	X	X		
Controls must be viewed, be in field of vision, and be suitably illuminated					X	X

[a]*If appropriately trans-illuminated.*
Source: Modified from Kroemer 1988b.

TABLE 12.20 Maximal Number of Stimuli Useful for Coding

Stimuli	Maximum number
Shape	5
Size	3–10
Color of lights (lamps)	3–5
Color of surfaces	5–9
Flash rate of lights (lamps)	2
Light intensity, lamp brightness	2
Intensity (loudness) of sounds	3
Frequencies of sounds	4
Duration of sounds	2

Source: Data from Cushman and Rosenberg (1991) and MIL STD 1472F; Cushman and Rosenberg (1991), MIL-STD-1472F.

Keying performance may be more accurate and slightly faster with the telephone keyset than with the calculator arrangement, but the performance differences are slight.

12.2.2 Computer Keyboards

Typewriters

Throughout the 19th century, many inventors tried to replace manual penmanship with mechanisms that could print on paper: "typewriting machines." They commonly followed the traditional practice of composing text from single letters, numerals, signs, and spaces. Their efforts were primarily directed at designing and constructing machinery that could generate the imprint as a consequence of an operator pushing a button ("key"). Given the technology at hand at that time, this proved to be a formidable task, and few of the proposed mechanisms were workable; into the 1870s, none was commercially successful.

The various typewriting machines featured a multitude of keyboards: some similar to the long black and white keys on the piano, some featuring double or triple rows of button-like keys, and some showing keys arranged in concave or convex circle segments.[5] While they were busy constructing the actual typing mechanisms, the would-be inventors did not worry much about how to assign letters, numbers, or other signs to specific keys, nor were they concerned with how to arrange them for easy use. "Human engineering" was not part of this design task in the 19th century.

QWERTY

Christopher Latham Sholes was the first inventor to successfully design, produce, and market a typewriter, including a special keyboard. His typewriting machine, patented on August 27, 1878 (US patent 207,559), shows a keyboard with four rows of altogether 44 keys. The keys on the third row are labeled, from the left, QWERTY. This arrangement of those six keys is often used as a short AQ5 label for the arrangement of all the letter ("alpha" or "alphabetic") keys. Remarkably, the current English–language QWERTY keyboard is still essentially true to Sholes's original design, with only the positions of the letters C and X exchanged from the original and the letter M moved one row. Sholes had previously obtained a series of patents on typing machines (in 1868, 79,265 and 79,868; in 1876, 182,511; and in 1878, 199,382, 200,351, 207,557, and 207,558), several of which were coauthored with other inventors. His first 1868 patent (79,265) showed 2 rows of keys, "similar to the keyboard of a piano." His next 1868 patent (79,868) also had 2 rows of keys, with 13 keys each, but now all in the same flat plane. The keys alternated in length and were lettered in numerical and alphabetic order. His 1876 patent and his first three patents of 1878 all exhibit similar arrangements, each with 3 rows of button-like keys affixed to lever-type bars. Those buttons show no assignments to letters or numbers.

Why Sholes decided on any of these key arrangements is now unknown. In his final patent 207,559, Sholes made 14 specific technical claims, but none refers to the key selection or layout. One drawing in this patent shows a frontal view of the invention with four staggered, horizontal rows of keys, with the row farthest from the operator the highest. Another drawing depicts a top view of the 4 straight rows, each with 11 keys, shown in Fig. 12.21.

Sholes's key layout

The 1878 QWERTY layout contains some remnants of an alphabetic arrangement. Sholes was a printer by trade, so we can surmise that the placement showed similarities to the arrangement of the printer's "type case," in which pieces were assorted according

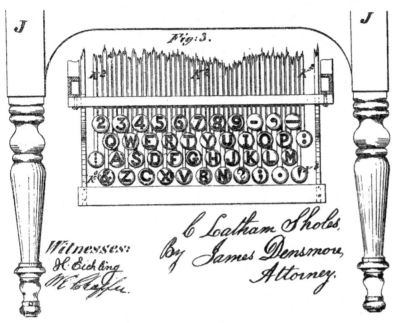

FIGURE 12.21   QWERTY Keyboard in Sholes's 1878 US Patent 207,559.

to convenience of use and not according to the alphabet. Another possible reason for the arrangement of the letters and keys may have been an effort to avoid having typebars (the moving lever-arms of the mechanical typewriter) collide when neighboring typebars were activated in a quick sequence. However, there is no contemporaneous evidence for either of these speculations.

Sholes's typographic machine became the predominant device to type text and numbers on paper. With millions of mechanical typewriters employed privately and in offices, use problems inherent to the design soon became apparent. The work required of the typists' hands and arms as they punched keys on the keyboard was tiresome. The keys had large displacement and stiff resistance, and typists struck them thousands of times in the course of the workday. Such strenuous effort overloaded many typists' hands and wrists: "myalgia" and related over-exertions, previously reported by telegraphists and pianists, now frequently afflicted typists as well.

From 1909 on, other patents for improved key locations were filed, but they maintained Sholes's original layout with its bent columns and straight rows. However, the keyboard's shortcomings were becoming apparent. In 1915, Fritz Heidner advocated substantial changes in the basic keyboard layout. He obtained US patent 1,138,474: on its first page, he announced that he had "invented certain new and useful improvements" in keyboards "to enable the operator to obtain a better view of the keys and to write with greater ease, in a less cramped position than ordinarily. With this object in view, I divide the keyboard into halves and locate the two groups of keys thus formed in such manner that the forearms of the operator ... instead of converging, lie substantially parallel with each other. ... [T]he hands have not to be twisted outward ... and there

Improved layout?

being thus much less strain upon the abducent muscles, writing is rendered considerably less fatiguing."

In addition to splitting the keyboard into left and right halves and placing them at a slant angle to allow better forearm posture, Heidner also arranged the keys in curved rows "in accordance with the natural form of the hand … [C]onvergence of the key groups further facilitates operation of the keys by the fingers in their natural position in the extended axis of the forearm." Fig. 12.22 is taken from his 1915 patent and shows his design recommendations.

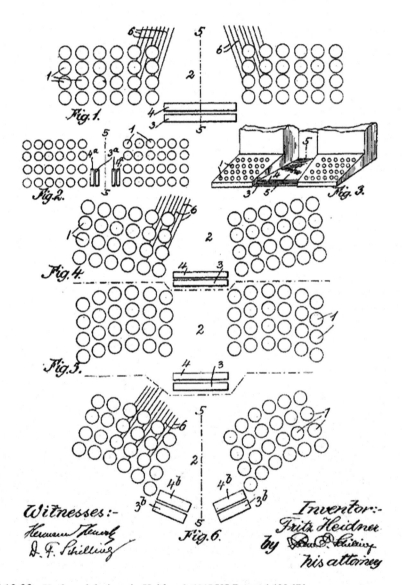

FIGURE 12.22   Keyboard designs in Heidner's 1915 US Patent 1,138,474.

However, no typewriter manufacturer adopted Heidner's improvements; in fact, his recommendations were seemingly forgotten[6] since they were not even mentioned in the literature or in subsequent related patents for about 70 years. Heidner's astute observations preceded by at least a decade the scientific research to measure effort and record fatigue associated with typing.[7] Proposals similar to Heidner's work reappeared later (without referring to him); examples include propositions in the 1940s by Griffith, in the 1960s by Kroemer, and thereafter by many other inventors and manufacturers of improved keyboards in the last decades of the 20th century.[8]

The mechanical nature of the early typewriters made it impractical to change the basic lever system and hence the design of the keyboard. Around 1950, electric auxiliary power reduced the amount of energy that the operator had to apply in each keystroke. Around 1960, electronics began to replace mechanics. The new technology would have allowed new and ergonomic designs of keyboards, but even in 1988 the ANSI/HFS 100 explicitly proscribed a conventional QWERTY keyboard. In the 1980s and 1990s, more keys were added on the sides and on top of Sholes's design so the total number of keys roughly doubled. Standardization perpetuated the convention of keys arranged in straight horizontal rows with zigzag columns on the QWERTY part but in straight columns on all other keysets.[9]

From manual to electronic

12.2.2.1 Problems with Current Keyboards

Certainly, the most serious problems with the use of keyboards stem from the excessive motions of the wrist and hand digits to accomplish the needed key activations. Typing at 50 words a minute, 5 letters per word, for 6 hours a day, means bending your fingers 90,000 times to press down keys, followed by the same number of finger elevations to release the keys. Many hands simply cannot take such frequent mechanical strain, and they respond with irritation and possibly painful overuse disorders (see Chapter 9).

Other problems that commonly arise are also related to design details of the keyboard. Often leading to mental and physical strain and fatigue, they include:

- Zigzag key columns on the QWERTY key pad, but straight columns on other keypads.
- Straight rows of keys, which are not aligned to fingertips.
- Horizontal rows of keys that enforce extreme inward rotation (pronation) of the forearms.
- Large numbers of keys in some keyboards (up to 130 on "full-size" keyboards) that require extreme digit and wrist motions and sideways stretch of the little finger.
- Arms and hands held over the keyboard with no support.
- QWERTY-type layout poorly suited for tablets and other touch screen devices when typing with thumbs.

12.2.2.2 Rethinking the Keyboard

With current technology, most of the (language-based) information we need to transfer to our computer is done by hand; the common interface is the conventional flat keyboard, often accompanied by other input means such as mouse, trackball, joystick, or light pen. The design of these interfaces affects the workload and the motor activities of the operator and dictates the layout of the computer workstation (see also Chapter 10).

Conventional keyboard

The keyboard that is still typically used in computing today is the one derived from a long-ago typewriter. There are many drawbacks to the conventional QWERTY keyboard due to its ergonomically contraindicated features. Associated with the widespread use of this keyboard are overuse disorders,[10] likely related to the frequency of key operation combined with awkward forearm and wrist postures and supports (see also Chapter 9).

Modified keyboards

Many proposed improvements to the QWERTY keyboard have been suggested, categorized into several broad categories. Some relocate letters on the keyboard and change the geometries of the keyboard, such as arranging keys in curved rows and columns. Others divide the keyboard into halves, one for each hand, arranged so that the center sections are higher than the outsides, so hands do not need to pronate as required on the flat keyboard. Still others involve activating two or more keys simultaneously—called chording—to generate one character or whole words, or chunks of words, such as those used by court reporters. And still others recognize that keys do not have to be of the conventional binary (ON–OFF) tap-down type but may be toggled or turned and have three or more different contact positions.

> Chord keyboards involve activating two or more keys simultaneously to generate one character or whole words, or chunks of words, such as those used by court reporters. And still others utilize keys that are toggled or turned and have three or more different contact positions, such as the ternary chord keyboard as by Lawrence Langley in his 1988 US patent 4,775,255, which has only four keys for each hand (Fig. 12.23). This allows the hand to rest on built-in wrist pads, and the keys can support the fingers because the keys are toggled rather than tapped down.

Adjusting to new design

In the past, new designs have been rejected with the argument that it would be too difficult and time-consuming to learn working on a different keyset—even if increases in typing speed and accuracy could ultimately be achieved. Anecdotal experience and research[11] indicate that we can learn to operate new input devices with surprising ease

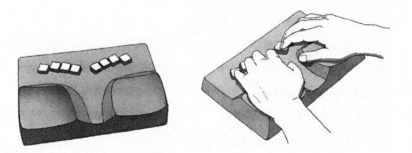

FIGURE 12.23   Ternary chord keyboard as in Langley's 1988 US Patent 4,775,255.

and speed. Who in 1990 would have predicted that, within two decades, mini keyboards would become ubiquitous on millions of pocket-sized mobile phones?

Many of the key arrangements on mobile phones continue follow the traditional QWERTY system; however, the tiny size of the keys disregards specifications in ANSI/ HFS 100 and design guidelines regarding suitable spatial location, dimensions of the overall keyboard and of each key, and appropriate key characteristics in displacement and resistance. Users have proven to be willing and able to accept new designs and are motivated to become proficient with them.

12.2.3 Other Input Devices

People vary in size, shape, ability, and preference (including language), so there is no one keyboard that is best for every person. The conventional large QWERTY keyboard is likely to remain in common office use because it is entrenched, but it is not suited to all. A variety of ergonomic designs have been developed to address specific needs under specific work conditions, and some are viable alternatives for QWERTY. Of course, given that most of us are accustomed to a traditional keyboard, there is a learning curve in acquiring new skills on new keysets. With advances in technologies, sensors, and activation devices, we should expect that there will be a variety of ways to transfer input signals to computers other than keyboards.

In addition to the traditional keyboards already discussed, other input devices can be used:

- Mouse: A palm-sized, hand-contoured block with one or more finger-operated button(s) commonly slid on a surface (mouse pad), mostly used to move a cursor.
- Puck: Similar in shape to a mouse but typically has a reticular window used on a digitizing surface (tablet).
- Trackball: A ball mounted in an enclosure whose protruding surface is moved by palm or hand digits, usually to move a cursor.
- Joystick: A short lever, operated by a fingertip or, if larger, by the hand, typically for moving a cursor for pointing or tracking.
- Stylus: A pencil-shaped, handheld device often used for object selection, freehand drawing, and cursor movement, usually on a tabletop digitizing surface.
- Light pen: Similar to a stylus but commonly used on a display surface.
- Tablet: A flat, slate-like panel over which a stylus or puck or just the tip of a finger is moved, usually for cursor movement and object selection.
- Overlay: An opaque overlay of a tablet that provides graphics.
- Touch screen (touch sensitive panel): An empty frame, or overlay, mounted over the display screen that locates the position of a finger or pointing device, used for object selection, object movement, or drawing.

Typical uses of these devices include moving a cursor, inputting a single character or number, manipulating the screen content, digitizing information, and pointing to insert or retrieve information, but the tasks change with evolving technology, software, and needs.

The design of the keyboard or any other manipulated input device, especially of a mouse or puck, determines hand and arm posture. The essentially horizontal surface of conventional keyboards and of most mice requires that the palm also be kept approximately horizontal; this is uncomfortable because it requires strong pronation in the forearm, close to the anatomically possible extreme. A more convenient angle of forearm and wrist rotation is achieved by a sideways down tilt of the mouse surface that is in contact with the palm. This is similar to splitting a keyboard and tilting the sections down to the side.

Hands and wrists are not the only body parts affected by the designs of keys and other input devices. The location of hand-operated input devices of any kind naturally determines the position of the hand, of the forearm, and of the upper arm and, consequently, even affects the posture of the trunk. As a rule, the best position for the hand is in front of the body, at elbow height. If mouse, puck, stick, or other input instruments are used jointly with a keyboard, they should all be placed closely together; in fact, many keyboard housings already contain a trackball or trackpad.

Supporting arm and hand appropriately relieves shoulder and back muscles from their holding tasks. Various kinds of arm, wrist, and hand rests can be employed during work, or at least during breaks in extended work. Such rest periods are helpful but by themselves are not sufficient to overcome unsuitable equipment and excessive workloads.

12.2.4 New Solutions

Traditionally, we input data into our computers through manual interaction between our fingers and such devices as keys, mouse, trackball, or light pen. These entry techniques force us into a complex task sequence:

- *First, breakdown*: We must break down formulas, words, and sentences into letters, numerals, and symbols.
- *Second, association*: We must make an association between each letter, numeral, or symbol to a distinct key, another device, or a function element.
- *Third, separate entries*: We must manually operate keys and the like for separate entry of the numerals, symbols, and letters into the computer.

After all of this, we then use the computer program to reconstruct the original words, sentences, and formulas which we had just broken into components. This is a time-consuming, complicated use of the human mind, and becomes intensive manual work. In a sense, we are wasting efforts of both our bodies and minds.

> Sign language (as used by people with hearing impairments) and shorthand (as used by stenographers) shows that it is often unnecessary to dissect information into letters, as customarily done when typing characters into a computer: inputs could be by other means, and in chunks.

It would be much better to transmit the message from the human to the computer as a whole, at least in batches. For this, voice communication with the computer is an obvious and (now) technically feasible solution. Speech recognition converts speech into machine-readable text, meaning a string of character codes. Speech becomes an especially interesting alternative when we want to input data hands-free, perhaps while walking or otherwise physically taking a break, or if otherwise unable or unwilling to keyboard.

One can combine methods, such as supplementing voice recognition to keyboarding. Using a combination of voice and keyboard operations, and maybe adding in a mouse or stylus, could be an extremely efficient and effective way of entering information into a computer, speeding up work performance while resting the hands and wrists, and reducing the risk of a musculoskeletal injury.

Many other ways exist to generate computer inputs. Consider that we might use:

- Hands and fingers for pointing, gestures, sign language, and tapping
- Arms for gestures, making signs, and moving or pressing control devices
- Feet for motions and gestures and for moving and pressing devices
- Legs for gestures and moving and pressing devices.
- Torso, including the shoulders, for positioning and pressing
- Head, also for positioning and pressing
- Mouth for lip movements, tongue, or breathing such as through a blow/suck tube
- Face for grimaces and other facial expressions
- Eyes for tracking

Combinations and interactions of these different actions could be useful as inputs, similar to the way we utilize them to communicate face to face, to convey information and express meaning and mood. Of course, the input method selected must be clearly distinguishable from environmental clutter or other "loose energy" (e.g., jittery movements or tapping fingers) that interferes with sensor pickup. The ability of a sensor to detect the input signals depends on the type and intensity of the signal generated. As an example, it may be quite difficult to distinguish between different facial expressions. Thus, the use of other than conventional input methods depends on the state of technology, which includes the tolerance of the system to either missed or misinterpreted input signals.

In the United States, ANSI/HFS standards and HFES guidelines refer to specific design features such as key spacing, key size, key actuation force, and key displacement. They also provide design dimensions for mouse and puck input devices, track-balls, joysticks, styli and light pens, tablets and overlays, and touch sensitive panels. However, the ANSI standard merely specifies "acceptable" applications based on human factors engineering research and experience, and the standard does not apply to operator health considerations or work practices. The standard permits alternate computer workstation technologies so as to not impede the development and use of novel solutions.

12.3 DISPLAYS

Displays provide the operator with information about the status of equipment. Displays are either visual (often a light, scale, counter, CRT, flat panel), auditory (such as a bell, horn, beep, recorded voice), or sometimes tactile (such as shaped knobs, or Braille writing). Labels and instructions/warnings are also special kinds of displays. (Additional information on displays used in computerized office workspaces can be found in Chapter 10.)

The four cardinal rules for displays are:

1. Display only the information that is essential for adequate job performance.
2. Display information accurately to facilitate operator decision making and control actions.
3. Present information in the most direct, simple, understandable, and usable form possible.
4. Present information in such a way that failure or malfunction of the display itself will be immediately obvious.

12.3.1 Selecting the Display

Visual or auditory

Which kind of display to select depends on the existing conditions and circumstances. A *visual display* is appropriate if the environment is noisy, the operator stays in place, the message is long, complex, remains displayed for some time, conveys change, will be referred to later, deals with spatial location. An *auditory display* is appropriate if the environment must be kept dark, the operator moves around, the message is short, simple, requires immediate attention, deals with events in time.

Selecting either a visual or an auditory display also depends on the purpose. The objective may be to provide:

- Status information: about the current state of the system.
- Historical information: about the past state of the system, such as the path that a milling tool has cut.
- Predictive information: about the future state of a system, such as the expected course and arrival of an airplane in flight, given present steering settings.
- Instructional information: telling the operator what to do, and how to do it.
- Commanding information: giving directions or orders for a required action.

12.3.1.1 *Emergency Signals*

An *emergency alert* is best conveyed by a penetrating auditory warning signal (see below) together with flashing light, usually red. The environment may be very busy in terms of lights or noise, which can mask warning signals: consider the examples of automobile traffic, construction sites, the interior of trucks or other vehicles. This leads to special considerations in the design of the warning systems and of hearing protection devices because affected people must perceive the alarm above ambient light or noise levels.[12]

Visual means to signal an emergency should use a luminance contrast with the immediate background of at least 3:1. Flash rate should be between three and five pulses per second, with ON time about equal to OFF time. If the flasher device should fail, the light must remain ON steadily; warning indicators must never turn off merely because a flasher fails. A voice warning (such as DANGER–STOP) should be used, if feasible. The operator may acknowledge the emergency by turning off either the light or sound signal.

Visual emergency alerts

Audible means to signal an emergency include bells, horns, sirens, and voice announcements. Temporal and voice signals are the most effective means. Temporal coding is commonly accomplished by interrupting a steady sound in the sequence such as 0.5 seconds ON, 0.5 seconds OFF, 0.5 seconds ON, 0.5 seconds OFF, in a repeating cycle. A distinctive three-pulse temporal pattern signals an immediate emergency evacuation. This pattern, which is used for smoke alarms, for example, is also called the Temporal-Three alarm signal, or T-3 (ISO 8201 and ANSI/ASA S3.41 Temporal Pattern) and the pattern of ON–OFF sounds is followed by a 1.5 seconds pause, repeated for a minimum of 180 seconds.

Audible emergency alerts

12.3.2 Visual Displays

Visual displays are of two basic types: "check" displays for attribute data or "read" displays for variable data.

12.3.2.1 "Check" Displays

"Check" displays indicate whether or not a certain condition exists, such as is temperature "normal" (often green) or "hot" (red) or "cold" (blue). Fig. 12.24 shows an example of a simple check display.

Check

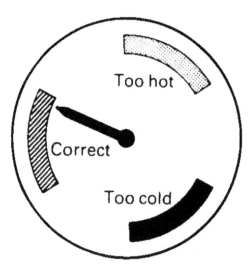

FIGURE 12.24 Typical "check" display.

LIGHT, OR COLOR, SIGNALS

"Check" displays often use lights (colors) to indicate the status of a system (such as ON or OFF) or to alert the operator that the system, or a portion thereof, is inoperative and that special action must be taken. Common light (color) coding systems are:

- A green signal indicates that the monitored equipment is in satisfactory condition and that it is all right to proceed. For example, a green display may provide such information as "go ahead," "in tolerance," "ready," "power on."
- A yellow signal advises that a marginal condition exists and that alertness is needed, that caution be exercised, that checking is necessary, or that an unexpected delay exists.
- A red signal alerts the operator that the system, or a portion thereof, is inoperative and that a successful operation is not possible until appropriate correcting or overriding action has been taken. Examples of a red-light signal include "malfunction," "failure," "error."
- A flashing red signal denotes an emergency condition that requires immediate action to avert impending personal injury, equipment damage, and other serious consequences.
- Blue generally has no special signal meaning but is used, often together with flashing red, on emergency vehicles.
- A white signal has no correct/wrong implications but may indicate that certain functions are on.

12.3.2.2 "Read" Displays

"Read"

"Read" displays convey quantitative information about details, such as providing exact information about a value, magnitude, direction, location, trend of changes, and the like.

Quantitative displays

Quantitative displays are more complex than those for simply checking a condition; they must be "read" with some care to extract quantitative data (temperature in degrees, for e.g.,). Traditionally, four types of displays are distinguished:

1. moving pointer (fixed scale),
2. moving scale (fixed pointer),
3. counters,
4. pictorial (icon).

These displays used to be mechanical, but now are often generated electronically. Table 12.21 lists relative advantages and disadvantages of these four kinds of displays. To ease the cognitive load of observers, it may be advisable to reduce the information content. For example, a quantitative indicator (be it numerical or a pointer) showing temperature in degrees may be simplified to a "check" display that just signals states, like "too cold," "correct", or "too hot."

Pointer and Scale

Many quantitative displays use a moving pointer over a fixed scale. The scale may be straight (either horizontal or vertical), curved, or circular. The scales should be simple

TABLE 12.21 Characteristics of Quantitative Displays

Used for	Moving pointer	Moving scale	Counter	Pictorial
Quantitative information	Fair Difficult to read while pointer is in motion	Fair Difficult to read while scale is in motion	Good Minimal time and error regarding exact numerical value but difficult to read when moving	Fair Direction of motion-scale may conflict causing ambiguity in interpretation
Qualitative information	Good Easy to locate pointer; numbers and scale need not be read; changes in position easily detected	Poor Difficult to judge direction and magnitude of deviation without reading numbers and scale	Poor Numbers must be read; changes in position not easily detected	Fair Easily associated with the real world situation
Setting	Good Simple and direct relation of pointer motion to motion of setting knob; position change aids monitoring	Fair Possibly ambiguous relation to motion of setting knob; no change in pointer position to aid change monitoring; not readable during rapid setting	Good Most accurate monitoring of numerical setting; yet, relation to motion of setting knob is less direct than with moving pointer; not readable during rapid setting	Fair Control/display relationship easy to observe
Tracking	Good Pointer position easily controlled and monitored; simplest relation to manual control motion	Fair No changes in position to aid monitoring; relation to control motion maybe ambiguous	Poor No gross changes in position to aid monitoring	Fair Control/display relationship easily observed
Difference estimation	Good Easy to determine negatively or positively by scanning scale	Fair Subject to reversal errors	Poor Requires mental calculation	Fair Control/display relationship easily observed
In general	Requires largest exposed and illuminated area on panel; scale length limited unless multiple pointers are used	Saves panel space; only a small section of scale need to be exposed and illuminated; allows long scale	Most economical use of space and illumination; scale length limited only by available number of digit positions	Pictures, symbols, icons need to be carefully designed and pretested

Source: Modified from MIL-HDBK 759, MIL-STD-1472F.

and uncluttered; graduation and numbering should be selected so that correct readings can be taken quickly. Numerals should be located outside the scale markings beyond the reach of the pointer so that they cannot be obscured. The pointer, on the other side of the scale, should end with its tip directly at the markings. Fig. 12.25 provides related information.

The scale mark should show only the level of detail the operator must read and not go into unnecessary detail. All major marks should be clearly numbered. Progressions of

Scale Design

A –Fixed scale–moving pointer preferred: three-level marking, numbered at each major mark. Pointer adjacent to graduation marks to preclude obscuration of either marks or numbers.

 B –For short, finite scale, every 5th graduation is marked; using only two-level marking.

 C –When scale crowding makes pointer-mark association difficult, scale may be graduated in units of two, with two-level scale marking and numbering at each major marking.

 D –When dial face is deeply inset within instrument case and visibility of numbers is more important than scale mark-pointer association, pointer may be located inside the graduations along with numbers at major markings. Pointer width should be narrowed at point at which it passes numbers.

 E –Moving scale against an index mark or pointer may be used when scale length precludes the fixed scale format (i.e., graduation marks would be too close together). Open window configuration helps operator focus on significant scale area.

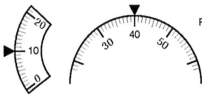 F –When open window configuration is oriented in vertical position, numbers should appear upright as each number passes the index mark or pointer. Total scale exposure is desirable when operator needs to refer to other portions of the scale.

FIGURE 12.25   **Alternatives for scale graduation, numbering, and pointer position.** *Source: Adapted from MIL-HDBK 759.*

either 1, 5, or 10 units between major marks are best. The largest admissible number of unlabeled minor graduations between major marks is nine, but only with a pronounced graduation at 5. Numbers should increase as follows: left-to-right; bottom-to-top; clockwise. Recommended dimensions for the scale markers are presented in Fig. 12.26.

12.3.2.3 *Locating and Arranging Displays*

Some guidelines on locating and arranging displays:

- Position displays within the normal viewing area of the operator, with their surfaces perpendicular to the line of sight. The more critical the display, the more centered it should be within the operator's central cone of vision.

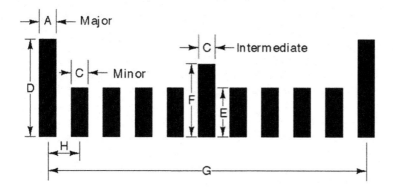

Minimum scale dimensions suitable even for low illumination

Dimension (in mm)	Viewing distance (in mm)		
	710	910	1,525
A (Major index width)	0.89	1.14	1.90
B (Minor index width)	0.64	0.81	1.37
C (Intermediate index width)	0.76	0.99	1.63
D (Major index height)	5.59	7.19	12.00
E (Minor index height)	2.54	3.28	5.44
F (Intermediate index height)	4.06	5.23	8.71
G (Major index separation between midpoints)	17.80	22.90	38.00
H (Minor index separation between midpoints)	1.78	2.29	.381

FIGURE 12.26 **Scale marker dimensions, suitable even for low illumination.** *Source: Adapted from MIL-HDBK 759.*

- Avoid glare on and from displays; a dark background can make a display act like a mirror.
- Group displays functionally or sequentially so that the operator can easily use them.
- Make sure that all displays are properly illuminated or self-luminant, coded, and labeled according to their function.

A group of pointer instruments can be arranged so that all pointers, under normal conditions, are aligned. If one of the pointers deviates from that normal pointer direction, its displacement from the aligned configuration is particularly obvious. Fig. 12.27 shows examples of such arrangements.

Aligned pointers

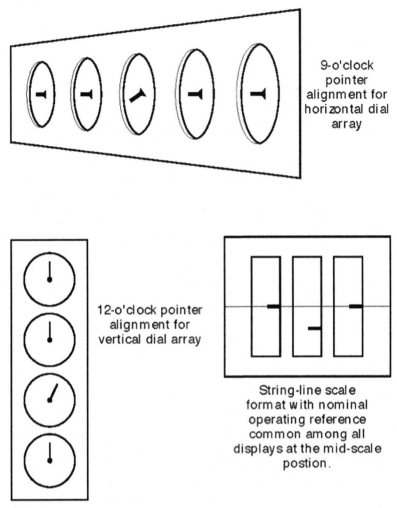

FIGURE 12.27   **Aligned pointers.** *Source: Adapted from MIL-HDBK 759.*

12.3.2.4 Display–Control Assignments

Display and
control

When a display shows the results of a control's settings, control and display should be located close to each other to allow for correct, convenient and fast adjustments. When the control is directly below or to the right of the display, their relationship usually is clear, as shown in Fig. 12.28A. Other expected spatial relationships exist or can be trained, but the stereotypes are often not strong and may depend on the user's background and experience. Fig. 12.28 shows examples of appropriate and of poor display–control arrangements.

Movement
relationships

Expected movement relationships are influenced by the types of controls and displays used. If both are linear or rotary, the expectation is that they move in corresponding and congruous directions: both up, or both clockwise. When the movements are

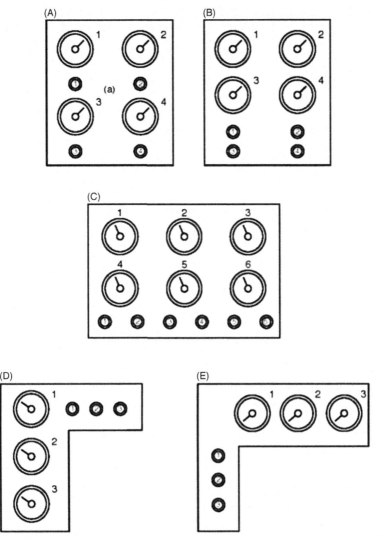

FIGURE 12.28 Acceptable (A) and difficult (B, C, D, E) arrangements of displays and their controls. *Source: Adapted from MIL-HDBK 759.*

incongruous, their preferred movement relationship can be taken from Fig. 12.29. The following rules generally apply:

- "Clockwise-for-increase": Turning the control clockwise should cause an increase in the displayed value.
- "Geared" (Warrick's) rule: A display (pointer) is expected to move in the same direction as the side of the control close to the display.

The *control/display* (C/D) *ratio* describes how much a control must be moved to adjust a display. The C/D ratio is like a gear ratio: if a large control movement produces only

C/D ratio

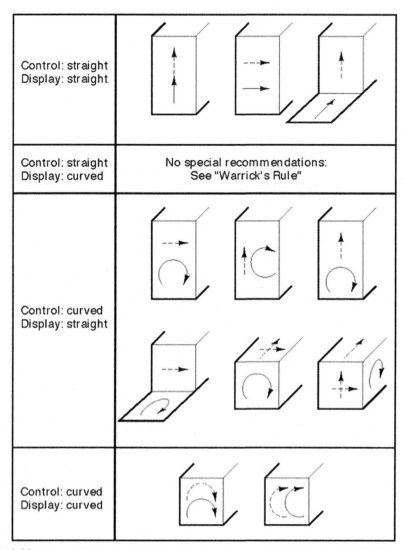

FIGURE 12.29   **Compatible control-display directions.** *Source: Adapted from Loveless (1962).*

a small display motion, the C/D ratio is high, and the control is said to have low sensitivity. The inverse ratio is often called gain.

Usually, two distinct movements are involved when an operator attempts a setting: first a fast motion to an approximate location, then a fine adjustment to the exact setting. The optimal C/D ratio minimizes the sum of the two movements. For continuous rotary controls, the C/D ratio is usually 0.08–0.3, for joysticks 2.5–4; however, the most appropriate ratios depend heavily on the given circumstance and must be determined for each application.

II. DESIGN APPLICATIONS

12.3.3 Auditory Displays

As a general rule, auditory signals are better suited than visual displays when the message must attract attention even in a busy environment. Therefore, auditory displays are predominantly used as warning devices, especially when the message is short or simple, often together with a flashing light.

Auditory signals may be single tones, sounds (mixture of tones), or spoken messages. Tones and sounds may be continuous, periodic, or unevenly timed. They usually come from horns, bells, sirens, whistles, buzzers, or loudspeakers. Use of tonal signals is recommended for qualitative information, such as for indications of status, or for warnings; speech may be appropriate for all types of messages.

Auditory signals

Tones, sounds, messages

- Tonal signals should be at least 10 dB louder than the ambient noise (see Chapter 5).
- Signal frequencies should differ from those dominating background noise.
- Signal frequency range should be within 200 and 5000 Hz, best between 500 and 1500 Hz. If the signal undulates or warbles, 500–1000 Hz are advised.
- Use frequencies below 1000 Hz when signals must travel long distances (more than 300 m).
- Use frequencies below 500 Hz when signals must "bend around" sound barriers or penetrate them.
- Buzzers may have frequencies as low as 150 Hz and horns as high as 4000 Hz.

The conspicuousness of a tonal signal can be elevated by increasing its intensity, by interrupting it repeatedly, and by changing its frequency. An example: increase from 700–1700 Hz in 0.85 s, be silent for 0.15 s (cycle time 1 second), then start over.

Conspicuous signals

Word messages may be pre-recorded or produced as digitized or synthesized speech. The most important aspects are good intelligibility together with proper intensity to penetrate background noise. Pre-recorded messages can sound natural and personal whereas synthesized speech usually does not sound natural and therefor may startle the listener—which draws attention to the message and can be effective in making people respond as needed.

Spoken messages

12.4 LABELS AND WARNINGS

12.4.1 Labels

Ideally, no label should be required on equipment, display or control to identify it and explain its use. However, if recognition and proper use are not intuitive or self-explanatory, labeling must be used for clear and accurate information. The following guidelines are recommended for labels:

Labelling

- Orientation: A label and the information printed on it should be oriented horizontally (assuming that is how the operator's accustomed language is written) so that it can be read quickly and easily.
- Location: A label should be placed on or very near the item that it identifies.

- Standardization: Placement of all labels should be consistent throughout the equipment and the system.
- Equipment functions: A label should primarily describe the function ("what does it do?") of the labelled item.
- Abbreviations: Common abbreviations may be used; if a new abbreviation is necessary, its meaning must be obvious to the reader. The same abbreviation should be used for all tenses and for the singular and plural forms of a word. Capital letters are fine and generally recommended; periods normally omitted.
- Brevity: The label inscription should be as concise as possible without distorting the intended meaning or information. The text should be clear, unambiguous, with minimal redundancy.
- Familiarity: Words familiar to the operator are best.
- Visibility and legibility: The operator must be able to easily and accurately read the label at expected actual reading distances, at the expected worst illumination level and within the expected environment vis-à-vis vibration and motion. Important considerations include the contrast between the lettering and its background; the height, width, stroke width, spacing, and style of letters (font); and the specular reflection of the background, cover, and other components.
- Font and size: The legibility of written information is determined by text typography: style, font, arrangement and appearance. The font chosen should be simple, bold, and vertical; examples include Calibri, Futura, Helvetica, Tempo, Vega.
- Character height recommendations depend on the viewing distance:
  - Viewing distance 35 cm, suggested height 22 mm
  - Viewing distance 70 cm, suggested height 50 mm
  - Viewing distance 1 m, suggested height 70 mm
  - Viewing distance 1.5 m, suggested height at least 1 cm
- The ratio of stroke width to character height should be between 1:8–1:6 for black letters on white background, and 1:10–1:8 for white letters on black background.
- The ratio of character width to character height should be about 3:5.
- The space between letters should be at least one stroke width.
- The space between words should be at least one character width.
- For continuous text, upper- and lower-case letters should be mixed.
- For short labels, upper-case only letters are best.

Note that any labels with words need to be understandable by the operators; this means that appropriate language(s) need to be represented.

12.4.2 Warnings

Safe use

All devices should be *safe to use*—this is the first priority in safety engineering. In reality, complete safety cannot always be achieved through design. Then the next priority is to *remove the dangerous interface*; for example, eliminate all level road–rail crossings. If impossible, then the next step is *guarding*. If a hazard still remains, one must *warn users of danger* associated with product use and *provide instructions for safe use* to prevent injury or damage.

Any known or knowable potential danger of injury from normal use of a product means a warning must be issued.

Reasons to warn

- The more serious the potential injury, the greater the duty to worn.
- The less obvious the danger, the greater the duty to worn.
- The more insidious the onset of injury, the greater the duty to worn.
- The larger the number of people at risk, the greater the duty to worn.

People should be warned about various concerns, including:

What to warn about

- The importance of proper use of the product or device
- Improper or excessive use may cause serious injury
- Certain people are at particularly elevated risk and must take particular care
- Seek medical attention immediately if any symptoms appear

A product manufacturer has the duty to warn:

Whom to warn

- Potential product users, so that they may learn to protect themselves
- Purchasers of the product who might be at risk so that work practices and work stations (where the product is used) may be properly modified
- Sales, marketing, and service people in contact with users and customers
- The general public

Guidelines on how (on what materials and surfaces) to warn include:

How to warn

- Warnings on the product itself (active is better than passive, see below)
- Warnings in instructions in product manual or use directions
- Warnings in promotional literature and advertising
- Warnings through sales and service staff's instructions to customers and users

12.4.2.1 Active Versus Passive Warning

An active warning usually consists of a sensor that notices inappropriate use, and of an alerting device that warns the human of an impending danger. While active warnings are preferable, in most cases passive warnings are used, which usually consist of a label attached to the product and of instructions for safe use in the user manual. Such passive warnings rely completely on the human to recognize an existing or potential dangerous situation, to remember the warning, and to behave prudently.

Active or passive

Factors that influence the effectiveness of product warning information have been compiled[13] and are listed in Table 12.22.

12.4.3 Design of Warnings

Labels and signs for passive warnings must be carefully designed by following the most recent government laws and regulations, national and international standards, and best-practice human engineering information.[14] Most warnings are in writing, but auditory warnings (see above) or tactile means are generally better.

TABLE 12.22   Factors that Influence the Effectiveness of Warnings on Products

	Low effect	High effect
User is familiar with product	X	
User has never used product before		Yes
High accident rate associated with product		Yes
Probability of accident is low	X	
Consequences of accident are likely to be severe		Yes
User is in a hurry	X	
User is poorly motivated	X	
User is fatigued or intoxicated	X	
User has been injured by product previously		Yes
User knows good safety practices		Yes
Warning label is adjacent to the hazard		Yes
Warning label is easily legible and easy to understand		Yes
Active warnings alert user only when some action is necessary		Yes
Active warnings frequently give false alarms	X	
Product carries warning labels that seem inappropriate	X	
Warning contains only essential information		Yes
Source of information on the warning is credible		Yes

Source: Adapted from Cushman and Rosenberg (1991).

Visual warnings   Warning labels and placards may contain text, graphics, and pictures; often in redundant form. Graphics, particularly pictures, pictograms and icons, appeal to people with different cultural and language backgrounds, if these depictions are selected carefully. One must remember, however, that users of different ages and experiences, and users of different national and educational backgrounds, may have rather different perceptions of dangers and warnings.

Symbols and icons   Symbols or icons are simplified drawings of objects or abstract signs, meant to identify an object, warn of a hazard, or indicate an action. They are common in public spaces, automobiles, computer displays, and maintenance manuals. The Society of Automotive Engineers and the ISO, for example, have developed extensive sets of symbols and established guidelines for developing new ones. Some of the symbols for use in vehicles, construction machinery, cranes, and airport handling equipment are reproduced in Fig. 12.30. Note that abstract symbols (a "line" for ON) and simplified pictorials may require learning or the viewer's familiarity with the object (i.e., that an hourglass represents "elapsed hours"). When selecting symbols or developing new icons, the cultural and educational background of the viewer must be carefully considered: many symbols have ancient cultural roots and may invoke unexpected and unwanted reactions.[15]

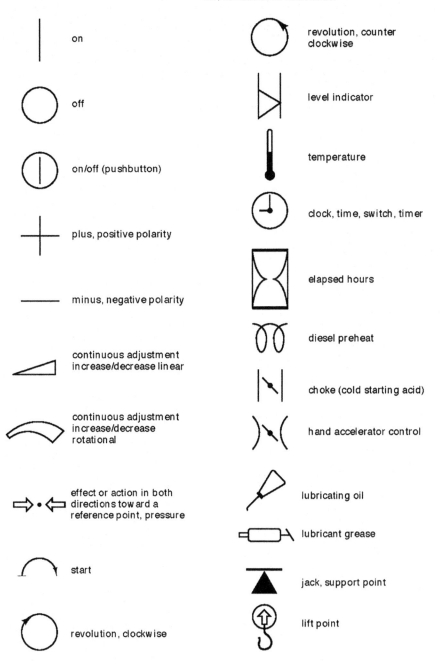

FIGURE 12.30 Sample of ISO symbols.

12.5 AUTONOMOUS SYSTEMS

With increasing sophistication of sensors, data processing, and communication systems, autonomous systems are becoming more common. Although automated systems do not themselves require ergonomics, the human factor remains: human designers have a responsibility to ensure human control is maintained using ergonomic means, considering both physical and psychological human factors.

12.5.1 Autonomous Ships

Ships

Ships are increasingly controlled by electronic systems: The bridge of a modern vessel is more likely to have computer screens and joysticks than a giant ship's wheel. Autonomous and unmanned vessels are already in use in some arenas, such as for clearing mines, underwater surveying and monitoring, and ocean data collecting. The newest ships serving the offshore oil and gas industry in the North Sea, for example, use dynamic positioning systems which collect data from satellites, gyrocompasses, and wind and motion sensors to automatically hold their position when transferring cargo—which is also done by remote control—to and from platforms, even in the heaviest of swells.

The shipping industry is as keen as its automotive counterpart to get autonomous ships running commercially. The advantages of unmanned ships are manifold, but primarily center on keeping people safe by removing them from the hazards of at-sea operations (in particular, accidents due to human error), and reducing the cost of these operations.

12.5.2 Autonomous Aircraft

Airplanes

Commercial airplanes have come a long way from when pilots set courses, manned controls, and switched navigation radio frequencies; the latest airliner models feature automatically updated flight-management systems, known as "fly-by-wire" systems. Modern commercial aircraft are generally flown by a computer autopilot that tracks its position using motion sensors and dead reckoning, corrected as necessary by a global positioning system (GPS). Software systems are also used to land commercial aircraft. The pilot and copilot aboard no longer need do much; on a 2.5-hour domestic flight, autopilots and flight-management systems do around 95% of the work.

Drones

Drones and unmanned aerial vehicles (UAVs) have been used when manned missions were too dangerous or otherwise not recommended. Today, drones are widely used for non-military uses as well: aerial photography, package deliveries, policing and surveillance. These UAVs are also becoming increasingly automated.

Spacecraft

Often we think of spaceships as being manned, such as the Apollo missions to the moon or flights to the International Space Station. However, most spacecraft are actually crewless. Space probes, lunar probes, interstellar probes, and satellites are all examples of robotic or unmanned spacecraft. Spaceships are generally autonomous due to the complexity of having astronauts on board: it is expensive and risky to include all the supplies that astronauts would require (such as food, breathable air, and medical supplies). Further, human astronauts may not live long enough to travel the vast distances of space.

12.5.3 Autonomous Vehicles

Connected vehicles (CV), which are available right now (in 2018), are those that allow vehicles to communicate with each other and the world around them. For example, a navigation system may provide dynamic route guidance, including information on congestion or construction on the road ahead. The goal of the connected vehicle is to supply useful information to a driver or a car so that the driver can make informed decisions. The CV is not making any choices for the driver; it is just supplying information.

Connected vehicles

A fully autonomous vehicle (AV) can drive itself; it does not require a human driver. Although AVs do not need CV technology to function, they are safer and more efficient with the technology. For example, AVs may perform better if they have the latest information on road conditions and congestion. AVs also require some form of connectivity to keep up-to-date on software.

Autonomous vehicles

Depending on the level of autonomy, more or less involvement of the human driver is required. At low levels of automation, most (or all) of the functions of the vehicle are controlled by the driver, such as steering and speed (acceleration and braking). As more automation is included, the driver serves less of a function while the AV takes more control.

Levels of autonomy

There are many (potential) benefits associated with AVs, including:

Benefits of AV

- Safety: With no human driver in control, collisions involving driver error would be virtually eliminated.
- Heightened productivity: Drivers in fully automated vehicles can spend their time working or in other pursuits when they no longer need to pay attention to driving tasks.
- On-demand transportation services: Ride sharing companies could provide door-to-door, on-demand, autonomous pick-up and drop-off, offering convenience to those who are unwilling or unable to drive (including children, seniors, and the physically disabled).
- Eco-friendliness: With less need for individual car ownership and reduced congestion, AVs could lead to less emission-related pollution.

Although it is still early, there are many potential implications regarding AVs to be considered.

Implications of AV

- Regulation and insurance: Governments need to consider how to regulate the emerging technology, for example, how to license AVs, who is responsible for maintaining AV software, who is responsible in the case of a failure or collision.
- Urban planning: Congestion may ease in urban areas, parking lots may become available for other uses, and more convenient commutes may impact where people choose to live.
- Security and policing: Although AVs may make emergency response times quicker (e.g., automated reporting of emergencies, or prioritization of emergency response

vehicles in traffic), they may also cause more emergencies if AVs are used for criminal or terroristic acts.
- Privacy Concerns: AVs are a risk for cyber hacking and privacy breaches are a concern.

12.5.4 Failures of Autonomous Systems

Large ships, large aircraft, spacecraft and autonomous road vehicles present increasingly common examples of systems which, under normal conditions, can operate under computer control, without human interference. However, if external conditions become unforeseeably abnormal, or if critical components of the craft and of its guidance system become non-functional, humans must interrupt the autonomy of the system and take over control of the craft.

A small number of events of this kind have taken place so far. Consequently, there is limited experience with recognition and analysis of problems and solutions. The following examples raise more issues than they present solutions. However, the designers of autonomous systems must be prepared to address the ergonomic aspects of the product: both the physical aspects of the controls and displays as well as the psychological abilities of the human operator. Physical ability to intervene requires, at a minimum, readily accessible controls that provide clear information on the status and/or required interventions. Psychological ability to intervene requires an operator who is engaged, alert, and trained.

DITCHED PASSENGER AIRPLANES

On August 21, 1963, a lack of fuel shut down the engines on a passenger plane near Leningrad. The pilot glided to the Neva river and ditched the plane close to a tugboat. While the plane began to flood, the tugboat captain broke the aircraft's windshield, tied a cable to the cockpit wheel and towed the craft onto a river bank. The 59 passengers and crew then evacuated the plane via a roof hatch.

On January 15, 2009, after about three minutes of regular climb-out from New York's LaGuardia Airport, at about 860 m altitude, US Airways Flight 549—helmed by pilot "Sully" Sullenberger—struck a flock of flying geese. This caused both jet engines to quickly lose power. The captain glided the plane for nearly 3 min over the Hudson river, finally ditching near some watercraft. Flight attendants on board compared the ditching to a "hard landing" with one impact, no bounce, then a gradual deceleration. The impact with the water ripped a hole in the underside of the airplane and twisted the fuselage, causing cargo doors to open; the plane began to fill with water. Initially, the plane floated and all 155 people aboard got out and were picked up by boats. Several passengers suffered injuries but only two required overnight hospitalization.

In both instances, the pilots observed and recognized the kind of system failure in time. They had enough time to choose appropriate action. They skillfully operated their equipment to successful recovery.

DOOMED FLIGHT 447

At night on June 1, 2009, Flight 447 from Rio de Janeiro to Paris crashed into the Atlantic after about 4 h of flying time. All 228 people aboard perished. The transcript from the cockpit recorders, gleaned from the "black box" flight recorder recovered nearly 2 years later, details what really happened:

> Cruising in severe weather at about 10,600 m, the captain was not in the cockpit, and the autopilot disengaged (probably because of the speed-sensor failure). The junior of the two copilots pulled on his side-stick controller, causing the aircraft to climb steeply in spite of a stall alarm. He continued to pull up, making the plane lose forward speed and consequently begin to fall. Additionally, severe turbulence buffeted the plane, making it nearly impossible to control, and the cockpit instruments showed conflicting information. When the more senior of the copilots took control of the airplane, also seemingly unaware that the plane was stalling, he pulled back on the stick as well.
>
> Finally, the captain returned to the cockpit. He was not initially told that the junior pilots continually kept the stalling plane nose-up. When he noticed, he made them put the plane's nose down. The plane began to regain speed but was still descending precipitously. When it had fallen to about 460 m, the aircraft's sensors detected the fast-approaching water surface and triggered a new alarm. Unfortunately, at this low altitude, the plane could not be pushed into a dive to gain airspeed and control. Without telling his superiors, the junior copilot once again pulled his side stick all the way back. The aircrew then realized that the plane was doomed to crash.

Here are some illustrative samples of the many published reports[16]:

Instrument failure caused uncertainty about the flying conditions. The reaction of the junior copilot to make the plane climb followed, in principle, the manufacturer's emergency drill, which, at that time, instructed pilots to climb at a five-degree pitch-up angle if airspeed readings became unreliable when flying above about 3000 m. However, the junior copilot kept the plane pitched up at 14 or 15 degrees in spite of the stall alarm; thus the plane was falling instead of climbing.

The other more experienced copilot did not realize that his junior colleague (without saying so) continually attempted to make the aircraft climb: the side sticks on the pilot seats moved independently, so when one pilot pulled back on his control, the other pilot did not feel this on his stick.

Communication between the two copilots was insufficient when they were alone in the cockpit and when the captain showed up.

As a result of this tragedy, airlines updated their training programs to teach appropriate actions: not making the plane climb steeply at high altitudes; understanding what it means and what to do when the stall alarm sounds; and practicing hand-flying the airplane during all phases of flight.

APOLLO 13

On April 13, 1970, an oxygen tank exploded on the US spacecraft Apollo 13 on its way to the Moon. This caused severe power reduction on board, loss of cabin heat, and shortage of potable water, and the astronauts had to repair the carbon dioxide removal system. The crew left the main Command Module and used the Lunar Landing Module as a "lifeboat" to return to Earth. NASA's ground personnel plotted a course that allowed the crippled space craft to re-enter earth atmosphere and make a successful splashdown four days later. The three astronauts on board were not injured; only one person suffered a urinary tract infection due to insufficient water intake.

The crew was alerted to the system failure by sound and feel. They recognized which actions were necessary and executed them, doubtlessly benefiting from previous training and from advice by their peers on the ground. Emergency evacuation—in this case, selection and use of the return vessel—had been carefully considered as an escape option, leading to a successful outcome.

SELF-DRIVING CAR CRASHES

In 2016, at least two self-driving cars crashed into stationary obstacles on the road; in both cases, the occupant was killed. The cars were equipped with crash avoidance systems. Reportedly, under the prevalent conditions, the computerized traffic-aware cruise control in the vehicle could not detect the obstacle and hence did not activate the automated brakes; neither did the person in the driver's seat.

The human occupants completely relied on the autopilot; they were not prepared to take action because the guidance system did not warn them about an impending failure, nor did they realize on their own that they should take over.

12.5.5 Taking Back Control From Autonomous Systems

Consequences of automation

For shipping, avionics, space travel, and now for autonomous road vehicles, computer-based control systems have been developed, and will be developed, with the intent of reducing uncertainty, danger, cost, and human work load. But automation also withholds important information from the human, including data on location, speed, and traffic conditions. If and when problems suddenly appear and the computer can no longer cope, the humans in the vessel might find themselves unexpectedly in serious trouble with an incomplete picture of what is happening. They need to understand what to do first, which threat is most pressing, which instruments and displays are providing reliable and actionable information, and which "instinctive" actions are appropriate and which are contraindicated.

Many experts feel it is imperative to keep a driver alert and engaged behind the wheel of "driverless" vehicles, ready to seize control of the car if an emergency arises.

This is because AVs, at least in the shorter term, will not be perfect: even if a thousand car-related emergency scenarios are tested, there will always be that additional scenario no one anticipated. Current drivers are still generally skeptical about the safety of AVs, and systems design should consider ergonomic features to ensure the driver remains engaged and ready to intervene rather than be a completely passive (and unprepared) passenger.

With little training for emergency circumstances, it is likely that many humans would not act appropriately or in a timely manner. Consequently, with any AV or vessel, human operators should be trained when and how to intervene. *Training for takeover*

The designer of any autonomous system must consider what kind of malfunctions could occur in the guidance system or in the vehicle itself. The design must also address how the human passengers are warned about a malfunction, how much time will they might have to react, and how they would know what to do—especially given that they probably have very little experience with such a situation. Ergonomic interventions can be effective contributors. *Warnings of system failure*

Switching to a road traffic system with autonomous automobiles provides an opportunity to redesign the old human-driven vehicles. Such updated vehicle technology[17] should address significantly different and better controllers of vehicle direction and speed: for example, removing pedals, replacing the steering wheel, using novel instrumentation. However, the designer must consider events when humans must intervene with the computerized system: how quickly can the responsible human passenger to move into the appropriate driver position and use appropriate controls to drive? *AVs*

To be prepared, the operator must be familiar with proper interventions. The cases of a successful airplane ditching, reported above, show significant similarities: the maneuvers occurred in daylight, instruments and displays were functioning correctly, the flight crews were well experienced and therefore could keep the vessel in control. Compare this with the actions of a confused and inexperienced airline crew, as in the doomed flight described above. The conduct of the Apollo 13 astronauts shows the importance of crew selection, training, skill, and outside support. Among the most important features is that the responsible person assesses the vessel's conditions quickly and correctly; decides on appropriate actions; and is able to execute these appropriate actions[18] even if the vehicle employs new technology and related new layouts of displays and controls. *Know what needs to be done*

12.6 CHAPTER SUMMARY

Although only few psychological rules are known about how we humans best perceive displayed information and operate controls, there are many well-proven design recommendations for traditional controls and dial-type displays. These are presented, in much detail, in this chapter.

With the widespread use of computers and electronics, new issues with input devices have emerged, some of which still beg for good solutions.

Warning signals, instructions on how to use equipment, and warnings to avoid misuse and danger are important, both ergonomically and juristically. Good design must consider the physical aspects of the controls and displays as well as the psychological abilities of the human operator.

Warnings, alerts, and instructions on what to do have become urgent issues as our transportation vessels become more and more autonomous. When automated control systems fail or malfunction, and humans need to assume control, we must ensure that the human operators have been trained and given the information they need to take control appropriately, effectively, and in a timely manner.

12.7 CHALLENGES

What are the implications for future control and display developments, if so few general rules exist that explain current usages?

Should one try to generate the same "population stereotypes" throughout the world?

Are there better means to control the direction of an automobile than by the conventional steering wheel?

Is it worthwhile to look for different keyboard and key arrangements if voice activation might become available soon?

Does user acceptance of keyboards that violate good ergonomic design guidelines mean we don't need to create better keyboards?

Why is it difficult to compare the usability of computer displays with the usability of information displayed on paper?

What is a good arrangement of controls for the burners on a stove top? Consider intuitiveness of the controls, safety, and ease of cleaning. How might this be different for different countries?

Why must a manufacturer warn about risks associated with false or excessive use of a product?

What are the potential upsides and downsides of autonomous vehicles in your or another country?

When is the risk of failure of autonomous vehicles sufficiently low to allow their use?

How many levels of failure must the designer of autonomous vehicles consider?

In what ways can ergonomic design reduce the risk of severe consequences of autonomous vehicle failures?

NOTES

1   For example, Van Cott and Kinkade (1972); Woodson (1954); Kroemer (1988); NASA STD 3000, 3001; SAE J 1138, 1139, 1048; [US] Department of Defense 2000; [US] MIL-STD-1472 and MIL-HDBK 759.
2   For example, HFS/ANSI 100; NASA STD 3000, 3001; SAE J 1138, 1139, 1048; [US] Department of Defense 2000; [US] MIL-STD-1472 and MIL-HDBK 759.
3   Courtney (1994).
4   The given values are measured edge-to-edge with the controls in their closest positions. For arrays of controls and operations with gloves, larger distances may be recommended; see Boff and Lincoln (1988).
5   See, for example, Adler (1997); Herkimer County Historical Society (1923); Martin (1949); Kroemer (2001).
6   We note without further speculation that Fritz Heidner was German, filing a US patent during the time of the first World War.
7   Such as that published by Schroetter (1925) and Klockenberg (1926).

8   See publications and reviews by, for example, Klemmer (1958); Kincaid and Gonzalez (1969); Kroemer (1972c, 2001, 2010); Seibel (1972); Norman and Fisher (1982); Noyes (1983); Gilad and Pollatschek (1986); Gopher and Raij (1988).

9   We discuss QWERTY to reflect the keyboard labels in English—certainly keyboards are also used with other languages. Converters allow a QWERTY keyboard to type in other languages such as Hindi or Arabic, or allow additional letters such as for Danish or French (which may also use a AZERTY keyboard layout) or Turkish (which may use a Turkish F-board). Most complex may be the Chinese keyboards, which need to accommodate many thousands of characters and often use either a phonetic or shape-based method of input.

10 See, for example, Arndt and Putz-Anderson (2006); Gerr, Monteilh, and Marcus (2006); Dennerlein (2006); Kumar (2001); Lewis, Potosnak, and Magyar (1997); Rempel (2008); [US] National Research Council 1999.

11 Such as by Anderson, Mirka, and Joines (2009), Marklin and Simoneau (2004), McMulkin and Kroemer (1994).

12 See, for example, Robinson, Casali, and Lee (1997), Seshagiri (1998), Van den Heever and Roets (1996).

13 By Cushman and Rosenberg (1991).

14 ANSI Z535.4; Brauer (2006); Peters and Peters (2013); Wogaltera, Conzolaa, and Smith-Jackson (2002).

15 Fruitiger (1998).

16 Final Report On the accident on 1st June 2009 to the Airbus A330-203 registered F-GZCP operated by Air France flight AF 447 Rio de Janeiro, Paris, published July 2012 by BEA (Bureau d'Enquêtes et d'Analyses pour la sécurité de l'aviation civile), https://www.bea.aero/docspa/2009/f-cp090601.en/pdf/f-cp090601.en.pdf. Accessed August 23, 2016.

17 Kroemer (2017).

18 Casner, Geven, Recker, and Schooler (2014); Gold, Koerber, Lechner, and Bengler (2016).

Handling Loads

Ergonomics
http://dx.doi.org/10.1016/B978-0-12-813296-8.00013-X

Overview

We all "handle" loads daily. We lift, hold, carry, push, pull, lower—while moving, packing, and storing objects. The material may be soft or solid, bulky or small, smooth or with corners and edges; it may come as bags, boxes, containers; with or without handles. We may handle objects occasionally or repeatedly; during leisure activities; and often as part of our occupational work. On the job, ergonomic design of material, containers, and workstations can help to reduce the risk of over-exertions and injuries; this is augmented by the use of instructions and training on how to "lift properly." For some jobs, selecting people who are physically capable of strenuous material handling may be appropriate.

13.1 STRAINS ASSOCIATED WITH LOAD HANDLING

Manipulation of even lightweight and small objects can strain us because we have to stretch, move, bend, or straighten our hands, arms, trunk, and legs. Heavy loads put additional strain on the body owing to their mass or bulk and/or lack of handles. Awkward body postures make load handling even more demanding.

Low back injuries

Material handling is among the most frequent and the most severe causes of injury all over the world, with strains in the low back area most frequent.[1] Chronic pain affects approximately 100 million adults in the United States, where low-back pain is the most common cause of job-related disability and a leading contributor to missed work. Back pain is the second most common neurological ailment in the United States—only headache is more common. Estimates are that low-back pain affects nearly 4 out of 10 people on Earth.[2]

To move an object with our hands, we must exert force and energy, and this strains hands, arms, shoulders, the trunk, and (since one usually stands) the legs. Holding or lifting/lowering a mass, pushing or pulling or dragging an object all put the musculo-skeletal system under stress, but the directions and magnitudes of the external and internal force and torque vectors differ depending on the particular task. The primary area of physiological and biomechanical concern for lifting an external load is the low back, particularly the discs of the lumbar spine. Thus, the operative words in the literature have been "low-back pain[3] related to lifting." Low-back pain (LBP) is a common general indicator of an over-exertion, probably due to improper work design or practice.

Sources of LBP

Neither the facets of the apophyseal joints nor the intervertebral discs seem to have pain-sensitive nerves. Thus, in each spinal unit, the three load-bearing elements (the two facet joints and one disc) can be injured without pain sensation. If the discs or facet joints are repeatedly injured, degenerative changes may set in; however these degenerative changes appear to be as frequent in persons who have LBP as in those who do not suffer from it. To complicate the matter even further, mechanical derangements of the intervertebral joint, such as a decrease in disc height, or a change in the positions of the components of the facet joints, may produce clinical symptoms that appear days after the acute phase of the injury is over. While it was long believed that a large portion of all chronic LBP is discogenic, muscular or ligament tissue problems near the spine may be just as frequent and important. Whether and to what extent these conditions contribute, and how they can be diagnosed and treated, have been a topic of often heated discussions, particularly between physicians and chiropractors. The same tasks will load individuals differently, so the actual reasons for back pain are quite often not clear.

The loading of the body may be dynamic and/or static, of fast or slow onset, and of short or long duration. There may be single or several events. If the same or similar strains reappear, that repetition may be at regular or irregular timing, and the strains may be of similar or dissimilar magnitudes. So, there is a great variety of stressful and possibly damaging events. This complexity means it is difficult to analyze why and how load handling causes an injury, how to treat the injury, and how to prevent injury in the future.

Types of strains

Large musculoskeletal strain may be experienced in sports—for example, in weight lifting. Other sources of strains are leisure and occupational activities. If done on the job, the activities are often labeled *manual material(s) handling*, abbreviated as MMH.[4]

"MMH"

In the course of such material handling, one may either exert energy intentionally toward an outside object, or the body may be subjected to unexpected energies. These unexpected events include catching a falling object as well as accidents, such as slipping and falling when we try to "catch ourselves." Aside from sports, the literature has dwelt primarily on lifting (moving an object by hand from a lower position to a higher one); lowering (the opposite of lifting); and same-level activities such as pushing and pulling, dragging and carrying, and holding. Studying the handling of materials allows us to perform ergonomic interventions to reduce the risk of over-exertions and injury.

Material handling

13.2 ASSESSING BODY CAPABILITIES FOR MATERIAL HANDLING

Handling material requires exerting energy to lift, lower, push, pull, drag, carry, or hold objects. The energy needed to do these tasks must be generated within the body and exerted in terms of force or torque over time to the outside object. Past research initially was concerned with static forces applied to the handled object, mostly in singular efforts or a small number of repeated efforts (see Chapter 2). Into the 1980s, these "isometric lift" measurements were the basis for rather simplistic (and in some cases surprising) statements about loads which, supposedly, men, women and even children could lift safely.[5]

13.2.1 Trunk Strains

Of all the musculoskeletal structures within the trunk—see Chapter 2—a variety of components may be strained: the spinal vertebrae (primarily the disc or facet joints), connective tissue such as ligaments and cartilage, and muscles with their tendons. These all may experience sprains or trauma. Compression strain of discs and vertebrae has been primarily studied, but tension is the primary loading of elastic elements such as muscles and connective tissues. Tension strains can be in form of linear elongation, or of bending movements, or of twisting torque. All structures may be subject to shear.

Newton's second law

Moving one's own body requires that muscles be tensed. Obviously, the faster the onset of a motion, especially if against an external resistance, the larger the internal forces. This is described in Newton's second law, where force equals mass multiplied by acceleration, $f = m \times a$. Since we are, on Earth, always under gravitational acceleration, we need to tense muscles even just to maintain a posture (unless we are fully supported, such as when lying down.)

Trunk loading

Several strong muscles connect the inferior and superior parts of the trunk; these are primarily the right and left latissimus dorsi, right and left erector spinae, right and left rectus abdominus, and the right and left internal and external obliques, as sketched in Fig. 13.1. They would "pull the shoulders onto the hips" (so to speak) if the spinal column would not keep them apart. Longitudinal contraction of any of the trunk muscles compresses the spinal column, which is (possibly with some relief internal forces from intra-abdominal pressure; see Fig. 2.12 in Chapter 2) the only solid load-bearing structure of the trunk.

Axial compression of the spine

Summation of all tensions in the longitudinal trunk muscles generates axial compression of the spinal column, as Fig. 13.1 shows. That large magnitude is augmented even further if an external load must be lifted. Fig. 13.2 illustrates, in simple

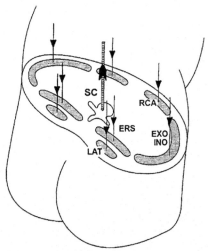

FIGURE 13.1 Sketch indicating the pull forces generated by the trunk muscles (RCA, rectus abdominus; EXO, INO, external and internal obliques; ERS, erector spinae; LAT, latissimus dorsi) which cause spinal compression (SC). *Source: Adapted from Schulz and Andersson (1981).*

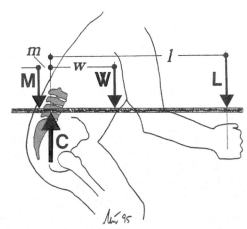

FIGURE 13.2 Anthromechanical model for calculating musculoskeletal back loading due to lifting.

biomechanical terms, how lift conditions determine spinal compression, as in the following example.

The external load L pulls downward. The (upper) body weight W acts downward as well. The distances l and w are the lever arms of load and weight to the point of support (the fulcrum of the system) at a disc of the spinal column. The forces L and W exert the moments ($L \times l$ and $W \times w$). Counteraction comes mostly from muscles (ERS and LAT in Fig. 13.1) that attach to the spinous processes of vertebrae: together they produce force M. It uses its lever arm m to the fulcrum to generate the moment ($M \times m$—note that the minus sign indicates torque in direction opposite to the torques due to W and L.) The upright force C at the spinal column keeps the system in place.

The system is in balance when these descriptive equations apply:

Sum of moments: $m\,M = l\,L + w\,W$; therefore $M = (l\,L + w\,W)/m$

Sum of forces: $C = L + W + M$

Assume the external load (L) to be 100 N with a lever arm (l) of 25 cm; the upper body W at 500 N with a lever (w) of 15 cm; the lever arm (m) of the muscle force is 5 cm. Then, the muscle force $M = 2000$ N, and

the column compression force $C = 2600$ N.

In the example above, the relatively large muscle force M (20 times the external load) and the huge column compression force C (26 times the external load) primarily result from the disadvantageously short lever arm m of the muscle M. That lever is anatomically determined, so we cannot change it—but the external load L is under our control, both in magnitude and in location relative to the body.

Both back muscle force M and column compression force C diminish as L and l get smaller—therefore the ergonomist recommends to reduce the external load and keep it close. M and C get even smaller if we keep the body more upright (which reduces l and w) and if we decrease upper body weight W—so, slim down that belly!

These anthromechanical[6] considerations demonstrate that the following admonitions can decrease the risks to the spinal column:

Simple advice for lifting

- Keep the external load small
- Keep the external load close to the body
- Keep the upper body upright
- Reduce excess upper body weight, including belly protrusion[7]

Around 1970, it became possible to actually measure the pressure inside intervertebral discs as affected by body postures and holding loads. Several early studies demonstrated that a bent or twisted back causes higher (and possibly dangerous) pressure on the intervertebral discs[8] than a straight back. Bending the back forward leads to heavy pressure and shear force on the front edge of the disc. Straightening the back eliminates shear, makes the facet joints carry some of the compressive load, and distributes the compression forces more evenly over the disc surface—all of which reduce the risk of damage.

Intervertebral disc loading

13.2.2 Muscle Strains

If a very strong force, such as for lifting a heavy object, must be exerted just once or occasionally, then the body's ability to generate such force is the limiting factor. As discussed in Chapter 2, that capability is determined by both individual and situational factors including the persons muscle strength and body position. This is a compelling reason not to design load handling tasks based on maximal voluntary muscle strength, as had been attempted prior to the 1980s.

13.2.3 Metabolic and Circulatory Strains

If the human must move (lift, lower, push, pull, drag, or carry) material over hours in ways that involve much of the body or even the whole body, then metabolic and circulatory capabilities (see Chapter 3) are likely to set limits to such activities. This fact became obvious in the development of the 1981 "NIOSH Lifting Guide"[9] (see below): Material handling that is repeated several times a minute, over hours, strains mostly metabolic and circulatory functions. This provides another reason to not design load handling tasks based on maximal voluntary muscle strength.

13.2.4 Psycho-physiologic Strains

Compared to static tests, designing and controlling experiments with realistic dynamic strength exertions is much more complicated.[10] In dynamic tests, the subject decides how much effort to exert; this includes the complex task of judging[11] (often unconsciously) how strenuous the test is. This judgment relies partly on experience and mostly on integrating internal feedback about various physiological (muscular, circulatory, metabolic, etc.) body functions involved in the effort. Based on this assessment, the subject decides what amount of strength or energy she/he is willing to exercise over a given time, whether 10 minutes or 8 hours, without endangering or exhausting her/himself. Such "psychophysical" experiments[12] have been conducted to assess tasks done particularly in industrial material handling and by firefighters, soldiers, and nurses.

13.3 CURRENT GUIDELINES FOR MATERIAL HANDLING

The two extremes for material handling are single maximal efforts and handling light loads over long periods of time. Assessing human abilities to move material has primarily been conducted through anthromechanical and psychophysical methods, each with some use of metabolic and circulatory measurements as appropriate. These methods allow us to generate useful guidelines for material handling.

13.3.1 Lifting and Lowering Guidelines (NIOSH)

Starting in the late 1970s, the United States, National Institute of Occupational Safety and Health (NIOSH) began developing guidelines for lifting and lowering. NIOSH combined several disciplinary approaches, including anthromechanics, physiology, and psychophysiology.

The predominant anthromechanical or biomechanical assessments involve load bearing capabilities of spinal joints (see Chapter 2) and of static and dynamic muscle strength (see Chapter 3). Among the physical criteria used to develop guidelines, most prominent was the highest acceptable disc compression in the lumbar spine (see example above with Fig. 13.2). NIOSH used 3.4 kN as the maximally allowable disc compression force.

Anthromechanics

Human perception appears able to integrate anthromechanical and physiological demands. Based on psychophysical studies conducted in the United States, NIOSH recommends that the job should be acceptable to most of the exposed population.

Psychophysiology

NIOSH used physiological criteria primarily to prevent whole body strain and fatigue. Important task characteristics concern the location of the external load and the duration of the work, especially frequency of work and work-rest pattern. NIOSH used the following limits for energy expenditure, where VO_2max is used as the measure of the worker's maximum aerobic capacity: 50% of VO_2max for up to 1 hour of continued work; 40% of VO_2max for 1–2 hours of continued work; 33% of VO_2max for 2–8 hours of continued work. Put another way, and in practical terms, this limits energy expenditure for continuous work to 33% of the worker's maximum capacity for an 8-hour shift.

Physiology

In summary, the limiting criteria used by NIOSH are:

- *Biomechanical* for high loads and low handling frequency
- *Psychophysiological* for moderate loads and low to moderate handling frequency
- *Physiological* for low loads and high handling frequency

The NIOSH guidelines apply to:

NIOSH weight limits

- Smooth lift (no jerking)
- Two-handed lift
- Load width below 75 cm
- Unrestricted posture
- Good foot traction
- Suitable environment with low humidity, appropriate temperature, good lighting.

The latest NIOSH guidelines[13] contain *Recommend Weight Limits* (RWLs) that 90% of US industrial workers, male or female, may lift or lower. The maximal weight is 23 kg (51 lb), under optimal conditions. However, optimal conditions are rare; therefore the load should usually be lower, as determined by several factors which include:

RWL

- The start and end points of the paths of the hands
- Whether or not the action is in front of the body (the asymmetry of the task)
- The frequency of the lifting and lowering actions
- The quality of coupling between hand and load.

These and some other factors become multipliers in an equation by which one calculates the RWL that applies for prevalent conditions.[14]

The relative risk of the level of physical stress from a given lifting task is estimated using a *Lifting Index* (LI). The LI is the ratio of load (weight) to RWL. The Composite LI is a

LI

modification of the LI that is used to assess multiple lifting tasks. The LI or Composite LI should be targeted to be at or below 1.0 for single or multiple lifting tasks; a lifting index over 1.0 is considered to pose a (theoretical) risk of low-back pain to most workers.

13.3.2 Pushing and Pulling, Lifting and Lowering Guidelines (Liberty Mutual)

Around the same time when the NIOSH developed its guidelines, researchers at the Liberty Mutual Insurance Company conducted psychophysical tests with North American workers to determine the efforts that they were willing to exert in pushing and pulling, in carrying, and in lifting and lowering loads.

Acceptable forces and weights

The experiments primarily relied on psychophysical methodology but also included measurements of oxygen consumption, heart rate, and anthropometric characteristics. The subjects controlled either weight of objects or the force that they applied; all other task variables such as frequency, size, height, and distance were selected by the experimenter. During the experiments, the subjects monitored their own feelings of exertion or fatigue; accordingly, they adjusted the weight or force so that they would "work as hard as they could without straining themselves, without becoming unusually tired, weakened, overheated, or out of breath." The results of the tests were compiled into tables of maximal acceptable weights and forces to 10%, 25%, 50%, 75%, and 90% of the male and female US population.[15]

Female and male load handlers

The "Liberty Mutual Tables" on lifting, lowering, pushing, pulling and carrying provide data about capability and limitations of US workers are grouped by females and males. Note that these tables do not apply to catching or throwing items, nor to one-handed tasks—by their nature, these tasks place uneven loads on the back and present a greater physical stress than two-handed lifts.

Some specific findings are:

- Body bending: Any task that begins or ends with the hands below knuckle height presents some risk. The deeper the bending motion, the greater the physical stress on the low back. Regardless of weight, frequent bending is not recommended.
- Body twisting: This motion puts uneven forces on the back generating additional physical stress. The greater the twist, the more physically stressful the task.
- Body reaching: The distance away from the body that a load is held greatly affects the forces on the back, shoulders, and arms. The farther the reach, the more physically stressful the task.

13.3.3 Comparing Guidelines

Altogether, there is good overlap and fair agreement between the recommendations for lifting and lowering. If there a difference, a prudent ergonomic choice is to use the lower value. The NIOSH values are unisex while Liberty Mutual considered female and male workers separately. Both sets of recommendations indicate, as to be expected, that high loads, far reaches, repetition, and deep bending or twisting the body reduce the acceptable efforts. Both sets of recommendations indicate that the manner of coupling between hand and load co-determines how much people are willing to exert. Missing handles, sharp edges, or objects that are so wide that they are difficult to grasp[16] reduce the acceptable load values.

Careful use of the guidelines should generate working conditions that are suitable for most able-bodied individuals. However, the recommendations compiled by NIOSH and by the Liberty Mutual Insurance Company were generated from and thus apply to North Americans. Other populations may have different body builds and strength characteristics; their work tasks, their customs and procedures at work, and their working conditions are possibly quite different. Special circumstances and local/regional/national codes, standards, or regulations must be considered.

13.3.4 Use of Lifting Belts

While preparing to lift or lower a load, our body automatically pressurizes the abdominal cavity. This intra-abdominal pressure (IAP) is believed to assist in supporting the curvature of the spinal column and to reduce some of the spinal compression force generated by the outside load and by the weight of the upper body (see Chapter 2). Wrapping a side brace around the abdominal region is meant to make the "walls" of the IAP column stiffer, which should make it easier to maintain or even increase the internal pressure. That idea may be a reason why porters and workers in Nepal traditionally wear a cloth wound tightly around the waist, called *patuka*, and weight lifters commonly use fairly stiff, wide, and contoured belts.

Intra-abdominal pressure

It has been advocated that people who handle material, even those who just bend and twist their body frequently, should wear such abdominal wraps, called variously lift(ing) belts, back belts, back braces or back supports. Studies have been performed to assess the effects of wearing such braces[17]: their conclusions neither summarily support nor condemn the wearing of supports. Certain material handlers, especially those who have suffered a back injury, may benefit from a suitable support which reduces back bending; however, candidates for belt wearing should be screened for cardiovascular risk which may be increased by belt pressure. Belt wearers should be aware that presence of the brace may generate a false sense of security. Even competitive weight lifters who wear belts suffer back injuries. Belts are not a substitute for ergonomic design of work task, work place, and work equipment.

Belt use?

13.4 TRAINING INDIVIDUALS

There are three major ergonomic approaches for safer and more efficient manual material handling: personnel training, personnel selection, and job design. The first two fit the person to the job, whereas the third fits the job to the person. For each approach one needs knowledge of the operator's capabilities and limitations. Thus, the research and knowledge base overlaps in all procedures.

Since lifting of loads is associated with a large portion of all back injuries, training in "safe lifting" procedures has been advocated and conducted for decades. In North America, approximately one-half of compensable low-back injuries are associated with lifting. Industrial experience has also identified lifting as a major cause of back injury. Hence, training on how to "lift safely" has been targeted to people who perform load handling activities in industry, and to groupings of industries and jobs. (See also Chapter 4 for the psychological aspects of training.)

The three essential components of a training program are:

Essential components

- Knowledge (of the capabilities of the worker, of the requirements of the task)
- Instruction (that transmits the required information to the worker)
- Practice (to help the worker learn and use information)

Benefits

Training should:

- Reduce injuries both in severity and frequency
- Develop specific material handling skills
- Further awareness and self-responsibility
- Improve specific physical fitness characteristics

When and whom to train?

Various instructional styles and media have been used. Training, either generic or customized, has been done in a single session or in various combinations of repetitions; at various times during employment; at the job site, in classrooms, or in off-site workshops. Participants in these training efforts have been selected at random, or chosen according to risk, previous injuries, age, etc.; they may be volunteers or they may include all employees.

13.4.1 Training in Proper Lifting Techniques

Instructions on "how to lift" are meant to improve lifting behavior and to reduce the likelihood of an over-exertion strain or injury. In the laboratory, it has been shown that proper training regimes can increase the capability for lifting. For example, after 4 weeks of training, initially inexperienced lifters showed increased work output through improved neuromuscular coordination while maintaining the energy expenditures through increases in muscular endurance. In another controlled study, 6 weeks of training improved muscular endurance, muscular strength, and cardiovascular endurance. However, few tightly controlled studies with large numbers of industrial material handlers have been performed,[18] and the validity of laboratory findings for industrial environments remains to be established.

> Many lifting instructions include the admonition "keep the back straight." However, what this means is that one should maintain the "natural curvature" of the "standing" spine, particularly its lumbar lordosis. Such imprecise use of words can lead to confusion.

ONE good lifting posture?

From the 1940s on, the most widely advocated method of lifting was the straight-back/bent-knees lift (*squat lift*), where workers lowered themselves to the load by bending the knees and then lifted it using their leg muscles. However, the leg muscles used in this procedure do not always have the needed strength. Further, awkward and stressful postures may be assumed if one tries to enforce this technique under unsuitable circumstances, such as when the object is bulky. Hence, the straight-back/bent-knees action evolved into the so-called *kinetic lift*, in which the back is kept mostly straight while the knees are straightening. Another variant was the *free-style lift* which may be better for some workers than the straight-back/bent-knees technique. In some situations, the *stoop lift* (with a strong bend at the waist and straight knees, sometimes called hip-hinge) may be superior to the squat posture (bent knees, flat back) which is often advocated; however, an intermediate *semi-squat lift* may be most versatile. If no single lifting method is best for all situations,[19] then training of proper lifting technique is an area of confusion: what method should be taught?

13.4.2 "Back Schools"

Education is used as preventive treatment for healthy participants and as conservative treatment for LBP patients instead of surgery. The schooling approach emphasizes knowledge, awareness, and attitude change by instructing the patient in anatomy, biomechanics, and injuries of the spine. The overall goal is to encourage the individual to take responsibility for his/her own health by means of proper nutrition, physical fitness, and awareness of the effects of posture and movement on the back. Such programs may also provide vocational guidance and information on stress management, and even on drug use and abuse.

"Back school"

Controlled research concerning back schools has been limited, but usually back school patients express increased understanding of their own back problem and a sense of better control of pain. Back school and physical therapy treatments appear about equally effective in reducing the number of days needed for pain relief, but the back school may be more economical because patients are treated in groups rather than individually.

Typical industrial case studies involve a training program together with instituting new safety rules, job redesign, and campaigns of posters, booklets, and cards (sometimes given out with the worker's paycheck to encourage reading). Often, reported results point to reductions in compensation costs together with fewer lost work days per injury; but, unfortunately, there is usually no control group. Other reports showed no significant differences, neither in the occurrence of back pain nor in the number of lost work days, between healthy employees who attended back school and a control group that did not. Occasionally, one may even see an increase in the number of reported pain incidents, but this might be attributed to a change in management attitude and a willingness on the part of employees to report back problems early—a positive step toward avoiding more serious injury.

Effectiveness

Key motivating factors in efforts to prevent back injuries include:

Motivators

- recognizing the problem and the employee's personal role in it
- gaining knowledge about the spine
- understanding the relevance of the problem by observing the working environment
- intending to alter the behavior by learning proper techniques
- practicing these new techniques
- following up the program at the workplace

"Hawthorne
effect"

The concept of the post-injury back school with its emphasis on attitude change and awareness to encourage compliance with proper procedures at work is an appealing approach that goes beyond traditional "safe lifting" teaching. Clearly, in many cases (especially the published cases), management is happy with the results and convinced of the program's effectiveness, even if its success may in reality be due to a "Hawthorne effect"[20] (the positive result of "treating" workers with actions that may or may not be effective, but to which they react positively because of the show of concern for and interest in them).

THE HAWTHORNE EFFECT — "BREAD AND BUTTER" FOR CONSULTANTS

Consultants hired by company management to improve unsatisfactory conditions of manual material activities are likely to be successful if their improvement strategies are well-directed and engage the material handlers and their supervisors. If after the campaign the level of performance declines, management is likely to call upon the same consultants again because they were previously successful.

13.4.3 "Fitness" Training

Fitness training

Another related approach to training workers in prevention of low-back injury relies on the idea of fitness training. Material handling is physical work, and it is reasonable to assume that many aspects of physical fitness, especially musculoskeletal strength, aerobic capacity, and flexibility are associated with the ability to perform load handling without injury. Flexibility, particularly of the trunk, appears to be needed for bending and lifting activities, which are part of material handling tasks. The well-known practices of Chinese and Japanese workers doing exercises before the workday, and of athletes training for better performance, follow that notion.

"Work
hardening"

The concept of "work hardening" includes purposefully designed exercises for improving specific body abilities deemed necessary to perform the job. Empirical evidence is mixed: some experiences indicate that the occurrence of musculoskeletal injuries in weaker workers is greater than for stronger workers, but there are also unexpected findings that stronger and fitter workers may be injured more often or more seriously than their weaker and less fit colleagues.[21]

13.4.4 Training: What? How? Whom?

Content

The basic question is what should be taught; the high-level purpose is to prevent over-exertion and injury, but the methods of how to achieve that depend on the specific aim. Previous efforts have generally focused on three areas:

1. Training of specific lifting techniques, skill improvement.
2. Improving awareness, attitude, and self-responsibility by explaining sequence and mechanics of possibly injurious events.
3. Making the body physically fit so that it is less susceptible to injury.

The concept of combining the three major topics has a basic, almost simplistic, appeal—both logically and practically. However, exactly what should be taught and trained, and in how much detail? How much knowledge of physics, kinesiology, anthromechanics, and physiology is needed? Who should be the trainer: engineer, psychologist, physiologist?

Having a physically fit body seems intuitive and rational for injury-free load handling. However, research regarding the role of physical training as a preventive method is still mostly lacking. For example, it remains unclear if training in body mobility or strength or endurance effectively reduce workers' back injuries, or if warm-up exercises before the work shift are beneficial.

Beyond these approach-specific research questions are the corollary problems of simultaneously implementing several training approaches. For example, how exactly does physical fitness interact with attitude change? In many cases, proposed programs contain elements of knowledge and attitude change, physical exercises, and specific load handling techniques. It is not clear what combinations are the best, or even if they are effective at all.

Once it has been determined what should be taught, attention must be paid to the formation of the course itself. Many courses are taught in a lecture format, with practice of techniques at some time during the session. Programmed interactive instruction, videos, posters, and handouts have been used: the (relative) effectiveness of these techniques is hard to ascertain.

How to train?

Another question is where the sessions should be held: would a classroom with desks and chairs, or even a gymnasium, be more appropriate than a lecture hall? The worksite appears to be the most practical location, but it may not be suitable for instructional purposes. Is it best to train employees working together as a group, or should the group be split up? Of course, in any situation, there are practical constraints that might limit the ability to implement training ideals, but currently information is not available on which to base sound judgment.

Where to train?

The majority of industrial back injuries has not been associated with objective pathological findings, and approximately half of back pain episodes cannot be linked to a specific incident. A variety of actions and events at work can lead to injuries. In the mining industry, over-exertion, slips/trips/falls, and jolts in vehicles have been the most frequently mentioned events. Back injuries have been directly associated with lifting (37–49% of cases), pulling (9–16%), pushing (6–9%), carrying (5–8%), lowering (4–7%), bending (12–14%), twisting (9–18%), and the percentages vary considerably among industries and occupations.[22] Industries with the largest incidence ratios for compensation claims for back injuries have been construction and mining. Other occupations with high incidence ratios include garbage collectors, truck drivers, health care providers, and, of course, material handlers.

Customized training

The traditional approach of training a specific lifting technique alone does not appear to be effective because there is no one technique that is appropriate for all lifts. It seems more appropriate to demonstrate and practice a variety of body postures and motions that are suitable under given conditions. Load handling specifics differ a great deal among industries and jobs (say, tire-making, mining, nursing) as well as within one industry or profession. They depend on the specific task, on available handling aids and equipment,

Task-specific training?

and on many other conditions. Therefore, it is a question of how training recommendations are applicable across settings. Even the group characteristics of material handlers in different industries might be important in designing a training program; for example, hospital workers might have higher educational skills than heavy-industry workers. Female employees may be predominant in a given industry or occupation, which might influence training, since on the whole, women are about two-thirds as strong as men.

Efficacy

In the United States, in 2014, the rate of over-exertion injuries across all industries was 33 per 10,000 full time workers.[23] Of the actually reported injuries, about every tenth one is serious, yet these serious injuries comprise by far the largest portion of the total cost. Hence, if one wanted to specifically prevent these serious injuries, about 4 of every thousand employees are the target sample, while the entire one-thousand should be in the educational program. This is a rather expensive approach, which may not appear cost-effective to an administrator.

Whom to train?

Obviously, groups of people who frequently and/or seriously suffer from injuries are of particular concern. However, at what lower levels of over-exertion should management or a public agency get involved? Since many material handling problems are related to work design, should job supervisors and possibly plant engineers and designers be trained as well?

Ethical considerations

People continue to get injured and experience pain, so it is unclear if everyone or just individuals at high risk should receive training. When conducting research, a no-treatment control group is often used. However, what are the responsibilities of the researcher in choosing whether to train or not train people in ways that might help prevent back injury? In the case of fitness training for injury prevention, there is the potential for injury from the training itself. Therefore, should training be mandatory or voluntary? If voluntary, will the training be given during working hours or on the worker's own time? Some individuals might feel that "mandatory" fitness training or nutrition guidelines, for example, decree a change in lifestyle that they may not want; it may feel overly intrusive to them. How far can—and should—employers and government agencies go?

Effectiveness

A major and basic question is how to judge the effectiveness of any given training. The most commonly used approach is to consider objective data derived from company records on cost and productivity, and from medical records to compare quantitative measures before and after training. The use of company loss data is common, but sometimes the exact meaning of these data is not clear, particularly if other actions take place during the period of data-recording that may have impacted the loss statistics. Apparently, there is not enough standardization among companies in the definition of terms, or in the actual derivation of the statistics, to allow reliable industry-wide comparisons. Subjective data result from asking trainees or managers questions such as whether the training seemed worthwhile. The value of such judgments is uncertain.

Training quality

It would not make sense to do evaluations of training if the students never learned the material in the first place. Thus, there is an interest in assessing whether they have actually absorbed what was presented in training sessions. Which test or measure of learning directly following the training could be used as an evaluative tool, and how well one must do on the evaluation to be considered trained or having a changed attitude? Is transfer easily made from the learning environment to the work environment? After training, for how long a time should one expect to observe effects of training?

Another unanswered question concerns the retention of information by the trainees after training has been completed. Was there sufficient original learning? It may have been good immediately following instruction, but why then is there usually an increase in injuries (or whatever measure is used) as time passes? Is this a reflection of good training that is forgotten over time, or was the training not "good" in the first place? Increasing retention of information is a common concern, so refresher courses (of some kind) probably should be offered, however, suitable time intervals between training need to be established.

Retention, refreshment

Studies on training outcomes reported in the literature are deficient when they do not utilize an experimental design that allows the assessment of reliability and validity of the results of the experimental treatment. Granted, it is difficult (because of work interference and the time and expense required) to conduct a field experiment in which one varies only the independent (training) variable and excludes confounding variables and uncontrollable interferences. However, including a control group in the experimental design is often feasible, and it allows for evaluation of the claimed effect of the experimental training treatment.

Evaluating training outcomes

Even with the large variety of past training approaches and of evaluations of their outcomes, a short review of cases[24] is of interest. Early reviews (1972, 1975, 1984, and 1989) of the effects of education to reduce back injuries in the United States detected no significant reductions. A 1980 comparison of the number of back injuries showed no significant difference between companies that conducted training programs in "safe lifting" with companies that did not have such programs. Several studies on nurses did not find effects on the incidence of low-back injury after receiving repeated instructions on lifting procedures. More than 200 physical therapists involved in back care education experienced the same incidence, prevalence, and recurrence of back pain as a control group.

Unsuccessful training

Few reports on the effectiveness of training include control groups; according to these, back school therapy was successfully applied to patients with back pain. In a study with nearly 4000 postal employees,[25] experienced physical therapists trained 2534 workers and 134 supervisors. Work groups of workers with their supervisors participated in a two-session back school (3 hours of training) followed by three to four reinforcement sessions over the next 5 years. The 75 people from either the intervention or the control groups who were injured during that time period, after their return to work, were randomly assigned to either receive (re)training or not.

During the observation period of more than 5 years, 360 workers reported back injuries: the rate was 21.2 injuries per 1000 worker-years of risk. The median work-time lost was 14 days.

A comparison of the intervention and control groups showed no effects of the education program on the rate of low-back injuries, the cost per injury, the time off work per injury, or the rate of injury repetition; however the trainees' knowledge of safe behavior increased.

13.4.5 Summary of Training

The preceding review of training covered about a hundred published sources. However, the literature coverage remains spotty: it neither addresses all topics of interest nor indicates many clear results. The first deficiency relates to the multifaceted magnitude of the problem, the second also to deficiencies in the experimental design and control of training approaches.

The experimental design should follow the sequence of

1. needs assessment
2. development or selection of a suitable training program
3. pre-training assessment,
4. doing the training
5. post-training assessment

This should be followed by program improvement and possible repetition of the training. Using such a systematic approach helps in the evaluation of reliability, significance, and validity of training for safe material handling.

Common sense suggests that "training" should be successful but hardly any specific training guidelines are well supported by controlled research. This leaves much room for speculation, guesswork, and charlatanry regarding the "best" way to train people for the prevention of back injuries related to load handling. Some, possibly many, training approaches are not effective in injury prevention, or their effects may be so uncertain and inconsistent that money and effort paid for training programs might be better spent on ergonomic job redesign. The legal responsibility of employers to provide training cannot be ignored: "…so long as it is a legal duty [in the United States] for employers to provide such training or for as long as the employer is liable to a claim of negligence for failing to train workers in safe methods of [material handling], the practice is likely to continue despite the lack of evidence to support it."[26] However: "In spite of more than 50 years of concerted effort to diminish task demand, the incidence of compensable back injuries has not wavered. … Rather than pursuing the 'right way to lift,' the more reasonable and humane quest might be for workplaces that are comfortable when we are well and accommodating when we are ill."[27]

If the job requirements are stressful, "doctoring the symptoms" (such as behavior modification) will not eliminate the inherent risk. Designing a safe job is fundamentally better than training people to behave safely. Yet it appears plausible to expect that at least certain training approaches should show positive results. Among these, training for

- work skills (body and load positioning and movement),
- awareness and attitude (physics and biomechanics associated with lifting, self-control), and
- fitness, strength and endurance

appeal to common sense and appear theoretically sound, even though none of these has yet proven successful according to the literature. Of course, the Hawthorne effect is likely to reduce incidents immediately after any reasonable training (which is, of course, a positive and desirable result), but after a fairly short time the injury statistics are likely to worsen again—which provides new reason to train.

Several general recommendations should help to reduce the risk of over-exertion injuries when lifting:

Leg-lift or back-lift: The "leg lift" has been heavily promoted in training, as opposed to the "back lift". It is indeed, normally, better to straighten the bent legs while lifting rather than unbending the back. But leg lifts can be done only with certain types of loads, either those of small size which fit between the legs, or those having "two handles" between which one can stand (two small boxes instead of one big case). Large and bulky loads usually cannot be lifted by unbending the knees without body contortion.

Rules for lifting and lowering: There are no comprehensive and definitive rules for risk-free lifting or lowering, which is a complex combination of moving body segments, changing joint angles, tightening muscles, and loading the spinal column. Here are some guidelines for proper lifting:

- Eliminate manual lifting (and lowering) from the task and workplace. If lifting or lowering needs to be done by a person, perform it between knuckle and shoulder height.
- Be in good physical shape. If you are not used to vigorous exercise, do not attempt to do difficult lifting or lowering tasks.
- Think before acting. Consider handling aids, place material conveniently, and make sure sufficient space is cleared.
- Get a good grip on the load. Test the weight before trying to move it. If it is too bulky or heavy, get a mechanical lifting aid, or somebody else to help, or both.
- Get the load close to the body. Place the feet close to the load. Stand in a stable position, and have the feet point in the direction of movement.
- Involve primarily straightening of the legs in lifting, bending the legs in lowering.

There are some actions to avoid:

- Do NOT twist the back, or bend sideways.
- Do NOT lift or lower, push or pull, awkwardly.
- Do NOT hesitate to get help, either mechanical or by another person.
- Do NOT lift or lower with arms extended.
- Do NOT continue lifting when the load is too heavy.

These rules and recommendations should not only be part of a worker training program but they should also inform how the job is designed. For example, the loads should be properly sized for easy moving without body contortion; placed at proper height to be handled in front of the trunk; and put in proper form and shape so that one can get a good grip.

13.5 SELECTING WORKERS BY PHYSICAL TESTING

While training is one approach to fit the person to the selected job, the other is to select the person to fit the job. This involves screening of individuals with the purpose of filling the strenuous jobs with those who can do them safely. This screening may be done either before employment, before placement on a new job, or during routine examinations during employment.

A basic premise of personnel selection by physical characteristics is that the risk of over-exertion injury from manual material handling decreases as the handler's capability to perform such activity increases. This means that a test should be designed that allows assessment of the match between a person's capabilities for load handling and the actual load handling demands of the job. Accordingly, this matching process requires that one knows, quantitatively, both the job requirements and the related capabilities of the individual. (Of course, if the job requirements are excessive, they should be mitigated before any matching is attempted.)

13.5.1 Selection by General Physical Capabilities

Central limitations

The human body must maintain a balance between external demands of the work and related internal capacities. The body is an energy "factory" that converts chemical energy derived from nutrients into externally useful physical energy. Final stages of this metabolic process take place at skeletal muscle, which needs oxygen transported from the lungs by the blood. The blood flow also removes byproducts generated in the energy conversion, such as carbon dioxide, water, and heat, which are dissipated in the lungs, where oxygen is absorbed into the blood. Heat and water are also dispelled through the skin (sweat). The blood circulation is powered by the heart. (Chapters 1, 2 and 3 of this book discuss these issues in some detail.) Thus, the pulmonary system, the circulatory system, and the metabolic system establish *central limitations* of a person's ability to perform strenuous work.

Local limitations

A person's capability may be limited also by muscular strength, by the ability for movement in body joints, and, in material handling, often by the stress responses of the spinal column and its supporting connective tissues. As discussed earlier, these are *local limitations* for the force or work that a person can exert.

The "weakest link"

Both central and local limitations may determine the individual performance capability. While handling material, the force or torque exerted with the hands must be transmitted through the body—that is, via the wrists, elbows, shoulders, trunk, hips, knees, ankles, and feet to the floor. In this chain of force vectors, the weakest link determines the whole body's capability to do the job. If muscles are weak, or if they have to pull under mechanical disadvantages, the available handling force is reduced. Often, the lumbar section of the spinal column is the weakest link in this kinematic chain: muscular or ligament strain or painful displacements of the vertebrae and/or of the intervertebral discs may limit a person's ability to handle material.

Assessment methods

Several methods of examination have been developed to assess an individual's capabilities for performing specified handling tasks.

The *medical examination* primarily identifies individuals with physical impairments or diseases. While x-ray examinations are now sparingly used, the physical/medical exam screens out the medically unfit on the basis of their physiological and orthopedic traits. Unless specific job requirements are known to the medical providers who perform the testing, they must evaluate the person's capabilities against generic job demands. Hence, the examination is often not a specific match of capabilities (more precisely, of limitations) with demands of the prospective job.

The *biomechanical examination* addresses mechanical functions of the body, primarily of the musculoskeletal type, such as load-bearing capacities of the spine, or muscle strength that must be exerted in certain postures or motions. Anthromechanical methods rely on explicit models of such body functions, which is both their strength and weakness: the results are only as good (that is, reliable and valid) as the underlying models.

The *physiological examination* usually identifies individual limitations in central capabilities, mostly of pulmonary, circulatory or metabolic functions. This provides essential criteria if these functions are indeed highly taxed by the material handling job, such as in frequent movements with heavy loads and substantial body involvement.

The *psychophysical examination* addresses all (central or local) functions strained in the test: hence, it may include all or many of the systems checked via medical, biomechanical, or physiological methods. It filters the strain experienced through the sensation of the subject who rates the perceived exertion. In tests of maximal voluntary exertions, the subject decides how much strain is acceptable under the given condition; for example, the maximum weight the subject is willing to lift.

13.5.2 Selection by Individual Strength Assessment

Several screening techniques exist that rely on assessing an individual's "strength" in terms of force or torque exerted to an instrument, or the weight (mass) that can be lifted—such as the biomechanical and psychophysical exams just discussed. These outcomes (should) serve to select workers who are able to perform defined material handling activities with no or acceptably small risks of over-exertion injuries; that is, selecting individuals whose capabilities match the job demands with a safety margin.

In the past, managers, foremen, or physicians had to rely on intuitive, experience-based guesses; but now a number of pre-employment or placement tests have been developed based on models and methods just discussed. Primarily, these tests differ in the techniques used to generate the external stresses that strain the local and central function capabilities to be measured.

Static tests rely on the subject exerting isometric muscle strength against an external measuring instrument. Since muscle length does not change, there is no displacement of body segments involved; hence, no time derivatives of displacement are considered. This establishes a mechanically and physiologically simple case which allows straightforward measurement of isometric muscle strength capability. The "Caldwell Regimen" (see Chapter 2) has been widely accepted (with some minor modifications) for such static strength measurement. **Static tests**

Dynamic tests appear more relevant to actual material handling activities but they are also more complex to perform, because of the large number of possible displacements and of their time derivatives (velocity, acceleration, and jerk). Hence, most current dynamic testing techniques employ either one of two ways to generate dynamic test stresses: **Dynamic tests**

1. In the *isokinematic* (or *isovelocity*, often incorrectly called "isokinetic") *technique*, the subject's body (limb) moves at constant angular velocity about a specific joint (usually knee, hip, shoulder, or elbow) while exerting maximal voluntary torque, which the equipment monitors throughout the angular displacement. Simpler equipment has a handle (or pedal) that is being moved at constant speed but, given the mechanics and geometry of human limbs, angular joint speed and muscle displacement are not controlled—see the discussion in Chapter 2, especially Table 2.4.

2. The *isoinertial technique* requires the subject to move a weight (constant mass) between two defined points. The maximal load which the subject can lift (or lower, push, pull, carry) is taken as the measure of that person's capability. Such tests do not prescribe the exact path of limbs, do not specify joint speeds or muscle motions, and do not assess the actual "strengths" of involved muscles. This rather realistic technique is part of many currently employed tests.

13.5.3 Pros and Cons of Screening Techniques

The main advantage of the static (isometric) techniques is the conceptual simplicity: putting a person into a few "frozen" positions and measuring the effort which she or he can generate in one given direction over a period of just a few seconds is indeed easily understood. Furthermore, since no displacement takes place (that is, joint angles and locations do not change), the physical conditions are uncomplicated as speed (or any time derivatives of displacement) is not a factor. This allows the use of rather simple instruments and permits relatively easy control of the experimental condition—which is why, in the 1980s, static assessments of individual capabilities supplanted the presumption of single "safe" loads for children, women, and men.

Unfortunately, performance on static tests has low predictive power for execution of dynamic tasks. This discrepancy triggered research to develop dynamic measuring techniques around 1980[28] which resemble actual material handling. A well-proven practical test of lifting capacity is to actually lift loads, starting with small loads and increasing the weight incrementally until the largest load that can be listed is determined. This incremental isoinertial test is easily understood, executed, and controlled; it is safe, reliable and valid; and has been used to test millions of US military recruits, male and female.[29] A strategy of increasing or decreasing the loads to determine a person's lifting capacity to the nearest 5 pounds (2.3 kg) is shown in Fig. 13.3. The equipment is inexpensive and robust. Such an isoinertial test is an efficient, reliable, and realistic approach to assess an individual's capability to lift.

13.6 ERGONOMIC DESIGN FOR LOAD HANDLING

How the job is designed determines the stress imposed on the human. Many factors affect whether a job is well-designed and whether it is safe, efficient, and agreeable to the operator: the size of the load, its weight, whether it has handles or not; the layout of the task, the kinds of body motions to be performed, the body forces and torques to be exerted; the frequency of these efforts; the organization of work and rest periods; and other engineering and managerial aspects. In this section, we provide further

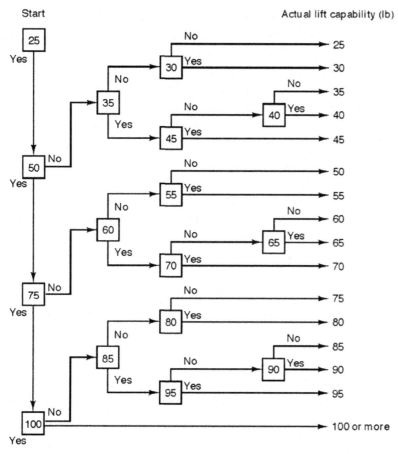

FIGURE 13.3 Sequence for determining an individual's lifting capability used in LIFTEST.
Source: Adapted from Kroemer (1982).

information on facility layout and the overall work environment, equipment, and work place and work task for ergonomic load handling.

13.6.1 Facility Layout

The layout of the overall work facility should ensure safe and efficient material transfer. Organizing the flow of material and work tasks determines the involvement of people and how they must handle the material.

In the real world, one generally encounters either one of two situations:

1. A facility exists, it must be used, and the building and its interior layout cannot be changed significantly. Through small changes, one must make the best use of what there is.
2. A new facility can be designed for the process at hand, to suit the workers and make work easy and efficient.

The opportunity to "do it right at the planning stage" is most desirable, allowing the closest approximation to the optimal solution. However even a modification of an existing facility should strive for the best (possible) solution. Therefore, the ideal case will be used here to guide us in this text, even when only modifications may be possible.

Material flow

A facility with well-planned material flow has short and few transportation lines. Reduction of material movement through proper facility layout can lower the cost of material transport considerably—which usually accounts to 30–75% of total operating cost, and may be even higher. Material movement adds no value to the product and can be dangerous. Hence, reduction of material handling is both a major cost and safety factor.

Process or product layout

There are two major design strategies: process layout and product layout. In *process layout*, all machines or processes of the same type are grouped together, such as all heat treating in one room, all production machines in another section, and all assembly work in a different division. Process layout has the advantage of keeping machines busy because different products or parts flow through the same workstations. It also allows for better control of any special conditions (such as climate control or clean room filtration) that a process strep or equipment requires. However, process layout requires a relatively large amount of material, has no fixed flow paths, and needs a great deal of floor space.

In *product layout*, all machines, processes, and activities needed for making a particular product are grouped together. This results in short throughput lines and requires relatively little floor space. However, the layout suits only the specific product, and breakdown of any single machine or of special transport equipment may stop the entire production line. Product layout is advantageous for material handling because routes of material flow can be predetermined and planned well in advance.

Flow charts and diagrams easily describe events and activities with simple sketches, symbols, and words. Even a cursory look at Fig. 13.4 shows waste of personnel, space and time: several workstations with 11 transports and 7 delays, all combined with many possible sources of injuries caused by manual material movement. Fig. 13.5 demonstrates an improved layout with one workstation, 5 transports and with only two delays. Even this set-up could still be improved.

13.6.2 Work Environment

A well-designed and maintained work environment (see also Chapter 6) can contribute to safe manual material activities.

Visual

The visual environment should be well lit, clean, and uncluttered; allowing good depth perception and discrimination of visual details, of differences in contrast, and of colors.

Thermal

The thermal environment should be within zones comfortable for the physical work, usually in the range of about 18–22°C. Thermal stress resulting from conditions that are too hot or too cold may contribute to material handling safety problems.

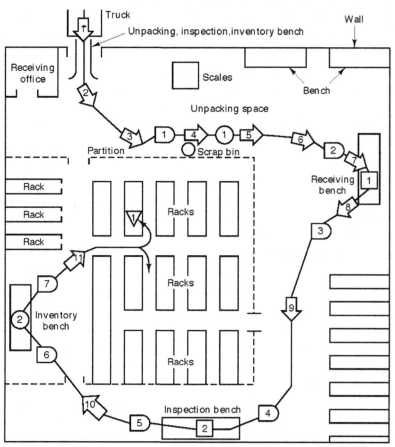

FIGURE 13.4 Flow diagram of the original setup for receiving, inspecting, inventorying and storing.
Source: Adapted from ILO 1974.

Acoustic

The acoustical environment should be agreeable, with sound levels preferably below 75 dBA. Warning sounds and signals indicating unusual conditions should be clearly perceptible by the operator. High noise conditions can contribute to overall straining of the operator, and hence affect safety of material handling.

Housekeeping

Good housekeeping practices help to avoid injuries. Safe gripping of shoes on the floor, or good support from the chair when seated, are necessary conditions for safe material handling. Poor coupling with the floor can result in slips, trips, or falls. Floor surfaces should be kept clean to provide a good coefficient of friction with the shoes. Clutter, loose objects on the floor, dirt, and spills can reduce friction and lead to slip, trip, and fall accidents—see Fig. 13.6.

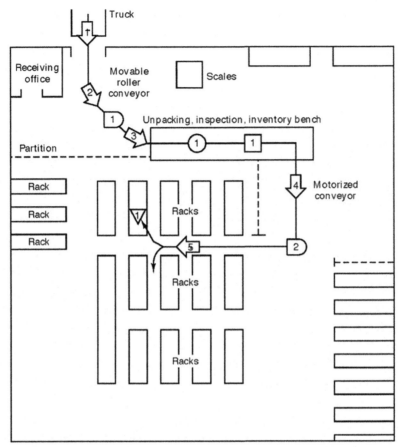

FIGURE 13.5 Flow diagram of improved setup for receiving, inspecting, inventorying and storing.
Source: Adapted from ILO 1974.

Some facilities follow the 5S methodology to organize a work space for efficiency; this can be a component of "Just in Time" and "Lean" manufacturing strategies. The 5S system is based on five Japanese concepts (*seiri, seiton, seiso, seiketsu,* and *shitsuke*), generally translated into English as:

- Sort (keep only those items you need)
- Set in order (arrange the items for optimal workflow)
- Shine (clean and inspect frequently)
- Standardize (share best practices)
- Sustain (maintain the good practices)

Sometimes a sixth S is added to represent Safety.

FIGURE 13.6 **Bad and good housekeeping.** *Source: Adapted from ILO, 1988.*

13.6.3 Work Equipment

Equipment may do the actual material handling or it may provide assistance at the workplace. This includes:

- Overhead cranes
- Mobile cranes, as in Fig. 13.7
- Hoists, as in Fig. 13.8
- Lift trucks, as in Fig. 13.9
- Lift tables, as in Fig. 13.10
- Conveyors, as in Fig. 13.11
- Carts, dollies, and hand trucks, as in Fig. 13.12
- Automated storage and retrieval systems

FIGURE 13.7   Crane on truck. *Source: Adapted from British Standard 40004.*

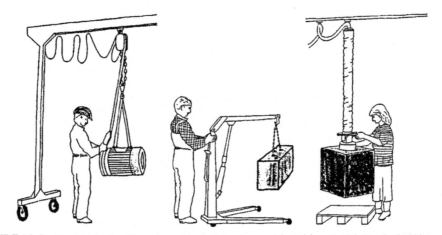

FIGURE 13.8   Electric, hydraulic and vacuum hoists. *Source: Adapted from British Standard 40004.*

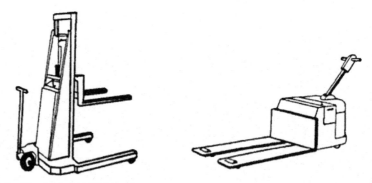

FIGURE 13.9   Hand-operated lift trucks.

II. DESIGN APPLICATIONS

FIGURE 13.10   Lift tables.

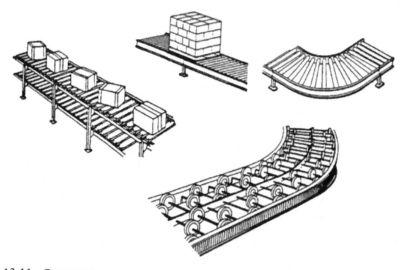

FIGURE 13.11   Conveyors.

FIGURE 13.12   Examples of a cart, dolly and hand truck for rolling materials instead of carrying them.

13.6.3.1 *Ergonomic Design of Equipment*

Will equipment
be used?

Equipment such as mentioned above can be designed to take over the lifting, lowering, pushing, pulling, carrying and holding of materials which would otherwise be performed manually by a person. But will it actually be used? Obviously, facility layout as well as workplace design must be suitable for the use of equipment. The equipment must be appropriate for the task. Furthermore, the operator must be convinced that it is worthwhile to go through the effort of getting, say, a hoist to use instead of heaving the material by hand; the operator must be sure that this behavior is not only tolerated but in fact encouraged by supervisors and high-level management.

We are likely to avoid employing machinery, such as a dolly to move material, if the operation is inconvenient, clumsy, time-consuming, or awkward. Equipment must not only be able to perform the material moving job, and to do so safely, but it must also "fit" the human operator. Convenience and ease of use are important. Unfortunately, some material movement equipment such as hoists and hand trucks show an alarming lack of consideration of human factors and safety principles in their design. The actual use of hand trucks depends on:

- Human force required to operate it
- Stability
- Ease of steering
- Handle design and location
- Ease of starting and stopping
- Ease of loading and unloading
- Security of the load

Older forklifts, as in Fig. 13.13, provide the appallingly worst example of bad design: the driver's seat is inconveniently tight, and transmits vibrations and impacts from the rolling wheels, and the driver's neck and trunk must be severely twisted and bent to see when driving backward—which is often necessary because the load obstructs the driver's forward view.

Handles

Control handles on equipment, containers or tools should be so designed and oriented that hand and forearm of the operator are aligned. Do not force the operator to work with a bent wrist: "Bend the handle, not the wrist." Carpal tunnel syndrome and other cumulative trauma problems (overuse disorders) are less likely to occur if a person can work with a straight wrist—see Chapter 9.

Chapter 12 contains detailed information about suitable design of hand controls. The handle should be of such shape and material that squeezing forces are distributed over the largest possible palm area. The handle diameter should be in the range of 2.5–5 centimeters. Its surface should be slip-resistant. Coupling with the hands is best done by protruding handles or by gripping-notch types; on containers, boxes and drawers, hand-hold cutouts or drawer-pull types may be acceptable.

Labels and
instructions

If equipment is designed for intuitive and obvious use, written instructions and warnings are not necessary. However, when instructions are needed, labeling should be done according to these rules (see also the appropriate section in Chapter 12):

- Write the instructions in the simplest and most direct manner possible.
- Give only the needed information.

FIGURE 13.13   **Typical forklift truck.**

- Describe the required action clearly. Never mix different instruction categories such as operation, maintenance, and warning.
- Use words (and language) familiar to the operator.
- Be brief, but not ambiguous.
- Locate labels in a consistent manner.
- Words should read horizontally, not vertically (note that this recommendation depends on the user's language and culture).
- Label color should contrast with equipment background.

13.6.4 Work-place and Work-task

As discussed at the beginning of this chapter, only a few decades ago it was believed that certain set weights could be lifted safely by men or women or children. Such a simplistic idea is no longer legitimate, nor acceptable. Epidemiological, medical, anthromechanical, psycho-physiologic, psychologic, and administrative knowledge has led to more reliable guidelines for manual material movement. They consider such variables as frequency, location, direction of handling activities and properties of the loads. Two major sets of recommendations are in the previous section on guidelines for material handling, but these still rely on assumptions and approaches that can be refined and further evaluated.

Ergonomic recommendations for load handling take for granted that the material handler is free to assume any body postures suitable for the job. However, there are conditions in which only limited room is available, such as in underground mines, or in aircraft cargo holds, or during emergency repair jobs. Stooped, kneeling, bent, sitting,

Constrained space and postures

supine, prone, reaching, stretching, contorted and otherwise restricted body positions reduce the operator's ability to handle objects, often severely, in terms of forcefulness, repetition, direction and distance. Additionally, the use of safety or environmental equipment (such as harnesses, belts, or even space suits) can further reduce mobility.

No people involved, no people at risk

The most successful solution is to "design out" manual material movement or, if unavoidable, assigning it to machines. If people must be involved, load weight and size should be kept small, together with ergonomic design of the work task, by selecting the proper type of body efforts (such as horizontal push instead of vertical lift) and a low frequency of efforts. Fig. 13.14 presents a flow chart of actions which can eliminate load handling or, at least, reduce the inherent risks.

Workplace

Design of work places and tasks should fit the human body and suit human capabilities. Naturally, the workplace itself must be well designed and maintained (nonslip floor, clean, orderly, uncluttered), and it should present an appropriate environment in terms of climate, lighting and sound—all conditions that contribute to avoiding accidents, mishaps, physical over-exertions and stress. Proper working height is a basic requirement. The location of the worker's hands (and hence the body posture) is determined by the location of the work object: it should be between hip and shoulder height, best between waist and chest height; and directly in front of the body so as to avoid contorting, twisting or bending the trunk (see Figs. 13.15 and 13.16).

Load

The object itself is important, of course, regarding its weight, bulk, pliability (for example, firm box vs. pliable bag), and ability to be grasped securely (shape, handles, surface texture). Whether it is light and small (as in Fig. 13.15) or large and heavy (as in Fig. 13.17), it should be delivered at about waist height. If grasping a load on the floor cannot be avoided, it should be done between the feet, not in front of them (see Fig. 13.18).

13.6.5 Simple Ergonomic Rules for Load Handling

Although general rules may not cover all specific cases, the following guidelines are useful.

GROUND RULE: ELIMINATE THE MANUAL TASK IF AT ALL POSSIBLE: "NO EXPOSURE, NO INJURY"

Rule 1: Reduce size, weight and forces. (Conversely: increase mass to make use of machinery economical.)
Rule 2: Provide material at the proper working height.
Rule 3: Keep the object close to the body, always in front; don't twist the body.
Rule 4: Minimize the distance through which the object must be moved.
Rule 5: All movement should be smooth and planned.
Rule 6: Move horizontally, not vertically: convert lifting and lowering to pushing, pulling, or carrying.
Rule 7: Don't lift (lower) anything that must be lowered (lifted) later.
Rule 8: For lifting, keep the trunk up and the knees bent.

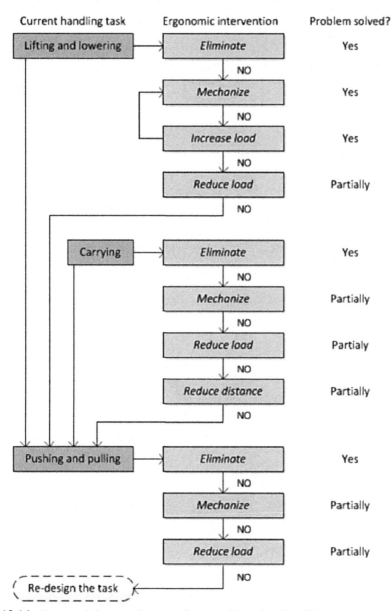

FIGURE 13.14 Ergonomic interventions to reduce or mitigate load handling.

If material handling cannot be avoided altogether, consider means to facilitate the task, making it less hazardous and strenuous. The following guidelines first address the (most hazardous) tasks of lifting and lowering, then carrying and holding, and finally pushing and pulling.

FIGURE 13.15   **Work in front of the body within easy reach.** *Source: Adapted from ILO, 1986.*

FIGURE 13.16   **Avoid body twisting.**

FIGURE 13.17   **Pick up loads from a shelf, not from the floor.** *Source: Adapted from ILO, 1986.*

II. DESIGN APPLICATIONS

FIGURE 13.18   **Grasp a load on the floor between your feet, not in front of them.**

Eliminate the need to lift or lower manually by supplying the material at the working height. If elimination of lifting/lowering is not feasible, consider use of

- lift table, lift platform, elevated pallet, lift truck
- crane, hoist, elevating conveyor
- work dispenser, gravity dump, gravity chute

Eliminate the need to carry by rearranging the workplace. If elimination of carrying is not feasible, convert it to pushing or pulling by use of:

- conveyor, cart, dolly, truck (Figs. 13.11 and 13.12)
- tables or slides between workstations
- shortening the carry distance
- reducing the load weight by
  - using lighter material for the object
  - making the object smaller
  - making the container lighter
  - making the container smaller
  - installing a conveyor between work stations

If you must carry a load, hold it close and in front of the trunk; get somebody to help you carrying long, heavy, awkward objects (Figs. 13.19 and 13.20).

Eliminate the need to push/pull by use of crane, lift truck, conveyor, slide, chute, etc. If elimination of pushing/pulling is not feasible, reduce the effort required by:

- reducing weight or size of the load
- using ramps, conveyor, dolly or truck, wheels and casters, air bearings, good maintenance of equipment and floor surface.

If a material handling job is associated with injuries, if people shy away from it, if it just looks awkward (as shown in Figs. 13.21 and 13.22) then it needs ergonomic attention and intervention.

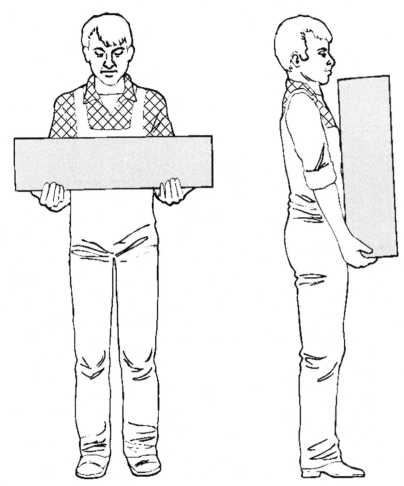

FIGURE 13.19   **Keep a load close to the body.**

13.7 CARRYING TECHNIQUES

We usually assume that carrying a load is done by hand, but there are other carrying techniques that are rarely mentioned in the literature yet are commonly used. One is clasping a load and pressing it against the trunk. This everyday technique is not easily described in biomechanical terms, and the associated muscular effort and energy expenditure depend very much on the actual clasping manner used, and on the load carried. Another common but seldom described technique is to support a load on the hips, such as a parent does when carrying a small child. Sometimes devices are attached to the body to facilitate carrying, such as when using a backpack or a yoke (as in Fig. 13.23).

Obviously, loads can be carried in many different ways. The technique that is most appropriate depends on many variables: the amount (weight) of load, its shape and size, its rigidity or pliability, the provision of handholds or other means of attachment,

FIGURE 13.20 **Get somebody to help you.** *Source: Adapted from ILO, 1988.*

the bulkiness or compactness of the load. The "best" technique also depends on the distance of carry, on whether the path is straight or curvy, flat or inclined, with or without obstacles; whether one can walk freely or must duck (because of limited space) or hide (such as a soldier who wants not to be detected). Some aspects of carrying loads on the body have been formally investigated.[30] The existing results and experiences may be summarized as follows:

- In general, the load should be carried near the mid-axis of the body, near waist height. The farther away from there, either toward the feet or toward one side, the more demanding the carrying becomes.
- Carrying a medium load, up to 30 kg, distributed on chest and back is the least energy-consuming method to carry. However, loads on the front must be kept small in size and very close to the body as not to hinder movements.

FIGURE 13.21   **An exceedingly awkward material handling task.** *Source: Adapted from NIOSH, 1997.*

- Carrying the load on the back, or well distributed across both shoulders and the neck, also consumes relatively little energy.
- Carrying the load in one hand is quite fatiguing and stressful, particularly for the muscles of the hand, shoulder, and back. Nevertheless, this is often done because of the convenience of quickly grasping and securely holding and releasing an object, and of carrying it over short distances while the other hand remains free for other tasks.

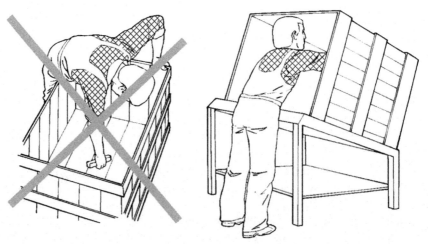

FIGURE 13.22   **Better than having a deep container on the floor is putting it on an inclined stand—but the person must still bend and reach far to retrieve material.** *Source: Adapted from ILO, 1988.*

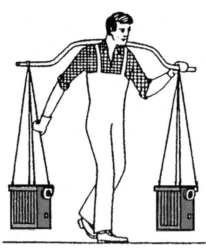

FIGURE 13.23 **A yoke across the shoulders facilitates carrying two heavy loads.** *Source: Adapted from ILO, 1988.*

- Carrying a load of proper size and weight on the head is also suitable, if one is used to doing so. While the load is quite a distance away from the center of body mass, it is right on top of it. Thus, carrying a load on the head requires balancing skills and a healthy spine, but does not demand much energy beyond that needed to move the body.

Table 13.1 provides an overview of the different ways of carrying a 30 kg load. For very small loads, the location probably has little effect on energy demands, while convenience and availability of space may be the determining factors: one might place items into pockets along the thighs, around the waist, on the chest, or on the upper sleeves. Carrying the load evenly distributed on chest and back is least demanding. Avoiding pressure points and allowing good body mobility are important considerations.

13.8 MOVING PATIENTS

Although the previously presented guidelines for load handling apply, in principle, patients are special and sensitive loads, often fragile, who need to be handled with care—which puts particular requirements on their caregivers and, especially, their low backs.[31]

People who take care of patients are at high risk for low-back injuries. Nurses and other caregivers must often lift, reposition, and support patients who are difficult to move because they may be weak and limp, often bandaged and connected to medical devices, and sometimes reluctant to be moved or uncooperative because of discomfort and pain. These individuals came in varying sizes and weights, they have irregular and deformable shapes, and offer no handholds.

TABLE 13.1 Considerations of Different Techniques of Carrying a Load of up to 30 kg on the Body, on a Straight, Level Path

Carrying 30 kg on straight, flat path:	Estimated energy expenditure (kcal/min)	Estimated muscular fatigue	Local pressure and ischemia	Stability of loaded body	Special considerations	
In one hand	?	Very high	Very high	Very poor	Load easily manipulated and released	Suitable for quick pick-up and release; for short-term carriage even of heavy loads.
In both hands, equal weights	Very high: about 7	High	High	Poor		
Clasped between arms and trunk	?	?	?	?	Compromise between hand and trunk use	
On head, supported with one hand	Fairly low: about 5	High if hand guidance needed	?	Very poor	May free hand(s); strongly limits body mobility; determines posture; pad is needed	If accustomed to this technique, suitable for heavy and bulky loads.
On neck, often with strap around forehead	Medium: about 5.5	?	?	Poor	May free hand(s); affects posture	
On one shoulder	?	High	Very high	Very poor	May free hand; strongly affects posture	Suitable for short-term transport of heavy and bulky loads.
Across both shoulders by yoke, held with one hand	High: about 6.2.	?	High	Poor	May free hand(s); affects posture	Suitable for bulky and heavy loads; pads and means of attachment must be carefully provided.
On back	Medium: 5.3 with backpack; 5.9 with bag held in place with hands	Low	?	Poor	Usually frees hands; forces forward trunk bend; skin-cooling problem	Suitable for large loads and long duration carriage. Packaging must be done carefully; attachment means shall not generate areas of high pressure on body.
On chest	?	Low	?	Poor	Frees hands; easy hand access; reduces trunk mobility; skin-cooling problem	Highly advantageous for several small loads that must be accessible.
Distributed on chest and back	4.8, lowest	Lowest	?	Good	Frees hands; may reduce trunk mobility; skin-cooling problem	Highly advantageous for loads that can be divided or distributed; suitable for long durations.
At waist, on buttocks	?	Low	?	Very good	Frees hands; may reduce trunk mobility	Around waist for smaller items, distributed in pockets or by special attachments; superior surface of buttocks often used to partially support backpacks.
On hip	?	Low	?	Very good	Frees hands; may affect mobility	Often used to prop up large loads temporarily.
On legs	?	High	?	Good	Easily reached with hands; may affect walking	Requires pockets in garments or special attachments.
On foot	Highest	Highest	?	Poor	Usually not useful	

Moving patients in, from, or to beds is a frequent task. This often forces the caregiver to assume awkward positions with high risk of over-exertion, particularly of back injury. Therefore, it is often advisable to call helpers; a second person might help to stabilize and balance a patient who otherwise is difficult to manipulate. However, two people are not twice as strong as one, partly because the timing of efforts is not exactly the same, particularly if the load is difficult to grasp and handle. Even under favorable conditions, such as good coordination and suitable placement of hands and feet, at best 90% but usually only about 80% of the sum of lift strengths of two or three individuals can be generated in isometric exertion. If the effort is dynamic instead of static, two people can together generate only about two-thirds of their combined single strengths; even less if three individuals co-operate.[32]

Helping each other

These problems have given rise to numerous attempts to redesign hospital beds and gurneys, especially to provide means for raising, lowering, or tilting; and to lift and transfer patients. One solution is to use equipment which is, in underlying mechanical design, similar to material handling devices discussed earlier in this chapter. However, acceptance in the medical workplace has been slow; in part because of the equipment's awkwardness, the need to move it from bed to bed, and its unpleasant appearance and high price. Ergonomic design of equipment and facilities, combined with suitable procedures and organization, should lead to great improvements both for patients and caregivers, as discussed in Chapter 14.

13.9 CHAPTER SUMMARY

Avoidance of unnecessary strain, over-exertions, and injuries in manual material movement is important for ethical and economic reasons. In essence, three major approaches exist: fit the person to the job through training, or through selection, or fit the job to the person through ergonomic design.

Training for "safe" manual material handling has been attempted in many different ways. Training relies on the assumption that there are safe procedures that can be identified, taught, and followed. Unfortunately, no single material handling technique, or any one training procedure, have yet been proven enduringly successful. Nevertheless, many claims of (often short-lived) improvements are made. Three essential components of a training program are knowledge, instruction, and practice.

Personnel selection relies on the assumption that fitter, stronger workers would be less susceptible to over-exertion than weaker colleagues. While static strength measurements of an individual are relatively easy to do, actual job demands of material handling generally are better reflected in dynamic tests. Isoinertial strength tests have been developed and applied to large numbers of military personnel. These techniques are available for application in industry.

Ergonomic design of load, task, equipment, and workstation, including the work environment and entire facility, appears to be the most successful approach. Human engineering solutions can minimize or mitigate causes of possible over-exertion injuries, thus generating fundamentally safer, more efficient, and more agreeable working conditions: ergonomic design goes beyond just addressing the symptoms, as worker training and selection do.

Developing successful personnel training and selection procedures and combining them with ergonomic design of task, workplace, and equipment reduces the problem of over-exertions and injuries of material handlers. In fact, the combination of all three approaches should provide the highest probability of success, and the best efficiency; this comprehensive approach has been reported to be successful in health care facilities where various and difficult manipulations have to be done on precious loads.

Both workers and managers are important players: one cannot do without the other. While material handlers are the direct recipients of proper ergonomic measures, the manager must fully understand and support them. Physical work, such as manual material movement, is accompanied by physical exertion of the body, energy expenditure, force generation, and accompanying fatigue and aches. Painful low-back symptoms, for example, are likely to appear in nearly everybody's life, whether one works in the shop, in an office, or at home. When workers experience LBP during work, management should not allow adversary situations to develop which are likely to result in prolonged disability. Instead, understanding and acceptance of low-back problems, early interventions, good follow-up and communication, and return-to-work-early programs may prevent, alleviate, or shorten disability.

13.10 CHALLENGES

What determines the time-dependent forces and impulses that must be applied to a load? How do these forces and impulses strain the body?

Why is the lower back so frequently over-exerted?

What are the relations between force and torque in the spinal column? What relationships exist among compressive forces, shear forces, bending, and twisting torques? How are these transmitted by joints, bones, and discs, and by the ligaments and muscles that attach to vertebrae?

What is the role of skill development (experience) in load handling? Which traits can be specifically assessed using the biomechanical approach?

Under what conditions would physiological functions (such as metabolism and muscular efforts) establish limitations for a load handler's capacity for handling loads?

Why does contracting longitudinal trunk muscles load the spinal column?

How does "asymmetric lifting" influence the use of different muscle groups and the loading of the spinal column?

Use the NIOSH lifting equations to determine the load (RWL) and relative risk (LI) for specific lifting tasks: for example, calculate the RWL and LI for a worker who lifts a reel of diameter 70 cm, width 30 cm, weight 15 kg from the floor to a tabletop at 1 m-height. What if the worker lifts this more than once per shift? If the worker wears gloves? If the worker needs to carefully place the reel after lifting it to the table? Is the ergonomic assessment different if you use the Liberty Mutual (Snook) tables?

What is the role of intra-abdominal pressure in lifting tasks, and in pushing and pulling as well as carrying?

What effects might the use of "lifting belts" have on one's attitude and ability in material handling?

Would "staying in practice" keep an aging worker safe from a lifting injury?

What ergonomic assessments might be needed to allow injured workers to return to their jobs?

What needs to be done to make "training for safe material handling" more effective?

Which lifting techniques are appropriate for all tasks and conditions?

How can the success of training be assessed?

Are pre-work exercises effective in reducing worker's back injuries? What kind of exercise is effective: mobility, strength, endurance, flexibility? Is the evidence convincing enough that these exercises should be required?

How much intrusion on one's lifestyle should one accept to become "fit" for material handling on the job?

What are the responsibilities of the employer and the employee to promote non-injurious material handling on the job?

Should training for proper material handling be done during working hours?

Should an employer provide exercise facilities for the employees?

Is there anything wrong with use of the Hawthorne effect to reduce over-exertion injuries?

Why is there not one safe load?

Which are appropriate means to select workers who are capable of handling heavy loads, or handling loads frequently?

How well can static strength testing predict a person's ability to perform load handling?

Which are the major ways for ergonomic design of the work process to reduce manual labor?

How can one redesign the forklift truck to provide better vision and riding conditions for the operator?

Which are the major procedural differences underlying the recommendations by Liberty Mutual (Snook) and by NIOSH?

Which loads are particularly suitable to be carried close to the chest and/or on the back?

Which ergonomic means can be devised to facilitate the moving of patients in hospitals?

NOTES

1 Leading causes of workplace injuries, Liberty Mutual Insurance Company Newsletter Winter 2013/14, www.libertymutual.com. Accessed July 31, 2017.
2 National Institutes of Health (NIH) (December 2014), https://www.nih.gov/news-events/news-releases/chronic-low-back-pain-research-standards-announced-nih-task-force. Accessed December 30, 2016.
3 Deyo and Weinstein (2001); Garg and Marras (2014); Marras (2008); Marras and Karwowski (2006); Snook (2000); Violante, Armstrong, & Kilbom (2000).
4 Since the Latin word "manus" means hand, the terminology "manual material handling" is a triple tautology.
5 ILO (1988).
6 Kroemer (2008).
7 Marras (2008).
8 Chaffin, Andersson, & Martin (2006); Kumar (2004); Marras and Karwowski (2006b).

9 NIOSH (1981) contains an extensive discussions of then current assessment methods; see also the 2003 edition of this book; Kroemer (1985, 1997c); Waters and Putz-Anderson (1998).

10 Astrand, Rodahl et al. (2004); Kroemer (1974, 1999); Marras and Karwowski (2006a); Winter (2009).

11 Dempsey (1998, 2006).

12 Snook (2005); Snook and Ciriello (1991).

13 Available at https://www.cdc.gov/niosh/docs/81-123/default.html. See also Colombini, Occhipinti et al. (2012); Lu (2012); Waters (2008); Waters and Putz-Anderson (1999), Lu, Putz-Anderson, Garg, & Davis (2016).

14 Since the guidelines are subject to updating, check for the newest NIOSH and OSHA information. There are also useful calculators available.

15 Ciriello (2001, 2007); Ciriello, McGorry, Martin, & Bezverkhny (1999); Snook (2005); Snook and Ciriello (1991).

16 Ciriello (2001, 2007); Kroemer (1997c).

17 See the 2003 edition of this book; Summary of NIOSH Back Belt Studies (October 2013) https://www.cdc.gov/niosh/topics/ergonomics/beltsumm.html, accessed 31 July 2017.

18 Genaidy, Asfour et al. (1990); Sharp and Legg (1988); see also the 2003 edition of this book.

19 Jones (1972); Sedgwick and Gormley, 1998.

20 Parsons (1974); Roethlisberger and Dickson (1939).

21 See the 2003 edition of this book.

22 Example data from the 2003 edition of this book.

23 NIOSH (August 2013), Safe Patient Handling and Mobility. https://www.cdc.gov/niosh/topics/safepatient/default.html. Accessed January 2, 2017.

24 See the 2003 edition of this book.

25 Daltrov et al. (1997).

26 NIOSH (1981, p. 99).

27 Hadler (1997, p. 935).

28 Ayoub (1982); Kroemer (1978, 1982, 1983); Kroemer, Marras et al. (1990).

29 McDaniel, Skandis, & Madole (1983); Stevenson, Greenhorn, Bryant, Deakin, & Smith (1996); Rice (1999); Teves, Wright, & Vogel (1985).

30 Listed by Kroemer, Kroemer, Kroemer, & Kroemer-Elbert (2003).

31 NIOSH (2013).

32 Kroemer (1999, 1997c, 1974); Kroemer, Kroemer and Kroemer (2006, 2003); National Research Council (1999).

14

Designing for Special Populations

Overview

Too often, people who design furniture, machinery, vehicles, safety equipment, devices, and the like design for a "regular adult" population in the age range of about 20–50 years. Anthropometry, biomechanics, physiology, psychology, attitudes and behavior of this population are fairly well known. However, there are other large population groups of specific concern: pregnant women, children, the aging, the significantly overweight, the disabled, and the ailing. These people constitute at least one third of the total population[1]; they need special ergonomic attention, but a great deal of information is missing or incomplete.

14.1 BACKGROUND

As discussed in Chapter 1, we have relatively complete anthropometric information for military populations, and we use these body dimensions to estimate dimensions for adult civilians since the latter are seldom measured as a large group. Fortunately, their body dimensions as well as their physical and psychological capabilities and traits are not very different; consequently, the ergonomic principles that help us design for civilian adults also apply to the military, and vice versa.

There are differences among adult men and women—for instance, in body sizes and in physical capabilities. However, in general, we can design nearly any workstation, piece of equipment, or tool so that it is usable by either women or men. In some cases, adjustability is needed or we may need to provide objects in different dimension ranges. These adjustments and ranges, however, generally are not gender specific but are simply needed to fit different people.

While healthy adults in their 20s, 30s, and 40s are commonly considered "the norm", and ergonomic information is widely available about this group (and most ergonomic efforts are made for this group), other groups need ergonomic focus as well. One group in particular needs specific attention: pregnant females. Another group that draws everyone's attention is infants and children. The younger they are, the more different they are from our ergonomic norm, the adult. Specific ergonomic information is available for small children, but little is systematically known about teenagers.

Aging people comprise another large group that needs special ergonomic attention. Body size and posture, physical capabilities, and psychological traits change, for some quickly and over a short time, for some slowly and over decades. To accommodate their special traits during their later working years, their retirement, and their waning period poses challenging yet ethically satisfying tasks to ergonomists.

Another group of people, most but not all of them adults, also need special ergonomic consideration: the impaired and disabled. They differ from their peers in size, posture, and abilities. In many cases, proper ergonomic design of their environment and their equipment can help them to overcome their disabilities, at least to a specific degree, and live a satisfactory life. Obesity also provides additional ergonomic challenges.

Injury or disease can quickly change an independent person into somebody who needs medical help, assistive devices/accommodations, or nursing care. Health care has become a major business with very special demands on organization and procedures to provide well-being and safety to patients and their caregivers.

14.2 DESIGNING FOR WOMEN OR MEN

Do we need to design special workstations and tools to fit each gender? We are accustomed to making general statements which are based on group averages and ranges, findings which may not be true for individuals: there are tall and strong women, and small men. Overall, large overlaps exist between capabilities and traits of males and females, for example, in body size or in muscle strength.

Gender differences

- In body dimensions, men are generally taller, with women having relatively shorter legs but wider hips. In muscle strength and work output, women are generally weaker than men,[2] usually between 60% and 90% of the male values. However, women's leg strength is only marginally lower, and muscle tension developed per cross-sectional unit is equal to men's. Mobility in body joints is generally but only marginally greater in women.
- In audition (hearing), girls and adult females have lower absolute thresholds for pure tones than boys and men. Aging men have larger hearing losses than aging women.
- In vision, both static and dynamic, boys and men have better acuity than girls and women. With increasing age, acuity declines earlier in females than in males. Females have more vision deficiencies.
- In gustation (taste), women may detect sweet, sour, salty, and bitter stimuli at lower concentrations than men; however, some studies have not substantiated these findings.
- In olfaction (smell), women can detect some substances more easily than men. Both taste and smell capabilities and preferences may change with the female menstrual cycle and during pregnancy.

- In cutaneous senses, little difference has been shown between sexes in the threshold for temperature sensation. Females usually feel less warm (less comfortable) than males in environments of 19–38°C initially, but adapt to the surrounding temperature more rapidly than males. Females often begin to sweat at higher temperatures than males, and acclimatize to work in severely hot conditions somewhat more slowly. Men may find it a bit easier to adjust to very hot and very cold conditions, but the differences are minor and conditional.[3]
- Regarding touch, sensitivity to vibration is about the same in males and females. However, females are more sensitive to pressure stimuli on their body, except on the nose. Pain sensations (a complex and difficult topic for physiological and behavioral reasons) seem to be about the same in males and females.[4]

Thus, there appear to be some differences in sensory functioning between the sexes, but differences usually are small. This is also true for many of the gender difference findings discussed below. This is somewhat surprising as male and female brains show anatomical, functional, and biochemical differences throughout life. Additionally, many studies have demonstrated that estrogen and testosterone accentuate cognitive functions, while it is unclear if progesterone affects cognitive function. However, not only do these differences tend to be small but also many appear to be getting smaller over time due to a variety of sociological and educational factors. Some other differences do seem to remain fairly robust, as discussed below.

Cognitive differences

On average, men usually perform better in mathematical tests than women, and women outperform men in verbal and social skills. In the United States, the results of the 2016 Scholastic Aptitude Test (SAT, taken generally by 16- and 17-year-old high school students intending to attend colleges and universities) reflect a pattern that has persisted for decades for all ethnic groups: high school boys outperform girls when it comes to SAT mathematics. However, these girls had better overall academic high school records than boys; on average, they earned better grades than their male peers, were more likely to graduate at the top of their graduating class, and were more likely to graduate from college.[5] In adults, women generally perform better in memory and language skills while men generally outperform women in visual-spatial ability and mathematical reasoning.[6]

Motor skill differences

In motor skills, girls and women are more skillful than boys and men on perceptual and psychomotor tests such as color perception, aiming and dotting, finger dexterity, inverted alphabet printing, and card sorting. Many of these are fine motor skills in which women tend to outperform men. Males perform better in speed-related tasks, on the rotary pursuit apparatus, and in other simple rhythmic eye–hand skills.[7] One of the most robust findings in gender differences is that boys and men are significantly better than girls and women in throwing accuracy.[8]

Coping differences

Regarding environmental stress, we note that the concept of stress encompasses a wide variety of situations, including job pressures, marital and family tensions, and physical aspects of the environment such as climate and noise. Among these stressors, few gender-specific differences have been proven. While females may be slightly more sensitive to sound and noise, the impact on performance, on health, and on social behavior seems to be similar for both sexes. Women appear to be better able to cope with reduced personal space than men: when forced into crowded quarters, men tend to maintain greater interpersonal distance and tend to react more negatively when people invade their personal space.[9] Likely due to socialization factors, women may be more prone to communicating less assertively when stressed, and to prefer working in a group rather than in isolation in challenging conditions.

Research on gender differences in response to monotonous work conditions or higher arousal conditions ("high pressure jobs") has shown inconclusive findings.[10] In a series of tests in which performance and subjective workloads were assessed, women tended to perform slightly better than men[11] on the majority of tasks, including grammatical reasoning, linguistic processing, mathematical processing, and memory search.

Regarding biological circadian rhythms (see Chapter 7), men and women react essentially the same. Regarding the menstrual cycle, of course, men do not have one; therefore it is of separate interest whether changes in physiological capabilities during the menstrual cycle impact performance. Women show distinct (but small) changes in basal temperature, in daily temperature variations, and of course in internal hormone production during the menstrual cycle. While this may affect some nervous system activity and sensory functioning, which possibly could impact performance on specific sensitive tasks, there is no evidence of significant variation in overall performance or in higher cognitive functioning.[12] Women in general maintain a similar performance level throughout compared to men, irrespective of premenstrual tension or pain. In a work task that requires continual attention, some minor changes of little practical importance may occur with the menstrual cycle. Women may be more vulnerable to work-induced fatigue during the premenstrual phase; in a study on hand steadiness, normally cycling women outperformed men consistently except during the week preceding their menses, while women taking oral contraception medication maintained a steady performance level.[13]

Cyclical differences

In neuropsychological assessments, there is some evidence that women and men perform equally when men are compared with women in their preovulatory phase, while women may experience a relative weakness in visual-reaction time, and short-term verbal and visual memory tasks (attentional tasks) and relative executive strengths in the *Stroop test* (reading ink color in place of written words) in the postovulatory phase.[14] This is consistent with many studies finding that estrogen may have a protective effect in preserving executive functions during the aging process. Overall, any such minor effects of the menstrual cycle on performance require no particular ergonomic interventions.

The stereotypical question whether females or males "perform better" is, put in such a general way, unanswerable, because of wide individual differences and many different performance measures. Task performance depends on various attributes: the overall nature of the task and any requirements for specific capabilities, the conditions under which the task must be performed, and the attitudes of the person performing the task and the subjective value the person places on it. Gender differences are often of far less importance than these attributes.[15]

General performance differences

Few statistics indicate differences in actual on-the-job performance. For example, analyzing US aviation records for pilot-error accidents during the years 1972–1981 indicated that female pilots had fewer accidents, and much less serious accidents, than their male colleagues.[16] A subsequent review of pilot-error-related accident records for the years 1986–1992 showed no significant difference between male and female pilots of comparable experience level.[17]

Regarding specific activities, only in those that require very large muscular strength and power do men (on average) develop higher output than women. However, even here individual capabilities, selection of teams, training and skill, age, or motivation, often contradict statements based on "averages."

No other genetically based sex differences produce essential differences in performance of males and females. While there are some other subtle variations, such as in sensory capabilities, or in reactions to climatic stress, the differences between genders are by far overshadowed by interindividual differences. Even within areas where differences—though small—have been found, such as in reaction to crowded conditions, or in fine manual skills, the question is unanswered as to whether these reflect biologically or genetically based differences, or are merely (or mostly) the results of sex roles acquired through group traditions in cultural contexts and under social pressure.

In general, one can design nearly any workstation and any piece of equipment or tool so that it is usable by either women or men. In some cases, adjustability is needed, or one may have to provide objects in different dimension ranges. These adjustments and ranges, however, often are not gender specific but are simply needed to fit different people.[18]

As this review of capabilities and performance has revealed, there are no gender-specific traits that require, or justify, design of workstations or work tools especially for men or women. Proper adjustment ranges will fit all people, whether they happen to be female or male. Although certain jobs may be preferred by one group or the other, using solely gender to restrict access to jobs is not supported by evidence.

14.3 DESIGNING FOR PREGNANT WOMEN

While pregnancy is one of the most common life events, surprisingly little systematic and scientific information is available in the literature for ergonomic purposes. Many brochures and books in anatomy and obstetrics contain normative tables about changes in body dimensions with pregnancy, but few studies have been published with large population samples in recent years.[19]

Size changes

Changes in body dimensions during pregnancy become apparent after 2 or 3 months, and most dimensions increase throughout the course of pregnancy. The most obvious increases are in body weight, abdominal protrusion, and circumference. Fig. 14.1 illustrates some of these changes. From the fourth month of pregnancy onwards, one study showed that waist circumference increased by 27%, weight by 17%, chest circumference

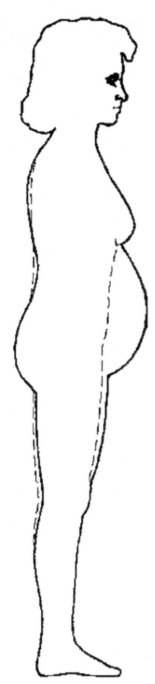

FIGURE 14.1   Changes in body dimensions and posture with pregnancy.

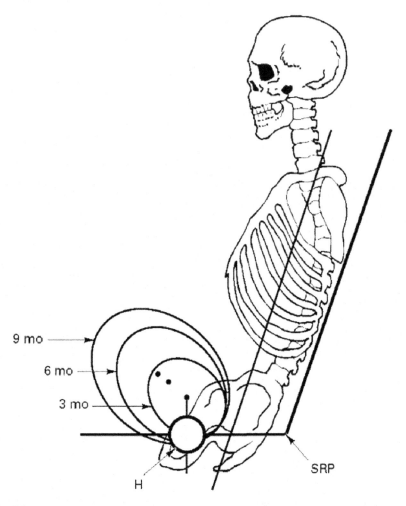

FIGURE 14.2   **Model of changes in body dimensions with pregnancy related to hip joint (H) and seat reference point (SRP).** *Source: Adapted from Culver and Viano (1990) and reprinted with permission from Human Factors, Vol. 32 (6), 1990.*

by 6%, and hip circumference by 4%.[20] For American women,[21] similar increases have been reported, with average increases of 17% in body weight, 8% in chest circumference, and 4% in hip circumference compared with the values measured at 16 weeks of pregnancy. The increases in the abdominal region can be less in women who start their pregnancies with heavier prepregnant bodies.

Mass distribution changes   Not only does the overall mass change during pregnancy but also its distribution—in ways that impact design considerations. For example, designs for crash protection in automobiles should consider the biomechanically significant changes in abdominal depth and circumference, accompanied by a shift in the center of mass of the body, as depicted in Fig. 14.2. The ellipses representing the fetus are superimposed on abdominal and pelvic geometry of a seated women to allow estimates of contact zones between the woman's body and restraining devices, or automobile interior surfaces, in the case of an impact.[22]

The increasing abdominal protrusion during pregnancy makes it more difficult for women to get close to work objects. The working area of the hands becomes smaller, and manipulating objects that are now farther ahead of the spinal column generates an increased compression and bending strain on the spine and on ligaments and muscles in the back. Increasing mass of the abdomen and its increasing moment arm relative to the spinal column further strain the back. (See also Chapter 13.) Variations in abdominal shape also change the body posture, which increasingly assumes a backward pelvic rotation (anterior pelvic tilt) accompanied by changes in lumbar lordosis and thoracic kyphosis. These events explain, at least partly, the common complaints of back strain and pain with pregnancy.

Mobility changes

Changes in mood and cognition have been reported in a study of working women who were near their last third of pregnancy.[23] Comparing their test results with those of a matched control group of nonpregnant female workers showed no significant differences with the exception that the pregnant women felt less alert, vigorous and energetic, and meaning they were probably more easily fatigued.

Mood and cognition changes

Capabilities for physical performance change during pregnancy. Although there are great variations among individuals, in general, advancing pregnancy correlates with a sharp decrease in the ability to perform work that requires a substantial amount of energy, mobility, or far reaches, particularly if repeated or over long periods.

Performance changes

In a 1992 survey, 200 British women,[24] between 29 and 33 weeks pregnant, were asked about their current performance of certain tasks compared to before becoming pregnant. Women reported difficulties related to back pain, reduced reach and clearance, feeling unstable, being fatigued, having reduced mobility, and having difficulties seeing objects near the body. In general, (suitable) sitting was found less straining than standing. Unsurprisingly, some tasks were found to be significantly more difficult to perform during pregnancy than in pre-pregnancy, such as:

- picking an object up from the floor and reaching high shelves
- working at a desk
- walking up stairs
- getting in and out of a car, using a seat belt, driving
- getting in and out of bed

Many everyday tasks become more difficult with pregnancy. However, differences among individuals are striking, with some women finding only a few tasks more difficult and others finding nearly all activities harder to do.

To accommodate pregnant women at the workplace, in transportation, and at home, consider offering these ergonomic adjustments:

- Keep work areas close to the body and ideally slightly elevated from their normal height.
- Avoid the lifting of heavy objects.
- Provide adjustable seating.
- Allow appropriate rest periods.
- Provide more space than usual for moving around, and avoid obstacles that may be difficult to see.

14.4 DESIGNING FOR CHILDREN

Growth

The 18-year time span between birth and early adulthood is characterized by substantial changes in body dimensions, body strength, skill, and other physical and psychological variables. At birth, we weigh about 3½ kg and are around 50 cm in length; the trunk represents about 70% of that length. In the two decades that follow, body length increases three- to four-fold, weight increases about 20-fold, and body proportions change dramatically. Our trunk accounts for just over 50% of our stature when we are grown—see Fig. 14.3.

Timing of growth

While most humans grow roughly in the manner just described, the timing of these changes is individually quite different, and appears to be related not only to genetic factors but also to environmental variables. In general, the growth rate in boys is rapid during infancy (up to 2 years) and then slows until the onset of puberty (at about 11 years) when growth rate increases again, peaks around age 14 years, and then again slows. Although final stature is attained by the early 20s, some boys have almost completed their growth by age 14 or 15 while others are just beginning a strong growth phase. In girls, development during infancy is also rapid; however the puberty growth spurt begins earlier, at about 9 years of age, and is fastest at about 12 years. Full adult stature is often complete at age 16. Hence, at about 11–13 years, many girls are taller than boys of the same age.

Secular trends

Several secular trends have been observed in recent decades:

- Increase in the growth rate of children: children seem to grow faster in their earlier years.
- Earlier onset of puberty, as indicated by menarche in girls and the adolescent growth spurt in both boys and girls.
- Increase in the individually achieved, final adult stature.

Such information is often more anecdotal than based on large scientific surveys. Moreover, information about children is not only specific to age groups but also to areas of origin. For example, a survey of more than 600 children[25] in the Netherlands indicated significant differences among the body dimensions of children growing up in different provinces of Holland.

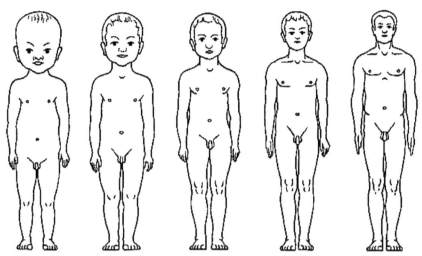

FIGURE 14.3 **Changes in body proportions from infancy to adulthood.** *Source: Adapted from Fluegel et al. (1986).*

Measurements on children are difficult to take, particularly if they are very young and unable or unwilling to follow instructions with a short attention span. For example, the body length (stature) of babies and infants, up to the age of 2 or 3 years, is customarily measured with the child lying on his/her back; later, stature is taken when he/she stands. The recumbent length is several centimeters larger than stature. Considering the importance given by pediatricians and parents to anthropometric information about the growth of children, it is surprising how sparse and outdated the available information is. Compilations of child data[26] were published in the early 2000s but they contain mostly information that even then was several decades old. The only current information in the United States appears in the annually updated Centers for Disease Control and Prevention (CDC) growth charts[27] but they refer only to stature and weight. Therefore, the data, such as compiled in Table 14.1, must be seen and used only as relative "examples in time and location" of anthropometric information on children.

Anthropometry of children

14.4.1 Designing for Body Size

Due to size and mobility, young children are sometimes injured by head, neck, and hand entrapment. Thus, in the 1970s the US Consumer Product Safety Commission sponsored anthropometric studies of American children.[28] The resulting reports not only provide related body dimensions (even though now dated) but also describe in admirable detail the relevant sampling and measuring strategies. Critical measurements include head breadth, chest depth, and hand clearance diameter. Based on the reports, recommendations and regulations were put forth for the primary risk factors for entrapment, such as the distance between the stakes of railings. The data showed that up to the age of 12 years, girls have narrower heads, shallower chests, and smaller hand clearance diameters than boys. Therefore, if openings are kept small enough to not let girls' bodies pass, boys should be safe as well.

Body size

TABLE 14.1 Body Mass and Location of the Center of Mass for US Girls and Boys: Averages (and Standard Deviations)

			Height of the center of mass of the body (in percent of stature)			
	Body mass (kg)		Standing (above floor)		Seated (above seat)	
Age (years)	Girls	Boys	Girls	Boys	Girls	Boys
0	4.6 (1.1)	4.8 (1.2)	59.4 (1.9)	58.5 (2.1)	50.2 (3.3)	48.0 (4.4)
0.5	6.7 (0.9)	7.4 (0.9)	58.1 (2.5)	59.1 (2.3)	47.1 (2.8)	46.6 (3.7)
1	8.9 (1.3)	9.5 (0.8)	58.1 (1.8)	58.5 (2.4)	44.6 (2.3)	45.6 (2.7)
2	11.2 (1.1)	12.2 (1.2)	57.5 (1.0)	57.5 (1.1)	41.3 (2.4)	39.3 (2.6)
3	12.8 (1.1)	14.2 (1.5)	59.3 (2.5)	58.9 (1.0)	39.1 (1.9)	37.6 (1.2)
4	15.4 (1.8)	15.8 (1.8)	58.8 (2.0)	59.7 (1.7)	37.9 (3.3)	37.2 (2.2)
5	17.7 (2.3)	18.3 (2.1)	59.3 (2.0)	58.9 (1.7)	35.3 (2.6)	36.6 (2.6)
6	19.3 (2.7)	20.8 (3.0)	59.3 (1.5)	59.1 (1.3)	34.0 (2.3)	35.0 (1.9)
7	21.8 (2.7)	23.2 (3.1)	58.6 (1.1)	58.7 (1.3)	33.3 (1.8)	33.1 (2.1)
8	24.2 (4.0)	25.3 (4.4)	58.0 (1.7)	58.6 (1.1)	32.2 (2.4)	32.3 (2.2)
9	27.7 (5.2)	27.7 (4.6)	58.0 (1.4)	57.9 (1.1)	30.7 (1.6)	32.1 (1.8)
10	30.6 (5.8)	30.4 (5.2)	57.5 (0.9)	58.0 (1.1)	30.2 (2.0)	31.1 (1.9)
11	34.4 (7.2)	35.4 (5.8)	57.4 (0.7)	57.7 (1.0)	29.5 (1.6)	30.0 (1.6)
12	38.1 (7.3)	38.8 (6.4)	57.4 (1.1)	57.8 (1.0)	29.4 (1.4)	30.1 (2.1)
13	48.0 (8.1)	40.7 (7.0)	57.4 (1.3)	58.0 (1.5)	29.2 (1.3)	29.7 (1.5)

Source: Adapted from Snyder et al. (1975).

Children's body dimensions are critical for the design of furniture, particularly furniture used in schools. Complicating the matter is that children with different body sizes are generally combined in the same rooms from preschool or kindergarten on up. Ideally, tables and chairs in various size ranges should be available to fit the children, but in practice, this is seldom possible. Adjustable furniture would offer a (potentially expensive) solution, but children, particularly very young ones, might have problems adjusting that furniture to their size and liking.

14.4.2 Designing for Body Mass

Body mass

For biomechanical design purposes, such as for restraining devices in automobiles, we need certain critical information on changes in body mass with age as well as on the location of the center of a child's mass when standing and sitting—as compiled in Table 14.1. These 1975 data indicate that the center of mass of a standing body, expressed in percent of body height above the floor, does not change much as we age, and is very similar for girls and boys. For seated children, the relative height of the center of mass above the seat decreases with age when expressed in percent of stature, again in a similar fashion for girls and boys. This information suggests that we do not need to make a distinction between girls and boys when it comes to designing body restraint systems.

14.4.3 Designing for Body Strength

Infants are still fairly uncoordinated and do not show great body strength, however their body strength develops quickly during early and middle childhood. Hand strength shows a strong positive correlation with age, but in those early years, there is little connection between gender and hand strength and dominance. Tables 14.2 and 14.3 list the torque and force capabilities measured in 1975 on American children between 3 and 10 years of age. We show the tables because they reflect how (average) strength increases with age, and they show the very large interindividual differences: the coefficients of variations range from a whopping 30%–60%. The tables combine the measurements

Strength

TABLE 14.2 Average Torques (with Standard Deviations) around Wrist, Elbow, and Knee Exerted by US Children (Girls and Boys Combined)

Age (years)	Wrist flexion (N-cm)	Wrist extension (N-cm)	Elbow flexion (N-cm)	Elbow extension (N-cm)	Knee flexion (N-cm)	Knee extension (N-cm)
3	84 (47)	63 (22)	606 (156)	616 (111)	500 (197)	1673 (616)
4	122 (61)	61 (28)	731 (233)	724 (259)	468 (194)	1866 (710)
5	152 (79)	69 (30)	932 (319)	901 (285)	706 (3510)	2301 (738)
6	224 (85)	90 (40)	1192 (299)	1034 (373)	956 (386)	2717 (961)
7	268 (105)	113 (47)	1687 (415)	1332 (441)	1775 (334)	3788 (1165)
8	352 (128)	122 (44)	2114 (506)	1612 (437)	1371 (564)	4762 (1391)
9	453 (188)	167 (74)	2248 (674)	1676 (527)	1986 (638)	5648 (1386)
10	434 (166)	164 (41)	2362 (603)	1569 (446)	2084 (842)	5553 (1826)
	$n = 211$	$n = 205$	$n = 495$	$n = 496$	$n = 267$	$n = 496$

Source: Data from Owings et al. (1975a,b)

TABLE 14.3 Average Side Pinch and Grasp Forces (with Standard Deviations) Exerted by US Children (Girls and Boys Combined)

Age in years	Thumb-finger side pinch[a] (in newton)	Power grasp (grip strength) (in newton)
3	18.6 (4.9)	45.1 (14.7)
4	26.5 (5.9)	57.9 (17.7)
5	31.4 (7.8)	71.9 (18.6)
6	38.3 (5.9)	89.3 (22.6)
7	41.2 (9.8)	105.0 (32.4)
8	47.1 (9.8)	124.6 (33.4)
9	52.0 (9.8)	145.2 (35.3)
10	51.0 (8.8)	163.8 (37.3)

[a] *Pinch surfaces 20 mm apart.*
Source: Data from Owings et al. (1975a,b)

taken on girls and boys because at these ages, no systematic differences between the genders were observed. Hand preference, as determined both by handwriting and ball throwing, was not associated with strength up to the age of 12 years. Among teenagers, however, on average, males were stronger than females.

14.5 DESIGNING FOR THE AGING

Age categories

Anthropometric, demographic, and capability-related information is usually collected in 5-year intervals until the age of 65, and then just combined for the remaining years, with only occasionally a time marker set at 75 years of age.[29]

Life expectancy

The number of years that humans can expect to live has increased dramatically. The average life span was less than 20 years until about 1000 BCE and increased into the low 20s in Ancient Greece around the year 0. In the Middle Ages, about 1000 CE, life expectancy increased in Western Europe to the 30s and reached the low 40s in the 19th century. In colonial America, until about 1700, the average life span was 35 years. It increased to about 50 years by 1900 and to about 75 around 1990 in the United States. Now, it approaches 80 years. Life expectancies depend on genetic heritage, gender, climate, hygiene, nutrition, diseases, wars, and accidents.[30]

Roles of the aging in society

The position of an older person in society has been quite variable in different eras and in different regions.[31] The aged person might be considered a wise and experienced leader or advisor or a useless and expensive appendix that is removed from societal life. Intermediate positions may be prevalent: consider, for example, the typical retirement age of about 60 to 65 for many occupations around the world, compared with the longevity of many research, entertainment, or political careers.

What does a person owe to society or society to a person? How much care can be expected from relatives and friends, how much from society, and what can the aging person do in return? These general and many specific concerns have been addressed. However, these remain largely unanswered questions, and are beyond the scope of this ergonomics text.

14.5.1 Changes in Anthropometry

Age brackets

Body dimensions are generally assessed in a cross-sectional approach: researchers measure all available people, and then combine those measurements within a defined age bracket. This does not create a big problem in the young adult population, because dimensions do not change very much in the 20–40-year age span. However, this is a major problem in the description of the aging (as well as of children) for several reasons:

- Among the aging, dimensions can change rapidly in some people over just a few years. For example, stature changes due to changing posture and shrinking thickness of spinal discs; weight changes due to shifts in nutrition, metabolism, and health; and musculature and strength change due to activity levels, habits, and health.
- The age brackets used for surveys are often broad, usually encompassing decades or even longer time spans, as opposed to the common 5-year bracket reported for younger cohorts. Hence, people with very different dimensions are contained in each reported sample.

Using solely chronological age is not a good way to classify changes of the aging (nor of children). They would be better described by a longitudinal procedure in which changes in body dimensions and capacities are observed within one individual over many years; unfortunately, few such data are available.[32]

Chronological age

Anthropometric information reported in the literature on sufficiently large samples is, as a rule, the result of cross-sectional surveys. Compilation of available data (as listed in Table 12.7 of the 2003 edition of this book) reveals current problems with anthropometric information: most of the samples are exceedingly small; surveys are done for a few age ranges only; there are often no distinctions between ethnic origin, region, socioeconomic status, health or other attributes that are codeterminers of anthropometry.

Available data

The apparent height loss with age (of about 1 cm per decade) starting in the 30s may be a result of:

Height and weight

- flattening of the cartilaginous disks between the vertebrae
- flattening or thinning of the bodies of the vertebrae
- general thinning of weight-carrying cartilage
- change in the S-shape of the spinal column in the side view, particularly an increased kyphosis in the thoracic area (hump back)
- scoliosis, a lateral deviation from the straight line displayed by the spinal column in the frontal view
- bowing of the legs and flattening of the feet

Contrast the loss in height with changes in weight: generally, American men have their largest body weight in their 30s, then lose weight with aging; American women are relatively light in their 20s, but then increase their weight with age, becoming heaviest, on average, at about 60 years of age.[33] (Note that the increasing prevalence of obesity since the early 2000s over much of the earth may have changed these weight data.)

14.5.2 Changes in Anthromechanics

In addition to the anthropometric changes that commonly occur with increasing age, there are numerous alterations in biomechanical features (discussed in Chapter 1).

The long bones become larger in outer diameter and larger in inner diameter (thus, more hollow), and larger pores appear in the remaining bone. Total bone mass decreases. Bones also become stiffer and more brittle because of a change in mineral content: this is the major effect in age-related osteoporosis. Women and those who rarely exercise are at higher risk of osteoporosis than men and active people. The changes in bone structure are associated with an increased risk of bone breakage, often as a result of (or the cause of) falls or other accidents in which sudden forces and impulses are applied. Injuries to the pelvic girdle, hip joint, femur, to the shoulder and arms are particularly frequent (and dangerous) in older women.

Bones

The lining of joints, the bony surfaces in joints, the supply of synovial fluids, and the elasticity and resilience of joint capsules and ligaments are all reduced with age. This leads to reduced mobility in the joints, often associated with pain.

Joints

While well-used muscles can retain their capabilities into advanced age, reduced use, often accompanied by decreased circulatory supply, generally leads to a loss of musculature and ensuing loss of strength.

Muscles

Skills

Although some—even many—skills can be retained through practice and with good health, others deteriorate for the reasons already discussed. In addition, reductions in the performance of the central and peripheral nervous systems, and diminished circulatory and metabolic capabilities, can impact skills. The reductions are most likely in activities that require exertion of large energy or forces, often combined with endurance requirements and controlled through perceptual information, particularly of the visual and vestibular systems.

Manipulation capability can require a group of skills: strength, mobility, and sensory control. Even though manipulation skills are critical for many tasks (on the job, at home, or during leisure), unified and standardized measuring techniques are not in use. There is little reliable information on manipulation skills for adults in general, and even less for aging individuals.

14.5.3 Changes in Respiration and Circulation

Diminished respiration and circulation

Respiratory capabilities decline with increasing age mostly because the alveoli in the lungs are less able to perform the exchanges of gases (oxygen and carbon dioxide). Furthermore, the intercostal muscles—groups of muscles that run between the ribs and help form and move the chest wall, expanding and shrinking the size of the chest cavity to facilitate breathing—and the chest diaphragm lose some of their ability to generate breathing space in the chest: hence vital capacity decreases. This is coupled with reduced blood flow, and may be complicated in the presence of breathing difficulties (emphysema) that often results from many years of smoking and pollution.

The elasticity of blood vessels seems to decrease with age. Resistance to blood passage in vessels may be increased due to lipid or calcified deposits along their walls. Blood cell production in the bone marrow is decreased. In particular, thin aging people may have reduced volumes of body fluids.

Heart functions also change. The heart tends to enlarge slightly with ageing, developing thicker walls and slightly larger chambers. The walls of the arteries become thicker and less elastic. Thus, higher blood pressure during systole and normal blood pressure during diastole is very common among older people. Cardiac output is lower. Heart rate takes longer to return to resting level after the rate has been increased. Neural control of the heart may be impaired.[34]

14.5.4 Changes in Somesthetic Functions

Somesthetic senses

The somesthetic senses include those related to touch, pressure, pain, vibration, temperature, and motion—see Chapters 5 and 6. These permit proprioception (sense of position and movement) and haptic perception.

Touch and taction

Skin, muscles, tendons, joints, and internal organs have nerve endings (receptors) that allow us to sense the type and amount of touch. With aging, absolute thresholds appear to increase, which may be associated with a loss of touch receptors. Aging may also lead to a reduction in blood supply to the spinal cord with ensuing damage to the nerve tracks, possibly a decline in the number of myelinated fibers in the spinal roots, and/or diminished blood flow to the peripheral structures of the body in general. Reduced ability to detect vibration, touch, and pressure increases the risk of injuries. Dietary deficiencies might also play a role.

Sensitivity for pain appears much reduced as one gets older. Possible explanations are reductions in the number of receptor organs in the skin, in the number of myelinated fibers in the peripheral nervous system, and in blood supply.

Pain

Sensitivity to temperature also seems to decline, but this may be partially offset by the often-observed desire of elderly people to be in warmer temperatures, outside or within a building. The observed differences in temperature behavior may be associated with a decline in the body's temperature regulation system.

Sensing temperature

Decrease in information from kinesthetic receptors, or decrease in the use of that information in the central nervous system, may contribute to the higher incidence of falls: with increasing age, we seem less able to perceive of being moved (such as in an automobile or airplane) or of moving one's own body parts.

Sensing motion

The ability to cope with the environment depends in large part on detecting, interpreting, and responding appropriately to sensory information. Both sensation (the reception of stimuli at sensors and the resulting neural impulses in the afferent part of the nervous system) and perception (the interpretation of the stimuli) change with age. The numbers of receptor cells in the skin (such as Meissner's and Pacinian corpuscles) decrease which, together with diminished arterial and venous flow in blood vessels, changes the stimulation of the nervous system. This may lead to increased variability in reception and integration of, and hence response to, external and internal stimuli.

Sensation and perception

With receptors becoming more sparse and nerve conduction velocity decreasing, especially with concurrent reduced synaptic function, performance may be diminished in perceiving, interpreting, and reacting to external stimuli. This may impact spontaneous decision-making and how appropriate and timely subsequent actions are. Malfunctioning of neurons and development of plaques and tangles in the brain may cause problems with body movements and with memory, thinking, and behavior. Peripheral nerve conduction may slow as a result of degeneration of the myelin sheath, reduced blood flow, bone overgrowth, and slower self-repair processes in axonal damage.

Reactions and actions

14.5.5 Changes in Vision

Changes in visual functions are tied to many concurrent anatomical, physiological, and psychological processes that develop with age. Using the analogy of the camera, the human eye as a photographic device loses precision. Structures that bend, guide, and transform light (see Chapter 5) change, which reduces the amount of light reaching the retina, and defocuses the image projected to the retina. It becomes more difficult to focus on near objects, particularly if they are elevated or moving quickly.

Eyes lose precision

One problem is the watering and tearing of eyes, due to accumulation of fluid on the outsideof the eyeball. Water can affect the properties of the eye as a lens, and it can simply be annoying.

Watery eyes

Sagging eyelids may reduce the amount of light reaching the cornea.

Droopy eyelids

Difficult to focus

The cornea flattens, which limits the ability to focus. Fatty deposits can reduce the transmission of light and scatter arriving light.

Smaller pupil

The opening of the pupil gets smaller, which further reduces the amount of light entering the eyeball. This disorder (*senile miosis*) is most noticeable in dim light. There is a possible benefit from the smaller diameter, however, similar to having a smaller aperture opening in a camera lens: the depth of field may be enhanced, meaning that objects both near and far are in better focus, although they appear dimmer.

Glaucoma?

The natural fluids produced in the eye may not drain well but collect inside the eyeball, and the ensuing pressure (*glaucoma*) may eventually destroy fibers of the optic nerve. Thus, regular medical checkups for glaucoma are recommended after the age of 40 to detect its occurrence before damage occurs.

Eye movements

Ocular motility, the ability to move the eyeball through muscular actions, becomes impaired, both in performing quick movements and in turning the eye to an extreme angle. Eye pursuit movements are impaired in most aging people, which limit both the ability to follow a target smoothly and to move the gaze to a target and to fixate it. Furthermore, the field of vision is reduced at its edges, particularly in the upward direction.

Contrast

The ability to detect contrast is reduced, which decreases one's ability to perceive details of an object or scene, particularly at twilight. Recognizing details from a cluttered background becomes difficult: picking out individual faces in a crowd at dim light is nearly impossible, even reading a book with large print may be difficult if inadequate white space exists between the black letters.

Light transmission

The vitreous humor may yellow, adding to problems in light transmission already caused by a yellowing lens. Liquid and gel portions may clump together, causing "floaters" (composed of cellular debris) or spots to appear in the field of vision. Pockets of liquid vitreous may form. Clumping and liquefaction together result in changing refraction, making the image formed on the retina less coherent. A sudden jolt or vibration may detach the posterior vitreous from the retina (probably as a result of macular edema), which brings about a severe vision impairment.

Less clarity

The cones and rods at the retina, the sensors of color and light, are reduced in number, particularly the cones at the fovea. This reduces the clarity of vision. Retinal pigment is reduced, a degenerative pigment (*lipofuscin*) begins to appear, and the retina becomes thinner, particularly at the periphery.

14.5.5.1 Changes in the Lens

Presbyopia

A common problem in people over 40 years of age is hardening of the crystalline lens, which reduces its ability to change shape. The small ciliary muscles that manipulate the lens may also atrophy. This makes it more difficult to accommodate (focus on) near objects, causing farsightedness or presbyopia (*presbyopia*, derived from the Greek *presbus*, meaning old man, and *opia*, meaning eye).

Perceived colors

The young eye has a slightly yellow-tinted lens, which acts as an ultraviolet filter for the retina. The aging eye becomes more yellow because of the development of fluorescent chromophores of yellow color. A yellower lens is a stronger light filter, absorbing more energy (hence raising the threshold for detection of light in general), and it absorbs some of the shorter (blue and violet) wavelengths of light. This changes perception of colors: white objects appear yellow, blue is harder to detect, and blue and green are difficult to distinguish. Color contrast and saturation is reduced.

With increasing age, water-insoluble dry proteins in the lens become more prevalent, and proteins aggregate into larger macromolecules. This decreases lens transparency and increases scattering of the light, affecting what is transmitted to the retina. Dispersion of light rays may act like a "luminous veil" through which one tries to see. | **Dimmed light and veiling**

The condition when the lens becomes opaque and "cloudy" enough to interfere with vision is called a cataract. This can occur at any age but is most common in the aging. Increased opacity distorts and decreases available light. The effects of a cataract on vision depend on its size, location, and density. A small cataract in the center of the lens is likely to affect vision far more than even a large cataract at the periphery. Cataracts can cause blurred or double vision, spots, difficulty in seeing at too little or too much illumination, change in the color of the pupil, and the sensation of having a film over the eye or looking through a waterfall (*cataract* is derived from the Latin for waterfall). | **Cataracts**

14.5.5.2 *Designing for Aging Vision*

Individuals develop visual impairments of different types and magnitudes, and at varying ages, but the most common difficulties are with near vision—especially in dim light, reading small print, and distinguishing similar colors. In addition to these basic deficits, complex functions generally deteriorate with age: visual perception, search, and processing; depth perception; coping with glare.

Most of the vision problems described above can be counteracted through relatively simple means: corrective lenses, brighter lighting, increased color contrast, larger characters with high contrast to the background, repositioned computer screen, shielded bright lights in the field of view. | **Simple everyday aids**

For work at computer workstations, separate optical lenses might be needed for viewing either the display or for associated reading and writing tasks. The display should be placed low rather than high with respect to the eyes. On the screen, bright reds and yellows are easier to distinguish instead of blues and greens. | **Computer work**

Extra lighting (increased illuminance) can improve the visual ability of aging people. However, since glare is a particular problem for this population, spotlights and other bright light sources must be carefully and thoughtfully placed. Aging people often find it more difficult to adapt to sudden changes in lighting, particularly to adapt from bright to dim conditions. | **More light**

14.5.6 Changes in Hearing

The ability to hear decreases quite dramatically in the course of a lifetime, initially at the high frequencies from 20 kHz to about 10 kHz. During adulthood, it diminishes further to about 8000 Hz, an impairment called age-related hearing loss (*presbycusis*, derived from the Greek *presbus*, meaning old man, and *akousis*, meaning hearing). Hearing difficulties may extend into the lower frequencies, and they are often coupled with noise-induced hearing loss (NIHL, see Chapter 5).

Biologic changes

The changes with age start at the outer ear (*pinna*), which becomes harder, less flexible, and may change in size and shape. Wax buildup is frequent in the ear canal. The Eustachian tube may become obstructed, leading to an accumulation of fluid in the middle ear. In the inner ear, atrophy and degeneration of hair cells in the basilar membrane of the cochlea can occur. Deficiencies in the bioelectric and biomechanical properties of the inner ear fluid and mechanical degeneration of the cochlear partition occur, often together with a loss of auditory neurons. These degenerations cause either frequency-specific or more general deficiencies in hearing capabilities.

Hearing sensitivity

While it is estimated that 70% of all individuals over 50 years of age have some kind of hearing loss, the changes are individually quite different. Typically, in populations that do not suffer from industry- or civilization-related noises, hearing sensitivity in the higher frequencies is less reduced than in people from so-called developed countries. Such changes related to the environment overlap with, and in some cases mask, age-dependent changes.

Understanding speech

Loss of hearing ability in the higher frequencies of the speech range reduces our ability to understand consonants that have high-frequency components. This explains why older people often are unable to discriminate between phonetically similar words, which may make it difficult to follow conversations in noisy environments. Severe hearing disorders may lead to speech disorders which, to some extent, may be psychologically founded: for example, if others have to speak loudly for you to hear, they may interact with you less. What's more, the person with the hearing loss may be hesitant to speak if he or she does not know what level of loudness to use. This can lead to the person suffering from hearing loss feeling isolated and alone, or appearing aloof or uninterested to others.

Hearing aids

Externally applied "hearing aids" can improve a person's ability to hear by amplifying sounds for which hearing deficiencies exist. This, however, is difficult to do if the deficient areas are not diagnosed carefully. Simple amplification of sounds usually also amplified unwanted background noise; however, selective suppression and amplification of sound is, fortunately, technically possible now. Improvement is very difficult if the hearing loss is due to destruction of structures of the middle and inner ear, and, at least with current technology, nearly impossible if the auditory nerves have been damaged. Further technical development may bring better help in the future; replacement of the cochlea, for example, by an electronic implant is already being performed.

Ergonomic interventions

The preferred or comfortable sound intensity level for speech increases with age: on average from about 55 dB at 20 years of age to more than 80 dB at 80 years for prose text. Furthermore, one can improve the clarity of the message by providing sound signals that are easily distinguishable and of sufficient intensity by avoiding masking background sounds ("noise"), and by increasing redundancy and reducing sentence complexity. Another solution is to augment information through other sensory channels, such as vision (e.g., adding illustrations and written texts to an auditory message) or tactile sense (e.g., using Braille).

14.5.7 Changes in Taste and Smell

Taste and smell lessen

The number of taste buds and the production of saliva in the mouth are often reduced with age, and fissuring of the tongue may occur. The rate of regeneration of taste buds (usually every 1–2 weeks), in particular after injury (say, contact with a too-hot beverage), may slow. Since taste is tied to smell, a loss of taste detection may actually be a loss in smell detection. The sense of smell may deteriorate with aging due to reduction in sensory cells and in receptor neurons, impacting both the olfactory sense itself and our ability to distinguish between odors. Although the senses of taste and smell are often said to be diminished, this finding is not uniform.[35]

14.5.8 Changes in Psychometric Performance

With aging, reaction and response times to stimuli typically increase. This may at least be partially explained by deficiencies in the sensory peripheral parts of the nervous system, delays in afferent provision of information to the central nervous system, and reduction of efficiency in the efferent part of the peripheral nervous system. However, successful performance involves perception, sensation, attention, short-term memory, decision-making, intelligence, and personality as well as motor behavior. In normal aging, short-term memory and new learning tend to be affected first. As soon as age 40, most people begin to demonstrate a gradual decline in recalling new information, including reading materials and visual information. Word-finding difficulties are common but not worrisome. Verbal abilities may begin to decline after about age 70. Studies show that by age 70, the amount of information recalled 30 min after hearing a story once is approximately 75% of that recalled at age 18.[36] Intellectual performance and most executive function are usually maintained until about age 80 in the absence of neurologic disorders. Visual-perceptual problems also decline after age 80. However, the effects of aging per se may be difficult to differentiate from the effects of various disease processes more common in the elderly, such as hypothyroidism, depression, diabetes, and cerebral-vascular incidents. Fortunately, many of these disease processes can be minimized with healthy lifestyles. Other protective factors include cognitive reserve (people who are more intelligent or cognitively skilled at a young age retain skills over time compared to those who are less capable at younger ages), social engagement, and a sense of purpose. There is tremendous variability in the aging populace.

Consequently, while diminished performance may be attributed to some of the physiological processes in aging, much depends on other processes of the mind and body which go beyond the scope of this text.[37]

14.5.9 Research Needs

Available research findings do not provide a complete picture of the changes with aging. Since many studies are performed as cross-sectional research, comparing the anthropometry or performance of people of similar age provides only that information. This yields little insight into the nature and effects of aging processes, and how to counteract them.

Chronological age is not a meaningful classifier. Better classification systems need to be established: the so-called biological age is one attempt in that direction, but its definition is rather difficult.

It is obvious that many or almost all physical, perceptual, cognitive, and decision-making capabilities decline with age. However, some of those losses are slow and are not easily observed. Some capabilities and facilities decline fast, or they may deteriorate slowly at first and then quickly at some point in time, perhaps again stabilizing for a while. Some of these changes are independent of each other, but many are linked, directly or indirectly. For example, failing physical health may have effects on attitude and intentness; failing eyesight might lead to a fall and serious injury with ensuing illness; failing hearing may lead to isolation and depression.

In addition, people develop different coping strategies. One's failing ability to recall names, for example, may be overcome to some extent by developing mnemonic strategies. An aging person may reduce activity boundaries by maintaining only those with which he or she is comfortable and can competently handle. Maintaining physical or mental activities, perhaps purposefully so, can counteract reductions in facility significantly and for long periods of time.

There is a great deal of evidence that skilled performance, both mental and physical, can be maintained by many individuals into advanced age for several reasons:

- In most of our daily tasks, on or off work, only a portion of our abilities is required. Thus, even with decreasing capacity, enough remains available to do what needs to be done.
- Certain tasks require particular experience and patience which may come with age.
- Continuing to use, even to train, one's skills and functions tends to maintain them into advanced age.

14.5.10 Designing for the Aging Driver

As we age, driving can help us stay connected, mobile, and independent. However, the risk of being hurt or even killed increases as we age, probably because of age-related declines in vision, hearing, and cognitive functioning. The pattern of accidents with age indicates increasingly missed or misinterpreted perceptual clues, slow or incorrect reactions to cues, and incorrect actions such as accelerating instead of braking.

To reduce the risk of accidents, many aging drivers may avoid driving at night, on unfamiliar roadways, and in congested areas and times. To help them, one should avoid "environmental clutter" of lights and colors such as experienced in many commercial streets, where illumination and advertising lights compete with traffic lights. (In some European countries, no red or green lamps may be used near traffic lights.)

Regarding the design of the interior of the automobile, the illuminance of instruments should be higher than the common $300\ cd/m^2$ (candela per square meter) to avoid "washout" in direct sunlight: $1200\ cd/m^2$ is more suitable. The commonly used minimal contrast ratio of 5:1 is sufficient for passive displays, but should be about 20:1 for moving displays. Regarding character size, a font size of 25 minutes of arc should be the minimum. Critical visual information should be displayed in the central area of view. Red versus blue or green is most easily distinguished, followed by yellow and white. These colors should be pure in saturation. Such design recommendations not only help older drivers but also are advantageous for younger eyes.

Other, more general recommendations include designing cars that are easier to get into and out of; making seat belt use easy and convenient; reducing complexity on the dash board by avoiding confusing or unneeded displays; and making controls easy to grasp and operate.

14.5.11 Designing for the Older Worker

The US Age Discrimination Acts have defined the "older" person as either age 40 or 45 and above. While we cannot set a particular age as the official onset of aging, there is no doubt that certain work tasks become more difficult as one approaches retirement. This includes strenuous physical exertions, such as moving heavy loads; tasks that require high mobility, particularly of the trunk and back; and work that requires high visual acuity and close focusing. In contrast, there are anecdotal indications that tasks which require patient and experience-generated skill may be performed better by at least some older workers than by young workers. Nearly all age-related difficulties can be mitigated or overcome by:

- workplace and tool design and selection,
- arrangement of work procedures and tasks,
- provision of working aids such as power-assisted tools or magnifying lenses, and
- managerial measures such as assigning appropriate work tasks and providing breaks.

Specific ergonomic measures are the logical results of intent to counteract known deficiencies. Recommendations include:

- Facilitating manipulation tasks by providing proper work height, arm and elbow supports, and adjustable seating.
- Counteracting vision deficiencies by providing corrective eye lenses, intense and well-directed illumination, and avoiding glare.
- If text must be read on paper or on a screen, ensuring that proper character size and contrast along with carefully selected color schemes are used.
- Controlling the auditory environment to keep background noise at a minimum and ensuring that auditory signals that must be heard are sufficiently loud and below frequencies of about 4 kHz.

> Use of proper ergonomic measures, carefully selected and applied, is just good human engineering practice. This helps all people of all age groups, not just the older individual.

14.5.12 Designing a Home for the Aging

A 50-year-old who purchases a home (or any durable products) will be a rather different user 10 or 20 or more years later. A person's ability to live at home, perhaps with

Problems/ manifestations	Senescence	Arteriosclerosis	Hypertension	Parkinson's disease	Peripheral neuropathy	Drowsiness	Cataracts/glaucoma	Arthritis	Paget's disease	Osteoporosis	Low back pain	Bronchitis/emphysema	Pneumonia	Diabetes	Senile dementia
General debility	X		X	X	X			X	X	X	X	X	X	X	
Mobility	X	X		X	X		X	X	X		X				
Posture	X			X			X	X	X	X	X				
Pain		X		X	X			X	X		X				
Incoordination		X		X	X		X	X							
Reduced sensory input	X	X			X	X	X								
Loss of balance	X	X		X			X								
Reduced joint mobility								X	X	X	X				
Weakness in muscles	X			X	X			X							
Auditory disorders	X					X		X						X	X
Locating body in space		X					X	X							
Shortening of breath		X		X								X	X		
Deformity								X	X	X					
Memory impairment		X													X
Visual problems	X	X					X							X	
Disorientation		X		X											X
Loss of sensation	X				X	X									
Cognition disturbance		X		X											
Incontinence		X													X
Speech disorders				X		X	X								
Touch disabilities	X				X			X						X	

FIGURE 14.4   **Problems commonly arising with aging.** *Source: Adapted from Kemmerling (1991).*

some outside help, or to be cared for at home, or to live in a care environment is a complex function of various abilities and disabilities. Good daily functioning of older people depends on their intellectual, mental, and physical capabilities. The elderly need a comprehensive and holistic approach to a variety of problems that occur with aging. Fig. 14.4 presents an overview of common disorders among the aged, and how these problems can be alleviated, at least to some degree, by proper ergonomic measures.

How much dependency?

Functional ability, or its opposite, dependency, is commonly assessed in terms of clusters of abilities, known as activities of daily living (ADL). ADLs describe basic self-care tasks; these are the kinds of skills that people usually learn in early childhood. They include feeding, toileting and maintaining continence, selecting and putting on appropriate clothing, grooming, bathing and washing, walking, and transferring (such as moving from bed to wheelchair). Instrumental activities of daily living (IADLs) are the more complex skills we typically learn as teenagers, needed to successfully live independently. These skills include managing finances, handling transportation (driving or navigating public transit), shopping, preparing meals, communicating with others by telephone and other devices, managing medication, and performing housework and basic home maintenance. Together, ADLs and IADLs represent the skills that people usually need to be able to manage in order to live as independent adults. These two scales have widely been used to assess people and classify them as being able to live independently or as requiring help and/or specifically designed environments.

While they are practical and self-explanatory, ADLs and IADLs lack specificity and objectivity and are difficult to scale. This deficiency can be alleviated by subdividing ADLs into more specific tasks[38] such as lifting/lowering, pushing/pulling, bending/stooping, and reaching. Further work in this area could result in a list of basic demands and activities similar to the "motion elements" used in industrial engineering for studies of work method. Nonetheless, these scales are still helpful, and widely in use.

Staying in our homes as we age has the major advantage of a "familiar" setting with all its physical and emotional implications. Unless designed—by happenstance or foresight—to be ergonomic, private homes usually need some adjustments to allow the aging inhabitant to perform all necessary activities, even with somewhat reduced sensory, motoric, and decision-making capabilities. In addition to stairs and other passage areas, kitchens and bathrooms are of particular concern.

Home, sweet home

In the kitchen, we store, prepare, and serve food. In many households, it is the de facto living room and message center. Early scientific study of kitchens was conducted by Lillian Gilbreth in the 1920s. Her quintessential study concerned mainly the flow of work, which she assessed using time and motion study methods that she pioneered with her husband. Her redesigned kitchen reduced motions by nearly 50%, and still influences kitchen design.

Kitchen

Seven principles, derived from time and motion studies and augmented by ergonomic findings, apply to the kitchen:

1. Design for a small "work triangle," the corners of which are the refrigerator, the sink, and the stove/range.
2. Kitchen components should facilitate the "work flow" for food preparation, which is to remove food from storage, mix or otherwise prepare ingredients near the sink, cook, and serve.
3. If there is traffic flow of people moving through the kitchen, this flow should not cut through the work triangle or work flow pattern.
4. Items should be stored at the point of first use, as determined by work triangle and flow.
5. The work space for the hands should be at or slightly below elbow height. This facilitates manipulation and visual control. Counter and sink heights are derived from elbow height. Note that walking aids, stools or chairs may also be used in the kitchen.
6. Reaches to items stored in the kitchen should be at or slightly below eye height to allow for visual control and easy arm and shoulder motions.
7. The motion and work space should not be reduced or interrupted by appliance or cabinet doors; consequently, they should open away from the person working in the kitchen.

Bathroom/toilet

The bathroom is one of the busiest and, unfortunately, most dangerous rooms in the house. Basic equipment includes a bathtub and/or shower, sink, and toilet, and often also storage facilities for toiletries, towels, etc. Bathtub and shower are the most common means for whole body cleansing—and they can be treacherous accident sites.[39] Their major danger stems from slipperiness between bare skin and floor or walls. The bathtub is particularly hazardous because of its slanted surfaces and the high sides; climbing in and out can be tricky for those who may have balance and mobility deficiencies. While resting in the tub, the angle of the backrest and its slipperiness comprise the most critical design aspects. Proper hand rails and grab bars, within easy reach both for sitting and getting in and out, are critical. The shower stall or walk-in tub may also have a slippery floor to step on, but its lower enclosure rim makes it easier to enter and exit. Use of the control handles for hot and cold water is quite often difficult for aged people, particularly when they are not at their own familiar home and have to cope with different handle designs and movement directions. Both better design principles and standardization would be helpful—for example, in the mode and direction of control movement to regulate water temperature. The sink may be difficult to use if it is set too deeply into a cabinet. The faucet often reduces the usable opening area of the wash basin. Proper height is important, as are the water controls. The toilet is, obviously, of great importance for eliminating body wastes and keeping the body openings and adjacent anatomic areas clean. Asking elderly users about their problems and for their solution suggestions leads to successful designs. For Western-style facilities, a variety of publications contains valuable ergonomic design recommendations, including grab bars, walk-in tubs, and raised toilet height; in other parts of the Earth quite different customs and conditions exist, for which at present only limited ergonomic information[40] is available.

14.5.13 Designing Nursing Homes

What many of us fear the most when we contemplate old age is suffering from severe functional disabilities. In the United States, currently only one of five people of age 65–74 report poor or only fair health, but there is the likelihood that health is declining: the number rises to one in three for those aged 85+.[41] As more health deficiencies, mental problems, or functional disabilities occur with further aging, the elderly person first needs more help in his or her own home.[42] Initially, that care may be secured privately, through relatives and friends or a hired person. For many, this is the beginning of a path that leads to a nursing home.

Therapeutic nihilism

Many aged individuals suffer from (often self-diagnosed) treatable disease conditions but have decided to accept aches, pains, and physical distress as something "normal" at advanced age. Further, since the ability to feel pain can lessen with age, they may neglect or even refuse to seek help or take prescribed medication. They may not want to appear weak and discouraged or to bother their caregiver, or they may fear the cure more than the illness. This makes it particularly difficult to provide them with the help and care that they need and deserve.

Nevertheless, aged people are the most frequent consumers of physicians' services. Usually, it is a physician's job to diagnose an illness, make a medical intervention, and seek a cure when possible. But this is often not the case with the elderly. The older we get, the more likely we are to suffer from a chronic illness that cannot be cured, although maybe alleviated (or ignored), at least for a while. Fighting disability requires strategies different from battling illness. These strategies include techniques specific to geriatric medicine and require collaboration with nurses, physical therapists, dietitians, psychologists, and ergonomists.

There are different kinds of institutions for the elderly who cannot stay at home. Some simply offer room, board, and personal care to the residents. Others are more like a hospital, offering intensive medical services to seriously ill people or hospice for end-of-life care. Some cater to certain religious or ethnic groups; some cater to people with dementia or Alzheimer's disease or those who are bedridden; others accept only occupants who are not severely physically or mentally impaired. Because of this diversity, we classify them by the intensity of care which they provide.

Room, board, care

Residents who need assistance in functioning but not intensive care are in "intermediate" care facilities. Here, the architectural and other ergonomic design recommendations mentioned earlier for the private home also apply to facilitate the residents' efforts to look after themselves. In addition, design and organizational means that facilitate the caregivers' activities must be considered, such as easy cleaning, awareness of immediate help requirements, and emergency access.

Intermediate care

While it is important that aged people have as much personal freedom as possible, being in a care facility limits their choices in the most basic aspects of life, such as where to live, when to get up or lie down, what to do, what meals to have. Home management should carefully provide various choices for the residents, prioritizing their interests over organizational ease.

Patients in "skilled nursing care facilities" (as called in the United States) usually have little ability to control themselves or their environment. The residents tend to be very old and in continual need of help and care. Owing to gender differences in longevity and types of illnesses and injuries, women are more likely than men to be disabled but not to have a life-threatening illness. Regarding the architecture and interior design of nursing homes for people who need intensive care, the earlier detailed recommendations for ergonomic designs still apply; but now the aspects of providing 24-hour supervision and care, and possibly intensive medical treatment, prevail. Unfortunately, ergonomic information on architecture and interior design for nursing care facilities and for rehabilitation equipment is piecemeal and incomplete, even at this publication date (2018). Of particular concern is preventing falls in nursing homes: they cause about 1800 deaths annually in the United States, and by 2020, the annual direct and indirect costs of fall injuries is expected to reach almost $55 billion.[43]

Skilled nursing care

14.6 DESIGNING FOR DISABLED INDIVIDUALS

We are all only temporarily able-bodied: as children, we lack the strength and skill that we hope to acquire during adulthood; and while aging, we lose some of the facilities that we previously enjoyed. Many of us suffer from injuries or illnesses that deprive us of certain capabilities, often only for some period of time but possibly lifelong. In the

United States, Census 2000 counted 49.7 million people with some type of long-lasting condition or disability—this represents nearly one person in five.[44]

Many of the design considerations discussed in the foregoing section as applying to aging people are also relevant to disabled individuals. This is particularly true for those design recommendations that help to alleviate, overcome, or sidestep impairments so the largest possible independence and everyday functioning capability are attainable.

> At pedestrian street crossings, it is beneficial to combine walk/stop lights with acoustic signals that indicate to people with impaired hearing and sight that they may or shall not cross the street. In addition to the familiar red and green lights, selected sounds help to indicate when it is safe to proceed. For example, a "cuckoo" sound may signal crossing in one direction and an electronic "chirp" indicate the orthogonal direction, or a simple verbal "walk" and "wait/stop" may be used. This combination of light and sound is beneficial not only for impaired and aging individuals but also for all other pedestrians.

Impairment, disability, handicap

Regarding the definition of terms: *impairment* is the abnormality itself; *disability* is the restriction that results from the impairment; *handicap* is the way that the impairment limits or prevents the functioning of the individual. Note that having a disability does not mean someone is handicapped.

In the United States, the 1990 Americans with Disabilities Act defines disability as "physical or mental impairment which substantially limits one or more of an individual's major ADL such as walking, hearing, speaking, learning, and performing manual tasks." The World Health Organization (WHO) defined disability as "any restriction or lack (resulting from an impairment) of ability to perform an activity in the manner, or in the range, considered normal." Impairments may be chronic, physiologic, psychological, or anatomical abnormality of bodily structure function caused by disease or injury. Since human activities are varied, there are many different kinds of disabilities resulting from impairments. More specifically, "work disability" is a dysfunction in the vocation for which a person is trained.

Types of impairment

Sorting and classifying impairments, as shown in Figs. 14.4 and 14.5, helps to understand that certain impairments (and ensuing disabilities) may result from very dissimilar conditions. For example, "poor balance" can be caused by conditions related to blood pressure, hemiplegia, paraplegia, amputation, multiple sclerosis, muscular dystrophy, cerebral palsy, Parkinson's disease, brain tumor, and other causes. The design matrix in Fig. 14.6 relates display design issues to impairments. Such charting identifies ergonomic challenges and indicates possible solutions.

Handicap

The disadvantage that results from impairment and disability is properly called a "handicap." In addition to overt impediments, it may entail loss of income, social status, or social contacts. Some handicaps can be alleviated, for example by providing a space to park a car next to a building entrance used by people with disabilities. Of course, it is much better to avoid a disability (and thus a handicap) by human engineering means, such as replacing a natural eye lens impaired by cataracts with a clear manufactured lens, or by replacing a damaged body joint with an artificial one.

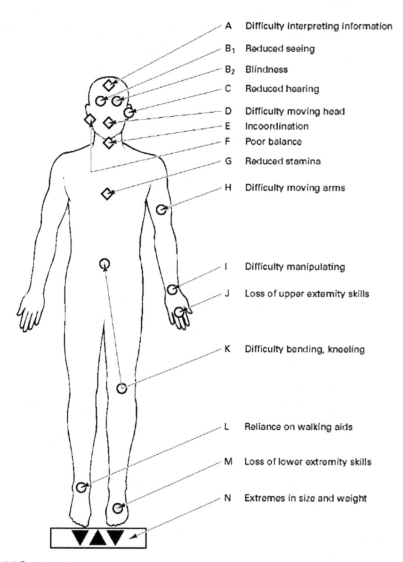

A	Difficulty interpreting information
B₁	Reduced seeing
B₂	Blindness
C	Reduced hearing
D	Difficulty moving head
E	Incoordination
F	Poor balance
G	Reduced stamina
H	Difficulty moving arms
I	Difficulty manipulating
J	Loss of upper extremity skills
K	Difficulty bending, kneeling
L	Reliance on walking aids
M	Loss of lower extremity skills
N	Extremes in size and weight

FIGURE 14.5 **Classification of impairments.** *Source: Adapted from Faste (1977).*

There is a wide range of judgment regarding the various impairments, disabilities, and handicaps, both with respect to their extent in relation to specific activities, and regarding specific age groups. For example, judged against "normal adults," children, pregnant women and many elderly people could be described as disabled. In spite of individual differences, we tend to consider juveniles, adults, and pregnant women as groups of people that have features that are sufficiently similar so as to allow general ergonomic design recommendations. However, disabilities are often very specific to that individual and, therefore, the person's special abilities (or lack of capabilities) must

Design-for-One

II. DESIGN APPLICATIONS

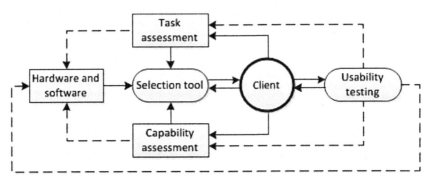

FIGURE 14.6 **Procedure for determining suitable aid for an impaired client.** *Source: Adapted from Casali and Williges (1990).*

be assessed on a case-by-case basis to help the individual with appropriate assistive devices or arrangements.

UCD and Design-for-All

Designing for the disabled user is best approached with the user at the forefront, and the objective becomes to improve a product's usability. This approach is called *User Centered Design* (USD), and it consists of five tenets: a detailed knowledge of the end user and his or her abilities and restrictions; active participation of the end user so his or her needs and requirements are heard and understood; a division of functions between users and technology; a tenacious approach that continues until the end user's requirements are met; and the involvement of a multidisciplinary team that includes engineers and physicians. Another approach is *Universal Design*, which involves designing products or environments so that they are both aesthetically pleasing and usable by everyone to the greatest extent possible, regardless of age, gender, ability, or status in life—this is also called *Design-for-All*.

Measuring residual abilities

Functional capacity evaluations (FCE) use testing methods and hardware to quantitatively measure the residual capabilities of impaired individuals. The Available Motions Inventory (AMI)[45], for example, has a number of panels with switches, knobs, and other devices to be reached, turned, and otherwise manipulated. These panels are put into standardized positions within the work area, and each person's ability or disability to operate these as required is recorded, together with the time needed and the strength exerted, as appropriate. Such systematic testing can generate a large database that allows the comparison of an individual's performance with that of others, able or disabled. This information facilitates a systematic approach to find work tasks and work conditions suitable for a given individual's capabilities, or to modify them to match the person's abilities.

Ergonomic interventions

Many disabilities can be mitigated by ergonomic means. The use of technology can be particularly helpful to address impairments. For example, for a paraplegic who is unable to do most physical tasks, a computer workstation can eliminate the need to handle papers or files, and can be designed to transcribe and communicate for the user with appropriate adaptive software and hardware such as voice activation, speech-to-text, and alternative input methods (say, a stick attached to the head, hand, or even held in the mouth).

Wheelchairs

Locomotion and transportation are serious problems for many disabled individuals. Providing technical aids in terms of crutches, walkers, and prostheses has been an age-old concern. Wheelchairs have been described for decades; they may be client-propelled, attendant-pushed, and motor-driven chairs for indoors and outside use, even for racing. In addition to movement, they should provide stable yet relaxed support, particularly avoiding pressure points and sores. Wheelchair design has come a long way from the hurtful folding chairs that added painful discomfort to the distress already felt by the user. Adjustments in height and angles of seat, backrest, and foot supports can stabilize patients. For long-term sitting, the user should be able to easily vary these settings to provide the necessary pressure relief to avoid bedsores caused by poor circulation due to prolonged pressure on body parts. Users of motorized wheelchairs appreciate such features as lowering the seat for fitting the thighs under a table, or as lifting the seat so that the person can reach for a high shelf, or tilting seat and back support to shift the user's body. Some even provide a stand-up design so the user is upright, promoting circulation and muscle tone as well as facilitating eye-level interactions with others while standing or walking/rolling.

The use of ergonomics in rehabilitation engineering is widespread; think of devices ranging from wrist splints to artificial limbs; from walking aids to specialized vehicles. Technology for people with disabilities has advanced to the systematic development and implementation of sophisticated devices. These include control actions stimulated by electromyography (EMG) signals, by movements of unimpaired body parts (such as the jaw), by voice control or even direction of a gaze. Some devices use the principal of sensory-substitution: translating visual information into sound, or auditory information into tactile sensations. The selection of assistive devices should consider a variety of criteria:

Helpful technology

- Affordability: Is the purchase, maintenance, and repair financially feasible for the user?
- Dependability and durability: Does the device operate with repeatable and predictable levels of accuracy for extended periods of time?
- Physical safety: Is there a low risk that the device causes physical harm to the user or to others?
- Portability: Is the device readily transportable to and operated in different locations?
- Learnability and usability: Can the user learn to use the newly received device in a timely way? Can he or she use it easily and safely for its intended purpose?
- Physical comfort and personal acceptability: Does the device provide comfort to the user? Does the person want to use it in public or privately?
- Flexibility and compatibility: Does the device interface with other devices that are currently used or will be used in the future?
- Effectiveness: Does the device improve the user's capabilities, independence, and both objective and subjective situation?

14.7 DESIGNING FOR OVERWEIGHT/OBESE WORKERS

About 34% of the United States work force is classified as "normal weight." This suggests that almost 2/3 of the American work force has body types that are not in the "normal" range, and most of these people are overweight rather than underweight. In fact, about 2/3 of all Americans are overweight and obese, including those not in the workforce. Other countries are catching up to these startling metrics: according to the WHO, obesity worldwide has more than doubled since 1980, and this trend has not yet been reversed. Increases in body size means that we have to consider that relevant workplace anthropometrics gathered many decades ago may be outdated.

14.7.1 Effects of Obesity in the Workplace

With increases in overweight and obesity among workers, there are also increases in prevalence of work limitations, reduced worker mobility, increased accidents and injuries, absenteeism, and premature mortality. About 20% of US health care costs have been attributed to obesity, and these costs are often borne by the employer. Currently, obese workers are associated with more excess health care cost than are smokers.[46] Overweight employees are significantly more likely to be injured in work environments that are not suited to their body size and mobility limitations. Excess body weight is a risk factor for many chronic illnesses, including hypertension, diabetes, respiratory illness, heart disease, osteoarthritis, carpal tunnel syndrome, low back pain, and various measures of psychological distress. Aside from chronic disease outcomes, obesity may also increase the risk of nonfatal traumatic injury. A BMI (body mass index, see Chapter 1) over 30 has been associated with an increased risk of injury, and obesity is an independent risk factor for certain work-related overuse disorders, including carpal tunnel syndrome.

14.7.2 Considerations for Overweight/Obese Workers

Injuries to the leg or knee appear to be especially prevalent among very obese workers in the manufacturing industry. Obese workers are likely to face psychosocial obstacles, including discrimination and the attributions of vice or weakness.

Adapting the workplace

Many relatively minor changes can be made in the workplace to provide increased safety, job retention, health, comfort, and productivity for the overweight/obese employee.

- Uniforms and safety gear need to be available in comfortable sizes and proportions. Body shape as well as body mass needs to considered, as overweight and obese workers vary in waist-to-hip ratios, upper body and abdominal obesity versus lower body obesity, and proportions of visceral fat versus subcutaneous fat within a given range of body mass.
- It must be considered that the larger the body size, the farther the load is from the spine. Thus, reaching, bending, squatting, lifting, balancing, stooping, and standing abilities are reduced, and telescopic tools and devices may be beneficial.
- Keyboard heights may need to be increased and/or seat height decreased, especially for womens who tend carry excess weight in a "pear" configuration (i.e., more weight around the hips than waist).
- For individuals with a very broad torso, a split and angled keyboard may be more suitable than a standard keyboard.
- Keyboards with a convex frontal curve can help accommodate the decreased reach capacity, and padded edges reduce tendon constriction of the wrist.

- For workers with an "apple" body shape configuration (tending to carry excess weight around the waist; often, men), work surfaces with cutouts or coves allow obese office workers to get closer to their work materials and computers.
- Seat pans in office chairs could be increased from the standard (in the United States) 19.25inch (49 cm) width to 23inch (58 cm) width. Not only are width requirements important but also sturdy armrests or other boosting devices should be present to assist in rising from the chair. Armrests of chairs may need to be lifted, tilted, or padded.
- Adaptations for reduced stamina and stability may be required, as the static force to maintain stability increases the heavier the body (or a limb) is. Anti-fatigue matting, sitting rather than standing workstations, reduced walking requirements are possible accommodations.
- Job rotation could be considered if engineering controls are cost prohibitive or impractical.
- Workplace safety personnel might consider adding policies or programs that address weight reduction and weight loss maintenance as part of ongoing comprehensive workplace safety and wellness strategies.

14.8 DESIGNING FOR PATIENTS AND HEALTHCARE PROVIDERS

Health care differs from other businesses and occupations[47]: its purpose is helping people in need, often under catastrophic conditions such as war, crime, accident, life-threatening injury, or disease. Such ethical concern can convince care providers that they know best what is good for others regardless of cost, and may result in the attitude that achieving these goals justifies work practices for care providers which exceed or disregard elsewhere-accepted roles and rules.[48] However, during the last decade it became increasingly recognized that even physicians and nurses should limit their working hours for their own sake and for the good of their patients; use checklists to make certain that consistent best practices are followed; and employ management procedures commonplace in other professions. Consequently, ergonomists and systems engineers contribute to improvement of patients' quality of care and their safety, as well as to improvement of care providers' working conditions and their safety at work.

Senior management of healthcare operations encounters classic business topics such as predictive modeling and data mining, logistics and supply; yet middle management and senior physicians encounter tasks specific to the emergency nature of many care giving tasks, such as making medical decisions under uncertainty. Other issues include design details, such as of emergency facility layout, operation of medical devices, even of written instructions for equipment use. All of these issues need to be seen as serving the purpose of helping patients.

In each aspect of patient care, care providers are at work—so their safety and working conditions need consideration as well as the provision of safe, efficient, effective, and compassionate care for patients. Several of these topics are treated in chapters in this book: making decisions and job stress in Chapter 4; work schedules in Chapter 7;

workplace design in Chapter 9; office design and computer use in Chapter 10; organizational and management issues as well as boredom and fatigue in Chapter 11; displays and controls in Chapter 12; load handling in Chapter 13. Specific conditions associated with health care may require special considerations and applications that exceed the discussions in previous chapters.

Emergency services

Emergency medical services are an often life-saving component of public safety, such as when accidents occur in transport and industry or when catastrophic events such as earthquakes or brush fires happen. The victims' survival or impact of injury depends on the time at which emergency helpers arrive, on their ability to judge what to do, and on their skill to administer in situ mechanical and/or medical treatment. The working conditions are unusually complex, fast paced, high stress, and often physically dangerous for all involved.

Ergonomic assistance to emergency services is at several levels. Proper size, design, and content of the paramedics' "response bag"[49] with its medications, bandages, instruments, and tools is essential; and so is the training of first aid responders in correct and efficient treatment of victims, and in assuring their own safety because of high risk for both acute and chronic injuries. Human factors engineering can also help in selecting and equipping ambulances and other vehicles needed to get aid emergency personnel to the accident site. One major point of concern is the intensity and quality of emergency treatment: Is an experienced physician needed to perform life-saving intervention on the spot? What does that mean for the equipment of the ambulance? What kinds of apparatus for life support of the patient are needed during transfer to the nearest hospital?

Workplace stress

Patient care routinely requires the staff to cope with some of the most stressful situations found in any kind of work. The staff must care for dependent and often demanding patients, they must deal with life-threatening injuries and illnesses of their patients and, often, with patient deaths. At the same time, their work is complicated by intricate equipment, tight schedules, understaffing, overwork, and complex hierarchies of authority and skills. All of these contribute to stress. Stress can affect caregivers' attitudes and behavior: frequently reported consequences of stress include difficulties in judging the seriousness of a potential emergency, communicating with very ill patients (and their families), and maintaining pleasant relations with patients and coworkers. Stressed caregivers are more likely to experience changes in appetite, ulcers, mental disorders, migraines, difficulty in sleeping, emotional instability, disruption of social and family life, and increased use of tobacco, alcohol, and drugs.

Hospital beds

Hospital beds have many available adjustments, controlled manually or assisted by a motor. The adjustment features are meant to allow the patients to assume different body postures, to relieve pressure points, and to make it easier for caregivers to turn patients and help them get in and out of the bed.

14.8.1 Moving Patients

Many tasks in caring for patients at home, in hospitals, and in nursing homes require considerable strenuous physical labor on the part of the caregivers, especially when transferring, lifting, sliding, and otherwise moving patients who cannot actively cooperate. Healthcare providers who handle patients suffer a high number of work-related back pain cases; overall, the injury rate[50] for US caregivers is about double the rate for all full-time workers in other occupations. Beyond the pain to the person, the cost of lost work days is of concern to management because it impacts the quality and expense of health care.

Ergonomic interventions help to reduce the risk of work-related strain injuries by designing patient handling devices and work procedures that can be performed safely. Certainly, patients are not "loads," but most of the biomechanical discussions in Chapter 13 regarding proper "load handling" help to avoid overexertions and injuries to the caregiver. Back injuries and shoulder strains often occur while lifting, transferring, and repositioning patients.

Several risk factors relate to different aspects of patient handling.[51]

Risks in patient handling

- Risks related to the task:
  - Force: amount of physical effort required to perform the task (such as heavy lifting, pulling, and pushing) or to maintain control of equipment and tools.
  - Repetition: performing the same motion or series of motions continually or frequently during the work shift.
  - Awkward positions: assuming positions that place stress on the body, such as leaning over a bed, kneeling, or twisting the trunk while lifting.
- Risks related to the patient:
  - Patients may be medically fragile, have wounds such as surgical sites, and be attached to tubing and equipment.
  - Patients have no handles.
  - Patient may not be able or willing to cooperate.
  - Patient response may be unpredictable when the patient is on medication, sedated, or in pain.
- Risks related to the environment:
  - Slip, trip, and fall hazards.
  - Uneven and soft support surfaces.
  - Space limitations (small rooms, equipment in the way).
  - Inadequate equipment.
  - Inadequate footwear and clothing.
  - Lack of knowledge or training.

It is useful to distinguish three ways of patient handling:

Transfer of patients

1. Manual transfer done by one or more caregivers using their own muscular strength and, wherever possible, any residual movement capacity of the patient involved.
2. Transfer using basic patient-handling aids such as low-friction fabric sheets, ergonomic belts, rotatable footboards, "trapeze bar" attached above the bed.
3. Transfer using sophisticated patient handling machinery like electro-mechanical lifting equipment.

Selection of the proper handling technique involves an assessment of the needs and abilities of both the patient and of the caregiver(s):

- A patient who is able and willing to partially support his own weight may be able to move from the bed to a chair using a standing assist device, whereas a person who is noncooperative may need a mechanical lift.

- A patient may be too large or too heavy for caregiver(s) to lift without mechanical assistance.
- Medical conditions such as pregnancy, wounds, tubes, or bandages may influence the choice of methods for lifting or repositioning.
- The physical environment may necessitate awkward positions and postures and create unsafe conditions for patient and caregiver(s).

Sole caregivers

If you are taking care of a patient[52] by yourself, you are at greatest risk for back over-exertion when pulling a person who is reclining in bed into a sitting position; transferring a person from a bed to a chair; and leaning over a person for long periods of time. The risk of back injury can be reduced by:

- maintaining proper alignment of your head and neck with your spine
- maintaining the natural curve of your spine and not bending at the waist
- not twisting your body when carrying a person
- keeping the person who is being moved close to your body
- having your feet shoulder-width apart to maintain your balance while standing
- using the muscles in your legs to lift and/or pull

If the task is too straining, seek help.

14.8.2 Electronic Health and Personal Records

Records

Electronic personal health records are now widely used in ambulatory practices, in health care facilities, and in nursing homes. These can be valuable tools but only if the records contain complete and correct information; incomplete and incorrect records can be dangerously misleading. Before admission to treatment, new patients (and often their companions) are usually asked to complete medical forms: designing these forms requires careful consideration of the users' ability to understand the questions they are being asked. People in acute distress, with low literacy, and/or with little knowledge about health-related medical terms may find it difficult to furnish the requested information[53] and, especially if they do not obtain needed assistance, may provide false or misleadingly incomplete data.

EHR

Proper setup of electronic health records (EHRs) can produce alerts[54] when prescribing a particular medication that may produce a health risk to a patient. Several conditions can trigger alerts:

- drug–drug alerts, when one drug alters the effects of another drug if they are administered together;
- drug-allergy alerts, when the patient has an allergy to a drug;
- drug-disease alerts, when a drug is already taken for another medical condition;
- drug-food alerts, when the effects of a drug are altered by food if they are taken together;
- overdose alerts, when the dose exceeds standard dosing values.

Medication alerts

It is a promising concept to automatically generate an alert of a risk when prescribing a particular medication to a particular patient. However, to be usable and be used, the software systems must be reliable and not generate an overwhelming number of alerts which are complex, unexplained, even cryptic to the prescriber. Too many alerts can cause "alert fatigue" which makes caregivers overlook or ignore clinically useful warnings; this can lead to serious consequences for patients.

14.8.3 Safety Guidelines, Standards and Laws

A plethora of safety-related guidelines, requirements, and regulations exists in many countries and in specific localities to govern ergonomics in health care. These documents are quite different from place to place; however, they are all meant to achieve a high level of safety and health for patients and caregivers. Relevant standards are being continually reviewed and updated: ISO health and safety standard 45001 addresses the global importance of occupational health and safety management systems, and is being aligned to OHSAS 18001, ISO 9001 (quality management), and ISO 14001 (environmental management). Those responsible for setting up and managing care facilities have to understand and comply with the relevant safety guidelines and laws.

14.8.4 Medical Devices

It is unacceptable to still encounter medical instruments with badly designed interfaces[55] that are difficult to use, even prone to misuse because of complicated or counterintuitive mechanics, controls, and displays. Guidelines, instructions, warnings, and labels need to be well-written and easy to understand: useful information on design and wording of written documents[56] is readily available. Given that patients undergoing any diagnosis or treatment can be considered impaired or disabled, we will provide some basic information on the use of ergonomics in the design of medical devices used to provide this diagnosis or treatment.

Medical devices are defined by regulatory agencies as any apparatus, appliance, software, material, or other article—whether used alone or in combination, and including the software intended by its manufacturer to be used specifically for diagnostic and/or therapeutic purposes and necessary for its proper application—intended by the manufacturer to be used for human beings for the purpose of:

- Diagnosis, prevention, monitoring, treatment, or alleviation of disease
- Diagnosis, monitoring, treatment, alleviation, or compensation for an injury or handicap
- Investigation, replacement, or modification of the anatomy or of a physiological process
- Control of conception

> Medical devices range in complexity from 3D magnetic resonance imaging machine, implantable replacement for the hip joint, suture used to close a wound, to a simple tongue depressor.

Standards   Regulations and guidance documents for the use of human factors engineering in the design of medical devices have been issued by regulatory agencies as well as professional societies and standards committees.[57] As defined by the US Food and Drug Administration (FDA) in 2016, these are intended to help industry follow "appropriate human factors and usability engineering processes to maximize the likelihood that new medical devices will be safe and effective for the intended users, uses, and use environments." Reducing the risk of use errors improves medical devices. The IEC 62366:2007 standard defines the process by which the manufacturer of the medical product must "analyze, specify, design, verify, and validate usability, as it relates to safety of a medical device. This usability engineering process assesses and mitigates risks caused by usability problems associated with correct use and use errors, that is, normal use. It can be used to identify but does not assess or mitigate risks associated with abnormal use."

Use errors   Usability needs to be considered throughout the design and development of the device, be it a simple mesh implant or a sophisticated heart–lung bypass machine. *Use error* is defined as an action or inaction that results in a different medical device response than intended by the manufacturer or expected by the user that can result in harm to the user(s) or patient. These use errors can be intended or unintended actions, including a slip (failure in attention), lapse (failure in memory), mistake (misapplication of a good rule or application of a bad rule), or ignorance. Note that "use" errors are not called "user" errors—this is to ensure that the focus (or blame) is placed on the task and not the operator.

User   Understanding user needs and minimizing use errors implies that the designer has identified who the user is. Different aspects of the device have a different "customer"— for example, the procurement analyst who negotiates contracts and orders the device; the radiologist who sets up and controls the ultrasound equipment used to image and diagnose; the supply manager who stocks product and reads the label to identify the product; the nurse who dispenses a sterile implant from its package and provides it to the surgical assistant; the surgeon who uses the electrocautery tool during surgery; the patient who receives the implant; the postoperative nurse who sets up the recovery monitoring equipment.

Ethnographic studies   By adapting the techniques of anthropologists and immersing themselves in the culture of users of medical devices, ergonomists can develop *ethnographic* approaches to design. *Contextual inquiries* (interviews done during the ethnography with the device users at the time of use) provide rich information on the user's experience. One way ethnography is used involves mapping out an entire surgical procedure: which tool is being manipulated by which person, for how long, for what purpose, requiring what level of physical and mental effort. Using a validated tool to measure workload during use of the device (such as the Surgery Task Load Index (SURG-TLX) approach to assess the impact of mental demands, physical demands, temporal demands, task complexity, situational stress, and distractions on surgeons in the operating room[58]) can add objective rigor to the ergonomic analysis. These techniques are more effective than simply asking users about their experiences in the abstract, and allow objective measurement of how the use of the device impacted the user. This then allows the ergonomist to adapt the design of the device to better suit the needs of the patient and the procedure.[59]

> **The plural of anecdote is not data.** Studies must include sufficient number of users that represent the entire intended population—whether in experience level, previous training, size, strength, cognitive or physical limitations, culture, or any other relevant aspect.[60]

Robotics

Robotic surgery (more accurately, robotically assisted surgery) has great promise, and is showing some clinical evidence of improved outcomes for many patients. The addition of detailed and multimodal imaging, fine movement control without tremors (augmented by motors), multiple arms directly under the surgeon's control, and additional mobility through robotic joints (not limited by human elbow or wrist flexibility) can allow precise surgical manipulation of tissue. There is the potential to include enhancements to human senses for haptic feedback, imaging beyond the visual spectrum (i.e., ultrasound for fluid flow in vessels, infrared for heat, ionizing radiation for markers of cellular activity), and even machine learning (artificial intelligence) where the robotic surgical assistant suggests surgical maneuvers to the physician or identifies potentially dangerous situations (such as unanticipated bleeding or cancerous regions). Adding all this complexity makes this a rich field for human factors design to ensure the equipment is suitably designed: with useful controls and ergonomic interfaces.

Usability studies

Medical device designers rely on ergonomics to help design against use errors. This includes performing usability studies to help set design requirements during the development of the device as well as to validate the final device; usability studies are required by regulatory agencies before a medical device can be cleared for sale. It is important to use good engineering practice to set up appropriate studies, analyze the results, perform root cause analyses to learn from the observations and results of the usability studies, and document[61] the process.

Usability studies must realistically simulate the conditions under which the proposed device will be used: this means that stress conditions should also be considered in the usability studies. For example, if the device is to be used in an emergency situation, then the usability studies should be realistic in setting and environment—potentially including all the background noise and distractions the eventual user might have in, say, a busy hospital emergency room. The study could subject the user to cognitive stress (such as adding timed tasks to the study or diverting the user's focus with additional and unnecessary information), all to better understand how the device might function in the real world. Note that subjecting the user to any physical or extreme stress may be unethical, and should be undertaken only after careful review and appropriate consent.

Color can aid usability: for example, by coding controls, or grouping them by function, sequence, or importance. Color influences cleanliness: white or light colors show dirt more readily (prompting cleaning), while the use of darker colors on foot operated controls may conceal scuff marks better. Color also influences the emotional response: black may make a hand-held device look elegant and more like a consumer product (thus less intimidating), whereas black on a large piece of hospital equipment may appear threatening to a nervous patient. Warm colors evoke feelings of youth and energy, and may be used for pediatric devices. More saturated colors may feel more energetic, whereas more muted colors may seem more tranquil. Contrasting colors can draw attention to important features and controls. Red is generally reserved for warning alerts and stop/off controls.[62]

Risk management

No device can be completely without risk to user or patient; the key is for designers to understand the use and users well enough to be able to reduce the probability of usability issues. Starting the process of iterative usability testing early in the development cycle allows for updates to the design when it is still easy to make changes. The goal is to minimize use errors and minimize use-associated risks: the usability process is linked with a risk management process. In order of preference, risk may be managed by creating inherent safety by design (e.g., using connectors that cannot be connected to the wrong component, automating device functions that are prone to manual use error); include protective measures in the device (e.g., adding shielding or safety guards, using alarms or alerts for hazardous conditions like "low battery"); and provide written information such as instructions and training (this is the least preferred because it requires the user to remember the information or to have time to find the manual and read it during an emergency).

For example, the designer of an infusion pump must consider if the user needs to be prompted to prime the pump before beginning an infusion, or if this setup step can be automated. If manual data entry is allowed, the interface should not allow incorrect programming to avoid dangerous medication dosages. The design of the pump may be different if it will only be used in a hospital by trained nurses, or if it will be used at home by family members.

It is critically important (as well as required by regulations) to include human factors consideration in the design of medical devices. Understanding that a user may omit steps or perform them in the wrong order, may try to connect the wrong components, or may incorrectly read or enter data, allow us to create interfaces that mitigate the risk of misuse. As with any application of ergonomics, we should not blame the user for being human.

14.9 CHAPTER SUMMARY

There are typical differences between adult men and women in body sizes and physical capabilities. However, in general, one can design nearly any workstation, any piece of equipment, or tool so that it can be used by either women or men. In some cases, particular adjustability is needed, or one may have to provide objects in different dimension ranges; yet, these adjustments and ranges are generally not gender specific but are simply needed to fit different people.

There are large population groups that especially need ergonomic design. Among them are pregnant women, who during their pregnancy experience significant changes in body dimensions and in capabilities to perform demanding physical work. Infants, children, and adolescents are other very large groups different from "regular" ergonomic customers. From birth to adulthood, body dimensions, physical characteristics, skill, intelligence, and attitude change enormously. Statistical information on these developments is usually ordered by age, which can be an ineffective classification because very large differences are likely to exist among juveniles at the same age. This lack of data is counter to the strong economic interest in providing goods to the young population, be it in clothing, furniture, or toys. There also is a need for safeguarding and protecting children—for example, by properly designing cribs, railings, and toys or by providing restraint devices to protect them in case of an automobile accident.

Like children, aging people are also usually grouped within chronological age brackets, which is an often misleading classification scheme because physical characteristics, capabilities, and attitudes vary widely among individuals of similar age. The change of capabilities (here usually a decline) can also rapidly or slowly alter within one individual over time, or may remain on about the same level for years. There is great societal and individual interest in providing an ergonomically properly designed and maintained environment to the older population, be it at the workplace, in the home, or in a care facility.

Helping impaired people to overcome their handicaps through assistive devices is another major concern of ergonomics and rehabilitation engineering. Systematic procedures are at hand to determine the kind and extent of disabilities for specific activities, and often help can be provided either with mechanical devices or computers. Electronic devices require relatively little physical capabilities from the user, but can control and operate rather complex machinery and provide information storage as well as transfer to others and to the impaired person.

The increasing prevalence of overweight and obesity in the human population is impacting the workplace as well. Many relatively minor changes can be made in the workplace to provide increased safety, job retention, health, comfort, and productivity for the overweight/obese employee.

Patients in acute care in hospitals as well as residents in nursing homes depend, often completely, on care staff members to provide often vital assistance and medications; they may also need help in daily living tasks such as dressing, bathing, feeding, and toileting. Each of these activities involves multiple person-to-person interactions, often with handling or transferring of patients/residents which can result in injuries to both parties. Possible solutions include the appropriate design of facilities, installations, and devices according to human factors principles and practices; and the training of staff in the proper use of these implements.

The shift from paper to electronic records in health care has brought similar human factors problems as seen earlier when other businesses made that transition. Imperfect questionnaires and data files lead to incomplete and incorrect decision-making; for example, upon initial admission, patients may not supply sufficient descriptions of their medical history. The following step, data management, needs appropriate software algorithms in order to detect, for example, medication interactions to which the prescribing physician should be alerted. If that purpose is not met, vital information may be overlooked, with potentially dire consequences.

Medical devices, be they diagnostic equipment, implants, surgical tools, or fracture splints, require ergonomics to be designed properly. If the use of these devices is overly complicated or not user-friendly, the safety and efficacy of the device can be reduced. Regulatory agencies now require the documented application of usability assessments before any medical device can be sold. The careful application of ergonomic principals and subsequent testing reduces the risk of use errors and improves safety for both users and patients.

Designing for special populations benefits everybody. For example, in public transportation (airports, subways, elevators, moving walkways, etc.), many people—not just the aging or impaired—report confusion about how to use the systems. Many of these problems can be overcome by applying common "human engineering principles" such as providing proper illuminance and avoidance of glare, using signs with legible content, supplying intuitive cues such as colored arrows with directional symbols, and employing auditory announcements that are timely, understandable, and redundant. An example is displaying the floor number in an elevator in large numerals, indicating what is located at this floor, and announcing it early and clearly over a public address system. Another example is entrance and exit passages and steps in buildings and transport systems, which are often difficult to negotiate for those disabled by pregnancy, age, or physical impairment. Uneven floors and damaged or misplaced floor coverings often pose problems in vehicles or hallways. Handholds in buses and trams include a variety of loops, columns, and hand grips—useful for every passenger.

Ergonomic knowledge and ergonomic procedures can be of help from childhood on. The ergonomic principles and techniques for architectural design and interior layout, for workplace, equipment, tools, and devices are the same for "anybody," they are just more critical and important for those with reduced abilities.

14.10 CHALLENGES

Most design is for "normal adults" in the age range of 20–40 years. Is this a homogeneous group? Are there developments within a person during this age span?

Are there physiological reasons to presume that the musculature of women is per se weaker than that of men?

Should we expect that a summation of small differences in sensory functions between the sexes should make a difference for professional task performance?

Is it true that females show better finger dexterity than males?

Are there specific occupations that appear to be better suited for women, or men?

What kind of seats would particularly accommodate pregnant women?

Would an air bag in an automobile promise to provide better forward protection to a pregnant women, or a small child, or an obese passenger, compared to a seat-shoulder belt?

What are the biomechanic problems related to side air bags for small individuals (including children) as compared to regular adults?

How would the proper design of restraint devices in automobiles be influenced by the age-related development of children?

Why is it so difficult to take anthropometric and biomechanical measurements on small children?

Is decades-old information about body size, proportions, and mass distribution of children still useful?

What kind of adjustment features might be suitable for furniture used in kindergarten and elementary school?

Which professional activities appear to be particularly affected by changes related to aging?

Are there any professions and activities for which aging people appear particularly suited?

Why are there so few longitudinal studies of people? What would it take to do such longitudinal studies?

Which physiological functions are likely to be maintained into old age by physical exercises?

What is the problem associated with sound amplification in most hearing devices?

Which would be suitable approaches to assess changes in somesthetic sensitivities with age?

Which tests may be included, possibly for all and not only for aging people, to qualify them for a "driver's license"?

To what extent would the architectural layout of a house for the aging differ from a layout suitable for younger people?

Consider classification schemes for the various "disabilities" with the intent to use these to develop aids and enablers.

Why are computers so particularly helpful for disabled people?

What can be done to make computers more acceptable to older people?

Propose replacements for ADLs and IADLs.

Consider the various uses of wheelchairs, temporary or long-term, by individuals with different impairments, and for different purposes.

Considering the potential costs of overweight and obesity in the workplace, does the employer have the right to try to influence an employee's weight? If so, to what extent?

Can certain classification schemes be developed with regard to specific design solutions?

We stated that the plural of anecdote is not data. Although anecdotes do not lend themselves to statistical analyses, they can point to information to be pursued—so, how many anecdotes does it take before we have a data set?

When designing a medical device, how much risk is acceptable in return for the benefit of improving quality of life a little or a lot? For the benefit of extending life for 1 month, 1 year, or longer?

NOTES

1   Kroemer (2006c).
2   Kumar (2004); Peebles and Norris (1998, 2000, 2003).
3   Burse (1979); Shapiro, Pandolf, Avellini, Pimental, and Goldman (1981).
4   Baker (1987).
5   There is great controversy regarding young adult gender differences in academic preparedness, grades, and test scores, as discussed by (among many others) MJ Perry http://www.aei.org/publication/2016-sat-test-results-confirm-pattern-thats-persisted-for-45-years-high-school-boys-are-better-at-math-than-girls/. Accessed July 31, 2017. Given high variability (and thus standard deviations) in SAT scores, comparisons should be undertaken cautiously. Further, SAT scores may not be an important predictor for college or other measure of success.
6   Downing (2008); Zaidi (2010).
7   Noble (1978).
8   Moreno-Briseno et al. (2010); Barnett, van Beurden, Morgan, Brooks, and Beard (2010).
9   Greene and Bell (1987); Redgrove (1976).
10 Hancock and Meshkati (1988); McWright (1988).
11 Schlegel, Schlegel, and Gilliland (1988).
12 Asso (1987); Hudgens, Fatkin, Billingsley, and Mazuraczak (1988); Matthews and Ryan (1994); Nakatani, Sato, Matsui, Matsunami, and Kumashiro (1997); Patkai (1985).
13 Hudgens et al. (1988).
14 Upadhayay and Guragain (2014); Souza, Ramos, Hara, Stumpf, and Rocha (2012).
15 Interest in gender-related differences in performance is not new. One of the first unbiased collections of facts about behavior of the sexes was H. Ellis' book *Man and Woman*, published by Black in London in six editions between 1894 and 1930. Any (presumed) differences are topics with emotional, social, and political overtones, often biased by the intention to establish or dismiss superiority of one or the other gender.
16 Vail (1986).
17 McFadden (1997).
18 In 2011, the BIC company introduced the BIC Cristal For Her pen with an "elegant design—just for her!" and a "thin barrel designed to fit a woman's hand" that was "diamond engraved … for an elegant and unique feminine style." Consumers did not agree (nor would ergonomists) that mass-market disposable pens needed to be designed differently for men and women; the product was soundly mocked and is no longer available.
19 When compiling anthropometric and ergonomic data on pregnant British women in 1986, Pheasant had to rely on body dimensions measured on pregnant Japanese women (Pheasant, 1986, pp 178–179).
20 As measured on 198 German women by Fluegel, Greil, and Sommer (1986).
21 As measured on 105 white women with an average age of 26 years by Rutter et al. (1984).
22 Culver and Viano (1990).
23 In a small study by Morris, Toms, Easthope, and Biddulph (1998).
24 Nicholls and Grieve (1992).
25 Steenbekkers and Molenbroek (1990).
26 Kroemer, Kroemer, and Kroemer (2003); Kroemer (2006a,b); Lueder and Rice (2007). Snyder, Spencer, Owings, and Schneider (1975) measured American children over 40 years ago, and the oft-used growth charges by Hamill et al. (1979) rely on data measured from 1963 to 1979 and then statistically smoothed.
27 http://www.cdc.gov/growthcharts/; the World Health Organization (WHO) also publishes growth charts.
28 Schneider, Lehman, Pflug, and Owings (1986); Snyder et al. (1975, 1977).
29 In the United States, there is a curious use of terms: a "middle-aged" person becomes "older" at 45 years of age; "elderly" at 65 years; "old" as he/she reaches 75 years; and "very old" or "old-old" if he/she lives beyond 85 years of age.
30 Aging of friends and relatives, and one's own aging, are of perpetual interest and concern. This book cannot provide a complete discussion of current knowledge and research regarding age-related changes. The literature abounds with anecdotal observations and case studies, but systematic research is scarce.

The recent phenomenon of large numbers of older people, many still politically and economically powerful, is now prompting study.

31 As portrayed in various writings by Margaret Mead (1901–1978); Belsky (2016); Blaikie (1993); in the United States by the Committee on an Aging Society (1998) and by Czaja (1990), both under the aegis of the (United States) National Research Council, and by Shepherd (1995) and Fisk and Rogers (1997).

32 Annis, Case, Clauser, and Bradtmiller (1991).

33 Annis et al. (1991); Barlow et al. (1990); Stoudt (1981).

34 Spence (1989).

35 Belsky (1990, 2016); Hayslip, Hicks Patrick, and Panek (2011); Kermis (1984).

36 Psychcorp (2009).

37 For more information, see the literature in geriatric psychology, physiology, and sociology; the Committee on an Aging Society (1988); Belsky (2016); Czaja (1990); Hayslip et al. (2011).

38 Clark, Czaja, and Weber (1990); Rogers, Meyer, Walker, and Fisk (1998); Czaja (1991, 1997).

39 Kira (1976).

40 Cai and You (1998); Ogawa and Arai (1995); Peloquin (1994); Singer and Graeff (1988); van Hoof et al. (2017).

41 As per 2014 CDC Data, https://agingstats.gov. Accessed July 31, 2017.

42 National Research Council (2011).

43 Li and Ali (2015).

44 The percentage varies depending on how "disability" is defined, see also Gardner-Bonneau (1990), National Research Council (2011).

45 Bronzino and Peterson (2015);Dryden and Kemmerling (1990);Kondraske (1988);Smith and Leslie (1990).

46 Moriarty et al. (2012).

47 This text uses material from Kroemer (2017), Healthcare for Patients and Providers, Chapter 22; In Fitting the Human; 7th ed., CRC: Boca Raton, FL.

48 Brill (2013); Carayon (2011); Charney (2010); Dillon (2008); Gawande (2015); Waterson (2014).

49 Bitan, Ramey, Philp, and Uukkivi (2015).

50 Data from OHSA, https://www.osha.gov/dsg/hospitals/documents/1.2_Factbook_508.pdf accessed July 31, 2017; Kurowski et al. (2014); Waters (2010).

51 Guidelines from the European Agency for Safety and Health at Work (http://osha.europa.eu); American Nurses Association "Handle with Care" Campaign Fact Sheet (http://www.nursingworld.org/handle-withcare/); US Occupational and Safety Health Agency (OSHA) (http://www.osha.gov/ergonomics/guidelines/nursinghome/index.html).

52 From the American Academy of Orthopaedic Surgeons, http://orthoinfo.aaos.org/topic.cfm?topic=A00096. Accessed December 12, 1016.

53 Czaja et al. (2015).

54 Mauney, Furlough, and Barnes (2015); Patterson et al. (2015).

55 Ali, Li, and Lisboa (2015); Weinger, Wiklund, and Gardner-Bonneau (2010).

56 Robinson (2009).

57 Before using these standards and regulations, be sure to check for the most current versions, for example with the US Food and Drug Administration (FDA) (https://www.fda.gov/MedicalDevices/), International Electrotechnical Commission (IEC) (https://www.iecee.org/documents/refdocs/), American National Standards Institute (ANSI) and Association for the Advancement of Medical Instrumentation (AAMI) (http://www.aami.org/standards/index.aspx), International Organization for Standardization (ISO) (https://www.iso.org/standards.html). US FDA (2016) "Applying Human Factors and Usability Engineering to Medical Devices. Guidance for Industry and Food and Drug Administration Staff" February 3, 2016. IEC 60601-1-6:2007 Medical electrical equipment—General requirements for basic safety and essential performance Collateral Standard: Usability. IEC 62366:2007 Medical devices—Application of usability engineering to medical devices. ANSI/AAMI HE74:2001 Human factors design process for medical devices. AAMI HE75: 2009 Human Factors Engineering—Design of Medical Devices. ISO14971:2007 Medical devices—Application of risk management to medical devices.

58 Wilson et al. (2011).

59 For further reading, see, for example, Wiklund and Wilcox (2005).

60 The 2016 FDA Guidance Document on "Applying Human Factors and Usability Engineering to Medical

Devices" includes suggestions for sample size for human factors validation testing, including reference to a study that concluded that using 15 people can find at least 90% (and an average 97%) of problems with a particular software, and using 20 people can find at least 95% (and an average 98%) of the problems.

61 Document your work properly so that it can be referred to in the future—and so that you can remember what you did. Good documentation practices are worth the effort.

62 C. Park, "Delivering Positive Medical Experience through Color on the Medical Device" Published on MDDI Medical Device and Diagnostic Industry News Products and Suppliers. http://www.mddionline.com/print/14367?cid=nl.x.qmed02.edt.aud.qmed.20170111, accessed January 11, 2017.

Why and How to Do Ergonomics

Overview

The study of ergonomics has been evolving with humans as our needs have evolved. The reasons to do ergonomics remain: we have a moral imperative to design for ease and efficiency, to reflect our progress in knowledge and technology, to provide economic advantages to all. Ergonomics can be applied on a micro- as well as a macro-scale. Ergonomics needs to be applied in both new designs and as an intervention in existing designs, and its efficacy can be judged. Finally, ergonomics allows us to ever strive to improve human and system performance, health, safety, comfort, and the quality of life.

Ergonomics
http://dx.doi.org/10.1016/B978-0-12-813296-8.00015-3

15.1 EVOLUTION OF THE QUEST FOR EASE AND EFFICIENCY

When the first pre-human selected a stone blade, was that a pre-ergonomics feat? Was affixing a handle to the blade an example of early human engineering? Was the micro- and macro-organization mastered by the Romans to train and supply, equip and lead their legions systems ergonomics? Or did the discipline start in the 19th century with the recognition that, for physiologic and psychologic and sociologic reasons, laborers needed rest and recuperation to restore their capabilities for work? How about the rules of "scientific management" and of "work efficiency" in the early 20th century; the role of organized labor, or of unions; or, during the second World War, aircraft and automobile design and manufacturing, computerization, space exploration?

Evidently, humans have always shaped tools and practices to their liking and for efficacy on the individual and system levels—with no one event or time as the starting point. And what we generally began to call ergonomics in 1950 or human factors engineering in 1956 (to echo the introduction to this book) rests on millennia of experience, on intuitive and intentional development. During the second half of the 20th century, that wisdom became more organized, more scientific, much more comprehensive, and its application better incorporated in designs of work, of implements, of human-technology systems.[1]

15.2 REASONS TO APPLY ERGONOMICS

There are three major reasons to apply ergonomics, related to ethics, progress, and economic advantages.

1. Moral imperative: to improve the human condition and quality of life at work, especially in regards to health, safety, comfort, outcome, and enjoyment. Certainly, work must be safe and healthy, and should be comfortable. As we have all experienced, it can and should be enjoyable to achieve results at work and through work that we value personally.
2. Progress in knowledge and technology: to follow the human quest to learn more about ourselves, our desires, capabilities and limitations, and to develop and apply new theories and practices. Historically, thinkers and tinkerers, scientists and engineers have led the innate urge for progress. Stand-still seems unacceptable; we strive to improve objects and conditions.
3. Economic advantages: to reduce effort and cost in work systems that includes humans as doers, users, and beneficiaries.

In many new designs of things and systems, the "human factor" already has been incorporated during the concept stage. Examples are meat cutters and computer mice, hip joint replacements and portable phones, spacecraft and modern automobiles, contemporary manufacturing and assembly plants. Some of these would be excessively dangerous, or could not function at all, without preplanning and design in human engineering.

Yet, there are many older objects and work systems that were designed with little consideration of the human as user or worker, such as the typewriter or computer keyboard, or the traditional beverage delivery industry. Here, replacement, retrofit and redesign are necessary to reduce injury or strain to humans and to remedy inefficiency.

15.3 MICRO- AND MACRO-ERGONOMICS

Ergonomics can make simple designs (such as toothbrushes, plier-handles, light switches) user-friendly, and can render complex systems (such as subways) safe and friendly and easy to use: human factors are important for nearly any item that we use, simple or complex, at work, at home, anytime and anywhere. Of course, designing a public transport system, or a manufacturing plant, is a much more complicated enterprise than designing a warning label.

Our transport and manufacturing systems comprise many people at different levels, all interacting with each other and with different technologies, having different tasks and responsibilities. To design such sociotechnical systems in terms of "human-organization interface technology" requires "macro-ergonomics"[2] which goes beyond the traditional "micro-ergonomic" interface technologies that concern specifics of hardware, software, job and environment. Of course, even designing a toothbrush or a warning sign involves more than just considering hand size or reading acuity; and thus to categorize tasks as micro- or macro-ergonomics can be a matter of judgment. The design of a work system through consideration of relevant technical, environmental, and social variables in their interactions is a common ergonomic task; if organizational design and management aspects, even economical and regional-political implications are prominently involved, the task is indeed in the realm of macro-ergonomics.[3]

15.4 HOW TO DO ERGONOMICS

15.4.1 Ergonomics in New Designs

Obviously, it is best to design ergonomically correctly from the beginning of the concept. This principle of "do it right from the start" applies to simple new items as well as to complex new human-technology systems. And if we cannot find the time to design it right the first time, all too often we must find the time to design it over again—or suffer the consequences (see Fig. 15.1).

The concept team must be "use-centered": this implies a goal (use for what?), an instrument (use what?), a process (use how?) and a user (use by whom?). This use (and user)-centered concept has been central to human engineering but too often was not applied thoroughly: surgery is but one example[4] of the necessary integration of all aspects.

The concept and design team must include expertise in all related aspects; in today's industry, this commonly includes marketing and sales. Human factors design guidance

FIGURE 15.1　**Finding time to design.** *Source: Courtesy of Richard Elbert.*

is amply available both in terms of experience with past similar products and processes and in form of new information and data. Since 1994, in the United States alone, on average, each year about two major books on human factors/ergonomics came in the market besides the one you are reading right now.[5] In addition, many more specialized books have appeared, such as on biomechanics or material handling.

Use of new information depends on its how easily it is to access, interpret, and apply. The designer will utilize what is viable, useful and usable, but stay away from information that is perceived as "difficult."[6]

- *Accessibility* refers to the ease, or difficulty, of searching for information and obtaining it for use. Where do we look for the needed information (with the need often not yet clearly defined)? How do we obtain it, such as from an obscure thesis or industry or military report? How do we use it, if it has been declared confidential or proprietary?
- *Interpretability* refers to the ability of the non-specialist to understand, evaluate, and decide to apply the information to the design project. Making the judgment about applying the data, or not, often requires considerable understanding of underlying theory and of complicated language on the part of the designer. How can we understand information if it is written in a foreign language, uses technical jargon, or is presented in a difficult format?
- *Applicability* involves risky judgment by the designer regarding the relevance, reliability, and generalizability of (or ability to reduce and specify) theoretical information to the conditions of the application case. The problem is three-fold, at least. First, research may address problems that have only some overlap with the practical question. Second, most experimental data are obtained under laboratory conditions which rigorously control experimental variables and exclude confounding influences in such a manner that the laboratory work becomes "unrealistic." Third, theoretical data may just be too complex to use.

Computer models can greatly facilitate the use of new information in the design of new products and processes, or in the retrofit of older designs. However, as discussed in Chapter 8, using an inadequate model, inputting misleading information, or using the output incorrectly all result in bad design.

Major steps in the development of a new product (meaning anything that we can design—shovel or heart by-pass machine or air traffic control system) are:

1. Recognition of a need
2. Statement of the need
3. Concept of the product's utilization
4. Exploration of candidate product designs
5. Selection of the specific design of the product
6. Technical development
7. Production and distribution
8. Use and maintenance
9. Retirement and recycling of the product

Of these nine steps, six involve human factors as major determiners of either the design of the product (Steps 1, 3, 4, 5, and 8) or its production (Step 7). The phases of selecting the product design features, and the use details (Steps 3, 4, and 5) are iterative and involve intensive ergonomic considerations, including:[7]

- Human factors problems expected and eliminated
- Skills required from the user and maintainer
- Training required of user and maintainer

These ergonomic concerns and problems must be addressed early, comprehensively, and thoroughly: they "make or break" the usability and usefulness of the product (and its commercial chances). Skill and training requirements should be as low as humanly possible to avoid use and maintenance problems. Their recognition (and subsequent elimination) can be made easier by comparisons with similar existing products but usually realistic use of prototypes or real-life employment of a preproduction specimen is necessary. Fig. 15.2 helps in preempting ergonomic problems in either new or existing systems.

15.4.2 Ergonomic Interventions in Existing Designs

Most of the products and systems that we use daily have already been around for a while, and we may have good reasons to keep them; but consider that even the familiar may be in urgent need of improvement. That need may come from injuries, safety concerns, difficult use, costly maintenance, insufficient productivity, and/or lack of competitiveness.

Fig. 15.2 presents a stepwise sequence for ergonomic improvement. This procedure applies not only to existing conditions but can also be used for new designs.

- We become aware of the problem(s) (#1) either because the "ergonomic expert" (often the user) knows better, or because of deficiencies in performance (quality or quantity), or people may simply not like what is going on.

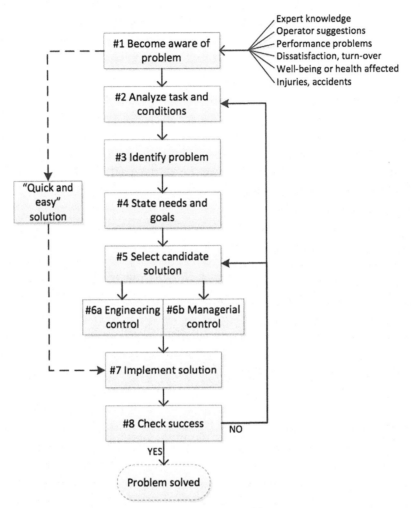

FIGURE 15.2   Steps to identify and preempt ergonomic problems.

- Identifying the true underlying problem is the most important Step (#3): missing the true reason invariably leads to selecting a candidate solution that misses the goal (#5). Selecting the solution (either technical or managerial, and often both) can be a difficult task because several approaches with different consequences may be feasible. For example: should one load packages on a truck by a forklift using pallets (which would require previous palletizing and may not fill the vehicle to capacity), or should it be done by hand (and by how many people, and may require a conveyer)? As another example, should one send astronauts or robots on a first visit to our moon, or to Mars?

A large variety of ergonomic procedures and techniques exist to improve working conditions. Taking computer workstations as an example (as discussed in Chapter 10), we can improve the layout of the room and workstation; improve the furniture (chair, support for the display, support for the input device, support for the source document); improve the lighting conditions and the physical climate; reduce unwanted sound; and consider the need for privacy. For "experimental cleanliness," we would like to introduce each intervention separately and then observe how it changes the person's feeling of well-being and performance. Yet, in reality, a comprehensive ergonomic approach combines these multiple measures. Therefore, a resulting improvement in attitude and work output is not easily traceable to any one single intervention.

15.4.3 The Hawthorne Effect

Improvements observed immediately after the intervention may be due to the so-called Hawthorne Effect. This was observed in the 1920s during experiments in the Hawthorne Works near Chicago: the ergonomic treatment consisted of improving the lighting conditions in a manufacturing/assembly task. Each rise in the lighting level was followed by an improved output; but when the illumination level was finally lowered, performance still improved. In somewhat simplified terms, one may conclude that paying attention to the workers, taking their comments and activities seriously into account, listening to them—in short, treating the workers as important—led to improved output regardless of the magnitude, even the direction, of the overt ergonomic measure taken.[8]

15.5 JUDGING THE EFFICACY OF ERGONOMIC INTERVENTIONS

As mentioned, often decisions must be made during the design phases of a new product about human engineering avenues to be taken, or about interventions in an existing system (see #5, #6, and #7 in Fig. 15.2) on the basis of subjective estimates, guesstimates, and gut feelings—often based on experiences with related products—about the resulting outcomes.

There are systematic ways to elicit choices from panels of experts or users. This involves:

- Defining the candidate solutions
- Establishing criteria by which to judge the candidates
- Laying out a score sheet according to the criteria
- Obtaining paired-comparisons scores from the participants
- Determining weightings for the criteria values
- Aggregating the data, accept the candidate solution(s) with the best overall score

The technique is straight-forward, fast, transparent, gives every participant the same voice, identifies the criteria, and shows how a change in their importance (weighting) influences the outcome in terms of selected candidate solution. The procedure works well for a small number of candidate solutions and criteria, but the technique can be used even in complex cases.[9]

COOLING A TRAILER, PARKED IN THE SUN, THAT MUST BE LOADED BY HAND WITH PACKAGES

Candidate Solutions: Cool trailer by spraying water on roof and sides (water); cool trailer by installing air conditioning (AC); cool trailer by forced air flow (fan); no cooling of trailer, but loading person takes a 2-min rest break in a cool room after every 5 min of loading (breaks).

Criteria: Effectiveness, availability, maintenance effort.

Scoring: 1–10, with 10 best.

Raw Scores Given by Raters:

	Water	AC	Fan	Breaks
Effectiveness	8	10	5	1
Availability	8	1	8	10
Maintenance	4	2	5	10
Total	20	13	18	**21**

With all criteria weighted the same, according to the raw scores, simply providing rest breaks would be the chosen solution.

With weight factors (1–10, with 10 highest) applied to the criteria, the scores are as follows:

	Weight factor	Water	AC	Fan	Breaks
Effectiveness	× 10	80	100	50	10
Availability	× 2	16	2	16	20
Maintenance	× 5	20	10	25	50
Total		**116**	112	91	80

With weighted criteria, cooling the trailer by spraying water on its roof and sides is the preferred solution.

15.5.1 Optimal Versus Good Solution

Seldom is one able to truly optimize a design: "optimizing" literally means to obtain the most favorable solution, the greatest degree of perfection.[10] But in the real world, there are constraints, often in terms of available time, resources, money, that allow the ergonomist to find a good (or good-enough) solution, the best one within the practical limitations, but not often the "optimal" solution. With respect to performance expectations, this usually means a compromise, such as between system cost and quality. In the example given above, cooling a trailer may be the chosen solution to prevent the worker's overheating, although loading and unloading by mechanical means would avoid human labor altogether, but appears too costly.

Another case altogether is the necessity to design for the extreme emergency: the proverbial airline pilot flies for hours in boring monotony which may be interrupted by seconds or minutes of highly exciting demands. We must "optimize" the cockpit design for those seconds of panicky terror and make it "good-enough" for routine operation.

At the beginning of this chapter, three reasons for ergonomics were given: moral obligation, implementing progress, gaining competitive and economic advantage. However, the decision still must be made to take which of the possible human factors approaches. This is often determined by weighing the gains to be made and the efforts needed.

15.5.2 Measuring the Results of Ergonomics

There are many criteria by which to measure outcomes and efforts. The result can be judged objectively by mission success or failure (often done by the military), by system performance in terms of quantity (productivity) or quality, or by subjective attitudes such as loyalty, feelings of well-being and satisfaction, demonstration of cooperative spirit and of social interactions.

15.6.2.1 Cost–Benefit Analysis

In western industry and business, the effects of ergonomic interventions have been commonly measured through monetary values, with dollar signs (or Euros or Rupees) affixed both to the human factors inputs and to the system outputs such as performance, productivity, accidents, injuries.

This can be a fairly simple task. Take the cost of providing the workers with human engineered tools which allowed better production, divide it by the money made by producing (and selling) more gadgets. But such cost/benefit calculation is more difficult in many work situations, when the output depends not only on the individuals, but also on circumstances over which they little or no control, such as working at a speed that is determined by a machine, by the movement of an assembly line, or by the number of customers waiting. In such conditions, ergonomic improvements of the working situation are not likely to result in increased performance; however, the improvements may result in improved quality of work, even if the tempo is machine-paced, or in less strain on the person and in better attitude and well-being.

Cost/benefit analyses may be beset by serious methodological difficulties, if they require the analyst to make value-laden assumptions. Costs of interventions are easier to count in money than in such benefits as better work attitude, loyalty and good will towards the employer, improved health, quality of life, and other positive economic side effects that may defy accurate estimation. For example, the art of estimating the number of occupational diseases prevented, or of work injuries avoided, is imperfect. Health benefits may accrue far into the future, and a human life or a lost limb do not have an established market value. Nevertheless, cost/benefit, economic efficiency, and cost-effectiveness analyses are attempts at rational decision-making, even if they are much more complex than is often naively suggested.

The literature brims with case studies of the monetary and other successes of human-engineering new products, and of ergonomic interventions to improve existing work systems.[11]

15.6 IMPROVING HUMAN AND SYSTEM PERFORMANCE, HEALTH, SAFETY, COMFORT, AND QUALITY OF LIFE

As succinctly stated and demonstrated in Fig. 15.3, human-centered design of simple objects as well of complex systems is, as a rule, better than what marketing and sales departments request and what engineering and production departments tend to deliver.

We acknowledge that there is a selfish aspect to the ultimate reason on how and why to do ergonomics: by improving the lives of those around us, we also get to improve our own.

We close this chapter and this book with a final, Top-10 list, and our best wishes that you go forth and design for ease and efficiency.

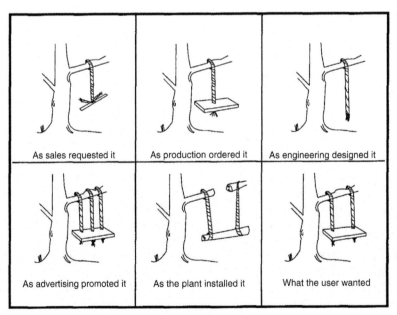

FIGURE 15.3   **Problems in design.** *Source: Rouse (1991).*

TOP TEN PRACTICES OF ERGONOMICS/HUMAN FACTORS ENGINEERING:[12]

#10: Design task, equipment and working conditions for real people. User selection and training cannot substitute for good design.

#9: Get into the design process early.

#8: Pick problems to work on that you can solve, and that will make a difference.

#7: Remember that simple is better: if it's difficult to explain or use, it won't be used.

#6: Bury complexity inside the system to make its every-day use simple.

#5: Good design appeals to common sense and is intuitively usable.

#4: Design to be forgiving of error and misuse: don't blame the user for being human.

#3: Design so that an operator can manage overload without performance consequences.

#2: Design to provide multiple ways of accomplishing the same thing.

#1: Design to promote human dignity and well-being, even pleasure.

NOTES

1 See publications by McFarland (1946), Lehmann (1953), Floyd and Welford (1954), McCormick (1957), Grandjean (1963), with many more thereafter (Kroemer 1993b).

2 As defined by Hendrick and Kleiner (2002).

3 Kleiner and Drury (1999).

4 As stated by Flach and Dominguez (1995).

5 See, for example, Bridger (1995), Bailey (1996), Bhattacharya and McGlothlin (1996), Chapanis (1996), Tayyari and Smith (1997), Karwowski and Salvendy (1998), Karwowski and Marras (1999), Hancock (1999), Mital, Kilbom, and Kumar (2000), Wickens, Lee, Liu, and Gordon-Becker (2004), Konz and Johnson (2007), Proctor and Van Zandt (2008), Weinger, Wiklund and Gardner-Bonneau (2010), Dekker (2014).

6 If the engineer or manager believes that research data are of value, then they will be sought, considered, and probably used. If they are hard to find, to understand, and to apply, they will probably not be used: as stated by many, including Burns and coworkers (1997).

7 Chapanis (1995, 1996).

8 The Hawthorne effect is the delight of the ergonomic consultant, who can be fairly well assured that almost any reasonable measure taken in the client's facility will have positive effects. Even if these effects wear off after the intervention has been completed, one can come back a while later and do, in essence, the same thing again, with similar positive effects. And, of course, to have positive effects, even if they are only indirectly related to the actual measures taken, is good for all involved. (Roethlisberger and Dickson, 1939; Parsons, 1974)

9 As discussed, for example, by Mitchell (1992).

10 Sheridan, on page 4 of his letter in the June 1988 issue of the Human Factors Society Bulletin.

11 See, for example, Alexander (1995), Oxenburgh (1997), Beevis (2003), Goggins et al. (2008), Puleio and Zhao (2015), AEP Return on Investment for Ergonomics Interventions, accessed July 19, 2017 https://www.humanscale.com/userfiles/file/return-on-investement_03272015.pdf www.humanscale.com

12 Based on Pew (1995) "Top-ten list of principles of human factors practice" and on Kroemer (1996) "Ten ergonomics principles in engineering."

Further Information

References

Aaras, A., & Ro, O. (1998a). Supporting the forearms at the table top doing VDU work. In S. Kumar (Ed.), *Advances in occupational ergonomics and safety* (pp. 549–552). Amsterdam, The Netherlands: IOS Press.

Aaras, A., Horgen, G., Bjorset, H. H., Ro, O., & Thoresen, M. (1998b). Musculoskeletal, visual and psychosocial stress in VDU operators before and after multidisciplinary ergonomic interventions. *Applied Ergonomics, 29*, 335–354.

Aaras, A., Ro, O., & Horgen, G. (1998). Back pain. A prospective three years epidemiological study of VDU workers. In S. Kumar (Ed.), *Advances in occupational ergonomics and safety* (pp. 553–555). Amsterdam, The Netherlands: IOS Press.

ACGIH. (1989). *Threshold limit values for 1980–90*. Cincinnati, OH: American Conference of Governmental Industrial Hygienists.

Ackerman, D. (1990). *A natural history of the senses*. New York, NY: Random House.

Adams, J. S. (1963). Toward an understanding of inequity. *Journal of Abnormal and Social Psychology, 67*, 422–436.

Adams, M. A., & Hutton, W. C. (1986). Has the lumbar spine a margin of safety in forward bending? *Clinical Biomechanics, 1*, 3–6.

Adams, E., Kaneko, S. T., & Rutter, B. (1993). Making sure it's right: Three phases of ergonomics research in the design of a pointing device. *Proceedings of the Human Factors and Ergonomics Society 37th annual meeting* (pp. 443-447). Santa Monica, CA: Human Factors and Ergonomics Society.

Adler, M. (1997). *Antique typewriters, from creed to QWERTY*. Atglen, PA: Schiffer.

Akerblom, B. (1948). *Standing and sitting posture*. Stockholm, Sweden: Nordiska Bokhandeln.

Åkerstedt, T., & Wright, K. P. (2009). Sleep loss and fatigue in shift work and shift work disorder. *Sleep Medicine Clinics, 4*(2), 257–271.

Albery, W. B., & Woolford, B. (1997). Design for macrogravity and microgravity environments. In G. Salvendy (Ed.), *Handbook of human factors and ergonomics* (2nd ed., pp. 935–963). New York, NY: Wiley (Chapter 29).

Alden, D. G., Daniels, R. W., & Kanarick, A. F. (1972). Keyboard design and operation: A review of the major issues. *Human Factors, 14*, 275–293.

Alderfer, C. P. (1969). An empirical test of a new theory of human need. *Organizational Behavior & Human Performance, 4*(2), 142–175.

Alderfer, C. P. (1972). *Existence, relatedness and growth: Human needs in organizational settings*. New York, NY: Free Press.

Alexander, D. C. (1995). The economics of ergonomics: Part II. *Proceedings of the Human Factors and Ergonomics Society 39th annual meeting* (pp. 1025-1027). Santa Monica, CA: Human Factors and Ergonomics Society.

Alfredsson, L., Akerstedt, T., Mattsson, M., & Wilborg, B. (1991). Self-reported health and well-being amongst night security guards: A comparison with the working population. *Ergonomics, 34*, 525–530.

Ali, H., Li, H., & Lisboa, P. (2015). Usability analysis and redesign of infusion pump user interface. *Proceedings of the International Symposium on Human Factors and Ergonomics in Health Care, 4*, 129–133.

Allread, W. G., Marras, W. S., & Parnianpour, M. (1996). Trunk kinematics of one-handed lifting, and the effects of asymmetry and load weight. *Ergonomics, 39*, 322–334.

American Heart Association (AHA). (2015). *Alcohol and heart health*. Available from: http://www.heart.org/en/healthy-living/healthy-eating/eat-smart/nutrition-basics/alcohol-and-heart-health.

American National Standards Institute. (2018). *ANSI Z535.4-2011 (R2017), Product Safety Signs and Labels*.

American Psychiatric Association. (2013). *Diagnostic and statistical manual of mental disorders* (5th ed.). Washington, DC: American Psychiatric Publishing.

Amoore, J. (1970). *Molecular basis of odor*. Springfield, IL: Thomas.

Anderson, A. M., Mirka, G. A., Joines, S. M. B., & Kaber, D. B. (2009). Analysis of alternative keyboards using learning curves. *Human Factors, 51*, 35–45.

Anderson, C. K., & Catterall, M. J. (1987). The impact of physical ability testing on incidence rate, severity rate, and productivity. In S. S. Asfour (Ed.), *Trends in ergonomics/human factors IV* (pp. 577–584). Amsterdam, The Netherlands: Elsevier.

Anderson, N. S. (1987). Cognition, learning and memory. In M. A. Baker (Ed.), *Sex differences in human performance* (pp. 37–54). Chichester, UK: Wiley (Chapter 3).

Andersson, B. J. G., Oertengren, R., Nachemson, A., & Elfstroem, G. (1974). Lumbar disc pressure and myoelectric back muscle activity during sitting – I studies on an experimental chair. *Scandinavian Journal of Rehabilitation Medicine, 6*, 104–114.

Andersson, B. J. G., Schultz, A. B., & Oertengren, R. (1986). Trunk muscle forces during desk work. *Ergonomics, 29,* 1113–1127.

Andersson, G. B. J. (1991). Low back pain. *Proceedings, Occupational Ergonomics: Work related upper limb and back disorders (not paginated).* San Diego, CA: American Industrial Hygiene Association San Diego Section.

Andersson, G. B. J. (1999). Epidemiology of back pain in industry. In W. Karwowski & W. S. Marras (Eds.), *The occupational ergonomics handbook* (pp. 913–932). Boca Raton, FL: CRC Press (Chapter 51).

Andersson, K., Karlehagen, S., & Jonsson, B. (1987). The importance of variations in questionnaire administration. *Applied Ergonomics, 18,* 229–232.

Andres, R. (1984). Mortality and obesity: The rationale for age-specific height-weight tables. In R. Andres, E. L. Bierman & W. R. Hazzard (Eds.), *Principles of geriatric medicine* (pp. 311–318). New York, NY: McGraw-Hill.

Andres, R. (1985). Impact of age on weight goals. *Proceedings, NIH consensus development conference* (pp. 77-81). Beshesda, MD: National Institutes of Health.

Andres, R. O. (2004). Pushing and pulling strength. In S. Kumar (Ed.), *Muscle strength.* Boca Raton, FL: CRC Press (Chapter 15).

Ankrum, D. R., & Nemeth, K. J. (1995). Posture, comfort, and monitor placement. *Ergonomics in Design, 3(2),* 7–9.

Annis, J. F., & McConville, J. T. (1996). Anthropometry. In A. Bhattacharya & J. D. McGlothlin (Eds.), *Occupational Ergonomics* (pp. 1–46). New York, NY: Dekker (Chapter 1).

Annis, J. F., Case, H. W., Clauser, C. E., & Bradtmiller, B. (1991). Anthropometry of an aging work force. *Experimental Aging Research, 17,* 157–176.

ANSI/HFES 100. (2000). *U.S. National Standard for Human Factors Engineering of Computer Workstations.* Santa Monica, CA: Human Factors and Ergonomics Society (In preparation; issued 2007).

Anson, J. G. (1982). Memory drum theory: Alternative tests and explanations for the complexity effects on simple reaction time. *Journal of Motor Behavior, 14(3),* 228–246.

Antonovsky, A. (1979). *Health, stress, and coping.* San Francisco, CA: Jossey-Bass.

Armstrong, T. J. (2006). The ACGIH TLV for hand activity work. In W. S. Marras & W. Karwowski (Eds.), *The occupational ergonomics handbook: Fundamentals and assessment tools for occupational ergonomics* (2nd ed.). Boca Raton, FL: CRC Press (Chapter 41).

Arnaut, L. Y., & Greenstein, J. S. (1990). Is display/control gain a useful metric for optimizing an interface? *Human Factors, 32,* 651–663.

Arndt, B., & Putz-Anderson, V. (2006). *Cumulative trauma disorders* (3rd ed.). Boca Raton, FL: CRC Press.

Arswell, C. M., & Stephens, E. C. (2001). Information processing. In W. Karwowski (Ed.), *International encyclopedia of ergonomics and human factors* (pp. 256–259). London, UK: Taylor & Francis.

Aschoff, H. (1981). In *Handbook of behavioral neurobiology* (4). New York, NY: Plenum.

Asfour, S. S., Genaidy, A. M., & Mital, A. (1988). Physiological guidelines for the design of manual lifting and lowering tasks: The state of the art. *American Industrial Hygiene Association Journal, 49,* 150–160.

ASHRAE (Ed.). (1985). *Physiological principles for comfort and health.* Atlanta, GA: American Society of Heating, Refrigerating, Air-Conditioning Engineers 1985 Fundamentals handbook (Chapter 8).

Asimov, I. (1963). *The human body: Its structure and organization.* New York, NY: New American Library/Signet.

Asimov, I. (1989). *Asimov's chronology of science and discovery.* New York, NY: Harper & Row.

Aspden, R. M. (1988). A new mathematical model of the spine and its relationship to spinal loading in the workplace. *Applied Ergonomics, 19,* 319–323.

Asso, D. (1987). Cyclical variations. In M. A. Baker (Ed.), *Sex differences in human performance.* Chichester, UK: Wiley (Chapter 4).

Astin, A. D. (1999). *Finger force capability: Measurement and prediction using anthropometric and myoelectric measures.* (Unpublished Master's thesis). Blacksburg, VA: Dept. of Systems Engineering, Virginia Tech.

Astrand, P. O., & Rodahl, K. (1977). *Textbook of work physiology* (2nd ed.). New York, NY: McGraw-Hill.

Astrand, P. O., & Rodahl, L. (1986). *Textbook of work physiology* (3rd ed.). New York, NY: McGraw-Hill.

Astrand, P. O., Rodahl, K., Dahl, H. A., & Stromme, S. B. (2004). Textbook of work physiology *Physiological bases of exercise.* (4th ed.). Champaign, IL: Human Kinetics.

Authier, M., Lortie, M., & Gagnon, M. (1994). Handling techniques: Impact of the context on the choice of grip and box movement in experts and novices. In F. Aghazadeh (Ed.), *Advances in industrial ergonomics and safety VI* (pp. 687–693). London, UK: Taylor & Francis.

Ayoub, M. M. (1982). The manual lifting problem: The illusive solution. *Journal of Occupational Accidents, 4,* 1–23.

Ayoub, M. M., & Dempsey, P. G. (1999). The psychophysical approach to manual materials handling task design. *Ergonomics, 42,* 17–31.

III. FURTHER INFORMATION

Ayoub, M. M., & Mital, A. (1989). *Manual materials handling.* London, UK: Taylor & Francis.

Ayoub, M. M., & Smith, J. L. (1988). Manual materials handling in unusual postures: Carrying of loads. *Proceedings of the Human Factors Society 32nd annual meeting* (pp. 675-679). Santa Monica, CA: Human Factors Society.

Ayoub, M. M., Selan, J. L., & Jiang, B. C. (1984). *A mini-guide for manual materials handling.* Lubbock, TX: Institute of Ergonomics Research, Texas Tech University.

Ayoub, M. M., Selan, J. L., & Jiang, B. C. (1986). Manual materials handling. In G. Salvendy (Ed.), *Handbook of human factors* (pp. 790–818). New York, NY: Wiley.

Ayres, J. (1998). Sensory integrations and praxis tests. *Manual.* Los Angeles, CA: WPS.

Bailey, R. W. (1996). *Human performance engineering* (3rd ed.). Upper Saddle River, NJ: Prentice Hall.

Bajraktarov, S., Novotni, A., Manusheva, N., Nikovska, D. G., Miceva-Velickovska, E., Zdraveska, N., et al. (2011). Main effects of sleep disorders related to shift work—Opportunities for preventive programs. *EPMA Journal, 2,* 365–370.

Baker, M. A. (Ed.). (1987). *Sex differences in human performance.* Chichester, UK: Wiley.

Baker, N. A., Cham, R., Cidboy, E. H., Cook, J., & Redfern, M. S. (2007). Kinematics of the fingers and hands during computer keyboard use. *Clinical Biomechanics, 22*(1), 34–43.

Balci, R., Aghazadeh, F., & Waly, S. M. (1998). Work-rest schedules for data entry operators. In S. Kumar (Ed.), *Advances in occupational ergonomics and safety* (pp. 155–158). Amsterdam, The Netherlands: IOS Press.

Ballard, B. (1995). How odor affects performance: A review. *Proceedings of ErgoCon '95, Silicon valley ergonomics conference and exposition* (pp. 191–200). San Jose, CA: San Jose State University.

Balogun, J. A. (1986). Optimal rate of work during load transportation on the head and by yoke. *Industrial Health, 24,* 75–86.

Barker, K. L., & Atha, J. (1994). Reducing the biomechanical stress of lifting by training. *Applied Ergonomics, 25,* 373–378.

Barlow, A. M., & Braid, S. J. (1990). Foot problems in the elderly. *Clinical Rehabilitation, 4,* 217–222.

Barnes, M. E., & Wells, W. (1994). If Hearing Aids Work, Why Don't People Use Them? *Ergonomics in Design, 2*(2), 19–24.

Barnett, L., van Beurden, E., Morgan, P., Brooks, L., & Beard, J. (2010). Gender differences in motor skill proficiency from childhood to adolescence: A longitudinal study. *Research quarterly for exercise and sport, 81,* 162–170. doi: 10.1080/0270136 7.2010.10599663.

Barron, G., & Yechiam, E. (2002). Private email requests and the diffusion of responsibility. *Computers in Human Behavior, 18*(5), 507–520.

Basbaum, A. I., & Julius, D. (2006). Toward better pain control. *Scientific American, 294*(6), 60–67.

Basmajian, J. V., & DeLuca, C. J. (1985). *Muscles alive* (5th ed.). Baltimore, MD: Williams & Wilkins.

Batavia, A. I., & Hammer, G. S. (1990). Toward the development of consumer-based criteria for the evaluation of assistive devices. *Journal of Rehabilitation Research and Development, 27,* 425–436.

Battie, M. C., Bigos, S. J., Fisher, L., Hansson, T. H., Jones, M. E., & Wortley, M. D. (1989). Isometric lifting strength as a predictor of industrial back pain reports. *Spine, 14,* 851–856.

Bauer, W., & Wittig, T. (1998). Influence of screen and copy holder positions on head posture muscle activity and user judgment. *Applied Ergonomics, 29,* 185–192.

Baumeister, R., Vohs, D., Aaker, J., & Garbinsky, E. (2013). Some key differences between a happy life and a meaningful life. *The Journal of Positive Psychology, 8*(6), 505–516.

Bedny, G. Z., & Zeglin, M. H. (1997). The use of pulse rate to evaluate physical work load in Russian ergonomics. *American Industrial Hygiene Association Journal, 58,* 375–379.

Beevis, D. (2003). Ergonomics—Costs and benefits revisited. *Applied Ergonomics, 34,* 491–496.

Bell, C. (1833). *The nervous system of the human body.* London, UK: Taylor & Francis.

Bellera, A., Foster, B. J., & Hanley, J. A. (2012). Calculating sample size in anthropometry. In V. R. Preedy (Ed.), *Handbook of anthropometry: Physical measures of human form in health and disease* (pp. 3–27). Heidelberg, Germany: Springer (Chapter 1).

Belsky, J. K. (1990). *The psychology of aging: Theory, research, and interventions.* Pacific Grove, CA: Brooks/Cole.

Belsky, J. (2016). *Experiencing the lifespan* (4th ed.). Duffield, UK: Worth Publishers.

Bendix, T., Poulsen, V., Klausen, K., & Jesnen, C. V. (1996). What does a backrest actually do to the lumbar spine? *Ergonomics, 39,* 533–542.

Bennett, P. B. & Elliott, D. H. (Eds.). (1993). *The physiology and medicine of diving* (4th ed.). Philadelphia, PA: Saunders.

Bennett, C. M., Baird, A. A., Miller, M. B., & Wolford, G. L. (2010). Neural correlates of interspecies perspective taking in the post-mortem Atlantic salmon: An argument for proper multiple comparisons correction. *Journal of Serendipitous and Unexpected Results, 1,*1–5.

Bensel, C. K., & Santee, W. R. (1997). Climate and clothing. In G. Salvendy (Ed.), *Handbook of human factors and ergonomics* (2nd ed., pp. 909–934). New York, NY: Wiley.

Ben-Shakhar, G., & Furedy, J. J. (1990). *Theories and applications in the detection of deception.* New York, NY: Springer.

III. FURTHER INFORMATION

Beranek, L., Blazier, W., & Figwer, J. (1971). Preferred noise criteria (PNC) curves and their application to rooms. *Journal Acoustical Society of America, 50*, 1223–1228.

Berger, E. H., & Casali, J. G. (1992). Hearing protection devices. In M. J. Crocker (Ed.), *Handbook of acoustics*. New York, NY: Wiley (Chapter 8).

Berger, E. H., Royster, L. H., Royster, J. D., Driscoll, D. P., & Layne, M. (2003). *The noise manual* (5th ed.). Fairfax, VA: American Industrial Hygiene Association.

Bergmann, G., Deuretzbacher, G., Heller, M., Graichen, F., Rohlmann, A., Strauss, J., et al. (2001). Hip contact forces and gait patterns from routine activities. *Journal of Biomechanics, 34*, 859–871.

Bergquist-Ullman, M., & Larsson, U. (1977). Acute low back pain in industry. *Acta Orthopadica Scandinavica, 170*, 1–117.

Bernard, B. P. (Ed.). (1997). *Musculoskeletal disorders and workplace factors. (DHHS NIOSH Publication No. 97-141)*. Washington, DC: U.S. Department of Health and Human Services.

Bernard, T. E. (2002). Thermal stress. In B. A. Plog (Ed.), *Fundamentals of industrial hygiene* (5th ed.). Itasca, IL: National Safety Council (Chapter 12).

Bernard, T. E., & Dukes-Dobos, F. N. (2002). *Heat stress* (2nd ed.). Atlanta, GA: AIHA.

Bernard, T. E., & Iheanacho, I. (2015). Heat index and adjusted temperature as surrogates for wet bulb globe temperature to screen for occupational heat stress. *Journal of Occupational and Environmental Hygiene, 12*(5), 323–333.

Bernhard, T. E., & Joseph, B. S. (1994). Estimation of metabolic rate using qualitative job descriptors. *American Industrial Hygiene Association Journal, 55*, 1021–1029.

Berns, T. A. R., & Milner, N. P. (1980). *TRAM: A technique for the recording and analysis of moving work posture (report 80: 23, pages 22–26)*. Stockholm, Sweden: Ergolab.

Best, J. B. (1997). *Cognitive psychology* (4th ed.). St. Paul, MN: West Publishing.

Bhattacharya, A., & McGlothlin, J. D. (1996). *Occupational ergonomics: Theory and applications*. New York, NY: Dekker.

Bhattacharya, A., & McGlothlin, J. D. (2012). *Occupational ergonomics: Theory and applications* (2nd ed.). Boca Raton, FL: CRC Press.

Bhise, V. D. (2011). *Ergonomics in the automotive design process*. Boca Raton, FL: CRC Press.

Bien, N., ten Oever, S., & Sack, A. (2012). The sound of size: Crossmodal binding in pitch-size synesthesia: A combined TMS, EEG, and psychophysics study. *NeuroImage, 59*(1), 663–672.

Bigos, S. J., Spengler, D. M., Martin, N. A., Zeh, J., Fisher, L., & Nachemson, A. (1986). Back injuries in industry: A retrospective study III. Employee-related factors. *Spine, 11*, 252–256.

Birkbeck, M. Q., & Beer, T. C. (1975). Occupation in relation to the carpal tunnel syndrome. *Rheumatology and Rehabilitation, 14*(4), 218–221.

Bishu, R. R., Wang, W., Hallbeck, M. S., & Cochran, D. J. (1992). Force distribution at hand/handle coupling: The effect of handle type. *Proceedings of the Human Factors Society 36th annual meeting* (pp. 816-820). Santa Monica, CA: Human Factors Society.

Bitan, Y., Ramey, S., Philp, G., & Uukkivi, T. (2015). Working with paramedics on implementing human factors improvements to their response bags. *Proceedings of the International Symposium on Human Factors and Ergonomics in Health Care, 4*, 179–181.

Bjoering, G., Johansson, L., & Haegg, G. M. (1998). The pressure distribution in the hand when holding powered drills. In S. Kumar (Ed.), *Advances in occupational ergonomics and safety* (pp. 436–438). Amsterdam, The Netherlands: IOS Press.

Bjoerkman, T. (1996). The rationalization movement in perspective and some ergonomic implications. *Applied Ergonomics, 27*, 111–117.

Black, T. R., Shah, S. M., Busch, A. J., Metcalf, J., & Lim, H. J. (2011). Effect of transfer, lifting, and repositioning injury prevention program on MSD injury among direct care workers. *Journal of Occupational and Environmental Hygiene, 8*(4), 226–235.

Blaikie, A. (1993). Images of age: A reflexive process. *Applied Ergonomics, 24*, 51–57.

Blanchard, B. S., & Fabrycky, W. J. (1996). *Systems engineering and analysis* (3rd ed.). Englewood Cliffs, NJ: Prentice Hall.

Blinkhorn, S. (1988). Lie detection as a psychometric procedure. In A. Gale (Ed.), *The polygraph test: Lies, truth, and science* (pp. 29–39). London, UK: Sage.

Bobick, T. G., & Gutman, S. H. (1989). Reducing musculo-skeletal injuries by using mechanical handling equipment. In K. H. E. Kroemer, J. D. McGlothlin & T. G. Bobick (Eds.), *Manual material handling: Understanding and preventing back trauma* (pp. 87–96). Akron, OH: American Industrial Hygiene Association.

Bobjer, O., Johansson, S., & Piguet, S. (1993). Friction between hand and handle. Effects of oil and lard on textured and non-textured surfaces; perception of discomfort. *Applied Ergonomics, 24*, 190–202.

Boff, K. R. & Lincoln, J. E. (Eds.). (1988). *Engineering data compendium: Human perception and performance*. Wright-Patterson AFB, OH: Armstrong Aerospace Medical Research Laboratory.

Boff, K. R., Kaufman, L. & Thomas, J. P. (Eds.). (1986). *Handbook of perception and human performance*. New York, NY: Wiley.

III. FURTHER INFORMATION

Bogner, M. S. (1998). An introduction to design, evaluation, and usability testing. In V. J. B. Rice (Ed.), *Ergonomics in health care and rehabilitation* (pp. 231–247). Boston, MA: Butterworth-Heinemann (Chapter 13).

Bogner, M. S., Weinger, M. B., Laugherey, K. R., Haas, E. C., & Admunson, D. O. (1998). Warnings in health care. *CSERIAC Gateway, 9*(3), 1–3.

Bohan, M., Chaparro, A., Fernandez, J. E., Kattel, B. P., & Choi, S. D. (1998). Cursor-control performance of older adults using two computer input devices. In S. Kumar (Ed.), *Advances in occupational ergonomics and safety* (pp. 541–544). Amsterdam, The Netherlands: IOS Press.

Bohle, P., & Tilley, A. J. (1989). The impact of night work on psychological well-being. *Ergonomics, 34*, 1089–1099.

Bonne, A. J. (1969). On the shape of the human vertebral column. *Acta Orthopaedica Belgica, 3–4*, 567–583 (35, Fasc).

Booher, H. R. (Ed.). (2003). *Handbook of human systems integration*. New York, NY: Wiley.

Booth-Jones, A. D., Lemasters, G. K., Succop, P., Atterbury, M. R., & Bhattacharya, A. (1998). Reliability of questionnaire information measuring musculoskeletal symptoms and work histories. *American Industrial Hygiene Association Journal, 59*, 20–24.

Borg, G. A. V. (1962). *Physical performance and perceived exertion*. Lund, Sweden: Gleerups.

Borg, G. A. V. (1982). Psychophysical bases of perceived exertion. *Medicine and Science in Sports and Exercise, 14*, 377–381.

Borg, G. (1990). Psychophysical scaling with applications in physical work and the perception of exertion. *Scandinavian Journal of Work, Environment and Health, 16*(Suppl. 1), 55–58.

Borg, G. (2001). Rating scales for perceived physical effort and exertion. In W. Karwowski (Ed.), *International encyclopedia of ergonomics and human* (pp. 358–541). London, UK: Taylor & Francis.

Borg, G. (2005). Scaling experiences during work: Perceived exertion and difficulty. In N. Stanton, A. Hedge, K. Brookhuis, E. Salas & H. Hendrick (Eds.), *Handbook of human factors and ergonomics methods* (pp. 11-1–11-7). Boca Raton, FL: CRC Press.

Borelli, G. A. (1680). *De motu animalium*. Romae: Bernabò.

Boudrifa, H., & Davies, B. T. (1984). The effect of backrest inclination. Lumbar support and thoracic support on the intra-abdominal pressure while lifting. *Ergonomics, 27*, 379–387.

Bowman, R. L., & Delucia, J. L. (1992). Accuracy of self-reported weight: A meta-analysis. *Behavior Therapy, 23*, 637–655.

Boyce, P. R. (1982). *Human factors in lighting*. New York, NY: MacMillan.

Boyce, P. R. (2014). *Human factors in lighting* (3rd ed.). Boca Raton, FL: CRC Press.

Bradtmiller, B. (2015). Anthropometry in human systems integration. In D. A. Boehm-Davis, T. D. Francis & J. D. Lee (Eds.), *APA handbook of human systems integration* (pp. 117–132). Washington, DC: American Psychological Association (Chapter 8).

Brauer, R. (2006). *Safety and health for engineers*. Hoboken, NJ: Wiley.

Braune, W., & Fischer, O. (1889). The center of gravity of the human body as related to the equipment of the German Infantryman. (In: German.). *Translation in human mechanics (AMRL-TDR-63-123)*. Wright-Patterson AFB, OH: Aerospace Medical Research Laboratory, 1963 (Abh. d. Math. Phys. Cl. d. k. Saechs. Gesell. d. Wissenschaften Leipzig 1889 15, pp 561.).

Brennan, R. B. (1987). Trencher operator seating positions. *Applied Ergonomics, 18*, 95–102.

Bridger, R. S. (1995). *Introduction to ergonomics*. New York, NY: McGraw-Hill.

Bridger, R. S. (2009). *Introduction to ergonomics* (3rd ed.). Boca Raton, FL: CRC Press.

Brill, S. (2013). Bitter pill. *Time, 181*(8), 16–24, 26, 28 passim.

Bringelson, L. S., Yeoh, C. H., & Wong, C. B. (1998). An empirical investigation of pointing devices for notebook computers. In S. Kumar (Ed.), *Advances in occupational ergonomics and safety* (pp. 545–548). Amsterdam, The Netherlands: IOS Press.

Bronzino, J. D., & Peterson, D. R. (2015). *Medical devices and human engineering*. Roca Baton, FL: CRC Press.

Brouha, L. (1960). *Physiology in industry*. Oxford, UK: Pergamon.

Brouha, L. (1967). *Physiology in industry* (2nd ed.). Riverside, NJ: Pergamon.

Brown, J. R. (1972). *Manual lifting and related fields: An annotated bibliography*. Toronto, ON: Labour Safety Council of Ontario, Ontario Ministry of Labour.

Brown, J. R. (1975). Factors contributing to the development of low-back pain in industrial workers. *American Industrial Hygiene Association Journal, 36*, 26–31.

Brown, I. D. (1994). Driver fatigue. *Human Factors, 36*, 298–314.

Brown, D. A., Coyle, I. R., & Beaumont, P. E. (1985). The automated hettinger test in the diagnosis and prevention of repetition strain injuries. *Applied Ergonomics, 16*, 113–118.

Brownell, K. D. (1995). Definition and classification of obesity. In K. D. Brownell & C. G. Fairburn (Eds.), *Eating disorders and obesity*. New York, NY: Guilford.

Brownell, K., Kersh, R., Ludwig, D., Post, R., Puhl, R., Schwartz, M., et al. (2010). Personal responsibility and obesity: A constructive approach to a controversial issue. *Health Affairs, 29*(3), 379–387.

III. FURTHER INFORMATION

Brunswick, E. (1956). *Perception and the representative design of experiments* (2nd ed.). Berkley, CA: University of California Press.

Buchholz, B., Paquet, V., Punnett, L., Lee, D., & Moir, S. (1996). PATH: A work-sampling based approach to ergonomic work analysis for construction and other non-repetitive work. *Applied Ergonomics, 27,* 177–187.

Buckle, P. W., David, G. C., & Kimber, A. C. (1990). Flight deck design and pilot selection: Anthropometric considerations. *Aviation, Space, and Environmental Medicine, 61,* 1079–1084.

Buis, N. (1990). Ergonomics, legislation and productivity in manual materials handling. *Ergonomics, 33,* 353–359.

Bullinger, H., Kern, P., & Braun, M. (1997). Controls. In G. Salvendy (Ed.), *Handbook of human factors and ergonomics* (2nd ed., pp. 697–728). New York, NY: Wiley (Chapter 21).

Burgdorf, A., van Riel, M., van Wingerden, J. P., van Wingerden, S., & Snijders, C. (1995). Isodynamic evaluation of trunk muscles and low-back pain among workers in a steel factory. *Ergonomics, 38,* 2107–2117.

Burgess-Limerick, R., & Abernethy, B. (1997). Toward a quantitative definition of manual lifting postures. *Human Factors, 39,* 141–148.

Burns, C. M., Vicente, K. J., Christoffersen, K., & Pawlak, W. S. (1997). Towards viable, useful and usable human factors design guidance. *Applied Ergonomics, 28,* 311–322.

Burrows, E., Thomas, G., & Rickards, J. (1998). A pre-intervention benefit/cost methodology – Refining the cost audit process. *Proceedings of the 30th annual conference of the Human Factors Association of Canada* (pp. 131-136). Windsor, ON: Human Factors Association of Canada.

Burse, R. L. (1979). Sex differences in human thermoregulatory responses to heat and cold stress. *Human Factors, 21,* 687–699.

Burt, S., & Punnett, L. (1999). Evaluation of interrater reliability for posture observations in a field study. *Applied Ergonomics, 30,* 121–135.

Burton, K. (1991). Measuring flexibility. *Applied Ergonomics, 22,* 303–307.

Bush-Joseph, C., Schipplein, O., Andersson, G. B. J., & Andriacchi, T. P. (1988). Influence of dynamic factors on the lumbar spine moment in lifting. *Ergonomics, 31,* 211–216.

Butler, D., Andersson, G. B. J., Trafimow, J., Schipplein, O. D., & Andriacchi, T. P. (1993). The influence of load knowledge on lifting technique. *Ergonomics, 36,* 1489–1493.

Cady, L. D., Bischoff, D. P., O'Connell, E., Thomas, P. C., & Allan, J. (1979a). Strength and fitness and subsequent back injuries in fire fighters. *Journal of Occupational Medicine, 21,* 269–272.

Cady, L. D., Bischoff, D. P., O'Connell, E. R., Thomas, P. C., & Allan, J. H. (1979b). Letters to editor: Authors' response. *Journal of Occupational Medicine, 21,* 720–725.

Cai, D., & You, M. (1998). An ergonomic approach to public squatting-type toilet design. *Applied Ergonomics, 29,* 147–153.

Caillet, R. (1981). *Low back pain* (3rd ed.). London, UK: Davis.

Cain, W. S., Leaderer, B. P., Cannon, L., Tosun, T., & Ismail, H. (1987). Odorization of inert gas for occupational safety: Psychophysical considerations. *American Industrial Hygiene Association Journal, 48,* 47–55.

Caldwell, L. S., Chaffin, D. B., Dukes-Dobos, F. N., Kroemer, K. H. E., Laubach, L. L., Snook, S. H., et al. (1974). A proposed standard procedure for static muscle strength testing. *American Industrial Hygiene Association Journal, 35,* 201–206.

Cannon, W. B. (1939). *The wisdom of the body.* New York, NY: Norton.

Caple, D. C., & Betts, N. J. (1991). RSI—Its rise and fall in telecom Australia 1981–1990. *Proceedings of the 11th congress of the International Ergonomics Association* (pp. 1037-1039). London, UK: Taylor & Francis.

Capodaglio, P., Capodaglio, E. M., & Bazzini, G. (1995). Tolerability to prolonged lifting tasks assessed by subjective perception and physiological responses. *Ergonomics, 38,* 2118–2128.

Capodaglio, P., Capodaglio, E. M., & Bazzini, G. (1997). A field methodology for ergonomic analysis in occupational manual materials handling. *Applied Ergonomics, 28,* 203–208.

Carayon, P., & Lim, S. Y. (2006). Psychosocial work factors. In W. S. Marras & W. Karwowski (Eds.), *The occupational ergonomics handbook* (2nd ed.). Boca Raton, FL: CRC Press Interventions, controls, and applications in occupational ergonomics (Chapter 5).

Carayon, P. (2011). *Handbook of human factors and ergonomics in healthcare and patient safety* (2nd ed.). Boca Raton, FL: CRC Press.

Carlton, R. S. (1987). The effects of body mechanics instruction on work performance. *The American Journal of Occupational Therapy, 41,* 16–20.

Carr, G., & Houtchens, C. J. (1989). Artificial gravity. *Final Frontier, 2*(3), 28-30, 61.

Carroll, A. E. (2015). "The Persistent Health Myth of 8 Glasses of Water a Day." *New York Times,* August 25, p. A3.

Carvalhais, A. B., Tepas, D. I., & Mahan, R. P. (1988). Sleep duration in shift workers. *Sleep Research, 17,* 109–124.

Casali, J. G., & Berger, E. H. (1996). Technology advancements in hearing protection circa 1995 active noise reduction, frequency-amplitude-sensitivity, and uniform attenuation. *American Industrial Hygiene Association Journal, 57,* 175–185.

III. FURTHER INFORMATION

Casali, J. G., & Gerges, S. N. Y. (2006). Protection and enhancement of hearing in noise. In R. C. Williges (Ed.), *Reviews of human factors and ergonomics* (2). Santa Monica, CA: Human Factors and Ergonomics Society (Chapter 7).

Casali, J. G., & Robinson, G. S. (2006). Noise in industry. In W. S. Marras & W. Karwowski (Eds.), *The occupational ergonomics handbook. Fundamentals and assessment tools for occupational ergonomics* (2nd ed.). Boca Raton, FL: CRC Press (Chapter 31).

Casali, J. G., & Wright, W. H. (1995). Do amplitude-sensitive hearing protectors improve detectability of vehicle backup alarms in noise? *Proceedings of the Human Factors and Ergonomics Society 39th annual meeting* (pp. 994–998). Santa Monica, CA: Human Factors and Ergonomics Society.

Casali, S. P., & Williges, R. C. (1990). Databases of accommodative aids for computer users with disabilities. *Human Factors, 32,* 407–422.

Casey, S. M. (1989). Anthropometry of farm equipment operators. *Human Factors Society Bulletin, 32,* 1–16.

Casner, S. M., Geven, R. W., Recker, M. P., & Schooler, J. W. (2014). The retention of manual flying skills in the automated cockpit. *Human Factors, 56*(8), 1506–1516.

Cavanagh, P. R. (1988). On "muscle action" vs "muscle contraction". *Biomechanics, 21,* 69.

Chaffin, D. B. (1981). Functional assessment for heavy physical labor. *Occupational health & safety, 50,* 24–32.

Chaffin, D. B. (1991). Occupational ergonomics. Proceedings. *Occupational ergonomics: Work-related upper limb and back disorders (not paginated).* San Diego, CA: American Industrial Hygiene Association, San Diego Section.

Chaffin, D. B., & Andersson, G. B. J. (1984). *Occupational biomechanics.* New York, NY: Wiley.

Chaffin, D. B., & Andersson, G. B. J. (1991). *Occupational biomechanics* (2nd ed.). New York, NY: Wiley.

Chaffin, D. B., Herrin, G. D., & Keyserling, W. M. (1978). Preemployment strength testing: An updated position. *Journal of Occupational Medicine, 20,* 403–408.

Chaffin, D. B., Andersson, G. B. J., & Martin, B. J. (1999). *Occupational biomechanics* (3rd ed.). New York, NY: Wiley.

Chaffin, D. B., Andersson, G. B. J., & Martin, B. J. (2006). *Occupational biomechanics* (4th ed.). New York, NY: Wiley.

Chandler, R. F., Clauser, C. E., McConville, J. R., Reynolds, H. M., & Young, J. W. (1975). *Investigation of inertial properties of the human body (AMRL-TR-74-137).* Wright-Patterson AFB, OH: Aerospace Medical Research Laboratory.

Chapanis, A. (Ed.). (1975). *Ethnic variables in human factors engineering.* Baltimore, MD: Johns Hopkins University Press.

Chapanis, A. (1995). Ergonomics in product development: A personal view. *Ergonomics, 38,* 1625–1638.

Chapanis, A. (1996). *Human factors in systems engineering.* New York, NY: Wiley.

Chapanis, A., Garner, W., & Morgan, C. (1949). *Applied experimental psychology: Human factors in equipment design.* New York, NY: Wiley.

Charney, W. (2010). *Handbook of modern hospital safety* (2nd ed.). Boca Raton, FL: CRC Press.

Chengular, S. N., Rodgers, S. H., & Bernard, T. E. (2003). *Kodak's ergonomic design for people at work* (2nd ed.). New York, NY: Wiley.

Chenoweth, D. (1983a). Fitness program evaluation: Results with muscle. *Occupational Health and Safety, 52,* 14–17 (40–42).

Chenoweth, D. (1983b). Health promotion: Benefit vs cost. *Occupational Health and Safety, 52,* 37–41.

Cherry, N. (1987). Physical demands of work and health complaints among women working late in pregnancy. *Ergonomics, 30,* 689–701.

Cheverud, J., Gordon, C. C., Walker, R. A., Jacquish, C., Kohn, L., Moore, A., et al. (1990a). *1988 anthropometric survey of US Army personnel: Correlation coefficients and regression equations NATICK/TR-90/032.* Natick, MA: United States Army Natick Research, Development and Engineering Center, Soldier Science Directorate.

Cheverud, J., Gordon, C. C., Walker, R. A., Jacquish, C., Kohn, L., Moore, A., et al. (1990b). *1988 anthropometric survey of US Army Personnel.* Natick, MA: US Army Natick Research, Development and Engineering Center (technical reports 90/031 through 036).

Chi, C. F., & Lin, F. T. (1998). A comparison of seven visual fatigue assessment techniques in three data-acquisition tasks. *Human Factors, 40,* 577–590.

Chiou, S., Bhattacharya, A., & Succop, P. A. (1996). Effect of workers' shoe wear on objective and subjective assessment of slipperiness. *American Industrial Hygiene Association Journal, 57,* 825–831.

Christensen, J. M. (1993). Forensic human factors psychology – Part 2 a model for the development of safer products. *CSERIAC Gateway, 4*(3), 1–5.

Christensen, J. M., & Talbot, J. M. (1986). Psychological aspects of space flight. *Aviation, Space, and Environmental Medicine, 57,* 203–212.

Christensen, J. M., Topmiller, D. A., & Gill, R. T. (1988). Human factors definitions revisited. *Human Factors Society Bulletin, 31,* 7–8.

CIE. (1951). *CIE Proceedings.* Paris: Bureau Central de la Commission Internationale de l'Eclairage 3.

III. FURTHER INFORMATION

Ciriello, V. M., & Snook, S. H. (1983). A study of size, distance, height, and frequency effects on manual handling tasks. *Human Factors, 25,* 473–483.

Ciriello, V. M. (2001). The effects of box size, vertical distance, and height on lowering tasks. *International Journal of Industrial Ergonomics, 28,* 61–67.

Ciriello, V. M. (2007). The effects of container size, frequency, and extended horizontal reach on maximum acceptable weights of lifting for female industrial workers. *Applied Ergonomics, 38*(1), 1–5.

Ciriello, V. M., McGorry, R. W., Martin, S., & Bezverkhny, I. B. (1999). Maximum acceptable forces of dynamic pushing: Comparison of two techniques. *Ergonomics, 42*(1), 32–39.

Ciriello, V. M., Snook, S. H., & Hughes, G. J. (1993). Further studies of psychophysically determined maximum acceptable weights and forces. *Human Factors, 35,* 175–186.

Clark, A., Fleche, E., Layard, R., Powdthavee, N., & Ward, G. (2017). *Origins of happiness: The science of well-being over the life course.* Princeton, NJ: Princeton University Press.

Clark, M. C., Czaja, S. J., & Weber, R. A. (1990). Older adults and daily living task profiles. *Human Factors, 32,* 537–549.

Clauser, C. E., McConville, J. T., & Young, J. W. (1969). *Weight, volume, and center of mass of segments of the human body AMRL-TR-69-70.* Wright-Patterson AFB, OH: Aerospace Medical Research Laboratory.

Cohen, S., Frank, E., Doyle, W. J., Skoner, D. P., Rabin, B. S., & Gwaltney, J. M. (1998). Types of stressors that increase susceptibility to the common cold in healthy adults. *Health Psychology, 17,* 214–223.

Colapinto, J. (2015). Lighting the brain. *The New Yorker, 18,* 74–83.

Colle, H. A., & Reid, G. B. (1998). Context effects in subjective mental workload ratings. *Human Factors, 40,* 591–600.

Colligan, M. J., & Tepas, D. I. (1986). The stress of hours of work. *American Industrial Hygiene Association Journal, 47,* 686–695.

Collins, M., Brown, B., Bowman, K., & Carkeet, A. (1990). Workstation variables and visual discomfort associated with VDTs. *Applied Ergonomics, 21,* 157–161.

Colombini, D., Occhipinti, E., Alvarez-Casado, E., & Waters, T. R. (2012). *Manual lifting.* Boca Raton, FL: CRC Press.

Colquhoun, W. P. (1985). Hours of work at sea: Watch-keeping schedules, circadian rhythms, and efficiency. *Ergonomics, 28,* 637–653.

Cometto-Muniz, J. E., & Cain, W. S. (1994). Perception of odor and nasal pungency from homologous series of volatile organic compounds. *Indoor Air, 4*(3), 140–145.

Committee on an Aging Society. (1988). In *The social and built environment in an older society.* Washington, DC: National Academy Press.

Committee on Vision. (1987). In *Work, aging, and vision.* Washington, DC: National Academy Press.

Congleton, J. J., Ayoub, M. M., & Smith, J. L. (1985). The design and evaluation of the neutral posture chair for surgeons. *Human Factors, 27,* 589–600.

Coniglio, I., Fubini, E., Masali, M., Masiero, C., Pierlorenzi, G., & Sagone, G. (1991). Anthropometric survey of Italian population for standardization in ergonomics. *Proceedings of the 11th congress of the International Ergonomics Association* (pp. 894-896). London, UK: Taylor, Francis.

Conrad, R., & Hull, A. J. (1968). The preferred layout for numerical data entry key sets. *Ergonomics, 11,* 165–173.

Conway, F. T., & Smith, M. J. (1997). Psychosocial aspects of computerized office work. In M. Helander, T. K. Landauer & P. Prabhu (Eds.), *Handbook of human–computer interaction* (pp. 1497–1517). Amsterdam, The Netherlands: Elsevier.

Conway, K., & Unger, R. (1991). Ergonomic guidelines for designing and maintaining underground coal mining equipment. In A. Mital & W. Karwowski (Eds.), *Workspace, equipment, and tool design* (pp. 279–302). Amsterdam, The Netherlands: Elsevier.

Cook, T. D., & Campbell, D. T. (1979). *Quasi-experimentation: Design and analysis issues for field settings.* Chicago, IL: Rand McNally.

Cooper, R., & Sawaf, A. (1998). *Executive EQ: Emotional intelligence in leadership and organization.* New York, NY: Perigree.

Coren, S. (1994). Most comfortable listening level as a function of age. *Ergonomics, 37,* 1269–1274.

Corlett, E. N. (2005). The evaluation of industrial seating. In J. R. Wilson & N. Corlett (Eds.), *Evaluation of human work* (3rd ed.). London, UK: Taylor & Francis (Chapter 27).

Corlett, E. N., & Bishop, R. P. (1976). A technique for accessing postural discomfort. *Ergonomics, 19,* 175–182.

Corlett, E. N., Madeley, S. J., & Manenica, I. (1979). Postural targeting: A technique for recording working postures. *Ergonomics, 22,* 357–366.

Costa, G. (1996). The impact of shift and night work on health. *Applied Ergonomics, 27,* 9–16.

Costa, G. (2010). Shift work and health: Current problems and preventive actions. *Safe Health Work, 1*(2), 112–123.

Courtney, A. J. (1994). The effect of scale-side, indicator type, and control plane on direction-of-turn stereotypes for Hong Kong Chinese subjects. *Ergonomics, 37,* 865–877.

III. FURTHER INFORMATION

Cowen, R. (1988). *Eyes on the workplace*. Washington, DC: National Academy Press.

Cox, T., & Griffiths, A. (2005). The nature and measurement of work-related stress: Theory and practice. In J. R. Wilson & N. Corlett (Eds.), *Evaluation of human work* (3rd ed.). London, UK: Taylor & Francis (Chapter 19).

Craig, B. N., Congleton, J. J., Kerk, C. J., Lawler, J. M., & McSweeney, K. P. (1998). Correlation of injury occurrence data with estimated maximal aerobic capacity and body composition in a high-frequency manual materials handling task. *American Industrial Hygiene Association Journal, 59*, 25–33.

Cramer, S., Sur, M., Dobkin, B., & O'Brien, C. (2011). Harvesting neuroplasticity for clinical applications: Review. *Brain, 134*, 1591–1609.

Cronk, C. E., & Roche, A. F. (1982). Race- and sex-specific reference data for triceps and subscapular skinfolds and weight/stature. *American Journal of Clinical Nutrition, 35*, 347–354.

Crowell, H. P. (1995). Human engineering design guidelines for a powered. *Full body exoskeleton (report ARL-TN-60)*. Aberdeen Proving Ground, MD: U.S. Army Research Laboratory.

Csikszentmihalyi, M. (1990). *Flow: The psychology of optimal experience*. New York, NY: Harper & Row.

Culver, C. C., & Viano, D. C. (1990). Anthropometry of seated women during pregnancy: Defining a fetal region for crash protection research. *Human Factors, 32*, 625–636.

Cushman, W. H., & Rosenberg, D. J. (1991). *Human factors in product design*. Amsterdam, The Netherlands: Elsevier.

Cymerman, A. Institute of Medicine (US) Committee on Military Nutrition Research. (1996). The physiology of high-altitude exposure. In B. M. Marriott & S. J. Carlson (Eds.), *Nutritional needs cold and in high-altitude environments: Applications for military personnel in field operations*. Washington, DC: National Academies Press (US).

Czaja, S. J. (Ed.). (1990a). *Human Factors: special issue on Aging, 32*, 505–622.

Czaja, S. J. (Ed.). (1990b). *Human factors research needs for an aging population*. Washington, DC: National Academy Press.

Czaja, S. J. (1991). Work design for older adults. In A. Mital & W. Karwowski (Eds.), *Workspace, equipment, and tool design* (pp. 345–369). Amsterdam: Elsevier.

Czaja, S. J. (1997). Using technology to aid in the performance of home tasks. In A. D. Fisk & W. A. Rogers (Eds.), *Handbook of human factors and the older adult* (pp. 311–334). San Diego, CA: Academic Press.

Czaja, S. J., Zarcadoolas, C., Vaughon, W. L., Lee, C. C., Rockoff, M. L., & Levy, J. (2015). The usability of electronic personal health record systems for an underserved adult population. *Human Factors, 57*, 491–506.

Czeisler, C. A., & Gooley, J. J. (2007). Sleep and circadian rhythms in humans. *Cold Spring Harbor Symposia on Quantitative Biology, 72*, 579–597.

Czeisler, C. A., Kronauer, R. E., & Allan, J. S. (1989). Bright light induction of strong (type 0) resetting of the human circadian pacemaker. *Science, 244*, 1328–1332.

Czeisler, C. A., Dumont, M., & Richards, G. S. (1990a). Disorders of circadian function: Clinical consequences and treatment. *NIH Consensus Development Conference on the Treatment of Sleep Disorders of Older People*. Bethesda, MD: National Institutes of Health.

Czeisler, C. A., Johnson, M. P., & Duffy, J. F. (1990b). Exposure to bright light and darkness to treat physiologic maladaptation to night work. *The New England Journal of Medicine, 322*, 1253–1259.

d'Ambrosio Alfano, F., Malchaire, J., Palella, B., & Riccio, G. (2014). WBGT index revisited after 60 years of use. *Annals of Occupational Hygiene, 58*, 955–970.

Daams, B. J. (1993). Static force exertion in postures with different degrees of freedom. *Ergonomics, 36*, 397–406.

Daams, B. J. (1994). Human force exertion in user-product interaction. *Background for design*. Delft, The Netherlands: Delft University Press, IOS Press.

Daams, B. J. (2001). Push and pull data. In W. Karwowski (Ed.), *International encyclopedia of ergonomics and human*. London, UK: Taylor & Francis (299–316; torque data, 334–342).

Daltrov, L. H., Iversen, M. D., Larson, M. G., Lew, R., Wright, E., Ryan, J., et al. (1997). A controlled trial of an educational program to prevent low back injuries. *New England Journal of Medicine, 337*, 322–328.

Daniels, G. S. (1952). *The "Average" Man? (technical note WCRD 53-7)*. Wright-Patterson AFB, OH: Wright Air Development Center.

Datta, S. R., & Ramanathan, N. L. (1971). Ergonomic comparison of seven modes of carrying loads on the horizontal plane. *Ergonomics, 14*, 269–278.

Davids, R. (1991). Multi-user human computer modeling—A mixed blessing. *CSERIAC Gateway, 2*(3), 11–12.

Davies, B. T. (1978). Training in manual handling and lifting. In C. G. Drury (Ed.), *Safety in manual materials handling* (pp. 175–178). Cincinnati, OH: NIOSH (DHEW (NIOSH) Publication No. 78–185).

Davis, J. R., & Mirka, G. A. (1997). A transverse contour model of distributed muscle forces and spinal loads during lifting and twisting. *Proceedings of the Human Factors and Ergonomics Society 41st annual meeting* (pp. 675-679). Santa Monica, CA: Human Factors and Ergonomics Society.

III. FURTHER INFORMATION

de Bruijn, I., Engels, J. A., & Van der Gulden, J. W. J. (1998). A simple method to evaluate the reliability of OWAS observations. *Applied Ergonomics, 29,* 281–283.

Deeb, J. M., Drury, C. G., & Pizatella, P. (1987). Handle placement on containers in manual materials handling. *Proceedings of the 9th International Conference on Production Research* (pp. 417-423). Amsterdam, The Netherlands: Elsevier.

Dekker, S. (2014). *The field guide to understanding human error.* Williston, VT: Ashgate.

Delleman, N. J., Haslegrave, C. M. & Chaffin, D. B. (Eds.). (2004). *Working postures and movements.* Boca Raton, FL: CRC Press.

Dempsey, P. G. (1998). A critical review of biomechanical, epidemiological, physiological and psychophysical criteria for designing manual materials handling tasks. *Ergonomics, 41,* 73–88.

Dempsey, P. G. (1999). Prevention of musculoskeletal disorders: Psychophysical basis. In W. Karwowski & W. S. Marras (Eds.), *The occupational ergonomics handbook.* Boca Raton, FL: CRC Press Chapter 60 (1101–1126).

Dempsey, P. G. (2006). Psychophysical approach to task analysis. In W. S. Marras & W. Karwowski (Eds.), *The occupational ergonomics handbook. Fundamentals and assessment tools for occupational ergonomics* (2nd ed.). Boca Raton, FL: CRC Press (Chapter 47).

Dempster, W. T., Sherr, L. A., Priest, J. G. (1964). Conversion scales for estimating humeral and femoral lengths and the lengths of functional segments in the limbs of American Caucasoid Males. *Human Biology, 36*(3), 246–262.

Dennerlein, J. (2006). The computer keyboard system design. In W. S. Marras & W. Karwowski (Eds.), *The occupational ergonomics handbook* (2nd ed.). Boca Raton, FL: CRC Press Interventions, controls, and applications in occupational ergonomics (Chapter 39).

Dennerlein, J. T. (2005). Finger flexor tendon forces are a complex function of finger joint motions and fingertip forces. *Journal of Hand Therapy, 18,* 120–127.

Dennerlein, J. T., Mote, C. D., & Rempel, D. M. (1998). Control strategies for finger movement during touch-typing. The role of the extrinsic muscles during a keystroke. *Experimental Brain Research, 121,* 1–16.

DePalma, G., Collins, S., Bercik, P., & Verdu, E. (2014). The microbiota-gut-brain axis in gastrointestinal disorders: Stressed bugs, stressed brain, or both? *The Journal of Physiology, 592*(14), 2989–2997.

Department of Defense. (2000). *Human engineering design data digest. Human Factors Engineering Technical Advisory Group.* Washington, DC: Federal Aviation Administration.

Deutsch, E. S., Dong, Y., Halamek, L. P., Rosen, M. A., Taekman, J. M., & Rice, J. (2016). Leveraging health care simulation technology for human factors research: Closing the gap between lab and bedside. *Human Factors, 58*(7), 1082–1095.

Deyo, R. A., & Weinstein, J. N. (2001). Low back pain. *New England Journal of Medicine, 344,* 363–370.

Dickinson, C. E., Campion, K., Foster, A. F., Newman, S. J., O'Rourke, A. M. T., & Thomas, P. G. (1992). Questionnaire development: An examination of the nordic musculoskeletal questionnaire. *Applied Ergonomics, 23,* 197–201.

Dillon, B. S. (2008). *Reliability, technology, human error, and quality in healthcare.* Boca Raton, FL: CRC Press.

DiNardi, S. R. (Ed.). (1997). *The occupational environment—Its evaluation and control.* Fairfax, VA: American Industrial Hygiene Association.

Doherty, E. T. (1991). Speech analysis techniques for detecting stress. *Proceedings of the Human Factors Society 35th annual meeting* (pp. 689-693). Santa Monica, CA: Human Factors Society.

Dohrenwend, B. P., Raphael, K. G., Schwartz, S., Stueve, A., & Skodol, F. (1993). The structured event probe and narrative rating method for measuring stressful life events *Handbook of stress: Theoretical clinical aspects.* (2nd ed.). New York, NY: Free Press (pp. 174-196).

Doidge, N. (2007). *The brain that changes itself.* New York, NY: Penguin.

Dowell, W. R., Price, J. M., & Gscheidle, G. M. (1997). The effect of VDT screen distance on seated posture. *Proceedings of the Human Factors and Ergonomics Society 41st annual meeting* (pp. 505-508). Santa Monica, CA: Human Factors and Ergonomics Society.

Downing, K., Chan, S., Downing, W., Kwong, T., & Lam, T. (2008). Measuring gender differences in cognitive functioning. *Multicultural Education and Technology Journal, 2*(1), 4–18.

Drillis, R. J. (1963). Folk norms and biomechanics. *Human Factors, 5,* 427–441.

Drillis, R., & Contini, R. (1966). *Body segment parameters (report 1166-03).* Office of Vocational Rehabilitation. Department of Health Education and Welfare. New York, NY: University School of Engineering and Science.

Driskell, J. E., & Mullen, B. (2005). The efficacy of naps as a fatigue countermeasure: A meta-analytic integration. *Human Factors, 47,* 360–377.

Druckman, D. & Bjork, R. A. (Eds.). (1990). *The mind's eye: Enhancing human performance.* Washington, DC: National Academy Press.

Druckman, D. & Swets, J. A. (Eds.). (1988). *Enhancing human performance: Issues, theories, and techniques.* Washington, DC: National Academy Press.

III. FURTHER INFORMATION

Drury, C. G., Deeb, J. M., Hartman, B., Woolley, S., Drury, C. E., & Gallagher, S. (1989). Symmetric and asymmetric manual materials handling; Part 1 physiology and psychophysics. *Ergonomics, 32*, 467–489.

Dryden, R. D., & Kemmerling, P. T. (1990). Engineering assessment. In S. P. Sheer (Ed.), *Vocational assessment of impaired workers* (pp. 107–129). Aspen, CO: Aspen Press.

Duchon, J., Wagner, J., & Keran, C. (1989). Forward versus backward shift rotation. *Proceedings of the Human Factors Society 33rd annual meeting* (pp. 806-810). Santa Monica, CA: Human Factors Society.

Duecker, J. A., Ritchie, S. M., Knox, T. J., & Rose, S. J. (1994). Isokinetic trunk testing and employment. *Journal of Occupational Medicine JOM, 36*, 42–48.

Dunham, R. B., Pierce, J. L., & Castaneda, M. B. (1987). Alternative work schedules: Two field quasi experiments. *Personnel Psychology, 40*, 215–242.

Dunn, W. (2001). The sensations of everyday life: Theoretical, conceptual, and pragmatic considerations. *American Journal of Occupational Therapy, 55*, 608–620.

Dupuis, H., & Zerlett, G. (1986). Whole-body vibration and disorders of the spine. *International Archives of Occupational and Environmental Health, 59*, 323–336.

Duquette, J., Lortie, M., & Rossignol, M. (1997). Perception of difficulties related to assembly work: General findings and impact of back health. *Applied Ergonomics, 28*, 386–396.

Durkin, J. L., & Dowling, J. J. (2003). Analysis of body segment parameter differences between four human populations and the estimation errors of four popular mathematical models. *Journal of Biomechanical Engineering, 125*(4), 515–522.

Dvorak, A. (1936). Typewriter keyboard. US Patent 2,040,248. Washington, DC: US Patent and Trademark Office.

Dvorak, A. (1943). There is a better typewriter keyboard. *The National Business Education Quarterly, 12*(2), 51–58, 66.

Easterby, R., Kroemer, K. H. E. & Chaffin, D. B. (Eds.). (1982). *Anthropometry and biomechanics—Theory and application*. New York, NY: Plenum.

Easterlin, R. (1974). Does economic growth improve the human lot? In R. David & M. Reder (Eds.), *Nations and households in economic growth: Essays in honor of Moses Abramovitz* (pp. 89–1125). New York, NY: Academic Press.

Eastman Kodak Company. (1983). *Ergonomic design for people at work* (Vol. 1). New York, NY: Van Nostrand Reinhold.

Eastman Kodak Company. (1986). *Ergonomic design for people at work* (Vol. 2). New York, NY: Van Nostrand Reinhold.

Eberhard, J. W., Barr, R. A. (1992). Safety and mobility of elderly drivers Part 2. *Special Issue of Human Factors, 34*, 1–65.

Edholm, O. G., & Murrell, K. H. F. (1974). *The ergonomics research society: A history 1949 to 1970*. Washington, DC: The Council of the Ergonomics Research Society.

Edmonds, C., Lowry, C., & Pennefather, J. (1994). *Diving and subaquatic medicine*. Oxford, UK: Butterworth-Heinemann.

Edworthy, J., & Adams, A. (1996). *Warning design*. London, UK: Taylor & Francis.

Edworthy, J., & Stanton, N. (1995). A user-centred approach to the design and evaluation of auditory warning signal: 1. Methodology. *Ergonomics, 38*, 2262–2280.

Elbert, K. E. K. (1991). *Analysis of polyethylene in total joint replacement* (Unpublished Doctoral Dissertion). Ithaca, NY: Cornell University.

Elkins, A., Derrick, D., Nunamaker, J., & Burgoon, J. (2017). *AVATAR—Automated virtual agent for truth assessments in real-time*. Tucson, AZ: University of Arizona/Homeland Security University Programs.

Enoka, R. M. (1988). *Neuromechanical basis of kinesiology*. Champaign, IL: Human Kinetics.

Esser, M., Hedden, S., Kanny, D., Brewer, R., Gfoerer, J., & Naimi, T. (2014). Prevalence of alcohol dependence among US adult drinkers 2009–2011. *Preventing Chronic Disease, 11*, 140329.

Estill, C. F., & Kroemer, K. H. E. (1998). Evaluation of supermarket bagging using a wrist motion monitor. *Human Factors, 40*, 624–632.

Ettema, A. M., Zhao, C., Amadio, P. C., O'Byrne, M. M., & An, K. N. (2007). Gliding characteristics of flexor tendon and teno-synovium in carpal tunnel syndrome: A pilot study. *Clinical Anatomy, 20*(3), 292–299.

Ewing, J. (1984). Detecting alcoholism: The CAGE questionnaire. *Journal of the American Medical Association, 252*(14), 1905–1907.

Fagarasanu, M., & Kumar, S. (2004). Hand strength. In S. Kumar (Ed.), *Muscle strength*. Boca Raton, FL: CRC Press (Chapter 10).

Fahrini, W. H. (1975). Conservative treatment of lumbar disc degeneration our primary responsibility. *Orthopedic Clinics of North America, 6*, 93–103.

Fallon, A. E., & Rozin, P. (1985). Sex differences in perceptions of desirable body shape. *Journal of Abnormal Psychology, 94*, 102–105.

Fallon, E. F., Dillon, A., Sweeney, M., & Herring, V. (1991). An Investigation of the concept of designer style and its implications for the design of CAD man–machine interfaces. In W. Karwowski & J. W. Yates (Eds.), *Advances in industrial ergonomics and safety III* (pp. 873–880). London, UK: Taylor & Francis.

III. FURTHER INFORMATION

Farris, B. A., Landwehr, H. R., Fernandez, J. E., & Agarwal, R. K. (1998). Physiological evaluation of mouse pad placement. In S. Kumar (Ed.), *Advances in occupational ergonomics and safety* (pp. 487–490). Amsterdam, The Netherlands: IOS Press.

Fast, A., Shapiro, M. D., & Edmond, J. (1987). Low back pain in pregnancy. *Spine, 12*, 368–371.

Faste, R. A. (1977). New system propels design for the handicapped. *Industrial Design, 2*, 51–55.

Feathers, D., D'Souza, C., & Paquet, V. (2015). Anthropometry in ergonomic design. In J. R. Wilson & S. Sharples (Eds.), *Evaluation of human work* (4th ed., pp. 725–749). Boca Raton, FL: CRC Press (Chapter 27).

Fernandez, J. E., Ayoub, M. M., & Smith, J. L. (1991). Psychophysical lifting capacity over extended periods. *Ergonomics, 34*, 23–32.

Fernie, G. (1997). Assistive devices. In A. D. Fisk & W. A. Rogers (Eds.), *Handbook of human factors and the older adult* (pp. 289–310). San Diego, CA: Academic Press (Chapter 12).

Fernstroem, E., & Ericson, M. O. (1997). Computer mouse or trackpoint—Effects on muscular load and operator experience. *Applied Ergonomics, 28*, 347–354.

Finomore, V. S., Shaw, T. H., Warm, J. S., Matthews, G., & Boles, D. B. (2013). Viewing the workload of vigilance through the lenses of the NASA-TLX and the MRQ. *Human Factors, 55*, 1044–1063.

Fisher, W., & Tarbutt, V. (1988). Some issues in collecting data on working postures. *Proceedings of the Human Factors Society 32nd annual meeting* (pp. 627-631). Santa Monica, CA: Human Factors Society.

Fisk, A. D. (1999). Human factors and the older adult. *Ergonomics in Design, 7*, 8–13.

Fisk, A. D. & Rogers, W. A. (Eds.). (1997). *Handbook of human factors and the older adult*. San Diego, CA: Academic Press.

Fisk, A. D., Rogers, W. A., Charness, N., Czaja, S. J. & Sharit, J. (Eds.). (2009). *Designing for older adults* (2nd ed.). Boca Raton, FL: CRC Press.

Fitts, P. M. (1954). The information capacity of the human motor system in controlling the amplitude of movement. *Journal of Experimental Psychology, 47*(6), 381–391.

Flach, J. M. (1989). An ecological alternative to egg-sucking. *Human Factors Society Bulletin, 32*, 4–6.

Flach, J. M., & Dominguez, C. O. (1995). USE - Centered Design: Integrating the User, Instrument, and Goal. *Ergonomics in Design, 3*, 19–24.

Flaherty, A. W. (2005). Frontotemporal and dopaminergic control of idea generation and creative drive. *The Journal of Comparative Neurology, 493*(1), 147–153.

Fletcher, H., & Munson, W. A. (1933). Loudness, its definition, measurement and calculation. *Journal of the Acoustic Society of America, 5*, 82–108.

Flier, J. F., & Maratos-Flier, E. (2007). What fuels fat. *Scientific American, 297*(3), 72–81.

Floyd, W. F., & Welford, A. T. (1954). *Human factors in equipment design*. London, UK: Taylor & Francis.

Fluegel, B., Greil, H., & Sommer, K. (1986). *Anthropologischer Atlas*. Berlin, Germany: Tribuene.

Folkard, S. & Monk, T. H. (Eds.). (1985). *Hours of work*. Chichester, UK: Wiley.

Folkard, S., & Tucker, P. (2003). Shift work, safety and productivity. *Occupational Medicine, 53*, 95–101.

Folkard, S., Lombardi, D. A., & Tucker, P. T. (2005). Shiftwork, sleepiness and sleep. *Industrial Health, 43*, 20–23.

Folkman, S., & Lazarus, R. S. (1988). The relationship between coping and emotion: Implications for theory and research. *Social Science and Medicine, 26*, 309–317.

Ford, E. (2006). Lie detection: Historical, neuropsychiatric, and legal dimensions. *International Journal of Law and Psychiatry, 29*(3), 159–177.

Foster, R., & Kreitzman, L. (2004). Rhythms of life. *The biological clocks that control the daily lives of every living thing*. London, UK: Profile.

Foster, G. D., Wadden, T. A., & Vogt, R. A. (1997). Body image in obese women before, during, and after weight loss treatment. *Health Psychology, 16*, 226–229.

Fox, J. G. (1983). Industrial music. In D. J. Oborne & M. M. Gruneberg (Eds.), *The physical environment at work* (pp. 221–226). New York, NY: Wiley.

Fransson-Hall, C., & Kilbom, A. (1993). Sensitivity of the hand to surface pressure. *Applied Ergonomics, 24*, 181–189.

Fransson-Hall, C., Gloria, R., Kilbom, A., Winkel, J., Karlqvist, L., & Wiktorin, C. (1995). A portable ergonomic observation method (PEO) for computerized on-line recording of postures and manual handling. *Applied Ergonomics, 26*, 93–100.

Fraser, T. M. (1980). *Ergonomic principles in the design of hand tools*. Geneva, Switzerland: International labour office (Occupational Safety and Health Series, No. 44).

Frazer, L. (1991). Sex in space. *Ad Astra, 3*, 42–45.

Fredericks, T. K., Gunn, E., Rozek, G., & Beert, T. (1998). Maximum acceptable weight of lift for an asymmetrical lowering task commonly observed in the parcel delivery industry. In S. Kumar (Ed.), *Advances in industrial ergonomics and safety* (pp. 361–364). Amsterdam, The Netherlands: IOS Press.

III. FURTHER INFORMATION

Freivalds, A. (1989). Understanding and preventing back trauma: Comparison of U.S. and European approaches. In K. H. E. Kroemer, J. D. McGlothlin & T. G. Bobick (Eds.), *Manual material handling: Understanding and preventing back trauma* (pp. 55–63). Akron, OH: American Industrial Hygiene Association.

Freivalds, A. (1999). Ergonomics of hand controls. In W. Karwowski & W. S. Marras (Eds.), *The occupational ergonomics handbook* (pp. 461–478). Boca Raton, FL: CRC Press (Chapter 27).

Freivalds, A. (2006). Upper extremity analysis of the wrist. In W. S. Marras & W. Karwowski (Eds.), *The occupational ergonomics handbook, fundamentals and assessment tools for occupational ergonomics* (2nd ed.). Boca Raton, FL: CRC Press (Chapter 45).

Freivalds, A. (2011). *Biomechanics of the upper limbs: Mechanics, Modeling and musculoskeletal injuries.* (2nd ed.). Boca Raton, FL: CRC Press.

Fries, R. C. (2012). *Reliable design of medical devices.* Boca Raton, FL: CRC Press.

Friis, R., & Sellers, T. (1999). *Epidemiology for public health practice* (2nd ed.). Gaithersburg, MD: Aspen Publishers.

Froeberg, J. E. (1985). Sleep deprivation and prolonged working hours. In S. Folkard & T. H. Monk (Eds.), *Hours of work* (pp. 67–76). Chichester, UK: Wiley (Chapter 6).

Froufe, T., Ferreira, F., & Rebelo, F. (2002). Collection of anthropometric data from primary schoolchildren. *Proceedings of CybErg 2002, the 3rd International cyberspace conference on ergonomics* (pp. 166–171). London, UK: International Ergonomics Association Press.

Fruitiger, A. (1989). *Signs and symbols.* New York, NY: Van Nostrand Reinhold.

Fry, H. J. (1986a). Overuse syndrome in musicians: Prevention and management. *Lancet, 2(8509),* 728–731.

Fry, H. J. (1986b). Overuse syndrome in musicians—100 years ago. An historical review. *The Medical journal of Australia,* 145(11–12), 620–625.

Fukuda, S. (2016). *Emotional engineering* (4). Heidelberg, Germany: Springer.

Fullenkamp, A. M., Robinette, K. M., & Daanen, H. A. M. (2008). *Gender differences in NATO anthropometry and the implication for protective equipment.* Wright-Patterson AFB, OH: Air Force Research Laboratory Human Effectiveness Directorate, Biosciences and Protection Division, Biomechanics Branch (AFRL-RH-WP-JA-2008-0014).

Gagnon, M., & Smyth, G. (1992). Biomechanical exploration of dynamic modes of lifting. *Ergonomics, 35,* 329–345.

Gagnon, M. (1997). Box tilt and knee motions in manual lifting: Two differential factors in expert and novice workers. *Clinical Biomechanics, 12,* 418–428.

Gagnon, M., Plamondon, A., & Gravel, D. (1995). Effects of symmetry and load absorption of a falling load on 3D trunk muscular movements. *Ergonomics, 38,* 1156–1171.

Gale, A. (Ed.). (1998). *The polygraph test: Lies, truth, and science.* London, UK: Sage.

Gallagher, S. (1999). Ergonomics issues in mining. In W. Karwowski & W. S. Marras (Eds.), *The occupational ergonomics handbook* (pp. 1893–1915). Boca Raton, FL: CRC Press (Chapter 106).

Gallagher, S., & Moore, J. S. (1999). Worker strength evaluation: Job design and worker selection. In W. Karwowski & W. S. Marras (Eds.), *The occupational ergonomics handbook* (pp. 371–386). Boca Raton, FL: CRC Press (Chapter 21).

Gallagher, S., Marras, W. S., & Bobick, T. G. (1988). Lifting in stooped and kneeling postures: Effects on lifting capacity, metabolic costs, and electromyography of eight trunk muscles. *International Journal of Industrial Ergonomics, 3,* 65–76.

Gallagher, S., Moore, J. S., & Stobbe, T. J. (2004). Isometric, isoinertial and psychophysical strength testing: Devices and protocols. In S. Kumar (Ed.), *Muscle strength.* Boca Raton, FL: CRC Press (Chapter 8).

Gallwey, T. J., & Fitzgibbon, M. J. (1991). Some anthropometric measures on an Irish population. *Applied Ergonomics, 22,* 9–12.

Garcia, D. T., Wong, S. L., Fernandez, J. E., & Agarwal, R. K. (1998). The effect of arm supports on muscle activity in shoulder and neck muscles. In S. Kumar (Ed.), *Advances in occupational ergonomics and safety* (pp. 483–486). Amsterdam, The Netherlands: IOS Press.

Gardner-Bonneau, D. J. (Ed.). (1990). Assisting people with functional impairments. *Special Issue of Human Factors, 32,* 379–475.

Gardner-Bonneau, D., & Gosbee, J. (1997). Health care and rehabilitation. In A. D. Fisk & W. A. Rogers (Eds.), *Handbook of human factors and the older adult* (pp. 231–255). San Diego, CA: Academic Press (Chapter 10).

Garg, A., & Ayoub, M. M. (1980). What criteria exist for determining how much load can be safely lifted? *Human Factors, 22,* 475–486.

Garg, A., & Marras, W. S. (2014). Epidemiological studies of workplace musculoskeletal disorders. *Special Issue of Human Factors, 56,* 5.

Garrett, J. W., & Kennedy, K. W. (1971). *A collation of anthropometry.* Wright-Patterson Air Force Base, OH: Aerospace Medical Research Laboratories (AMRL-TR-68-1).

Gavande, A. (2008). The itch. *The New Yorker,* 58–65.

Gaver, W. W. (1997). Auditory interfaces. In M. Helander, T. K. Landauer & P. Prabhu (Eds.), *Handbook of human–computer interaction* (2nd ed., pp. 1003–1041). Amsterdam, The Netherlands: Elsevier (Chapter 42).

III. FURTHER INFORMATION

Gawande, A. (1998). The pain perplex. *The New Yorker*, 86–94.

Gawande, A. (2015). Overkill. *The New Yorker*, 42–53.

Gawron, V. J. (1997). High-g environments and the pilot. *Ergonomics in Design*, 5, 18–23.

Genaidy, A. M., Asfour, S. S., Mital, A., & Waly, S. M. (1990a). Psychophysical models for manual lifting tasks. *Applied Ergonomics*, 21, 295–303.

Genaidy, A. M., Gupta, T., & Alshedi, A. (1990b). Improving human capabilities for combined manual handling tasks through a short and intensive physical training program. *American Industrial Hygiene Association Journal*, 51, 610–614.

Genaidy, A. M., Al-Shedi, A. A., & Karwowski, W. (1994). Postural stress in industry. *Applied Ergonomics*, 25, 77–87.

Genaidy, A., Barkawi, H., & Christensen, D. (1995). Ranking of static non-neutral postures around the joints of the upper extremity and the spine. *Ergonomics*, 38, 1851–1858.

Gerard, M. J., Armstrong, T. J., Foulke, J. A., & Martin, B. J. (1996). Effects of key stiffness on force and the development of fatigue while typing. *American Industrial Hygiene Association Journal*, 57, 849–854.

Gerr, F., Monteilh, C. P., & Marcus, M. (2006). Keyboard use and musculoskeletal outcomes among computer users. *Journal of Occupational Rehabilitation*, 16(3), 265–277.

Giacomin, J., & Quattrocolo, S. (1997). An analysis of human comfort when entering and exiting the rear seat of an automobile. *Applied Ergonomics*, 28, 397–406.

Giarmatzis, G., Jonkers, I., Wesseling, M., Van Rossom, S., & Verschueren, S. (2015). Loading of hip measured by hip contact forces at different speeds of walking and running. *Journal of Bone and Mineral Research*, 30(8), 1431–1440.

Gibbons, J. D. (1997). *Nonparametric methods for quantitative analysis* (3rd ed.). Columbus, OH: American Sciences.

Gibson, J. J. (1966). *The senses considered as perceptual systems*. Boston, MA: Houghton Mifflin.

Gil, H. J. C., & Tunes, D. B. (1977). Posture recording: A model for sitting posture. *Applied Ergonomics*, 20, 53–57.

Gilad, I., & Pollatschek, M. A. (1986). Layout simulation for keyboards. *Behaviour and Information Technology*, 5, 273–281.

Gilliland, K., & Schlegel, R. E. (1994). Tactile stimulation of the human head for information display. *Human Factors*, 36, 700–717.

Gils, H. J., & Tunes, E. (1989). Posture recording: A model for sitting posture. *Applied Ergonomics*, 20(1), 53–57.

Godin, G., & Gionet, N. (1991). Determinants of an intention to exercise of electric power commission's employees. *Ergonomics*, 34, 1221–1230.

Goggins, R. W., Spielholz, P., & Nothstein, G. L. (2008). Estimating the effectiveness of ergonomics interventions through case studies: Implications for predictive cost-benefit analysis. *Journal of Safety Research*, 39, 339–344.

Gold, C., Koerber, M., Lechner, D., & Bengler, K. (2016). Taking over control from highly automated vehicles in complex traffic situations. *Human Factors*, 58(4), 642–652.

Goldenberg, M., Ishak, W., & Danovitch, I. (2017). Quality of life and recreational cannabis use. *The American Journal on Addictions*, 26, 8–25.

Goleman, D. (1995). *Emotional intelligence: Why it can matter more than IQ*. New York, NY: Bantam Books.

Goleman, D. (2006). *Emotional intelligence: Why it can matter more than IQ*. New York, NY: Bantam Books.

Goleman, D. (2011). *Leadership: The Power of Emotional Intelligence*. Northampton, MA: More Than Sound LLC.

Goodman, H. J., & Choueka, J. (2005). Biomechanics of the flexor tendons. *Hand Clinics*, 21, 129–149.

Goodman, L. S., de Yang, L., Kelso, B., & Liu, P. (1995). Cardiovascular effects of varying g-suit pressure and coverage during +g_z positive pressure breathing. *Aviation Space and Environmental Medicine*, 66, 829–836.

Gopher, D., & Raij, D. (1988). Typing with a two-hand chord keyboard: Will the QWERTY become obsolete? *IEEE Transactions on Systems, Man, and Cybernetics*, 18, 601–609.

Gordon, C. C., Churchill, T., Clauser, C. E., Bradtmiller, B., McConville, J. T., Tebbetts, I., et al. (1989). *1988 anthropometric survey of U.S. Army personnel: Summary statistics interim report*. Natick, MA: United States Army Natick Research, Development and Engineering Center (Technical Report NATICK/TR-89-027).

Gordon, C. C., Corner, B. D., & Brantley, J. D. (1997). *Defining extreme sizes and shapes for body armor and load-bearing systems design: Multivariate analysis of U.S. Army Torso Dimensions*. Natick, MA: U.S. Army Natick RD&E Center (TR-97/012).

Gordon, C., Blackwell, C. L., Bradtmiller, B., Parham, J. L., Barrientos, P., Paquette, S. P., et al. (2014). *2012 anthropometric survey of US army personnel: Methods and summary statistics*. Natick, MA: US Army Natick Soldier Research, Development and Engineering Center (TR-15/007).

Gould, S. J. (1981). *The mismeasure of man*. New York, NY: Norton.

Gould, S. J. (1988). A novel notion of Neanderthal. *Natural History*, 97(6), 16–21.

Gowers W.R. (1886). *Manual: Diseases of the nervous system*. Philadelphia, PA: Blakiston (1888).

Graeber, R. C. (1988). Aircrew fatigue and circadian rhythmicity. In E. L. Wiener & D. C. Nagel (Eds.), *Human factors in aviation* (pp. 305–344). San Diego, CA: Academic Press (Chapter 10).

Grafman, J. (2000). Conceptualizing functional neuroplasticity. *Journal of Communication Disorders, 33*(4), 345–356.

Grandjean, E. (1963). *Physiological design of work.* Thun, Switzerland: Ott (in German).

Grandjean, E. (1987). *Ergonomics in computerized offices.* London, UK: Taylor & Francis.

Grandjean, E. (Ed.). (1969). *Sitting posture.* London, UK: Taylor & Francis.

Grandjean, E., Huenting, W., & Nishiyama, K. (1984). Preferred VDT workstation settings body postures and physical impairments. *Applied Ergonomics, 15,* 99–104.

Grant, A. (1990). Homo-quintadus computers and rooms (repetitive orcular orthopedic motion stress). *Optometry and Vision Science, 67,* 297–305.

Green, B. F., & Anderson, L. K. (1955). The tactual identification of shapes for coding switch handles. *Journal of Applied Psychology, 39,* 219–226.

Greenberg, J. (2010). *Behavior in organizations: Understanding and managing the human side of work* (10th ed.). Upper Saddle River, NJ: Prentice-Hall.

Greene, T. C., & Bell, P. A. (1987). Environment stress. In M. A. Baker (Ed.), *Sex differences in human performance* (pp. 81–106). Chichester, UK: Wiley (Chapter 5).

Greenstein, J. S. (1997). Pointing devices. In M. Helander, T. K. Landauer & P. Prabhu (Eds.), *Handbook of human–computer interaction* (2nd ed., pp. 1317–1348). Amsterdam, The Netherlands: Elsevier (Chapter 55).

Greenwood, K. M. (1991). Psychometric properties of the diurnal type scale of torsvall and akerstedt (1980). *Ergonomics, 34,* 435–443.

Greiner, T. M. (1991). *Hand anthropometry of U.S. Army personnel.* Natick, MA: U.S. Army Natick Research, Development and Engineering Center (Technical report TR-92/011).

Greiner, T. M., & Gordon, C. C. (1990). *An assessment of long-term changes in anthropometric dimensions: Secular trends of U.S. Army Males.* Natick, MA: U.S. Army Natick Research, Development and Engineering Center (TR-91/006).

Grieco, A. (1986). Sitting posture: An old problem and a new one. *Ergonomics, 29,* 345–362.

Griffin, M. J. (1990). *Handbook of human vibration.* San Diego, CA: Academic Press.

Griffin, M. J. (1997). Vibration and motion. In G. Salvendy (Ed.), *Handbook of human factors and ergonomics* (2nd ed., pp. 828–857). New York, NY: Wiley (Chapter 25).

Griffin, M. J. (1998). A comparison of standardized methods for predicting the hazards of whole-body vibration and repeated shocks. *Journal of Sound and Vibration, 215*(4), 883–914.

Griffin, M. J. (2004). Minimum health and safety requirements for workers exposed to hand-transmitted vibration and whole-body vibration in the European Union: A review. *Occupational and Environmental Medicine, 61,* 387–397.

Griffin, M. J. (2012). *Handbook of human vibration* (3rd ed.). London, UK: Academic Press (1st ed. 1990, 2nd ed. 1996).

Griffin, M. D., & French, J. R. (1991). *Space vehicle design.* Washington, DC: American Institute of Aeronautics and Astronautics, Inc.

Groenquist, R., & Hirvonen, M. (1994). Pedestrian safety on icy surfaces: Anti-slip properties of footwear. In F. Aghazade (Ed.), *Advances in industrial ergonomics and safety VI* (pp. 315–322). London, UK: Taylor & Francis.

Grubin, D. (2010). The polygraph and forensic psychiatry. *Journal of the American Academy of Psychiatry and the Law, 38*(4), 446–451.

Gudjonsson, G. H. (1988). How to defeat the polygraph tests. In A. Gale (Ed.), *The polygraph test: Lies, truth, and science* (pp. 126–136). London, UK: Sage.

Guignard, J. C. (1985). Vibration. In L. V. Cralley & L. J. Cralley (Eds.), *Patty's industrial hygiene and toxicology* (pp. 635–724). New York, NY: Wiley (Chapter 15).

Guyll, M., & Contrada, R. J. (1998). Trait hostility and ambulatory cardiovascular activity: Responses to social interaction. *Health Psychology, 17,* 30–39.

Guyton, A. C. (1979). *Physiology of the human body* (5th ed.). Philadelphia, PA: Saunders.

Gyi, D. E., & Porter, J. M. (1999). Interface pressure and the prediction of car sear comfort. *Applied Ergonomics, 30,* 99–107.

Haas, E. C., & Casali, J. G. (1995). Perceived urgency of and response time to multi-tone and frequency-modulated warning signals in broadband noise. *Ergonomics, 38,* 2313–2326.

Hackett, T. P., Rosenbaum, J. F., & Tesar, G. E. (1988). Emotion, psychiactric disorders, and the heart. In E. Braunwald (Ed.), *Heart disease—A textbook of cardiovascular medicine* (pp. 1883–1900). Philadelphia, PA: Saunders.

Hackman, J. R., & Oldham, G. R. (1980). *Work redesign.* Reading, MA: Addison-Wesle.

Hadler, N. M. (1997). Back pain in the workplace. *Spine, 22,* 935–940.

Hagberg, M., & Rempel, D. (1997). Work-related disorders and the operation of computer VDT's. In M. Helander, T. K. Landauer & P. Prabhu (Eds.), *Handbook of human–computer interaction* (2nd ed., pp. 1415–1429). Amsterdam, The Netherlands: Elsevier.

III. FURTHER INFORMATION

Hahn, H. A., & Price, D. L. (1994). Assessment of the relative effects of alcohol on different types of work behavior. *Ergonomics, 37*, 435–448.

Haisman, M. F. (1988). Determinants of load carrying ability. *Applied Ergonomics, 19*, 111–121.

Hall, H. W. (1973). "Clean" versus "Dirty" lifting, an academic subject for youth. *American Society of Safety Engineers Journal, 18*, 20–25.

Hall, J. E. (2016). *Guyton and Hall textbook of medical physiology* (13th ed.). Amsterdam, The Netherlands: Elsevier.

Hall, K. D., Heymsfield, S. B., Kemnitz, J. W., Klein, S., Schoeller, D. A., & Speakman, J. R. (2012). Energy balance and its components: Implications for body weight regulation. *The American Journal of Clinical Nutrition, 95*(4), 989–994.

Hamill, J., & Hardin, E. C. (1997). Biomechanics. In S. R. DiNardi (Ed.), *The occupational environment—Its evaluation and control* (26). Fairfax, VA: American Industrial Hygiene Association Page 699 in chapter.

Hamill, P. V. V., Drizd, T. A., Johnson, C. L., Reed, R. B., Roche, A. F., & Moore, W. M. (1979). Physical growth: National Center for Health Statistics Percentiles. *American Journal of Clinical Nutrition, 32*, 607–629.

Hammer, W., & Price, D. (2000). *Occupational safety management and engineering* (5th ed.). Englewood Cliffs, NJ: Prentice-Hall.

Han, S. H., Williges, B. H., & Williges, R. C. (1997). A Paradigm for sequential experimentation. *Ergonomics, 40*, 737–760.

Hancock, P. (1999). *Human performance and ergonomics: Perceptual and cognitive principles*. Cambridge, MA: Academic Press.

Hancock, P. A. (2017). Whither workload? Mapping a path for its future development. In L. Longo & M. C. Leva (Eds.), *Human mental workload: Models and applications* (pp. 3–17). Cham, Switzerland: Springer.

Hancock, P. A. & Meshkati, N. (Eds.). (1988). *Human mental workload*. Amsterdam, The Netherlands: Elsevier.

Hangartner, M. (1987). Standardization in olfactometry with respect to odor pollution control, and assessment of odor annoyance in the community. *80th annual meeting of the APCA*. New York, NY: Air Pollution Control Association. Presentations 87-75A.1 and 87-75B.3.

Hansen, L., Winkel, J., & Jorgensen, K. (1998). Significance of mat and shoe softness during prolonged work in upright position: Based on measurement of low back muscle EMG, foot volume changes. Discomfort and ground force reactions. *Applied Ergonomics, 29*, 217–224.

Harrison, A. A., Clearwater, Y. A., & McKay, C. P. (1991). *From antarctica to outer space: Life in isolation and confinement*. New York, NY: Springer.

Hart, S. G., & Staveland, I. E. (1988). Development of NASA-TLX (Task Load Index): Results of experimental and theoretical research. In P. A. Hancock & N. Meshkati (Eds.), *Human mental workload* (pp. 185–218). Amsterdam, The Netherlands: Elsevier.

Harvey, P., Stoner, J., Hochwarter, W., & Kacmar C (2007). Coping with abusive supervision: The neutralizing effects of ingratiation and positive affect on negative employee outcomes. *The Leadership Quarterly, 18*(3), 264–280.

Harvey, R., & Peper, E. (1997). Surface electromyography and mouse use position. *Ergonomics, 40*, 781–789.

Hashemi, L., & Dempsey, P. G. (1997). Body parts and nature of injuries associated with manual materials handling workers' compensation claims. *Proceedings of the Human Factors and Ergonomics Society 41st annual meeting* (pp. 619-623). Santa Monica, CA: Human Factors and Ergonomics Society.

Havenith, G. (2005). Thermal conditions measurement. In N. Stanton, A. Hedge, K. Brookhuis, E. Salas & H. Hendrick (Eds.), *Handbook of human factors and ergonomics methods*. Boca Raton, FL: CRC Press (Chapter 60).

Hay, J. G. (1973). *The center of gravity of the human body*. Washington, DC: American Association for Health, Physical education, Recreation Kinesiology III (2044).

Hayne, C. R. (1981). Lifting and handling. *Health and Safety at Work, 3*, 18–21.

Hayslip, B., & Panek, P. (1989). *Adult development and aging*. New York, NY: Harper & Row.

Hayslip, B., Hicks Patrick, J., & Panek, P. E. (2011). *Adult development and aging* (5th ed.). Malabar, FL: Krieger.

Heacock, H., Koehoorn, M., & Tan, J. (1997). Applying epidemiological principles to ergonomics: A checklist for incorporation sound design and interpretation of studies. *Applied Ergonomics, 26*, 165–172.

Hedge, A. (2017). *Ergonomic workplace design for health, wellness, and productivity*. Boca Raton, FL: CRC Press.

Hedge, A., & James, T. (2012). Ergonomic issues of computer use in a major healthcare system. In *Proceedings of the fourth international conference on applied human factors and ergonomics* (pp. 2630–2639). San Francisco, CA: Applied Human Factors and Ergonomics.

Heidner, F. (1915). Type-writing machine. US Patent 1,138,474. Washington, DC: US Patent and Trademark Office.

Helander, M. G. (1981). *Human factors/ergonomics for building and construction*. New York, NY: Wiley.

Helander, M. G. (1982). *Ergonomic design of office environments for visual display terminals*. Blacksburg, VA: Virginia Tech (VPI & SU) (Report for NIOSH).

Helander, M. G. (Ed.). (1988). *Handbook of human–computer interaction*. Amsterdam, The Netherlands: North-Holland.

III. FURTHER INFORMATION

Helander, M. G. (2003). Forget about ergonomics in chair design? Focus on aesthetics and comfort! *Ergonomics*, 46, 1306–1319.

Helander, M. G. (2006). *A guide to human factors and ergonomics* (2nd ed.). Boca Raton, FL: CRC Press.

Helander, M. G., & Khalid, H. M. (2006). Affective and pleasurable design. In G. Salvendy (Ed.), *Handbook of human factors and ergonomics* (3rd ed., pp. 543–572). Hoboken, NJ: Wiley (Chapter 21).

Helander, M., & Nagamachi, N. (Eds.). (1992). *Design for manufacturabilty: A systems approach to concurrent engineering and ergonomics.* London, UK: Taylor & Francis.

Helander, M. G., & Zhang, L. (1997). Field studies of comfort and discomfort in sitting. *Ergonomics*, 401, 895–915.

Helmers, K. F., Posluszny, D. M., & Krantz, D. S. (1994). Associations of hostility and coronary artery disease: A review of studies. In A. W. Siegman & T. W. Smith (Eds.), *Anger, hostility and the heart* (pp. 67–96). Hillsdale, NJ: Erlbaum.

Hendrick, H. W. (1996). The ergonomics of economics is the economics of ergonomics. In *Proceedings of the Human Factors and Ergonomics Society 40th annual meeting* (pp. 1–10). Santa Monica, CA: Human Factors and Ergonomics Society.

Hendrick, H. W., & Kleiner, B. M. (1999). *Macroergonomics: An introduction to work system analysis and design.* Santa Monica, CA: Human Factors and Ergonomics Society.

Hendrick, H. W., & Kleiner, B. M. (2002). *Macroergonomics: Theory, methods, and applications* (2nd ed.). Boca Raton, FL: CRC Press (2005).

Herkimer County Historical Society. (1923). *The story of the typewriter.* Published in commemoration of the fiftieth anniversary of the invention of the writing machine (pp. 1873–1923). Herkimer, NY: Herkimer County Historical Society.

Hertzberg, H. T. E. (1968). The conference on standardization of anthropometric techniques and terminology. *American Journal of Physical Anthropology*, 28, 1–16.

Herzberg, F. (1966). *Work and the nature of man.* New York, NY: Thomas.

Herzberg, F. (1968). One more time: How do you motivate employees? *Harvard business review*, Sept–Oct 1987 issue, 5–16. (Reprint 87507).

Herzog, W. (2008). Determinants of muscle strength. In S. Kumar (Ed.), *Biomechanics in ergonomics* (2nd ed.). Boca Raton, FL: CRC Press (Chapter 7).

Heuer, H., & Owens, D. A. (1989). Vertical gaze direction and the resting posture of the eyes. *Perception*, 18, 353–377.

Heuer, H., Bruewer, M., Roemer, T., Kroeger, H., & Knapp, H. (1991). Preferred vertical gaze direction and observation distance. *Ergonomics*, 34(3), 379–392.

Heymsfield, S. B., Allison, D. B., Heshka, S., & Pierson, R. N. (1995). Assessment of human body composition. In D. B. Allison (Ed.), *Handbook of assessment methods for eating behaviors and weight-related problems.* Thousand Oaks, CA: Sage Publications.

Human Factors Ergonomics Society (H.F.E.S.) 300 Committee. (2004). *Guidelines for using anthropometric data in product design.* Santa Monica, CA: Human Factors and Ergonomics Society.

Hidalgo, J., Genaidy, A., Karwowski, W., Christensen, D., Huston, R., & Stambough, J. (1995). A cross-validation of the NIOSH limits for manual lifting. *Ergonomics*, 38, 2455–2464.

Hidalgo, J., Genaidy, A., Karwowski, W., Christensen, D., Huston, R., & Stambough, J. (1997). A comprehensive lifting model: Beyond the NIOSH lifting equation. *Ergonomics*, 40, 916–927.

Hill, S. G., & Kroemer, K. H. E. (1986). Preferred declination and the line of sight. *Human Factors*, 28, 127–134.

Hill, S. G., Iavecchia, H. P., Byers, J. C., Bittner, A. C., Zaklad, A. L., & Christ, R. E. (1992). Comparison of four subjective workload rating scales. *Human Factors*, 34, 429–439.

Himmelskin, J. S., & Andersson, G. B. J. (1988). Low back pain: Risk evaluation preplacement screening. *Journal of Occupational Medicine*, 3, 255–269.

Hinkelmann, K., & Kempthorne, O. (1994). *Design and analysis of experiments, vol. 1: Introduction to experimental design.* New York, NY: Wiley.

Hinkelmann, K., & Kempthorne, O. (2005). *Design and analysis of experiments, vol. 2: Advanced experimental design.* New York, NY: Wiley-Interscience.

Hirsch, R. S. (1970). Effect of standard vs. alphabetical formats in typing performance. *Journal of Applied Psychology*, 54, 484–490.

Hocking, B. (1987). Epidemiological aspects of "Repetition Strain Injury" in telecom Australia. *The Medical Journal of Australia*, 147, 218–222 (Sept.).

Hoffman, R. G., & Pozos, R. S. (1989). Experimental hypothermia and cold perception. *Aviation Space and Environmental Medicine*, 60, 964–969.

Holewijn, M., & Lotens, W. A. (1993). The influence of backpack design on physical performance. *Ergonomics*, 35, 149–157.

Holland, D. A. (1991). *Systems and human factors concerns for long-duration space flight.* (Unpublished Master's Thesis). Blacksburg, VA: Virginia Tech.

Holmes, T. H., & Rahe, R. H. (1967). The social readjustment rating scale. *Journal of Psychosomatic Research*, 11, 213–218.

Holzmann, P. (1981). ARBAN—A new method of analysis of ergonomic effort. *Applied Ergonomics*, 13, 82–86.

III. FURTHER INFORMATION

Hood, D. C., & Finkelstein, M. A. (1986). Sensitivity to light. In K. R. Boff, L. Kaufman & J. P. Thomas (Eds.), *Handbook of perception and human performance* (pp. 5.1–5.66). New York, NY: Wiley.

Hornberger, S., Knauth, P., Costa, G. & Folkard, S. (Eds.). (2000). *Shiftwork in the 21st century*. Frankfurt, Germany: Lang.

Horne, J. A. (1985). Sleep loss: Underlying mechanisms and tiredness. In S. Folkard & T. H. Monk (Eds.), *Hours of work* (pp. 53–65). Chichester, UK: Wiley.

Horne, J. A. (1988). *Why we sleep—The functions of sleep in humans and other mammals*. Oxford, UK: Oxford University Press.

Horne, J. A. (2006). *Sleepfaring: A journey through the science of sleep*. Oxford, UK: Oxford University Press.

Horowitz, A. (2013). *On looking: Eleven walks with expert eyes*. New York, NY: Scribner.

Horscroft, J. A., Kotwica, A. O., Laner, V., & West, J. A. (2017). Metabolic basis to Sherpa altitude adaptation. *Proceedings of the National Academy of Sciences of the United States of America*, *114*(24), 6382–6387.

Hotzman, J., Gordon, C. C., Bradtmiller, B., Corner, B. D., Mucher, M., Kristensen, S., et al. (2011). *Measurer's handbook, US Army and Marine Corps anthropometric surveys, 2010–2011*. Natick, MA: Army Natick Soldier Research, Development and Engineering Center (TR-11/017).

House, L. H., & Pansky, B. (1967). *A functional approach to neuroanatomy* (2nd ed.). New York, NY: McGraw-Hill.

Houy, D. A. (1983). Range of joint motion in college males. *Proceedings of the Human Factors Society 27th annual meeting* (pp. 374–378). Santa Monica, CA: Human Factors Society.

Howard, I. P. (1986). The perception of posture, self-motion, and the visual vertical. In K. R. Boff, L. Kaufman & P. Thomas (Eds.), *Handbook of perception and human performance: Vol I. Sensory processes and perception* (pp. 18-1–18-62). New York, NY: Wiley.

Howard-Jones, P. (2014). Neuroscience and education: Myths and messages. *Nature Reviews Neuroscience*, *15*, 817–824.

Howarth, P. A. (2005). Assessment of the visual environment. In J. R. Wilson & N. Corlett (Eds.), *Evaluation of human work* (3rd ed.). London, UK: Taylor & Francis (Chapter 24).

Hsiao, H., Long, D., & Snyder, K. (2002). Anthropometric differences among occupational groups. *Ergonomics*, *45*, 136–152.

Hsiao, H., Whitestone, J., Kau, T. Y., & Hildreth, B. (2015). Firefighter hand anthropometry and structural glove sizing. *Human Factors*, *58*(8), 1359–1377.

Hudgens, G. A., Fatkin, L. T., Billingsley, P. A., & Mazuraczak, J. (1988). Hand Steadiness: Effects of sex, menstrual phase, oral contraceptives, practice, and handgun weight. *Human Factors*, *30*, 51–60.

Huenting, W., Granjean, E., & Maeda, K. (1980). Constrained postures in accounting machine operators. *Applied Ergonomics*, *14*, 145–149.

Huffman, J. A., & Lehman, K. R. (1997). Pointing devices in the retail environment. *Proceedings of the Human Factors Ergonomics Society 41st annual meeting* (pp. 415–419). Santa Monica, CA: Human Factors and Ergonomics Society.

Hughes, R. E., & An, K. N. (2008). Biomechanics models of the hand, wrist, and elbow in ergonomics. In S. S. Kumar (Ed.), *Biomechanics in ergonomics* (2nd ed.). Boca Raton, FL: CRC Press (Chapter 14).

Hughes, R. E. (1995). Choice of optimization models for predicting spinal forces in a three-dimensional analysis of heavy work. *Ergonomics*, *38*, 2476–2484.

Hukins, D. W. L., & Meakin, J. R. (2000). Relationship between structure and mechanical function of the tissues of the intervertebral joint. *American Zoology*, *40*(1), 42–52.

Human Factors and Ergonomics Society. (1998). Human Factors and Ergonomics Society strategic plan. *Human Factors and Ergonomics Society directory yearbook*. Santa Monica, CA: Human Factors, Ergonomics Society 388, (1998–1999).

Hunt, P., & Craig, D. R. (1954). *The relative discriminability of thirty-one differently shaped knobs (WADC-TR-54-108)*. Wright-Patterson AFB, OH: Wright Air Development Center.

Hutchinson, G. (1995). Taking the guesswork out of medical device design. *Ergonomics in design*, *3*, 21–26.

Ignazi, G., Mollard, R., & Coblentz, A. (1982). Progress and prospect in human biometry. In R. Easterby, K. H. E. Kroemer & D. B. Chaffin DB (Eds.), *Anthropometry and biomechanics—Theory and application* (pp. 71–98). New York, NY: Plenum.

Ignazi, G., Martel, A., Mollard, R., & Coblentz, A. (1996). Anthropometric measurements evolution of a french school children and adolescent population aged four to eighteen. *Proceedings of the 4th Pan Pacific conference on occupational ergonomics* (pp. 111-114). Hsinchu, ROC: Ergonomics Society of Taiwan.

ILO. (1974). *Introduction to work study*. Geneva, CH: International Labour Office.

ILO. (1986). *Introduction to work study* (3rd ed.). Geneva, CH: International Labour Office.

ILO. (1988). *Maximum weights in load lifting and carrying (occupational safety and health series, no. 59)*. Geneva, CH: International Labour Office.

Imran, N. (2008). Hand grasping, finger pinching and squeezing. In S. Kumar (Ed.), *Biomechanics in ergonomics* (2nd ed.). Boca Raton, FL: CRC Press (Chapter 9).

Imrhan, S. N. (1998). Manual torquing: A review of empirical studies. In S. Kumar (Ed.), *Advances in occupational ergonomics and safety* (pp. 439–442). Amsterdam: IOS Press.

III. FURTHER INFORMATION

Imrhan, S. N. (1999). Push–pull force limits. In W. Karwowski & W. S. Marras (Eds.), *The occupational ergonomics handbook* (pp. 407–420). Boca Raton, FL: CRC Press.

Inciardi, J. (1999). *Harm reduction: National and international perspectives.* Thousand Oaks, CA: Sage Publications.

ISO. (1985). *Evaluation of human exposure to whole-body vibration (ISO standard 2631).* Geneva, CH: International Organization for Standardization.

ISO. (1987). *Mechanical vibration and shock: Mechanical transmissibility in the human body in the Z direction (ISO standard 7962).* Geneva, CH: International Organization for Standardization.

ISO. (1989). *Hot environments (ISO standard 7243).* Geneva, CH: International Organization for Standardization.

ISO. (1995). *Ergonomics of the thermal environment—Estimation of the thermal insulation and evaporative resistance of a clothing ensemble (ISO standard 9920).* Geneva, CH: International Organization for Standardization.

ISO. (1997). *Mechanical vibration and shock, biodynamic coordinate systems (ISO standard 8727).* Geneva, CH: International Organization for Standardization.

ISO. (2004). *Mechanical vibration and shock, evaluation of human exposure to whole-body vibration, Part 5: Method for evaluation of vibration containing multiple shocks (ISO standard 2631-5:2004).* Geneva, CH: International Organization for Standardization.

ISO. (2010a). *3D scanning methodologies for internationally compatible anthropometric databases (ISO 20685).* Geneva, CH: International Organization for Standardization.

ISO. (2010b). *Basic human body measurements for technological design (ISO 7250).* Geneva, CH: International Organization for Standardization.

ISO. (2012). *General requirements for establishing anthropometric databases. (ISO 15535).* Geneva, CH: International Organization for Standardization.

ISO. (2016). *Ergonomic principles related to mental work-load. (ISO/DIS 10075).* Geneva, CH: International Organization for Standardization.

Jaeger, M. (1987). Biomechanical human model for analysis and evaluation of the strain in the spinal column while manipulating loads (in German). *Biotechnik Series, 17*(33). Duesseldorf, Germany: VDI Verlag.

Jaeger, M., & Luttmann, A. (1986). Biomechanical model calculations of spinal stress for different working postures in various workload situations. In N. Corlett, J. Wilson & I. Manenica (Eds.), *The ergonomics of working postures: Models, methods and cases* (pp. 144–423). London, UK: Taylor & Francis (Chapter 15).

Janis, I. (1972). *Victims of groupthink: A psychological study of foreign-policy decisions and fiascoes.* Boston, MA: Houghton Mifflin.

Jarrett, A. (Ed.). (1973). *The physiology and pathology of the skin.* London, UK: Academic Press.

Jaschinski-Kruza, W. (1991). Eyestrain in VDU users: Viewing distance and the resting position of ocular muscles. *Human Factors, 33,* 69–83.

Jenkins, J., & Rickards, J. (1997). Can i benefit from the cost of ergonomics? Exploring a pre-intervention methodology. *Proceedings of the 29th annual conference of the Human Factors Association of Canada* (pp. 139–147). Windsor, ON: Human Factors Association of Canada.

Jenkins, W. L. (1953a). *Design factors in knobs and levers for making settings on scales and scopes WADC-TR-53-2.* Wright-Patterson AFB, OH: Aero Medical Laboratory.

Jenkins, W. L. (1953b). *Design factors in knobs and levers for making settings on scales and scopes (WADC-TR-53-2).* Wright-Patterson AFB, OH: Aero Medical Laboratory.

Jenkins, J. P. (Ed.). (1991). *Human performance for long duration space missions (final report, NASA-SSTAC Ad Hoc Committee).* Washington, DC: NASA.

Jensen, M. C., Brant-Zawadzki, M. N., Obuchowski, N., Modic, M. T., Malkasian, D., & Ross, J. S. (1994). Magnetic resonance imaging of the lumbar spine in people without back pain. *New England Journal of Medicine, 331,* 69–73.

Jensen, R. C. (1985). A model of the training process devised from human factors and safety literature. In R. E. Eberts & C. G. Eberts (Eds.), *Trends in ergonomics/human factors II* (pp. 501–509). Amsterdam, The Netherlands: Elsevier.

Johansson, A., Johansson, G., Lundquist, P., Akesson, I., Odenrick, P., & Akelsson, R. (1998). Evaluation of a workplace redesign of a grocery checkout system. *Applied Ergonomics, 29,* 261–266.

Johnson, D. A. (1998a). New stairway—Old problems. *Ergonomics in Design, 6,* 7–10.

Johnson, S. L. (1998b). Selecting computer-based ergo tools. *IIE solutions,* 40–45.

Johnson, S. L., & Lewis, D. M. (1989). A psychophysical study of two-person manual material handling tasks. *Proceedings of the Human Factors Society 33rd annual meeting* (pp. 651-653). Santa Monica, CA: Human Factors Society.

Johnson, L. C., Tepas, D. I., Colquhoun, W. P. & Colligan, M. J. (Eds.). (1981). *Biological rhythms, sleep and shift work.* New York, NY: Spectrum.

Johnson, R. C., Doan, J. B., Stevenson, J. M., & Bryant, J. T. (1998). An analysis of subjective responses to varying a load centre of gravity in a backpack. In S. Kumar (Ed.), *Advances in industrial ergonomics and safety* (pp. 248–251). Amsterdam, The Netherlands: IOS Press.

III. FURTHER INFORMATION

Jones, D. F. (1972). Back injury research. *American Industrial Hygiene Association Journal, 33*, 596–602.

Jones, R. G. (1990). Worker independence and output: The Hawthorne studies reevaluated. *American Sociological Review, 55*, 176–190.

Jones, C., Manning, D. P., & Bruce, M. (1995). Detecting and eliminating slippery footwear. *Ergonomics, 38*, 242–249.

Jorna, P. G. A. M. (1993). Heart rate and workload variations in actual and simulated flight. *Ergonomics, 36*, 1043–1054.

Jorna, G. C., Mohageg, M. F., & Snyder, H. L. (1989). Performance, perceived safety and comfort of the alternating tread stair. *Applied Ergonomics, 20*(1), 26–32.

Juergens, H. W., Aune, I. A., & Pieper, U. (1990). *International data on anthropometry (occupational safety and health series no. 65)*. Geneva, Switzerland: International Labour Office.

Kahn, J. F., & Monod, H. (1989). Fatigue induced by static work. *Ergonomics, 32*, 839–846.

Kahn, R., & Rowe, L. J. (1998). *Successful aging*. New York, NY: Pantheon.

Kahneman, D., & Riis, J. (2005). Living, and thinking about it: Two perspectives on life. In N. Baylis, F. A. Huppert & B. Keverne (Eds.), *The science of well-being* (pp. 285–301). Oxford, UK: Oxford University Press.

Kahneman, D., & Deaton, A. (2010). High income improves evaluation of life but not emotional well-being. *Proceedings of the National Academy of Sciences, 107*, 16489–16493.

Kajimoto, H., Saga, S. & Konyo, M. (Eds.). (2016). *Pervasive haptics*. Heidelberg, Germany: Springer.

Kaleps, I., Clauser, C. E., Young, J. W., Chandler, R. F., Zehner, G. F., & McConville, J. (1984). Investigation into the mass distribution properties of the human body and its segments. *Ergonomics, 27*, 1225–1237.

Kamarck, T., & Jennings, J. R. (1991). Behavioral factors in sudden cardiac death. *Psychological Bulletin, 109*, 42–75.

Kamarck, T. W., Shiffman, S. M., Smithline, L., Goodie, J. L., Paty, J. A., Gnys, M., et al. (1998). Effects of task strain, social conflict, and emotional activation on ambulatory cardiovascular activity: Daily life consequences of recurring stress in multiethnic sample. *Health Psychology, 17*, 17–29.

Kanner, A. D., Coyne, J. C., Schaefer, C., & Lazarus, R. S. (1981). Comparison of two modes of stress measurement: Daily hassles and uplifts vs. major life events. *Journal of Behavioral Medicine, 4*, 1–39.

Kantowitz, B. H., & Sorkin, R. D. (1983). *Human factors: Understanding people-system relationships*. New York, NY: Wiley.

Karasek, R. A., Baker, D., Marxer, F., Ahlborn, A., & Theorell, T. (1981). Job decision latitude, job demands, and cardiovascular disease: A prospective study of Swedish men. *American Journal of Public Health, 71*, 694–705.

Karasek, R. A., Theorell, T., Schwartz, J. E., Schnall, P. L., Pieper, C. F., & Michela, J. L. (1988). Job characteristics in relation to the prevalence of myocardial infarction in the US Health Examination Survey (HES) and the Health and Nutrition Examination Survey (HANES). *American Journal of Public Health, 78*, 910–919.

Karhu, O., Karkonen, R., Sorvali, P., & Vepsalainen, P. (1981). Observing working postures in industry: Examples of OWAS application. *Applied Ergonomics, 12*, 13–17.

Karpandji, I. A. (1988). *The physiology of the joints*. Edinburgh, UK: Churchill Livingstone.

Karwowski, W. (1988). Maximum load lifting capacities of males and females in teamwork. *Proceedings of the Human Factors Society 32nd annual meeting* (pp. 680-682). Santa Monica, CA: Human Factors Society.

Karwowski, W. (1991). Psychophysical acceptability and perception of load heaviness by females. *Ergonomics, 34*, 487–496.

Karwowski, W. (Ed.). (2001). *International encyclopedia of ergonomics and human factors*. London, UK: Taylor & Francis.

Karwowski, W. (2006a). From past to future. *Human Factors and Ergonomics Society Bulletin, 11*, 1–3 (49).

Karwowski, W. (Ed.). (2006b). *International encyclopedia of ergonomics and human factors* (2nd ed.). Boca Raton, FL: CRC Press.

Karwowski, W. & Marras, W. S. (Eds.). (1999). *The occupational ergonomics handbook*. Boca Raton, FL: CRC Press.

Karwowski, W., & Pongpatanasuegsa, N. (1988). Testing of isometric and isokinetic lifting strengths of untrained females in teamwork. *Ergonomics, 31*, 291–301.

Karwowski, W. & Salvendy, G. (Eds.). (1998). *Ergonomics in manufacturing: Raising productivity through workplace improvement*. Dearborn, MI: Society of Manufacturing Engineers.

Karwowski, W., Soares, M. M. & Stanton, N. A. (Eds.). (2011). *Handbook of human factors and ergonomics in consumer product design*. Boca Raton, FL: CRC Press.

Kaufman, J. E. & Haynes, H. (Eds.). (1981). *IES lighting handbook, 1981 Application Volume*. New York, NY: Illuminating Engineering Society of North America.

Kazarian, L., & Graves, G. A. (1977). Compressive strength characteristics of the human vertebral centrum. *Spine, 2*, 1–14.

Keegan, J. J. (1952). Alterations to the lumbar curve related to posture and sitting. *Journal of Bone and Joint Surgery, 35*, 589–603.

Keele, S. W. (1986). Motor Control. In K. R. Boff, L. Kaufman & J. P. Thomas (Eds.), *Handbook of human perception and human performance* (pp. 30.1–30.60). New York, NY: Wiley.

Keesey, R. E. (1995). A set-point model of body weight regulation. In K. D. Brownell & C. G. Fairburn (Eds.), *Eating disorders and obesity* (pp. 46–50). New York, NY: Gilford.

III. FURTHER INFORMATION

Keesey, R. E., & Powley, T. L. (1986). The regulation of body weight. *Annual Review of Psychology, 37*, 109–133.

Keir, P. J., Bach, J. M., Engstrom, J. W., & Rempel, D. M. (1996). Carpal tunnel pressure: Effects of wrist flexion/extension. *Proceedings of the American Society of Biomechanics 20th annual meeting* (pp. 169-170). Atlanta, GA: American Society of Biomechanics.

Keller, E., Becker, E., & Strasser, H. (1991). An objective assessment of learning behavior with a single-hand chord keyboard for text inputs (in German). *Zeitschrift Fur Arbeitswissenschaft, 45*, 1–10.

Kelly, P. L., & Kroemer, K. H. E. (1990). Anthropometry of the elderly: Status and recommendations. *Human Factors, 32*, 571–595.

Kemmerling, P. T. (1991). Human factors engineering for the disabled and aging. In *Course information ISE* (5654). Blacksburg, VA: Virginia Tech (Fall Semester, 1991).

Kenny, W. L., Wilmore, J., & Costill, D. (2015). *Physiology of sport and exercise* (6th ed.). Champaign, IL: Human Kinetics.

Kermis, M. D. (1984). *Psychology of human aging.* Boston, MA: Allyn and Bacon.

Keyserling, W. M. (1986a). Postural analysis of the trunk and shoulder in simulated real time. *Ergonomics, 29*, 569–583.

Keyserling, W. M. (1986b). A computer-aided system to evaluate postural stress in the workplace. *American Industrial Hygiene Association Journal, 47*, 641–649.

Keyserling, W. M. (1990). Computer-aided posture analysis of the trunk, neck, shoulders, and lower extremities. In W. Karwowski, A. M. Genaidy & S. S. Asfour (Eds.), *Computer-aided ergonomics* (pp. 261–272). London, UK: Taylor & Francis.

Keyserling, W. M. (1998). Methods for evaluating postural work load. In W. Karwowski & G. Salvendy (Eds.), *Ergonomics in manufacturing* (pp. 167–187). Dearborn, MI: Society of Manufacturing Engineers.

Kiiski, J., Heinonen, A., Järvinen, T. L., Kannus, P., & Sievänen, H. (2008). Transmission of vertical whole body vibration to the human body. *Journal of Bone and Mineral Research, 23*(8), 1318–1325.

Kincaid, R. D., & Gonzalez, B. K. (1969). *Human factors design considerations for touch-operated keyboards. Final Report 12091-FR.* St. Paul, MN: Honeywell, Inc.

King, K. B. (1997). Psychologic and social aspects of cardiovascular disease. *Annals of Behavioral Medicine, 19*, 264–270.

Kinney, J. M. (Ed.). (1980). *Assessment of energy metabolism in health and disease.* Columbus, OH: Ross Laboratories.

Kinney, J. S. & Huey, B. M. (Eds.). (1990). *Application principles for multicolored displays.* Washington, DC: National Academy Press.

Kira, A. (1976). *The bathroom.* New York, NY: Viking.

Kirk, J., & Schneider, D. A. (1990). *Physiological and perceptual responses to load carrying in female subjects using internal and external frame backpacks.* (Technical Report TR-91/023). Natick, MA: United States Army Natick Research, Development and Engineering Center.

Kitzes, W. F. (1996). Forensic safety analysis: Investigation and evaluation. In Wiley Editorial Staff (Ed.), *1996 Wiley expert witness update—New developments in personal injury litigation* (pp. 85–119). New York, NY: Wiley (Chapter 3).

Kivimäki, M., Jokela, M., Nyberg, S. T., & Singh-Manoux, A. (2015). Long working hours and risk of coronary heart disease and stroke: A systematic review and meta-analysis of published and unpublished data for 603838 individuals. *Lancet, 386*, 1739–1746.

Kleiner, B. M., & Drury, C. G. (1999). Large-scale regional economic development: Macroergonomics in theory and practice. *Human Factors and Ergonomics in Manufacturing, 9*, 151–163.

Klemmer, E. T., & Lockhead, G. R. (1962). Productivity and errors in two keying tasks: A field study. *Journal of Applied Psychology, 46*(6), 401–408.

Klemmer, E. T. (1958). *A ten-key typewriter.* (Research Memorandum RC-65). Yorktown Heights, NY: IBM Research Center.

Klesges, R. C. (1995). Cigarette smoking and body weight. In K. D. Brownell & C. G. Fairburn (Eds.), *Eating disorders and obesity* (pp. 61–64). New York, NY: Gilford.

Kline, D. W., & Scialfa, C. T. (1997). Sensory and perceptual functioning: Basic research and human factors implications. In A. D. Fisk & W. A. Rogers (Eds.), *Handbook of human factors and the older adult* (pp. 27–54). San Diego, CA: Academic Press.

Klockenberg, E. A. (1926). *Rationalization of the typewriter and of its use (in German).* Berlin, Germany: Springer.

Knapik, J. (1989). *Loads carried by soldiers: A review of historical, physiological, biomechanical and medical aspects.* (Technical Report T19-89). Natick, MA: United States Army Natick Research, Development and Engineering Center.

Knapick, J., Harman, E., & Reynolds, K. (1996). Load carriage using packs: A critical review of physiological, biomechanical and medical aspects. *Applied Ergonomics, 27*, 207–216.

Knapik, J., Hickey, C., Ortega, S., Nagel, J., & de Pontbriand, R. (1997). Energy cost of walking in four types of snowshoes. *Proceedings of the Human Factors Ergonomics Society 41st annual meeting* (pp. 702–706). Santa Monica, CA: Human Factors Ergonomics Society.

Knauth, P. (1996). Designing better shift systems. *Applied Ergonomics, 27*, 39–44.

III. FURTHER INFORMATION

Knauth, P. (2006). Workday length and shiftwork issues. In W. S. Marras & W. Karwowski (Eds.), *The occupational ergonomics handbook: Interventions, controls, and applications in occupational ergonomics* (2nd ed.). Boca Raton, FL: CRC Press (Chapter 29).

Knauth, P. (2007a). Extended work periods. *Industrial Health, 45,* 125–136.

Knauth, P. (2007b). Schicht- und Nachtarbeit, shift work and night work (in German). In K. Landau (Ed.), *Lexikon Arbeitsgestaltung*. Stuttgart, Germany: Gentner.

Kogi, K. (1985). Introduction to the problems of shiftwork. In S. Folkard & T. H. Monk (Eds.), *Hours of work* (pp. 115–184). Chichester, UK: Wiley.

Kogi, K. (1991). Job content and working time: The scope for joint change. *Ergonomics, 34,* 757–773.

Kondraske, G. (1988). Rehabilitation engineering: Towards a systematic process. *IEEE Engineering in Medicine and Biology Magazine, 10,* 11–15.

Konz, S. A. (1990). *Work design: Industrial ergonomics* (3rd ed.). Scottsdale, AZ: Publishing Horizons.

Konz, S. (1991). Japanese children. *Ergonomics, 34,* 971.

Konz, S. (1995). *Work design: Industrial ergonomics* (5th ed.). Scottsdale, AZ: Publishing Horizons.

Konz, S., & Johnson, S. (2000). *Work design: Industrial ergonomics* (5th ed.). Scottsdale, AZ: Holcomb Hataway.

Konz, S., & Johnson, S. (2007). *Work design: Industrial ergonomics* (6th ed.). Scottsdale, AR: Holcomb Hataway.

Koradecka, D. (Ed.). (2000). *Wojciech jastrzebowski*. Warsaw, Poland: Cental Institute for Labour Protection.

Kragt, H. (Ed.). (1992). *Enhancing industrial performance: Experiences of integrating the human factor*. London, UK: Taylor & Francis.

Kramer, A. F., Coyne, J. T., & Strayer, D. L. (1993). Cognitive function at high altitude. *Human Factors, 35,* 329–344.

Kroemer, K. H. E. (1964). On the effect of the spatial position of keyboards on typing performance (in German). *Intrnationale Zeitschrift Angewandte Physiologie, Einschliesslich Arbeitsphysiologie, 20,* 240–251.

Kroemer, K. H. E. (1965). Ergonomic aspects of control operation (in German). *Doctoral dissertation*. Hannover: Technical University Hannover.

Kroemer, K. H. E. (1971). Foot operation of controls. *Ergonomics, 14,* 333–361.

Kroemer, K. H. E. (1972a). *Pedal operation by the seated operator*. New York, NY: Society of Automotive Engineers (SAE Paper 72004).

Kroemer, K. H. E. (1972b). Human engineering the keyboard. *Human Factors, 14,* 51–63.

Kroemer, K. H. E. (1973). COMBIMAN—COMputerized BIomechanical MAN-model. *Proceedings IfU. Colloquium space technology—A model for safety techniques accident prevention?* (pp. 73–88). Cologne, Germany: Verlag TUV.

Kroemer, K. H. E. (1974). *Designing for muscular strength of various populations*. AMRL-Technical Report 72-46. Wright-Patterson AFB, OH: Aerospace Medical Research Laboratory.

Kroemer, K. H. E. (1978). The assessment of human strength. In C. G. Drury (Ed.), *Safety in manual materials handling, DHEW (NIOSH) publication 78–185* (pp. 39–45). Washington, DC: US Government Printing Office.

Kroemer, K. H. E. (1979). A new model of muscle strength regulation. *Proceedings of the annual conference of the Human Factors Society* (pp. 19–20). Santa Monica, CA: Human Factors Society.

Kroemer, K. H. E. (1981). Engineering anthropometry: Designing the work place to fit the human. *Proceedings of the annual conference of the American Institute of Industrial Engineers* (pp. 119–126). Norcross, GA: AIIE.

Kroemer, K. H. E. (1982). *Development of LIFTEST, a dynamic technique to assess the individual capability to lift material, final report*. NIOSH contract 210-79-0041. Blacksburg, VA: Ergonomics Laboratory, IEOR Department, Virginia Tech.

Kroemer, K. H. E. (1983). An isoinertial technique to assess individual lifting capability. *Human Factors, 25*(5), 493–506.

Kroemer, K. H. E. (1985a). Office ergonomics: Work station dimensions. In D. C. Alexander & B. M. Pulat (Eds.), *Industrial ergonomics* (pp. 187–201). Norcross, GA: Institute of Industrial Engineers.

Kroemer, K. H. E. (1985b). Testing individual capability to lift material: Repeatability of a dynamic test compared with static testing. *Journal of Safety Research, 16,* 1–7.

Kroemer, K. H. E. (1986). Coupling the hand with the handle. *Human Factors, 28*(3), 337–339.

Kroemer, K. H. E. (1988a). VDT workstation design. In M. Helander (Ed.), *Handbook of human computer interaction* (pp. 521–539). Amsterdam, The Netherlands: Elsevier (Chapter 23).

Kroemer, K. H. E. (1988b). Ergonomics. In A. Plog (Ed.), *Fundamentals of industrial hygiene* (pp. 283–334). Chicago, IL: National Safety Council.

Kroemer, K. H. E. (1989a). Engineering anthropometry. *Ergonomics, 32,* 767–784.

Kroemer, K. H. E. (1989b). Cumulative trauma disorders: Their recognition and ergonomic measures to avoid them. *Applied Ergonomics, 20,* 274–280.

III. FURTHER INFORMATION

Kroemer, K. H. E. (1991a). Sitting at work: Recording and assessing body postures, designing furniture for computer workstations. In A. Mital & W. Karwowski (Eds.), *Work space, equipment and tool design* (pp. 93–109). Amsterdam, The Netherlands: Elsevier.

Kroemer, K. H. E. (1991b). Experiments with the TCK—A keyboard with built-in wrist rest and only eight keys. In W. Karwowski & J. W. Yates (Eds.), *Advances in industrial ergonomics and safety III* (pp. 537–542). London, UK: Taylor & Francis.

Kroemer, K. H. E. (1992). Avoiding cumulative trauma disorders in shop and office. *American Industrial Hygiene Association Journal, 53,* 596–604.

Kroemer, K. H. E. (1993a). Operation of ternary chorded keys. *International Journal of Human–Computer Interaction, 5,* 267–288.

Kroemer, K. H. E. (1993b). Psychology plus physiology plus biomechanics equal ergonomics? *Proceedings of the 1993 International Industrial Engineering conference* (pp. 278-284). Norcross, GA: Institute of Industrial Engineers.

Kroemer, K. H. E. (1994). Locating the computer screen: How high, how far? *Ergonomics in Design, 1,* 40 (Original article 1993. *Ergonomics in Design, 4,* 7–8).

Kroemer, K. H. E. (1996). Ten ergonomic principles in engineering. *Ergonomia, 19,* 65–76.

Kroemer, K. H. E. (1997a). Design of the computer workstation. In M. G. Helander, T. K. Landauer & P. V. Prabhu (Eds.), *Handbook of human–computer interaction* (2nd ed., pp. 1395–1414). Amsterdam, The Netherlands: Elsevier.

Kroemer, K. H. E. (1997b). *Ergonomic design of material handling systems.* Boca Raton, FL: CRC Press/Lewis Publishers.

Kroemer, K. H. E. (1997c). *Ergonomic design of material handling systems.* Boca Raton, FL: CRC Press.

Kroemer, K. H. E. (1998a). *Reviews of publications related to keyboarding—In chronological order. (Report, 15 February 1998).* Radford, VA: K.H.E. Kroemer Ergonomics Research Institute, Inc.

Kroemer, K. H. E. (1998b). Relating muscle strength and its internal transmission to design data. In S. Kumar (Ed.), *Advances in occupational ergonomics and safety* (pp. 349–352). Amsterdam, The Netherlands: IOS.

Kroemer, K. H. E. (1999). Assessment of human muscle strength for engineering purposes: Basics and definitions. *Ergonomics, 42,* 74–93.

Kroemer, K. H. E. (1999a). Engineering anthropometry. In W. Karwowski & W. S. Marras (Eds.), *The industrial ergonomics handbook* (pp. 139–165). Boca Raton, FL: CRC Press.

Kroemer, K. H. E. (1999b). Human strength evaluation. In W. Karwowski & W. S. Marras (Eds.), *The industrial ergonomics handbook* (pp. 205–227). Boca Raton, FL: CRC Press.

Kroemer, K. H. E. (2001). Keyboards and keying: An annotated bibliography of the literature from 1878 to 1999. *International Journal Universal Access in the Information Society* UAIS 1/2, 99-160, http://www.springerlink.com/index/yp9u5phcq-pyg2k4b.pdf.

Kroemer, K. H. E. (2006a). Designing children's furniture and computers for school and home. *Ergonomics in Design, 3,* 8–16.

Kroemer, K. H. E. (2006b). Designing for older people. *Ergonomics in Design, 4,* 25–31.

Kroemer, K. H. E. (2006c). *"Extra-Ordinary" ergonomics: How to accommodate small and big persons, the disabled and elderly, expectant mothers and children.* Boca Raton, FL: CRC Press.

Kroemer, K. H. E. (2008). Anthropometry and biomechanics: Anthromechanics. In S. Kumar (Ed.), *Biomechanics in ergonomics* (2nd ed.). Boca Raton, FL: CRC Press (Chapter 2).

Kroemer, K. H. E. (2010). 40 Years of human engineering the keyboard. *Proceedings of the 54th annual meeting of the Human Factors and Ergonomics Society* (pp. 1134–1138). Santa Monica, CA: Human Factors, Ergonomics Society.

Kroemer, K. H. E. (2017a). *Fitting the human: Introduction to ergonomics/human factors engineering* (7th ed.). Boca Raton, FL: CRC Press.

Kroemer, K. H. E. (2017b). Healthcare for patients and providers. In *Fitting the human.* (7th ed.). Boca Raton, FL: CRC Press (Chapter 22).

Kroemer, K. H. E., & Kroemer, A. D. (2001). *Office ergonomics.* London, UK: Taylor & Francis.

Kroemer, K. H. E., & Kroemer, A. D. (2006). *Office ergonomics* (Korean edition). Seoul, Korea: Kokje Publishing.

Kroemer, A. D., & Kroemer, K. H. E. (2017). *Office ergonomics, ease and efficiency at work* (2nd ed.). Boca Raton, FL: CRC Press.

Kroemer, K. H. E., Kroemer, H. J., & Kroemer-Elbert, K. E. (1990a). *Engineering physiology: Bases of human factors/ergonomics* (2nd ed.). New York, NY: Van Nostrand Reinhold.

Kroemer, K. H. E., Kroemer, H. J., & Kroemer-Elbert, K. E. (1997). *Engineering physiology: Bases of human factors/ergonomics* (3rd ed.). New York, NY: Van Nostrand Reinhold—Wiley.

Kroemer, K. H. E., Kroemer, H. B., & Kroemer-Elbert, K. E. (2003). amended reprint of the 2001 *Ergonomics: How to design for ease and efficiency.* (2nd ed.). Upper Saddle River, NJ: Prentice Hall/Pearson.

Kroemer, K. H. E., Kroemer, H. J., & Kroemer-Elbert, K. E. (2010). *Engineering physiology: Bases of human factors/ergonomics* (4th ed.). Heidelberg, Germany: Springer.

III. FURTHER INFORMATION

Kroemer, K. H. E., Marras, W. S., McGlothlin, J. D., McIntyre, D. R., & Nordin, M. (1990b). Assessing human dynamic muscle strength. *International Journal of Industrial Ergonomics, 6*, 199–210.

Kroemer, K. H. E., McGlothlin, J. D. & Bobick, T. J. (Eds.). (1989). *Manual material handling: Understanding and preventing back trauma*. Akron, OH: American Industrial Hygiene Association.

Kroemer, K. H. E., & Robinette, J. C. (1968). *Ergonomics in the design of office furniture: A review of european literature* (AMRL-TR 68-90) (pp. 115–125). Wright-Patterson Air Force Base, OH: Aerospace Medical Research Laboratories. (Also published with shortlist of references (1969) in *International Journal of Industrial Medicine and Surgery*).

Kroemer, K. H. E., & Robinson, D. E. (1971). *Horizontal static forces exerted by men standing in common working positions on surfaces of various tractions* (AMRL-Technical Report 70–114). Wright-Patterson AFB, OH: Aerospace Medical Research Laboratory.

Kroemer, K. H. E., Snook, S. H., Meadows, S. K. & Deutsch, S. (Eds.). (1988). *Ergonomic models of anthropometry, human biomechanics, and operator–equipment interfaces*. Washington, DC: National Academy Press.

Krumwiede, D., Konz, S., & Hinnen, P. (1998). Floor mat comfort. In S. Kumar (Ed.), *Advances in occupational ergonomics and safety* (pp. 159–162). Amsterdam, The Netherlands: IOS Press.

Kulkarni, D. S. R., Chitodkar, V., Gurja, V., Chaisas, C. V., & Mannikar, A. V. (2012). SIZE-INDIA, anthropometric size measurement of indian driving population. *SAE technical paper 2011-260108*. Warrendale, PA: SAE International.

Kumar, S. (1997). The effect of sustained spinal load on intra-abdominal pressure and EMG characteristics of trunk muscles. *Ergonomics, 40*, 1312–1334.

Kumar, S. (2001). Theories of musculoskeletal injury causation. *Ergonomics, 44*(1), 17–47.

Kumar, S. (Ed.). (2004). *Muscle strength*. Boca Raton, FL: CRC Press.

Kumar, S. (2008a). Cumulative load and models. In S. Kumar (Ed.), *Biomechanics in ergonomics* (2nd ed.). Boca Raton, FL: CRC Press (Chapter 22).

Kumar, S. (Ed.). (2008b). *Biomechanics in ergonomics* (2nd ed.). Boca Raton, FL: CRC Press.

Kumar, S. & Mital, A. (Eds.). (1996). *Electromyography in ergonomics*. London, UK: Taylor & Francis.

Kumar, S., Chaffin, D. B., & Redfern, M. (1988). Isometric and isokinetic back and arm lifting strengths: Device and measurement. *Biomechanics, 21*, 35–44.

Kuorinka, I. & Forcier, L. (Eds.). (1995). *Work related musculoskeletal disorders: A reference book for prevention*. London, UK: Taylor & Francis.

Kuorinka, I., Jonsson, B., Kilbom, A., Vinterberg, H., Biering-Sorensen, F., Andersson, G., et al. (1987). Standardized Nordic questionnaires for the analysis of musculoskeletal symptoms. *Applied Ergonomics, 18*, 233–237.

Kurowski, A., Buchholz, B., & Punnett L (2014). A physical workload index to evaluate a safe resident handling program for nursing home personnel. *Human Factors, 56*, 669–683.

Kwallek, N., & Lewis, C. M. (1990). Effects of environmental colour on males and females: A red or white or green office. *Applied Ergonomics, 21*, 275–278.

Landau, K. (Ed.). (2000). *Ergonomic software tools in product and workplace design. A review of recent developments in human modeling and other design aids*. Stuttgart, Germany: Ergon.

Landy, F. J., & Conte, J. M. (2006). *Work in the 21st century: An introduction to industrial and organizational psychology* (2nd ed.). Malden, MA: Blackwell.

Langleben, D., & Moriarty, J. (2013). Using brain imaging for lie detection: Where science, law and research policy collide. *Psychol Public Policy Law, 19*(2), 222–234.

Langley, L. W. (1988). Ternary Chord-type Keyboard. US Patent 4,775,255. Washington, DC: US Patent and Trademark Office.

Lankhorst, G. J., van de Stadt, R. J., Vogelaar, T. W., van der Korst, J. K., & Prevo, A. J. H. (1983). The effect of the Swedish back school in chronic ideopathic low back pain. *Scandanavian Journal of Rehabilitation and Medicine, 15*, 141–145.

Lateck, J. C., & Foster, L. W. (1985). Implementation of compressed work schedules: Participation and job redesign as critical factors for employee acceptance. *Personnel Psychology, 38*, 75–92.

Latko, W. A., Armstrong, T. J., Foulke, J. A., Herrin, G. D., Rabourn, R. A., & Ulin, S. S. (1997). Development and evaluation of an observational method for assessing repetition in hand tasks. *American Industrial Hygiene Association Journal, 58*, 278–285.

Laughery, K. R., Jackson, A. S., & Fontenelle, G. A. (1988). Isometric strength tests: Predicting performance in physically demanding transport tasks. *Proceedings of the Human Factors Society 32nd annual meeting* (pp. 695–699). Santa Monica, CA: Human Factors Society.

Lavender, S. A. (2006). Training lifting techniques. In W. S. Marras & K. Karwowski (Eds.), *The occupational ergonomics handbook* (2nd ed.). Boca Raton, FL: CRC Press.

Lavender, S. A., Chen, S. H., Li, Y. C., & Andersson, G. B. J. (1998). Trunk muscle use during pulling tasks: Effects of a lifting belt and footing conditions. *Human Factors, 40*, 159–172.

III. FURTHER INFORMATION

Lavender, S. A., Oleske, D. M., Nicholson, L., Andersson, G. B. J., & Hahn, J. (1997). A comparison of four methods commonly used to determine low-back disorder risk in a manufacturing environment. *Proceedings of the Human Factors and Ergonomics Society 41st annual meeting* (pp. 657–660). Santa Monica, CA: Human Factors and Ergonomics Society.

Lazarus, R. S., & Cohen, J. B. (1977). Environmental stress. In L. Altman & J. F. Wohlwill (Eds.), *Human behavior and the environment: Current theory and research* (2). New York, NY: Plenum.

Lazarus, R. S., & Folkman, S. (1984). *Stress, appraisal, and coping.* New York, NY: Springer.

Le Bon, C., & Forrester, C. (1997). An ergonomic evaluation of a patient handling device: The elevate and transfer vehicle. *Applied Ergonomics, 28,* 365–374.

Leamon, T. B. (1994). Research to reality: A critical review of the validity of various criteria for the prevention of occupationally induced low back pain disability. *Ergonomics, 37,* 1959–1974.

Leclercq, S., Tisserand, M., & Saulnier, H. (1995). Tribological concepts involved in slipping accident analysis. *Ergonomics, 38,* 197–208.

Lee, Y. H., & Chen, Y. L. (1996). An isoinertial predictor for maximal acceptable lifting weights of Chinese subjects. *American Industrial Hygiene Association Journal, 57,* 456–463.

Lee, J. C., & Healy, J. C. (2005). Normal sonographic anatomy of the wrist and hand. *Radiographics, 25,* 1577–1590.

Lee, C. H., Hosni, Y. A., Guthrie, L. L., Barth, T., & Hill, C. (1991). Design and evaluation of a work seat for overhead tasks. In W. Karwowski & J. W. Yates (Eds.), *Advances in industrial ergonomics and safety III* (pp. 555–562). London, UK: Taylor & Francis.

Lee, N., Swanson, N., Sauter, S., Wickstrom, R., Waikar, A., & Mangum, M. (1992). A Review of physical exercises recommended for VDT operators. *Applied Ergonomics, 23,* 387–408.

Leekham, S., Nieto, C., Libby, S., Wing, L., & Gould, J. (2007). Describing the sensory abnormalities of children and adults with autism. *Journal of Autism and Developmental Disorders, 37*(5), 894–910.

Lehmann, G. (1953). *Praktische arbeitsphyiologie.* Stuttgart, Germany: Thieme.

Lehmann, G. (1962). *Praktische arbeitsphysiologie* (2nd ed.). Stuttgart, Germany: Thieme.

Lehto, M., & Landry, S. J. (2014). *Introduction to human factors and ergonomics for engineers* (2nd ed.). Boca Raton, FL: CRC Press.

Lehto, M., & Salvendy, G. (1995). Warnings: A supplement not a substitute for other approaches to safety. *Ergonomics, 38,* 2155–2163.

Lepore, J. (2009). Not so fast. *The New Yorker,* 114–122.

Lesiuk, T. (2005). The effect of music listening on work performance. *Psychology of Music, 33,* 173–191.

Levine, D. B., Zitter, M. & Ingram, L. (Eds.). (1990). *Disability statistics: An assessment. Committee on National Statistics, National Research Council.* Washington, DC: National Academy Press.

Lewis, J. R. (1994). Sample sizes for usability studies: Additional considerations. *Human Factors, 36,* 368–378.

Lewis, J. R., Potosnak, K. M., & Magyar, R. L. (1997). Keys and keyboards. Chapter. In M. G. Helander, T. K. Landauer & P. V. Prabhu (Eds.), *Handbook of human–computer interaction* (pp. 1285–1315). (54). Amsterdam, The Netherlands: Elsevier.

Leyk, D., Kuechmeister, G., & Juergen, H. W. (2006). Combined physiological and anthropometrical databases as ergonomic tools. *Journal of Physiological Anthropology, 25,* 363–369.

Li, H., & Ali, H. (2015). Human factors considerations in the design of falls prevention technologies for nursing homes: A case study. *Proceedings of the International Symposium on Human Factors and Ergonomics in Health Care, 4*(1), 97–102.

Li, C., Hwang, S., & Wang, M. (1990). Static anthropometry of civilian Chinese in Taiwan using computer-analyzed photography. *Human Factors, 32,* 359–370.

Liberty Mutual. (2018). *Manual materials handling guidelines.* Regularly updated on https://libertymmhtables.libertymutual.com/CM_LMTablesWeb/taskSelection.do?action=initTaskSelection.

Lieberman, H. R., Castellani, J. W., & Young, A. J. (2009). Cognitive function and mood during acute cold stress after extended military training and recovery. *Aviation, Space, and Environmental Medicine, 80,* 629–636.

Lin, L., & Cohen, H. H. (1995). Fall accident patterns involved in litigation. *Proceedings of the Silicon Valley ergonomics conference exposition Ergocon '95.* San Jose, CA: San Jose State University.

Litterick, I. (1981). QWERTYUIOP-dinosaur in a computer age. *New Scientist,* January 8, 66–68.

Liu, Q., Zhou, R. L., Zhao, X., Chen, X. P., & Chen, S. G. (2016). Acclimation during space flight: Effects on human emotion. *Military Medical Research, 3,* 15.

Locke, J. C. (1983). Stretching away from back pain injury. *Occupational Health and Safety, 52,* 8–13.

Locke, E. A., & Latham, G. P. (1990). *A theory of goal setting & task performance.* Englewood Cliffs, NJ: Prentice Hall.

Lockhead, G. R., & Klemmer, E. T. (1959). *An evaluation of an eight-key word-writing typewriter.* (Research report RC-180). Yorktown Heights, NY: IBM Research Center.

Lockwood, A. H. (1989). Medical problems of musicians. *New England Journal of Medicine, 320,* 221–227.

III. FURTHER INFORMATION

Lohman, T. G., Roche, A. F. & Martorel, R. (Eds.). (1988). *Anthropometric standardization reference manual*. Champaign, IL: Human Kinetics.

Longo, L. & Leva, M. C. (Eds.). (2017). *Human mental workload: Models and applications*. Cham, Switzerland: Springer.

Louhevaara, V., Ilmarinen, J., & Oja, P. (1985). Comparison of three field methods for measuring oxygen consumption. *Ergonomics, 28,* 463–470.

Loveless, N. E. (1962). Direction-of-motion stereotypes: A review. *Ergonomics, 5,* 357–383.

Lowe, B., & Freivalds, A. (1998). Design and evaluation of prototype tool handles. In S. Kumar (Ed.), *Advances in occupational ergonomics and safety* (pp. 417–420). Amsterdam, The Netherlands: IOS Press.

Lu, M. L. (2012). *A practical guide to the revised NIOSH lifting equation (NLE)*. Cincinnati, OH: National Institute for Occupational Safety and Health.

Lu, H., & Aghazadeh, F. (1998). Modeling of risk factors in the VDT workstation systems. *Occupational Ergonomics, 1,* 189–210.

Lu, M. L., Putz-Anderson, V., Garg, A., & Davis, L. (2016). Evaluation of the impact of the revised NIOSH lifting equation. *Human Factors, 58*(5), 667–682.

Lueder, R. & Rice, V. B. (Eds.). (2007). *Ergonomics for children*. Boca Raton, FL: CRC Press.

Lundervold, A. (1951a). Electromyographic investigations of position and manner of working in typewriting. *Acta Physiologica Scandinavica, 24,* 84–104.

Lundervold, A. J. S. (1951b). Electromyographic investigations of position and manner of working in typewriting. *Acta Physiologica Scandinavica, 24*(84), 1–171 (Supplement).

Lykken, D. T. (1988). The case against the polygraph test. In A. Gale (Ed.), *The polygraph test. Lies, truth, and science* (pp. 111–125). (9). London, UK: Sage. (Chapter 9).

MacDougall, D. (1907). Hypothesis concerning soul substance together with experimental evidence of the existence of such substance. *American medicine, 2,* 240–243.

Mace, W. M. (1977). James J. Gibson's strategy for perceiving: Ask not what's inside your head, but what your head's inside. In R. E. Shaw & J. Bransford (Eds.), *Perceiving, acting, and knowing*. Hillsdale, NJ: Erlbaum.

Mack, K., Haslegrave, C. M., & Gray, M. I. (1995). Usability of manual handling aids for transporting materials. *Applied Ergonomics, 26,* 353–364.

Maekinen, T. M., Palinkas, L. A., Reeves, D. L., Paeaekkoenen, T., Rintamaeki, H., Leppaeluoto, J., et al. (2006). Effect of repeated exposures to cold on cognitive performance in humans. *Physiology & Behavior, 87,* 166–176.

Malone, R. L. (1991). *Posture taxonomy*. (Unpublished Master's Thesis) Blacksburg, VA: Virginia Tech.

Manchikanti, L., Singh, V., Datta, S., Cohen, S., & Hirsh, J. (2009). Comprehensive review of epidemiology, scope, and impact of spinal pain. *Pain Physician, 12*(4), E35–E70.

Mandal, A. C. (1975). Work-chair with tilting seat. *Lancet, 305*(7907), 642–643.

Mandal, A. C. (1982). The correct height of school furniture. *Human Factors, 24,* 257–269.

Mansfield, N. J. (2005). *Human response to vibration*. Boca Raton, FL: CRC Press.

Marklin, R. W., & Simoneau, G. G. (2004). Design features of alternative computer keyboards: A review of experimental data. *Journal of Orthopaedic & Sports Physical Therapy, 34,* 638–649.

Marklin, R. W., Simoneau, G. G., & Monroe, J. F. (1997). The effect of split and vertically-inclined computer keyboards on wrist and forearm posture. *Proceedings of the Human Factors and Ergonomics Society 41st annual meeting* (pp. 642–646). Santa Monica, CA: Human Factors and Ergonomics Society.

Marklin, R. W., Simoneau, G. G., & Monroe, J. F. (1999). Wrist and forearm posture from typing on split and vertically-inclined computer keyboards. *Human Factors, 41,* 559–569.

Marlatt, G., & Donovan, D. (Eds.), (2005). *Relapse prevention: Maintenance strategies in the treatment of addictive behaviors*. (2nd ed.). New York, NY: Guilford Press.

Marquié, J., Tucker, P., Folkard, S., Gentil, C., & Ansiau, D. (2013). Chronic effects of shift work on cognition: Findings from the VISAT longitudinal study. *Occupational and Environmental Medicine, 72,* 258–264.

Marras, W. S. (2008). *The working back: A systems view*. New York, NY: Wiley.

Marras, W. S., & Karwowski, W. (Eds.), (2006a). *The occupational ergonomics handbook: Vol. 1. Fundamentals and assessment tools for occupational ergonomics* (2nd ed.). Boca Raton, FL: CRC Press.

Marras, W. S., & Karwowski, W. (Eds.), (2006b). *The occupational ergonomics handbook: Vol. 2. Interventions, controls, and applications in occupational ergonomics* (2nd ed.). Boca Raton, FL: CRC Press.

Marras, W. S., & Mirka, G. A. (1996). Intra-abdominal pressure during trunk extension motions. *Clinical Biomechanics, 11,* 267–274.

Marras, W. S., & Radwin, R. G. (2006). Biomechanical modeling. In R. S. Dickerson (Ed.), *Reviews of human factors and ergonomics* (1). Santa Monica, CA: Human Factors and Ergonomics Society (Chapter 1).

III. FURTHER INFORMATION

Marras, W. S., & Rangarajulu, S. L. (1987). Trunk force development during static and dynamic lifts. *Human Factors, 29*, 19–29.

Marras, W. S., & Reilly, C. H. (1988). Networks of internal trunk-loading activities under controlled trunk-motion conditions. *Spine, 13*, 661–667.

Marras, W. S., & Schoenmarklin, R. W. (1991). Wrist motions and CTD risk in industrial and service environments. *Proceedings of the 11th congress of the International Ergonomics Association* (pp. 36-38). London, UK: Taylor, Francis.

Marras, W. S., Lavender, S. A., Leurgans, S. E., Fathallah, F. A., Ferguson, S. A., Allread, W. G., et al. (1993a). Biomechanical risk factors for occupationally related low back disorders. *Ergonomics, 38*, 377–410.

Marras, W. S., Lavender, S. A., Leurgans, S. E., Rajalu, S. L., Allread, W. G., Fathallah, F. A., et al. (1993b). The role of dynamic three-dimensional trunk motion on occupationally-related low back disorders. *Spine, 18*, 617–628.

Marras, W. S., McGlothlin, J. D., McIntyre, D. R., Nordin, M., & Kroemer, K. H. E. (1993c). *Dynamic measures of low back performance.* Fairfax, VA: American Industrial Hygiene Association.

Marras, W. S., Walter, B. A., Purmessur, D., Mageswaran, P., & Wiet, M. G. (2016). The contribution of biomechanical–biological interactions of the spine to low back pain. *Human Factors, 58*(7), 965–975.

Marteniuk, R. G., Ivens, C. J., & Brown, B. E. (1996). Are there task specific performance effects for differently configured numeric key pads? *Applied Ergonomics, 27*, 321–325.

Martimo, K. P., Verbeek, J., Karppinen, J., Furlan, A. D., Takala, E. P., Kuijer, P. P. F., et al. (2008). Effect of training and lifting equipment for preventing back pain in lifting and handling: Systematic review. *British Medical Journal, 336*, 429–431.

Martin, R. (1914). *Lehrbuch der anthropology.* Jena, Germany: Fischer (in German).

Martin, E. (1949). *The typewriter and its historical development.* Aachen, Germany: Basten (in German).

Martin, B. J., Roll, J. P., & Gauthier, G. M. (1986). Inhibitory effects of combined agonist and antagonist muscle vibration on H-reflex in men. *Aviation, Space, and Environmental Medicine, 57*, 681–687.

Martin, B., Rempel, D., Sudarsan, P., Dennerlein, J., Jacobson, M., Gerard, M., et al. (1998). Reliability and sensitivity of methods to quantify muscle load during keyboard work. In S. Kumar (Ed.), *Advances in occupational ergonomics and safety* (pp. 485–489). Amsterdam, The Netherlands: IOS Press.

Martins, A., Ramalho, N., & Morin, E. (2010). A comprehensive meta-analysis of the relationship between emotional intelligence and health. *Personality and Individual Differences, 6*, 554–564.

Maslach, C., & Jackson, S. E. (1981). The measurement of experienced burnout. *Journal of Occupational Behaviour, 2*, 99–113.

Maslach, C., Schaufeli, W. B., & Leiter, M. P. (2001). Job burnout. *Annual Review of Psychology, 52*, 397–422.

Maslow, A. H. (1943). A theory of human motivation. *Psychological Review, 50*, 370–396.

Maslow, A. H. (1954). *Motivation and personality* (2nd ed.). New York, NY: Harper.

Matthews, G., & Ryan, H. (1994). The expression of the "pre-menstrual syndrome" in measures of mood and sustained attention. *Ergonomics, 37*, 1407–1417.

Matthews, G., Reinerman-Jones, L. E., Barber, D. J., & Abich, J. (2015). The psychometrics of mental workload: Multiple measures are sensitive but divergent. *Human Factors, 57*, 125–143.

Mattila, M., Karwowski, W., & Vilkki, M. (1993). Analysis of working postures on building construction sites: The computerized OWAS method. *Applied Ergonomics, 24*, 405–412.

Mauney, J., Furlough, C., & Barnes, J. (2015). Developing a better clinical alert system in EHRs. *Proceedings of the International Symposium on Human Factors and Ergonomics in Health Care, 4*, 29–36 (June).

Max-Planck-Gesellschaft zur Foerderung der Wissenschaften. (1998). Tastatur. German Patent 1,255,117. Munich, Germany: German Patent Office.

Mayer, E. (2011). Gut feelings: The emerging biology of gut–brain communication. *Nature Reviews Neuroscience, 12*(8), 453–466.

Mayer, E., Knight, R., Mazmanian, S., Cryan, J., & Tillisch, K. (2014). Gut microbes and the brain: Paradigm shift in neuroscience. *The Journal of Neuroscience, 34*, 15490–15496.

Mazure, C. M. (1998). Life stressors as risk factors in depression. *Clinical Psychology: Science and Practice, 5*, 291–313.

McAtamney, L., & Corlett, E. N. (1993). RULA: A survey method for the investigation of work-related upper limb disorders. *Applied Ergonomics, 24*, 91–99.

McCauley-Bush, P. (2011). *Ergonomics: Foundational principles, applications, and technologies.* Boca Raton, FL: CRC Press.

McClelland, I. L. (1982). The ergonomics of toilet seats. *Human Factors, 24*, 713–725.

McConville, J. T., Churchill, T., Kaleps, I., Clauser, C. E., & Cuzzi, J. (1980). *Anthropometric relationships of body and body segment moments of inertia.* Wright-Patterson AFB, OH: Aerospace Medical Research Laboratory (AFAMRL-TR-80-119).

McConville, J. T., Robinette, K. M., & Churchill, T. (1981). *An anthropometric data base for commercial design applications. Final report NSF DAR-80 09 861.* Yellow Springs, OH: Anthropology Research Project, Inc.

McDaniel, J. W. (1998). Human modeling: Yesterday, today, and tomorrow. Keynote address at the SAE international conference, digital human modeling for design and engineering, Dayton, OH.

III. FURTHER INFORMATION

McDaniel, J. W., Skandis, R. J., & Madole, S. W. (1983). *Weight lift capabilities of air force basic trainees.* Wright-Patterson Air Force Base, OH: Air Force Aerospace Medical Research Laboratory (AFAMRL-TR-83-0001).

McEwen, B. (2018). Physiology and neurobiology of stress and adaptation: Central role of the brain. *Physiological Reviews, 87,* 873–904.

McFadden, K. L. (1997). Predicting pilot-error incidents of US airline pilots using logistic regression. *Applied Ergonomics, 28,* 209–212.

McFarland, R. A. (1946). *Human factors in air transport design.* New York, NY: McGraw-Hill.

McGill, S. M. (1999a). Dynamic low back models: Theory and relevance in assisting the ergonomist to reduce the risk of low back injury. In W. Karwowski & W. S. Marras (Eds.), *The occupational ergonomics handbook (945–965).* Boca Raton, FL: CRC Press.

McGill, S. M. (1999b). Update on the use of back belts in industry: More data—Same conclusion. In W. Karwowski & W. S. Marras (Eds.), *The occupational ergonomics handbook (1353–1358).* Boca Raton, FL: CRC Press.

McGill, S. M. (2006). Back belts. In W. S. Marras & W. Karwowski (Eds.), *The occupational ergonomics handbook, 2nd ed, interventions, controls, and applications in occupational ergonomics.* Boca Raton, FL: CRC Press (Chapter 30).

McGill, S. M., & Norman, R. W. (1987). Reassessment of the role of intra-abdominal pressure in spinal compression. *Ergonomics, 30,* 1565–1588.

McGill, S. M., Sharratt, M. T., & Seguin, J. P. (1995). Loads on spinal tissues during simultaneous lifting and ventilatory challenge. *Ergonomics, 38,* 1772–1792.

McGill, S. M., Norman, R. W., & Cholewicki, J. (1996). A simple polynomial that predicts low back compression during complex 3-D tasks. *Ergonomics, 39,* 1107–1118.

McLeod, D., & Morris, A. (1996). Ergonomics cost benefits case study in a paper manufacturing company. *Proceedings of the Human Factors and Ergonomics Society 40th annual meeting* (pp. 698–701). Santa Monica, CA: Human Factors and Ergonomics Society.

McMillan, G. R. (2001). Brain and muscle signal-based control. In W. Karwowski (Ed.), *International encyclopedia of ergonomics and human factors* (pp. 379–381). London, UK: Taylor & Francis.

McMillan, G. R., & Calhoun, G. L. (2001). Gesture-based control. In W. Karwowski (Ed.), *International encyclopedia of ergonomics and human factors* (pp. 237–239). London, UK: Taylor & Francis.

McMin, T. (2013). "A-Weighting": Is it the metric you think it is? *Proceedings of acoustics* (pp. 1–4). Victor Harbor, Australia: Australian Acoustical Society.

McMulkin, M. L., & Kroemer, K. H. E. (1994). Usability of a one-hand ternary chord keyboard. *Applied Ergonomics, 25*(3), 177–181.

McMulkin, M. L., & Sivasubramanian, T. (1998). Hand forces in weaker and stronger members during two-person lifts. In S. Kumar (Ed.), *Advances in industrial ergonomics and safety (316–319).* Amsterdam, The Netherlands: IOS Press.

McVay, E. J., & Redfern, M. S. (1994). Rampway safety: Foot forces as a function of rampway angle. *American Industrial Hygiene Association Journal, 55,* 626–634.

McWright, A. (1988). Gender differences in the strain responses to job demand. *Proceedings of the Human Factors Society 32nd annual meeting* (pp. 853–856). Santa Monica, CA: Human Factors Society.

Meakin, J. R., Hukins, D. W. L., & Aspden, R. M. (1996). Euler buckling as a model for the curvature and flexion of the human lumbar spine. *Biological Sciences, 263,* 1383–1387 (1375).

Megaw, T. (2005). The definition and measurement of mental workload. In J. R. Wilson & N. Corlett (Eds.), *Evaluation of human work* (3rd ed.). London, UK: Taylor & Francis (Chapter 18).

Meindl, B. A., & Freivalds, A. (1992). Shape and placement of faucet handles for the elderly. *Proceedings of the Human Factors Society 36th annual meeting* (pp. 811–815). Santa Monica, CA: Human Factors Society.

Meister, D. (1989). *Conceptual aspects of human factors.* Baltimore, MD: Johns Hopkins University Press.

Meister, K. J. (1990). A few implications of an ecological approach to human factors. *Human Factors Society Bulletin, 33,* 1–4.

Meister, D. (1999). *The history of human factors and ergonomics.* Mahwah, NJ: Erlbaum.

Mellerowicz, H., & Smodlaka, V. N. (1981). *Ergometry.* Baltimore, MD: Urban and Schwarzenberg.

Mello, R. P., Damokosh, A. I., Reynolds, K. L., Witt, C. E., & Vogel, J. A. (1988). *The physiological determinants of load bearing performance at different march distances.* Natick, MA: U.S. Army Research Institute of Environmental Medicine (T15-88).

Merck Manual of Medical Information. (1997). *Home edition.* Whitehouse Station, NJ: Merck & Co., Inc.

Merletti, R., Farina, D., & Rainoldi, R. (2004). Myoelectric manifestations of muscle fatigue. In S. Kumar (Ed.), *Muscle strength.* Boca Raton, FL: CRC Press (Chapter 18).

Merrill, B. A. (1995). Contributions of poor movement strategies to CTD/solutions or faulty movements in the human house. *Proceedings of the Silicon Valley ergonomics conference exposition Ergocon '95* (pp. 222–228). San Jose, CA: San Jose State University.

III. FURTHER INFORMATION

Merryweather, A. S., & Bloswick, D. S. (2008). Biomechanics during ladder and stair climbing and walking on ramps and other irregular surfaces. In S. Kumar (Ed.), *Biomechanics in ergonomics*. Boca Raton, FL: CRC Press (Chapter 25).

Metropolitan Life Foundation. (1983). Comparison of 1959 and 1983 metropolitan height and weight tables. *Statistical Bulletin, 64*, 6–7.

Meyer, D. E., & Kieras, D. E. (1997). A computational theory of executive cognitive processes and multi-task performance: Part 1 basic mechanisms. *Psychological Review, 104*, 3–65.

Meyers, J. M., Miles, J. A., Faucett, J., Janowitz, I., Tejeda, D. G., & Kabashima, J. N. (1997). Ergonomics in agriculture: Workplace priority setting in the nursery industry. *American Industrial Hygiene Association Journal, 58*, 121–126.

Michaels, S. E. (1971). QWERTY vs alphabetic keyboards as a function of typing skills. *Human Factors, 13*, 419–426.

MIL-HDBK 759. (2018). Human factors engineering design for army material (Metric). *US Army missile command*. DLA document services. Redstone Arsenal, (700 Robbins Avenue, Philadelphia, PA 19111-5094).

Miller, R. L. (1977). Bend your knees. *National Safety News, 115*, 57–58.

Miller, R. J. (1990). Pitfalls in the conception, manipulation, and measurement of visual accommodation. *Human Factors, 32*, 27–44.

Miller, T. Q., Smith, T. W., Turner, C. W., Guijarro, M. L., & Hallet, A. J. (1996). A meta-analytic review of research on hostility and physical health. *Psychological Bulletin, 119*, 322–348.

Miller, T. D., Balady, G. J., & Fletcher, G. F. (1997). Exercise and its role in the prevention and rehabilitation of cardiovascular disease. *Annals of Behavioral Medicine, 19*(3), 220–229.

MIL-STD-1472. (2018). Human engineering design criteria for military systems, equipment, and facilities. *US Army missile command*. DLA document services. Redstone Arsenal (700 Robbins Avenue, Philadelphia, PA 19111-5094).

Minors, D. S., & Waterhouse, J. M. (1987). The role of naps in alleviating sleepiness during an irregular sleep-wake schedule. *Ergonomics, 30*, 1261–1273.

Mirka, G. A., & Baker, A. (1996). An investigation of the variability in human performance during sagittally symmetric lifting tasks. *IIE Transactions, 28*, 745–752.

Mirka, G. A., Kelaher, D., Baker, A., Harrison, A., & Davis, J. (1997). Selective activation of the external oblique musculature during axial torque production. *Clinical Biomechanics, 12*, 172–180.

Mirka, G. A., Baker, A., Harrison, A., & Kelaher, D. (1998). The interaction between load and coupling during dynamic manual materials handling. *Occupational Ergonomics, 1*, 3–11.

Mirmohammadi, S. J., Mehrparvar, A. H., Mostaghaci, M., Davari, M. H., Bahaloo, M., & Mashtizadeh, S. (2016). Anthropometric hand dimensions in a population of Iranian male workers in 2012. *International Journal of Occupational Safety and Ergonomics, 22*(1), 125–130.

Mishan, E. J., & Quah, E. (2007). *Cost–benefit analysis*. London, UK: Routledge.

Mital, A., Kilbom, A., & Kumar, S. (2000). *Ergonomics guidelines and problem solving*. Oxford, UK: Elsevier Science.

Mitchell, N. B. (1992). A simple method for hardware and software evaluation. *Applied Ergonomics, 23*, 277–280.

Mobbs, D., Petrovic, P., & Marchant, J. L. (2007). When fear is near: Threat imminence elicits prefrontal–periaqueductal gray shifts in humans. *Science, 317*, 1079–1083.

Mobbs, D., Marchant, J. L., & Hassabis, D. (2009). From threat to fear: The neural organization of defensive fear systems in humans. *The Journal of Neuroscience, 29*, 12236–12243.

Moffett, J. A. K., Chase, S. M., Portek, I., & Ennis, J. R. (1986). A controlled perspective study to evaluate the effectiveness of a back school in the relief of chronic low back pain. *Spine, 11*, 120–121.

Momtahan, K., Hetu, R., & Tansley, B. (1993). Audibility and identification of auditory alarms in the operating room and intensive care unit. *Ergonomics, 36*, 1159–1176.

Monk, T. H. (1989). Shift worker safety: Issues and solutions. In A. Mital (Ed.), *Advances in industrial ergonomics and safety I* (pp. 887–893). Philadelphia, PA: Taylor & Francis.

Monk, T. H. (2006). Shiftwork. In W. S. Marras & W. Karwowski (Eds.), *The occupational ergonomics handbook, fundamentals and assessment tools for occupational ergonomics* (2nd ed.). Boca Raton, FL: CRC Press (Chapter 32).

Monk, T. H., & Tepas, D. I. (1985). Shiftwork. In Cooper & M. J. Smith (Eds.), *Job stress and blue collar work* (pp. 65–84). New York, NY: Wiley (Chapter 5).

Monk, T. H., & Wagner, J. A. (1989). Social factors can outweigh biological ones in determining night shift safety. *Human Factors, 31*, 721–724.

Monk, T. H., Folkard, S., & Wedderburn, A. I. (1996). Maintaining safety and high performance on shiftwork. *Applied Ergonomics, 27*, 17–23.

Monod, H., & Valentin, M. (1979). The predecessors of ergonomy (in French). *Ergonomics, 22*, 673–680.

III. FURTHER INFORMATION

Moon, S. D., & Sauter, S. L. (Eds.), (1996). *Beyond biomechanics: Psychosocial aspects of musculoskeletal disorders in office work.* London, UK: Taylor & Francis.

Moon, R. E., Vann, R. D., & Bennett, P. B. (1995). The physiology of decompression illness. *Scientific American, 273:* 2, 70–77.

Moore, T. G. (1974). Tactile and kinesthetic aspects of pushbuttons. *Applied Ergonomics, 5,* 66–71.

Moore, J. S., & Garg, A. (1997). Participatory ergonomics in a meat packing plant Part I: Evidence of lnag-term effectiveness. *American Industrial Hygiene Association Journal, 58,* 127–131.

Moore, A., Wells, R., & Ranney, D. (1991). Quantifying exposure in occupational manual tasks with cumulative trauma potential. *Ergonomics, 34,* 1433.

Moray, N. (1988). Mental workload since 1979. *International Reviews of Ergonomics, 2,* 123–150.

Moreno-Briseseno, P., Diaz, R., Campos-Romo, A., & Fernandez-Ruis, J. (2010). Sex-related differences in motor learning and performance. *Behavioral and Brain Function, 6,* 74.

Morgan, S. (1991). Wrist factors contributing to CTS can be minimized, if not eliminated. *Occupational Health and Safety, 60,* 47–51.

Moriarty, J. P., Branda, M. E., Olsen, K. D., Shah, N. D., Borah, B. J., Wagie, A. E., et al. (2012). The effects of incremental costs of smoking and obesity on health care costs among adults: A y-year longitudinal study. *Journal of Occupational and Environmental Medicine, 54(3),* 286–291.

Morris, A. (1984). Program compliance key to preventing low back injuries. *Occupational Health and Safety, 53,* 44–47.

Morris, N., Toms, M., Easthope, Y., & Biddulph, J. (1998). Mood and cognition in pregnant workers. *Applied Ergonomics, 29,* 377–381 (See also the related 1999 Letter to the Editor by in *Applied Ergonomics, 30,* 177).

Morrow, D. Y., & Jerome, L. (1990). The influence of alcohol and aging on radio communication during flight. *Aviation, Space and Environmental Medicine, 61,* 12–20.

Motowidlo, S. J., Packard, J. S., & Manning, M. R. (1986). Occupational stress: Its causes and consequences for job performance. *Journal of Applied Psychology, 71,* 618–629.

Muchinsky, P. M. (1983). *Psychology applied to work.* Homewood, IL: Dorsey.

Muchinsky, P. M., & Culbertson, S. S. (2016). *Psychology applied to work* (11th ed.). Summerfield, NC: Hypergraphic Press.

Muckler, F. A., & Seven, S. A. (1992). Selecting performance measure: "Objective" versus "Subjective" measurement. *Human Factors, 34,* 441–455.

Mullick, A. (1997). Listening to people: Views about the bathroom. *Proceedings of the Human Factors and Ergonomics Society 41st annual meeting* (pp. 500–504). Santa Monica, CA: Human Factors and Ergonomics Society.

Munsel, A. H. (1942). *Book of color.* Baltimore, MD: Munsell Color Book Corp.

Musil, P. (1995). In B. Pike & S. Luft (Eds.), *Precision and soul: Essays and addresses.* Chicago, IL: University of Chicago Press (Musil, R 1890–1942, from Gesammelte Werke in Neun Bänden).

Myers, D. G., & Diener, E. (1996). The pursuit of happiness. *Scientific American, 1996,* 54–56.

Myung, R., & Smith, J. L. (1997). The effect of load carrying and floor contaminants on slip and fall parameters. *Ergonomics, 40,* 235–246.

Nachemson, A. (1989). Individual factors contributing to low back pain. *Presented at the Occupational Orthopaedics conference of the AAOS.* New York, NY: American Academy of Orthopaedic Surgeons.

Nachemson, A., & Elfstroem, G. (1970). Intervertebral dynamic pressure measurements in lumbar discs. *Scandinavian Journal of Rehabilitation Medicine Supplement, 1,* 1–40.

Nagamachi, M. (2001). Relationships among job design, macro ergonomics, and productivity. In H. W. Hendrick & B. M. Kleiner (Eds.), *Macro ergonomics. Theory, methods, and applications.* Mahwah, NJ: Erlbaum (Chapter 6).

Nagamachi, M. (2011). *Kansai engineering.* Boca Raton, FL: CRC Press (2 vols).

Nakanishi, Y., & Nethery, V. (1999). Anthropometric comparison between Japanese and Caucasian American male university students. *Applied Human Science, 18(1),* 9–11.

Nakashima, A., & Cheung, B. (2006). The effects of vibration frequencies on physical, perceptual and cognitive performance. *DRDC Toronto TR 2006-218.* Toronto, ON: Defence R&D Canada.

Nakatani, C., Sato, N., Matsui, M., Matsunami, M., & Kumashiro, M. (1997). Menstrual cycle effects on a VDT-based simulation task: Cognitive indices and subjective ratings. *Ergonomics, 36,* 311–339.

NASA/Webb. (1978). *Anthropometric Sourcebook.* Houston, TX: LBJ Space Center (3 volumes). (NASA reference publication 1024.).

NASA. (1989). *Man-systems integration standards (revision A).* Houston, TX: L.B.J. Space Center (NASA-STD 3000) SP 34-89-230.

NASA. (circa 2015). NASA-STD-3001 Space Flight Human-System Standard Vol. 1 (Crew Health); Vol. 2 (Human Factors, Habitability and Environmental Health); NASA/SP-2010-3407 Human Integration Design Handbook (HIDH).

III. FURTHER INFORMATION

National Academy of Sciences. (1980). Recommended standard procedures for the clinical measurement and specification of visual acuity (report of working group 39 committee on vision). *Archives of Opthalmology, 41*, 103–148.

National Institutes of Health. (1985). *Health implications of obesity.* NIH Consensus Statement, 5(9), (pp. 1–7). Washington, DC: US Government Printing Office.

National Institutes of Health. (1990). *Noise and hearing loss.* NIH Consensus Statement, 8(1), (pp. 1–24). Washington, DC: US Government Printing Office.

National Research Council. (1979). *Odors from stationary and mobile source.* Washington, DC: National Academy Press.

National Research Council. (1983a). Committee on human factors. *Research needs for human factors.* Washington, DC: National Academy Press.

National Research Council. (1983b). Committee on vision. *Video displays, work and vision.* Washington, DC: National Academy Press.

National Research Council. (1998). *Work-related musculoskeletal disorders: A review of the evidence.* Washington, DC: National Academy Press.

National Research Council. (1999). *Work-related musculoskeletal disorders: Report, workshop summary, and workshop papers.* Washington, DC: National Academy Press.

National Research Council. (2003). Committee to review the scientific evidence on the polygraph. *The polygraph and lie detection.* Washington, DC: National Academies Press.

National Research Council. (2011). Committee on the Role of Human Factors in Home Health Care, Board on Human-Systems Integration, Division of Behavioral and Social Sciences and Education, National Research Council of the National Academies. *Health care comes home, the human factors.* Washington, DC: National Academies Press.

Nayak, U. S. L. (1995). Elders-led design. *Ergonomics in Design, 3*, 8–13.

Nelson, N. A., & Silverstein, B. A. (1998). Workplace changes associated with a reduction in musculoskeletal symptoms in office workers. *Human Factors, 40*, 337–350.

Neuffer, M. B., Schulze, L. J. H., & Chen, J. (1997). Body part discomfort reported by legal secretaries and word processors before and after implementation of mandatory typing breaks. *Proceedings of the Human Factors and Ergonomics Society 41st annual meeting* (pp. 624–628). Santa Monica, CA: Human Factors and Ergonomics Society.

New, Y. Y., Toyama, S., Akagawa, M., Yamada, M., Sotta, K., Tanzawa, T., et al. (2012). Workload assessment with Ovako working posture analysis system (OWAS) in Japanese Vineyards with focus on pruning and berry thinning operations. *Journal of the Japanese Society for Horticultural Science, 81*(4), 320–326.

Newell, A. F., & Gregor, P. (1997). Human computer interfaces for people with disabilities. In M. G. Helander, T. K. Landauer & P. V. Prabhu (Eds.), *Handbook of human–computer interaction* (pp. 813–824). Amsterdam, The Netherlands: Elsevier (Chapter 35).

Nicholls, J. A., & Grieve, D. W. (1992). Performance of physical tasks in pregnancy. *Ergonomics, 35*, 301–311.

Nicholson, A. S. (1989). A Comparative study of methods for establishing load handling capabilities. *Ergonomics, 32*, 1125–1144.

Nicholson, A. S. (1991). Anthropometry and workspace design. In A. Mital & W. Karwowski (Eds.), *Workspace, equipment and tool design* (pp. 3–28). Amsterdam, The Netherlands: Elsevier.

Nicogossian, A. E., Huntoon, C. L., & Pool, S. L. (1989). *Space physiology and medicine* (2nd ed.). Philadelphia, PA: Lea and Febiger.

NIOSH. (1981). Work practices guide for manual lifting. In *DHHS (NIOSH) publication* (pp. 81–122). Washington, DC: US Government Printing Office.

NIOSH. (1994). Back belt working group. In *Workplace use of back belts.* Cincinnati, OH: NIOSH (DHHS NIOSH Report 94-122).

NIOSH. (1997). Alternative keyboards. In *DHHS NIOSH publication* (pp. 97–148). Washington, DC: U.S. Government Printing Office.

Noble, C. E. (1978). Age, race, and sex in the learning and performance of psycho-motor skills. In R. T. Osborne, C. E. Noble & N. Weyl (Eds.), *The biopsychology of age, race, and sex* (pp. 287–378). New York, NY: Academy Press.

Nojkov, B., Rubenstein, J. H., Chey, W. D., & Hoogerwerf, W. A. (2010). The impact of rotating shift work on the prevalence of irritable bowel syndrome in nurses. *The American Journal of Gastroenterology, 105*(4), 842–847.

Nordin, M. (1991). Worker training and conditioning. *Proceedings of occupational ergonomics: Work related upper limb and back disorders (not paginated).* San Diego, CA: American Industrial Hygiene Association, San Diego Section.

Nordin, M., & Frankel, V. H. (1989). *Basic biomechanics of the musculoskeletal system.* Philadelphia, PA: Lea & Febiger.

Nordin, M., Andersson, G. B. J., & Pope, M. H. (1997). *Musculoskeletal disorders in the workplace: Principles and practices.* St. Louis, MO: Mosby.

Norman, D. A. (1991). Cognitive science in the cockpit. *CSERIAC Gateway, 2*, 1–6.

Norman, D. A., & Fisher, D. (1982). Why alphabetic keyboards are not easy to use: Keyboard layout doesn't much matter. *Human Factors, 24*, 509–519.

III. FURTHER INFORMATION

Noyes, J. (1983a). The QWERTY keyboard: A review. *International Journal of Man–Machine Studies, 18*, 265–281.

Noyes, J. (1983b). Chord keyboards. *Applied Ergonomics, 14*, 55–59.

Nunnely, S. A., French, J., Vanderbeek, R. D., & Stranges, S. F. (1995). Thermal study of anti-g ensembles aboard F-16 aircraft in hot weather. *Aviation, Space and Environmental Medicine, 66*, 309–311.

Nussbaum, M. A., & Chaffin, D. B. (1996). Development and evaluation of a scaleable and deformable geometric model of the human torso. *Clinical Biomechanics, 11*, 25–34.

Nygren, T. W. (1991). Psychometric properties of subjective workload measurement techniques. *Human Factors, 33*, 17–33.

Oberman, L., & Ramachandran, V. (2008). Preliminary evidence for deficits in multisensory integration is autism spectrum disorders. *Social Neuroscience, 3*, 348–355.

Oborne, D. J. (1983). Vibration at Work. In D. J. Oborne & M. M. Gruneberg (Eds.), *The physical environment at work* (pp. 143–177). New York, NY: Wiley.

Occelli, V., Esposito, G., Venutie, P., Arduino, G., & Zampini, M. (2013). The Takete-Maluma phenomenon in autism spectrum disorders. *Perception, 42*(2), 233–241.

Occhipinti, E., & Colombini, D. (1998). Proposed precise index for assessment of exposure to repetitive movements of the upper limbs (OCRA Index). In S. Kumar (Ed.), *Advances in occupational ergonomics and safety* (pp. 467–470). Amsterdam, The Netherlands: IOS Press.

Occhipinti, E., Colombini, D., & Grieco, A. (1991). A procedure for the formulation of synthetic risk indices in the assessment of fixed working postures. *Proceedings of the 11th congress of the International Ergonomics Association (3–5)*. London, UK: Taylor, Francis.

Occhipinti, E., Colombini, D., Frigo, C., Pedotti, A., & Grieco, A. (1985). Sitting posture: Analysis of lumbar stresses with upper limbs supported. *Ergonomics, 28*, 1333–1346.

O'Donnell, R. D., & Egglemeyer, F. T. (1986). Workload assessment methodology. In K. R. Boff, L. Kaufman & J. P. Thomas (Eds.), *Handbook of perception and human performance* (pp. 42.1–42.49). (II). New York, NY: Wiley.

Oezkaya, N., Nordin, M., Goldsheyder, D., & Leger, D. (2012). *Fundamentals of biomechanics*. Heidelberg, Germany: Springer.

Ogawa, I., & Arai, K. (1995). Ergonomic considerations for western-style toilets used in Japan. *Proceedings of the International Ergonomics Association World Conference 1995* (pp. 745–748). Rio de Janeiro, Brazil: Associacao Brasiliera der Ergonomia Rio de Janeiro.

Ong, C. N., & Kogi, K. (1990). Shiftwork in developing countries: Current issues and trends. In A. J. Scott (Ed.), *Shiftwork* (pp. 417–428). Philadelphia, PA: Hanley and Belfus.

Oriet, L. P., & Dutta, S. P. (1989). Investigations of modelling two-worker lifting teams. In A. Mital (Ed.), *Advances in industrial ergonomics and safety I* (pp. 679–683). London, UK: Taylor & Francis.

Osczevski, R., & Bluestein, M. (2005). The new wind chill equivalent temperature chart. *Bulletin American Meteorological Society, 86*, 1453–1458.

Osler, W. (1892). The principles and practice of medicine, IX. *Professional spasms: Occupation neuroses*. New York, NY: Appleton.

Ostlere, S. J., & Gold, R. H. (1991). Osteoporosis and bone density measurement methods. *Clinical Orthopaedics, 271*, 149–163.

Owen, D. (2006). The soundtrack of your life: Muzak in the realm of retail theatre. *The New Yorker*, 66–71.

Owens, D. A., & Leibowitz, H. W. (1983). Perceptual and motor consequences of tonic vergence. In C. Schor & K. Ciuffreda (Eds.), *Vergence eye movements: Basic and clinical aspects* (pp. 25–74). Boston, MA: Butterworths.

Owings, C. L., Chaffin, D. B., Snyder, R. G., & Norcutt R (1975). *Strength characteristics of U.S. children for product safety design*. Report 011903-F. Ann Arbor, MI: The University of Michigan.

Oxenburgh, M. S. (1997). Cost–benefit analysis of ergonomics programs. *American Industrial Hygiene Association Journal, 58*, 150–156.

Palinkas, L. A. (2001). Mental and cognitive performance in the cold. *International Journal of Circumpolar Health, 60*(3), 430–439.

Panek, P. E. (1997). The older worker. In A. D. Fisk & W. A. Rogers (Eds.), *Handbook of human factors and the older adult* (pp. 363–394). San Diego, CA: Academic Press.

Panjabi, M. M., Goel, V., Oxland, T., Takata, K., Duranceau, J., Krag, M., et al. (1992). Human lumbar vertebrae quantitative three-dimensional anatomy. *Spine, 17*, 299–306.

Paquet, V., & Feathers, D. (2004). An anthropometric study of manual and powered wheelchair users. *International Journal of Industrial Ergonomics, 33*(3), 191–204.

Paquette, S., Gordon, C., & Bradtmiller, B. (2009). *Anthropometric survey (ANSUR) II Pilot study: Methods and summary statistics* (TR-09/014). U.S. Army Natick Soldier Research. Natick, MA: Development and Engineering Center.

Park, D. C., & Jones, T. R. (1997). Medication adherence and aging. In A. D. Fisk & W. A. Rogers (Eds.), *Handbook of human factors and the older adult* (pp. 257–287). San Diego, CA: Academic Press (Chapter 11).

Parsons, H. M. (1974). What happened at Hawthorne? *Science, 18*, 922–932.

Parsons, H. M. (1990). Assembly ergonomics in the hawthorne studies. *Proceedings of the International Ergonomics Association conference on human factors in design for manufacturability process planning* (pp. 299-305). Buffalo, NY: Helander Dept. of IE, SUNYAB.

Parsons, K. C. (1995). International heat stress standards: A review. *Ergonomics, 32*, 6–22.

Parsons, K. C. (2003). *Human thermal environments* (2nd ed.). London, UK: Taylor & Francis.

Parsons, K. C. (2005). Ergonomic assessments of thermal environments. In J. R. Wilson & N. Corlett (Eds.), *Evaluation of human work* (3rd ed.). London, UK: Taylor & Francis (Chapter 23).

Parsons, K. C. (2014). *Human thermal environments* (3rd ed.). Boca Raton, FL: CRC Press.

Patel, T., Sanjog, J., Chatterjee, A., & Karmakar, S. (2016). Statistical interpretation of collected anthropometric data of agricultural workers from northeast India and comparison with national and international databases. *IIE Transactions on Occupational Ergonomics and Human Factors, 4*(4), 197–210.

Patkai, P. (1985). The menstrual cycle. In S. Folkard & T. H. Monk (Eds.), *Hours of work* (pp. 87–96). Chichester, UK: Wiley.

Patterson, E. S., Latkany, P., Brick, D., Gibbons, M. C., Ramaiah, M., & Lowry, S. Z. (2015). Integrating electronic health records into clinical workflow. *Proceedings of the International Symposium on Human Factors and Ergonomics in Health Care June, 4*, 42–49.

Paul, J. A., & Douwes, M. (1993). Two-dimensional photographic posture recording and description: A validity study. *Applied Ergonomics, 24*, 83–90.

Paykel, E. S. (1997). The interview for recent life events. *Psychological Medicine, 27*, 301–310.

Peebles, L., & Norris, B. (1998). *ADULTDATA: The handbook of adult anthropometric and strength measurements—Data for design safety*. London, UK: Department of Trade and Industry DTI/Pub 2917/3k/6/98/NP.

Peebles, L., & Norris, B. (2000). *Strength data*. London, UK: Department of Trade and Industry DTI/URN 00/1070.

Peebles, L., & Norris, B. (2003). Filling "gaps" in strength data for design. *Applied Ergonomics, 34*, 73–88.

Peloquin, A. A. (1994). *Barrier-free residental design*. New York, NY: McGraw-Hill.

Pereira, A., Hsieh, C. M., Laroche, C., & Rempel, D. (2014). The effect of keyboard key spacing on typing speed, error, usability, and biomechanics Part 2. *Vertical Spacing Human Factors, 56*, 752–759.

Pereira, A., Wachs, J. P., Park, K., & Rempel, D. (2015). A user-developed 3-D hand gesture set for human–computer interaction. *Human Factors, 57*, 607–621.

Peres, N. J. V. (1961). Process work without strain. *Australian Factory, 1*, 1–12.

Permanen, J. (2012). Some reasons to revise the International Standard ISO 226 2003: Acoustics—Normal equal-loudness-level contours. *Open Journal of Acoustics, 2*, 143–149.

Perotto, A. O. (1994). *Anatomical guide for the electomyographer: The limbs and trunk*. Springfield, IL: Thomas.

Peters, G. A., & Peters, B. J. (2013). *Human safety*. CreateSpace Independent Publishing Platform.

Peterson, D. R., & Bronzino, J. G. (2015). *Biomechanics: Principles and practices*. Boca Raton, FL: CRC Press.

Pew, R. W. (1995). Pew's principles of human factors practice. *Proceedings of the Silicon Valley ergonomics conference exposition Ergocon '95 (pp. 3–4)*. San Jose, CA: San Jose State University.

Pew, P. W. & Mavor, A. S. (Eds.). (2007). *Human-system integration in the system development process*. Washington, DC: The National Academies.

Peyer, K. E., Morris, M., & Sellers, W. I. (2015). Subject-specific body segment parameter estimation using 3D photogrammetry with multiple cameras. *PeerJ, 3*, e831.

Phalen, G. S. (1966). The carpal-tunnel syndrome: Seventeen years' experience in diagnosis and treatment of six hundred fifty-four hands. *Journal of Bone & Joint Surgery, 48*(2), 211–228.

Phalen, G. S. (1972). The carpal-tunnel syndrome clinical evaluation of 598 hands. *Clinical Orthopaedics and Related Research, 83*, 29–40.

Pheasant, S. (1986). *Bodyspace: Anthropometry, ergonomics, and design*. London, UK: Taylor & Francis.

Pheasant, S. (1996). *Bodyspace: Anthropometry, ergonomics, and design* (2nd ed.). London, UK: Taylor & Francis.

Pheasant, S., & Haslegrave, C. M. (2006). *Anthropometry, ergonomics and the design of work* (3rd ed.). London, UK: Taylor & Francis.

Phillips, C. A. (2000). *Human factors engineering*. New York, NY: Wiley.

Plog, B. A. (Ed.). (2001). *Fundamentals of industrial hygiene* (5th ed.). Itasca, IL: National Safety Council.

Pokorny, J., & Smith, V. C. (1986). Colorimetry and color discrimination. In K. R. Boff, L. Kaufman & J. P. Thomas (Eds.), *Handbook of perception and human performance* (pp. 8.1–8.51). New York, NY: Wiley (Chapter 8).

Poore, G. V. (1872). Writer's cramp: Its pathology and treatment. *Practitioner, 40*, 341–350.

Poore, G. V. (1887). Clinical lecture on certain conditions of the hand and arm which interfere with the performance of professional acts, especially piano-playing. *The British Medical Journal, 26*, 441–444 February issue.

III. FURTHER INFORMATION

Popkin, S. M. (Ed.), (2015). *Reviews of human factors and ergonomics: Vol. 10. Worker fatigue and transportation safety.* Santa Monica, CA: Human Factors and Ergonomics Society.

Popkin, S. M., Howarth, H. D., & Tepas, D. I. (2006). Ergonomics of work systems. In G. Salvendy (Ed.), *Handbook of human factors and ergonomics* (3rd ed., pp. 761–800). Hoboken, NY: Wiley.

Porter, G. (1996). Organizational impact of workaholism: Suggestions for researching the negative outcomes of excessive work. *Journal of Occupational Health Psychology, 1,* 70–84.

Post, D. L. (1997). Color and human–computer interaction. In M. Helander, T. K. Landauer & P. Prabhu (Eds.), *Handbook of human–computer interaction* (2nd ed., pp. 573–615). Amsterdam, The Netherlands: Elsevier.

Potvin, J. R. (1997). Use of NIOSH equation inputs to calculate lumbosacral compression forces. *Ergonomics, 40,* 650–655.

Potvin, J. R., & Bent, L. R. (1997). NIOSH equation distances associated with the Liberty Mutual (Snook) lifting table box width. *Ergonomics, 40,* 691–707.

Prentice, A. M., & Jebb, S. A. (2001). Beyond body mass index. *Obesity Reviews, 2*(3), 141–147.

Price, D. L. (1988). Effects of alcohol and drugs. In G. A. Peters & B. J. Peters (Eds.), *Automotive engineering and litigation* (pp. 489–551). (2). New York, NY: Garland.

Price, D. L., Radwan, M. A. E., & Tergou, D. E. (1986). Gender, alcohol pacing and incentive effects on an electronics assembly task. *Ergonomics, 29,* 393–406.

Priel, V. Z. (1974). A numerical definition of posture. *Human Factors, 16,* 576–584.

Proctor, R. W., & Van Zandt, T. (1994). *Human factors in simple and complex systems.* Boston, MA: Allyn and Bacon.

Proctor, R. W., & Van Zandt, T. (2008). *Human factors in simple and complex systems* (2nd ed.). Boca Raton, FL: CRC Press.

PsychCorp. (2009). *Wechsler memory scale WMS-IV; technical and interpretive manual.* San Antonio, TX: Pearson.

Putz-Anderson, V. (1988). *Cumulative trauma disorders: A manual for musculoskeletal diseases of the upper limbs.* London, UK: Taylor & Francis.

Putz-Anderson, V., & Waters, T. (1991). *Revisions in NIOSH guide to manual lifting.* Paper presented at national conference entitled "A national strategy for occupational musculoskeletal injury prevention: Implementation issues and research needs." Ann Arbor, MI: University of Michigan.

Quick, J. C. (1999). Occupational health psychology: Historical roots and future directions. *Health Psychology, 18*(1), 82–88.

Rabbitt, P. (1991). Management of the working population. *Ergonomics, 34,* 775–790.

Rabbitt, P. (1997). Aging and human skill: A 40th anniversary. *Ergonomics, 40,* 962–981.

Radwin, R. G. (1997). Force dynamometers and accelerometers. In W. Karwowski & W. S. Marras (Eds.), *The occupational ergonomics handbook* (pp. 565–581). Boca Raton, FL: CRC Press.

Radwin, R. G., Beebe, D. J., Webster, J. G., & Yen, T. Y. (1996). Instrumentation for occupational ergonomics. In A. Bahttacharya & J. D. McGlothlin (Eds.), *Occupational ergonomics* (pp. 165–193). New York, NY: Dekker (Chapter 7).

Ramazzini, B. (1713). *De morbis artificum.* (Diseases of Workers, W. C. Wright. Trans. Thunder Bay, Canada: OH&S Press (1993)).

Ramsey, J. D. (1995). Task performance in heat: A review. *Ergonomics, 32,* 154–165.

Ramsey, T., Davis, K. M., Kotowski, S. E., Anderson, V. P., & Waters, T. (2014). Reduction of spinal loads through adjustable interventions at the origin and destination of palletizing tasks. *Human Factors, 56,* 1222–1234.

Raschke, U. (2005). Digital human modeling research and development user needs panel. *SAE Technical Paper Series, 01,* 2745.

Raskin, D. C. (1988). Does science support polygraph testing? In A. Gale (Ed.), *The polygraph test: Lies, truth, and science* (pp. 96–110). London, UK: Sage.

Rastakis, L. (1998). Human-centered design project revolutionizes air combat. *CSERIAC Gateway, 9*(1), 1–6.

Rea, M. S. (2005). Photometric characterization of the luminous environment. In N. Stanton, A. Hedge, K. Brookhuis, E. Salas & H. Hendrick (Eds.), *Handbook of human factors and ergonomics methods.* Boca Raton, FL: CRC Press (Chapter 68).

Reddy, N. P., & Gupta, V. (2007). Toward direct biocontrol using surface EMG signals: Control of finger and wrist joint models. *Medical Engineering & Physics, 29*(3), 398–403.

Redfern, M. S., & Bidanda, B. (1994). Slip resistance of the shoe-floor interface under biomechanically relevant conditions. *Ergonomics, 37,* 511–524.

Redgrove, J. A. (1976). Sex differences in information processing: A theory and its consequences. *Journal of Occupational Psychology, 49,* 29–37.

Refinetti, R. (2016). *Circadian physiology* (3rd ed.). Boca Raton, FL: CRC Press.

Reid, G. B., & Nygren, T. E. (1988). The subjective workload assessment techniques: A scaling procedure for measuring mental workload. In P. A. Hancock & N. Meshkati (Eds.), *Human mental workload* (pp. 185–218). Amsterdam, The Netherlands: Elsevier.

III. FURTHER INFORMATION

Reiss, B. (2017). *Wild nights: How taming sleep created our restless world*. New York, NY: Basic Books.

Reith, M. S. (1982). *The relationship of isometric grip strength, optimal dynamometer settings, and certain anthropometric factors.* (Unpublished Master's Thesis). Richmond, VA: Virginia Commonwealth University.

Remington, R. J., & Rogers, M. (1969). *Keyboard literature survey phase 1: Bibliography* (TR 29.0042). Research Triangle Park, IBM Systems Development Division.

Rempel, D. (2008). The split keyboard: An ergonomics success story. *Human Factors, 50*, 385–392.

Rempel, D., Serina, E., & Klinenberg, E. (1997). The effects of keyboard keyswitch make force on applied force and finger flexor muscle activity. *Ergonomics, 40*, 800–808.

Rempel, D., Tittiranonda, P., Burastero, S., Hudes, M., & So, Y. (1999). Effect of keyboard keyswitch design on hand pain. *Journal of Occupational and Environmental Medicine, 41*, 111–119.

Rempel, D., Willems, K., Anshel, J., Jaschinski, W., & Sheedy, J. (2007). The effects of visual display distance on eye accommodation, head posture, and vision and neck symptoms. *Human Factors, 49*, 830–838.

Rice, V. J. B. (Ed.). (1998). *Ergonomics in health care and rehabilitation*. Boston, MA: Butterworth-Heinemann.

Rice, V. J. (1999). Preplacement strength screening. In W. Karwowski & W. S. Marras (Eds.), *The occupational ergonomics handbook* (pp. 1299–1319). Boca Raton, FL: CRC Press.

Ridder, C. A. (1959). *Basic design measurements for sitting*. Fayetteville, AR: University of Arkansas (Bulletin 616, Agricultural Experiment Station).

Ripple, P. H. (1952). Accommodative amplitude and direction of gaze. *American Journal of Ophthalmology, 35*, 1630–1634.

Robbins, H. (1963). Anatomical study of the median nerve in the carpal tunnel and etiologies of the carpal-tunnel syndrome. *The Journal of Bone & Joint Surgery, 45*, 953–966.

Robinette, K. M., & Hudson (2006). Anthropometry. In G. Salvendy (Ed.), *Handbook of human factors and ergonomics* (3rd ed., pp. 322–339). New York, NY: Wiley.

Robinette, K. M., & Vetch, D. (2016). Sustainable sizing. *Human Factors, 58*(5), 657–664.

Robinson, P. A. (2009). *Writing and designing manuals and warnings* (4th ed.). Boca Raton, FL: CRC Press.

Robinson, G. S., & Casali, J. G. (1995). Audibility of reverse alarms under hearing protectors for normal and hearing-impaired listeners. *Ergonomics, 38*, 2281–2299.

Robinson, D. W., & Dadson, R. S. (1956). A re-determination of the equal-loudness relations for pure tones. *British Journal of Applied Physics, 7*, 166–181.

Robinson, G. S., Casali, J. G., & Lee, S. E. (1997). *Role of driver hearing in commercial motor vehicle operations: An evaluation of the FHWA hearing requirement—Final report*. Washington, DC: Federal Highway Administration (Report DTFH 61-95-C-00172; NTIS PB98-114606).

Rodahl, K. (1989). *The physiology of work*. London, UK: Taylor & Francis.

Rodgers, S. H. (1997). Work physiology—Fatigue and recovery. In G. Salvendy (Ed.), *Handbook of human factors and ergonomics* (2nd ed., pp. 268–297). New York, NY: Wiley (Chapter 10).

Rodin, J. (1993). Cultural and psychosocial determinants of weight concerns. *Annals of Internal Medicine, 119*, 643–645.

Roebuck, J. A., Kroemer, K. H. E., & Thomson, W. G. (1975). *Engineering anthropometry methods*. New York, NY: Wiley.

Roebuck, J., Smith, K., & Raggio, L. (1988). *Forecasting crew anthropometry for shuttle and space station (STS 88-0717)*. Downey, CA: Rockwell International.

Roebuck, J. A. (1995). Anthropometric methods. *Designing to fit the human body*. Santa Monica, CA: Human Factors and Ergonomics Society.

Roethlisberger, F. J., & Dickson, W. J. (1939). *Management and the worker*. Cambridge, MA: Harvard University Press.

Rogers, A. S., Spencer, M. B., Stone, B. M., & Nicholson, A. N. (1989). The influence of a 1-hr nap on performance overnight. *Ergonomics, 32*, 1193–1205.

Rogers, W. A., Meyer, B., Walker, N., & Fisk, A. D. (1998). Functional limitations to daily living tasks in the aged: A focus group analysis. *Human Factors, 40*, 111–125.

Rohmert, W. & Rutenfranz, J. (Eds.). (1983). *Practical work physiology (in German)* (3rd ed.). Stuttgart, Germany: Thieme.

Rosa, R. R., & Colligan, M. J. (1988). Long workdays versus rest days: Assessing fatigue and alertness with a portable performance battery. *Human Factors, 30*, 305–317.

Roscoe, A. H. (1993). Heart rate as a psychophysical measure for in-flight work assessment. *Ergonomics, 36*, 1055–1062.

Ross, L. E., & Mundt, J. C. (1988). Multi-attribute modeling analyses of the effects of a low blood alcohol level on pilot performance. *Human Factors, 30*, 293–304.

Rouse, W. B. (1991). Human-centered design: Creating successful products, systems, and organisations. *CSERIAC Gateway, 2*(4), 1–3.

Rowe, M. L. (1983). *Backache at work*. Fairport, NY: Perinton.

III. FURTHER INFORMATION

Rubio, S., Diaz, E., Martin, J., & Puente, J. M. (2004). Evaluation of subjective mental workload: A comparison of SWAT, NASA-TLX, and workload profile methods. *Applied Psychology an International Review, 53*(1), 61–86.

Rutter, B. G., Haager, J. A., Daigle, G. C., Smith, S., McFarland, N., & Kelsey, N. (1984). Dimensional changes throughout pregnancy: A preliminary report. *Carle Select Papers, 36*, 44–52.

Sachs, M., Ellis, R., Schlaug, G., & Psyche, L. (2016). Brain connectivity reflects human aesthetic responses to music. *Social Cognitive and Affective Neuroscience, 11*(6), 884–891.

Sacks, O. (1990). *The man who mistook his wife for a hat*. New York, NY: Harper Perennial.

Sacks, O. (2007). *Musicophilia, tales of music and the brain*. New York, NY: Knopf. (Chapter 22, pp. 264–275). (In particular, Chapter 22: *Athletes of the small muscles: Musician's dystonia*).

SAE. (1973). *Subjective rating scale for evaluation of noise and ride comfort characteristics related to motor vehicle tires*. Detroit, MI: Society of Automotive Engineers (SAE J1060, Recommended Practice 29.11).

Salvendy, G. (Ed.). (2012). *Handbook of human factors and ergonomics* (4th ed.). New York, NY: Wiley.

SAMSHA. (2016). *Key substance use and mental health indicators in the United States: Results from the 2015 National Survey on Drug Use and Health*. [US] Substance Abuse and Mental Health Services Administration (SAMSHA). (HHS Publication No. SMA 16-4984).

Sanders, M. S., & McCormick, E. J. (1987). *Human factors in engineering and design* (6). New York, NY: McGraw-Hill.

Sauter, S. L., Schleifer, L. M., & Knutson, S. J. (1991). Work posture, workstation design, and musculoskeletal discomfort in a VDT data entry task. *Human Factors, 32*, 151–167.

Saval, N. (2014). Cubed. *A secret history of the workplace*. New York, NY: Doubleday.

Sayer, J. R., Sebok, A. L., & Snyder, H. L. (1990). Color-difference metrics: Task performance prediction for multichromatic CRT applications as determined by color legibility. *SID Digest* (pp. 265–268). Playa del Rey, CA: Society for Information Display.

Scerbo, M. W. (1995). Usability Testing. In J. Weimer (Ed.), *Research techniques in human engineering* (pp. 72–111). Englewood Cliffs, NJ: Prentice Hall.

Scheier, M. G., & Bridges, M. W. (1995). Person variables and health: Personality predispositions and acute psychological states as shared determinants for disease. *Psychosomatic Medicine, 57*, 255–268.

Scherrer, J. (1967). *Physiologie de travail*. Paris, France: Masson.

Schlegel, B., Schlegel, R. E., & Gilliland, K. (1988). Gender differences in criterion task set performance and subjective rating. *Proceedings of the Human Factors Society 32nd annual meeting* (pp. 848-852). Santa Monica, CA: Human Factors Society.

Schmidt, R. A. (1988). *Motor control and learning* (2nd ed.). Champaign, IL: Human Kinetics.

Schnall, P. L., Schwartz, J. E., Landsbergis, P. A., Warren, K., & Pickering, T. G. (1992). Relation between job strain, alcohol, and ambulatory blood pressure. *Hypertension, 19*, 488–494.

Schneck, D. J. (1992). *Mechanics of muscle* (2nd ed.). New York, NY: New York University Press.

Schneider, S., & Susi, P. (1994). Ergonomics and construction: A review of potential hazards in new construction. *American Industrial Hygiene Association Journal, 55*, 635–649.

Schneider, L. W., Lehman, R. J., Pflug, M. A., & Owings, C. L. (1986). *Size and shape of the head and neck from birth to four years*. Ann Arbor, MI: The University of Michigan, Transportation Research Institute (Final Report CPSC-C-83-1250).

Scholey, M., & Hair, M. (1989). Back pain in physiotherapists involved in back care education. *Ergonomics, 32*, 179–190.

Schroetter, H. (1925). Knowledge of the energy consumption of typewriting (in German). *Pflüger's Archiv für die Gesamte Physiosologie des Menschen und der Tiere, 207*(4), 323–342.

Schultz, A. B., & Andersson, G. B. J. (1981). Analysis of loads on the lumbar spine. *Spine, 6*, 76–82.

Scott, D., & Marcus, S. (1991). Hand impairment assessment: Some suggestions. *Applied Ergonomics, 22*, 263–269.

Scripture, E. W. (1899). *The new psychology*. New York, NY: Charles Scribner's Sons.

Sedgwick, W. (1888). *Studies from the biological laboratory*. Baltimore, MD: N. Murray, Johns Hopkins University 399.

Sedgwick, A. W., & Gormley, J. T. (1998). Training for lifting: An unresolved ergonomic issue? *Applied Ergonomics, 29*, 395–398.

Seibel, R. (1972). Data entry devices and procedures. In H. P. Van Cott & R. G. Kinkade (Eds.), *Human engineering guide to equipment design* (pp. 312–344). Washington, DC: U.S. Government Printing Office (Chapter 7).

Seidel, H. (1988). Myoelectric reactions to ultra-low frequency. Whole-body vibration. *European Journal of Applied Physiology, 57*, 558–562.

Seidel, H., & Heide, R. (1986). Long-term effects of whole-body vibration: A critical survey of the literature. *International Archives of Occupational and Environmental Health, 58*, 1–26.

Seligman, M. (2002). *Authentic happiness*. New York, NY: Free Press.

Selye, H. (1956). *The stress of life*. New York, NY: McGraw-Hill. (Revised edition 1978).

Selye, H. (1974). *Stress without distress*. Philadelphia, PA: Lippincott.

III. FURTHER INFORMATION

Serina, E. R., & Rempel, D. M. (1996). Fingertip pulp response during keystrikes. *Proceedings of the American Society of Biomechanics 20th annual meeting* (pp. 237–238). Atlanta, GA: American Society of Biomechanics.

Serow, W. J., & Sly, D. F. (1988). The demography of current and future aging cohorts. *Committee on an Aging Society America's Aging—The social and built environment in an older society* (pp. 42–102). Washington, DC: National Academy Press.

Seshagiri, B. (1998). Occupational noise exposure of operators of heavy trucks. *American Industrial Hygiene Association Journal, 59*, 205–213.

Seshagiri, B., & Stewart, B. (1992). Investigation of the audibility of locomotive horns. *American Industrial Hygiene Association Journal, 53*, 726–735.

Seven, S. A. (1989). Workload measurement reconsidered. *Human Factors Society Bulletin, 32*, 5–7.

Shackel, B., Chidsey, K. D., & Shipley, P. (1969). The assessment of chair comfort. *Ergonomics, 12*, 169–306.

Shah, R. K. (1993). A pilot survey of the traditional use of the patuka round the waist for the prevention of back pain in Nepal. *Applied Ergonomics, 24*, 337–344.

Shapiro, Y., Pandolf, K. B., Avellini, B. A., Pimental, N. A., & Goldman, R. F. (1981). Heat balance and transfer in men and women exercising in hot-dry and hot-wet conditions. *Ergonomics, 24*, 375–386.

Sharp, M. A., & Legg, S. J. (1988). Effects of psychophysical lifting training on maximal repetitive lifting capacity. *American Industrial Hygiene Association Journal, 49*, 639–644.

Sharp, M., Rice, V., Nindle, B., & Williamson, T. (1993). Maximum lifting capacity in single and mixed gender three-person teams. *Proceedings of the Human Factors and Ergonomics Society 37th annual meeting* (pp. 725–729). Santa Monica, CA: Human Factors and Ergonomics Society.

Sheedy, J. (2006). Vision and work. In W. S. Marras & W. Karwowski (Eds.), *The occupational ergonomics handbook: Fundamentals and assessment tools for occupational ergonomics* (2nd ed.). Boca Raton, FL: CRC Press (Chapter 18).

Shepherd, R. J. (1995). Exercise behavior of the elderly. Introduction theories of exercise behavior. In S. Harris, E. Heikkinen & W. S. Harris (Eds.), *Physical activity, aging and sports* (pp. 97–105). (IV). Albany, NY: Center for the Study of Aging (Chapter 18).

Sherrick, C. E., & Cholewiak, R. W. (1986). Cutaneous sensitivity. In K. R. Boff, L. Kaufmann & J. P. Thomas (Eds.), *Handbook of perception and human performance* (pp. 12.1–12.58). New York, NY: Wiley.

Sholes, C. L. (1878). *Improvement in type-writing machines*. US Patent 207,559. Washington, DC: US Patent and Trademark Office.

Siegel, J. M. (2003). Why we sleep. *Scientific American, 289*(5), 92–97.

Siekmann, H. (1990). Recommended maximum temperatures for touchable surfaces. *Ergonomics, 21*, 69–73.

Siervogel, R. M., Roche, A. F., Guo, S., Mukherjee, D., & Chumlea, W. C. (1991). Patterns of change in weight/stature from 2 to 18 years: Findings from long-term serial data for children in the Fels longitudinal growth study. *International Journal of Obesity, 15*, 479–485.

Silverstein, B. A. (1985). *The prevalence of upper extremity cumulative trauma disorders in industry*. (Unpublished Doctoral Dissertation). Ann Arbor, MI: University of Michigan.

Silverstein, B. A., Armstrong, T. J., Longmate, A., & Woody, D. (1988). Can in-plant exercise control musculoskeletal symptoms? *JOM Journal of Occupational Medicine, 30*, 922–927.

Simmons, J. P., Nelson, L. D., & Simonsohn, U. (2011). False-positive psychology: Undisclosed flexibility in data collection and analysis allows presenting anything as significant. *Psychological Science, 22*(11), 1359–1366.

Simoneau, G. G., Marklin, R. W., & Monroe, J. F. (1999). Wrist and forearm postures of users of conventional computer keyboards. *Human Factors, 41*, 413–424.

Sinclair, D. C. (1973). Mapping of spinal nerve roots. In A. Jarrett (Ed.), *The physiology and pathophysiology of the skin* (vol. 2, pp. 349). London, UK: Academic Press.

Singer, L. D., & Graeff, R. F. (1988). *A bathroom for the elderly. Report on grants 230-11-110H-150-8903081 and CAE-86-005-01*. Blacksburg, VA: College of Architecture and Urban Studies, Virginia Tech.

Sinke, C., Neufeld, J., Wiswede, D., Emrich, H., Bleich, S., Münte, T., et al. (2014). N1 enhancement in synesthesia during visual and audio-visual perception in semantic cross-modal conflict situations: An ERP study. *Frontiers in Neuroscience, 8*, 21.

Skottke, E. M., Debus, G., Wang, L., & Huestegge, L. (2014). Carryover effects of highly automated convoy driving on subsequent manual driving performance. *Human Factors, 56*, 1272–1283.

Smith, P. C. (1976). Behaviors results and organizational effectiveness: The problem of criteria. In M. D. Dunnette (Ed.), *Handbook of industrial and organizational psychology* (pp. 745–775). Chicago, IL: Rand McNally.

Smith, R. V. & Leslie, J. H. (Eds.). (1990). *Rehabilitation engineering*. Boca Raton, FL: CRC Press.

Smith, L., Macdonald, I., Folkard, S., & Tucker, P. (1998a). Industrial shift systems. *Applied Ergonomics, 29*, 273–280.

III. FURTHER INFORMATION

Smith, M. J., Karsh, B. T., Conway, F. T., Cohen, W. J., James, C. A., Morgan, J. J., et al. (1998b). Effects of a split keyboard design and wrist rest on performance, posture, and comfort. *Human Factors, 40*, 324–336.

Snook, S. H. (1987). Approaches to preplacement testing and selection of workers. *Ergonomics, 30*, 241–247.

Snook, S. H. (1988a). The control of low back disability: The role of management. Paper presented at the American Industrial Hygiene Conference, San Francisco, CA.

Snook, S. H. (1988b). *Low back pain.* (Available from Dr. S. H. Snook, 10472 SE Amberjack Court, Hobe Sound, FL 33455-3262.).

Snook, S. H. (1991). Low back disorders in industry. *Proceedings of the Human Factors Society 35th annual meeting* (pp. 830-833). Santa Monica, CA: Human Factors Society.

Snook, S. H. (2000). Back risk factors: An overview. In F. Violante, T. Armstrong & A. Kilbom (Eds.), *Occupational ergonomics. Work related musculoskeletal disorders of the upper limb and back.* London, UK: Taylor & Francis (Chapter 11).

Snook, S. H. (2005). Psychophysical tables: Lifting, lowering, pushing, pulling, and carrying. In N. Stanton, A. Hedge, K. Brookhuis, E. Salas & H. Hendrick (Eds.), *Handbook of human factors and ergonomics methods.* Boca Raton, FL: CRC Press (Chapter 13).

Snook, S. H., & Ciriello, V. M. (1991). The design of manual handling tasks: Revised tables of maximum acceptable weights and forces. *Ergonomics, 34*, 1197–1213.

Snyder, R. G. (1975). Impact. In J. F. Parker & V. R. West (Eds.), *Bioastronautics data book, NASA SP-3006* (pp. 221–295). Washington, DC: US Government Printing Office.

Snyder, R. G., Schneider, L. W., Owings, C. L., Reynolds, H. M., Golomb, D. H., & Schork, M. A. (1977). *Anthropometry of infants, children and youths to age 18 for product safety design (final report UM-HSRI-88-17).* Ann Arbor, MI: The University of Michigan, Highway Safety Research Institute.

Snyder, H. L. (1985a). The visual system: Capabilities and limitations. In L. E. Tannas (Ed.), *Flat-panel displays and CRTs* (pp. 54–69). New York, NY: Van Nostrand Reinhold.

Snyder, H. L. (1985b). Image quality: Measures and visual performance. In L. E. Tannas (Ed.), *Flat panel displays and CRTs* (pp. 71–90). New York, NY: Van Nostrand Reinhold.

Snyder, R. G., Spencer, M. L., Owings, C. L., & Schneider, L. W. (1975). *Physical characteristics of children as related to death and injury for consumer product safety design.* Ann Arbor, MI: The University of Michigan, Highway Safety Research Institute (Final Report, UM-HSRI-BI-75-5).

Soderberg, G. L. (Ed.). (1992). *Selected topics in surface electromyography for use in the occupational setting: Expert perspectives (DDHS-NIOSH Publication 91-100).* Washington, DC: US Department of Health and Human Service.

Sokal, R. R., & Rohlf, F. J. (2011). *Biometry: The principles and practices of statistics in biological research* (4th ed.). New York, NY: Freeman.

Soldo, B. J., & Longino, C. F. (1988). Social and physical environments for the vulnerable aged committee on an aging society. *America's aging—The Social and Built Environment in an Older Society* (pp. 103–133). Washington, DC: National Academy Press.

Solomon, S. S., & King, J. G. (1997). Fire truck visibility: Red may not be the most visible color. *Ergonomics in Design, 5*(2), 4–10 .

Sommerich, C. M., & Marras, W. S. (1992). Temporal patterns of trunk muscle activity throughout a dynamic asymmetric lifting motion. *Human Factors, 34*, 215–230.

Sommerich, C. M., & Marras, W. S. (2004). Electromyography and muscle force. In S. Kumar (Ed.), *Muscle strength.* Boca Raton, FL: CRC Press (Chapter 17).

Souza, E. G. V., Ramos, M. G., Hara, C., Stumpf, B. P., & Rocha, F. L. (2012). Neuropsychological performance and menstrual cycle: A literature review. *Trends Psychiatry Psychother, 34*(1), 5–12.

Spain, W. H., Ewing, W. M., & Clay, E. M. (1985). Knowledge of causes, controls aids prevention of heat stress. *Occup Health Safety, 54*(4), 27–33.

Spelt, P. F. (1991). Introduction to artificial neural networks for human factors. *Human Factors Society Bulletin, 34*, 1–4.

Spence, A. P. (1989). *Biology of human aging.* Englewood Cliffs, NJ: Prentice Hall.

Sprent, P. (2000). *Applied nonparametric statistical methods.* Boca Raton, FL: Chapman & Hall/ CRC Press.

Staff, K. R. (1983). *A comparison of range of joint mobility in college females and males.* (Unpublished Master's Thesis). College Station, TX: Texas A&M University.

Staffel, F. (1884). On the hygiene of sitting (in German). *Zbl. Allgemeine Gesundheitspflege, 3*, 403–421.

Staffel, F. (1889). *The types of human postures and their relations to deformations of the spine.* Wiesbaden, Germany: Bergmann (in German).

Stahre, M., Roeber, J., Kanny, D., Brewer, R., & Xingyou, Z. (2014). Contribution of excessive alcohol consumption to death and years of potential life lost in the United States. *Preventing Chronic Disease, 26*(11), E109.

Stanton, N., Hedge, A., Brookhuis, K., Salas, E. & Hendrick, H. (Eds.). (2005). *Handbook of human factors and ergonomics methods.* Boca Raton, FL: CRC Press.

III. FURTHER INFORMATION

Steenbekkers, L. P. A., & Molenbroek, J. F. M. (1990). Anthropometric data of children for non-specialist users. *Ergonomics, 33,* 421–429.

Stegemann, J. (1984). *Physiology of performance* (3rd ed.). Stuttgart, Germany: Thieme (in German).

Stekelenburg, M. (1982). Noise at work: Tolerable limits and medical control. *American Industrial Hygiene Association Journal, 43,* 402–410.

Sterling, P., & Eyer, J. (1988). Allostasis: A new paradigm to explain arousal pathology. In S. Fisher & J. Reason (Eds.), *Handbook of life stress, cognition and health* (pp. 629–649). Oxford, England: John Wiley.

Stevenson, J. M., Andrew, G. M., Bryant, J. T., Greenhorn, D. R., & Thomson, J. M. (1989). Isoinertial tests to predict lifting performance. *Ergonomics, 32,* 157–166.

Stevenson, J. M., Greenhorn, D. R., Bryant, J. T., Deakin, J. M., & Smith, J. T. (1996). Gender differences in performance of a selection test using the incremental lifting machine. *Applied Ergonomics, 27,* 45–52. Rebuttal and Reply. *Applied Ergonomics, 27,* 133–137.

Stobbe, T. J., & Plummer, R. W. (1988). Sudden-movement/unexpected loading as a factor in back injuries. In F. Aghazadeh (Ed.), *Trends in ergonomics/human factors V* (pp. 713–720). Amsterdam, The Netherlands: Elsevier.

Stockbridge, H. C. W. (1957). *Micro-shape-coded knobs for post office keys*. London, UK: Ministry of Supply (Techn. Memo No. 67).

Stoudt, H. W. (1981). The anthropometry of the elderly. *Human Factors, 23,* 29–37.

Straker, L. M., Stevenson, M. G., & Twomey, L. T. (1996). A comparison of risk assessment of single and combination handling tasks: 1 Maximum acceptable weight measures. *Ergonomics, 39,* 121–140.

Straker, L. M., Stevenson, M. G., Twomey, L. T., & Smith, L. M. (1997). A comparison of risk assessment of single and combination handling tasks: 3 Biomechanical measures. *Ergonomics, 40,* 708–728.

Strokina, A. N., & Pakhomova, B. A. (1999). *Anthropo-ergonomic atlas*. Moscow, Russia: Moscow State University Publishing House (in Russian, ISBN 5-211-04102-X).

Strong, E. P. (1956). *A comparative experiment in simplified keyboard retraining and standard keyboard supplementary training*. Washington, DC: General Services Administration.

Stuart-Buttle, C., Marras, W. S., & Kim, J. Y. (1993). The influence of anti-fatigue mats on back and leg fatigue. *Proceedings of the Human Factors and Ergonomics Society 37th annual meeting* (pp. 769–773). Santa Barbara, CA: Human Factors and Ergonomics Society.

Sullivan, E. (2001). *The concise book of lying*. New York, NY: Farrar, Straus, and Giroux Press.

Swanson, N. G., Galinsky, T. L., Cole, L. L., Pan, C. S., & Sauter, S. L. (1997). The impact of keyboard design on comfort and productivity in a text-entry task. *Applied Ergonomics, 28,* 9–16.

Swets, J. A., & Bjork, R. A. (1990). Enhancing human performance: An evaluation of "New Age" techniques considered by the US. Army. *Psychological Science, 1,* 85–96.

Swink, J. R. (1966). Intersensory comparisons of reaction time using an electro-pulse tactile stimulus. *Human Factors, 8,* 143–145.

Tache, J., & Selye, H. (1986). On stress and coping mechanisms. In C. D. Spielberger & I. G. Sarason (Eds.), *Stress and anxiety: A sourcebook of theory and research* (vol. 10, pp. 3–24). (10). Washington, DC: Hemisphere.

Tamhane, A. C. (2009). *Statistical analysis of designed experiments: Theory and applications*. Hoboken, NJ: Wiley.

Tanzer, R. C. (1959). The carpal-tunnel syndrome: A clinical and anatomical study. *Journal of Bone and Joint Surgery, 41-A(4),* 626–634.

Tattersall, A. J., & Foord, P. S. (1996). An experimental evaluation of instantaneous self-assessment as a measure of workload. *Ergonomics, 39,* 740–748.

Taylor, F. (1911). *The principles of scientific management*. New York, NY: Norton.

Tayyari, F., & Smith, J. L. (1997). *Occupational ergonomics—Principles and applications*. New York, NY: Chapman and Hall.

Tennant, C., & Andrews, G. (1976). A scale to measure the stress of life events. *Australian and New Zealand Journal of Psychiatry, 10,* 27–32.

Tepas, D. I. (1985). Flexitime compressed work weeks and other alternative work schedules. In S. Folkard & T. H. Monk (Eds.), *Hours of work* (pp. 147–164). Chichester, UK: Wiley (Chapter 13).

Tepas, D. I., Paley, M. J., & Popkin, S. M. (1997). Work schedules and sustained performance. In G. Salvendy (Ed.), *Handbook of human factors and ergonomics* (2nd ed., pp. 1021–1058). New York, NY: Wiley.

Teves, M. A., Wright, J. E., & Vogel, J. A. (1985). *Performance on selected candidate screening test procedures before and after army basic and advanced individual training*. Natick, MA: United States Army Research Institute of Environmental Medicine (T13/85).

Thompson, J. K. (1990). *Body image disturbance: Assessment and treatment*. Elmsford, NY: Pergamon.

Thompson, J. K. (1995). Assessment of body image. In D. B. Allison (Ed.), *Handbook of assessment methods for eating behaviors and weight-related problems* (pp. 119–148). Thousand Oaks, CA: Sage.

III. FURTHER INFORMATION

Thoumie, P., Drape, J. L., Aymard, C., & Bedoiseau, M. (1998). Effects of a lumbar support on spine posture and motion assessed by electrogoniometer and continuous recording. *Clinical Biomechanics, 13*, 18–26.

Tichauer, E. R. (1973). *The biomechanical basis of ergonomics*. New York, NY: Wiley.

Tittiranonda, P., Burastero, S., & Rempel, D. (1999). Risk factors for musculoskeletal disorders among keyboard users. *Occupational Medicine: State of the Art Reviews, 14*, 17–38.

Toomingas, A., Mathijssenm, S. E., & Tornquist, E. W. (2011). *Occupational physiology*. Boca Raton, FL: CRC Press.

Tougas, G., & Nordin, M. C. (1987). Seat features recommendations for workstations. *Applied Ergonomics, 18*, 207–210.

Troy, B. S., Cooper, R. A., Robertson, R. N., & Grey, T. L. (1995). Analysis of work postures of manual wheelchair users. *Proceedings of the Silicon Valley ergonomics conference exposition Ergocon'95* (pp. 166–171). San Jose, CA: San Jose State University.

Tschöp, M., & Morrison, K. (2001). Weight loss at high altitude. *Advances in Experimental Medicine and Biology, 502*, 237–247.

Tsunawake, N., Tahara, Y., Yukawa, K., Katsuura, T., Harada, A., Iwanaga, K., et al. (1995). Changes in body shape of young individuals from the aspect of adult physique model by factor analysis. *Applied Human Science, 14*, 227–234.

Turek, F. W. (1989). Effects of stimulated physical activity on the circadian pacemaker of vertebrates. In S. Daan & E. Gwinner (Eds.), *Biological clocks and environmental time* (pp. 135–147). New York, NY: Guilford.

Turner, C. (1990). How much alcohol is in a standard drink? An analysis of 125 studies. *British Journal of Addiction, 85*(9), 1171–1175.

Turville, K. L., Psihogios, J. P., & Mirka, G. A. (1998). The effects of video display terminal height on the operator: A comparison of the 15° and 40° recommendations. *Applied Ergonomics, 29*, 239–246.

Tyrrell, R. A., & Leibovitz, H. W. (1990). The relation of vergence effort to reports of visual fatigue following prolonged near work. *Human Factors, 32*, 341–357.

Ugbolue, U. C., Hsu, W. H., Goitz, R. J., & Li, Z. M. (2005). Tendon and nerve displacement at the wrist during finger movements. *Clinical Biomechanics, 20*(1), 50–56.

Ungar, E., & Stroud, K. (2010). *A new approach to defining human touch temperature standards*. Houston, TX: NASA/Johnson Space Center.

University of Nottingham. (2002). *Strength data for design safety, phase 2*. London, UK: Department of Trade and Industry DTI URN 01/1433.

Upadhayay, N., & Guragain, S. (2014). Comparison of cognitive functions between male and female medical students: A pilot study. *Journal of Clinical and Diagnostic Research, 8*(6), BC12–15.

Use, J. (2015). Are bosses necessary? *The Atlantic*, 28–32.

Vagg, P., & Spielberger, C. (1999). The job stress survey: Assessing perceived severity and frequency of occurrence of generic sources of stress in the workplace. *Journal of Occupational Health Psychology, 4*(3), 288–292.

Vail, G. J. (1988). A gender profile: U.S general aviation pilot-error accidents. *Proceedings of the Human Factors Society 32nd annual meeting* (pp. 862–866). Santa Monica, CA: Human Factors Society.

Vail, G. J., & Ekman, L. (1986). Pilot-error accidents: Male vs. female. *Applied Ergonomics, 17*, 297–303.

Van Cott, H. P. & Kinkade, R. G. (Eds.). (1972). *Human engineering guide to equipment design (rev. ed)*. Washington, DC: U.S. Government Printing Office.

Van den Heever, D. J., & Roets, F. J. (1996). Noise exposure of truck drivers: A comparative study. *American Industrial Hygiene Association Journal, 57*, 564–566.

Van der Grinten, M. P. (1991). Test–retest reliability of a practical method for measuring body part discomfort. *Proceedings of the 11th congress of the International Ergonomics Association* (pp. 54-56). London, UK: Taylor, Francis.

Van der Grinten, M. P., & Smitt, P. (1992). Development of a practical method for measuring body part discomfort. In S. Kumar (Ed.), *Advances in industrial ergonomics and safety IV* (pp. 311–318). London, UK: Taylor & Francis.

Van Hoof, J., Demiris, G. & Wouters, E. J. M. (Eds.). (2017). *Handbook of smart homes, health care and well-being*. Heidelberg, Germany: Springer.

Van Hulle, C., Schmidt, N., & Goldsmith, H. (2012). Is sensory over-responsivity distinguishable from childhood behavior problems? A phenotypic and genetic analysis. *Journal of Child Psychology and Psychiatry, 53*, 64–72.

Vercruyssen, M., & Hendrick, H. L. (2011). *Behavioral research and analysis* (4th ed.). Boca Raton, FL: CRC Press.

Vicente, K. J. (2002). Ecological interface design: Progress and challenges. *Human Factors, 43*, 62–78.

Vicente, K. J., & Harwood, K. (1990). A few implications of an ecological approach to human factors. *Human Factors Society Bulletin, 33*, 1–4.

Vicianova, M. (2015). Historical techniques of lie detection. *European Journal of Social Psychology, 11*(3), 522–534.

III. FURTHER INFORMATION

Victor, T. W., Lee, J., & Regan, M. A. (2013). *Driver distraction and Inattention.* Abingdon, UK: Ashgate.

Vidulich, M. A., & Tsang, P. S. (1985). Assessing subjective workload assessment: A comparison of SWAT and the NASA-bipolar methods. *Proceedings of the Human Factors Society 29th annual meeting* (pp. 71-75). Santa Monica, CA: Human Factors Society.

Vieira, E. R. (2008). Low back disorders. In S. Kumar (Ed.), *Biomechanics in ergonomics* (2nd ed.). Boca Raton, FL: CRC Press (Chapter 18).

Village, J., & Morrison, J. B. (2008). Whole body vibration. In S. Kumar (Ed.), *Biomechanics in ergonomics.* Boca Raton, FL: CRC Press (Chapter 21).

Villanueva, M. B. G., Sotoyama, M., Jonai, H., Takeuchi, Y., & Saito, S. (1996). Adjustments of posture and viewing parameters of the eye to changes in the screen height of the visual display terminal. *Ergonomics, 39,* 933–945.

Vink, P. (2005). *Comfort and design: Principles and good practice.* Boca Raton, FL: CRC Press.

Vink, P., & Kompier, M. A. J. (1997). Improving office work: A participatory ergonomics experiment in a naturalistic setting. *Ergonomics, 40,* 435–449.

Vink, P., Douwes, M., & Van Woensel, W. (1994). Evaluation of a sitting aid: The back-up. *Applied Ergonomics, 25,* 170–176.

Violante, F., Armstrong, T., & Kilbom, A. (Eds), (2000). *Occupational ergonomics: Work related musculoskeletal disorders of the upper limb and back.* London, UK: Taylor & Francis.

Virzi, R. A. (1992). Refining the test phase of usability evaluation: How many subjects is enough? *Human Factors, 34,* 457–468.

Vivoli, G., Bergomi, M., Rovesti, S., Carrozzi, G., & Vezzosi, A. (1993). Biochemical and haemodynamic indicators of stress in truck drivers. *Ergonomics, 36,* 1089–1097.

Von Noorden, G. K. (1985). *Binocular vision and ocular motility* (3rd ed.). St. Louis, MO: Mosby.

Vredenburgh, A. G., Saifer, A. G., & Cohen, H. H. (1995). You're going to do what with that thing? *Ergonomics in Design, 2,* 16–20.

Wadden, T. A., Foster, G. D., & Letizia, K. A. (1994). One-year behavioral treatment of obesity: Comparison of moderate and severe caloric restriction and the effects of weight maintenance therapy. *Journal of Consulting and Clinical Psychology, 62,* 165–171.

Wagner, C. (1974). Determination of finger flexibility. *European Journal of Applied Physiology, 32,* 259–278.

Wagner, C. (1988). The Pianist's hand: Anthropometry and biomechanics. *Ergonomics, 31,* 97–131.

Waikar, A., Lee, K., Aghazadeh, F., & Parks, C. (1991). Evaluating lifting tasks using subjective and biomechanical estimates of stress at the lower back. *Ergonomics, 34,* 33–47.

Walji, A. H. (2008). Functional anatomy of the upper limb (extremity). In S. Kumar (Ed.), *Biomechanics in ergonomics* (2nd ed.). Boca Raton, FL: CRC Press (Chapter 8).

Walker, G., Stanton, N. A., & Salmon, P. (2015). *Human factors in automotive engineering and technology.* Abingdon, UK: Ashgate.

Wang, J., Pierson, R. N., Jr., & Heymsfield, S. B. (1992). The five level model: A new approach to organizing body composition research. *American Journal of Clinical Nutrition, 56,* 19–28.

Wang, M. J. J., Wang, E. M. Y., & Lin, Y. C. (2002). *Anthropometric data book of the Chinese people in Taiwan.* Hsinchu: The Ergonomics Society of Taiwan.

Wargo, M. J. (1967). Human operator response speed frequency and flexibility: A review and analysis. *Human Factors, 9,* 221–238.

Wasserman, D. E. (1987). *Human aspects of occupational vibrations.* Amsterdam, The Netherlands: Elsevier.

Wasserman, D. E. (1998). Vibration-induced cumulative trauma disorders. In W. Karwowski & G. Salvendy (Eds.), *Ergonomics in manufacturing (369–379).* Dearborn, MI: Society of Manufacturing Engineers.

Wasserman, D. E., Phillips, C. A., & Petrofsky, J. S. (1986). The potential therapeutic effects of segmental vibration on osteoporosis. *Proceedings of the 12th international congress on acoustics.* Toronto, Canada. Paper F2-1.

Wasserman, D. E., Wilder, D. G., Pope, M. H., Magnusson, M., Aleksiev, A. R., & Wasserman, J. F. (1997). Whole-body vibration exposure and occupational work-hardening. *Journal of Occupational and Environmental Medicine, 39*(5), 403–407.

Waters, T. R., & Putz-Anderson, V. (1998). Assessment of manual lifting—The NIOSH approach. In W. Karwowski & G. Salvendy (Eds.), *Ergonomics in manufactoring (205–241).* Dearborn, MI: Society of Manufacturing Engineers.

Waters, T. R., & Putz-Anderson, V. (1999). Revised NIOSH lifting equation. In W. Karwowski & W. S. Marras (Eds.), *The occupational ergonomics handbook* (pp. 1037–1061). Boca Raton, FL: CRC Press.

Waters, T. R. (2008). Revised NIOSH lifting equation. In S. Kumar (Ed.), *Biomechanics in ergonomics.* Boca Raton, FL: CRC Press (Chapter 20).

Waters, T. R. (2010). Introduction to ergonomics for healthcare workers. *Rehabilitation Nursing, 35,* 91–185.

III. FURTHER INFORMATION

Waterson, P. (2014). *Patient safety culture*. Abingdon, UK: Ashgate.

Webb, P. (1985). *Human calorimeters*. New York, NY: Praeger.

Webb, R. D. G., & Tack, D. W. (1988). Ergonomics, human rights and placement tests for physically demanding work. In F. Aghazadeh (Ed.), *Trends in ergonomics/human factors V* (pp. 751–758). Amsterdam, The Netherlands: Elsevier.

Weimer, J. (1995). Developing a research project. In I. J. Weimer (Ed.), *Research techniques in human engineering* (pp. 20–48). Englewood Cliffs, NJ: Prentice Hall.

Weinger, M. B., Wiklund, M. E., & Gardner-Bonneau, D. J. (2010). *Handbook of human factors in medical device design*. Boca Raton, FL: CRC Press.

Weinsier, R. L. (1995). Clinical assessment of obese patients. In K. D. Brownell & C. G. Fairburn (Eds.), *Eating disorders and obesity (463–468)*. New York, NY: Gilford.

Whinnery, J. E., & Murray, D. G. (1990). *Enhancing tolerance to acceleration (g_z) stress: The "Hook" maneuver. (NADC-90088-60)*. Warminster, PA: Naval Air Development Center.

Whitcome, K. K., Shapiro, L. J., & Lieberman, D. E. (2007). How women bend over backwards for baby. *Nature, 450*(7172), 1075–1078.

White, H. A., & Lee Kirby, R. (2003). Folding and unfolding manual wheelchairs: An ergonomic evaluation of health-care workers. *Applied Ergonomics, 34*(6), 571–579.

Wickens, C. D. (2017). Mental workload: Assessment, prediction and consequences. In L. Longo & M. C. Leva (Eds.), *Human mental workload: Models and applications* (pp. 18–29). Cham, Switzerland: Springer.

Wickens, C. D., & Carswell, C. M. (1997). Information processing. In G. Salvendy (Ed.), *Handbook of human factors and ergonomics* (2nd ed., pp. 89–129). New York, NY: Wiley.

Wickens, C. D., Gordon, S. E., & Liu, Y. (1998). *Introduction to human factors engineering*. New York, NY: Longman.

Wickens, C. D., Lee, J., Liu, Y., & Gordon-Becker, S. (2004). *An introduction to human factors engineering* (2nd Ed.). Upper Saddle River, NJ: Prentice-Hall/Pearson Education.

Wierwille, W. W. (1992). Visual and manual demands of in-car controls and displays. In J. B. Peacock & W. Karwowski (Eds.), *Automotive ergonomics: Human factors in the design and use of the automobile (299–320)*. London, UK: Taylor & Francis.

Wierwille, W. W., & Casali, J. G. (1983). A validated rating scale for global mental workload. *Proceedings of the Human Factors Society 27th annual meeting* (pp. 129–133). Santa Monica, CA: Human Factors Society.

Wierwille, W. W., & Eggemeyer, F. T. (1993). Recommendations for mental workload measurement in a test and evaluation environment. *Human Factors, 35*, 263–281.

Wierwille, W. W., Rahimi, M., & Casali, J. (1985). Evaluation of 16 measures of mental workload using a simulated flight task emphasizing mediational activity. *Human Factors, 27*, 489–502.

Wiker, S. F., Chaffin, D. B., & Langolf, G. D. (1989). Shoulder posture and localized muscle fatigue and discomfort. *Ergonomics, 32*, 211–237.

Wiklund, M. E., & Wilcox, S. B. (2005). *Designing usability into medical products*. Boca Raton, FL: Taylor & Francis/CRC Press.

Wiktorin, C., Selin, K., Ekenvall, L., Kilbom, A., & Alfredsson, L. (1996). Evaluation of perceived and self-reported manual forces exerted in manual materials handling. *Applied Ergonomics, 27*, 231–239.

Williams, V. H. (1988). *Isometric forces transmitted by the fingers: Data collection using a standardized protocol*. (Unpublished Master's Thesis). Blacksburg, VA: Virginia Tech.

Williams, R. B., Barefoot, J. C., & Shekelle, R. B. (1985). The health consequences of hostility. In M. A. Chesney & R. H. Rosenman (Eds.), *Anger and hostility in cardiovascular and behavioral disorders, (173–185)*. Washington, DC: Hemisphere.

Williamson, D. F. (1995). Prevalence and demographics of obesity. In K. D. Brownell & C. G. Fairburn (Eds.), *Eating disorders and obesity*. New York, NY: Guilford.

Williges, R. C. (1995). Review of experimental design. In J. Weimer (Ed.), *Research techniques in human engineering* (pp. 49–71). Englewood Cliffs, NJ: Prentice Hall.

Williges, R. C. (2007). *CADRE: Computer-aided design reference for experiments. Electronic Book CD-ROM-07-01*. Blacksburg, VA: Virginia Tech.

Wilmore, J. H., Costill, D., & Kenney, W. L. (2008). *Physiology of sport and exercise* (4th ed.). Champaign, IL: Human Kinetics.

Wilson, G. F., & Eggemeyer, F. T. (1994). Mental workload assessment. *CSERIAC Gateway, 5*(2), 1–3.

Wilson, J. R., & Corlett, E. N. (1990). *Evaluation of human work*. London, UK: Taylor & Francis.

Wilson, J. R. & Corlett, E. N. (Eds.). (1995). *Evaluation of human work. A practical ergonomics methodology*. London, UK: Taylor & Francis.

Wilson, J. R. & Corlett, N. (Eds.). (2005). *Evaluation of human work* (3rd ed.). London, UK: Taylor & Francis.

Wilson, J. R. & Sharples, S. (Eds.). (2015). *Evaluation of human work* (4th ed.). Boca Raton, FL: CRC Press.

III. FURTHER INFORMATION

Wilson, M. R., Poolton, J. M., Malhotra, N., Ngo, K., Bright, E., & Masters, R. S. (2011). Development and validation of a surgical workload measure: The surgery task load index (SURG-TLX). *World Journal of Surgery*, 35, 1961–1969.

Wing, A. M. (1983). Crossman & Goodeve (1963) twenty years on. *Quarterly Journal of Experimental Psychology*, 35, 245–249.

Winkelmolen, G. H. M., Landeweerd, J. A., & Drost, M. R. (1994). An evaluation of patient lifting techniques. *Ergonomics*, 37, 921–932.

Winter, D. A. (1990). *Biomechanics and motor control of human movement* (2nd ed.). New York, NY: Wiley.

Winter, D. A. (2009). *Biomechanics and motor control of human movement* (4th ed.). New York, NY: Wiley.

Wofford, J. C., & Daly, P. S. (1997). A cognitive-affective approach to understanding individual differences in stress propensity and resultant strain. *Journal of Occupational Health Psychology*, 2, 134–147.

Wogaltera, M. S., Conzolaa, V. C., & Smith-Jackson, T. L. (2002). Research-based guidelines for warning design and evaluation. *Applied Ergonomics*, 33, 219–230.

Wood, E. H., Code, C. F., & Baldes, E. J. (1990). Partial supination versus Gz protection. *Aviation, Space, and Environmental Medicine*, 61, 850–858.

Woodhouse, M. L., McCoy, R. W., Redondo, D. R., & Shall, L. M. (1995). Effects of back support on intra-abdominal pressure and lumbar kinetics during heavy lifting. *Human Factors*, 37(3), 582–590.

Woodson, W. E., & Conover, D. W. (1964). *Human engineering guide for equipment designers* (2nd Ed.). Berkeley, CA: University of California Press.

Woodson, W. E. (1954). *Human engineering guide for equipment designers*. Berkely, CA: University of California Press.

Woodson, W. E. (1981). *Human factors design handbook*. New York, NY: McGraw-Hill.

Woodson, W. E., Tillman, B., & Tillman, P. (1991). *Human factors design handbook* (2nd ed.). New York. NY: McGraw-Hill.

World Health Organization, Department of Nutrition for Health and Development. (2006). *Length/height-for-age, weight-for-age, weight-for-length, weight-for-height and body mass index-for age: Methods and development*. Geneva, Switzerland: WHO Press.

Wright, W. C. (1993). Diseases of workers. *Translation of Bernadino Ramazzini's 1713 De Morbis Articum*. Thunder Bay, ON: OH&S Press.

Wu, G., Siegler, S., Allard, P., Kirtley, C., Leardini, A., Rosenbaum, D., et al. (2002). ISB recommendation on definitions of joint coordinate system of various joints for the reporting of human joint motion Part I: Ankle, hip, and spine. *Journal of Applied Biomechanics*, 35, 543–555.

Wu, J., Yang, J., & Yoshitake, M. (2014). Pedal errors among younger and older individuals during different pedal operating conditions. *Human Factors*, 56, 621–630.

Wyszecki, G. (1986). Color appearance. In K. R. Boff, L. Kaufman & J. P. Thomas (Eds.), *Handbook of perception and human performance (9.1–9.57)*. New York, NY: Wiley.

Xing, L. (1988). Furniture took ages to grow legs. *China Daily October*, 31(5).

Yettram, A. L., & Jackman, J. (1981). Equilibrium analysis for the forces in the human spinal column and its musculature. *Spine*, 5, 402–411.

Youle, A. (2005). The thermal environment. In K. Gardiner & M. Harrington (Eds.), *Occupational hygiene* (3rd ed.). Oxford, UK: Blackwell.

Youle, A. (Ed.). (1990). *The thermal environment (technical guide no 8 British Occupational Hygiene Association)*. Leeds, UK: Science Reviews Ltd. and H and H Scientific Consultants Ltd.

Young, S. L., Brogmus, G. E., & Bezverkhny, I. (1997). The forces required to pull loads up stairs with different handtrucks. *Proceedings of the Human Factors and Ergonomics Society 41st annual meeting* (pp. 697-701). Santa Monica, CA: Human Factors and Ergonomics Society.

Yu, T., Roht, L. H., Wise, R. A., Kilian, D. J., & Weir, F. W. (1984). Low-back pain in industry: An old problem revisited. *Journal of Occupational Medicine*, 26, 517–524.

Yun, M. H., & Freivalds, A. (1995). Analysis of hand tool grips. *Proceedings of the Human Factors and Ergonomics Society 39th annual meeting* (pp. 553-557). Santa Monica, CA: Human Factors and Ergonomics Society.

Zacharkow, D. (1988). Posture: Sitting, Standing, Chair Design and Exercise. Springfield, IL: Thomas.

Zaidi, Z. (2010). Gender differences in the human brain: A review. *The Open Anatomy Journal*, 2, 37–55.

Zar, J. H. (1998). *Biostatistical analysis* (4th ed.). Englewood Cliffs, NJ: Prentice-Hall.

Zar, J. H. (2010). *Biostatistical analysis* (5th ed.). Englewood Cliffs, NJ: Prentice-Hall.

Zellers, K. K., & Hallbeck, M. S. (1995). The effects of gender, wrist and forearm position on maximum isometric power grasp force, wrist force, and their interactions. *Proceedings of the Human Factors and Ergonomics Society 39th annual meeting* (pp. 543–547). Santa Monica, CA: Human Factors and Ergonomics Society.

III. FURTHER INFORMATION

Zhang, X., & Chaffin, D. B. (2006). Digital human modeling for computer-aided design. In W. Karwowski & W. S. Marras (Eds.), *Interventions, controls, and applications in occupational ergonomics*. Boca Raton, FL: CRC Press (Chapter 10).

Zhao, C., Ettema, A. M., Osamura, N., Berglund, L. J., An, K. N., & Amadio, P. C. (2007). Gliding characteristics between flexor tendons and surrounding tissues in the carpal tunnel: A biomechanical cadaver study. *Journal of Orthopaedic Research, 25*(2), 185–190.

Ziobro, E. (1991). A contactless method for measuring postural strain. In W. Karwowski & J. W. Yates (Eds.), *Advances in industrial ergonomics and safety III* (pp. 421–425). London, UK: Taylor & Francis.

Zipperer, L. (2014). *Patient safety*. Abingdon, UK: Ashgate.

III. FURTHER INFORMATION

Glossary

The following list contains descriptions of terms used in this book. In other contexts, terms may have additional meanings—see, for example, admittance.

A

abdomen the part of the human body between the diaphragm and the pelvis that contains the digestive organs; same as belly, gut; not the same as stomach.

abduct to pivot away from the body or one of its parts; opposed to adduct.

absolute threshold the amount of energy necessary to just detect a stimulus. Also called the detection threshold or merely threshold. Often taken as the value at which some specified probability of detection exists, such as 0.50 or 0.75. See differential threshold.

absorbed power the power dissipated in a mechanical system as a result of an applied force.

accelerance the ratio of acceleration to force during simple harmonic motion. (Accelerance is the inverse of apparent mass or effective mass.)

acceleration a vector quantity that specifies the rate of change of velocity.

accommodation in vision, an adjustment of the curvature (thickness) of the lens of the eye (which change the eye's focal length) to bring the image of an object into proper focus on the retina.

achromatic lacking chroma or color.

acoustics the science of sound.

acromion the most lateral point of the lateral edge of the scapula. Acromial height is usually equated with shoulder height.

action, activation (of muscle) see contraction.

activator control(ler).

acuity the visual ability to discriminate fine detail. (Visual acuity is often expressed in terms of the angle subtended, or physical size, of the smallest recognizable object.) See contrast sensitivity; fovea; Landolt C; Snellen chart.

adaptation a change in sensitivity to the intensity or quality of stimulation over time. May be an increase in sensitivity (such as in dark adaptation) or a decrease in sensitivity (such as with continued exposure to a constant stimulus).

adduct to pivot toward the body or one of its part; opposed to abduct.

ADL activities of daily living.

admittance the ratio of displacement to force in a vibrating mechanical system. The displacement and force may be taken at the same point or at different points in the same system during simple harmonic motion.

ADP adenosine diphosphate. See mitochondrion.

aetiology a part of medical science concerned with the causes of disease. Also spelled etiology.

afferent conducting nerve impulses from sense organs to the central nervous system. See efferent.

agonist a contracting muscle, usually counteracted by an opposing muscle; same as protagonist. See antagonist.

amplitude the maximum value of a quantity (often a sinusoidal quantity). (Also called peak amplitude or single amplitude.)

anastomosis a connection between two blood vessels.

anatomy the science of the structure of the body.

angular frequency the product of the frequency of a sinusoidal quantity with 2π (rad/s).

ankylosis stiffening or immobility of a joint as a result of disease, together with a fibrous or bony union across the joint.

annulus fibrosus a ring of fiber that forms the circumference of an intervertebral disc.

antagonistic muscles pairs of muscles that act in opposition to each other, such as extensor and flexor muscles.

anterior at the front of the body; opposed to posterior.

anthromechanics biomechanics applied to the human.

anthropometric dummy a physical model constructed to reproduce the dimensions and ranges of movement of the human body for a specified percentile of an identified population.

anthropometry the measurement of human physical form (e.g., height or reach).

apparent mass the ratio of force to acceleration during simple harmonic motion.

arithmetic mean see mean.

arousal a general term indicating the extent of readiness, or activation, of the body. See autonomic nervous system.

Ergonomics.
http://dx.doi.org/10.1016/B978-0-12-813296-8.00021-9

arteriole a minute artery with a muscular wall.

artery a blood vessel carrying blood away from the heart. See vein.

arthritis a degenerated condition of joints characterized by inflammation. See osteoarthritis.

arthrosis degeneration of a joint.

articular relating to a joint.

ATP adenosine triphosphate. See mitochondrion.

atrophy wasting of body tissue.

audio frequency any normally audible frequency of a sound wave (e.g., 20–20,000 Hz).

audio frequency sound sound with a spectrum lying mainly at audio frequencies.

audiogravic illusion the apparent tilt of the body related to an auditory stimulus when an observer is exposed either to linear translational acceleration or deceleration or to centripetal acceleration. See oculogravic illusion.

audiogyral illusion apparent movement of a source of sound that is stationary with respect to an observer when the observer is rotated. See oculogyral illusion.

autonomic nervous system a principal part of the nervous system that is mainly self-regulating. It includes the sympathetic nervous system (involved in arousal) and the parasympathetic nervous system (involved in digestion and the maintenance of functions that protect the body). See nervous system.

average in statistical sense, same as mean. See mean.

axilla the armpit.

axis one of three mutually perpendicular straight lines passing through the origin of a coordinate system.

axon a nerve fiber that normally conducts nervous impulses away from a nerve cell and its dendrites to another nerve cell or to effector cells. Axons vary from about 0.25–100 μm in width and can be many centimeters long. Axons more than 0.5 μm thick are usually covered by a myelin sheath. See neuron; synapse.

B

ballistic relating to the parabolic motion characteristic of a projectile; in this book, the initial phase of such a motion.

bandpass filter a filter that has a single transmission band extending from a (nonzero) lower cutoff frequency to a (finite) upper cutoff frequency. Used to remove unwanted low- and high-frequency oscillations.

bandwidth for a filter, this is the difference between the nominal upper and lower cutoff frequencies, generally expressed in hertz, or as a percentage of the passband center frequency, or in octaves.

bel a unit of level when the base of the logarithm is 10. Used with quantities proportional to power. See decibel.

biceps brachii the large muscle on the anterior surface of the upper arm, connecting the scapula with the radius.

biceps femoris a large posterior muscle of the thigh.

biodynamics the science of (human) body motions in terms of a mechanical system; a subdivision of biomechanics.

biomechanics the science of the (human) body in terms of a mechanical system.

biopsy removal of tissue from a living person for diagnostic purposes.

bite bar a bar, plate, or mount held between the teeth of the upper and lower jaws so that there is no relative motion between the bar and the skull (head). Used to indicate the position of the head.

blood pressure pressure exerted on the artery walls by the blood. Systolic blood pressure is the maximal pressure produced by contraction of the left ventricle; diastolic blood pressure is the minimal pressure when the heart muscle is relaxed.

brachialis the muscle connecting the shoulder and upper arm with the ulna, crossing both shoulder and elbow joints.

brain that part of the central nervous system enclosed within the skull. See nervous system; electroencephalogram.

breakthrough the error appearing in the output of a system controlled by a human operator as a result of vibration transmitted via the limb of the operator. (Also called vibration-related error or feed-through.)

brightness individual perception of the intensity of a given visual stimulus.

bump a mild form of mechanical shock.

bursitis inflammation of a bursa. Bursae provide slippery membranes for tendons and ligaments to slide over bones. Bursitis may arise from infection or repeated pressure, friction, or other trauma.

buttock protrusion the maximal posterior protrusion of the right buttock.

C

CAD computer-aided design.

cadaver a dead body.

candela (cd) the SI base unit of the luminous intensity of a light source.

canthus juncture of the eyelids.

capillary minute blood vessel. Capillaries connect the arterial system to the venous system.

cardiovascular relating to the heart and the blood vessels or the circulation of blood.

carpal bones the scaphoid, lunate, triquetral, pisiform, trapezium, trapezoid, capitate, and hamate bones located in the proximal section of the hand.

carpal tunnel syndrome pressure of swollen tissue on the median nerve at the point where it passes through the carpal tunnel in the hand formed by the carpal bones and the transverse ligament.

cartilage the tough, smooth, white tissue (gristle) covering the moving surfaces of joint.

case–control study in epidemiology, a study in which individual cases of disease are matched with individuals from a control group. See cohort.

Celsius scale of temperature on which the melting point of ice is 0° and the boiling point of water is 100°, also sometimes still called centigrade.

center frequency the geometric mean of the nominal cutoff frequencies of a passband filter. The center frequency $= (f_1 f_2)^{-2}$, where f_1 and f_2 are the cutoff frequencies.

centigrade former name for the Celsius scale of temperature.

central nervous system part of the nervous system consisting of the brain and the spinal cord, often abbreviated as CNS. See cerebellum; cerebral cortex; cerebrum; nervous system.

central tendency a measure of a distribution of values as given by a typical value, such as the mean, geometric mean, median, or mode.

cerebellum the posterior part of the brain consisting of two connected lateral hemispheres. The cerebellum is involved in muscle coordination and the maintenance of body equilibrium. See central nervous system.

cerebral cortex the outermost layer of the cerebrum consisting of "gray matter." The cerebral cortex is involved in sensory functions, motor control, and some higher processes. See central nervous system.

cerebrum the largest part of the brain, consisting of two interconnected hemispheres with an inner core of "white matter" (myelinated fibers) and an outer covering of "gray matter" (unmyelinated fibers—the cerebral cortex). The cerebrum is involved in higher mental activities, including the interpretation of sensory signals, reasoning, thinking, decision-making, and the control of voluntary actions. See nervous system; central nervous system; myelin.

cervical relating to the neck. See vertebra.

cervicale the posterior protrusion of the spinal column at the base of the neck caused by the tip of the spinous process of the seventh cervical vertebra.

chroma attribute of color perception that determines to what degree a chromatic color differs from an achromatic color of the same lightness.

chromatic or achromatic induction a visual process that occurs when two color stimuli are viewed side by side and each stimulus alters the color perception of the other. The effect of chromatic or achromatic induction is usually called simultaneous contrast or spatial contrast.

chronic of long duration with slow progress; not acute.

circadian circadian rhythms are physiological cycles with a length of 24 h. See infradian; ultradian.

clavicle the collarbone linking the scapular with the sternum.

clinical related to the consideration of symptoms of a disease as opposed to the scientific observation of changes.

closed system a system isolated from all inputs from outside. See open system.

closed-loop system a system whose output is fed back and used to manipulate the input quantity. Closed-loop systems are frequently called feedback control systems. See open-loop system.

CNS central nervous system.

coccys a small triangular bone forming the lowest segment of the vertebral column (the tailbone).

cochlea coiled, snail-shaped structure in the inner ear. The cochlea has a spiral canal making two and a half turns around a central core. The canal consists of three parallel fluid-filled subcanals separated from each other by Reissner's membrane and the basilar membrane. See Corti organ.

cognitive task a task involving mental processes.

cohort a defined population group. The term is used in epidemiology to refer to a group followed prospectively in a cohort study.

collimation the process of making rays of light parallel as if they came from an object at infinite distance. Collimation may be achieved with an optical device such as a lens or mirror.

color illuminant color perceived as belonging to an area that emits light as primary source.

color object color perceived as belonging to an object.

color surface color perceived as belonging to a surface from which the light appears to be reflected or radiated.

color, related color seen in direct relation to other colors in the field of view.

color, unrelated color perceived to belong to an area in isolation from other colors.

compliance the reciprocal of stiffness.

concentric (muscle effort) shortening of a muscle against a resistance.

conditioned response a response that anticipates an event. May refer to the response to a normally neutral stimulus that has been elicited by "conditioning."

condyle articular prominence of a bone.

cone receptor of light in the retina. Cones, which mediate color vision, are the only receptors located in the fovea. The density of cones decreases toward the periphery of the retina. Cones function in daylight (photopic) viewing conditions. See rod.

confidence interval for a normal distribution of measured data points, the range within which a value will lie with a given degree of probability.

confounded said of the results of an experiment when the observed effect may be caused by more than one variable.

contraction (of muscle) literally, "pulling together" the Z lines (that delineate the length of a sarcomere), caused by the sliding action of actin and myosin filaments. Contraction develops muscle tension only if external resistance against the shortening exists. The term "isometric contraction" (a contradiction in terms!) assumes (falsely) that no change in sarcomere length occurs; in an "eccentric contraction" the sarcomere is actually lengthened. Therefore, it is often better to use the term action, effort, or exertion instead of contraction. See muscle; isometric.

contralateral relating to the opposite side. See ipsilateral.

contrast sensitivity a measure of the ability of the visual system to detect variations in contrast. Contrast sensitivity is dependent on the angular size subtended by test objects. It is the dependent variable used in measuring visual performance as a function of spatial frequency. See acuity.

contrast the difference in luminance between two areas.

control group a group of persons with characteristics similar to those of an experimental group that is not exposed to the conditions under investigation.

coordinate system orthogonal system of axes to indicate the directions of motions or forces. By convention, biodynamic coordinate systems follow the "right-hand rule": the positive directions of the x-, y-, and z-axes are designated by the directions of the first finger, the second finger, and the thumb, respectively, of the right hand.

Coriolis force force that arises when a body that is undergoing rotation also undergoes translation. The force arises from a cross-coupling of the motions; the resultant motion is called Coriolis acceleration.

Coriolis oculogyral illusion the visual and postural illusion that occurs within a rotating environment when one moves the head about an axis orthogonal to the axis of rotation. A small light at the center of rotation will appear to rise diagonally forward to the right if the head is tilted to the left while undergoing clockwise rotation.

cornea transparent outer layer of the anterior portion of the eye.

coronal plane same as frontal plane.

corpuscle an encapsulated nerve ending. See Golgi-Mazzoni corpuscle; Golgi (tendon) organ; Krause's end bulb; Meissner's corpuscle; Merkel's disc; Pacinian corpuscle; Ruffini ending.

correlation coefficient a number that expresses the degree and direction of a relationship between two (or more) variables. A correlation coefficient of −1.00 indicates perfect negative correlation; +1.00 indicates perfect positive correlation; 0 indicates no correlation. See product-moment correlation.

correlation in statistics, a relationship between two (or more) variables such that increases in one variable are accompanied by systematic increases or decreases in the other variable. See correlation coefficient; multiple correlation; product-moment correlation; rank-order correlation.

cortex see cerebral cortex.

Corti organ complex structure in the cochlea of the inner ear associated with hearing. Includes the basilar membrane and attached hair cells; sounds impinging on the tympanic membrane (the eardrum) cause a vibration of the ossicles (three small bones), which is then transmitted to the basilar membrane, causing movement and firing of the hair cells. See cochlea; ear.

Coulomb damping the dissipation of energy that occurs when a particle in a vibrating system is resisted by a force, the magnitude of which is a constant that is independent of displacement and velocity and the direction of which is opposite to the direction of the velocity of the particle. Also called dry friction damping.

covariance a measure of the extent to which changes in one variable are accompanied by changes in a second variable.

crest factor the ratio of the peak value to the root-mean-square value of a quantity over a specified time interval.

criterion a characteristic by which something may be judged (plural: criteria).

III. FURTHER INFORMATION

cross-axis coupling the motion occurring in one axis due to excitation in an orthogonal axis.

cross-sectional study an investigation in which a group of persons is studied at one point in time. (Also called prevalence study.) See longitudinal study.

CRT cathode-ray tube.

cumulative injury see repetitive strain injury.

cutaneous sensory system sensory system with receptors in or near the skin. (Receptors include those responsible for the senses of touch, pressure, warmth, cold, pain, taste, and smell.) See proprioception; corpuscle.

cycle the complete range of values through which a periodic function passes before repeating itself.

D

dactylion the tip of the middle finger.

damper a device for reducing the magnitude of a shock or vibration by dissipation of energy. Also called an absorber.

damping the dissipation of energy with time or distance. See Coulomb damping.

danger an unreasonable and unacceptable combination of hazard and risk. See hazard, risk.

dB(A) A-weighted sound pressure level.

dB see decibel.

decay time for a shock pulse, the interval of time required for the value of the pulse to drop from some specified large fraction of the maximal value to some specified small fraction thereof.

decibel one-tenth of a bel. A level in decibels is 10 times the logarithm (base 10) of the ratio of powerlike quantities. The power level in decibels is $L_P = 10 \log_{10} (p^2/p_0) = 20 \log_{10} (p/p_0)$.

degeneration deterioration of physical tissues in the condition or function. Degeneration may be caused by injury or disease processes. The function may be impaired or destroyed.

degrees of freedom the number of independent variables in an estimate of some quantity.

dendrite a branching treelike process involved in the reception of neural impulses from neurons and receptors. Dendrites are rarely more than 1.5 mm in length. See axon.

dependent variable any variable, the values of which are the result of changes in an independent variable. See also experimental design.

dermis layers of skin between the epidermis and subcutaneous tissue. The dermis contains blood and lymphatic vessels, nerves and nerve endings, and glands and hair follicles.

deterministic function a function whose value can be predicted from knowledge of its behavior at previous times.

diastole the dilation and relaxation of the heart during which its cavities fill with blood. See systole.

differential threshold the difference in value of two stimuli that is just sufficient to be detected. Also called difference threshold. See absolute threshold.

digits the fingers (not the thumb) and toes.

diopter a unit of measurement of the refractive power of lenses, equal to the reciprocal of the focal depth in meters. See Snellen chart.

disability loss of function or ability.

disease an interruption, cessation, or disorder of a body function. A disease is identified by at least two of the following: an identifiable cause, a recognizable group of signs and symptoms, and consistent anatomical alterations. See sign, syndrome.

displacement control(ler) a control, often hand operated, that is operated by moving the control. The output is proportional to the displacement of the control. See also force; isometric control; isotonic control.

displacement a vector quantity that specifies the change of position of a body with respect to a reference position.

distal the end of a body segment farthest from the center of the body. Opposed to proximal.

dominant frequency a frequency at which a maximum value occurs in a spectral-density curve.

dorsal at the back; also, at the top of hand or foot; opposed to palmar, plantar; and ventral.

dorsum the back or posterior surface of a part.

double amplitude see peak-to-peak value.

double-blind said of an experimental procedure in which neither the subject nor the person administering the experiment knows its crucial aspects (often, whether a substance being administered is a placebo or a drug).

driving-point impedance the ratio of force to velocity taken at the same point in a mechanical system during simple harmonic motion.

dummy test device or physical model simulating one or more of the anthropometric or dynamic characteristics of the human or animal body for experimental or test purposes.

duodenum the first part of the small intestine following the stomach.

dynamic stiffness (1) the ratio of change of force to change of displacement under dynamic conditions; (2) the ratio of force to displacement during simple harmonic motion.

dynamic relating to the existence of motion due to forces; not static, not at equilibrium. See mechanics.

dynamics a subdiscipline of mechanics that deals with forces and bodies in motion.

E

ear the organ of hearing. The ear consists of: (1) the external ear (the pinna and ear canal to the tympanic membrane or eardrum); (2) the middle ear (the cavity beyond the eardrum and including the ossicles, that is, the malleus, incus, and stapes); and (3), the inner ear (the cochlea, semicircular canals, utricle, and saccule). See Corti organ; vestibular system.

ear–eye plane a standard plane for orientation of the head (as observed in sagittal view of the right side), especially for defining the angle of the line of sight. The ear–eye plane is established by a line passing through the right auditory meatus (ear hole) and the right external canthus (juncture of the eyelids), with both eyes on the same level. The ear–eye plane is about 11 degrees more inclined than the Frankfurt plane. See frontal plane, medial plane, transverse plane.

eccentric (muscle effort) lengthening of a resisting muscle by external force.

ECG see electrocardiogram.

EEG see electroencephalogram.

effective mass see apparent mass.

effector a muscle at the terminal end of an efferent nerve that produces an intended response; often the hand or foot in human–machine systems; the actuator in robotics.

efferent conducting of nerve impulses from the central nervous system toward the peripheral nervous system (e.g., to the muscles). See afferent.

effort see contraction.

EKG see electrocardiogram.

electrocardiogram (ECG; or German, elektokardiogramm EKG) the graphical presentation of the time-varying electrical potential produced by heart muscle.

electroencephalogram (EEG) the graphical presentation of the time-varying electrical potentials of the brain.

electromyogram (EMG) the graphical presentation of the time-varying electrical potentials of muscle(s).

electrooculogram (EOG) the graphical presentation of the time-varying electrical potentials of the eye muscle(s).

embolism obstruction or occlusion of a vessel.

EMG see electromyogram.

empirical based on observation and experiment rather than theory.

end organ term used to refer to a sensory receptor. See corpuscle.

endocrine relating to internal secretion of hormones by glands. The hormones are usually distributed through the body via the bloodstream.

endolymph a clear fluid in the membranous semicircular canals of the vestibular system. The flow of endolymph relative to the canals causes movement of the cupula and the firing of hair cells consistent with rotation of the head.

EOG see electrooculogram.

epicondyle the bony eminence at the distal end of the humerus, radius, and femur.

epidemiology the study of the prevalence and spread of disease in a community.

epidermis the outer layer of the skin.

equivalent comfort contour outline of the magnitudes of vibration, expressed as a function of frequency, that produce broadly similar degrees of discomfort.

equivalent continuous value of the A-weighted sound pressure level of a continuous, steady sound that, within a specified time interval T, has the same mean-square sound pressure as a sound under consideration.

ergometer an apparatus for measuring the work performed by a person using it.

ergometry the measuring of the amount of physical work done by the body.

ergonomics the discipline that examines human characteristics for the appropriate design of the living and work environment. Also called human factors or human engineering, mostly in the United States.

etiology a part of medical science concerned with the causes of disease. Also spelled aetiology.

ex vivo outside the living body. See in vivo.

exertion see contraction.

III. FURTHER INFORMATION

experimental design the plan of an experimental investigation. Experimental designs are aimed at maximizing the sensitivity and ease of interpretation of experimental outcomes (measured on the "dependent" variable as a result of manipulating the "independent variable"), while minimizing the influence of unwanted effects. The design is usually associated with some statistical procedures used to test for significant results.

extend to move adjacent body segments so that the angle between them is increased, as when the bent leg is straightened; opposed to flex.

extensor a muscle whose contraction tends to straighten a limb or other body part. An extensor muscle is the antagonist of the corresponding flexor muscle.

external away from the central long axis of the body; the outer portion of a body segment.

exteroception the perception of information about the world outside the body. Involves the cutaneous sensory system and the senses of vision, hearing, taste, and smell. See proprioception; interoception.

exteroceptor any sensory receptor mediating exteroception.

extrinsic muscle muscle located outside the performing body segment, such as the hand.

extrinsic variable a variable external to a subject, the properties of which are not directly under the control of the subject.

eye movements see saccade.

eye see cone; fovea; retina; rod.

F

facet a small, smooth area on a bone, especially the superior and inferior articular facets of a vertebra. These form two small posterior joints that connect adjoining vertebrae (in addition to the larger anterior disc joint) and restrict the relative movement between the vertebrae.

false negative a test result which falsely indicates that a particular condition or attribute is absent. See false positive; type-II error.

false positive a test result which falsely indicates that a particular condition or attribute is present. See false negative; type-I error.

farsightedness an error of refraction, when accommodation is relaxed, in which the parallel rays of light from an object at infinity are brought to focus behind the retina. Also called hyperopia or hypermetropia.

fatigue weariness resulting from bodily (or mental) exertion, reducing the performance. The condition can be removed and performance restored by rest.

Fechner's law law stating that the psychological sensation P produced by a physical stimulus of magnitude I increases in proportion to the logarithm of the intensity of the stimulus: $P = k \log I$. See Stevens' (power) law; Weber's law.

feedback control system see closed-loop system.

feedback input of some information to a system about the output of the system.

feedthrough see breakthrough.

femur the thigh bone.

fiber see muscle.

fibril see muscle.

filament see muscle.

filter a device for separating oscillations on the basis of their frequency. A filter, which may be mechanical, acoustical, electrical, analog, or digital, attenuates oscillations at some frequencies more than those at other frequencies.

Fitts's law the motion time MT to a target depends on the length D of the path and the size W of the target: $\mathrm{MT} = a + b \log_2 (2D/W)$.

flex to move a joint in such a direction as to bring together the two parts it connects, as when the elbow is being bent; opposed to extend.

flexor a muscle whose contraction tends to flex. A flexor muscle is the antagonist of the corresponding extensor muscle.

force control(ler) a control, often hand operated, that is operated by applying force to the control which does not move. The output is proportional to the force applied to the control. See also displacement; isometric control; isotonic control.

fovea the fovea (centralis) of the eye is a small pit in the center of the retina that contains cones, but no rods. When a person looks directly at a point, its image falls on the fovea, which covers an angle of about 2 degrees. Visual acuity is normally greatest for images on the fovea.

Frankfurt plane (occasionally falsely spelled Frankfort plane) a standard plane for orientation of a skull (head). The plane is established by a line passing through the right tragion (approximately the ear hole) and the lowest point of the right orbit (eye socket), with both eyes on the same level. See ear–eye plane; frontal plane; medial plane; transverse plane.

III. FURTHER INFORMATION

free dynamic in the context of muscle strength, an experimental condition in which neither displacement nor its time derivatives nor force is manipulated as independent variables.

frequency weighting a transfer function used to modify a signal according to a required dependence on vibration frequency.

frequency the reciprocal of the fundamental period. Frequency is expressed in hertz (Hz), one unit of which is equals to one cycle per second.

frontal plane any plane at a right angle to the medial (midsagittal) plane dividing the body into anterior and posterior portions. Also called the coronal plane. See ear–eye plane; Frankfurt plane; medial plane; transverse plane.

functional disorder any disorder for which there is no known organic pathology.

G

G, acceleration due to gravity the acceleration produced by the force of gravity at the surface of the earth, g = 9.80665 m/s². Denoted G (or g) if used as a unit.

gain the amplification provided by a system.

galvanic skin response a measure of the electrical characteristics of the skin, usually the electrical resistance, which varies with emotional tension and other factors. A polygraph, or lie detector, uses the galvanic skin response as an indicator of a person's emotional state.

Gaussian distribution items occur in a bell-shaped (binomial) distribution that can be described fully by mean and standard deviation. Same as normal distribution.

geometric mean the geometric mean of two quantities is the square root of the product of the two quantities.

glabella the most anterior point of the forehead between the brow ridges in the midsagittal plane.

glenoid cavity the depression in the scapula below the acromion into which fits the head of the humerus, forming the "shoulder joint."

gluteal furrow the furrow at the juncture of the buttock and the thigh.

Golgi (tendon) organ a proprioceptive sensory nerve whose ending is mainly embedded within fibers of tendons at their junction with muscles. The Golgi organ is a "stretch" receptor activated by a change in tension between muscles and bone. See corpuscle.

Golgi-Mazzoni corpuscle an encapsulated nerve ending found in the dermis of the skin.

gustation tasting.

H

habituation reduction in human response to a stimulus as a result of cumulative exposure to the stimulus. Habituation is often assumed to involve activity of the central nervous system. See adaptation.

haemoglobin the red respiratory protein of erythrocytes. Haemoglobin consists of globin, which is a protein, and haem, which is an iron compound. Also spelled hemoglobin.

handicap the manner by which an impairment or abnormality limits or prevents the functioning of an individual.

hand-transmitted vibration (or shock) mechanical vibration (or shock) applied or transmitted directly to the hand–arm system, usually through the hand or its digits. A common example is the vibration from the handles of power tools.

haptic relating to the combined feeling from sensors in the skin (tactile sensation) and motion and displacement sensors (kinesthetic, proprioceptive sensation).

harmonic a sinusoidal oscillation whose frequency is an integral multiple of the fundamental frequency. The second harmonic is twice the frequency of the fundamental, etc.

hazard condition or circumstance that presents a potential for injury. See danger, risk.

head-up display a fixed display presenting information in the normal line of sight of the observer, such as in a windshield of an automobile or aircraft.

helmet-mounted display a visual display mounted on a helmet and moving with the head.

hemoglobin see haemoglobin.

hernia protrusion of a part of the body through tissue normally or muscle containing it.

herniated disc see prolapsed disc.

high-pass filter a filter that has a single transmission band extending from some critical cutoff frequency of interest. Used to remove unwanted oscillations at low frequencies.

H-point the pivot point of the torso and thigh on two-dimensional and three-dimensional devices used to measure a vehicle's seating accommodations.

III. FURTHER INFORMATION

hue attribute of color perception that uses color names and combinations thereof, such as bluish purple, yellowish green. The four unique hues are red, green, yellow, and blue, none of which contains any of the others.

human analog (or surrogate) in anthromechanics, a body that has biodynamic properties representative of those of the human body. See dummy.

human engineering see ergonomics.

human-factors (engineering) see ergonomics.

humerus the bone of the upper arm.

hyper over, above, exceeding, excessive.

hyperopia farsightedness.

hypertrophy increase in bulk of some tissue in the body without an increase in the number of cells.

hypo under, below, diminished.

hypothesis testing statistical procedure to determine whether an experimental outcome is likely to be the effect (1) of random events (not rejecting the null hypothesis) or (2) of the experimental treatment of the independent variable (rejecting the null hypothesis).

hypothesis a supposition made as a starting point for reasoning or investigation, possibly without an assumption as to its truth. Hypotheses should be formulated so that they can be evaluated (tested). See hypothesis testing; statistical significance; theory.

I

IADL instrumental activities of daily living.

iliac crest the superior rim of the pelvic bone.

ilium see pelvis.

impact a single collision of one mass with a second mass.

impedance the ratio of a harmonic excitation of a system to its response (in consistent units), both of whose arguments increase linearly with time at the same rate. See mechanical impedance.

impulse the integral with respect to time of a force taken over a time during which the force is applied; often simply the product of the force and the time during which the force is applied.

in toto as a whole.

in vitro in an artificial environment, such as in a test tube.

in vivo in the living body.

independent variable any variable whose values are independent of changes in the values of other variables; in an experiment, the variable that is manipulated so that its effect on one or more dependent variables can be observed. See also experimental design.

inferior below, lower, in relation to another structure.

infra below, low, under, inferior, after.

infradian infradian rhythms are physiological cycles that have periods longer than 24 h. See also circadian; ultradian.

infrasonic frequency any frequency lower than normally audible sound waves (below 20 Hz).

infrasound sound with a spectrum lying mainly at infrasonic frequencies.

injury damage to body tissue due to trauma.

inner ear the innermost part of the ear, containing the vestibular system and cochlea. Also called the labyrinth, a name derived from the complex maze-like structure within the bone.

innervation (1) provision of an organ with nerves; (2) nervous stimulation or activation of an organ.

inseam a term used in tailoring to indicate the inside length of a sleeve or trouser leg. The inseam is measured on the medial side of the arm or leg.

intelligence quotient (IQ) a total score obtained on specific standardized tests measuring human intelligence. Median IQ score is 100, and approximately two-thirds of the population scores between 85 and 115.

intelligibility the ability to understand, depending on acoustical conditions, the meanings of words, phrases, sentences, and entire speeches.

inter between, among.

internal near the central long axis of the body; inside a body segment.

interoception the perception of information about the interior functioning of the body. Interoception involves receptors in the viscera, glands, and blood vessels and the senses of hunger, thirst, nausea, etc. See proprioception, exteroception.

interoceptor any sensory receptor mediating interoception.

intersubject variability variability among subjects. See intrasubject variability.

interval scale a scale in which differences between intervals have quantitative significance; although the intervals between values are significant, the absolute values are not. Allowable arithmetic operations are addition and subtraction, but not multiplication or division; thus, one cannot use proportions. The valid statistical operations on an interval scale include calculation of the mean value and of the standard deviation. See nominal scale; ordinal scale; ratio scale.

intervertebral disc flexible pad between the main bodies of vertebrae. Intervertebral discs have a soft jellylike core, the nucleus pulposus, enclosed by hard fibrous tissue that is attached to the bodies of the adjacent vertebrae. Discs makeup about 25% of the length of the vertebral column. See annulus fibrosus; prolapsed disc; facet; vertebra.

intra within, on the inside.

intrasubject variability variability within a subject. See intersubject variability.

intrinsic muscle muscle located inside the performing body segment, such as the hand.

involuntary action an action not under voluntary control. Either a reflex action, a very well-learned action, or a normal action that proceeds from the genetic makeup of the body.

ipsilateral relating to the same side. See contralateral.

ischaemia see ischemia.

ischemia a deficiency in blood supply to a part of the body, often as a result of the narrowing or complete blockage of an artery or arteriole.

ischial tuberosity bony projection at the lower and posterior section of the coxal bone (also called pelvic bone or hip bone) that is part of the pelvic girdle. When a person is seated on a flat, rigid surface, the contact pressure is usually greatest beneath the ischial tuberosities.

ischium the dorsal and posterior of the three principal bones that compose either half of the pelvis.

iso equal, the same.

isoforce a theoretical construct in which muscular force (tension) is assumed to be constant. The term is synonymous with isotonic, see there.

isoinertial a condition in which muscle moves a constant mass.

isokinematic a theoretical construct in which the velocity of muscle shortening (or lengthening) is assumed to be constant. See isokinetic; contraction.

isokinetic a theoretical construct in which the force (or tension) within muscle is assumed to be constant. Note that some ergometers said to be isokinetic are in fact isokinematic. See isokinematic; contraction.

isometric contraction a muscular effort that causes tension, but no movement. See contraction.

isometric control a control, often hand operated, that can be operated by the isometric contraction of muscles. The control does not move, but responds to the applied force or torque. See also displacement, isometric, isotonic control.

isometric a theoretical construct in which the length of the muscle is assumed to be constant. See static.

isotonic control a control, often hand operated, that can be operated by isotonic contraction of muscles. The control moves, but offers the same resistance to the applied force or torque at all positions. See also isometric control.

isotonic a theoretical construct in which muscle tension (force) is assumed to be constant, see isoforce. (In the past, isotonic was occasionally erroneously applied to any condition other than isometric.)

J

jerk a vector quantity that specifies the rate of change of acceleration.

JND just-noticeable difference.

joule (J) the work done by a force of 1 N acting over a distance of 1 m. The joule is the SI unit of work. It is equivalent to 10^7 ergs and is the energy dissipated by 1 W in 1 s.

just noticeable difference (JND) the difference between two stimuli that is just barely perceived under some defined condition. See Weber's law.

K

kinematics a subdivision of dynamics that deals with the motions of bodies, but not with the forces that cause them.

kinesthetic the feeling of motion, especially from the muscles, tendons, and joints. Also called somatosensory. See haptic, tactile.

kinetics a subdivision of dynamics that deals with forces that cause bodies to move.

knuckle the joint formed by the meeting of a finger bone (phalanx) with a palm bone (metacarpal).

Krause's end bulb encapsulated corpuscle in parts of the skin, mouth, and other locations generally believed to be sensitive to cold.

kyphosis backward (convex) curvature of the spine. See lordosis, scoliosis.

L

labyrinth see inner ear.

Landolt C the letter C presented at various orientations as a test of visual acuity. (Also called Landolt ring.) See acuity; Snellen chart.

latency the period of apparent inactivity between the time a stimulus is presented and the moment that a specified response occurs.

lateral near or toward the side of the body; opposed to medial.

LBP low-back pain.

ligament fibrous band between two bones at a joint. Ligaments are flexible, but inelastic.

lightness visual perception of how much more or less light a stimulus emits in comparison to a "white" stimulus also contained in the field of view.

line of sight the line connecting the point of fixation in the visual field with the center of the entrance pupil and the center of the fovea in the fixating eye.

linear function one variable is said to be a linear function of another if changes in the first variable are directly proportional to changes in the second.

linear system a system in which the response is proportional to the magnitude of the excitation.

longitudinal study an investigation in which a person, or a group of persons, is studied over the course of time. See cross-sectional study.

longitudinal wave a wave in which the direction of displacement caused by the wave motion is in the direction of propagation of the wave.

lordosis forward (concave) curvature of the lumbar spine. See kyphosis; scoliosis.

low-pass filter a filter that has a single transmission band extending from zero frequency up to a finite frequency. (Used to remove unwanted high-frequency oscillations from a signal.)

lumbago a term used to describe pain in the middle and lower back. May be associated with sciatica, some combination of pulled muscles and sprained ligaments, or an unknown cause. See vertebral column; sciatica; prolapsed disc.

lumbar part of the back and sides between the ribs and the pelvis. See vertebra.

lumen (lm) the luminous flux emitted from a point source of uniform intensity of 1 cd into a unit solid angle. The lumen is the SI unit by which luminous flux is evaluated in terms of its visual effect.

lumen the opening (cross section) of a blood vessel.

lux the SI unit of illumination equal to $1 \ lm/m^2$.

M

macula lutea a small yellowish area (near the center of the retina of the eye) that contains the fovea where visual perception is most acute. See eye, fovea, retina.

macula the maculae acousticae are the two patches of sensory cells in the utricle and saccule of the vestibular system of the inner ear. See vestibular system; otoliths.

magnitude measure of largeness, size, or importance.

malinger to feign an illness, possibly for compensation, sympathy, or the avoidance of work.

malleolus a rounded bony projection in the ankle region. The tibia has such a protrusion on the medial side, the fibula one on the lateral side.

masking a phenomenon in which the perception of a normally detectable stimulus is impeded by a second stimulus. The second stimulus may be presented at a different point in the same sensory system, causing "lateral masking." Alternatively, it may be presented at a different time, either before, in "forward masking," or after, in "backward masking," of the test stimulus.

matched groups an experimental procedure in which groups of subjects are matched for variables that may affect results, but that are not studied in the experiment.

maximal value the value of a function when any small change in the independent variable causes a decrease in the value of the function.

maximum the maximal value.

mean (1) the mean value of a number of discrete quantities is the algebraic sum of the quantities, divided by the number of quantities; (2) the mean value of a function $x(t)$ over an interval between t_1 and t_2 is given by $\int x(t)\,dt$. The mean is a measure of central tendency. Also called the arithmetic mean or average. See geometric mean; median; mode.

meatus the juncture of the eyelids.

mechanical advantage in a biomechanical context, the lever arm (moment arm, leverage) at which a muscle works around a bony articulation.

mechanical impedance the ratio of force to velocity, where the force and velocity, may be taken at the same or different points in the same system during simple harmonic motion.

mechanical system an aggregate of matter comprising a defined configuration of mass, stiffness, and damping.

mechanics the branch of physics that deals with forces applied to bodies and their ensuing motions.

mechanoreceptor a receptor that responds to mechanical pressures, such as a touch receptor in the skin. See cutaneous sensory system; corpuscle.

medial plane a vertical plane through the middle of the body (in the anatomical position) that divides it into right and left halves. Also called the midsagittal plane. See ear–eye plane; Frankfurt plane; frontal plane; transverse plane.

medial near, toward, or in the midline of the body; opposed to lateral.

median the value, in a series of observed values, that has exactly as many observed values above it as below it; the middle score in a distribution of scores ordered according to their magnitude. The median is a primary measure of central tendency for skewed distributions. See mean; mode.

Meissner's corpuscle specialized encapsulated nerve ending found in the papillae of the skin of the hand and foot, the front of the forearm, the lips, and the tip of the tongue. Meissner's corpuscle responds to pressure and vibration within a small area and it adapts rapidly. It is often a principal means of sensing vibration of the skin in the range from 5 to 60 Hz, depending on various conditions. Meissner's corpuscles are located in glabrous skin and are orientated perpendicular to the skin surface. Also called a tactile corpuscle. See Pacinian corpuscle.

Merkel's disc specialized free nerve ending (corpuscle) found immediately below the epidermis and around the ends of some hair follicles. The corpuscle is believed to respond to pressure applied perpendicularly to the skin at frequencies below about 5 Hz. See Ruffini ending.

mesopic a condition in which illumination is between 0.1 and 0.01 lux. See photopic; scotopic.

metacarpal one of the long bones of the hand between the carpus and the phalanges.

metacarpus collectively, the five bones of the hand between the carpus and the phalanges.

midsagittal plane same as medial plane. See frontal plane; transverse plane.

minute (of arc) (′) unit of angular measure equal to 1/60 of a degree.

mitochondrion (plural mitochondria) a small granular body floating inside a cell's cytoplasm. Contains a maze of tightly folded membranes within which oxygen and nutrients (brought into the cell by the blood via the circulatory system) are processed by enzymes to reform adenosine triphosphate, ATP, from adenosine diphosphate, ADP.

MMH manual material handling. (Note the triple tautology.)

modality in psychology and physiology, an avenue of sensation (e.g., the visual modality).

mode in statistics, the value, of a series of values, that is the most frequently observed. A measure of central tendency. See mean; median.

model a representation of some aspect of an idea, an item, or a functioning.

modulation the variation in the value of some parameter that characterizes a periodic oscillation.

monotonic a relationship between two variables in which, for every value of each variable, there is only one corresponding value of the other variable. The graphical representation of a monotonic function is a steadily rising or steadily falling curve.

morbidity the prevalence of disease in a population.

motion sickness vomiting (emesis), nausea, or malaise provoked by actual or perceived motion of the body or its surroundings.

motivation a desire, or an incentive, to achieve.

motor unit the set of all muscle filaments under the control of one efferent nerve axon.

motor(ic) in life sciences, a term used to refer to processes or anatomical areas associated with muscular action.

multiple correlation (R) the relation between a dependent variable and two (or more) independent variables. See correlation.

multivariate any procedure in which more than one variable is considered simultaneously.

muscle contraction shortening of muscle tissue as a result of contractions of motor units. Usually, tension develops between the origin and insertion of the muscle as it contracts. See contraction.

muscle fibers elements of muscle, containing fibrils.

muscle fibrils elements or muscle fibrils, containing filaments.

muscle filaments elements of muscle fibrils (polymerized protein molecules) capable of sliding along each other, thus shortening the muscle and, if doing so against resistance, generating tension. See contraction.

muscle spindle an end organ in skeletal muscle that is sensitive to stretching of the muscle in which it is enclosed. Muscle spindles taper at both ends and lie parallel to regular muscle bundles.

muscle strength the capability of a muscle to generate force (tension) between its proximal (origin) and distal (insertion) ends.

muscle a tissue bundle of fibers, able to contract or lengthen. Specifically, striated muscle (skeletal muscle) that tends to move body segments about each other under voluntary control. There are also smooth and cardiac muscles.

myelin fatty substance that encloses some nerve fibers. Myelinated fibers propagate nerve impulses faster than unmyelinated fibers.

myo a prefix referring to muscle.

myopia see nearsightedness.

N

narrowband filter a bandpass filter for which the passband width is relatively narrow (e.g., one-third octave or less).

natural frequency a frequency of free vibration resulting from only elastic and inertial forces of a mechanical system. See resonance.

nearsightedness inability to see distant objects distinctly, owing to an error of refraction, if accommodation is relaxed, in which parallel rays of light from an object at infinity are brought to focus in front of the retina. Also called myopia.

nerve cell see neuron.

nerve a whitish cord made up of myelinated or unmyelinated neural fibers (or both) and held together by a connective tissue sheath. Afferent nerves transmit stimuli from sensors (extero- and interoceptors) to the central nervous system CNS. Efferent nerves transmit stimuli in the reverse direction, from the CNS to effectors. See myelin; axon; dendrite; synapse; effector.

nervous system a system composed of neural tissue controlling the human body's structures and organs. May be subdivided in many ways (e.g., central nervous system and peripheral nervous system, or autonomic nervous system and somatic nervous system).

neuromuscular relating to the relationship between nerve and muscle, especially the motor innervation of skeletal muscle.

neuron the functional unit of the nervous system, consisting of cell body, dendrite, and axon. See synapse.

nociceptor specialized nerve endings involved in the sensation of pain.

noise (1) any disagreeable or undesired sound; (2) sound, generally of a random nature, the spectrum of which does not exhibit clearly defined frequency components; (3) electrical (or mechanical) oscillations of an undesired or random nature.

nominal scale the simplest form of a scale, qualitative and consisting of a set of categories or labels. A nominal scale facilitates the identification of items, to determine whether they are equivalent and to count them. See interval; ordinal; ratio scale.

normal distribution items occur in a bell-shaped (binomial) distribution that can be described fully by mean and standard deviation. Same as Gaussian distribution.

nucleus pulposus the soft fibrocartilage central portion of the intervertebral disc.

null hypothesis the assumption, used to test statistical significance, that there is no difference between sets of data. See hypothesis testing; statistical significance; theory; type-I error; type-II error.

nyctalopia *night "blindness" (actually, the condition of a persona having reduced vision in dim light).*

nystagmus eye movements consisting of a slow drift of the eyeball, followed by a rapid return to the original position.

O

obesity *BMI (body mass index) of 30.0 or higher. See overweight.*

objective in life sciences, possessing the property of being real and measurable in physical units. Often used in contrast to subjective.

occipital the back of the head, skull, or brain.

occular relating to the eye.

III. FURTHER INFORMATION

octave bandwidth filter a bandpass filter for which the passband is one octave, that is, the difference between the upper and lower cutoff frequencies is one octave.

octave the interval between two frequencies that have a frequency ratio of 2.

oculogravic illusion the apparent tilt of the body relative to the visual scene when an observer is exposed either to linear translational acceleration or deceleration or to centripetal acceleration. See audiogravic illusion.

oculogyral illusion apparent movement of a point of light that is stationary with respect to an observer when the observer is rotated. See audiogyral illusion.

oculomotor related to eye movements and their muscular control.

OD overuse disorder. See repetitive strain injury. (Can also refer to overdose of drugs.)

olecranon the proximal end of the ulna.

olfaction smelling.

omphalion the center point of the navel.

one-half octave the interval between two frequencies that have a frequency ratio of $2^{1/2}$ (= 1.4142).

one-third octave the interval between two frequencies that have a frequency ratio of $2^{1/3}$ (= 1.2599).

open system a system that is influenced by inputs from the outside. See closed system.

open-loop system a system with no feedback. See closed-loop system.

orbit the eye socket.

ordinal scale a scale in which items are placed in order according to a characteristic. An ordinal scale is the simplest quantitative scale: items can be ranked, and large values imply that there is more of the characteristic, but do not indicate how much more. Nonparametric statistics is based on ordinal scales. The only other valid statistical operations on ordinal data are the determination of medians, percentiles, interquartile ranges, and similar procedures. See interval; nominal; ratio scale.

orthogonal (1) at right angles; (2) said of variables that are independent of each other.

os bone.

oscillation the variation, usually with time, of the magnitude of a quantity with respect to a specified reference when the magnitude is alternately greater and smaller than some mean value.

osteoarthritis, osteoarthrosis degeneration of the joints, with some loss of the low-friction cartilage linings and the formation of rough deposits of bone. The term osteoarthrosis is preferred when there is no inflammation.

osteoporosis the wasting, or atrophy, of bone. Bone becomes more porous, grows more brittle, and changes its geometry. Osteoporosis is often associated with aging, but also can arise from immobilization of the limbs and nutritional deficiencies.

otoliths crystalline particles of calcium carbonate and a protein adhering to the gelatinous membrane of the maculae of the utricle and saccule.

overuse injury see repetitive strain injury.

overweight a body mass index (BMI) of 25.0–<30.0. See obesity.

P

Pacinian corpuscle specialized oval-shaped encapsulated nerve ending. The Pacinian corpuscle responds to pressure over a diffuse area and adapts rapidly. It is a principal means of sensing vibration (45–400 Hz, depending on various conditions). Pacinian corpuscles are located in the palmar skin of the hands, in the plantar skin of the feet, on some tendons and ligaments, and in other locations. They are the largest nerve endings found in the skin (up to about 2 mm in length) and consist of concentric layers that make them look like a minute onion. Also called lamellated corpuscle. See Meissner's corpuscle.

palmar relating to the palm of the hand or sole of the foot. Also called plantar or volar. Opposed to dorsal.

palpation examination by feeling with the hands.

parasympathetic nervous system see autonomic nervous system.

patella the kneecap.

pathology (1) the science of diseases, (2) the structural and functional deviations from normal that constitute or characterize a particular disease.

peak value the maximal value of a quantity during a given interval. The peak value is usually taken as the maximum deviation of the quantity from the mean value.

peak-to-peak value the algebraic difference between the extreme values of a quantity.

Pearson product-moment correlation see product-moment correlation.

pelvis the basin-shaped ring of bone consisting of the ilium, pubic arch, and ischium.

perception awareness of some event; process by which the mind refers its sensations to external objects as their cause.

peripheral nervous system network of afferent and effluent nerves. Often abbreviated as PNS.

phalanges (singular: phalanx) bones of the fingers or toes. There are 14 on each hand, three for each finger and two for the thumb.

phase the fractional part of a period (in radians or degrees) through which a sinusoidal motion has advanced from some reference time.

phon unit of perceived loudness level. The loudness level of a sound is the sound pressure level of a 1000-Hz pure tone judged by the listener to be equally loud.

photopic a condition in which illumination is bright (above 0.1 lux). See mesopic; scotopic.

physiology the biological science of the normal functions and phenomena of living organisms.

pink vibration a random vibration that has an equal mean-square acceleration for any frequency band having a bandwidth proportioned to the center frequency of the band. The energy spectrum of pink random vibration, as determined by octave, one-third-octave, etc., filters have a constant value. See white vibration.

pitch (1) perceived sound, depending on the frequency; (2) rotational motion about the y-axis. See coordinate system.

pivot to rotate a body joint; to abduct or adduct the joint.

placebo a substance with no intended or known medicinal effect.

plane see ear–eye plane; Frankfurt plane; frontal plane; medial plane; transverse plane.

plantar relating to the sole of the foot. See palmar.

plasma the fluid portion of blood or lymph.

plethysmograph a device for the measurement and graphical representation of the changes in volume (often blood volume), of some part of the body.

PNS peripheral nervous system. See there.

polygraph See galvanic skin response.

popliteal pertaining to the ligament behind the knee or to the part of the leg behind the knee.

posterior at the back of the body; opposed to anterior.

predisposition a condition of having special susceptibility to a condition or disease.

probability (p) an expression of the likelihood of occurrence of an event; usually expressed as a ratio of the number of occurrences of a given event of interest to the number of occurrences of all types of events considered.

procedure words ("pro-words") *specific and standardized words and word-pairs that are used in radio transmissions.*

product-moment correlation (r) the linear correlation between two variables, based on the calculation of the mean of the products of the deviations of each score from the mean value of each variable. Also called the Pearson product-moment correlation.

prolapse slipped out of the normal position.

pronation twisting the forearm about its long axis counterclockwise when one looks down one's own right arm, clockwise when one looks down one's left arm. Opposite of supination.

prone the position of the body when lying face downward. Opposite of supine.

proprioception the perception of information about the position, orientation, and movement of the body and its parts. See interoception; exteroception.

proprioceptor any sensory receptor that mediates proprioception.

protagonist same as agonist.

proximal the end of a body segment nearest the center of the body; opposed to distal.

psychology the science concerned with the human behavior and mental processes.

psychophysics science concerned with the quantitative relations between the perception of stimuli and their physical characteristics.

pure tone a sound whose characteristic (such as pressure) varies sinusoidally with time.

R

radian (rad) unit of angular measure equal to the angle subtended at the center of a circle by an arc the length of which is equal to the radius of the circle. (There are 2π radians in 360 degrees; 1 rad $\approx$ 57.3 degrees.)

radius the bone of the forearm on its thumb side.

random process a set (ensemble) of time functions that have no specific pattern or purpose but that can be characterized through statistical properties. Also called a stochastic process.

range the interval between the highest and lowest scores in a distribution.

III. FURTHER INFORMATION

rank in statistics, the position of a score relative to all other scores ordered according to their value. When ranked, scores form an ordinal scale.

rank-order correlation a correlation between two variables based on the differences in the rank orders of the two variables. Also called Spearman rank-order correlation.

rate coding the time sequence in which efferent signals arrive at a motor unit to cause contractions.

rating scale a scale of words or numbers along some dimension that may be used to report subjective reactions. Many types of rating scale may be devised. Scales may be bipolar, giving the options "good" and "bad," or unipolar. The number of steps on the scale is often in the range from 5 to 9. See scaling.

ratio scale a scale in which the ratios of values on the scale have quantitative significance. A ratio scale represents the relative order of units as well as the exact value between units, and includes a true zero value. See interval; nominal; ordinal scale.

reaction time (RT) the time between the presentation of a stimulus and the beginning of an observer's response. Simple reaction time is the time to make a single simple response to a single stimulus. Choice reaction time is the time involved when there are two or more stimuli and two or more corresponding possible responses to every stimulus.

reclining posture a body posture at some angle between seated upright and being supine.

recruitment coding the time sequence in which efferent signals arrive at different motor units to cause them to contract.

reflex an involuntary reaction in response to a stimulus applied peripherally and transmitted to the nervous centers of the brain or spinal cord. A reflex involves a receptor that is sensitive to the stimulus, an effector that responds to the stimulus, and a reflex arc between receptor and effector.

reliability the reliability of a test or measurement method is given by the degree to which it produces similar values when applied repeatedly under the same test conditions. See sensitivity; specificity; validity.

repeatability the extent to which the outcome of an experiment would recur if the experiment was repeated.

repetition performance of the same activity more than once.

repetitive strain injury an injury that arises from repeated strains (traumata) when a single strain (a trauma) gives no observable signs of injury. Often abbreviated as RSI. Also called cumulative injury or overuse injury.

residual in statistics, the part of the variance that cannot be attributed to the factors which have been considered.

resonance the increase in amplitude of oscillation of an electrical or mechanical system exposed to a forced oscillation whose frequency of excitation is equal or very close to the natural undamped frequency of the system.

retina the light-sensitive membrane covering the inner rear surface of the eye.

rhythmic the same action repeated in equal intervals.

risk probability of injury. See danger, hazard.

rms see root-mean-square value.

rod receptor of light in the retina. Rods function under conditions of low illumination (scotopic) vision. They do not mediate color vision or render perception of fine detail. There are no rods in the fovea. See cone.

roll rotational motion about the x-axis. See coordinate system.

root-mean-square (rms) value the square root of the average of the squares of a set of numbers.

RSI repetitive strain injury.

Ruffini ending specialized encapsulated nerve ending (corpuscle) found in the dermis of hairy skin. The Ruffini ending responds to lateral stretching of the skin (and possibly warmth) and adapts slowly. See Merkel's disc.

S

saccade a quick jump of the eyes from one fixation point to another.

saccule the smaller of the two vestibular sacs in the membranous labyrinth of the inner ear. Like the utricle, the saccule contains a layer of receptors that are sensitive to translational forces arising from translational acceleration of the head or rotation of the head in an acceleration field. See vestibular system; macula; utricle.

sacrum the large wedge-shaped base for the lumbar vertebrae. It consists of five fused sacral vertebrae and forms the posterior section of the pelvis where it articulates with the hip bones.

sagittal ("in the line of an arrow shot from the bow") pertaining to the medial (midsagittal) plane of the body or to a parallel plane.

saturation attribute of color perception that determines the degree to which a chromatic color differs from an achromatic color, regardless of lightness.

scalar any quantity that is completely defined by its magnitude. See vector.

scale a procedure for placing items, individuals, events, or sensations in some series. Four types of scale are used: a nominal scale, an ordinal scale, an interval scale, and a ratio scale. See these terms.

scaling in psychology, the determination of (a point on) a scale, along a psychological dimension (such as effort) that has a continuous mathematical relation to some physical dimension (such as energy consumption). See rating scale; scale.

scapula the shoulder blade.

sciatica pain in the area of the sciatic nerve.

scoliosis a sideways curvature of the spine. See kyphosis; lordosis.

scotopic a condition in which illumination is dim (below 0.01 lux). See mesopic; photopic.

semantic scale a set of words describing various degrees of a characteristic, formed in a nominal or interval scale. See interval scale, nominal scale, rating scale, scale.

semicircular canals three small membranous ducts (tubes) that form loops of about two-thirds of a circle within each inner ear.

sensation the reception of stimuli at the sensors and their translation into neural impulses; followed by perception. See perception.

sensitivity (1) as applied to a transducer, the ratio of a specified output quantity (e.g., electrical charge) to a specified input quantity (e.g., acceleration); (2) in screening, the proportion of individuals with a positive test result for the effect that the test is intended to reveal. See reliability; specificity; validity.

sensorimotor relating to a signal transmitted via a neural circuit from a receptor to the central nervous system and back to a muscle.

shock mechanical shock exists when a force, a position, a velocity, or an acceleration is suddenly changed so as to excite transient disturbances in a system.

sign in medicine, any abnormality that is discovered by a physician during an examination of a patient. (A sign is an objective symptom of disease.) See symptom.

skeleton in humans, all or part of the (about) 206 separate bones in the body.

skin the outer covering of the body, consisting of epidermis, dermis, and subcutaneous tissues.

slipped disc see prolapsed disc.

Snellen acuity visual acuity measured using a standard chart containing rows or letters of graduated sizes, expressed as the distance in which a given row of letters is correctly read, compared to the distance at which the letters can be read by a person with clinically normal eyesight. For example, an acuity score of 20/50 indicates that the individual who was tested can read letters at the distance of 20 ft that a normally sighted person can read at 50 ft.

Snellen chart chart used to obtain approximate measures of visual acuity.

somatic relating to the body.

somatosensory see kinesthetic.

sound (1) acoustic oscillation capable of exciting the sensation of hearing; (2) the sensation of hearing excited by an acoustic oscillation.

spasm an involuntary muscular contraction.

Spearman rank-order correlation see rank-order correlation.

specificity in screening, the proportion of individuals with a negative test result for what the test is intended to reveal (such as true negative results as a proportion of the total of true-negative and false-positive results). See reliability; sensitivity; validity.

spectrum a description of a quantity as a function of frequency or wavelength.

sphyrion the most distal extension of the tibia on the medial side under the malleolus.

spinal cord the column of neural tissue running the length of the vertebral column from the brainstem to the bottom of the lumbar vertebrae. Along that length, 31 sets of nerve roots emerge on each side to form 31 pairs of spinal nerves.

spinal nerve in humans, one of the 31 pairs of nerves coming from the spinal cord: 8 cervical, 12 thoracic, 5 lumbar, 5 sacral, and 1 coccygeal.

spine the stack of vertebrae. See vertebral column.

spinous (or spinal) process of a vertebra the posterior prominence.

spondylosis degeneration of the spinal column causing, for example, osteoarthritis of vertebrae.

static at rest or in equilibrium; not dynamic. With regard to muscle effort, often used as equivalent to isometric because no body motion results. See isometric; contraction.

statics a subdivision of mechanics that deals with bodies at rest.

statistical significance the probability that a result would have occurred if only chance factors were operating; hence, the degree to which the result may be attributed to some systematic effect. A probability less than 5% is often chosen as the significance level, expressed as a significance of 5%, or $p = .05$. A lower probability is required ($p < .05$) if a result is not assumed to be a chance finding. The significance of a set of test results is determined by statistical tests. See type-I error, type-II error.

sternum the breastbone.

Stevens' (power) law the relationship between the magnitude P of a psychological sensation produced by a stimulus of magnitude I given by $p = kI^n$. The value of n is a characteristic of the physical stimulus, for example, noise or vibration, while k depends on the units of measurement. See Fechner's law; Weber's law.

stiffness the ratio of change of force (or torque) to the corresponding change in translational (or rotational) displacement of an elastic element.

stochastic process see random process. (Derived from the Greek word for "guess.")

stomach the internal organ between esophagus and small intestine where digestion begins; occasionally falsely used instead of belly or abdomen.

strain (1) in engineering, a dimensionless value given by the ratio of the deformation caused by a stress to the size of the material stressed; (2) in life sciences, the experienced result of stress generated by physical or mental demands. See stress; stressor.

stress (1) in engineering, the load applied to a material that causes a change in its dimensions (strain); (2) in psychology, either the cause of strain or the resulting strain itself. See stressor.

stressor a cause of strain or stress. The term helps to remove the confusing use of the word "stress": the cause is a stressor and the effect may be called either the stress (psychological use) or the strain (engineering use).

stylion the most distal point on the styloid process of the radius.

styloid process a long, spinelike projection of a bone.

sub- a prefix designating below or under.

subjective in life sciences, something that is dependent on an individual. See objective; scale; scaling.

superior above, in relation to another structure; higher.

supination twisting the forearm about its long axis clockwise when one looks down one's own right arm, counterclockwise on one's left arm. Opposite of pronation.

supine the position of the body when lying face upward. Opposite of prone.

supra- prefix denoting above, superior, or in very large quantities.

sympathetic nervous system see autonomic nervous system.

symptom in medicine, an abnormality in function, appearance, or sensation.

synapse the junction of one neuron with another, where nerve impulses are transmitted (chemically or electrically) from the axon of one neuron to the dendrites of the other.

syndrome in medicine, a combination of signs and symptoms that collectively indicates a disease.

synergistic effect the effect of a combination of two or more stressors. A synergistic effect is greater than the arithmetic sum of the effects of the individual stressors.

systole the contraction of the heart by which blood is driven from the ventricles through the aorta and pulmonary artery. See diastole.

T

tactile relating to touch or the sense of touch perceived through the skin. See corpuscle; haptic; kinesthetic; Meissner's corpuscle; Merkel's disc; Pacinian corpuscle; Ruffini ending.

tarsus the collection of bones in the ankle joint.

tendon reflex a reflex muscular contraction elicited by a sudden stretching of a tendon. Sensors in tendons sense stretching and trigger a reflexive contraction of the muscle to oppose movement and maintain posture. The "knee jerk" caused by percussion of the tendon at the knee is a tendon reflex.

tendon fibrous cord joining a muscle to a bone.

tendonitis inflammation of a tendon.

tenosynovitis inflammation of a tendon and its sheath.

theory a reasoned explanation of how something occurs, or may occur, but without absolute proof. A theory is more securely established than a hypothesis.

thoracic outlet syndrome compression of fifth cervical and first thoracic nerves and the subclavian artery by muscles in the region of the first rib and clavicle.

threshold see absolute threshold; differential threshold.

tibia the main bone of the lower leg (shinbone).

tibiale the uppermost point of the medial margin of the tibia.

tinnitus a subjective sensation of noises in the ear (ringing, whistling, booming, etc.) experienced in the absence of an external acoustic stimulus.

III. FURTHER INFORMATION

tissue a collection of similar cells and their surrounding structures.

tolerance the ability to endure a stimulus without harm.

tone (1) in muscles, a state in which there is normal muscular tension with slight stretching of the muscles maintained by proprioceptive reflexes; (2) in acoustics, a sound at one given frequency.

torsion the act of twisting by the application of forces at right angles to the axis of rotation.

touch the act of, or the sensation that arises from, contact between the body and an object.

tracking task a task that involves continuously following a target. The two principal types of tracking task are pursuit tracking, in which movements of the target are directly indicated, and compensatory tracking, in which the difference between the actual and the desired location of the target is indicated.

tragion the point located at the notch just above the tragus of the ear.

tragus the conical eminence of the auricle (pinna or external ear) in front of the ear hole.

trait a characteristic of a person.

transducer a device designed to receive energy from one system and supply energy, of either the same or a different kind, to another in such a manner that the desired characteristics of the input energy appear at the output.

transfer function a mathematical relation between the output (response) and the input (excitation) of a system.

transfer impedance in a mechanical sense, the ratio of the force taken at one point in a mechanical system to the velocity taken at another point in the same system during simple harmonic motion.

transfer mobility the ratio of the velocity taken at one point in a mechanical system to the force taken at another point in the same system during simple harmonic motion.

translation the (linear, straight-line) movement of an object so that all its parts follow the same direction.

transmissibility the nondimensional ratio of the response amplitude of a system in steady-state forced vibration to the excitation amplitude, expressed as a function of the vibration frequency. The ratio may be of forces, displacements, velocities, or accelerations.

transmission the sending of muscle force across one or more body joints.

transverse plane a plane across the body at right angles to the frontal plane and the medial (midsagittal) plane. See ear–eye plane, Frankfurt plane.

trauma an injury caused by harsh contact with an object. (From the Greek word for "wound.")

triceps the muscle of the posterior upper arm crossing the elbow.

trochanterion the tip of the bony lateral protrusion of the proximal end of the femur.

tuberosity a (large) rounded prominence on a bone.

Type-I error the erroneous rejection of a true hypothesis ("false positive"). A Type-I error is an error that arises in statistical tests and that is more likely when a low level of significance is required, such as $p = .05$, before rejecting the null hypothesis. See hypothesis testing; null hypothesis; statistical significance.

Type-II error the failure to reject a false hypothesis ("false negative"). A Type-II error is an error that arises in statistical tests and that is more likely when a high level of significant is required, such as $p < .001$, before rejecting the null hypothesis. See hypothesis testing; null hypothesis; statistical significance.

U

ulna the bone of the forearm on the side of the little finger.

ultra- prefix meaning beyond, extreme, or excessive.

ultradian ultradian rhythms are physiological cycles repeated throughout a 24-h time period. See also circadian; infradian.

ultrasonic frequency any frequency higher than normally audible sound waves (such as above 20,000 Hz).

umbilicus depression in the abdominal wall where the umbilical cord was attached to the embryo.

use error an aspect of a task that results in a different response than intended.

user error an error by an operator that results in a different response than intended.

utricle the larger of the two vestibular sacs in the membranous labyrinth of the inner ear.

V

validity in tests, a measure of how well the test assesses what it purports to assess. See reliability; sensitivity; specificity.

Valsalva maneuver forced expiratory effort with either closed glottis or closed nose and mouth. The Valsalva maneuver may be used to inflate the eustachian tube; it may also increase one's tolerance to large downward acceleration by preventing blood from returning to the heart from the head.

variable see dependent variable; independent variable.

III. FURTHER INFORMATION

variance the square of the standard deviation.

vascular relating to, or containing, blood vessels.

vaso- combining form denoting blood vessel. Also vas- and vasculo-.

vasoconstriction narrowing of a blood vessel.

vasodilation enlargement of a blood vessel.

vasomotor relating to the nerves that control the muscular walls of blood vessels.

vasospasm contraction of the muscular walls of the blood vessels.

vector a quantity that is completely determined by its magnitude and direction. See scalar.

vein a blood vessel carrying blood toward the heart. See artery.

ventral pertaining to the anterior side of the trunk.

venule a minute vein.

vertebra one of the bones of the spinal column. In humans there are 33 vertebrae: 7 cervical vertebrae, 12 thoracic vertebrae, 5 lumbar vertebrae, 5 sacral vertebrae (fused into one bone, the sacrum), and 4 coccygeal vertebrae (fused into one bone, the coccyx).

vertebral column the series of 33 vertebrae that extend from the coccyx to the cranium; the backbone or spine.

vertex the top of the head.

vertigo an inappropriate sensation of movement of the body or the visual field caused by a disturbance of the mechanisms responsible for equilibrium.

vessel any duct, tube, or canal conveying body liquid, such as blood.

vestibular system collective term for the three semicircular canals and the two vestibular sacs (utricle and saccule) within the labyrinth of the inner ear.

vestibule any small cavity at the front of a canal; especially the middle part of the inner ear containing the utricle and saccule.

vibration the variation with time of the magnitude of a quantity that is descriptive of the motion or position of a mechanical system when the magnitude is alternately greater and smaller than some average value.

viscera the internal organs of the body, including the digestive, respiratory, urogenital, and endocrine systems and the heart, spleen, etc.

viscosity the resistance of a fluid to shear forces and, therefore, to flow.

volar relating to the palm of the hand or sole of the foot. See palmar, plantar.

volition conscious, voluntary action. See reflex.

W

Weber's law law stating that the just-noticeable difference JND in the stimulus magnitude I is proportional to the magnitude of the stimulus. The relation may be expressed as $\Delta I/I$ = constant. See Fechner's law; Stevens' law.

Weber–Fechner law the combination of Weber's law and Fechner's law.

white vibration a random vibration that has equal mean-square acceleration for any frequency band of constant width over the spectrum of interest. See pink vibration.

X

x-axis see coordinate system.

Y

yaw rotational motion about a z-axis. See coordinate system.

y-axis see coordinate system.

Z

z-axis see coordinate system.

Zeitgebers time markers or cues given by the environment to reset the body's internal clock.

III. FURTHER INFORMATION

Index

Printed in the United States
By Bookmasters